DATE DUE

Coordination Chemistry of Macrocyclic Compounds

Coordination Chemistry of Macrocyclic Compounds

Edited by

Gordon A. Melson

Virginia Commonwealth University
Richmond, Virginia

PLENUM PRESS • NEW YORK AND LONDON

Library of Congress Cataloging in Publication Data

Main entry under title:

Coordination chemistry of macrocyclic compounds.

Includes index.
CONTENTS: Melson, G. A. General introduction. – Melson, G. A. Synthesis of macrocyclic complexes. – Thermodynamics and kinetics of cation-macrocycle interaction, by J. D. Lamb and others. [etc.]
1. Coordination compounds–Addresses, essays, lectures. 2. Cyclic compounds–Addresses, essays, lectures. I. Melson, Gordon A., 1937-
QD474.C68   541'.2242   78-27023
ISBN 0-306-40140-1

©1979 Plenum Press, New York
A Division of Plenum Publishing Corporation
227 West 17th Street, New York, N.Y. 10011

All rights reserved

Printed in the United States of America

Contributors

Lawrence J. Boucher, Department of Chemistry, Western Kentucky University, Bowling Green, Kentucky

James J. Christensen, Departments of Chemistry and Chemical Engineering and the Thermochemical Institute, Brigham Young University, Provo, Utah

Neil F. Curtis, Department of Chemistry, Victoria University of Wellington, Wellington, New Zealand

Bill Durham, Department of Chemistry, Wayne State University, Detroit, Michigan

Delbert J. Eatough, Departments of Chemistry and Chemical Engineering and the Thermochemical Institute, Brigham Young University, Provo, Utah

John F. Endicott, Department of Chemistry, Wayne State University, Detroit, Michigan

Virgil L. Goedken, Department of Chemistry, Florida State University, Tallahassee, Florida

Reed M. Izatt, Departments of Chemistry and Chemical Engineering and the Thermochemical Institute, Brigham Young University, Provo, Utah

John D. Lamb, Departments of Chemistry and Chemical Engineering and the Thermochemical Institute, Brigham Young University, Provo, Utah

Jean-Marie Lehn, Department of Chemistry, Université Louis Pasteur de Strasbourg, Strasbourg, France

Gordon A. Melson, Department of Chemistry, Virginia Commonwealth University, Richmond, Virginia

Alexander I. Popov, Department of Chemistry, Michigan State University, East Lansing, Michigan

F. L. Urbach, Department of Chemistry, Case Western Reserve University, Cleveland, Ohio

Preface

Chemists have been aware of the existence of coordination compounds containing organic macrocyclic ligands since the first part of this century; however, only during the past few years have they expanded research into the chemistry of these compounds. The expansion was initiated in the early 1960s by the synthesis and characterization of compounds containing some new macrocyclic ligands. The synthesis of compounds which may serve as model systems for some natural products containing large rings as ligands provided the main goal for the early expansion of research effort; indeed, a recurrent theme behind much of the reported chemistry has been the analogy between synthetic macrocyclic compounds and many natural-product systems.

More recently, the emphases of reported research have ranged over the whole spectrum of chemistry, and the number of publications that discuss macrocyclic chemistry has increased at a dramatic rate. The completed research has been reported in a variety of journals throughout the world but there has been no previous attempt to bring the major developments together under one cover. This book, therefore, attempts to satisfy the need for a single source in which there is both a collection and a correlation of information concerning the coordination chemistry of macrocyclic compounds. The chapters in this book discuss various aspects of macrocyclic chemistry, and while these chapters as a whole constitute an in-depth survey of the state-of-the-art of the field, each chapter is written as a complete unit. The chapter layout is designed so that the presented information begins with aspects of the synthesis of macrocyclic compounds and progresses through discussions of physical properties and reactivity to the relationship of these compounds to natural products. Although the chemistry of natural-product macrocyclic compounds is not discussed in depth, the relationship between synthetic macrocyclic compounds and natural products is a continuous theme throughout the book.

It is hoped that two major goals will be realized by the publication of this book. The first goal is to provide a source of information for scientists working in the field of macrocyclic chemistry, while the second is to stimulate further research into this fascinating and increasingly important area of chemistry.

Virginia Commonwealth University   Gordon A. Melson

Contents

4.   Structural Aspects

Neil F. Curtis

5. Ligand Field Spectra and Magnetic Properties of Synthetic Macrocyclic Complexes

F. L. Urbach

6. Chemical Reactivity in Constrained Systems

John F. Endicott and Bill Durham

7. Metal Complexes of Phthalocyanines

Lawrence J. Boucher

8.   Coordination Chemistry of Porphyrins
Lawrence J. Boucher

9.   Physicochemical Studies of Crown and Cryptate Complexes
Alexander I. Popov and Jean-Marie Lehn

10.   Natural-Product Model Systems
Virgil L. Goedken

General Introduction

Gordon A. Melson

1. Introductory Comments

The field of coordination chemistry of macrocyclic compounds has undergone spectacular growth during the past 15 years. This growth has largely been due to the synthesis of a great number and variety of synthetic macrocycles which behave as coordinating ligands for metal ions. The development of the field of bioinorganic chemistry has also been an important factor in spurring the growth of interest in complexes of macrocyclic compounds since it has been recognized that many complexes containing synthetic macrocyclic ligands may serve as models for biologically important species which contain metal ions in macrocyclic ligand environments.

Although several reviews dealing with different aspects of the coordination chemistry of macrocyclic compounds have been published over the past 15 years, no attempt to bring the developments in the field together in one source has been made. This book is an attempt to rectify this deficiency in the chemical literature; it is anticipated that scientists actively working in the field of macrocyclic compound chemistry and those who work in related fields will be able to obtain information they require from the data contained herein. A difficulty which is encountered in compiling a collection and making correlations of data in a rapidly expanding field, such as the field of macrocyclic compound chemistry, is that by the date of publication the material reported no longer reflects the current state of knowledge. This is inevitable; however, every attempt has been made to ensure that the data included in the different areas under discussion accurately reflect the "state of the art" in that area at the date of submission.

Gordon A. Melson • Department of Chemistry, Virginia Commonwealth University, Richmond, Virginia 23284.

2. General Comments

2.1. Definition of a Macrocyclic Compound

A macrocyclic compound is defined throughout the following material as a cyclic compound with nine or more members (including all hetero atoms) and with three or more donor (ligating) atoms.

2.2. Historical Background

Coordination compounds containing macrocyclic ligands have been known and studied since the beginning of this century; however, until quite recently, the number and variety of these compounds was limited. Complexes of porphyrins, corrins, and phthalocyanines have been investigated because of their relation to important naturally occurring species containing macrocycles such as heme, cytochromes, or chlorophyll, or because of their potential as dyestuffs or pigments. Several books and review articles which summarize the early research or porphyrins and phthalocyanines have been published; however, it is not appropriate to discuss this material in this introduction; Chapters 7 and 8 review this early research in more detail.

A few scattered reports of "synthetic macrocycles" were made prior to 1960; for example, Linstead, Elvidge, and co-workers[1-4] in a series of papers in the 1950s reported the synthesis of a variety of macrocyclic compounds, some of which are potential tri- and tetradentate ligands and are related to porphyrins and phthalocyanines. Some complexes of these new macrocyclic compounds with copper, nickel, cobalt, etc., were reported. The synthesis of 1,4,8,11-tetraazacyclotetradecane was first reported in 1936,[5] although cobalt and nickel complexes of this macrocycle were not prepared until the mid-1960s.[6,7] The above examples are not the only references relating to synthetic macrocycles which were reported prior to 1960, but they do serve to illustrate the status of the field of macrocyclic chemistry and particularly the area dealing with the coordination chemistry of macrocyclic compounds prior to the recent development of interest in the field. The synthetic macrocycles that were reported were not usually prepared with a view to studying their coordination chemistry; indeed the synthesis of coordination compounds containing macrocyclic ligands was very often incidental to the research rather than the prime motive for conducting the research. Thus prior to 1960 there appeared to be little interest in developing the coordination chemistry of macrocyclic compounds.

In the early 1960s several groups, working independently, synthesized a variety of coordination compounds containing synthetic macrocyclic ligands. Curtis,[8] working in New Zealand, described the reaction between tris-ethylenediaminenickel(II) perchlorate and acetone, and later[9] assigned to the

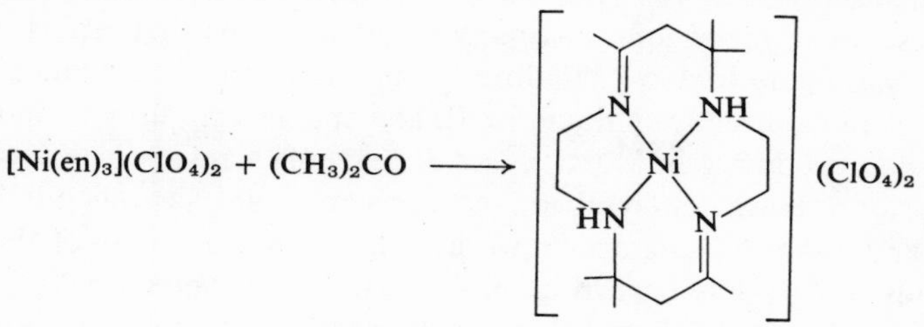

Scheme 1

product the macrocyclic structure shown in Scheme 1. Following these reports, the generality of the above type of reaction was demonstrated, and a new series of synthetic macrocycles was developed.[10] At approximately the same time that Curtis discovered the above reaction, Thompson and Busch, working in the U.S.A., achieved the first deliberate synthesis of a compound containing a new synthetic macrocyclic ligand by the reaction described in Scheme 2.[11,12]

Also in 1962, Schrauzer[13] and Umland and Thierig[14] in Germany reported the formation of macrocyclic complexes by reactions outlined in Scheme 3. Several metal ions and R and X groups were involved in the

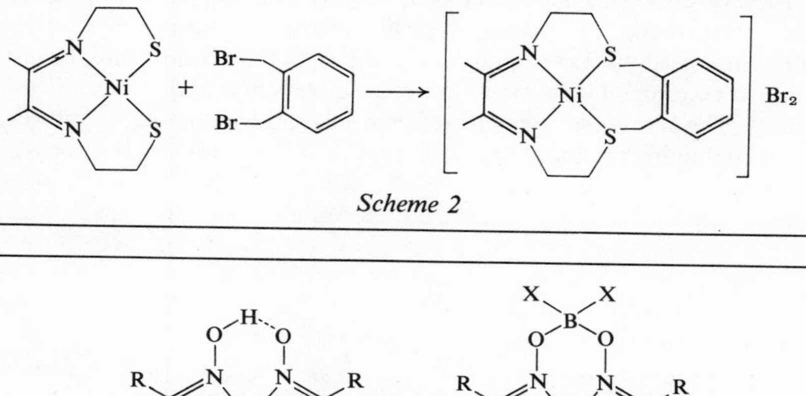

Scheme 2

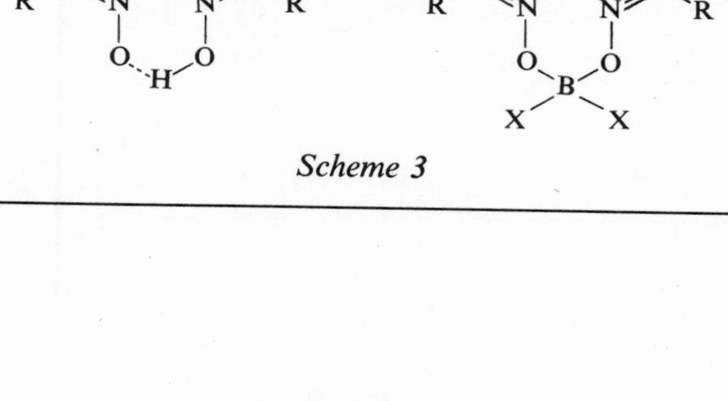

Scheme 3

individual reactions. In 1963 the self-condensation of *o*-aminobenzaldehyde in the presence of metal ions was reinvestigated[15] (the first report of these reactions was made in 1954,[16] although at that time the products of the reactions were incorrectly characterized), and species containing a tetrameric condensation product were isolated (see Scheme 4). A review article covering the syntheses of these new synthetic macrocyclic complexes was published in 1964,[17] and attention was drawn to the relationship of many of these new compounds to naturally occurring, biologically important species.

After 1964 the field of coordination chemistry of macrocyclic compounds expanded rapidly, although the scope and variety of the compounds studied was limited. The realization that macrocyclic compounds may behave as models for important natural products provided the stimulus for the expansion; however, this realization also resulted in a narrow and somewhat restricted expansion.

In 1967 a new series of macrocyclic compounds which have the ability to function as complexing agents was reported by Pedersen,[18] who synthesized a number of cyclic polyethers or "crown" compounds with a variety of ring sizes, a number of ether oxygens, and substituent groups. The ability of these compounds to strongly coordinate to alkali metal and alkaline earth ions resulted in a broadening of the field of macrocyclic chemistry to include metal ions not previously investigated. Following the initial reports, different potential donor atoms were introduced into the crowns and a new era of macrocyclic chemistry was developed. Shortly after the crown compounds were reported, the first macropolycyclic complexing agents, or cryptands, were synthesized by Lehn and co-workers.[19] These compounds have the ability to accommodate a metal ion of suitable size and form an inclusion complex. The cryptands originally synthesized were bicyclic species; however, species containing three and four macrocycles have recently been reported.[20]

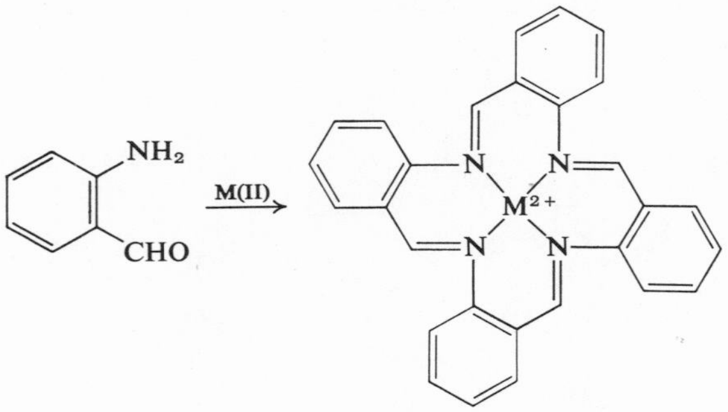

Scheme 4

Several review articles covering the chemistry of crowns and cryptands have been published; Chapter 9 deals with the coordination chemistry of these compounds and should be consulted for more details.

The development of any area of chemistry is usually accompanied by the introduction of new terminology; the area dealing with the coordination chemistry of macrocyclic compounds is no exception. Early in the 1960s it was recognized that the syntheses of many macrocyclic ligands were promoted by the presence of metal ions. The term "template effect," which reflects the controlling influence of the metal ion in a particular synthesis, was introduced by Busch.[12,17,21] Two template effects have been noted. In the kinetic template effect, it is the directive influence of the metal ion which controls the steric course of a sequence of stepwise reactions, whereas in the thermodynamic template effect, the metal ion perturbs an existing equilibrium in an organic system and the required product is produced, often in high yield, as its metal complex. Considerable use has been made of these effects in *in situ* reactions of macrocyclic complexes (see Chapter 2). The term "macrocyclic effect" was introduced by Cabbiness and Margerum in 1969[22] to account for the greater thermodynamic stability of complexes containing macrocyclic ligands when compared with those of nonmacrocyclic ligands of similar structure. Although the greater stability is an experimental fact, a great deal of discussion and controversy has arisen over its thermodynamic origin. The macrocyclic effect and an extension known as the "cryptate effect"[23] are discussed in more detail in Chapter 3. The phrase "multiple juxtapositional fixedness" was first introduced in an article by Busch and co-workers[24] to account for the enhanced stability of macrocyclic complexes in terms of their inertness toward substitution, even in the presence of strong acids. A macrocyclic ligand cannot dissociate from a metal ion by a replacement (or substitution) process similar to that which occurs for noncyclic ligands since the structure of the ring does not provide an "end" at which the successive, stepwise removal process may be initiated. Thus the increased stability of macrocyclic complexes compared with similar nonmacrocyclic complexes must be explained by consideration of both thermodynamic and kinetic effects (see reference 24 and Chapter 3).

In concluding this short section on the historical aspects of the coordination chemistry of macrocyclic compounds it should be noted that the research conducted from the mid-1960s to date has concentrated on the synthesis, characterization, and reactions of compounds containing macrocyclic ligands and metal ions which are similar to, and thus may serve as models for, compounds which are known to have significant biological roles (see Chapter 10 for a discussion of natural product model systems). It is anticipated that this emphasis on model systems will continue during the forthcoming years; however, it is also anticipated that new uses and applications of coordination compounds containing macrocyclic ligands will be developed, and that the whole field of macrocyclic chemistry will continue to grow at an increasing pace.

2.3. Abbreviations of Macrocyclic Compounds

Since the development of the field of chemistry dealing with macrocycles and coordination compounds containing macrocyclic ligands, authors have sought simple abbreviations for the macrocycles which may be used in place of their lengthy and cumbersome systematic names. Originally, trivial names, often related to the name of the investigator or the research group who first synthesized the compound, or nonsystematic abbreviations based on the structure or complete systematic name of the macrocycle, were adopted (see, for example, reference 25). However, as the number and complexity of the synthesized macrocycles increased, it became apparent that a systematic abbreviation scheme was necessary. Curtis[10] introduced a systematic approach for the determination of a macrocycle abbreviation based on the ring size and degree of unsaturation. Busch[26,27] modified and extended the scheme of Curtis: the principle modification was the use of locants for the positions of unsaturated linkages; the extensions involved the inclusion of the type and position of donor atoms and substituents into the abbreviation, and also considered fully saturated macrocycles. However, systems with fused rings, for example, macrocycles derived from 2,6-diacetylpyridine or from the self-condensation of o-aminobenzaldehyde, were not included in the modified scheme and were still referred to by trivial names.

During subsequent years, the use of abbreviations not based on the systematic approach continued, and although the modified scheme of Busch was further extended to include anionic macrocycles and those containing fused rings, ambiguities still existed, largely because a set of rules for the systematic derivation of a macrocycle abbreviation had not been drawn up. The abbreviations used in this book are derived by the application of a set of rules based on extensions of Busch's modified scheme; it is hoped that if these rules are adopted by scientists working in the field of macrocyclic chemistry and are applied to the derivation of a macrocycle abbreviation, then the confusion which has been prevalent in this area can be eliminated. The rules which follow concern abbreviations for "two-dimensional" macrocycles. The cryptands are not included, although it seems possible to extend the principles of this approach to include at least some of these "three-dimensional" macrocycles.

Rules for Derivation of an Abbreviation of a Macrocycle

Since the primary focus of interest in a macrocyclic ligand is the macrocyclic ring of atoms containing the ligating atoms, the abbreviation will be based on a macromonocycle, regardless of the type of nomenclature used to name the complete ring system. The following set of rules is proposed for the derivation of an abbreviation for a macrocycle.*

1. The size of the macromonocyclic ring is expressed by an Arabic number enclosed in square brackets, e.g., [14], [15], etc.

2. A term denoting unsaturation (if any) follows the size designation. Unsaturation is denoted by the usual nomenclature terms, e.g., "ene," "diene," "triene," "tetraene," etc., preceded by the appropriate locants (see rule 4 below). If no unsaturation is present, the term "ane" is used.

3. The ligating hetero atoms are identified after the unsaturation designation, and indicated by their element symbols; if more than one kind of ligating atom is present, the atoms are cited in alphabetic order. The number of each kind of ligating atom is indicated by a subscript; the position being indicated by a preceding locant (see rule 4 below). Nonligating hetero atoms are identified in parentheses after the ligating hetero atoms in the same manner as above.

4. The numbering of the macromonocycle begins with a hetero atom of highest priority, i.e., one that occurs earliest in the following list: O, S, Se, N, P, As, Sb, etc. [see *Chemical Abstracts Index Guide*, Vol. 76, p. 128 (1972)] and proceeds in *either* direction so that (i) the lowest set of locants for all hetero atoms is obtained, e.g., 1,2,6 is lower than 1,3,4; (ii) the hetero atoms of highest priority have lowest locants; and (iii) sites of unsaturation have lowest locants.

5. Simple substituents attached to a simple macromonocycle are cited in front of the abbreviation for the macrocycle, in order conforming to the numbering order, using standard abbreviations such as Me for methyl, Et for ethyl, Bzl for benzyl, Bz for benzoyl, Ph for phenyl, and Ac for acetyl where possible, preceded by appropriate locants (see rule 4 above). Other substituents may be denoted by usual formula representations or names, such as COOH, C_6H_{11}, oxo, etc.

6. Stereoisomers of a macromonocycle are identified by using the symbols *ms* for *meso* and *rac* for racemic, these terms being applied to the orientation of substituents in "equivalent" positions within the macrocycle. These equivalent positions are identified by the use of locants (in parentheses). Substituents (other than hydrogen) at these positions are identified as in rule 5 above. Although it is recognized that the use of the terms meso and racemic is not in agreement with their accepted definitions given in the IUPAC organic nomenclature rules, the practice of using them in the manner described above is continued throughout this text in order to make reference to the original literature as convenient as possible. Some comments and suggestions for modification of the rule concerning isomerism are offered on p. 8.

7. An anionic macrocyclic ligand is indicated by the addition of "ato"

---

* The author wishes to express his appreciation to Dr. Warren H. Powell of Chemical Abstracts Service for helpful comments and discussions concerning these rules.

to the abbreviation, followed by the charge designated by a Ewens–Bassett number. (The charge designation is often not used in metal complexes containing metal ions in common oxidation states and where the charge can readily be determined from the number and type of counterions.)

8. For macropolycyclic ligands, rings fused to the macromonocycle are described as "substituents" to the macromonocycle, even if attached at more than two points, using locant sets to describe the points of fusion. Primed locants can be used when necessary to differentiate between different "fused" rings. These "substituents" are designated by abbreviations such as bzo for benzo (or *o*-phenylene), pyo for pyridine, etc. Unsaturation which is common to both the fused ring and the macrocyclic ring is included in the abbreviation as described in rule 2 above.

The rules outlined above provide a systematic scheme for the derivation of a macrocycle abbreviation. However, the application of these rules often gives rise to a lengthy and cumbersome abbreviation for a particular macrocycle. This abbreviation may be simplified in many cases, if no ambiguity exists, by the elimination of locants for substituents, unsaturation and donor atoms, and the charge designation for anionic macrocycles. The examples labeled **1–30** show, for a particular macrocycle, the complete, systematic abbreviation which is obtained by application of the above set of rules, a "simplified" abbreviation, and in some cases a trivial name (in parentheses) which has been extensively used as an abbreviation.

It was noted in rule 6 (p. 7), that the terms meso and racemic when applied to macrocyclic compounds have not been used in the manner in which they are defined in the IUPAC organic nomenclature rules (see rules E-4.5 and E-4.7). It is possible that some molecules (as a whole) may be meso compounds according to the organic rules, and thus they may be identified by the prefix *meso*; for example, **25** is a *meso* compound and may be identified as *meso*-[2,6-Me_2-3,4,5-pyo-[14]-3-ene-1,4,7,11-N_4].

If stereochemical positions are to be identified in these abbreviations, they may be identified by the application of other IUPAC organic rules, e.g., rule E-2.3.2 states "When one substituent and one hydrogen atom are attached at each of two positions of a monocycle, the steric relations of the two substituents are expressed as *cis* or *trans*, followed by a hyphen and placed before the name of the compound." If this rule is applied to **25**, the abbreviation will be *meso*-[(*cis*-2,6)-Me_2-3,4,5-pyo[14]-3-ene-1,4,7,11-N_4], or more simply as (*cis*-2,6)-Me_2-3,4,5-pyo[14]-3-ene-1,4,7,11-N_4 since the *meso* term may be omitted if the steric relation of the methyl groups is denoted in the abbreviation. The situation is more difficult if the macromonocycle contains one substituent and one hydrogen atom attached at each of more than two positions; in this case, according to rule E-2.3.3 of the IUPAC organic nomenclature rules, the steric relations of the substituents are expressed by adding *r* (for reference substituent) followed by a hyphen, before the locant of

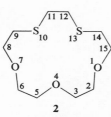

1

[14]ane-1,4,8,11-N₄
[14]aneN₄
(cyclam)

2

[15]ane-1,4,7-O₃-10,13-S₂
[15]aneO₂S₂

3

[11]ane-1,4,7-N₃
[11]aneN₃

4

[11]ane-1,4,8-N₃
[11]aneN₃

5

3-Me-[9]ane-1,4,7-N₃
Me[9]aneN₃

6

1,4,7,10-Bzl₄-[12]ane-1,4,7,10-N₄
N-bzl₄[12]aneN₄
(tbcyclen)

7

2,3,9,10-Me₄-[14]-1,3,8,10-tetraene-
1,4,8,11-N₄
Me₄[14]tetraeneN₄
(TIM)

8

2,3-Me₂-[14]-1,3-diene-
1,4,8,11-N₄
Me₂[14]dieneN₄
(DIM)

9

11,13,13-Me₃-[13]-10-ene-
1,4,7,10-N₄
Me₃[13]eneN₄

10

6,7,13,14-Me₄-[14]-5,7,12,14-tetraene-
1,5,8,12-N₄-(2,4,9,11-N₄)
Me₄[14]tetraeneN₄(N₄)

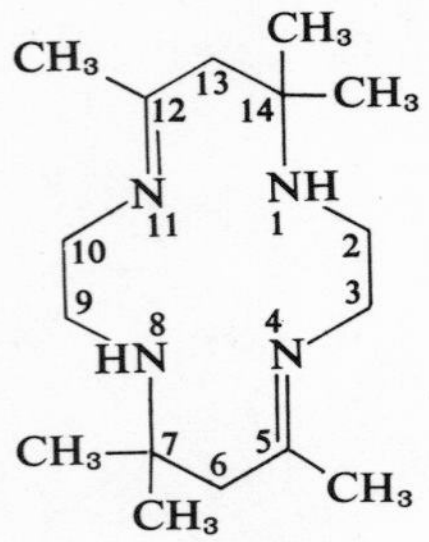

11

5,7,7,12,14,14-Me$_6$-[14]-4,11-diene-
1,4,8,11-N$_4$
Me$_6$[14]4,11-dieneN$_4$
(1,7-CT)

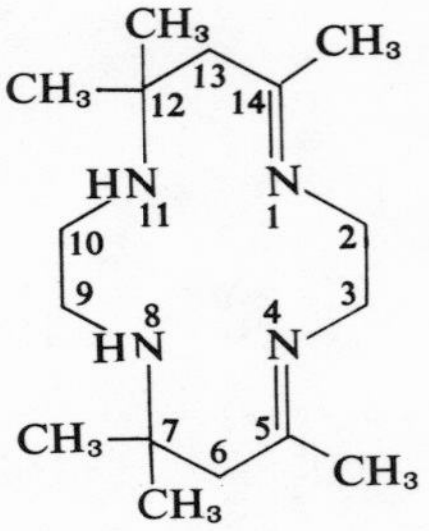

12

5,7,7,12,12,14-Me$_6$-[14]-4,14-diene-
1,4,8,11-N$_4$
Me$_6$[14]4,14-dieneN$_4$
(1,4-CT)

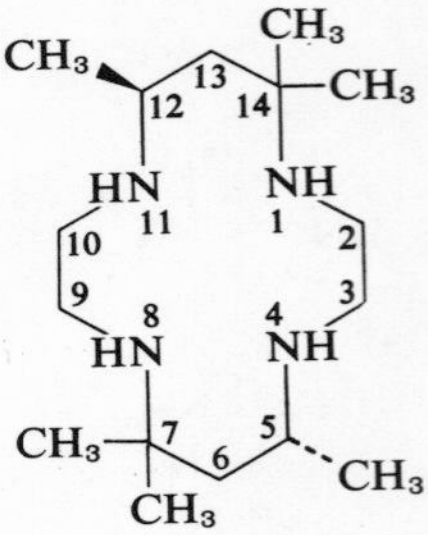

13

ms-(5,12)-7,7,14,14-Me$_6$-[14]ane-
1,4,8,11-N$_4$
ms-(5,12)-Me$_6$[14]aneN$_4$
(tet *a*)

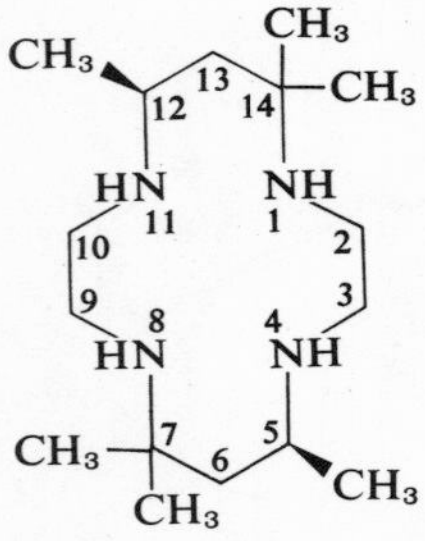

14

rac-(5,12)-7,7,14,14-Me$_6$-[14]ane-
1,4,8,11-N$_4$
rac-(5,12)-Me$_6$[14]aneN$_4$
(tet *b*)

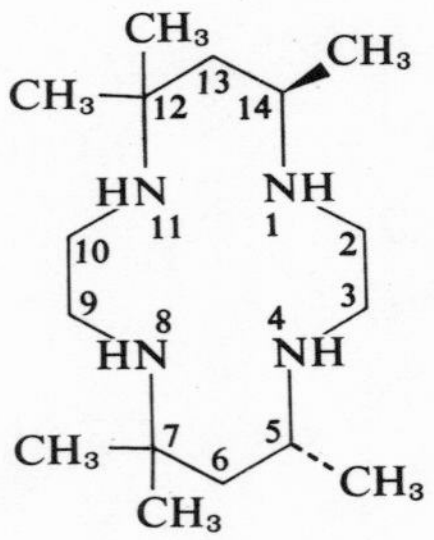

15

rac-(5,14)-7,7,12,12-Me$_6$-[14]ane-
1,4,8,11-N$_4$
rac-(5,14)-Me$_6$[14]aneN$_4$
(tet *c*)

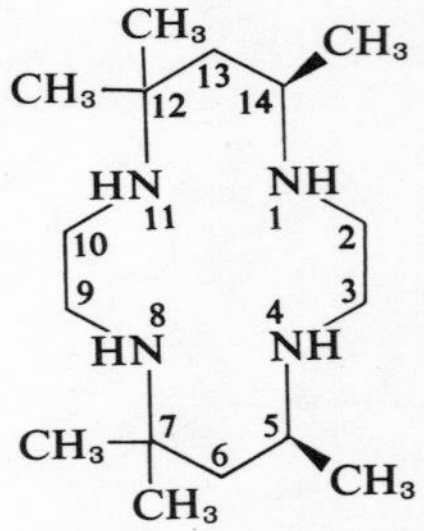

16

ms-(5,14)-7,7,12,12-Me$_6$-[14]ane-
1,4,8,11-N$_4$
ms-(5,14)-Me$_6$[14]aneN$_4$
(tet *d*)

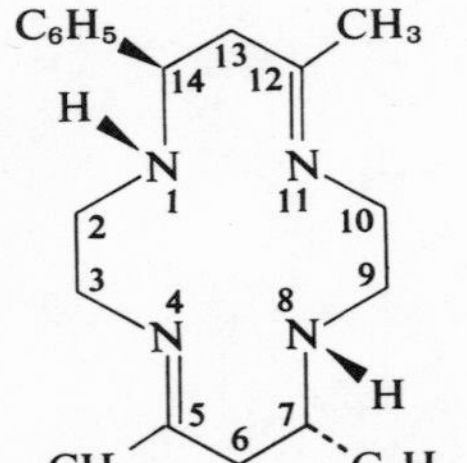

17

When coordinated:
rac-(1,8)-5,12-Me$_2$-*ms*-(7,14)Ph$_2$-[14]-
4,11-diene-1,4,8,11-N$_4$
rac-(1,8)-*ms*-(7,14)Ph$_2$Me$_2$[14]dieneN$_4$
(*C-meso-N-rac* isomer)

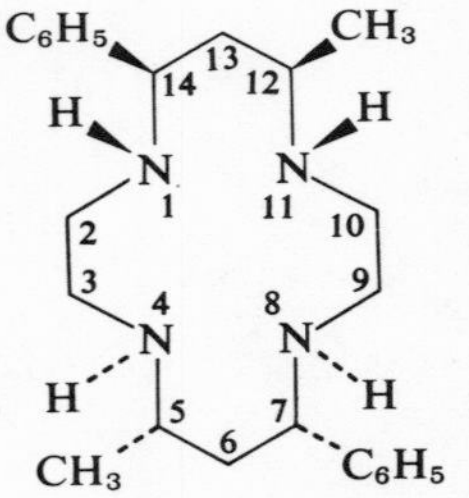

18

When coordinated:
ms-(1,8)-*ms*-(4,11)-*ms*-(5,12)Me$_2$-*ms*-(7,14)Ph$_2$-
[14]ane-1,4,8,11-N$_4$
ms-(1,8)-*ms*-(4,11)-*ms*-(5,12)Me$_2$-
ms-(7,14)Ph$_2$-[14]aneN$_4$
(*C-meso-C-meso-N-meso-
N-meso* isomer)

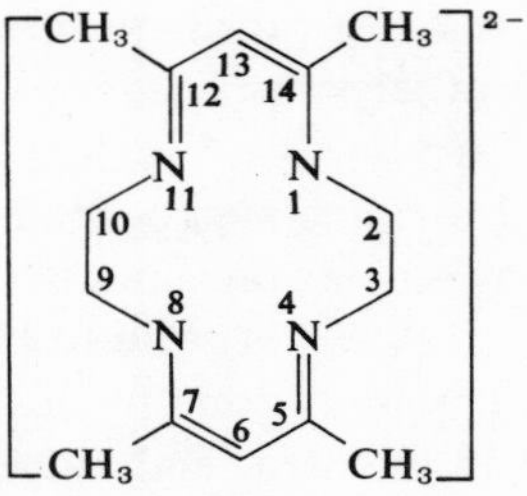

19

5,7,12,14-Me₄-[14]-4,6,11,13-tetraenato(2−)-
1,4,8,11-N₄
Me₄[14]tetraenato(2−)N₄ or
Me₄[14]tetraenatoN₄

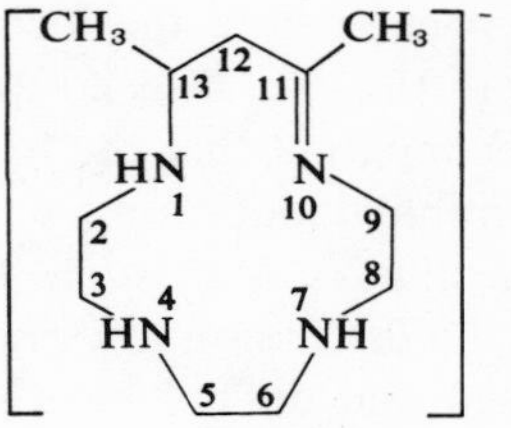

20

11,13-Me₂-[13]-10,12-dienato(1−)-
1,4,7,10-N₄
Me₂[13]dienato(1−)N₄ or
Me₂[13]dienatoN₄
(AT)

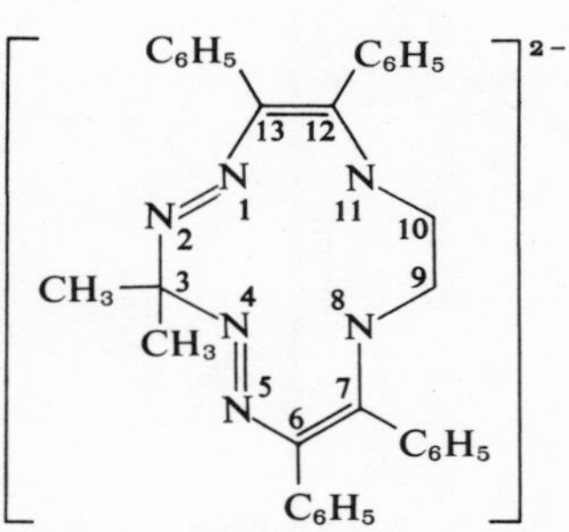

21

3,3-Me₂-6,7,12,13-Ph₄-[13]-1,4,6,12-tetraenato(2−)-
1,4,8,11-N₄-(2,5-N₂)
Me₂Ph₄[13]tetraenato(2−)N₄(N₂) or
Me₂Ph₄[13]tetraenatoN₄(N₂)

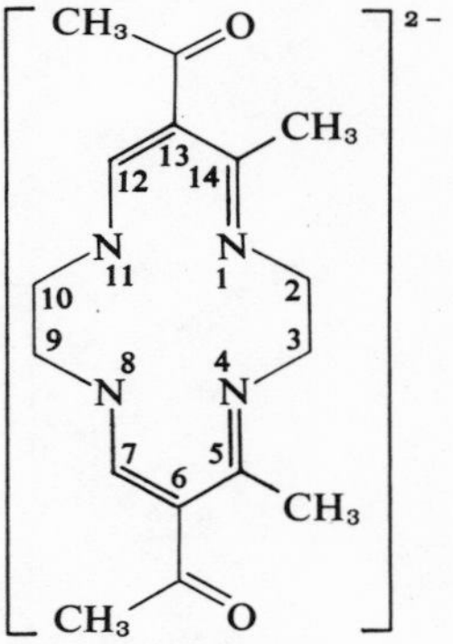

22

5,14-Me₂-6,13-Ac₂-[14]-4,6,12,14-tetraenato(2−)-
1,4,8,11-N₄
Me₂Ac₂[14]tetraenato(2−)N₄ or
Me₂Ac₂[14]tetraenatoN₄

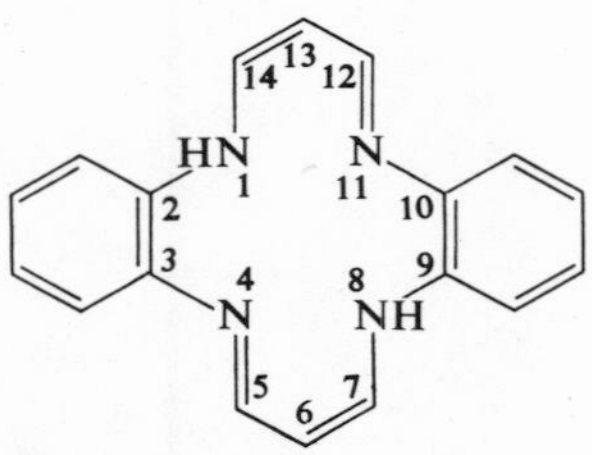

23

2,3:9,10-Bzo₂-[14]-2,4,6,9,11,13-hexaene-
1,4,8,11-N₄
Bzo₂[14]hexaeneN₄

In dianionic form:
2,3:9,10-Bzo₂-[14]-2,4,6,9,11,13-hexaenato(2−)-
1,4,8,11-N₄
Bzo₂[14]hexaenato(2−)N₄ or
Bzo₂[14]hexaenatoN₄

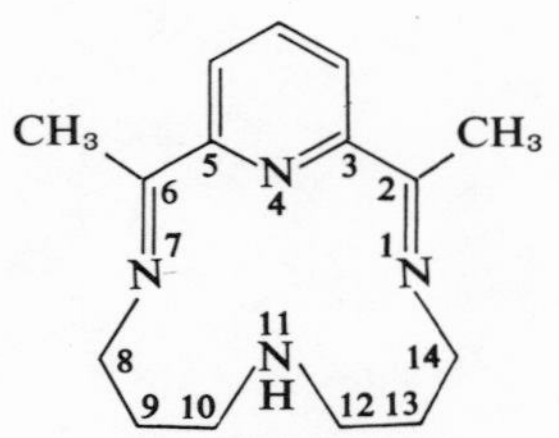

24

2,6-Me₂-3,4,5-pyo-[14]-1,3,6-
triene-1,4,7,11-N₄
Me₂pyo[14]trieneN₄
(CR)

the lowest numbered of the substituents and *c* or *t* (as appropriate), followed by a hyphen, before the locants of the other substituents to express their relation to the reference substituent.

It is apparent that the problem of describing stereochemical configurations in abbreviations for macrocycles is a complex one, and one that requires a great deal of further study. At this time it does not seem to be appropriate to take the discussion any further than has been done in this section; it is recognized that difficulties are prevalent in the area, and indeed an IUPAC committee has been established to consider the area of nomenclature and abbreviations of macrocyclic compounds. The rules outlined earlier in this section are applicable to most systems and have been used by and communi-

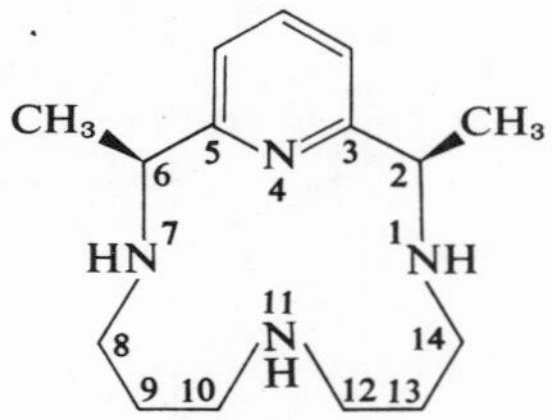

25

ms-(2,6)-Me$_2$-3,4,5-pyo-[14]-3-ene-
1,4,7,11-N$_4$
ms-Me$_2$pyo[14]eneN$_4$
(*ms*-CRH)

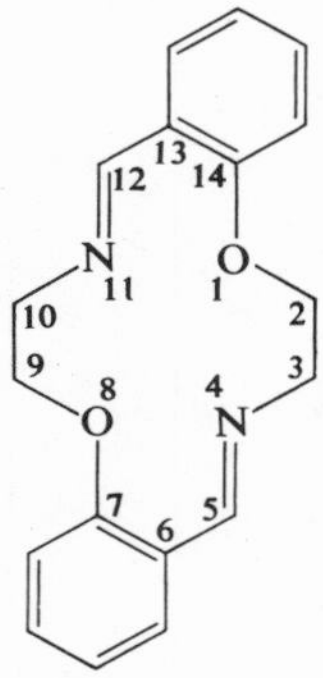

26

6,7:13,14-Bzo$_2$-[14]-4,6,11,13-tetraene-
4,11-N$_2$-1,8-O$_2$
Bzo$_2$[14]tetraene-4,11-N$_2$-1,8-O$_2$ or
Bzo$_2$[14]tetraeneN$_2$O$_2$

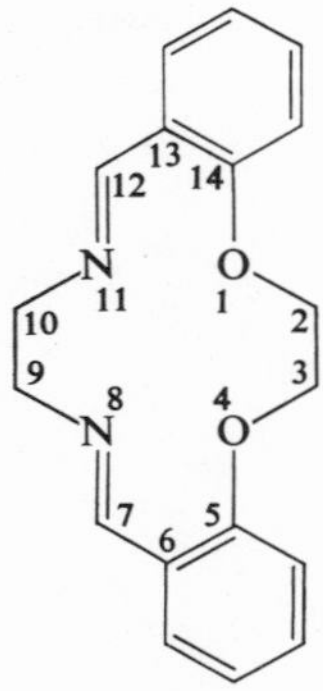

27

5,6:13,14-Bzo$_2$-[14]-5,7,11,13-tetraene-
8,11-N$_2$-1,4-O$_2$
Bzo$_2$[14]tetraene-8,11-N$_2$-1,4-O$_2$ or
Bzo$_2$[14]tetraeneN$_2$O$_2$

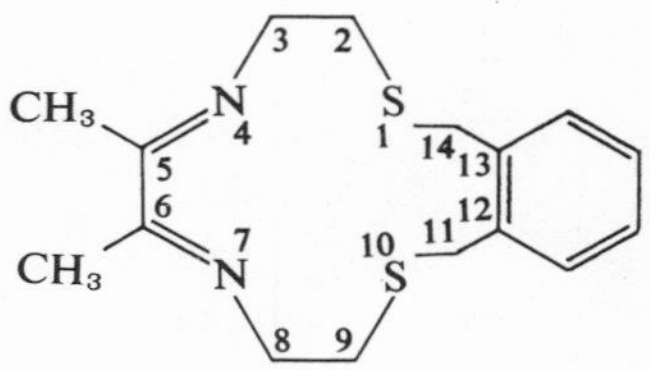

28

5,6-Me$_2$-12,13-bzo-[14]-4,6,12-triene-
4,7-N$_2$-1,10-S$_2$
Me$_2$bzo[14]trieneN$_2$S$_2$

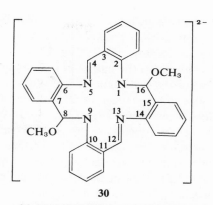

29

3,4:7,8:11,12:15,16-Bzo$_4$-[16]-1,3,5,7,9,11,13,15-
octaene-1,5,9,13-N$_4$
Bzo$_4$[16]octaeneN$_4$
(TAAB)

30

2,3:6,7:10,11:14,15-Bzo$_4$-8,16-(CH$_3$O)$_2$-[16]-
2,4,6,10,12,14-hexaenato(2−)-1,5,9,13-N$_4$
Bzo$_4$(CH$_3$O)$_2$[16]hexaenato(2−)N$_4$ or
Bzo$_4$(CH$_3$O)$_2$[16]hexaenatoN$_4$
(TAAB(OMe)$_2$)

cated between scientists working in the area. Thus the described rules have the advantage of familiarity even though they may have the disadvantage of some inaccuracy. It is anticipated that resolution of the problems encountered in deriving an abbreviation will take place during the next few years, although unfortunately this may be at the expense of simplicity. This author hopes that in resolving these problems we shall not lose sight of the goal of deriving and using an abbreviation, that is, to provide a simple and readily communicable reference for a complex molecule.

Abbreviations for macrocyclic polyethers may readily be obtained by application of the above set of rules, e.g., for **31**, the systematic abbreviation is 2,3:11,12-bzo$_2$-[18]-2,11-diene-1,4,7,10,13,16-O$_6$, or, in simplified form, bzo$_2$[18]dieneO$_6$. However, since the discovery of these compounds by Pedersen[18] they have been known as "crown ethers," and simple abbreviations based on the number and type of substituent groups, macrocyclic ring size, and number of potential donor (or hetero) atoms have been used for identification. For example, the commonly used abbreviation for **31** is dibenzo-18-crown-6, or, in shortened form, DB18C6. Chapter 9 lists some commonly used abbreviations for a variety of crown compounds and outlines a scheme for the derivation of these abbreviations. The scheme has many

31

shortcomings, however, and ambiguities may readily arise. For example, the scheme does not enable isomers such as those which may occur in mixed ether–thioether species (see **2**) to be distinguished, nor does it take into consideration the nature of the linkages between hetero atoms etc. Application of the systematic abbreviation scheme outlined above would provide a unique abbreviation for all compounds, and eliminate the ambiguities which arise by using the current scheme. In spite of this justification for the application of the systematic scheme to all macrocycles under discussion in this book, commonly used abbreviations rather than the systematic abbreviations will be used in chapters dealing with crown compounds. This decision was reached so that the material in these chapters may easily be compared with that in the original literature. It is this author's hope that scientists working in the area of crown compound chemistry will begin to use the systematic abbreviations, so that it will be necessary to use only one scheme in order to uniquely describe a macrocycle in an abbreviated form.

The systematic abbreviation scheme, at this time, does not include macropolycyclic compounds, such as cryptands, clathrochelates, etc. However, it does seem possible to extend the system so as to handle many of the "three-dimensional" macrocycles of this type. Trivial names, based on the number of cycles and the number and distribution of hetero atoms on each hydrocarbon strand, have been derived for the cryptands. For example, **32** is

32

described as a [2]-cryptand ([2] indicating a bicyclic compound) and is given the trivial name cryptand 222 or more simply C222, the numbers relating to the numbers of oxygen atoms on each strand of the bicyclic ligand. (Chapter 9 illustrates the derivation of these trivial names for a variety of cryptands.) Obviously the use of trivial names based solely on the criteria above will result in ambiguities. A systematic abbreviation can be derived for the cryptands and clathrochelates by considering a strand as a substituent (at two points) to a basic macromonocycle, an abbreviation for the latter being derived as described earlier in this chapter. A systematic abbreviation could also be derived for many cryptands by describing the system containing the ligating atoms as bicyclic or polycyclic. However, in order to facilitate easy reference to the original literature, trivial names for cryptands will be used in this book. It is hoped that a scheme for systematic derivation of a unique abbreviation for macropolycyclic compounds will shortly be developed and receive acceptance by scientists working in this most interesting area of macrocyclic chemistry.

2.4. Units

The International System (SI) of units is, with some exceptions, used throughout this book in an attempt to produce consistency of reported data. A great deal of physical data on macrocycles and coordination compounds containing macrocyclic ligands has been obtained during the development of the chemistry of these species. Much of the data in a particular area has been reported in a variety of units, and attempts to correlate the data are often impeded because of the different units employed. For example, wavelengths for UV-visible and near-infrared spectra have been reported in Å, kK, cm^{-1} $m\mu$, and nm; however, throughout the following chapters all wavelengths are quoted in nanometers (nm). Bond distances, etc., reported in Å and pm in the original literature are quoted herein in picometers, (pm). Data relating to vibrational spectra are reported in cm^{-1} since all the data are quoted in these units in the original literature. Thermochemical and thermodynamic data are reported in units related to calories (cal) or joules (J) in the original literature; in the following chapters calories will be used since these are the units employed in the vast majority of the original reports. Some conversion factors which may be useful for correlation of the data in this book with those in the chemical literature follow: $1 \text{ Å} = 100 \text{ pm}$ $(= 0.1 \text{ nm})$; $1 \text{ m}\mu = 1 \text{ nm}$; $1 \text{ kK} = 10^3 \text{ cm}^{-1} = 10^{-3} \text{ cm}$; $1 \text{ nm} = 10^{-9} \text{ m} (= 10^{-7} \text{ cm})$; $1 \text{ cal} = 4.1840$ joule (J).

2.5. Chapter Layout

The chapters which comprise this book deal with the coordination chemistry of macrocyclic compounds. Various aspects of this chemistry are discussed and, while the chapters as a whole comprise an in-depth survey and reflect the current "state of the art" of the field, each chapter is a complete "stand alone" unit. It was anticipated that a chapter would be read (or consulted) for information dealing with a particular aspect of macrocyclic chemistry, and the breakdown of the complete material into individual chapters was made in an attempt to make this information readily available without an extensive search. Some overlap between the chapters is inevitable; indeed it is desirable to provide continuity; however, this overlap has been reduced to a minimum. The chapter layout was designed so that the material proceeds from synthesis of macrocyclic compounds through physical properties and reactivity to the relation of these compounds to natural products. Although the chemistry of natural product macrocyclic compounds is not discussed in great detail (this has been the subject of several recent texts), the relevance of synthetic macrocyclic compounds to the natural products is a continuous theme throughout all of the chapters.

References

1. R. P. Linstead and M. Whalley, *J. Chem. Soc.* **1952**, 4839 (1952).
2. G. E. Ficken and R. P. Linstead, *J. Chem. Soc.* **1952**, 4846 (1952).
3. J. A. Elvidge and R. P. Linstead, *J. Chem. Soc.* **1952**, 5008 (1952).
4. J. A. Elvidge and J. H. Golden, *J. Chem. Soc.* **1957**, 700 (1952).
5. J. Van Alphen, *Rec. Trav. Chim. Pays-Bas* **55**, 835 (1936).
6. B. Bosnich, C. K. Poon, and M. L. Tobe, *Inorg. Chem.* **4**, 1102 (1965).
7. B. Bosnich, M. L. Tobe, and G. A. Webb, *Inorg. Chem.* **4**, 1109 (1965).
8. N. F. Curtis, *J. Chem. Soc.* **1960**, 4409 (1960).
9. N. F. Curtis and D. A. House, *Chem. Ind.* **42**, 1708 (1961).
10. N. F. Curtis, *Coord. Chem. Rev.* **3**, 3 (1968).
11. M. C. Thompson and D. H. Busch, *Chem. Eng. News* (Sept. 17), 57 (1962).
12. M. C. Thompson and D. H. Busch, *J. Am. Chem. Soc.* **86**, 3651 (1964).
13. G. N. Schrauzer, *Chem. Ber.* **95**, 1438 (1962).
14. F. Umland and D. Thierig, *Angew. Chem.* **74**, 388 (1962).
15. G. A. Melson and D. H. Busch, *Proc. Chem. Soc.* **1963**, 223 (1963).
16. G. L. Eichhorn and R. A. Latif, *J. Am. Chem. Soc.* **76**, 5180 (1954).
17. D. H. Busch, *Rec. Chem. Prog.* **25**, 107 (1964).
18. C. J. Pedersen, *J. Am. Chem. Soc.* **89**, 2495, 7017 (1967).
19. B. Dietrich, J.-M. Lehn, and J.-P. Sauvage, *Tetrahedron Lett.* **1969**, 2885, 2889 (1969).
20. J.-M. Lehn, J. Simon, and J. Wagner, *Nouv. J. Chim.* **1**, 77 (1977).
21. L. F. Lindoy and D. H. Busch, in *Preparative Inorganic Reactions* (W. Jolly, ed.), Vol. 6, p. 1, Interscience, New York (1971).
22. D. K. Cabbiness and D. W. Margerum, *J. Am. Chem. Soc.* **91**, 6540 (1969).
23. J.-M. Lehn and J. P. Sauvage, *J. Am. Chem. Soc.* **97**, 6700 (1975).
24. D. H. Busch, K. Farmery, V. Goedken, V. Katovic, A. C. Melnyk, C. R. Sperati, and N. Tokel, *Adv. Chem. Ser.* **100**, 44 (1971).
25. D. H. Busch, *Helv. Chim. Acta* **50**, 174 (1967).
26. V. L. Goedken, P. H. Merrell, and D. H. Busch, *J. Am. Chem. Soc.* **94**, 3397 (1972).
27. J. C. Dabrowiak, P. H. Merrell, and D. H. Busch, *Inorg. Chem.* **11**, 1979 (1972).

Synthesis of Macrocyclic Complexes

Gordon A. Melson

1. Introduction

Since the recognition of the importance of complexes containing macro-cyclic ligands, considerable effort has been directed toward the development of reliable syntheses for these interesting and important compounds. Several review articles which deal primarily with the synthesis of complexes containing synthetic macrocyclic ligands* have appeared.[1–13]

The procedures that have been described fall, generally, into three categories; first, syntheses involving complexation reactions; second, *in situ* syntheses; and third, syntheses involving modification of the macrocyclic ligand and/or the metal ion. It should be noted here that many compounds may be synthesized by procedures from more than one category. In this section, procedures will be discussed in general terms and both advantages and disadvantages of each approach will be outlined. Details concerning specific syntheses will follow in later sections, discussed according to ligand type.

The first category involves a complexation reaction between a presynthesized macrocyclic ligand and the metal ion in solution. Advantages of this approach are that the macrocyclic ligand may be isolated, purified, and characterized before the synthesis of the complex. Purification and characterization of an organic ligand is often more easily accomplished than purification and characterization of a metal complex. The usual volatility, diamagnetism, and solubility of the ligand in less polar solvents makes spectroscopic

---

* See also Chapters 7, 8, and 9 for review articles dealing with phthalocyanines, porphyrins, crowns, and cryptands.

Gordon A. Melson · Department of Chemistry, Virginia Commonwealth University, Richmond, Virginia 23284.

techniques such as mass spectrometry and NMR spectroscopy, which are commonly used for characterization, more amenable to the organic ligand than to the complex. Thus, after purification and characterization, the complexation reaction should be relatively free of side reactions which may lead to impurities in the desired product. Another advantage of this approach is that changes in the macrocyclic ligand upon complexation may be readily detected if the physical properties of the ligand can be compared with those of the complex. Conclusions concerning the mode of bonding of the ligand to the metal ion may also be drawn. Interest in complexes containing tetraaza macrocycles led Holm and co-workers[14-18] and Honeybourne[19] to develop general syntheses for macrocycles of varying ring size (14–16 members) and degree of unsaturation. Reaction between the ligands and metal acetates in ethanolic solution enabled a wide variety of macrocyclic complexes to be synthesized. One potential disadvantage of the complexation reaction approach is that the synthesis of the organic macrocycle often results in low yields of the desired product. In attempts to increase the yield of the macrocycle and reduce side reactions such as polymerizations, the final ring closure reaction in a stepwise synthesis is often carried out under conditions of high dilution. The tetramine 1,4,8,11-tetraazacyclotetradecane, cyclam, [14]aneN$_4$ was, until recently, most conveniently prepared by the condensation of 1,3-dibromopropane with 1,4,8,11-tetraazaundecane in ethanolic potassium hydroxide solution under conditions of high dilution.[20,21] In spite of the conditions, the yield of [14]aneN$_4$ was very low ($\sim 5\%$). The more general synthetic procedure developed for cyclic amines by Richman and Atkins[22] produces higher yields of [14]aneN$_4$ via the tosylated derivative, and an *in situ* synthesis of the ligand has recently been developed by Barefield *et al.*[23,24] The yield of [14]aneS$_4$, the sulfur analog of [14]aneN$_4$, may be significantly improved by carrying out the condensation between 1,3-dibromopropane and 1,4,8,11-tetrathioundecane in high dilution.[25,26]

In spite of problems which may be encountered during the synthesis of the macrocyclic ligands, the complexation reaction approach has been extensively employed for the synthesis of complexes containing synthetic macrocycles, in particular, cyclic tetramines, cyclic polyethers (crowns), and macrobicyclic ligands such as cryptands (see Chapter 9 for a discussion of complexes containing crowns and cryptands). Perhaps the greatest advantage of this approach to synthesis is the generality of the process. In principle, the number of complexes which may be synthesized from a ligand and a metal ion is restricted only by the nature of the reactants. Admittedly this is a gross simplification of the process; however, this approach is considerably more general in nature than the *in situ* process, which is more metal-ion specific and thus may be more limited in scope. For example, the isolation of Me$_6$[14]-dieneN$_4$ ligands[27-29] has enabled many more complexes containing these ligands and their reduced derivatives[30] to be synthesized than could have been obtained via *in situ* processes.

The second category involves the synthesis of the macrocyclic ligand in the presence of the metal ion. Such *in situ* reactions have been widely used in syntheses of many synthetic and "classical" macrocycles such as porphyrins, corrins, etc. The presence of a metal ion in certain cyclization reactions has been shown to markedly increase the yield of the cyclic product. The *in situ* synthesis of [14]aneN$_4$ described by Barefield[23,24] involves, in the first step, the use of nickel(II) ions in a ring closure reaction between 1,5,8,12-tetraazadodecane and glyoxal. After reduction of the condensation product with hydrogen, a yield of the free ligand [14]aneN$_4$ in excess of 65% may be obtained. This should be compared with the low yield of this ligand obtained from procedures reported earlier.[20,21] The importance of the metal ion in the ring closure reaction is demonstrated by allowing the amine and glyoxal to react together before addition of the nickel ions. No cyclic product is obtained even after several days. The generality of the above approach for the synthesis of macrocyclic tetramines has recently been demonstrated.[31] The function of the metal ion in *in situ* reactions has been the source of much discussion since the development of macrocyclic syntheses of this type.[1-6,12] There is no doubt that the metal ion plays an important role in directing the steric course of the reaction. Lindoy and Busch[6] have discussed this role in terms of coordination template effects. If the directive influence of the metal ion controls the steric course of a sequence of stepwise reactions, the kinetic template effect is considered to be operative. If the metal ion perturbs an existing equilibrium in the organic system by preferential coordination with one of the components, and by this coordination the required product is obtained as its metal complex, then the thermodynamic (or equilibrium) template effect is considered to be operative. Most *in situ* syntheses of macrocyclic ligands probably involve the kinetic template effect; however, it is not always possible to be definite about the role of the metal ion and to conclude which of the two effects is operating (or more important) in a specific synthesis. The first example of the deliberate synthesis of a new synthetic macrocycle was an *in situ* process described by Thompson and Busch.[32] In the complex 2,3-pentanedione-*bis*(mercaptoethylimino)nickel(II), **1**, the mercaptide groups are coordinated to the nickel ion in such positions that reaction with a suitable difunctional alkyl halide, e.g., α,α'-dibromo-*o*-xylene, causes ring closure to occur. Thus the nickel(II) complex **2** containing a new macrocyclic ligand is obtained. A study of the kinetics of the above reaction has been reported by Blinn and Busch.[33] This study showed that the reaction with α,α'-dibromo-*o*-xylene proceeds via a slow first step involving attack at one coordinated mercaptide followed by a second rapid, ring closure step, in contrast to the two slow consecutive steps observed for the reaction of **1** with benzyl bromide. After the first mercaptide group has reacted with α,α'-dibromo-*o*-xylene, the second mercaptide and halide are oriented such that the cyclization reaction proceeds rapidly. This control of the steric course of the reaction by the metal ion is the basis for the kinetic template effect. Many

1   **2**

macrocycle formation reactions involve condensations between amines and carbonyl-containing compounds. The anticipated synthesis of tetraaza macrocycles that may serve as models for biological molecules has provided much of the impetus to research in this area. Jäger has described a variety of reactions between complexes containing tetradentate ligands with *cis*-oriented CO groups and diamines.[34-39] Macrocyclic complexes of general

3

formula **3** are obtained: $M = Ni, Cu$; $R_1 = H$; $R_2 = COCH_3, COC_6H_5$ $COOC_2H_5$; $R_3 = CH_3, C_6H_5$; $X, Y = -(CH_2)_2-, -(CH_2)_3-$. Bamfield[40] extended the work of Jäger to systems where R_2 and R_3 are part of a cyclo-

4   **5**

hexane ring. If the cyclohexane ring contains a CO group, cyclization takes place with ethylenediamine to give a macrocyclic ligand with X = Y = —$(CH_2)_2$—. More recently, Melson and co-workers[41,42] reported ring closure reactions between **4** (R = C_6H_5, CH_3) and ethylenediamine; the macrocyclic complexes **5** were isolated. Reactions between complexes containing bidentate or tetradentate amines and carbonyl compounds have also been shown to produce tetraaza macrocyclic ligands. The reaction between *tris*-ethylenediaminenickel(II) perchlorate and acetone was first studied by Curtis,[43] who later showed that complexes containing the isomeric ligands $Me_6[14]4,11$-dieneN_4, **6**, and $Me_6[14]4,14$-dieneN_4, **7**, were produced.[44,45]

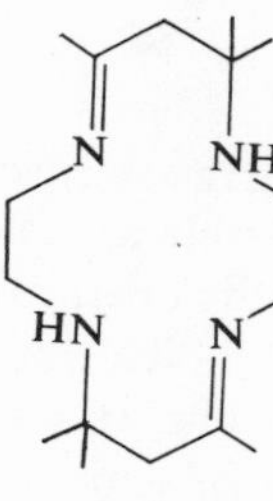

6 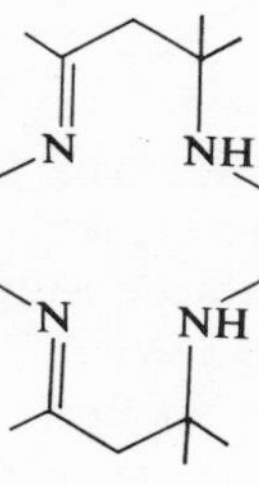   **7**

When triethylenetetramine complexes are employed in a similar reaction with acetone, a single ring closure step produces the macrocyclic ligand $Me_3[13]$-eneN_4, **8**, coordinated to the metal ion. The synthesis and characterization of complexes containing macrocyclic ligands of the above type obtained by reactions between amines and ketones have been discussed by Curtis[3] and will be described in more detail later in the chapter.

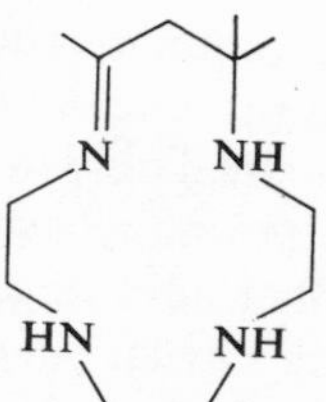

8

The above reactions are examples of macrocycle syntheses involving the isolation of intermediate complexes (usually tetradentate) followed by ring closure reactions. However, in the majority of *in situ* syntheses, the macrocyclic ligand is produced by condensation reactions between the reactants in the presence of a metal ion without the isolation of any intermediate species. A typical synthesis of this type involves the reaction between 1,3-propanediamine and 2,3-butanedione in the presence of iron(II), nickel(II), or cobalt(II) salts.[46,47] Complexes containing the tetraaza macrocycle $Me_4[14]$tetraeneN_4, **9**, can be isolated from the reaction medium. Many examples of "direct *in situ*" syntheses will be described later in the chapter. It is apparent that the

9

metal ion plays an important role in directing the steric course of reactions of this type, although in most cases it is not possible to tell conclusively whether the equilibrium or the kinetic template effect is operative or more important in the synthesis. Polarization effects, kinetic lability, thermodynamic stability, and the stereochemistry of both metal ion and reactants are some of the important factors that relate to the course of a particular reaction. More research in the area of mechanisms of macrocycle formation in the presence of metal ions should help to elucidate those factors that are important in controlling the course of the reaction and should lead to the design of new and improved synthetic procedures.

There are both advantages and disadvantages of the *in situ* approach to macrocycle synthesis. One advantage of this approach is that in many cases the yield of a particular macrocycle can be improved by the addition of a metal ion during the course of a reaction. The improved yield is usually attained by the elimination, or substantial reduction, of side reactions such as polymerizations or reactions to produce nonmacrocyclic products. Often a particular macrocycle can be obtained only in the presence of a metal ion and isolated as its metal complex, the macrocycle itself being unstable in the absence of the metal ion. Another advantage of this approach is the possibility of selectivity. By employing a metal ion with certain steric requirements, or preferences, a reaction may be controlled, and thereby a particular macrocyclic product may be obtained at the expense of an undesired macrocyclic product. (Such control of the macrocyclic product has been demonstrated in the metal-ion-controlled self-condensation of *o*-aminobenzaldehyde; see later.) Although the steric requirements of a metal ion may be used to advantage in some cases, they may be a disadvantage in others; for example, they may preclude the formation of a desired macrocyclic product. Another potential disadvantage of the *in situ* approach is that the macrocyclic product is usually coordinated to the metal ion, and its removal from the metal ion is not always possible. Thus *in situ* syntheses are less general and more metal-ion specific than syntheses involving the metal ion and a preformed macrocyclic ligand. In spite of these drawbacks and the possibility that the macrocyclic complex may be contaminated with other metal-complex impurities, the *in situ* synthesis approach has gained widespread acceptance as a powerful technique for the synthesis of macrocyclic ligands. Often the *in situ* and complexation approaches to synthesis are combined; the macrocyclic ligand is first synthesized by an *in situ* reaction, and then separated from the metal ion by a displacement process. Reactions between the free ligand and other metal ions may then be carried out in solution.

The third approach to the synthesis of macrocyclic complexes involves the modification of the ligand and/or the metal ion in the macrocyclic complex. Since this subject is dealt with in detail in Chapter 6 only a few general comments will be made here. Several macrocyclic ligands have been chemically modified while remaining coordinated to a metal ion. Most macrocyclic complexes are kinetically inert to ligand substitution, and thus the metal ion plays an important role in these "coordinated-ligand" reactions. The most extensively studied reactions are those involving oxidative dehydrogenation. Curtis[3] demonstrated the ease of interconversion of a series of complexes containing differing degrees of unsaturation. Since this demonstration, many oxidative dehydrogenations of coordinated macrocyclic ligands have been reported and a variety of reagents have been employed. The products are often metal-ion specific, a dramatic demonstration of the importance of the metal ion in controlling the reactions of a coordinated ligand. Coordination of a macrocyclic ligand to a metal ion may change the acid–base properties of the ligand. Cummings and co-workers[48,49] have demonstrated the interconversion of **10** and **11** (M = Cu, Ni) by adjustment of the pH of the solution containing these species. The relative acidity of the neutral ligand depends markedly on the metal ion and is 1000-fold less for the copper(II) complex than the nickel(II) complex. The change in basicity of

10   **11**

the amine groups in some coordinated macrocyclic ligands has enabled substituted species to be readily obtained. It has been shown that the secondary amines of [14]aneN$_4$ may be deprotonated with potassium hydroxide in dimethylsulfoxide. Reaction of the deprotonated species with methyl iodide leads to the tetra-*N*-methylated derivative.[50] It is interesting to note that the configurations of the methyl groups are different in the ligand produced by alkylation of the coordinated macrocycle from those in the ligand obtained by organic synthesis.[51] *N*-methylation, deacylation, and protonation reactions have been reported[52] for the nickel(II) complex of **3**, with $R_1 = H$, $R_2 = CH_3CO$, $R_3 = CH_3$, $X = -(CH_2)_2-$, and $Y = -(CH_2)_3-$, and electrophilic displacement reactions with several nitrogen electrophiles have been shown[53] to produce dinitro species with some coordinated macrocyclic ligands similar to **3**.

In addition to reactions involving modifications of the macrocyclic ligand system, reactions involving oxidation and reduction of the metal ion have been widely investigated.[54–60] The production of unusual oxidation states has been achieved by both chemical and electrochemical techniques, and macrocyclic ligands have been shown to be particularly effective in stabilizing a wide range of oxidation states. Chapter 6 deals with this area in detail.

The three categories into which the syntheses of complexes containing macrocyclic ligands may be conveniently placed contain many examples, and the general approaches have been widely applied. The procedure of choice for the synthesis of a particular macrocyclic complex will obviously depend upon many factors, and more than one approach to the synthesis may be applicable. The syntheses discussed in this chapter are arranged according to the number of donor atoms in the macrocyclic ligand and then subdivided by donor type.

2. Tridentate Ligands

There have been relatively few reports of complexes containing tridentate macrocyclic ligands when compared with the large number of papers dealing with complexes containing tetradentate macrocycles. However, tridentate macrocyclic ligands coordinate readily to metal ions producing stable complexes, many of which have unusual chemical and physical properties.

The syntheses of some cyclic triamines have been described[22,61] and a variety of metal complexes obtained from some of these ligands[61–65] (Table 1). Koyama and Yoshino[61] obtained the complex [Co([9]ane-1,4,7-N_3)$_2$]Br$_3$ by oxidation of a mixture of the ligand trihydrobromide and cobalt(II) chloride in basic solution. However, Barefield[62] isolated [Co([9]ane-1,4,7-N_3)(NO$_2$)$_3$] from an acetate-buffered medium containing NaNO$_2$. This complex was used as a starting material from which several cobalt(III) complexes containing one [9]ane-1,4,7-N_3 ligand were obtained. Koyama and Yoshino[61] also prepared cobalt(III) complexes with [10]ane-1,4,7-N_3 and [11]ane-1,4,8-N_3 but were unable to isolate a product from the aerial

TABLE 1. Complexes of Certain Macrocyclic Triamines, L

Name	l	m	n	Complex	Reference
([9]ane-1,4,7-N$_3$)	2	2	2	[CoL$_2$]Br$_3$	61
				[CoLBr$_3$]·H$_2$O	62
				[CoLX$_3$], X = NO$_2$, Cl	62
				[CoL(1-tmOH^a)(ClO$_4$)$_3$	62
				[CoL(1-tmOH^a)X]X$_2$·2H$_2$O, X = Cl, Br	62
				[CoL(2-tmOH^a)](ClO$_4$)$_3$	62
				[CoL(2-tmOH^a)Cl]Cl$_2$·2H$_2$O	62
				[CoL(dienOH)]Cl$_3$	62
				CuL$_2$Cl$_2$·4.5H$_2$O	63
				CuL$_2$(NO$_3$)$_2$	63
				NiL$_2$Cl$_2$·4H$_2$O	63
				NiL$_2$(NO$_3$)$_2$·H$_2$O	63
				NiL$_2$Cl$_2$·2.5H$_2$O	63
				NiL(NCS)$_2$	64
([10]ane-1,4,7-N$_3$)	2	2	3	[CoL$_2$]Br$_3$·H$_2$O	61
				NiL(NCS)$_2$	64
([11]ane-1,4,7-N$_3$)	2	2	4	NiL(NCS)$_2$	64
				CoL(NCS)$_2$	65
([12]ane-1,4,7-N$_3$)	2	2	5	NiL(NCS)$_2$	64
				[NiLCl]ClO$_4$	64
				Ni$_2$L$_3$(ClO$_4$)$_4$·4H$_2$O	64
				CoL(NCS)$_2$	65
([13]ane-1,4,7-N$_3$)	2	2	6	NiL(NCS)$_2$	64
				CoL(NCS)$_2$	65
([11]ane-1,4,8-N$_3$)	2	3	3	[CoLBr]Cl$_2$	61
				NiL(NCS)$_2$	65
([12]ane-1,4,8-N$_3$)	2	3	4	NiL(NCS)$_2$	64
				[NiLCl]ClO$_4$	64
				Ni$_2$L$_3$(ClO$_4$)$_4$·5H$_2$O	64
				CoL(NCS)$_2$	65
([13]ane-1,4,8-N$_3$)	2	3	5	NiL(NCS)$_2$	64
				[NiLCl]ClO$_4$	64
				CoL(NCS)$_2$	65
([14]ane-1,4,8-N$_3$)	2	3	6	NiL(NCS)$_2$	64
([12]ane-1,5,9-N$_3$)	3	3	3	NiL(NCS)$_2$	64
				[NiLCl]ClO$_4$	64
				Ni$_2$L$_3$(ClO$_4$)$_4$·5H$_2$O	64
				CoL(NCS)$_2$	65

—*continued overleaf*

TABLE 1 (continued)

Name	l	m	n	Complex	Reference
([13]ane-1,5,9-N_3)	3	3	4	NiL(NCS)$_2$	64
				[NiLCl]ClO$_4$	64
				CoL(NCS)$_2$	65
([14]ane-1,5,9-N_3)	3	3	5	NiL(NCS)$_2$	64
([15]ane-1,5,9-N_3)	3	3	6	NiL(NCS)$_2$	64
				CoL(NCS)$_2$	65

[a] 1-tmOH = 2,3-diamino-1-propanol; 2-tmOH = 1,3-diamino-2-propanol.

oxidation of cobalt(II) in a basic ethanolic solution containing [12]ane-1,5,9-$N_3 \cdot 3HBr$. Nonomaya[64,65] synthesized nickel(II) and cobalt(II) complexes of a variety of cyclic triamines (see Table 1); both five- and six-coordinated complexes were isolated. The syntheses of $R(-)Me[9]$ane-1,4,7-N_3 and $[Co(R(-)Me[9]ane-1,4,7-N_3)]I_3 \cdot 2H_2O$ have been described[66]; the circular dichroism spectrum of the complex ion has been obtained and discussed. Kinetic and equilibrium studies of complex formation between copper(II) and zinc(II) ions with [9]ane-1,4,7-N_3, [10]ane-1,4,7-N_3, [11]ane-1,4,8-N_3, and [12]ane-1,5,9-N_3 have been reported; although strong complexation was demonstrated, no complexes were isolated.[67–69] Reaction of 1,5,9-trithiacyclododecane, [12]ane-1,5,9-S_3, with nickel fluoborate in dry nitromethane results in a red–brown solution from which the complex Ni([12]ane-1,5,9-S_3)$_2$(BF$_4$)$_2$ can be obtained[70]; this complex behaves as a 2:1 electrolyte in nitromethane. Recently, the ligands **11A** and **11B**, R = C_6H_5, have been synthesized, and complexation reactions of **11A** with Mo(CO)$_6$ and Cr(CO)$_6$ and of **11B** with nickel(II) and cobalt(II) chlorides reported.[71] In all of the

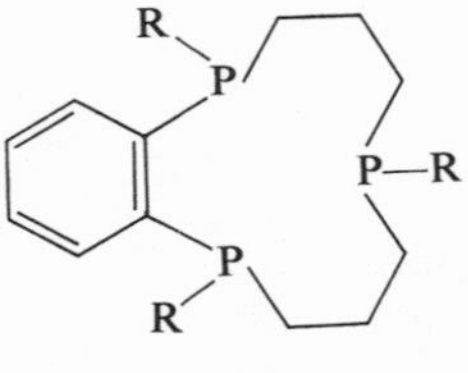

11A 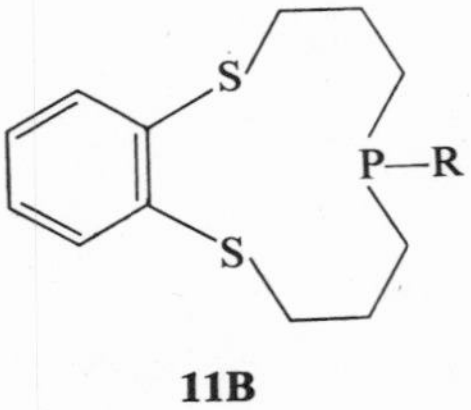　　　　　　**11B**

complexes discussed above, it is assumed that the macrocycle occupies a triangular face of the metal-ion coordination sphere. Thus in the *bis*-ligand complexes, a "sandwich" structure is suggested with distortion along the threefold axis of the [ML$_6$] (L = N, S) chromophore.

The self-condensation of *o*-aminobenzaldehyde has been studied both in the absence and in the presence of metal ions. McGeachin[72] has shown that in acidified aqueous solutions and in the absence of metal ions several cyclic products, including **12** and **13**, are obtained.* However, Melson and Busch[73] found that in ethanol and in the presence of nickel(II) ions, the condensation

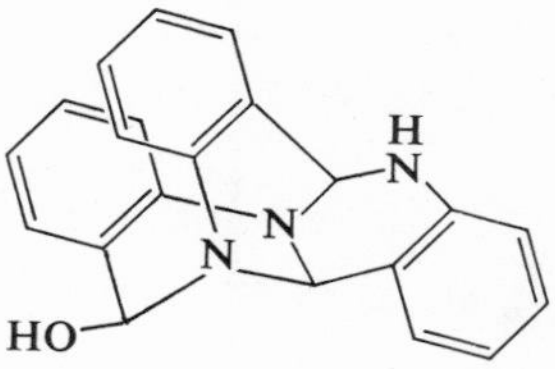

12

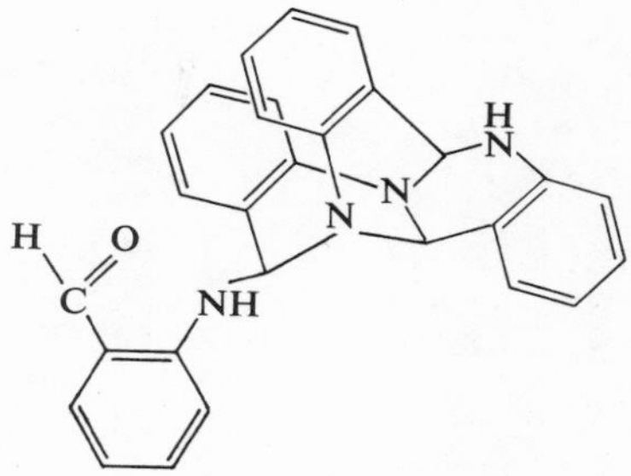

13

results in the formation of complexes containing the tridentate and tetra-dentate macrocyclic ligands **14**, $bzo_3[12]hexaeneN_3$, (TRI) and **15**, $bzo_4[16]-octaeneN_4$, (TAAB), respectively. Selective crystallization of the perchlorate

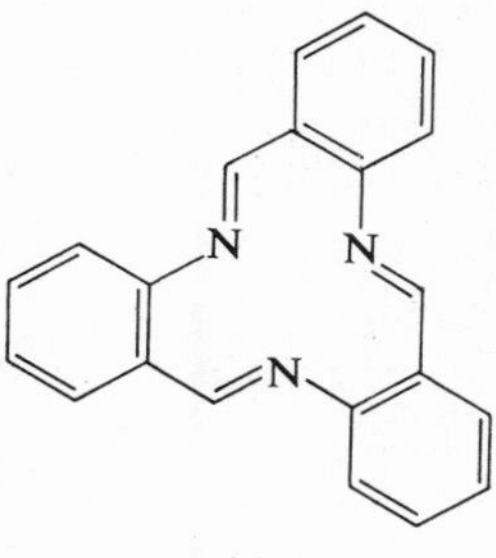

14

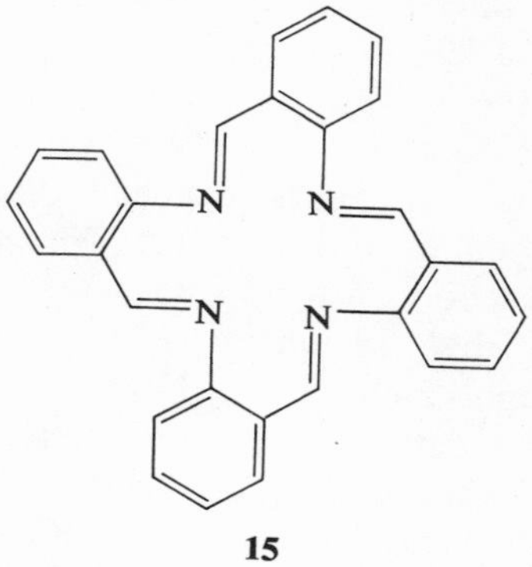

15

derivatives results in separation of the complexes containing **14** and **15**. A series of nickel(II) complexes containing **14** can readily be obtained by meta-thesis reactions from the perchlorate derivative (Table 2). The macrocycle structure and its mode of coordination as a tridentate ligand to the nickel(II) ion was confirmed by X-ray diffraction studies[74] on $Ni(TRI)(H_2O)_2(NO_3)_2$. Taylor and Busch[75] were able to resolve $[Ni(TRI)(H_2O)_3]^{2+}$ by column chromatography on microcrystalline cellulose and potato starch, since the complex ion does not possess a plane of symmetry, center of inversion, or an alternating rotation–reflection axis. The ion does not racemize in neutral or

---

* Busch and co-workers have recently identified a red diacid salt as the product of the self-condensation in the presence of strong or mineral acids (see p. 88).

Gordon A. Melson

TABLE 2. Complexes of Tribenzo[b, f, j][1, 5, 9]-triazacycloduodecine,
14, bzo$_3$[12]hexaeneN$_3$ (TRI)

Complex	X	Reference
Ni(TRI)X$_2$·H$_2$O	ClO$_4$, BF$_4$, Br, I, NO$_3$, NCS	73
Ni(TRI)X$_2$·(H$_2$O)$_3$	B(C$_6$H$_5$)$_4$	73
Ni(TRI)$_2$X$_2$·H$_2$O	NO$_3$, Cl, Br, I	76, 77
Ni(TRI)$_2$X$_2$	PF$_6$, ClO$_4$, NCS	77
Ni(5-Cl-TRI)X$_2$·S	NO$_3$; S = C$_2$H$_5$OH, H$_2$O	77
Ni(5-Cl-TRI)(TRI)X$_2$·2H$_2$O	NO$_3$	77
Co(TRI)$_2$X$_2$·nH$_2$O	Cl, BF$_4$, PF$_6$; $n = 1$ NO$_3$, Br, NCS, ClO$_4$; $n = 2$ Cl, I, NCS, BF$_4$; $n = 3$ Br; $n = 5$	78
[VO(TRI)OH]$_2$X$_4$·S	Cl; S = (CH$_3$)$_2$CO HgI$_4$; S = 4H$_2$O B(C$_6$H$_5$)$_4$	80

acidic aqueous solution. This inertness has been explained by the phenomenon referred to as "multiple juxtapositional fixedness" (MJF).[5] Reaction of 12 with nickel(II) ions in a 2:1 mole ratio in methanol[76,77] results in the formation of the complex Ni(TRI)$_2$(NO$_3$)$_2$·H$_2$O in which the ligands are coordinated to the nickel ion at opposite triangular faces of the octahedron, in a similar manner to that discussed earlier for complexes of tridentate amines and thioethers. No complexes containing 15 are produced in this reaction. Attempts to resolve [Ni(TRI)$_2$]$^{2+}$ were not successful, suggesting that the ion exists in the *meso* configuration, with the ligands having opposite chiralities. (Note that two geometrical isomers are possible for this ion depending on the relative clockwise or counterclockwise progression of the repeating unit, carbon, benzene, nitrogen or carbon, nitrogen, benzene.) Several derivatives of [Ni(TRI)$_2$]$^{2+}$ have been obtained (Table 2). Reaction between partially resolved [Ni(TRI)]$^{2+}$ and 12 produces [Ni(TRI)$_2$]$^{2+}$ with no observable rotation, suggesting that the reaction proceeds stereospecifically. Reaction of 2-amino-5-chlorobenzaldehyde with nickel(II) ions in ethanol does not produce any nickel-containing species; however, treatment of the acid-catalyzed self-condensation product 12 with nickel(II) nitrate results in the formation of Ni(5-Cl-TRI)(NO$_3$)$_2$·C$_2$H$_5$OH·(5-Cl-TRI is the macrocyclic tridentate ligand analogous to 14.) The mixed-ligand complex Ni(5-Cl-TRI)(TRI)(NO$_3$)$_2$·H$_2$O can be obtained[77] from the reaction of 12 with Ni(5-Cl-TRI)(NO$_3$)$_2$·S, S = C$_2$H$_5$OH, H$_2$O, but not from the reaction of [Ni(TRI)]$^{2+}$ with the acid-catalyzed self-condensation product from 2-amino-5-chlorobenzaldehyde. In the presence of cobalt(II) ions, the self-condensation of *o*-aminobenzaldehyde produces a mixture of products.[78] The mixture may be purified by washing with methanol followed by oxidation of the product with concentrated nitric acid. Fractional crystallization from acidic 1-propanol results in two distinct crystalline forms, and from these forms two series of complexes of type $Co(TRI)_2X_3 \cdot nH_2O$ may be produced by metathesis (Table 2). The two series of complexes are diastereomeric forms of the $[Co(TRI)_2]^{3+}$ ion. The racemic isomer (series B) has been resolved into its optical components via the *d*-dibenzoyltartaric acid salt and the absolute configuration of $(+)5461-[Co(TRI)_2]I_3 \cdot 3H_2O$. has been determined from an X-ray diffraction study.[79] The series A species have been assigned the *meso*-configuration similar to that for the $[Ni(TRI)_2]^{2+}$ ion. Hawley and Blinn[80] have investigated reactions between vanadyl salts and *o*-aminobenzaldehyde or **12** or the *bis*-hydroperchlorate of **13**. Salts of the hydroxy-bridged dimeric ion $[VO(TRI)OH]^{2+}$ were obtained. No species containing **15** were obtained from these reactions. It should be noted here that neither the self-condensation of *o*-aminobenzaldehyde nor the rearrangement of **12** in the presence of copper(II) ions produces a complex containing **14**, species containing **15** being obtained exclusively.[76,81,82] The formation of a unique condensation product is suggested to arise because of the stereochemistry of the copper(II) ion, and this result is important evidence in concluding that the nature of the metal ion plays an important role in controlling the steric course of condensation reactions. This topic will be discussed in more detail later in the chapter (see Section 3).

The reaction of **16** with a variety of metal ions, Cu(I), Cu(II), Ni(II), Co(II), Zn(II), and Cd(II), and **17** with Cu(II) ions results in the formation of neutral complexes containing one (presumably) tridentate dianionic macro-cyclic ligand.[83,84] Copper complexes with a similar ligand containing sulfur atoms in noncoordinating positions of the macrocyclic ring have also been reported.[85]

$R =$ **16**

$R =$ (with CH_3) **17**

3. Tetradentate Ligands

Most research on macrocyclic complexes has focused on species containing a first-row transition-metal ion and a tetradentate ligand. The majority of the ligands contain nitrogen donor atoms and are coordinated in a planar manner to the metal ion. The emphasis on these species is undoubtedly related to the existence of naturally occurring metal complexes such as metalloporphyrins, vitamin B_{12}, chlorophyll, etc. The development of effective, reliable syntheses for complexes containing synthetic macrocyclic ligands that may serve as models for the natural complexes is one reason for the rapid development of the chemistry of macrocyclic complexes.

The relationship between the size of the metal ion and the cavity or ring size of a tetradentate macrocyclic ligand has been examined.[5] Critical ring sizes occur at different values for different sets of donor atoms. With four sulfur donor atoms, a 14-membered ring is necessary to coordinate in a planar manner to a nickel(II) ion. However, if the macrocycle has four nitrogen donors in a fully saturated ring, a 13-membered ring will encircle a first-row transition-metal ion, whereas a 12-membered ring, such as [12]aneN$_4$[86] or N-bzl$_4$[12]aneN$_4$, **18**,[87] folds and does not coordinate in a planar manner (see p. 50). Lehn,[88] as part of his studies on complexes of various macrocyclic, macrobicyclic, and macrotricyclic ligands, estimated (from molecular models) the cavity diameter of the cyclic polyether **19** to be 120–150 pm;

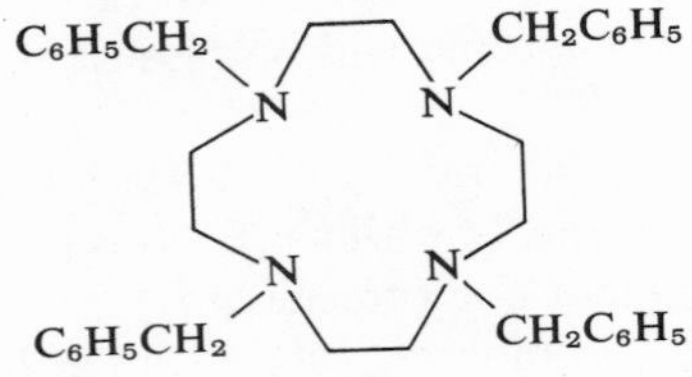

18

this ligand forms a stable complex only with the sodium ion of the alkali metal ions.

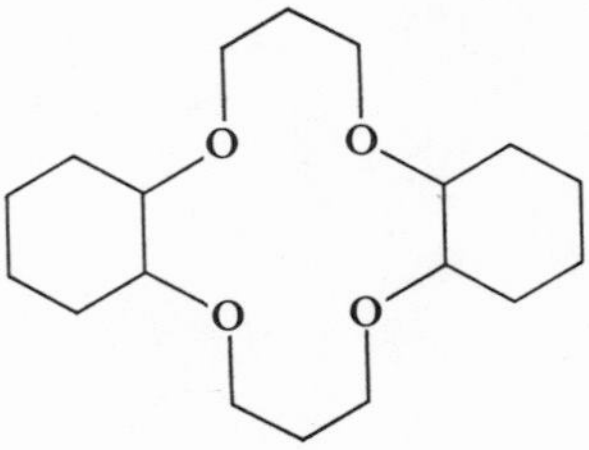

19

A quantitative assessment of the metal-ion–ring-size relationship has been made by Busch and co-workers[89] using cobalt and nickel(II) complexes of [13]aneN$_4$, [14]aneN$_4$, [15]aneN$_4$, and [16]aneN$_4$. A minimum strain energy was found for the 14-membered ring, with the greater strain energies of the 12- and 16-membered rings relating to the known resistance of these

TABLE 3. Ideal Metal–Nitrogen Bond Lengths and Planarity
of Macrocyclic Ligands[a]

Ligand ring size	Average ideal bond length (pm)	Average deviation from the ideal N_4 plane (pm)
12	183	41
13	192	12
14	207	0
15	222	14
16	238	0

[a] From reference 89.

rings to coordinate in a planar manner.[86,87,90] For a given size of tetraaza
macrocyclic ligand, the calculated idealized M–N distance represents the
hole size for which the strain energy in the ligand is at a minimum. The ideal
M–N bond distances are shown in Table 3. A metal ion giving rise to the
ideal M–N distance for a particular macrocycle would complex with minimum strain of the ligand. A larger or smaller metal ion would require a
change in the M–N distance with an accompanying increase in the strain
energy of the ligand. The calculations also indicate that if the 14-membered
ring provides the best "fit" for a cobalt(III) ion, the 15-membered ring will
fit the high-spin nickel(II) ion best. Most transition-metal–nitrogen bond
lengths are within the 180–240 pm range calculated for the idealized values
(Table 3), and many macrocyclic ligands of varying ring size coordinate
readily to transition-metal ions to produce stable compounds.

3.1. N_4 Donor Atoms

Most complexes containing macrocyclic ligands with four nitrogen
donor atoms have been synthesized via condensation reactions between
amines and carbonyl compounds. Both complexation reactions between a
metal ion and a presynthesized ligand and *in situ* syntheses have been employed. The first and most extensively studied series of synthetic tetraaza
macrocycles were initially prepared by Curtis, who reported that the reaction
of *tris*(ethylenediamine)nickel(II) perchlorate and acetone at room temperature produces a yellow crystalline product.[43] The compound is diamagnetic
and very resistant to hydrolysis. The tetra-*N*-isopropylidene structure **20** was
assigned to the complex ion, although the resistance to hydrolysis was in
marked contrast to that of **21**, which was reported to hydrolyze readily.[91]
The 14-membered macrocyclic structure **6** (Me₆[14]4,11-dieneN₄), which is
isomeric with **20**, was later proposed.[44] The cyclic structure is consistent

$$R = \text{(furan ring with } O \text{ and } CH_3 \text{)}$$

with the chemical inertness of the product, the infrared spectrum, the products of hydrolysis (ethylenediamine and mesityl oxide), and the uptake of 2 moles of hydrogen. Reactions of acetone with nickel(II) and copper (II) complexes of other 1,2-diamines[92,93] and 1,3-propanediamine[94] were reported. Complexes containing 14- and 16-membered[23] macrocyclic ligands can be isolated from the reaction media. Reactions between acetone and nickel(II) and copper(II) complexes of triethylenetetramine(trien) were reported to give rise to complexes containing the 13-membered macrocycle $Me_3[13]eneN_4$, **8**,[95] and recently[96] nickel(II) and copper(II) complexes of $Me_3[14]eneN_4$, **23**, have been prepared. Nickel(II) complexes of **23** are highly resistant to acid,

like other 14-membered macrocyclic complexes, and much more resistant to acid attack than complexes of **8**, suggesting that ring size rather than substitution determines the ease of hydrolysis. The reaction of $Ni_2(trien)_3Cl_4 \cdot 3H_2O$ with propionaldehyde, *n*-butyraldehyde, and *iso*-butyraldehyde forms com-

(a) $R_1 = R_2 = R_4 = H; R_3 = CH_3;$
$R_5 = CH_3CH_2$

(b) $R_1 = R_2 = R_4 = H; R_3 = CH_3CH_2$
$R_5 = CH_3CH_2CH_2$

(c) $R_1 = R_4 = R_5 = H;$
$R_2 = R_3 = CH_3$

24

plexes of general formula **24**, although no corresponding complexes of copper(II) are obtained from $Cu(trien)Cl_2 \cdot nH_2O$.[97] Nickel(II) complexes containing $Me_6[14]4,11\text{-dieneN}_4$ with oxalate[98] and the copper(II) complex $[Cu(Me_6[14]4,11\text{-dieneN}_4)]_2CN(ClO_4)_3$ have also been reported.[99] (A crystal structure of the latter complex has recently been reported.[100])

The product of the reaction of *tris*(ethylenediamine)nickel(II) salts with acetone occurs in two noninterconvertible isomeric forms, originally designated A and B by Curtis *et al.*,[45] with the A isomer occurring in two interconvertible forms, Aα and Aβ. The three isomers are similar in appearance and physical properties although some differences in their infrared spectra are noted. When $Ni(en)_3(ClO_4)_2$ reacts with acetone at room temperature, the first crystals of cyclic complex which separate from the reaction medium are isomerically pure $B(ClO_4)_2$. The $A\alpha(ClO_4)_2$ isomer usually begins to crystallize after about 25% of the total product has separated. The separation occurs due to the lower solubility of $B(ClO_4)_2$ in the reaction medium. The A–B isomerism was suggested to arise because of possible *cis* or *trans* arrangements of the imine linkages in the macrocyclic ring, with A corresponding to the *trans* isomer $Me_6[14]4,11\text{-dieneN}_4$, **6**, and B corresponding to the *cis* isomer $Me_6[14]4,14\text{-dieneN}_4$, **7**. In water or methanol, $A\alpha(ClO_4)_2$ is slowly converted into the Aβ form. If kept under the saturated aqueous solution, solid $A\alpha(ClO_4)_2$ slowly dissolves and recrystallizes as $A\beta(ClO_4)_2$. No apparent interconversion takes place in the solid state. In water and methanol solution, the two forms are in equilibrium, the β form being favored at low temperatures and the α form at high temperatures. In acetone solution, Aα is the stable form and Aβ is converted into Aα, slowly at room temperature, more rapidly on heating. Salts of these macrocyclic complex cations can be prepared from the perchlorates by methathesis, the species $A\alpha(ZnCl_4) \cdot H_2O$, $A\beta(ZnCl_4)$, $B(ZnCl_4) \cdot H_2O$, $A\alpha(NCS)_2 \cdot H_2O$, $A\alpha(NCS)_2$, $A\beta(NCS)_2$, $B(NCS)_2$, $A\alpha(NCS)(ClO_4) \cdot \frac{2}{3}H_2O$, and $A\alpha(NCS)(ClO_4)$ have been reported.[45,101] When cyanide ions are added to the orange aqueous solution of $A\alpha(ClO_4)_2$, a short-lived pink species, shown to be the pentacoordinate cyanide adduct $[A\alpha(CN)]^+$, is formed, which decomposes slowly to form the $[Ni(CN)_4]^{2-}$ ion.[102] When cyanide ions are added to aqueous solutions of Aβ or B forms, the color slowly fades to that of $[Ni(CN)_4]^{2-}$.

Bailey and Maxwell[103] confirmed from X-ray studies that the A isomers possess the *trans* configuration of the two imine groups in the macrocyclic ligand and that the Aα–Aβ isomerism arises from the asymmetry at the two tetrahedral nitrogen atoms in the diene complex, Aα corresponding to the racemic mixture of isomers and Aβ the *meso* isomer. Ryan *et al.*[104,105] reported the crystal and molecular structure of the perchlorate salt of the B form and confirmed the *cis* configuration for the imine groups. It was also concluded that the B form was a racemate and not the *meso* isomer. Warner, Rose, and Busch independently came to the same conclusions concerning the *cis–trans* isomerism from a PMR study and, working with the fluoborate and thiocyanate derivatives, assigned complete structures to all the isomers.[106,107] The orange isomer of $Ni(Me_6[14]dieneN_4)(NCS)_2$ was resolved into enantiomers and thus confirmed as the racemate ($A\alpha$ form). (The crystal and molecular structure of this compound has been reported.[108]) All attempts to resolve the yellow (*meso* or $A\beta$) isomer failed. By the use of column chromatography utilizing optically active absorbents, attempts were made to resolve the thiocyanate and fluoborate derivatives of $[Ni(Me_6[14]4,14-dieneN_4)]^{2+}$. Small rotations and ORD and CD curves were obtained.[107] It was concluded that equilibrium concentrations of the "missing" *meso* isomer of $[Ni(Me_6[14]4,14-dieneN_4)]^{2+}$ were below the detection limits of the physical methods employed. Equimolar mixtures of $Ni(en)_3(ClO_4)_2$ and $Ni(1,2-pn)_3(ClO_4)_2$ or $Ni(1,3-pn)_3(ClO_4)_2$ react with acetone to yield as major products nickel(II) complexes of $Me_7[14]dieneN_4$, and $Me_6[15]dieneN_4$, with smaller amounts of the complexes with $Me_6[14]dieneN_4$ and $Me_6[16]$-dieneN$_4$.[109] The complexes containing the 14-membered macrocyclic ligands are extremely resistant to chemical attack, being unaffected by boiling acid or basic solutions, or by ammoniacal dimethylglyoxime or sulfide, whereas the nickel(II) complex with the 15-membered ligand is slowly attacked and the complex with the 16-membered ligand is rapidly attacked. Attempts to prepare the copper(II) complex of $Me_6[15]dieneN_4$ were not successful. MacDermott and Busch[110] showed that, under various conditions, mesityl oxide, diacetone alcohol, and acetone all react with *tris*-(ethylenediamine)nickel(II) salts to yield complexes containing either $Me_6[14]4,11-dieneN_4$ or the nonmacrocyclic tetradentate ligand 1,9-diamino-4,6,6-trimethyl-3,7-diazanona-3-ene. However, when *tris*(1,3-propanediamine)nickel(II) salts were used, no reaction was reported to occur with diacetone alcohol. Black and Greenland[111] carried out the reaction under slightly different conditions and obtained the nickel(II) complex of $Me_6[16]dieneN_4$ (assigned the *trans* configuration of imine linkages). Reactions between $[Ni(1,3pn)_3]^{2+}$ and a variety of α-hydroxy ketones were then investigated, and macrocyclic nickel(II) complexes were obtained from benzoin and anisoin. No macrocyclic product was obtained when acetoin was used. Sadasvian and Endicott[28] reported that the reaction of $Fe(en)_3(ClO_4)_2$ with dry acetone at room temperature results in the formation of small transparent crystals on the walls of the container. After a period of approximately 36 hr, the crystals were removed by filtration, washed with acetone, and separated from a black precipitate by decanting in the presence of ethanol. A 63% yield of $Me_6[14]4,11-dieneN_4 \cdot 2HClO_4$ was obtained. The complexes $[NiL](ClO_4)_2$, $[CuL](ClO_4)_2$, $[ZnL(H_2O)Cl]ClO_4$, $[CoLCO_3]ClO_4$, and $[CoL](ClO_4)_2$ (L = $Me_6[14]4,11-dieneN_4$) were prepared from the dihydroperchlorate salt by treatment with the appropriate metal ion in basic solution. Although the iron(II) undoubtedly plays an important role in the synthesis of the macrocyclic ligand, no complexes containing iron(II) were obtained. Curtis and Hay[27] found that the macrocycle dihydroperchlorate could also be formed from the reaction of the monohydrogenperchlorate salt of ethylenediamine and mesityl oxide or acetone. Since the macrocycle is formed in the absence of a metal ion, these authors suggested that the iron(II) in the previous reaction serves to block polymerization reactions and then generates perchloric acid to give the insoluble dihydrogenperchlorate. $Me_8[14]4,11$-diene$N_4 \cdot 2HClO_4$ can be obtained from the reaction of 1,2-pn $\cdot HClO_4$ with acetone, diacetone alcohol, or mesityl oxide.[112] Hideg and Lloyd[113] investigated the generality of the reaction between ethylene-diamine and α,β-unsaturated ketones and reported that with benzilidene-acetones, 14-membered macrocyclic compounds are obtained, which are reduced by sodium borohydride to give tetraazacyclotetradecanes. The reduction of $Me_6[14]4,11$-diene$N_4 \cdot 2HClO_4$ and $Me_8[14]4,11$-diene$N_4 \cdot 2HClO_4$ by sodium borohydride to the fully saturated macrocycles was also reported by Tait and Busch.[30] Recently, reactions between en $\cdot HClO_4$ and α,β-unsaturated ketones have been studied in more detail.[114] Macrocyclic ligand dihydrogenperchlorates of general formula $(R_1R_2)_3[14]4,11$-diene$N_4 \cdot 2HClO_4$ ($R_1 = R_2 = H$; $R_1 = H$, $R_2 = CH_3$; $R_1 = H$, $R_2 = C_2H_5$; $R_1 = H$, $R_2 = i\text{-}C_3H_7$; $R_1 = R_2 = CH_3$) were prepared and nickel(II) complexes of the macrocyclic ligands were obtained from the ligand salts and nickel(II) acetate or from the free ligand and anhydrous nickel(II) perchlorate (see p. 76). Love and Powell,[29,115] isolated the free ligand $Me_6[14]4,14$-dieneN_4 (L) by treatment of the nickel perchlorate complex with cyanide ions in non-aqueous solution. After removal of $K_2Ni(CN)_4$, the complexes $[CuL](ClO_4)_2$, $[ZnL]ZnCl_4 \cdot CH_3OH$, and $[FeLCl_2]FeCl_4 \cdot H_2O$ were prepared from the ligand solution. For the cobalt series of complexes a dimeric system was proposed; however, the determination[116] of the crystal structure of one of these compounds showed it to be a cobalt(III) species, $[CoL(CN)_2]ClO_4$. Reactions of the iron(III) or zinc(II) complex with dilute $HClO_4$ or addition of a methanolic solution of $Me_6[14]4,14$-dieneN_4 to dilute $HClO_4$ leads to the isolation of the dihydrogenperchlorate salt of **25** in 80% yield. Copper(II) and nickel(II) complexes of **26** and a copper(II) complex of **27** can be

25

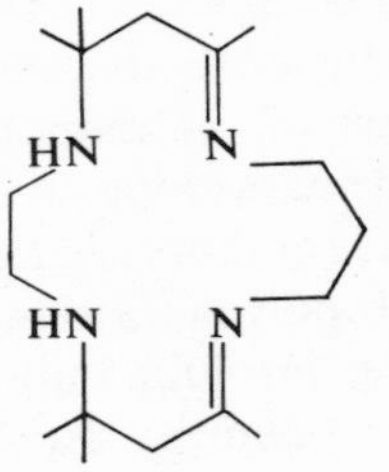

26 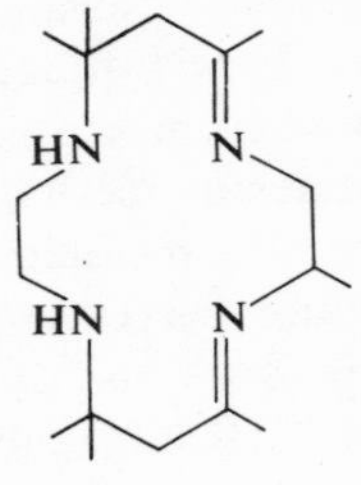   **27**

obtained in 20–50% yield by condensation of **25** with the appropriate diamine in the presence of the metal ion in methanol or aqueous solution. Blight
and Curtis[92] first investigated the reaction of *tris*(1,2-propanediamine)-
nickel(II) perchlorate and acetone and observed that over a period of days,
approximately equal amounts of two noninterconvertible isomeric cations
were formed as orange and yellow perchlorate salts. Reinvestigations of this
reaction by Curtis *et al.*[112] have enabled conclusions concerning the nature
of the isomerism in these species to be drawn. Both species have the *trans*
arrangement of the imines in the macrocyclic ligand and thus may be formulated as isomers of **28**, $[Ni(Me_8[14]4,11\text{-}dieneN_4)]^{2+}$. The orange isomer possesses a centrosymmetric cation and is thus the *meso* isomer, whereas the
yellow isomer is a racemic mixture, the isomerism being due to the different
configurations of the methyl groups in the 1,2-propanediamine backbone.

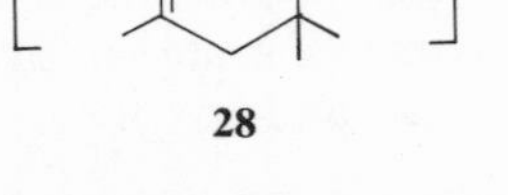

28

The crystal and molecular structure of the yellow isomer has been reported.[117]
Reaction of $Me_8[14]4,11\text{-}dieneN_4 \cdot 2HClO_4$ with nickel(II) results in the
formation of the *meso* isomer, and thus the ligand salt is also assigned the
meso configuration. Ito and Busch[118] have studied the stereochemistry of
the 3,10 *meso* and 3R, 10R isomers of **28** by analysis of their PMR spectra.
The reaction of $Cu(1,2\text{-}pn)_2(ClO_4)_2$ with acetone at 110°C was reported[92] to
form a copper(II) complex analogous to **28**, in low yield. At room temperature, $Cu(1,2\text{-}pn)_3(ClO_4)_2$ reacts slowly with acetone to form the same compound and an isomeric product.[112] The previously reported compound is

also formed from $Me_8[14]4,11$-diene$N_4 \cdot 2HClO_4$ and is therefore assigned the *meso* configuration; the new isomer is assigned the racemic configuration. The reaction of $Cu(en)_2Cl_2$ with acetone and diacetone alcohol in the presence of monoethanolamine has been reported to give rise to $Cu(Me_8[14]4,11$-diene$N_4)Cl_2$ and a variety of nonmacrocyclic products.[119]

The syntheses of iron(II) complexes containing $Me_6[14]4,11$-dieneN_4 were reported by Goedken *et al.*[120] These authors found that the addition, under N_2, of $Me_6[14]4,11$-diene$N_4 \cdot 2HClO_4$ to a solution of $Fe(ClO_4)_2$ in acetonitrile and triethyl orthoformate followed by neutralization by the addition of triethylamine causes a deep red solution to be formed, from which a pink precipitate is produced on heating. The precipitate is $[Fe(Me_6[14]4,11$-diene$N_4)(CH_3CN)_2](ClO_4)_2$, isomer A. The addition of ether to the filtrate from above causes a second crop of pink crystals to be obtained of identical formula, isomer B. The less-soluble isomer (A) was assigned the *meso* configuration and the more-soluble isomer (B) the racemic isomer. This isomerism is due to the possible configurations of the asymmetric nitrogens in the complex. A variety of complexes with other anions and axial ligands was also obtained. It was found that the coordination number and spin state of the metal ion are dependent on the axial ligands. High-spin, five-coordinate[121] iron(II) complexes $[FeLX]ClO_4$ ($X = Cl, Br, I;\ L = Me_6[14]4,11$-diene$N_4$) were isolated when weak axial ligands were present; low-spin, six-coordinate complexes were isolated with strong axial ligands, $[FeLY_2](ClO_4)_2$ (Y a neutral ligand) or $[FeLX_2]$ (X a uninegative ligand). (Nitrogen-donor ligands such as NCS^-, CH_3CN, and imidazole all yield low-spin complexes.) The high-spin complex $FeLI_2$ can be prepared from the low-spin $[FeL(CH_3CN)_2]I_2$ by heating for 24 hr at 100°C under vacuum. $[FeL(phen)](ClO_4)_2$ exists in spin-state equilibrium. Low-spin iron (III) complexes can be prepared from $[FeL(CH_3CN)_2](ClO_4)_2$ by using oxidants such as halogens or molecular oxygen, e.g., $[FeLX_2]Y$ ($X = Cl, Y = ClO_4$; $X = NCS, Y = BPh_4$) and $[FeL(CH_3CN)_2](ClO_4)_3$. The oxo-bridged iron(III) species $[(FeLH_2O)_2O](ClO_4)_4$ and the unstable iron(II) complex $[(FeL)_2OH](ClO_4)_3 \cdot H_2O$ can be obtained in the absence of acid in oxygen-donor solvents. The species $[FeLCl]ClO_4$ and $[FeL(NCS)_2]$ have also been reported by Rillema *et al.*,[57] and the five-coordinate structure of the complex $[FeLCl]I$ has been confirmed by Goedken *et al.*[122] The compounds $[FeL(imid)_2](BPh_4)_2$, $FeL(CN)_2$, and $[FeL(CH_3CN)_2](SO_3CF_3)_2$ have also been reported.[123,124]

Sadasvian and Endicott[28] reported an air- and water-sensitive reaction between $Co(en)_2(ClO_4)_2$ and acetone; a readily hydrolyzed crystalline product was obtained which, after treatment with 1,1,1-trifluoroacetone under N_2, yielded $Co(Me_6[14]4,11$-diene$N_4)(ClO_4)_2$. Reaction between $Me_6[14]4,11$-diene$N_4 \cdot 2HClO_4$ ($L \cdot 2HClO_4$) and cobalt(II) acetate in acetonitrile results in the formation of a mixture of $[CoL](ClO_4)_2$ and $[CoL](ClO_4)_2 \cdot H_2O$, which may be separated by fractional crystallization.[125] In a similar reaction,[57] between cobalt(II) chloride and L·2HClO$_4$ in dimethylformamide, [CoL] CoCl$_4$ was formed, which when treated with excess NH$_4$ClO$_4$ produced [CoL] (ClO$_4$)$_2$. Recrystallization of this compound from hot water results in the hydrated species CoL(ClO$_4$)$_2$·2H$_2$O, and treatment with pyridine (py) produces CoL(ClO$_4$)$_2$·2py. The complexes [Co(Me$_6$[14]4,11-dieneN$_4$)ClO$_4$]ClO$_4$ and [Co(Me$_6$[14]4,14-dieneN$_4$)ClO$_4$]ClO$_4$ were prepared from cobalt(II) acetate and L·2HClO$_4$ or Me$_6$[14]4,14-dieneN$_4$, respectively, in MeOH.[126] Dissolution of these low-spin, five-coordinate complexes in water results in the displacement of the coordinated perchlorate giving complexes of the form [CoL(H$_2$O)$_2$]$^{2+}$ in aqueous solution. The compound Co(Me$_8$[14]4,11-dieneN$_4$)(ClO$_4$)$_2$ can be prepared from the ligand dihydroperchlorate and cobalt(II) chloride in DMF.[57]

Attempted condensation reactions between *bis*- and *tris*(ethylenediamine)cobalt(II) complexes and acetone were unsuccessful[3]; however, the reaction between an excess of Me$_6$[14]4,11-dieneN$_4$·2HClO$_4$ and Na$_3$[Co(CO$_3$)$_3$]·n-H$_2$O in aqueous solution produces the red cobalt(III) complex [Co(Me$_6$[14]4,11-dieneN$_4$)CO$_3$]ClO$_4$.[28] Treatment of this complex with concentrated HCl in aqueous methanol followed by evaporation on the steam bath[127] results in the isolation of green crystals of [Co(Me$_6$[14]4,11-dieneN$_4$)Cl$_2$]ClO$_4$, isomer a. Isomer b of this complex can be prepared by the addition of a hot methanolic solution of cobalt(II) acetate to the ligand dihydroperchlorate salt. After filtration, a stream of O$_2$ was passed through the filtrate for 24 hr and methanol was allowed to evaporate. Treatment of the red glassy residue with concentrated HCl results in the formation of green crystals. Recrystallization from methanol produces two distinctly different kinds of crystal, dark green plates, approximately 20%, isomer a, and fine green needles, isomer b. Isomer a is less soluble than isomer b in methanol, and the isomeric products can be readily separated by fractional crystallization from MeOH. It was assumed that the isomerism in these cobalt(III) species is analogous to that noted[106,107] for nickel(II) complexes of Me$_6$[14]4,11-dieneN$_4$, viz., different geometrical arrangements of the protons on the secondary nitrogen atoms of the coordinated ligand. Metathetical reactions between [CoLCl$_2$]ClO$_4$, isomer a (L = Me$_6$[14]4,11-dieneN$_4$) and a variety of salts has led to the isolation of the complexes [CoLX$_2$]ClO$_4$ (X = CN, N$_3$, NO$_2$, Br) and [CoL(NCS)$_2$]ClO$_4$·SCN. When the latter is recrystallized from water containing NaSCN, [CoL(NCS)$_2$]SCN is obtained. Allowing [CoLCl$_2$]ClO$_4$ to stand in dilute HClO$_4$ for about one week leads to crystallization from solution of [CoL(H$_2$O)Cl](ClO$_4$)$_2$, and treatment with liquid ammonia[57] gives [CoL(NH$_3$)$_2$]Cl$_2$·ClO$_4$·H$_2$O. In all these complexes, the additional ligands are *trans* to one another. Lawrance[128] has described the synthesis of [CoL(NH$_3$)$_2$](ClO$_4$)$_3$·$\frac{1}{2}$H$_2$O from [CoLCl$_2$]ClO$_4$. Poon and co-workers[129,130] described syntheses of [CoLX$_2$]ClO$_4$ (X = Cl,

Br) which were modified from those reported earlier[126] and prepared a series of complexes containing the $[CoLAX]^{n+}$ ion ($A = H_2O$, $X = Cl$, Br, CN, $n = 2$; $A = NCS$, $X = Cl$, Br, $n = 1$; $A = N_3$, $X = Cl$, Br, $n = 1$; $A = CN$, $X = Cl$, Br, $n = 1$) in connection with studies on kinetics and mechanisms of substitution reactions of octahedral macrocyclic amine complexes[131,132] (see Chapter 6). Liteplo and Endicott[133] have prepared $[CoLOH(C_2H_3O_2)]$ ClO_4 from $L \cdot 2HClO_4$ and cobalt(II) acetate in methanol. When concentrated $HClO_4$ was added to solid samples or concentrated solutions of this complex, or to $CoLCO_3{}^+$, the species $CoL(H_2O)_2(ClO_4)_3$ was obtained; at pH 5–7, this species is converted to $[CoLH_2O(OH)](ClO_4)_2$. The addition of an excess of freshly prepared Ag_2O to $L \cdot 2HClO_4$ in water has resulted in the isolation of a small amount of the complex $AgL(ClO_4)_2$.[134]

The chromium(II) complex $[Cr(Me_6[14]4,11\text{-}dieneN_4)(py)](PF_6)_2$ (py = pyridine) can be isolated from the reaction of $Cr(CF_3SO_3)_2 \cdot 4py$ and $Me_6[16]4,11\text{-}dieneN_4 \cdot 2CF_3SO_3H$ in the presence of triethylamine.[135] Treatment of the complex with sodium nitrite in acetonitrile/ethanol gives rise to the chromium(I) complex, $[Cr(Me_6[14]4,11\text{-}dieneN_4)(NO_2)(NO)]PF_6$. The crystal and molecular structure of the chromium(I) complex has been determined.[135]

Recently, Curtis and co-workers have reinvestigated the reactions of some nickel(II) and copper(II) complexes of ethylenediamine and 1,2- and 1,3-propanediamine with acetone[136–138] and other aliphatic carbonyl compounds. By working with *bis*- and *tris*-diamine complexes of both metal ions and isolating intermediates in the formation of the macrocyclic complexes, these studies have enabled some conclusions concerning the mechanisms of these *in situ* condensation reactions to be reached. The formation of the three-carbon bridge in the macrocyclic complexes has been the source of considerable discussion since these complexes were first reported.[3,6] Curtis proposes (see Scheme 1) that in the condensation reactions, the initial product is an isopropylideneamine species, which undergoes an aldol-type reaction between the imino group and an acetone molecule to yield a β-amino ketone. This may be stabilized by coordination of the keto group, or may undergo an imine formation reaction with a primary amine to yield a compound with a ligand which has two amine residues linked by a three-carbon atom "amine–imine bridge." The formation of two such bridges for diamines results in a tetraaza macrocyclic complex. *Bis*(ethylenediamine)nickel(II) perchlorate reacts with methyl ethyl ketone at room temperature to give a *bis*(β-amino ketone)nickel(II) complex, which in pyridine undergoes intramolecular imine formation to yield the nickel(II) complex of $Me_2Et_4[14]4,11\text{-}dieneN_4$, **29**.[139] The salt $Me_2Et_4[14]4,11\text{-}dieneN_4 \cdot 2HClO_4$ crystallizes when ethylenediamine hydroperchlorate reacts with methyl ethyl ketone and can be used to form nickel(II) and copper(II) complexes by reaction with the appropriate metal acetate. *Tris*(ethylenediamine)nickel(II) perchlorate reacts very slowly with

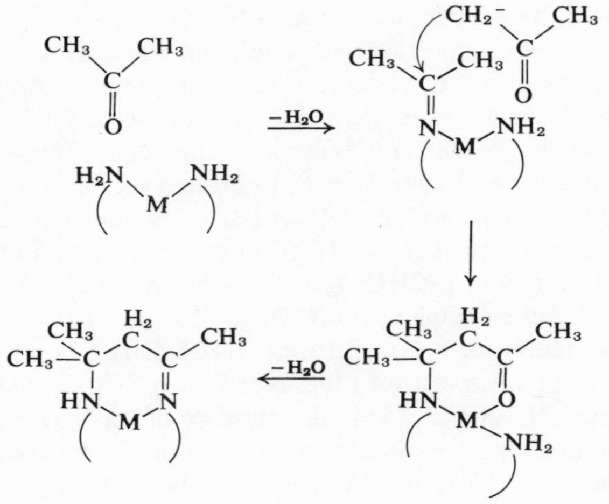

Scheme 1

methyl ethyl ketone to give the complex of $Me_2Et_4[14]4,14$-dieneN$_4$, the "*cis* imine" isomer of **29**. Other nickel(II) and some cobalt(III) complexes of **29** and nickel(II) complexes of the *cis* isomer have been synthesized, their

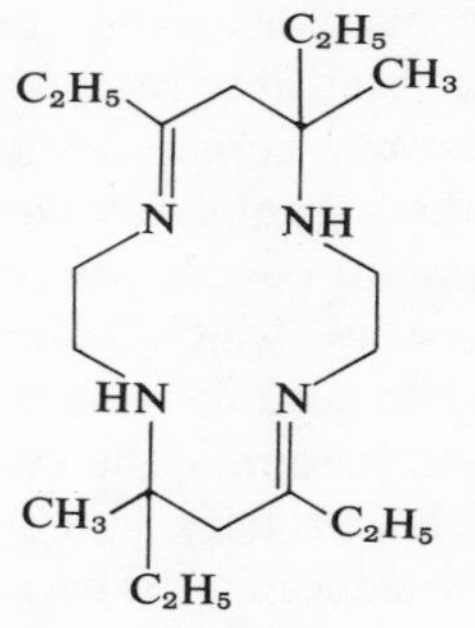

29

PMR spectra studied, and isomerism discussed.[140] The reactions of *bis*-ligand complexes of nickel(II) containing a variety of diamines with diacetone alcohol have been described.[141] Complexes of β-hydroxyimines are isolated. MacDermott and Busch[110] have previously studied some of these reactions and shown that under more forcing conditions complexes containing macro-cyclic ligands are obtained. Related complexes formed from some α-hydroxy ketones have previously been described.[142]

Reactions of $Me_6[14]dieneN_4$ complexes will be discussed in detail in Chapter 6; however, syntheses of complexes containing the fully saturated ligands obtained from the above complexes will be described here. Reduction of the imine groups in nickel(II) complexes of $Me_6[14]4,11$-dieneN$_4$ and $Me_6[14]4,14$-dieneN$_4$ is readily achieved catalytically by hydrogen over Pt or Ni, Ni/Al alloy in alkaline solution, by electrolysis, and by sodium borohydride in aqueous solution.[143–145] Complexes containing the isomeric ligands **30** and **31** obtained from $Me_6[14]4,11$-dieneN$_4$ and $Me_6[14]4,14$-

30   **31**

dieneN$_4$ complexes, respectively, are obtained. For each ligand system, two isomeric complexes are produced on reduction; these isomers, which may be separated by fractional crystallization of the oxalate derivatives, arise due to the introduction of two asymmetric carbon centers into the ligand system upon reduction of the two imine groups. The $Ni(Me_6[14]aneN_4)^{2+}$ complexes are resistant to attack by acid, but react with cyanide ions to produce the $[Ni(CN)_4]^{2-}$ ion and liberate the free macrocyclic ligands, from which the original nickel(II) complexes can be reformed. For **30**, the isomeric ligands have been referred to as tet *a* for the *C-meso* isomer and tet *b* for the *C-racemic* isomer. For **31**, tet *c* refers to the *C-racemic* isomer and tet *d* to the *C-meso* isomer. Tait and Busch[30] have reported that reaction of $Me_6[14]4,11$-dieneN$_4 \cdot 2HClO_4$ with an excess of NaBH$_4$ in the absence of metal ions also produces the fully reduced ligand in 85–90% yield. Fractional reprecipitation of the amine from the minimum of hot methanol by the addition of water produces the *meso* (least soluble) and racemic isomers. Attempts to reduce copper(II) complexes of $Me_6[14]dieneN_4$ ligands were not successful; however, copper(II) complexes of **30** and **31** may be readily obtained from the ligands and copper(II) salts.[55,143] Bauer *et al.*[146] have recently reported that copper(II) salts react with *rac*-(5,12)-$Me_6[14]aneN_4$ (tet *b*) in basic solution to form a blue complex which readily converts to a more stable red complex. The blue to red conversion can be prevented by quenching the basic solution with acid a few seconds after mixing the reactants. In the presence of chloride and perchlorate ions, a binuclear, five-coordinate compound of the blue complex $[Cu(rac$-(5,12)-$Me_6[14]aneN_4)]_2Cl(ClO_4)_3$ can be isolated. The bromide and iodide derivatives are not binuclear species, although the mono-meric complexes [Cu(rac-(5,12)-Me$_6$[14]aneN$_4$)X]ClO$_4$ (X = Br, I) are obtained. The binuclear chloro complex has been shown to contain the ligand in a folded form. A variety of other complexes containing the Me$_6$[14]aneN$_4$ ligands have been synthesized. These include nickel(II) complexes of ms-(5,12)-Me$_6$[14]aneN$_4$, tet *a*, and rac-(5,12)-Me$_6$[14]aneN$_4$, tet *b*, with borohydride,[144] nitrate,[147] thiocyanate,[148] acetate and formate,[149] oxalate,[150] and acetylacetonate[151] ions, and nickel(II) and copper(II) complexes of ms-(5,12)-Me$_6$[14]aneN$_4$ with cyanide.[99,102] The physical properties of the complexes indicate that the nickel(II) ion is six coordinated in all the species. With the exception of the acetylacetonate derivative, the ms-(5,12)-Me$_6$[14]aneN$_4$ macrocyclic ligand is coordinated in a planar manner with the anions in *trans* positions (in [Ni(ms-(5,12)-Me$_6$[14]aneN$_4$)acac]ClO$_4$, the ligand folds to accommodate the acetylacetonate ion[151]). However, the ligand rac-(5,12)-Me$_6$[14]aneN$_4$ is folded in all the above complexes with the anions coordinated in *cis* positions. The structure of [Ni(rac-(5,12)-Me$_6$-[14]aneN$_4$)OAc]ClO$_4$·$\frac{1}{2}$H$_2$O has confirmed the folded conformation for the macrocyclic ligand with the four secondary amine groups coordinated to adjacent sites of the nickel(II) ion and the acetate ion occupying the other two sites.[152] With anions that show little tendency to coordinate, such as ClO$_4^-$, BF$_4^-$, square-planar nickel(II) complexes are formed with rac-(5,12)-Me$_6$[14]aneN$_4$. The synthesis and structures of two perchlorate derivatives and the tetrachlorozincate(II) derivative of nickel(II) complexes with rac-(5,12)-Me$_6$[14]aneN$_4$ have been reported.[153] The planar conformation of the ligand was confirmed and ring conformations were discussed. Bryan and Dabrowiak[154] resolved the ligand rac-(5,12)-Me$_6$[14]aneN$_4$ via its nickel(II) complex. The optically active ligand was removed from the nickel(II) ion and used in the formation of copper(II) and manganese(III) complexes.

It should be noted here, that each cyclic tetramine, *C-meso* and *C-rac* isomers of **30** and **31**, has four secondary amino groups that give rise to two pairs of equivalent asymmetric centers when coordinated to a metal ion. The combination of two equivalent asymmetric carbon centers and two pairs of equivalent asymmetric nitrogen centers leads to a large number of possible configurational isomers for the complexes of **30** or **31**. For example, the complex ion derived from nickel(II) and **30** can exist in 20 theoretically possible isomeric forms.[155] The isomers associated with the carbon centers should be noninterconvertible and the isomerism should persist in the free amines, as observed for the tet-*a*–tet-*b* and the tet-*c*–tet-*d* pairs. Isomerism arising from the coordinated asymmetric nitrogen centers should be lost when the ligand is displaced from the complex. Warner and Busch[155] have investigated the isomerism of some of these species through stereochemical analysis and PMR and infrared spectral data. Factors determining the stabilities of configurations and conformations are discussed.

Reduction of $[Ni(Me_8[14]4,11\text{-}dieneN_4)]^{2+}$ by sodium borohydride in water was reported by Rillema *et al.*[57] Attempts to separate the isomers by techniques used by Curtis[143] for separation of isomers of **30** produced only a small amount of "$Ni((CH_3)tet\ b)(C_2O_4)_{1/2}ClO_4$" (presumably the *C*-racemic isomer); thus it was concluded that the compound formed is predominantly the $Ni((CH_3)_2tet\ a)^{2+}$ (*C-meso*) isomer. Tait and Busch[30] reported that $Me_8[14]4,11\text{-}dieneN_4 \cdot 2HClO_4$ can be reduced by sodium borohydride and the product separated into *meso* and racemic isomeric tetramine ligands. The ligands can then be used in complexation reactions with metal ions. Catalytic reduction of $[Ni(Me_6[16]4,12\text{-}dieneN_4)](ClO_4)_2$ was achieved by using hydrogen and a platinum oxide catalyst in methanol.[156] The low-spin five-coordinate complexes $[Ni(Me_6[16]aneN_4)X]X$ (X = Cl, Br, I) were obtained from the perchlorate derivative. Proton magnetic resonance spectra indicate that only one ligand isomer is formed in the reduction, although the identity of this isomer was not stated.

Iron(II) complexes of $ms\text{-}(5,12)\text{-}Me_6[14]aneN_4$ (derived from **30**, tet *a*) have been synthesized. Rillema *et al.*[57] prepared $Fe(ms\text{-}(5,12)\text{-}Me_6[14]\text{-}aneN_4)Cl_2$ by treatment of the ligand dihydrate with $FeCl_2$ in DMF. Dabrowiak *et al.*[157] reported the synthesis of $[Fe(ms\text{-}(5,12)\text{-}Me_6[14]aneN_4)\text{-}(CH_3CN)_2](BF_4)_2 \cdot 1\frac{1}{2}CH_3CN$ by the reaction of anhydrous iron(II) acetate with the ligand in dry, degassed acetonitrile in an inert atmosphere, and Olson and Vasilevskis[56] reported the isolation of $[Fe(ms\text{-}(5,12)\text{-}Me_6[14]\text{-}aneN_4)(CH_3CN)_2](BF_4)_2 \cdot 2CH_3CN$ from the reaction of $Fe(CH_3CN)_6(BF_4)_2$ and the anhydrous macrocyclic amine in acetonitrile (both coordinated and solvated CH_3CN molecules can be removed by heating *in vacuo*). The acetato derivative $Fe(ms\text{-}(5,12)\text{-}Me_6[14]aneN_4)(OAc)_2$ can be isolated as an intermediate in the synthesis employing iron(II) acetate. The *bis*-acetonitrile adduct is used as the starting material in the methathetical syntheses of $Fe(ms\text{-}(5,12)\text{-}Me_6[14]aneN_4)X_2$ (X = NO_2, SCN, Cl, Br, I). The dichloro species may be oxidized (with caution) by 70% $HClO_4$ to $[Fe(ms\text{-}(5,12)\text{-}Me_6[14]aneN_4)Cl_2]ClO_4$. In all complexes the macrocyclic ligand is assumed to be coordinated in the plane with the unidentate ligands occupying *trans* axial sites. When these unidentate ligands are CH_3CN, CN, NO_2, and NCS, the complexes are low spin, whereas when the ligands are Cl, Br, I, CH_3CO_2, BF_4, the complexes are high spin.

Most cobalt complexes containing $Me_6[14]aneN_4$ ligands contain the metal in the $+3$ oxidation state, although the reaction between $Co(ClO_4)_2 \cdot 6H_2O$ and $Me_6[14]aneN_4 \cdot 2H_2O$ (**30**) in ethanol has been shown[57,125] to produce $Co(Me_6[14]aneN_4)(ClO_4)_2$. Whimp and Curtis[158] described the syntheses of $[Co(ms\text{-}(5,12)\text{-}Me_6[14]aneN_4)Cl_2]ClO_4$ and $[Co(rac\text{-}(5,12)\text{-}Me_6[14]aneN_4)Cl_2]ClO_4$ by reaction of the ligands (tet *a* and tet *b*, respectively) and cobalt(II) chloride in a buffered ethanol–water solution followed by oxidation with hydrogen peroxide and the addition of $NaClO_4$. The compounds may also be prepared by the reaction between $Na_3[Co(CO_3)_3]$ · $3H_2O$ and the tetramines in aqueous solution acidified with HCl. The derivatives $[Co(ms\text{-}(5,12)\text{-}Me_6[14]aneN_4)(H_2O)_2]^{3+}$, $[Co(ms\text{-}(5,12)\text{-}Me_6[14]aneN_4)(H_2O)_2(NO_2)]^{2+}$, $[Co(ms\text{-}(5,12)\text{-}Me_6[14]aneN_4)XY]^+$ (X = Y = OH, Cl, Br, NCS, NO_3, CN; X = NO_2, Y = OH; X = Cl, Y = N_3) and $[Co(rac\text{-}(5,12)\text{-}Me_6[14]aneN_4)X_2]^+$ (X = Cl, CN, N_3, NO_2) have also been described. In these complexes the macrocycle coordinates in a square-planar manner to the cobalt(III) ion with the unidentate ligands in axial positions. Some of these complexes have also been synthesized by Sadasivan *et al.*,[127] and recently the optically active complex $(+)\text{-}trans[Co(+)\text{-}(5,12)\text{-}Me_6[14]ane\text{-}N_4Cl_2]BF_4$ has been reported.[159] This complex may be prepared from the optically active ligand, resolved via the nickel(II) complex,[154] and cobalt(II) chloride in methanol. Poon and co-workers have described the syntheses of $[Co(ms\text{-}(5,12)\text{-}Me_6[14]aneN_4),(A)X]^+$ (X = Cl, A = NO_2, N_3, NCS, CN;[160] X = Br, A = N_3, $NCS^{[129]}$). Complexes of cobalt(III) with $rac\text{-}(5,12)\text{-}Me_6[14]aneN_4$ (tet *b*) and a bidentate ligand have also been reported.[151,161,162] These complexes, $[Co(rac\text{-}(5,12)\text{-}Me_6[14]aneN_4)(bident)]^{n+}$ (bident = CO_3, C_2O_4, $n = 1$; bident = acac, $n = 2$; bident = en, $n = 3$) are assumed to contain the ligand in a folded conformation, with the bidentate ligands occupying *cis* positions. This difference in mode of coordination of *ms*- and *rac*-$Me_6[14]aneN_4$ is similar to that found in the nickel(II) complexes of these ligands described earlier. Whimp and Curtis[163] have described the syntheses of cobalt(III) complexes with the isomeric forms of **31**, viz., $ms\text{-}(5,14)\text{-}Me_6[14]aneN_4$ (tet *d*) and $rac\text{-}(5,14)\text{-}Me_6[14]aneN_4$ (tet *c*). Both isomers form *trans*-cobalt(III) complexes with unidentate ligands occupying axial coordination sites: $[Co(rac\text{-}(5,14)\text{-}Me_6[14]aneN_4)XY]^+$ (X = Y = Cl, CN, NCS, NO_2; X = Cl, Y = OAc), $[Co(rac\text{-}(5,14)\text{-}Me_6[14]aneN_4)(H_2O)_2]^{3+}$, and $[Co(ms\text{-}(5,14)\text{-}Me_6[14]aneN_4)Cl_2]^+$. In addition, the *meso* form of the ligand forms a *cis* derivative $[Co(ms\text{-}(5,14)\text{-}Me_6[14]aneN_4)C_2O_4]^+$. Attempts to prepare *cis* derivatives of the racemic isomer were unsuccessful. Kane-Maguire *et al.*[162] have reported the synthesis of $[Co(ms\text{-}(5,14)\text{-}Me_6[14]aneN_4)CO_3]ClO_4$ from the ligand and $Na_3[Co(CO_3)_3]$ in aqueous solution.

Curtis and Cook[164] reported that when rhodium(III) chloride was heated with the dihydrogenperchlorate salts of $ms\text{-}(5,12)\text{-}Me_6[14]aneN_4$ or $rac\text{-}(5,12)\text{-}Me_6[14]aneN_4$ in aqueous solution, reaction occurred slowly to yield the *trans*-dichloro complexes. When the reaction with $rac\text{-}(5,12)$-$Me_6[14]aneN_4$ was carried out in the presence of additional chloride ion, a sparingly soluble complex of stoichiometry $Rh(rac\text{-}(5,12)\text{-}Me_6[14]aneN_4)Cl_3$ was formed in approximately 40% yield together with the *trans*-dichloro complex. The compound $Rh(rac\text{-}(5,12)\text{-}Me_6[14]aneN_4)Cl_3$ was readily converted into, and reformed from, unambiguously *cis* compounds and is considered to have a *cis*-folded amine structure. The complex ions *trans*-$[Rh(ms\text{-}(5,12)\text{-}Me_6[14]aneN_4)X_2]^+$ (X = Cl, Br, I, NCS, OAc) and *trans*-

[Rh(rac-(5,12)-Me$_6$[14]aneN$_4$)X$_2$]$^+$ (X = Cl, Br, I, NCS) have been prepared. It is suggested that the compound Rh(rac-(5,12)-Me$_6$[14]aneN$_4$)Cl$_3$ and the analogous bromide and iodide species are not simple *cis*-dihalo–halide species; however, the chloride is readily soluble in aqueous carbonate solution and a variety of *cis*-[Rh(rac-(5,12)-Me$_6$[14]aneN$_4$)XY]$^{n+}$ species (XY = CO$_3$, NO$_3$, ClO$_4$, $n = 2$; XY = C$_2$O$_4$, $n = 1$; X = Y = NCS, $n = 1$; X = OAc, Y = OH, $n = 1$) may be obtained.

Manganese(II) and (III) complexes with *ms*-(5,12)-Me$_6$[14]aneN$_4$ have been synthesized by Bryan and Dabrowiak.[165] The complexes Mn(*ms*-(5,12)-Me$_6$[14]aneN$_4$)X$_2$ (X = CF$_3$SO$_3$, OAc) were prepared by reaction of the ligand with the manganese(II) salt in degassed, hot acetonitrile. The species [Mn(*ms*-(5,12)-Me$_6$[14]aneN$_4$)X$_2$]$^+$ (X = Cl, Br, NCS) were also described. Kestner and Allred[166] reported that addition of silver(I) perchlorate or nitrate to an aqueous suspension or 50% methanolic solution of *ms*-(5,12)-Me$_6$[14]aneN$_4$ resulted in the disproportionation of silver(I) to the free metal and the yellow complex Ag(*ms*-(5,12)-Me$_6$[14]aneN$_4$)(ClO$_4$)$_2$ (or nitrate*). However, in dry acetonitrile, the white compound Ag(*ms*-(5,12)-Me$_6$[14]aneN$_4$)ClO$_4$ precipitates. This species is insoluble in common solvents, but in water or methanol it disproportionates to give the silver(II) complex. Recently, Dei and Mani[168,169] described the syntheses of V(*ms*-(5,12)-Me$_6$[14]aneN$_4$)X$_2$ (X = Cl, Br, I) and both blue and purple forms of chromium(II) complexes Cr(*ms*-(5,12)-Me$_6$[14]aneN$_4$)X$_2$ (X = Cl, Br, I). It was proposed that in the purple chromium(II) complexes, the macrocyclic ligand adopts a planar conformation, whereas in the blue species, the ligand is folded.

The reaction between Ni$_2$(trien)$_3$Cl$_4$ (trien = triethylenetetramine) and *iso*-propylaldehyde in the presence of zinc chloride results in the formation of the nickel(II) complex of **32**, Me$_2$[13]eneN$_4$, isolated as the ZnCl$_4^{2-}$ salt.[170]

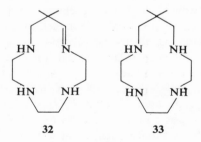

      32   **33**

The coordinated ligand may be reduced by NaBH$_4$ or hydrogen over platinum to give complexes of the fully saturated ligand **33**, Me$_2$[13]aneN$_4$. A crystal structure of [Ni(Me$_2$[13]aneN$_4$)](ClO$_4$)$_2$ has been reported.[171] Treatment of

* The crystal and molecular structure of this compound has recently been reported.[167]

[Ni(Me$_2$[13]aneN$_4$)]$^{2+}$ with cyanide results in displacement of the ligand **33**. Cobalt(III) complexes of **33** may be synthesized by reaction of the ligand with Na$_3$[Co(CO$_3$)$_3$]·3H$_2$O in water or methanol or from the ligand and a cobalt(II) salt followed by oxidation.[172] Both planar and folded conformations for the ligand have been observed, and complexes prepared with the amine in three configurations of the chiral nitrogen centers. Interconversion of these configurations occurs slowly under acidic conditions, rapidly under basic conditions. *N*-methylation of **33** has been achieved and complexation reactions of the substituted ligand with nickel(II) and copper(II) reported by Buxtorf *et al.*[173]; five-coordination in the complexes is indicated.

The ligands **30** and **31** are both derivatives of the unsubstituted cyclic amine 1,4,8,11-tetraazacyclotetradecane **34**, [14]aneN$_4$, often referred to by the trivial name cyclam. The amine was first synthesized by Van Alphen[21]

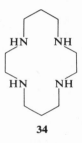

34

and later by Stetter and Mayer.[174] The yields of amine from these syntheses were low, and an *in situ* synthesis has been developed by Barefield *et al.*[23,24,31] In this synthesis, a condensation reaction between 1,5,8,12-tetraazadodecane and glyoxal in the presence of nickel(II) ions is followed by reduction with hydrogen using a Raney nickel catalyst, and then displacement of [14]aneN$_4$ by the addition of cyanide ions. Bosnich *et al.*[20] prepared the first complexes of [14]aneN$_4$ by reaction of the ligand, prepared by a modification of Van Alphen's synthesis, with cobalt(II) chloride in methanol followed by oxidation of the reaction medium by air in the presence of hydrochloric acid (see Table 4). The species *trans*-[Co[14]aneN$_4$)Cl$_2$]Cl may be isolated. This complex can be used as a starting material from which a variety of complexes of general formula [(Co[14]aneN$_4$)XY]$^{n+}$ (X = Y = Br, NO$_2$, NCS, N$_3$, n = 1; X = Y = NH$_3$, n = 3; X = Cl, Y = NCS, N$_3$, n = 1) may be obtained by metathetical reactions. Some of these species may also be obtained directly from the ligand and appropriate salts. Complexes of nickel(II) halides and nickel(II) perchlorate with [14]aneN$_4$ were also prepared by Bosnich *et al.*[175] The species Ni([14]aneN$_4$)X$_2$ (X = Cl, Br) are paramagnetic in the solid state, whereas when X = I or ClO$_4$, the complexes are diamagnetic. A crystal and molecular structure determination of the chloro species showed the ligand coordinated in a planar manner about the nickel(II) ion with the chloride

TABLE 4. Complexes of Certain Unsubstituted Tetramines, L

$$
\begin{array}{c}
-(CH_2)_n- \\
HN \quad\quad NH \\
(CH_2)_m \quad\quad (CH_2)_k \\
HN \quad\quad NH \\
-(CH_2)_l-
\end{array}
$$

Name	k	l	m	n	Complex	Metal Ion	Reference
([12]aneN$_4$)	2	2	2	2	*cis*-[CoLX$_2$]$^{n+}$	Co(III)	86, 196
					cis-[RhLX$_2$]$^{n+}$	Rh(III)	86
					NiLX$_2$	Ni(II)	204
([13]aneN$_4$)	2	2	2	3	*cis*-[FeL(CN)$_2$]	Fe(II)	193
					trans-[FeL(NCS)$_2$]	Fe(II)	193
					cis- and *trans*-[CoLX$_2$]$^+$	Co(III)	195
([14]aneN$_4$)	2	3	2	3	*cis*-[CoLX$_2$]$^+$	Co(III)	20, 181, 185, 186, 188
					trans-[CoLX$_2$]$^+$	Co(III)	20, 132, 178–184, 195
					[(CoLXO)$_2$]$^{n+}$	Co(III)	178
					trans-[NiLX$_2$]$^{n+}$	Ni(II)	175, 194
					trans-[NiLX$_2$]$^{n+}$	Ni(III)	191
					trans-[FeLX$_2$]$^{n+}$	Fe(II)	193
					cis- and *trans*-[FeLX$_2$]$^{n+}$	Fe(III)	191
					cis- and *trans*-[MnLX$_2$]$^{n+}$	Mn(III)	191, 198
					cis- and *trans*-[RhLXY]$^{n+}$	Rh(III)	189
					cis- and *trans*-[CrLX$_2$]$^{n+}$	Cr(III)	190
					LiL(ClO$_4$)	Li(I)	192
					(LiX)$_2$L	Li(I)	192
([15]aneN$_4$)	2	3	3	3	*cis*-[FeLNO$_2$]$^+$	Fe(II)	193
					trans-[FeLX$_2$]$^{n+}$	Fe(II)	193
					cis-[CoLCO$_3$]$^+$	Co(III)	195
					trans-[CoLX$_2$]$^{n+}$	Co(III)	195
					NiLX$_2$	Ni(II)	194
					[MnLBr$_2$]$^+$	Mn(III)	198
([16]aneN$_4$)	3	3	3	3	*trans*-[FeLX$_2$]$^{n+}$	Fe(II)	193
					[(FeLNO$_2$)$_2$NO$_2$]$^+$	Fe(II)	193
					cis-[CoLCO$_3$]$^+$	Co(III)	195
					trans-[CoLX$_2$]$^{n+}$	Co(III)	195
					NiLX$_2$	Ni(II)	194

ions in *trans* (axial) positions.[176] Hertli and Kaden[177] investigated the reaction of cyanide ions with nickel(II) complexes of [14]aneN$_4$. The adducts [Ni([14]aneN$_4$)CN]$^+$ and [Ni([14]aneN$_4$)(CN)$_2$] were detected as intermediates before displacement of the ligand from the nickel(II) ion. The addition of cobalt(II) perchlorate to an aqueous solution of [14]aneN$_4$ and a salt of the appropriate anion results in the crystallization of species of general formula $[Co([14]aneN_4)XO]_2^{n+}$ (X = Cl, NCS, NO_3, $n = 2$),[178] which contain a bridging peroxo group between the cobalt(III) ions. In the absence of coordinating anions, the complex *trans*-$[H_2OCo([14]aneN_4)O_2([14]aneN_4)-CoH_2O](ClO_4)_4$ is isolated. The water molecules may be replaced by other anions; this procedure may be used to prepare the species above, and also the azide derivative, which cannot be formed by the direct reaction procedure. The peroxo bridge is readily broken by acids HY, and mixed complexes *trans*-$[Co([14]aneN_4)XY]^+$ (X = NO_2, Y = Cl, Br; X = NCS, Y = Cl; X = N_3, Y = Cl) may be prepared. It is suggested that the peroxo species are the first stable species that can be isolated during the aerial oxidation of aqueous solutions of cobalt(II) salts and the tetramine. Poon and co-workers have carried out extensive studies of the kinetics and mechanisms of substitution reactions of octahedral macrocyclic amine complexes.[131] These reactions are discussed in detail in Chapter 6. During the course of these studies, several new cobalt (III) complexes have been reported: *trans*-$[Co([14]aneN_4)CNX]^+$ (X = Cl,[179] Br[180]), *trans*-$[Co([14]aneN_4)-N_3H_2O]^{2+}$,[180] and *trans*-$[Co([14]aneN_4)(H_2O)_2]^{3+}$,[181] and improved syntheses for some existing complexes have been described.[119,179,180,182–184] Complexes of general-formula *trans*-$[Co([14]aneN_4)(amH)_2](ClO_4)_3$ (amH = glycine, S-alanine, S-phenylalanine, S-leucine) have been reported.[185]

The isolation of small amounts of a material suggested to be *cis*-$[Co([14]aneN_4)Cl_2]Cl$ was first reported by Bosnich *et al.*[20] More recently, Poon and Tobe[181] and Poon[186] described the preparation and characterization of a series of complexes of the type *cis*-$[Co([14]aneN_4)X_2]Y$ (X = $\frac{1}{2}CO_3$, Cl, Y = Cl; X = NO_2, N_3, NCS, Y = NO_3 and $\frac{1}{2}PtCl_6$) and *cis*- and *trans*-$[Co([14]aneN_4)(H_2O)_2](ClO_4)_3$. During the preparative experiments, substitution takes place with retention of configuration although a slow isomerization to the more stable *trans* isomer can be observed. The *cis* to *trans* isomerization of *cis*-$[Co([14]aneN_4)Cl_2]^+$ has been shown to go by way of the labile *cis*-$[Co([14]aneN_4)OH \cdot H_2O)]^{2+}$ cation.[187] Treatment of a methanolic solution of *trans*-$[Co([14]aneN_4)Cl_2]Cl$ with an excess of ethylenediamine (en) results in the crystallization of pink crystals of *cis*-$[Co([14]aneN_4)en]Cl_3 \cdot 3H_2O$.[188] A crystal structure confirmed that the macrocyclic ligand is coordinated to four adjacent octahedral sites, with the ethylenediamine molecule occupying the remaining sites. The synthesis of *cis*-$[Co([14]aneN_4)-ox]ClO_4$ (ox = oxalate ion) from *trans*-$[Co([14]aneN_4)Cl_2]Cl_2$ and potassium oxalate has been described.[185] Series of complexes of the type *cis*- and *trans*-$[Rh([14]aneN_4)XY]^{n+}$ (X = Y = Cl, Br, I, N_3, NO_2, NCS, OH, $n = 1$; X = Y = H_2O, $n = 3$; X = OH, Y = H_2O, $n = 2$) and *trans*-$[Rh([14]-aneN_4)XY]^{n+}$ (X = Cl, Y = OH, Br, I, N_3, NCS, $n = 1$; X = Br, Y = OH, I, N_3, NCS, $n = 1$; X = I, Y = OH, N_3, NCS, $n = 1$; X = N_3, Y = OH, $n = 1$; X = H_2O, Y = OH, Cl, Br, I, N_3, $n = 2$) have been described.[189]

The *cis* and *trans* isomers were assigned on the basis of their absorption spectra, and *cis*-$[Rh([14]aneN_4)Cl_2]^+$ was resolved via the *d*-α-bromocamphor-π-sulfonate. Similar series of complexes containing chromium(III) have been reported.[190] These include *cis*-$[Cr([14]aneN_4)X_2]^+$ (X = Cl, Br, NCS, NO_2, N_3) and *trans*-$[Cr([14]aneN_4)Cl_2]^+$. The *cis*- and *trans*-$[Cr([14]aneN_4)(H_2O)_2]^{3+}$ cations, together with the corresponding hydroxyaquo and dihydroxo derivatives have also been characterized. Chan and Poon[191] described the synthesis and characterization of complexes of the type *cis*- and *trans*-$[Fe([14]aneN_4)X_2]^+$ (X = Cl, Br, NCS), *trans*-$[Mn([14]aneN_4)X_2]^+$ (X = Cl, Br, NCS, N_3), and *trans*-$[Ni([14]aneN_4)X_2]^+$ (X = Cl, Br). The assignment of geometrical configuration was made on the basis of infrared spectra in the 790–910 cm^{-1} region. Fenton *et al.*[192] showed that $[14]aneN_4$ reacts with lithium salts to give complexes of the type $Li([14]aneN_4)ClO_4$ and $(LiX)_2([14]aneN_4)$ (X = Br, I, ClO_4). Recently, Watkins *et al.*[193] reported the syntheses of some iron(II) complexes with [13]-, [14]-, [15]-, and [16]aneN$_4$. Great care is necessary to prevent contamination of the products with iron(III) or unsaturated species; consequently the compounds were not exposed to oxygen or water. The complexes $Fe([n]aneN_4)X_2$ (n = 13, X = NCS, CN; n = 14, X = NCS, CN, NO_2; n = 15, X = NCS, CN; n = 16, X = NCS), $[Fe([n]aneN_4)(CH_3CN)_2](PF_6)_2$ (n = 14, 15, 16), $[Fe([15]aneN_4)NO_2]PF_6$, and $[[Fe([16]aneN_4)NO_2]_2NO_2]PF_6$ were prepared. Species containing $[14]aneN_4$ were assigned a *trans* configuration similar to iron(II) complexes of *ms*-$Me_6[14]aneN_4$.[157] With $[13]aneN_4$, the NCS$^-$ derivative is assigned a *cis* configuration, whereas the CN$^-$ is *trans*; with $[15]aneN_4$ and the unidentate ligands CH_3CN, NCS$^-$, and CN$^-$, a *trans* configuration is assigned but the nitrite ion is found to be bidentate; thus the configuration must be *cis* in this compound. Both the acetonitrile and cyanide derivatives with $[16]aneN_4$ are assigned *trans* configurations.

Nickel(II) and cobalt(III) complexes with $[n]aneN_4$ ligands have also been prepared; the complexes $Ni([n]aneN_4)X_2$ (n = 14, X = N_3; n = 15, 16, X = Cl, Br, N_3, NCS), *cis*-$[Co([n]aneN_4)X_2]^+$ (n = 13, X = Cl, Br; n = 15, 16, X = CO_3), and *trans*-$[Co([n]aneN_4)X_2]^+$ (n = 13, X = N_3, NCS, NO_2, CN; n = 14, X = Cl; n = 15, 16, X = Cl, Br, N_3, NCS, NO_2, CN) were isolated.[194,195] Spectrochemical properties of the nickel and cobalt complexes and the kinetics of aquation of the dichloro cobalt macrocyclic ligand complexes have been reported.[194–196] The EPR spectra of the nickel(III) complex, *trans*-$[Ni([14]aneN_4)X_2]ClO_4$ (X = Cl, Br) and of the iron(III) species *trans*-$[Fe([14]aneN_4)X_2]^+$ (X = Cl, NCS, Br) have been obtained and discussed[197]; the nickel(III) complexes undergo dissociation in DMSO.*

---

* Bryan and Calvert[198] recently reported complexes of manganese(III) with [14]- and [15]aneN$_4$; these complexes were obtained by oxidation of manganese(II) chloride or bromide with chlorine or bromine in the presence of the ligand.

N-alkylation reactions of [14]aneN$_4$ and its metal complexes have been reported[50,51,173,199-201]; these reactions are discussed in Chapter 6. Nickel(II), copper(II), zinc(II), and iron(II) complexes with *N*-Me$_4$[14]aneN$_4$ have been synthesized.[50,199,200,202] Many of these complexes are suggested to be five-coordinate; an X-ray structure determination of [Ni(*N*-Me$_4$[14]-aneN$_4$)N$_3$]ClO$_4$ showed that five-coordination is forced on the metal ion by the stereochemistry of the nitrogen donor set.[199]

Complexes of 1,4,7,10-tetraazacyclododecane, cyclen, and [12]aneN$_4$ with cobalt(III) and rhodium(III) were first described by Collman and Schneider.[86] The ligand was synthesized by the method of Stetter and Mayer[174] and the complexes [Co([12]aneN$_4$)X$_2$]$^+$ (X = Cl, Br, NO$_2$, $\frac{1}{2}$CO$_3$, $\frac{1}{2}$C$_2$O$_4$) and [Rh([12]aneN$_4$)Cl$_2$]$^+$ obtained from it by reaction with cobalt(II) salts, followed by oxidation, or with rhodium(III) chloride, respectively. All complexes were assigned a *cis* configuration. The structure of [Co([12]aneN$_4$)(NO$_2$)$_2$]Cl has been reported.[203] The reaction of cyanide ions with nickel(II) complexes of [12]aneN$_4$ has been studied by Hertli and Kaden.[177] A five-coordinate adduct is proposed before metallation of the complex occurs. An improved synthesis of [12]aneN$_4$ has been reported.[22]

The complexes Ni([12]aneN$_4$)X$_2$ (X = Cl, Br, NO$_3$) have recently been prepared by reaction of the ligand with an appropriate nickel(II) salt.[204] The physical properties of these complexes suggest that in all cases the nickel(II) ion is six coordinated. An aqueous solution of Ni([12]aneN$_4$)$^{2+}$ is blue with an electronic spectrum typical of high-spin, octahedral nickel(II) in which two water molecules are coordinated in *cis* positions. Addition of large amounts of NaClO$_4$ results in a brown solution; if the solution is heated it becomes bright yellow.[205] These color changes are interpreted in terms of the presence in solution of a yellow, diamagnetic nickel(II) complex in which [12]aneN$_4$ coordinates in a square-planar manner. Thus, *under forcing conditions*, the 12-membered macrocycle [12]aneN$_4$ is able to encircle a nickel(II) ion (see p. 30). A study of the blue to yellow conversion of the nickel(II) complex of [14]aneN$_4$ has recently been completed; the yellow form is produced by either increasing the temperature or by increasing the ionic strength of the solution.[206]

Kalligeros and Blinn[87] have described syntheses of complexes containing 1,4,7,10-tetrabenzyl-1,4,7,10-tetraazacyclododecane (*N*-bzl$_4$[12]aneN$_4$, **18**). The species Ni(*N*-bzl$_4$[12]aneN$_4$)X$_2 \cdot n$H$_2$O (X = Cl, $n = 0.5$; X = NO$_3$, $n = 0.5$; X = Cl·ClO$_4$, $n = 1$) were isolated and the ligand shown to coordinate in a folded conformation. The structures of chloride and chloride–perchlorate derivatives were suggested to be trigonal bipyramidal, while the nitrate derivative was assigned a *cis* octahedral structure. Other *cis* octahedral species Ni(*N*-bzl$_4$[12]aneN$_4$)X (X = CO$_3$, oxalate) and Co(*N*-bzl$_4$[12]aneN$_4$)X$_2$ (X = NO$_3$) have been prepared.[267] The cobalt(II) complexes are not readily oxidized. A comparison of the physical properties of the Ni(II) and Cu(II) complexes of N-bzl$_4$[12]aneN$_4$ and [12]aneN$_4$ has recently been made.[204–206,208]

Reactions between a tetradentate ligand and a difunctional reagent in the presence of a metal ion have been extensively used for the synthesis of complexes containing tetradentate macrocyclic ligands. When the reactants involve amines and carbonyl-containing compounds, ionization of an amine hydrogen often occurs and the resultant macrocyclic ligand is thus anionic.

Jäger has studied reactions between diamines and β-ketoiminato complexes of the type **35**.[35] The square-planar macrocyclic complexes **36**,[34,39]

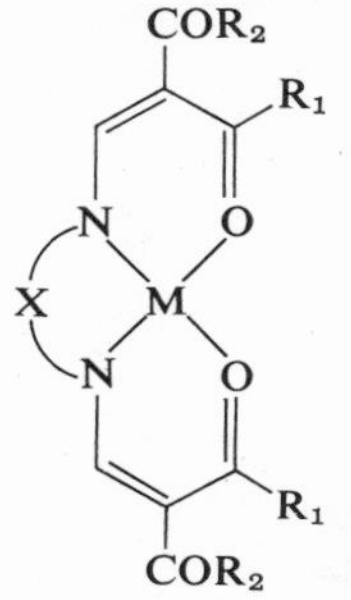

35

37,[36,38] and **38**[37] have been synthesized by reaction of the appropriate β-ketoiminato complex with *o*-phenylenediamine, ethylenediamine, and 1,3-propanediamine, respectively, in the absence of a solvent. Treatment of the copper(II) complexes of some of these ligands with H$_2$S results in the elimina-

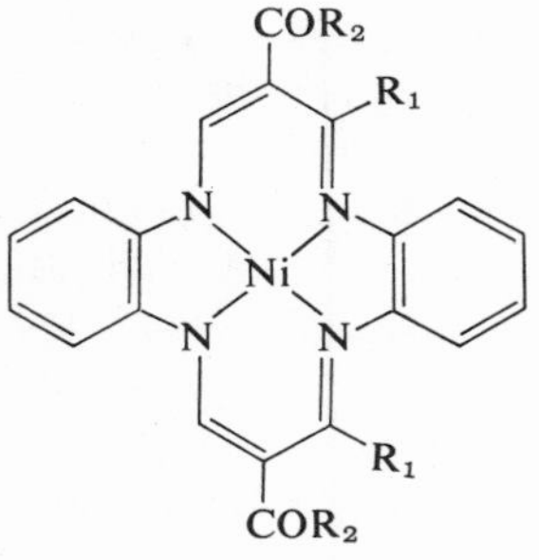

$R_1 = CH_3$; $R_2 = CH_3$, C_6H_5, OC_2H_5
$R_1 = C_6H_5$; $R_2 = OC_2H_5$

36

tion of CuS and the isolation of the free macrocyclic ligand.[38] It should be noted that the reactivity of the β-ketoiminato complexes is dependent on the nature of the substituent in the β (or *meso*) position. When a —COR or —COOR group is present, cyclization occurs although no reaction is observed between ethylenediamine and complexes that do not have these substituents present.[15] Cyclization does occur, however, when compounds having X = *o*-C$_6$H$_4$, with the substituent in the *meso* position being H or

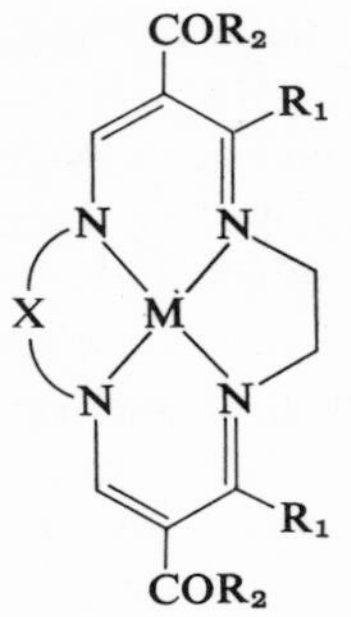

X = —(CH₂)₂—
R₁ = CH₃; R₂ = CH₃, C₆H₅, OC₂H₅; M = Cu, Ni
X = *o*-C₆H₄
R₁ = CH₃; R₂ = CH₃; M = Cu, Ni
R₂ = C₆H₅; M = Cu

37

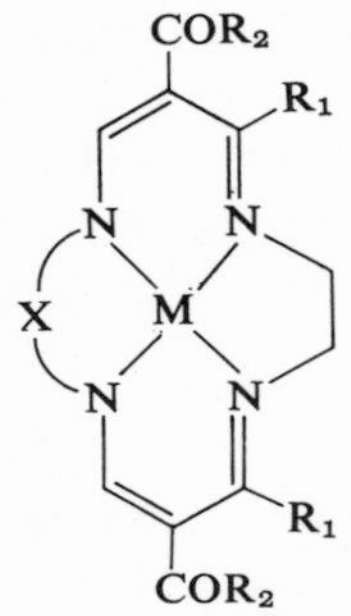

X = —(CH₂)₃—
R₁ = CH₃; R₂ = CH₃; M = Cu, Ni
X = —(CH₂)₃—
R₁ = CH₃; R₂ = CH₃; M = Ni

38

—COR, are heated in molten *o*-phenylenediamine.[39]* When the *meso* substituent and R₁ are part of a cyclohexane ring and X = *o*-C₆H₄, cyclization will not take place with either aliphatic or aromatic diamines.[40] If the cyclohexane ring contains a —CO group, cyclization does not take place with *o*-phenylenediamine, but with ethylenediamine; cyclization is observed, accompanied by amine exchange. Funke and Melson[210] studied the rate of interchange between "coordinated" and "uncoordinated" —COCH₃ groups in **35** (X = —(CH₂)₂—, M = Ni, R₁ = R₂ = CH₃). The preexchange lifetime τ, i.e., the lifetime of a particular configuration prior to isomerization taking place, was calculated and it was suggested that the carbon atom of the coordinated CO group is the preferred site for nucleophilic attack by amines in the ring closure process. These authors have also carried out kinetic and mechanistic studies on the ring closure reaction of the same complex with ethylenediamine in tetrahydrofuran/ethanol.[211] The reaction is second order in amine and first order in both complex and hydroxide ion and is solvent dependent, the rate of ring closure decreasing as the polarity of the solvent increases. A mechanism involving a preequilibrium between the complex and

---

* A crystal and molecular structure of the macrocyclic complex with hydrogen as the *meso* substituent has been reported.[209]

two molecules of diamine followed by base-catalyzed elimination of water and rapid ring closure has been proposed.[212] Knowledge of the factors controlling the formation of these "Jäger-type" macrocyclic complexes has enabled new, efficient syntheses to be developed.[210,211] These syntheses take place in tetrahydrofuran/ethanol solution in the presence of hydroxide ion and an excess of amine. (No macrocycle is produced in the absence of a metal ion under otherwise identical conditions, indicating that the template effect is operative in these syntheses.) The new procedures give higher yields of the complexes than the original processes described by Jäger which involve heating a β-ketoiminato complex with anhydrous liquid amine in the absence of solvent. Reactions of the macrocyclic complexes[52,53] will be discussed in Chapter 6.

Green and co-workers[213-215] have described reactions between the dialdehyde **39*** and amines in the presence of hydrated metal(II) acetates in methanol. Cobalt(II) complexes were prepared under N_2 to prevent oxidation. The neutral, square-planar macrocyclic complexes **40** are obtained, with R = —$(CH_2)_2$—, —$(CH_2)_3$—, —$(CH_2)_4$—, o-C_6H_4; M = Ni, Co, Cu.

39   40

It was originally reported[213] that, in the absence of metal ions, a high-melting polymeric material was obtained that does not react with nickel, cobalt, or copper ions. However, it was subsequently shown[214] that, under conditions of high dilution, the dialdehyde **39** and ethylenediamine react together to produce the macrocyclic ligand in about 50% yield, which reacts with nickel(II) acetate in a few minutes in boiling methanol to form the macrocyclic complex **40** (R = —$(CH_2)_2$—, M = Ni), Ni(bzo$_2$[14]tetraen-atoN$_4$). The polymeric material from ethylenediamine is converted to the macrocyclic complex after boiling with methanolic nickel(II) acetate for 7 days. Unlike ethylenediamine, 1,3-diaminopropane and 1,4-diaminobutane

* Improved syntheses of **39** and similar diamino dialdehydes have recently been reported.[216]

react with **39** in the absence of metal ions under conditions of "normal concentration" to form the corresponding macrocyclic ligands, bzo$_2$[15]-tetraeneN$_4$ and bzo$_2$[16]tetraeneN$_4$, with no polymeric material being detected. 1,2-diaminobenzene does not react with **39** in boiling methanol over a period of six days, although in the presence of zinc(II), AlCl$_3$, SbF$_3$, and the pyridinium ion, the free macrocyclic ligand, bzo$_3$[14]pentaeneN$_4$, with R = o-C$_6$H$_4$, is formed. Reaction between **39** and ethylenediamine in the presence of iron(II), zinc(II), and cadmium(II) produces a mixture of the metal-free macrocycle and polymeric material, but no macrocyclic complex analogous to **40**. Only polymer is formed in the presence of magnesium(II), manganese(II), silver(I), and aluminum(III) chloride. The significance of

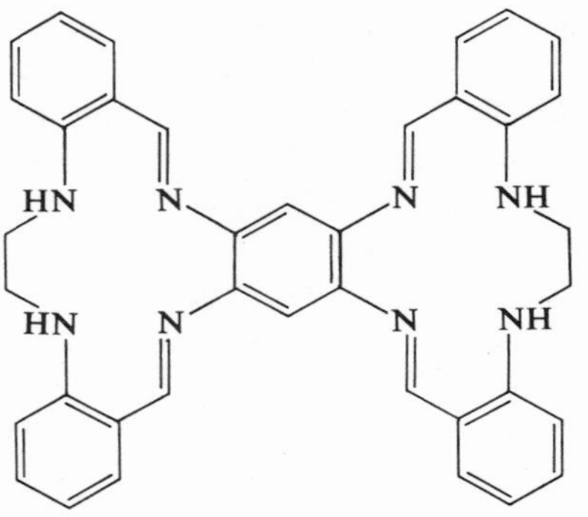

41

these results and their relation to proposed mechanisms for the formation of the various products has been discussed.[214] The ESR spectrum of **40** (R = —(CH$_2$)$_2$—; M = Cu, Cu(bzo$_2$[14]tetraenatoN$_4$) has been discussed,[217] and structural information on **40** (R = —(CH$_2$)$_4$—, M = Cu, and R = —(CH$_2$)$_2$—, M = Ni) has been reported.[218,219] Fleischer *et al.*[220] described the reaction between **39** and 1,2,4,5-tetraaminobenzene in methanol

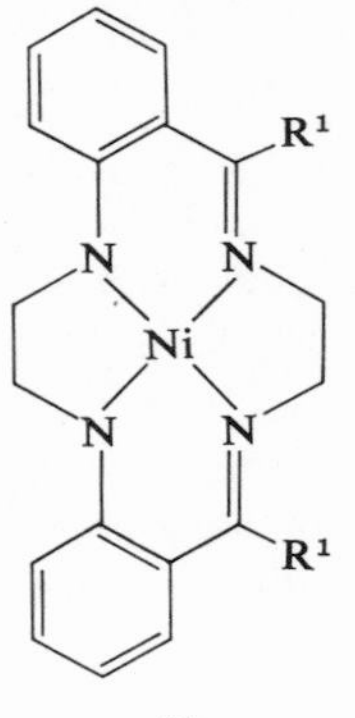

42

under N_2. The ligand **41** is obtained, which when treated with copper(II) or nickel(II) acetate in pyridine incorporates two metal ions, with identical N_4^{2-} donor sets and an intramolecular M–M distance estimated to be approximately 800 pm. Complexes of general formula **42** ($R^1 = H$, CH_3, C_6H_5) have been prepared by reaction of 1,2-dibromoethane with nickel(II) complexes of the tetradentate ligands obtained from *o*-aminobenzaldehyde derivatives and ethylenediamine. When $R^1 = H$, the complex is identical to **40** ($R = —(CH_2)_2—$, $M = Ni$).

Black and Lane[222] reported syntheses of the neutral, square-planar nickel(II) complexes **43** by reaction of the dinitro dialdehyde with ethylene-

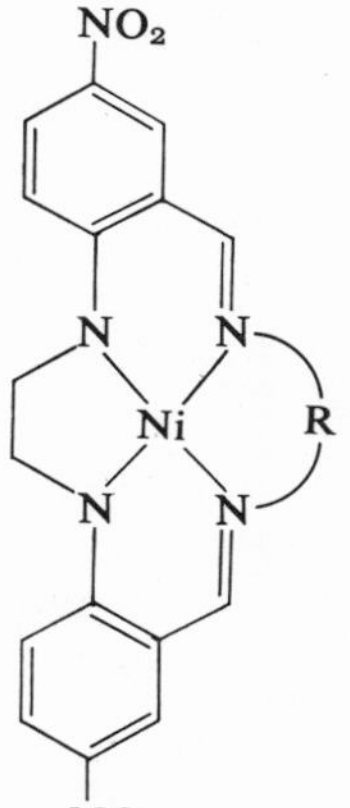

$R = —(CH_2)_2—,$

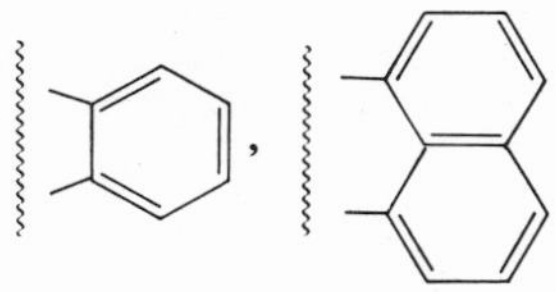

43

diamine, *o*-phenylenediamine, and 1,8-diaminonaphthalene in the presence of nickel(II) acetate in boiling DMF. The generality of this reaction was later

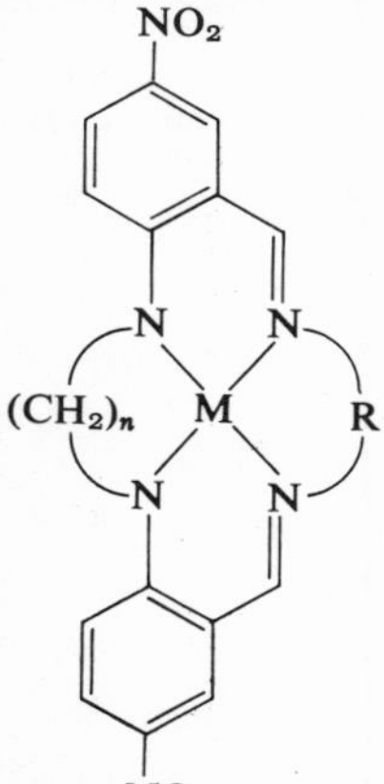

$n = 2; R = —(CH_2)_3—, —C(CH_3)_2CH_2—,$

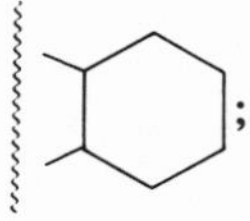

; $M = Ni, Co, Cu$

$n = 3; R = —(CH_2)_2—, —(CH_2)_3—, —C(CH_3)_2CH_2—,$

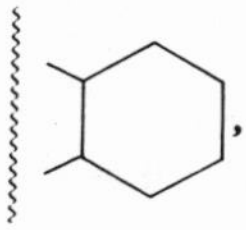

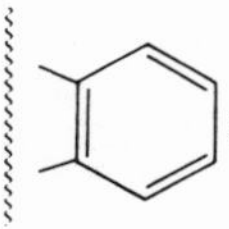

; $M = Ni, Co, Cu$

44

demonstrated[223] and complexes **44** were obtained. Similar macrocyclic complexes of manganese, magnesium, and zinc could not be prepared, and polymeric organic materials were obtained from reactions employing these metal ions. Reaction of **45** with anhydrous nickel(II) acetate in anhydrous MeOH, followed by addition of the appropriate diamine acid $Et_4N^+I^-$, produces complexes containing the macrocycle **46** $(R = -(CH_2)_2-,$

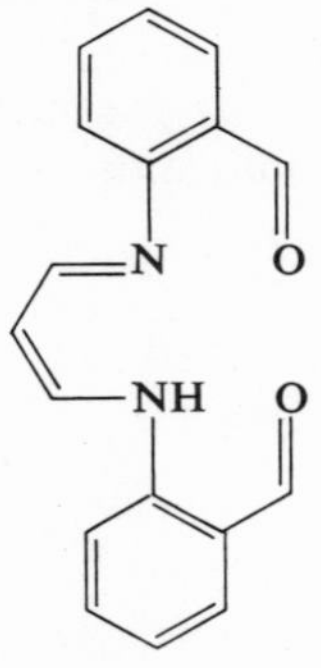

45 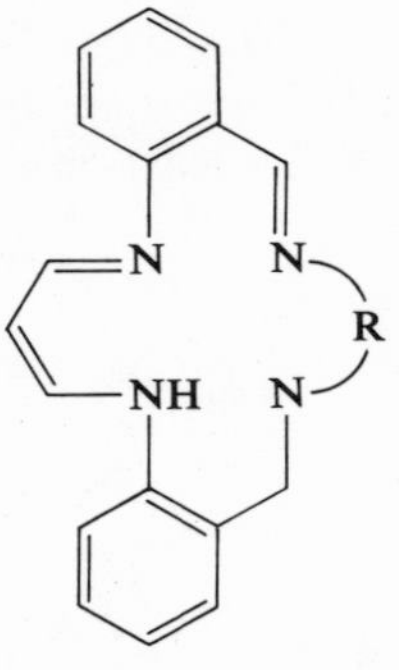   **46**

$-(CH_2)_3$, $-CH(CH_3)CH_2-$) and o-C_6H_4, which coordinates as a mono-anionic ligand.[224] Modifications of the procedure have led to the preparation

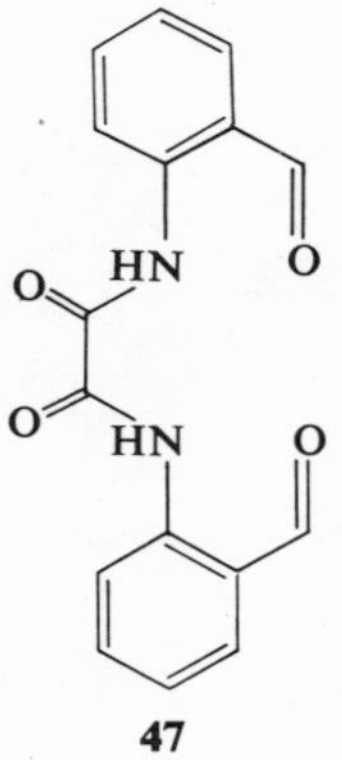

47 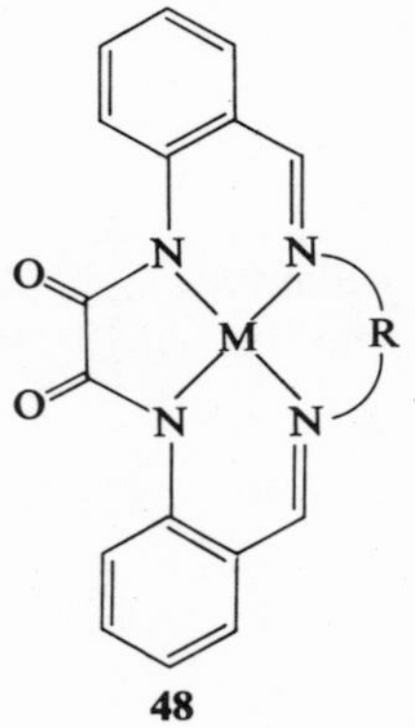   **48**

$R = -(CH_2)_2-, -(CH_2)_3-, M = Ni, Cu; R =$

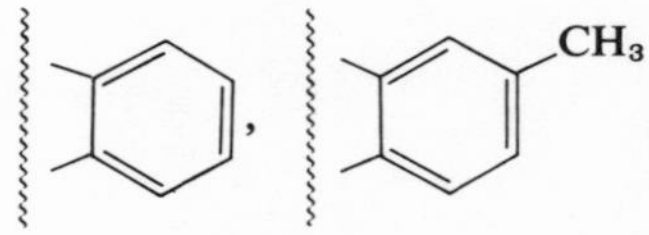

, M = Ni, Co

$R =$

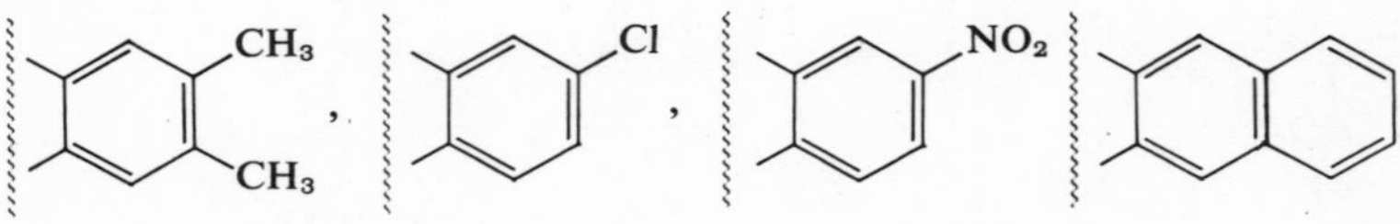

, M = Ni

of the copper(II) monoperchlorate and *bis*-thiocyanotocobalt(III) complexes of **46**, R = —(CH$_2$)$_2$—. Recently,[225] reactions of **47**, with amines and a hydrated metal acetate in DMF under nitrogen at 120°C have been reported; the neutral complexes of **48** crystallize out on cooling.

In the syntheses described above, the macrocyclic ligands were produced by *in situ* reactions between ligands with an N$_2$O$_2$ donor set and diamines. Reactions between a tetradentate amine and a dicarbonyl compound in the presence of a metal ion have also been shown to give rise to complexes containing tetradentate macrocyclic ligands with an N$_4$ donor set. Depending on the nature of the carbonyl compound, the resulting macrocycles may be anionic or neutral.

Cummings and Sievers[226,227] reported syntheses for the nickel(II) complexes, **49** (M = Ni, R = CH$_3$, $x = y = 2$, X = I, Br·$\frac{1}{2}$H$_2$O, SCN,

49   **50**

NO$_3$·$\frac{1}{2}$H$_2$O, BF$_4$, PF$_6$, ClO$_4$; R = CF$_3$, $x = y = 2$, X = I, Br, SCN, NO$_3$, BF$_4$, PF$_6$). In the *in situ* syntheses, triethylenetetramine, acetic acid, and the β-diketone (acetylacetone or trifluoroacetylacetone) are added to a refluxing aqueous solution of nickel(II) acetate. After addition of NaI and base the iodide species crystallize from solution. A second synthetic procedure involves the reaction of the presynthesized sexadentate ligand **51** (R = CH$_3$, CF$_3$) with nickel(II) acetate. The iodide derivatives of **49** (R = CH$_3$, CF$_3$) can be isolated as before. In a third procedure, the neutral nickel(II) complex of **51** (R = CF$_3$[228]) undergoes rearrangement by refluxing in aqueous solution of pH 5 for 6 hr. Addition of NaI to the concentrated solution at pH 10

51

gives **49** (R = CF$_3$, X = I, M = Ni). Attempts to prepare the cyclic ligands in the absence of nickel(II) ions were not successful, indicating the important role played by the metal ion in these macrocycle syntheses. The crystal and molecular structure of **49** (M = Ni, R = CH$_3$, X = ClO$_4$), [Ni(Me$_2$[14]-dienatoN$_4$)]·ClO$_4$, has been reported [229]; the square-planar coordination of the 13-membered tetradentate macrocycle was confirmed. Copper(II) analogs of **49** (M = Cu, R = CH$_3$, X = NO$_3$·H$_2$O, Br·H$_2$O, I·H$_2$O, SCN·H$_2$O, PF$_6$) have been synthesized [48]; however, under more acidic conditions copper(II) species containing **50** (M = Cu, R = CH$_3$, X = I, PF$_6$), in which the ligand is protonated in the γ position, have been isolated. The relative acidity of the coordinated ligand in the copper complex is 1000 times less than in the analogous nickel(II) complex.

The syntheses have been extended [49] to include complexes containing 14-membered macrocyclic ligands, **49** (M = Ni, Cu, R = CH$_3$, $x = 2$, $y = 3$, X = NO$_3$, Br, I, PF$_6$) and **50** (M = Ni, Cu, R = CH$_3$, $x = 2$, $y = 3$, X = I, PF$_6$). The nickel(II) complex containing a neutral 15-membered ligand, **50** (M = Ni, R = CH$_3$, $x = 3$, $y = 2$) has also been obtained, although no complex with the corresponding anionic ligand was formed. No complexes with 16-membered ligands could be obtained. The pK_a values for the coordinated 14-membered ligands are similar to those of the 13-membered analogs. Curtis and Milestone [230] prepared complexes [NiL(acac)]X (L = 1,4,8,11-tetraazaundecane) and triethylenetetramine, and showed that in hot ethanol/acetic acid the macrocyclic complexes **49** (M = Ni, R = CH$_3$, $x = 2$, $y = 3$; M = Ni, R = CH$_3$, $x = y = 2$, respectively) could be obtained. Cyclization does not take place on prolonged refluxing in ethanol or in ethanol containing pyridine. Elfring and Rose [237] have reported syntheses and some reactions of nickel(II) complexes of **49** (M = Ni, R = CH$_3$, $x = y = 2$, X = PF$_6$). The PMR spectra of all the complexes prepared have been discussed. Recently, Holtman and Cummings [232] showed that catalytic reduction of the diimine linkages of **50** (M = Ni, R = CH$_3$, $x = y = 2$; $x = 2$, $y = 3$) in aqueous solution over Raney nickel yields nickel(II) complexes containing the fully reduced dimethyl-substituted ligands, and Roberts *et al.* [223] reported syntheses of low-spin cobalt(II) complexes of **49** (M = Co, R = CH$_3$, $x = 2$, $y = 3$, X = I, PF$_6$) and **50** (M = Co, R = CH$_3$, $x = 2$, $y = 3$, X = I, PF$_6$, $\frac{1}{2}$[Co(NCS)$_4$]). The crystal and molecular structure of **50** (M = Co, R = CH$_3$, $n = 2$, $y = 3$, X = PF$_6$, [Co(12,14-Me$_2$[14]11,14-dieneN$_4$)·H$_2$O](PF$_6$)$_2$) was also reported.

Durham *et al.* [234] recently reported that the cobalt(II) complex [Co([14]4,7-dieneN$_4$)(H$_2$O)$_2$](PF$_6$)$_2$ reacts with oxygen to form a cobalt(II) complex containing the macrocyclic ligand **51A**, in which a ketonic oxygen has been introduced. It is interesting to note that although a cobalt(II) center appears to be necessary for this reaction to occur, it does not undergo oxidation in the process. The crystal and molecular structures of the diaquo and

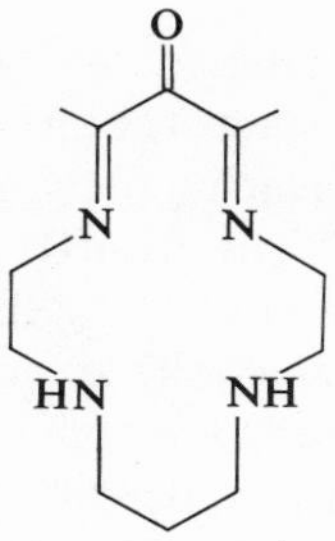

51A

dichloro cobalt(II) complexes with this ligand have been determined; the ketonic oxygen is bent up from the plane of the two imine bonds. No reactions of this keto group have been reported.

Condensation of 1,5,8,12-tetraazadodecane with biacetyl in methanol followed by the addition of cobalt(II) acetate, air oxidation, and perchloric acid results[47] in the isolation of cobalt(III) complexes containing **52**,

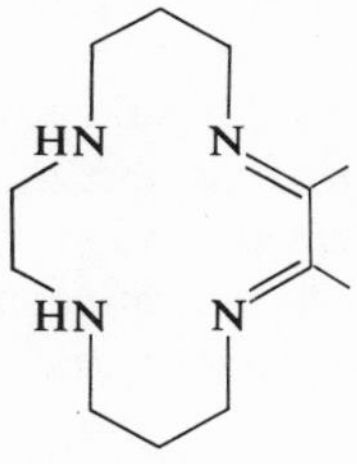

52

9,10-Me$_2$[14]8,10-dieneN$_4$. The complexes *trans*-[Co(Me$_2$[14]dieneN$_4$)X$_2$]Y (X = Br, Y = ClO$_4$, Br, PF$_6$; X = Cl, Y = ClO$_4$, PF$_6$; X = NO$_2$, Y = ClO$_4$) may also be isolated. Poon and Wong[184,235] prepared *trans*-[Co(Me$_2$[14]dieneN$_4$)N$_3$X]ClO$_4$ (X = Cl, Br), *trans*-[Co(Me$_2$[14]dieneN$_4$)(NO$_2$)X]ClO$_4$ (X = Cl, Br), and *trans*-[Co(Me$_2$[14]dieneN$_4$)(NCS)$_2$]ClO$_4$, and studied their acid hydrolysis. Hydrogenation of [Co(Me$_2$[14]dieneN$_4$)Br$_2$]ClO$_4$ using a Raney nickel catalyst gives rise to [Co(Me$_2$[14]aneN$_4$)Br$_2$]-ClO$_4$, containing the fully reduced ligand Me$_2$[14]aneN$_4$, **53**.[45] Cobalt(III)

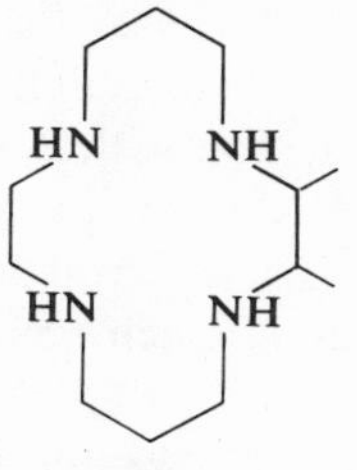

53

complexes of **53** can also be prepared by hydrogenation of [Ni(Me$_2$[14]-dieneN$_4$)](ClO$_4$)$_2$ in methanol using a Raney nickel catalyst followed by removal of the ligand by treatment of the complex with cyanide ions, and complexation with cobalt salts. The iron(II) complex [Fe(Me$_2$[14]aneN$_4$)-(CH$_3$CN)$_2$](ClO$_4$)$_2$ has been synthesized by the reaction of the ligand with anhydrous iron(II) acetate in acetonitrile under N$_2$.[123] The perchlorate derivative was obtained by the addition of perchloric acid to the solution. If oxygen gas is bubbled through a perchloric acid solution of the above complex, the iron(III) species [Fe(Me$_2$[14]aneN$_4$)(CH$_3$CN)$_2$](ClO$_4$)$_3$, crystallizes from solution. The nickel(III) complexes [Ni(Me$_2$[14]aneN$_4$)X$_2$]ClO$_4$ (X = Cl, Br, NCO, NO$_3$, $\frac{1}{2}$SO$_4$) have been prepared by oxidation of Ni(Me$_2$[14]aneN$_4$)(ClO$_4$)$_2$ with (NH$_4$)$_2$S$_2$O$_8$ in aqueous solution or concentrated nitric acid, followed by addition of the appropriate anion.[236]

The complex Ni(Me$_2$[14]aneN$_4$)(ClO$_4$)$_2$ can be isolated from the hydrogenation of Ni(Me$_2$[14]dieneN$_4$)(ClO$_4$)$_2$; the species Ni(Me$_2$[14]-aneN$_4$)X$_2$ (X = Cl, Br, N$_3$, NCS) can be prepared by metathesis from the perchlorate salt, or by reaction of the free ligand with the appropriate metal salt.[194]

Barefield and co-workers[23,24,31] have reported reactions between glyoxal and a variety of tetradentate amines as part of the development of synthetic procedures for completely saturated tetradentate macrocyclic ligands. The only "intermediate" containing the α-diimine group that was isolated[31] was **54**. The reaction product from the condensation of 1,5,8,12-tetraazadodecane

54

with glyoxal in the presence of nickel(II) ions was isolated and tentatively assigned as an imine–carbinolamine rather than a diimine. It was noted[31] that the metal-ion-assisted cyclization reaction employing glyoxal was more efficient than a corresponding reaction with 2,3-butanedione (biacetyl).

Black and Lane[237] isolated the complexes **55** and **56** from reactions involving biacetyl and a tetramine in the presence of nickel(II) salts. Generalization of this reaction type was not achieved because of the difficulty of purification of the products.

In situ reactions between α-diketones and diamines in the presence of a metal ion have been employed for the synthesis of complexes containing

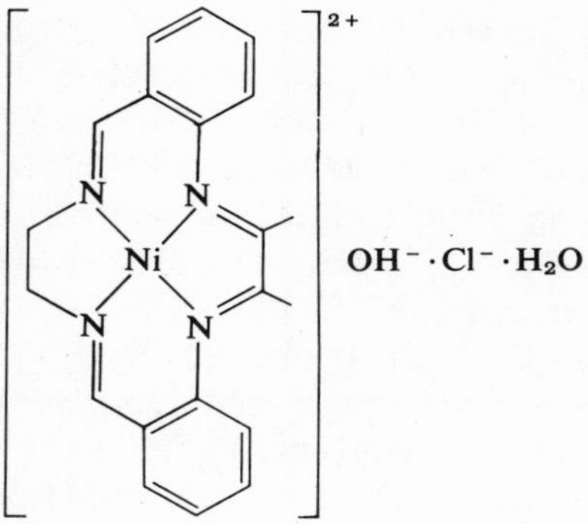

$OH^- \cdot Cl^- \cdot H_2O$

$ClO_4^- \cdot H_2O$

55 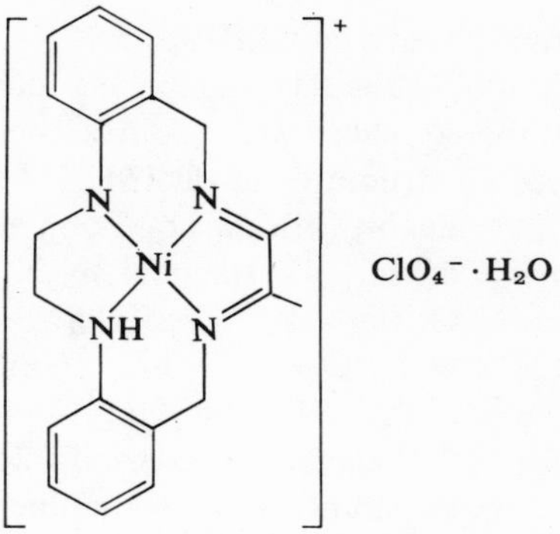   **56**

macrocyclic ligands. Baldwin and Rose[238] reported the synthesis of nickel(II) complexes of **9**, $Me_4[14]$tetraeneN_4,* by reaction of 1,3-propanediamine monohydrochloride with biacetyl in methanol in the presence of nickel(II)

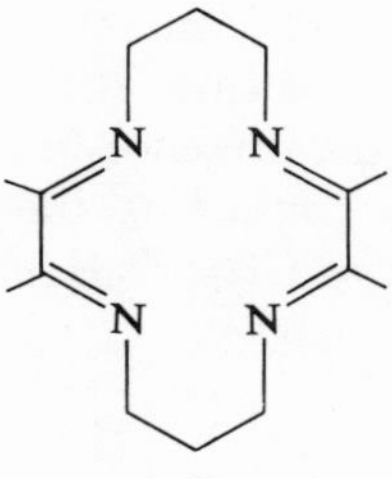

9

acetate. The complexes $Ni(Me_4[14]$tetraene$N_4)X_2$ ($X = \frac{1}{2}ZnCl_4$, PF_6, NCS) were isolated. Cobalt(III) complexes, $[Co(Me_4[14]$tetraene$N_4)X_2]Y$ ($X = Cl$, $Y = ClO_4$, PF_6; $X = Br$, $Y = PF_6$, Br, ClO_4; $X = Y = I$; $X = NO_2$, ClO_4), were prepared by a similar procedure to that outlined above using cobalt(II) acetate, with no attempt made to protect the reaction mixture from the atmosphere.[47] If the reaction is carried out under an N_2 atmosphere, the complexes $[Co(Me_4[14]$tetraene$N_4)X]BPh_4$ ($X = Cl$, Br, I) are obtained.[6] Compounds containing a cobalt–alkyl bond and the macrocyclic ligand $Me_4[14]$tetraeneN_4 have been obtained from the cobalt(II) and cobalt(III) complexes of the ligand by reduction with $NaBH_4$ followed by the addition of a large excess of the alkyl halide, or directly from the cobalt(II) complexes by reaction with alkyl halides.[239] The species $[RCo(Me_4[14]$tetraene$N_4)X]^{n+}$ ($R = CH_3$, $X = Cl$, Br, I, $n = 1$, $X = CH_3CN$, $n = 2$; $R = C_6H_5CH_2$, $X = Cl$, Br, I, $n = 1$, $X = CH_3CN$, $n = 2$) were synthesized and characterized. The syntheses involving reduction are proposed to proceed via the

* The complete abbreviation is $2,3,9,10$-$Me_4[14]$-$1,3,8,10$-tetraene-$1,4,8,11$-N_4.

formation of a cobalt(I) species. Recently, Tait *et al.*[126] generated, in aqueous solution, cobalt(I) complexes containing a variety of tetradentate 14-membered macrocyclic ligands, including $Me_4[14]$tetraeneN$_4$. Glick *et al.*[240] reported structures of $[Co(Me_4[14]$tetraeneN$_4)(H_2O)_2](ClO_4)_2$ and $[Co(Me_4[14]$tetraeneN$_4)(NH_3)_2]Br_3$. Poon and Wong[184,235] reported the syntheses of *trans*-$[Co(Me_4[14]$tetraeneN$_4)N_3X]^{n+}$ (X = N$_3$, NCS, Cl, Br, $n = 1$; X = H$_2$O, $n = 2$), *trans*-$[Co(Me_4[14]$tetraeneN$_4)NCSX]^{n+}$ (X = Cl, Br, NCS, $n = 1$; X = H$_2$O, $n = 2$), and *trans*-$[Co(Me_4[14]$tetraeneN$_4)NO_2X]^{n+}$ (X = Cl, Br, NCS, $n = 1$; X = H$_2$O, $n = 2$). Cragel and Douglas[185] described the complexes *trans*-$[Co(Me_4[14]$tetraeneN$_4)(amH)_2]^{3+}$ (amH = glycine, S-alanine, S-phenylalanine, and S-leucine) which were obtained from the reaction of the *trans*-diaquo species with the corresponding amino acid. The synthesis and characterization of the six-coordinate low-spin iron(II) complexes $[Fe(Me_4[14]$tetraeneN$_4)L_1L_2](PF_6)_2$ [L$_1$ = L$_2$ = CH$_3$CN (imidazole) and L$_1$ = CH$_3$CN, L$_2$ = CO] were reported by Baldwin *et al.*[46] The complexes were prepared under N$_2$ and in the presence of tin(II) chloride by an *in situ* reaction involving 1,3-propanediamine, biacetyl, and iron(II) chloride, and precipitated by addition of ammonium hexafluorophosphate.

The reaction of biacetyl or benzil with 1,3-propanediamine or 1,3-diamino-2-hydroxypropane in the presence of cobalt(II) has been shown to produce cobalt(III) complexes containing the hydroxy-substituted macrocyclic ligands **56A** and **56B** (R = CH$_3$, C$_6$H$_5$).[241] The hydroxyl groups on

56A   **56B**

the complexed ligands are unreactive toward a variety of reagents. When 2 moles of benzil react with slightly more than 1 mole of 1,3-propanediamine, the compound **56C** is produced, which will only react further with a second diamine if cobalt(II) ions are present, demonstrating the importance of the metal ion in the ring closure reaction. Substitution of 1,3-diamino-3-hydroxypropane for 1,3-propanediamine in the initial reaction does not give rise to a hydroxy-substituted analog of **56C**.

Until 1971, most complexes containing tetraaza macrocyclic ligands had been prepared by *in situ* reactions between amines and carbonyl compounds

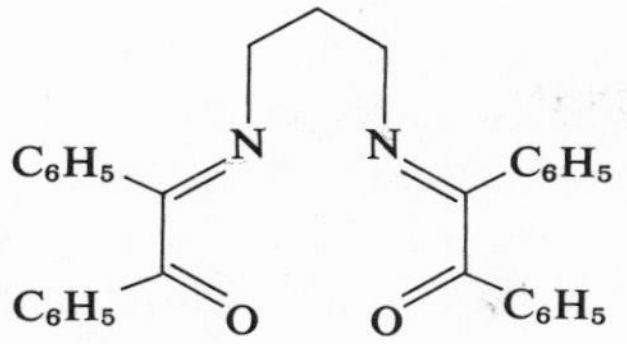

56C

in the presence of a metal ion. There were some notable exceptions, including the organic synthesis of [14]aneN$_4$ and Me$_6$[14]dieneN$_4$ ligands, although yields of these ligands are generally improved by *in situ* syntheses. Truex and Holm[14,15] reported a nontemplate synthesis of **57**, 5,7,12,14-Me$_4$[14]-4,6,11,13-tetraeneN$_4$ (or simply Me$_4$[14]tetraeneN$_4$). This ligand reacts with metal(II) acetates or with other metal(II) salts under conditions of non-

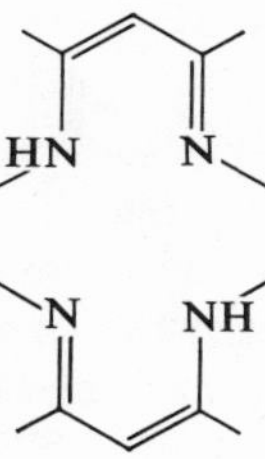

57

aqueous chelation to give the neutral macrocyclic complexes M(Me$_4$[14]-tetraenato(2−)N$_4$) (M = Fe, Co, Ni, Cu, Zn). In these complexes, the ligand coordinates as a dianion, two hydrogen atoms having been ionized. Attempts to prepare complexes derived from the neutral macrocycle **57** were not successful, although the species [Ni(Me$_4$[14]tetraenato(1−)N$_4$)]BF$_4$, in which the ligand coordinates as a −1 ion, was obtained. Recently, Olszanski and Melson[242] isolated the species Sc(Me$_4$[14]tetraeneN$_4$)(NCS)$_3$·$\frac{1}{2}$THF, from the reaction of an ethereal solution of scandium thiocyanate with the ligand in tetrahydrofuran. In this complex, the ligand is in its neutral form. An alternative synthesis of **57** was reported,[17] although attempts to

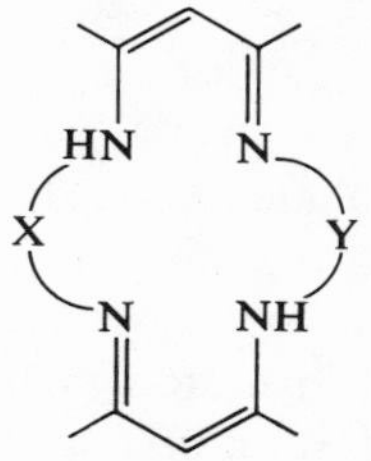

58

synthesize larger tetraaza ring systems by replacement of ethylenediamine with 1,3-propanediamine in these procedures were not successful. However, Busch *et al.*[243] were able to synthesize the *bis*-hexafluorophosphate salt of the 16-membered macrocyclic ligand **58** [X = Y = (CH₂)₃] by a procedure based on that of Holm *et al.*[17] Various authors have noted that **57** could not be obtained by reaction of *bis*-(acetylacetone)ethylenediiminemetal(II) complexes with ethylenediamine even under forcing conditions.

A series of ligands of general formula **59** (X = Y = C_2H_4; X = C_2H_4, Y = C_3H_6; X = $(CH_3)_2CCH_2$, Y = C_3H_6; X = $(CH_3)_2CC(CH_3)_2$, Y =

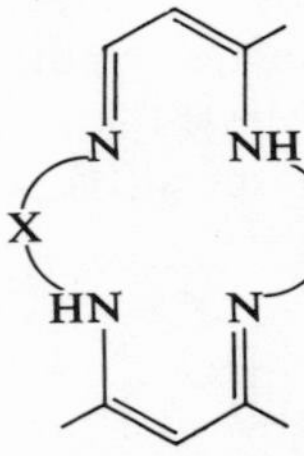

59

C_3H_6; X = Y = C_3H_6) has been synthesized by Busch and co-workers. The ligands are produced by deacylation[52] and demetallation of the corresponding "Jäger-type complex," **60**, with dry HCl gas in ethanol solvent. Addition

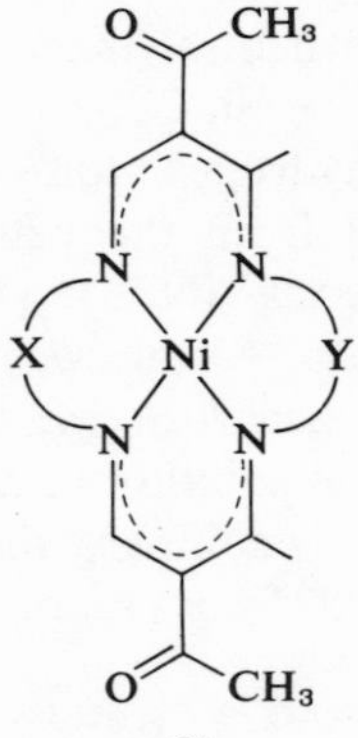

60

of ammonium hexafluorophosphate allows the dihydrohexafluorophosphate salts of **59** to crystallize from solution. The complex *trans*-[Fe(Me₂[14]-tetraeneN₄)(CH₃CN)₂](PF₆)₂ can be synthesized by reaction of [Fe(CH₃CN)₆]-(SO₃CF₃)₂ with the ligand salt in acetonitrile. This complex was used as a starting material for the preparation of *trans*-[Fe(Me₂[14]tetraeneN₄)(py)₂]-(PF₆)₂ (py = pyridine) and *trans*-[Fe(Me₂[14]tetraeneN₄)X₂] (X = NO₂, NCS). When [Fe(CH₃CN)₆]²⁺ is allowed to react with the Me₂[15]tetraeneN₄ and $Me_2[16]tetraeneN_4$ ligands, simple *bis*-acetonitrile adducts are not obtained[243,244]; the acetonitrile molecules add electrophilically to the γ carbons of the macrocycle, producing complexes of novel *cis*-hexaene sexadentate ligands (see p. 124 and Chapter 6). A similar species based on **58** ($X = Y = C_3H_6$, $Me_4[16]tetraeneN_4$) was also reported.[243] These novel complexes are the first examples of low-spin iron(II) coordinated exclusively to six nonconjugated imine donor groups. The crystal and molecular structure of the hexaene complex containing the sexadentate ligand derived from **59**, $X = Y = C_3H_6$ has been determined; the *cis* arrangement for the coordinated iminoethyl groups was confirmed.[245]

Recently, Busch and co-workers[246] discussed deprotonation reactions of iron(II) complexes with ligands derived from **58** and **59**; neutral, four-coordinate square-planar iron(II) complexes are obtained. Reprotonation in the presence of pyridine produces *trans–bis*(pyridine)*bis*(β-diimine) complexes for all species studied. In the presence of acetonitrile, protonation of the complex containing the dianionic ligand derived from **59** ($X = Y = C_3H_6$) regenerated the *trans–bis*(acetonitrile) adduct. However, for the complex containing the 15-membered dianionic ligand ($X = C_2H_4$, $Y = C_3H_6$) the *trans*-acetonitrilepentaene derivative is obtained. With the complex containing the 16-membered dianionic ligand ($X = Y = C_3H_6$) the sexadentate *trans*-hexaene ligand is produced. In these two complexes, acetonitrile electrophilically attacks the γ carbon of a charge-delocalized six-membered chelate ring. The mechanism of this reaction has been discussed; deuteration experiments indicate that the reaction proceeds via initial protonation of the macrocycle, followed by coordination and electrophilic activation of an acetonitrile molecule.

The chromium(III) complex, $[Cr(Me_2[14]tetraenatoN_4)(py)_2]PF_6$, containing the dianionic form of **59** ($X = Y = C_2H_4$) has been prepared[135]; treatment of a methanolic solution of this complex with sodium nitrite produces the chromium(I) complex, $Cr(Me_2[14]tetraenatoN_4)NO$.

A nontemplate synthesis of 14-, 15-, and 16-membered tetraaza macrocycles of general formula **61** ($X = Y = C_2H_4$; $X = C_2H_4$, $Y = C_3H_6$;

61

X = Y = C_3H_6) was devised by Holm and co-workers.[16,17] The method is based on the electrophilic reactivity of 1,2-dithiolium cations and vinylogous β-amino thiones with primary amines. Copper(II), nickel(II), and cobalt(II) complexes of the three macrocycles were prepared by reaction with the metal(II) acetates in DMF; the ligands coordinate as dianions. Reaction of the macrocycles with iron(II) and basic iron(III) acetate in hot DMF gives rise to neutral iron(II) complexes analogous to those above, and the iron(III) complexes **62** (X = Y = C_2H_4, Z = OOCCH$_3$, SC$_6$H$_5$, Br; X = C_2H_4, Y = C_3H_6, Z = OOCCH$_3$, SC$_6$H$_5$, Br; X = Y = C_3H_6, Z = OOCCH$_3$, SC$_6$H$_5$, SCH$_2$C$_6$H$_5$, Br, OOCC$_6$H$_5$).[247] In a later paper,[18] the iron(III)

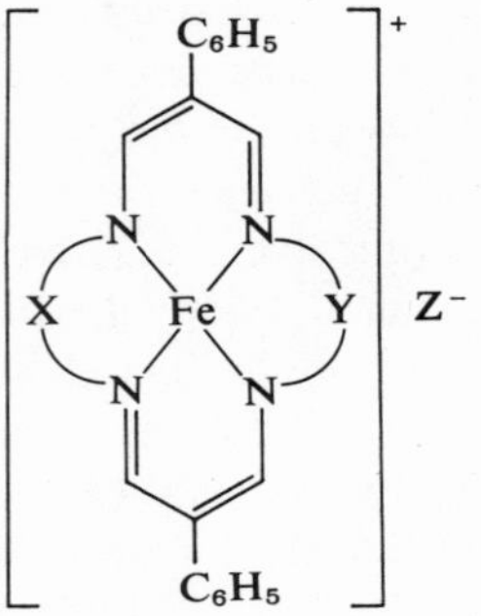

62

complexes **62** (X = Y = C_2H_4, Z = Cl, I, SC$_6$H$_5$·py; X = C_2H_4, Y = C_3H_6, Z = Cl, SC$_6$H$_5$·py; X = Y = C_3H_6, Z = $C_6H_5CH_2C_6H_5$) and [Fe(Ph$_2$[16]tetraenato(2−)N$_4$)]$_2$O were reported. The complexes FeLX (L = dianionic macrocyclic ligand, X = Cl, Br, I, OOCR, SC$_6$H$_5$) are five-coordinate. The magnetic susceptibilities, Mössbauer, and EPR parameters were obtained and discussed.

Oxidative dehydrogenation reactions of a variety of metal complexes containing the ligands **57** and **61** have been reported by Holm and co-workers[248–250]; these are discussed in detail in Chapter 6.

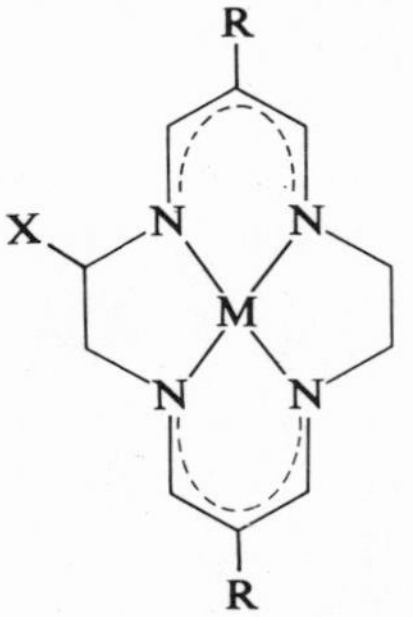

X = H, R = —N=N—C$_6$H$_5$, M = Ni, Cu;
X = H, R = —N=N—(2-CF$_3$)—C$_6$H$_4$, M = Ni, Cu;
X = H, R = —N=N—(4-Cl)-C$_6$H$_4$, M = Ni;
X = H, R = —(4-Cl)-C$_6$H$_4$, M = Ni;
X = CH$_3$, R = —N=N—C$_6$H$_5$

63

L'Eplattenier and Pugin[251] reported the syntheses of the neutral complexes **63** and **64**. These compounds were obtained in high yield by the condensation of aliphatic 1,2- or 1,3-diamines with phenylazo-malondialdehydes in the presence of divalent metal ions such as Ni(II) or Cu(II). Formation of complexes containing a N_2O_2 donor set followed by reaction in ethanol with diamines also leads to the complexes.

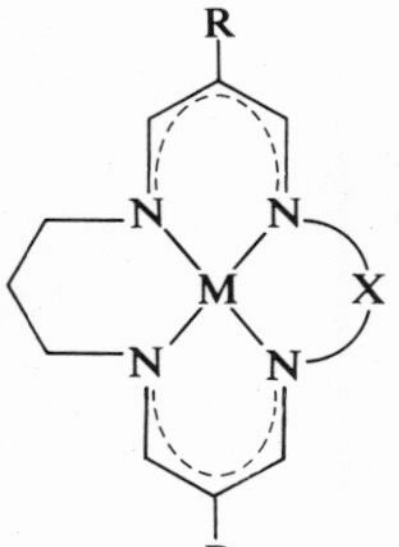

$X = C_3H_6$, $R = $ —N=N—C_6H_5, M = Ni, Cu;
$X = C_3H_6$, $R = $ —N=N—(2-Cl)—C_6H_4, M = Ni, Cu;
$X = C_2H_4$, $R = $ —N=N—C_6H_5, M = Ni, Cu;
$X = o$-C_6H_4, $R = $ —N=N—C_6H_5, M = Ni

64

Macrocyclic ligands derived from aromatic amines which may coordinate as dianions have been synthesized by a variety of synthetic techniques. Hiller *et al.*[252] synthesized **65**, $bzo_2[14]hexaeneN_4$, by the reaction of propargyl-

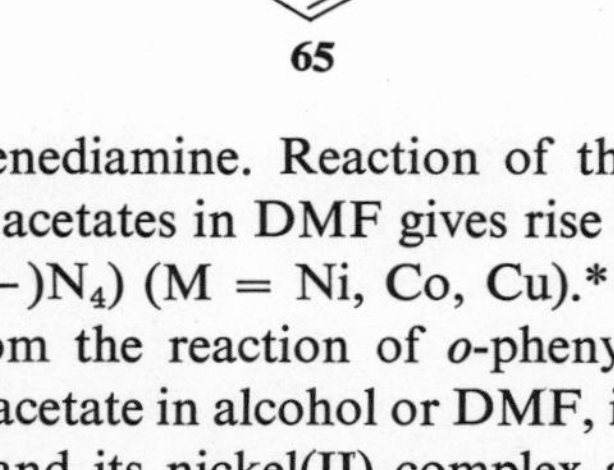

65

aldehyde with *o*-phenylenediamine. Reaction of the ligand with nickel(II) cobalt(II), or copper(II) acetates in DMF gives rise to the neutral complexes $M(bzo_2[14]hexaenato(2-)N_4)$ (M = Ni, Co, Cu).* The complexes can also be prepared directly from the reaction of *o*-phenylenediamine, propargylaldehyde, and the metal acetate in alcohol or DMF, in 50–60% yield. Various reactions of the ligand and its nickel(II) complex are reported. Chave and Honeybourne[254] synthesized the nickel(II) and copper(II) complexes of the neutral macrocycle $bzo_2[14]tetraeneN_4$ and reported the magnetic properties of the cobalt(II) and copper(II) complexes.[255] The synthetic approach used was extended to produce a variety of complexes containing the dianionic

---

* The nickel(II) and cobalt(II) complexes have been shown to be isostructural.[253]

form of macrocyclic ligands, **66**[19] (X = Y = o-C_6H_4; X = Y = C_2H_4; X = Y = 1,8-naphthalene; X = o-C_6H_4, Y = C_2H_4, C_3H_6). In a typical reaction, bromomalondialdehyde, CHO·CHBr·CHO, in cold ethanol reacts at room temperature with *bis*-(o-phenylenediamine)metal(II) acetate to produce the hydrated neutral complexes; yields range from 60% for the cobalt species to almost quantitative for the copper complexes.

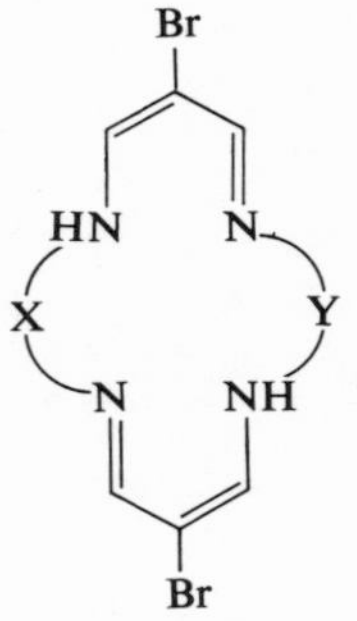

66

The macrocyclic ligand **67**, $Me_4bzo_2[14]hexaeneN_4$ has been synthesized[256] by a modification of the procedure reported earlier by Jäger.[39] The ligand was prepared by a template condensation between o-phenylenediamine

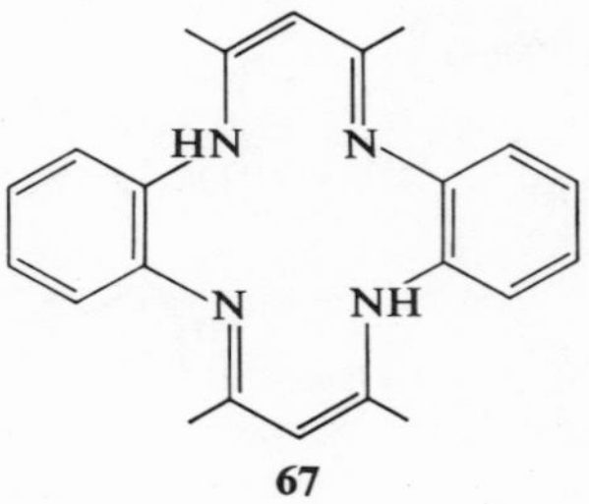

67

and acetylacetone in the presence of nickel(II) ions, stripped from the nickel(II) with anhydrous HCl in ethanol and isolated as the hydrochloride salt. Neutralization with base enables the neutral ligand to be isolated. Monocarbon monoxide complexes of the type [Fe($Me_4bzo_2[14]hexaenatoN_4$)-(base)CO] (base = acetonitrile, pyridine, 4-picoline, or hydrazine) were prepared by reaction of the free ligand with an anhydrous source of iron(II), usually as an amine complex, under an atmosphere of carbon monoxide. The structure of the hydrazine adduct was reported. Reaction of the ligand with *tris*-(o-phenylenediamine)iron(II) perchlorate in acetonitrile under nitrogen with triethylamine yields a high-spin five-coordinate iron(III) complex [Fe($Me_4bzo_2[14]hexaenatoN_4$)Cl], in which the perchlorate ion has been reduced to the chloride ion.[257] Use of the isothiocyanato derivative rather than the perchlorate produces the four-coordinate species [Fe(Me$_4$bzo$_2$-[14]hexaenatoN$_4$)], which readily adds one molecule of CO to yield a five-coordinate adduct. This five-coordinate adduct in turn readily adds a ligand such as pyridine in the sixth coordination position. The four-coordinate complex is sensitive to oxygen, forming the oxo-bridged dimer [Fe(Me$_4$bzo$_2$-[14]hexaenatoN$_4$)]$_2$O; it reacts with chloroform or carbon tetrachloride abstracting a chlorine atom to give the high-spin five-coordinate chloro species, with pyridine to give a *bis*-pyridine adduct, and with alkyl halides such as CH$_3$I, C$_2$H$_5$I, or C$_6$H$_5$CH$_2$Br to form a low-spin organo-iron(III) complex Fe(Me$_4$bzo$_2$[14]hexaenatoN$_4$)R] (R = CH$_3$, C$_2$H$_5$, C$_6$H$_5$CH$_2$), and the five-coordinate halide species. The organo-iron(III) complexes are identical to those prepared from [Fe(Me$_4$bzo$_2$[14]hexaenatoN$_4$)NCS] by reaction with RNHNH$_2$ (R = CH$_3$, C$_2$H$_5$, C$_6$H$_5$).[258] The reaction of the free ligand with cobalt(II) acetate in methanol–acetonitrile solution under N$_2$ yields the low-spin, four-coordinate cobalt(II) complex [Co(Me$_4$bzo$_2$[14]hexaenatoN$_4$)] in approximately 90% yield.[259] Oxidation of the reaction mixture with O$_2$ leads to the formation of a cobalt(III) complex in which one of the 2,4-pentanediiminato-like rings has been converted to a β-diimine function with an adjacent α,β-unsaturated carbonyl. High-spin, five-coordinate cobalt(III) complexes of the type [Co(Me$_4$bzo$_2$[14]hexaenatoX] (X = Cl, Br, I) can be prepared, and a variety of reactions involving the coordinated ligand have been reported. A *bis*-pyridine adduct can be isolated from the five-coordinate species. A crystal and molecular structure of the five-coordinate iodo complex has been completed.[260] Neves and Dabrowiak[261] isolated a number of manganese(II) and (III) and zinc(II) complexes of **67** in its dianionic form. The manganese(II) and zinc(II) complexes [M(Me$_4$bzo$_2$[14]hexaenatoN$_4$)·Amine] (Amine = Et$_3$N or (n-Pr)$_3$N) are five-coordinate with a distorted square-pyramidal environment about the metal ion. The high-spin manganese(III) complexes Mn(Me$_4$bzo$_2$[14]hexaenatoN$_4$)X (X = Cl, Br, SCN, N$_3$) can be isolated, the chloro and bromo species crystallizing from acetonitrile with one molecule of solvent. It is suggested that in the azide derivative, the manganese(III) ions are linked by an azide bridge in the solid state.* Bell and Dabrowiak[263] showed that the reaction of the ligand **67** with an equimolar quantity of M(CO)$_6$ (M = Mo, W) gives rise to the species M(Me$_4$-bzo$_2$-[14]hexaeneN$_4$)(CO)$_4$ (M = Mo, W) in which the macrocycle is concluded to coordinate as a neutral, bidentate ligand in *cis* positions of the octahedron around the metal atom. Olszanski and Melson[242] also found that the macrocycle coordinates as a neutral ligand in the complex obtained from the ligand and scandium thiocyanate, viz., Sc(Me$_4$bzo$_2$[14]hexaeneN$_4$)(NCS)$_3$; it is concluded that the macrocycle is tetradentate in this species. Conductivity

---

* A study of the redox behavior of these complexes has been reported.[262]

measurements and infrared spectral data suggest that the scandium(III) ion is five-coordinated, with a distorted square-pyramidal coordination environment.

The crystal and molecular structures of the ligand $Me_4bzo_2[14]hexaeneN_4$, and the complexes of $Me_4bzo_2[14]hexaenato(2-)N_4$ with Fe(II), Co(III), Fe(III), and Mn(II) have been reported; the iron(II) complex is four-coordinated, whereas the other species are five-coordinated.[264,265]

The condensation of aromatic diamines with 1,3-dicarbonyl compounds in the presence of a divalent metal ion such as Ni(II), Cu(II), or Co(II) yields complexes containing dianionic, tetradentate, noncyclic N_2O_2 ligands, or complexes with dianionic macrocyclic ligands of general formula **68**[266] (see Table 5). Some of the noncyclic complexes react with diamines to give

TABLE 5. Complexes of the Following General Formula[a]:

M	R^1	R^2	R^3	R^4	R^5
Ni	H	H	CH_3	H	CH_3
Ni	CH_3	CH_3	CH_3	H	CH_3
Ni	H	Cl	CH_3	H	CH_3
Ni	Cl	Cl	CH_3	H	CH_3
Ni	H	OC_2H_5	CH_3	H	CH_3
Ni	H	H	H	$-N=N-C_6H_5$	H
Cu	H	H	H	$-N=N-C_6H_5$	H
Ni	H	H	H	$-N=N-C_6H_4-NHCOCH_3$	H
Ni	$SO_2C_2H_5$	H	H	$-N=N-C_6H_5$	H
Ni	Cl	Cl	H	$-N=N-C_6H_5$	H
Cu	Cl	Cl	H	$-N=N-C_6H_5$	H
Co	Cl	Cl	H	$-N=N-C_6H_5$	H
Cu	Cl	Cl	H		H

[a] From reference 266.

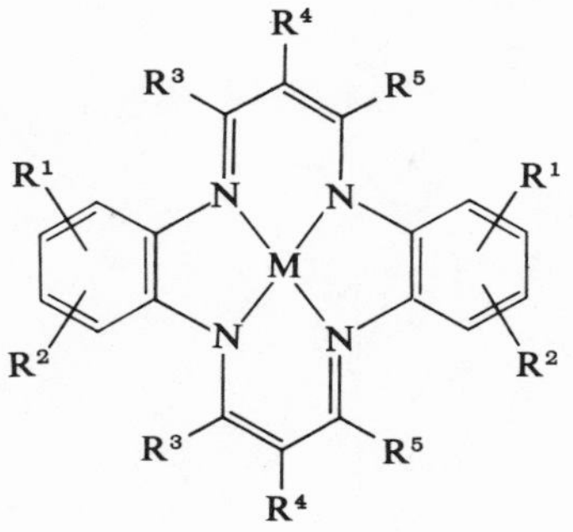

68

macrocyclic complexes,* for example **69**, with

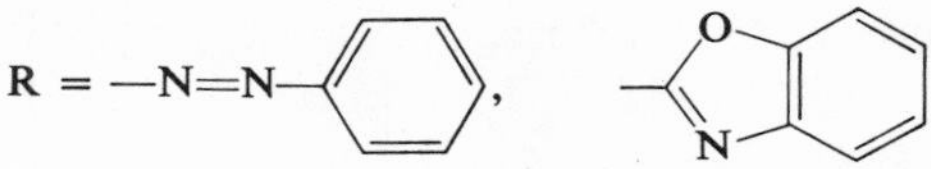

$$R = -N=N-\bigcirc,$$

reacts with an aromatic diamine to give **70**, and with ethylenediamine to give **71**. If an excess of ethylenediamine is used in the reaction, **72** is obtained.

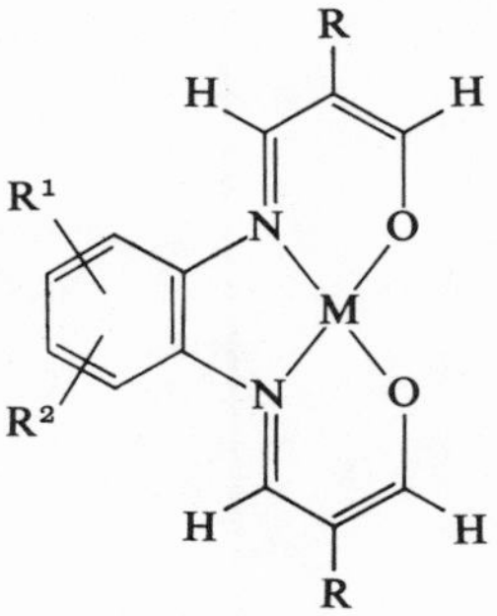

69 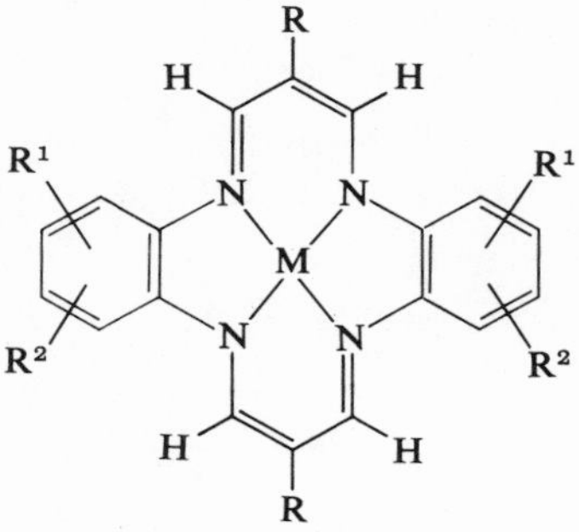   **70**

The macrocycle **67**, $Me_4bzo_2[14]hexaeneN_4$, reacts with aromatic diazonium salts, RN_2^+ (R = phenyl, *p*-nitrophenyl, and *p*-chlorophenyl) to give *bis*-azo macrocycles.[267] The synthesis and characterization of **73A** and **73B** and their nickel(II) complexes have been described[267,268]; the ligands coordinate as dianions.

The nickel(II) complex of **67** reacts with *p*-nitro and *p*-methoxybenzene-diazonium tetrafluoroborates to yield complexes of **73C** and **73D**, respectively.[269]

---

* The reactivity of these complexes is dependent upon the nature of the substituent in the γ position as observed by other researchers.

71   **72**

The reaction between benzylidene acetone ($C_6H_5CH\!\!=\!\!CHCOCH_3$) and ethylenediamine in the presence of anhydrous potassium carbonate in cyclohexane/ether produces a 14-membered tetraaza macrocyclic diene.[270] This

73

73A, R = $-N\!\!=\!\!N\!\!-$⟨ ⟩$-Cl$

73B, R = $-\overset{\overset{\textstyle O}{\|}}{C}-$⟨ ⟩$-N\diagdown_{N}$⟨ ⟩

73C, R = $-N\!\!=\!\!N\!\!-$⟨ ⟩$-NO_2$

73D, R = $-N\!\!=\!\!N\!\!-$⟨ ⟩$-OCH_3$

product may be either the *cis* isomer, **74**, or the *trans* isomer, **75**, or a mixture of both. The product of the reaction, obtained in 80% yield, appeared to be

74   **75**

homogeneous, and Hideg and Lloyd, who first reported the reaction, favored the *cis* isomer and later provided some chemical and mass-spectral evidence to support this structure.[271] Cook *et al.*[272] prepared nickel(II) and copper(II) complexes of the macrocyclic diimine, and from NMR spectroscopy and comparisons with other macrocyclic ligands of known structure suggested the ligand was the *trans-C-meso* diimine **75**, in which the two carbon centers have different chiralities. Reduction of the macrocyclic diimine with sodium borohydride in methanol gives only three tetramines, confirming that the diimine obtained in the original reaction was a pure diastereoisomer. A crystal structure of the copper(II) complex of one of the tetramines showed that the complex has the *C-meso-C-meso-N-meso-N-meso* configuration **76**, providing further evidence in support of the *trans-C-meso* structure for the original diimine macrocycle.[273] The macrocycle **75**, $Me_2Ph_2[14]dieneN_4$,

76

reacts rapidly with nickel(II) salts in methanol[272] to form the square-planar complexes $[Ni(Me_2Ph_2[14]dieneN_4)]X_2$ ($X = ClO_4$, $\frac{1}{2}ZnCl_4$). With unidentate anions, the species *trans*-$[Ni(Me_2Ph_2[14]dieneN_4)X_2]$ ($X = Cl$, NCS, NO_3, and $OOCCH_3$) are obtained; recrystallization of the perchlorate derivative from concentrated aqueous ammonia produces *trans*-$[Ni(Me_2Ph_2$-$dieneN_4)(NH_3)(H_2O)](ClO_4)_2$. In all these species, the macrocycle coordinates as a planar ligand (β configuration, *C-meso-N-meso*, **77**). However, $[Ni(Me_2Ph_2[14]dieneN_4)](ClO_4)_2$ reacts with ethylenediamine or acetylacetone plus base to yield the derivatives *cis*-$[Ni(Me_2Ph_2[14]dieneN_4)en]$-$(ClO_4)_2$ and *cis*-$[Ni(Me_2Ph_2[14]dieneN_4)(acac)]ClO_4$, respectively, in which

77   **78**

the macrocycle adopts a folded α configuration, *C-meso-N-rac*, **78**. The latter complex reacts with oxalic acid to give *cis*-[(Ni(Me$_2$Ph$_2$[14]dieneN$_4$))$_2$-C$_2$O$_4$](ClO$_4$)$_2$ with bridging oxalate ions. Treatment of these chelated complexes with acid removes the chelate by protonation and preserves the nitrogen configuration to give metastable perchlorate or tetrachlorozincate(II) salts with the macrocyclic ligand in the folded (α) configuration. These species are stable in the solid state, in aprotic solvents, or in strongly acidic protic media, but the cation isomerizes slowly in neutral or rapidly in basic solution to give the β configuration for the coordinated planar ligand. With copper(II) ions, the complexes Cu(Me$_2$Ph$_2$[14]dieneN$_4$)X$_2$ (X = ClO$_4$, NO$_3$, NCS, Cl·4H$_2$O) have been prepared with the ligand in the β configuration. Cobalt(III) complexes of **75** were prepared[274] by reaction of the ligand with Na$_3$Co(CO$_3$)$_2$·3H$_2$O in methanol or by the reaction of the ligand with a cobalt(II) salt in methanol followed by air or hydrogen peroxide oxidation and acidification; the species *trans*-[Co(Me$_2$Ph$_2$[14]dieneN$_4$)X$_2$]Y (X = Cl, Br, OOCCH$_3$, Y = ClO$_4$; X = NCS, Y = NCS·H$_2$O) were obtained in which the ligand has the β configuration. Reaction of the *trans*-dichloro perchlorate complex with acetylacetone in basic solution or ethylenediamine gives the complexes *cis*-[Co(Me$_2$Ph$_2$[14]dieneN$_4$)(acac)]ClO$_4$ or *cis*-[Co-(Me$_2$Ph$_2$[14]dieneN$_4$)en](ClO$_4$)$_3$ with the ligand in the α configuration. The, presumably *trans*, diamine [Co(Me$_2$Ph$_2$[14]dieneN$_4$)(NH$_3$)$_2$](ClO$_4$)$_3$·H$_2$O can also be obtained from the *trans*-dichloro complex. The compounds *trans*-[Co(Me$_2$Ph$_2$[14]dieneN$_4$)XY]Z (X = Y = Z = Cl; X = Cl, Y = NO$_2$, Z = ClO$_4$; X = Cl, Y = NO$_2$, Z = Cl; X = Br, Y = NO$_2$, Z = ClO$_4$, X = Y = NO$_2$, Z = ClO$_4$) and *trans*-[Rh(Me$_2$Ph$_2$[14]dieneN$_4$)Cl$_2$]ClO$_4$ have also been prepared[275]; in all these compounds the β configuration is proposed for the macrocyclic ligand. Reduction of the ligand **75** with NaBH$_4$ in methanol produces three tetramines; the *C-meso-C-meso* isomer was isolated from the reaction medium and in the complexes obtained from this ligand, viz., *trans*-[Co(Me$_2$Ph$_2$[14]aneN$_4$)X$_2$]Y (X = Cl, Y = Cl·H$_2$O; X = Cl, Y = ClO$_4$; X = Br, Y = Br·2H$_2$O), it is concluded[275] that the tetramine also has the *N-meso-N-meso* configuration as found in the copper(II) complex **76**. The complexes *trans*-[Rh(Me$_2$Ph$_2$[14]aneN$_4$)Cl$_2$]Cl·H$_2$O and *trans*-[Rh(Me$_2$Ph$_2$[14]aneN$_4$(H$_2$O)$_2$](ClO$_4$)$_3$·2H$_2$O have also been prepared. Recently, the crystal structure of the reduced ligand was reported; the *trans* structure was confirmed.[276] Caulkett *et al.* also provided evidence for the *trans* structure by the synthesis and characterization of **78A** from the reaction of the reduced tetramine with glyoxal in acetonitrile.[277]

The reaction of methyl vinyl ketone with 1,3-propanediamine in the presence of nickel(II) and hydrogen ions was reported to give rise to the nickel(II) hexafluorophosphate complex of **79**, Me$_2$[16]5,16-dieneN$_4$, in which the imines are *cis* to one another in the 16-membered macrocyclic ligand.[278,279] This complex represents only 13% of the total nickel(II)

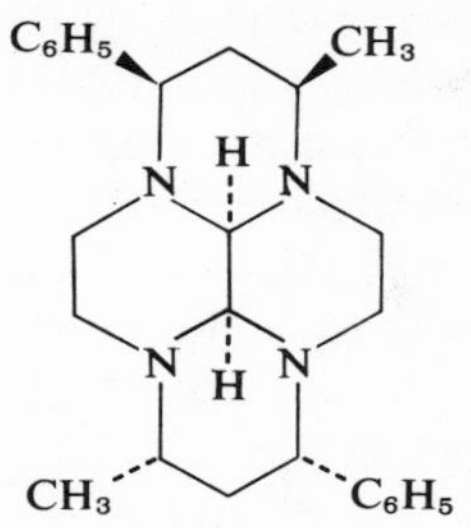

78A

present in the solution. Several derivatives of the nickel complex cation can be obtained from [Ni(Me$_2$[16]dieneN$_4$)](PF$_6$)$_2$ by the addition of the appropriate alkali metal salt. With unidentate anions or ligands, the nickel(II)

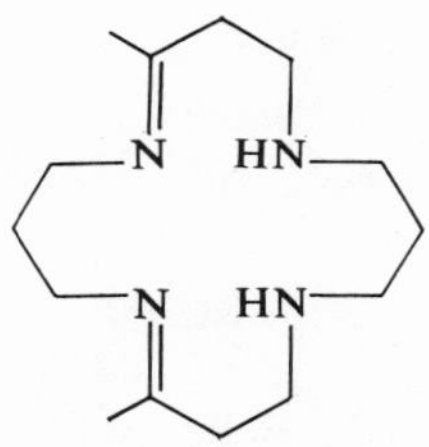

79

complexes are either square planar or tetragonally distorted six-coordinate species; however, in the pseudooctahedral species containing either acetylacetone or oxalate ions the macrocyclic ligand is assumed to coordinate in a folded configuration. Interconvertible (designated A and B) isomers of [Ni(Me$_2$[16]dieneN$_4$)]$^{2+}$ as dichloro and hexafluorophosphate derivatives have been separated and purified. It is concluded that the A and B isomers are the racemic and *meso* isomers of the cation. Myers and Rose[28] also studied the reaction of ethylenediamine with methyl vinyl ketone in the presence of nickel(II) ions; the complex [Ni(Me$_2$[14]dieneN$_4$)](PF$_6$)$_2$, containing **80**, Me$_2$[14]dieneN$_4$, in which the imines are *trans* to one another, is obtained.

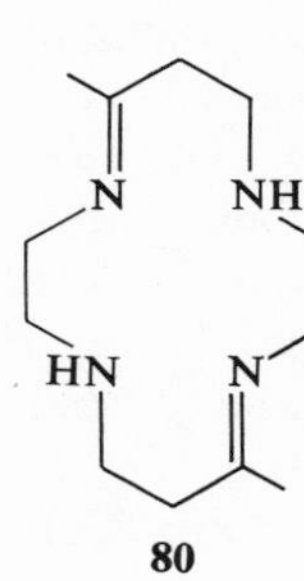

80

This ligand had previously been prepared by the reaction between ethylene-diamine monohydrogenperchlorate and methyl vinyl ketone,[280] and nickel(II) and copper(II) complexes isolated and studied.[281] The generality of the reaction between α,β-unsaturated ketones and ethylenediamine mono-hydrogenperchlorate was investigated by Kolinski and Korybut-Daszkie-wicz[114]; the ligands **81** ($R_1 = R_2 = R_3 = H$; $R_1 = R_3 = H$, $R_2 = CH_3$; $R_1 = R_3 = H$, $R_2 = C_2H_5$; $R_1 = R_3 = H$, $R_2 = i\text{-}C_3H_7$ and $R_1 = R_2 = CH_3$, $R_3 = H$) were prepared as the dihydrogenperchlorate salts. Isomeric

81

nickel(II) complexes with **81** ($R_1 = R_2 = R_3 = H$), $[Ni(Me_2[14]dieneN_4)]$-$(ClO_4)_2$, were prepared and studied; the isomerism arises from N-racemic or N-meso configurations for the coordinated N–H groups (α and β isomers, respectively). The α isomer has been partially resolved into optically active forms. Nickel(II) complexes of the ligand **81** ($R_1 = H$, $R_2 = CH_3$) were isolated and characterized by NMR spectroscopy; species containing the ligand in N-rac-C-rac[Bβ(e–e)], N-meso-C-rac[Bγ(a–e)], N-rac-C-rac[Bα(a–a)], N-rac-C-meso[Aα(a–e)], and N-meso-C-meso[Aβ(e–e)] were characterized. X-ray crystallography was carried out on the N-rac-C-rac stereoisomer[282] and on the N-rac-C-meso stereoisomer[283]; the former complex has both methyl groups axial and the latter has one axial and one equatorial methyl group. Hay and Jeragh[284] found that the reaction between methyl vinyl ketone and en·$HClO_4$ is not completely stereospecific and approximately 15% of the C-rac ligand is formed in conjunction with the C-meso ligand. A series of octahedral cobalt(III) complexes of general formula trans-$[CoLX_2]^{n+}$ [$X = Cl$, Br, NO_2, CN, N_3, CH_3COO, NH_3, or $Cl(NO_2)$] have been pre-pared by anion-metathesis reactions with the C-meso-N-meso(e–e) stereo-isomer of trans-$[CoLCl_2]^+$. NMR studies confirm the configuration in all the complexes. The μ-peroxo complex $[\{CoL(OH_2)\}_2O_2](ClO_4)_4 \cdot H_2O$ and the four-coordinate complex $[CuL](ClO_4)_2$ are also reported. A number of metal(II) complexes with **80** have been synthesized[285]: $[M(Me_2[14]dieneN_4)]$-$(ClO_4)_2$ (M = Ni, Cu, Co, Zn), trans-$[Co(Me_2[14]dieneN_4)X]^{n+}$ (X = Cl,

Br, NO_2, NCS, $n = 2$; X = NH_3, $n = 3$), and *cis*-[Co(Me$_2$[14]dieneN$_4$)-(bident)]$^{n+}$ (bident = CO_3, $n = 1$; bident = acetylacetonate ion, $n = 2$). Two isomers of *trans*-[Co(Me$_2$[14]dieneN$_4$)Cl$_2$]ClO$_4$ have been isolated, corresponding to *N-meso* (isomer a) and *N*-racemic (isomer b) diastereoisomers. The former is obtained almost exclusively by the reaction of the ligand dihydrogenperchlorate with Na$_3$[Co(CO$_3$)$_3$], whereas reaction with cobalt(II) acetate gives predominantly the latter isomer. Interconversion takes place in basic solution. The crystal and molecular structure [Co(Me$_2$[14]dieneN$_4$)(NCS)$_2$]SCN·H$_2$O has been determined[286]; the *trans* arrangement of the diimines was confirmed with the N–H bonds of the secondary amine groups in a *cis* configuration, corresponding to the *N*-racemic isomer. The ligand coordinates in a planar manner about the cobalt(III) ion with coordinated isothiocyanato groups. Lawrance[128] has reported the synthesis of *trans*-[Co(Me$_2$[14]dieneN$_4$)(NH$_3$)$_2$](ClO$_4$)$_3$·$\frac{1}{2}$H$_2$O from the *trans*-dichloro species, and Hay and Lawrance[287] used the same starting material for the synthesis of the *trans*-dibromo and *trans*-aquochloro species. Since the dichloro complex contains the *N-meso* ligand isomer, it is assumed that the complexes obtained from it retain the same configuration for the coordinated secondary amines. When the ligand dihydrogenperchlorate and cobalt(II) perchlorate in a 1:1 molar ratio are dissolved in water at room temperature and mixed, the solution darkens, and addition of NaClO$_4$ yields a brown precipitate.[288] This precipitate corresponds to *trans*-[(H$_2$O)–Co(Me$_2$[14]dieneN$_4$)–O$_2$–Co(Me$_2$[14]dieneN$_4$)–(H$_2$O)](ClO$_4$)$_4$·H$_2$O, which contains a bridging peroxo group between the cobalt(III) ions and is similar to that obtained from the reaction of cobalt(II) perchlorate and [14]aneN$_4$ under similar conditions.[178] Addition of a coordinating anion such as nitrite to a solution of the aquo complex leads to replacement of the labile aquo ligands by the anion.

When ethylenediamine monohydrogenperchlorate and 3-methylbut-3-en-2-one react in methanol, the product is the dihydrogenperchlorate of the *trans*-diimine Me$_4$[14]4,11-dieneN$_4$, **81** (R$_1$ = R$_2$ = H, R$_3$ = CH$_3$).[289] Fractional crystallization of the complex [Ni(Me$_4$[14]4,11-dieneN$_4$)](ClO$_4$)$_2$ from aqueous solution gives three isomers (a, b, and c) identified by ^{1}H NMR spectroscopy. Isomer b is the centrosymmetric *N-meso-C-meso* with both methyl groups axial; isomer c is the *N-rac-C-rac* isomer, which also has axial methyl groups. Isomer a, which has not (as yet) been obtained isomerically pure is suggested to be either the *N-meso-C-rac* (one methyl axial, one equatorial) or the *N-rac-C-meso* (one methyl axial, one equatorial) stereoisomer. This isomer is always found in association with b, and this a and b mixture in alkaline (CD$_3$)$_2$SO undergoes conversion to isomer d, which is the *N-meso-C-meso* isomer with equatorial methyl groups and the most thermodynamically stable diastereoisomer of the *C-meso* series of complexes. Isomer c does not undergo isomerization under similar conditions and is thus considered to be the most thermodynamically stable of the *C-rac* series of complexes. The compounds [M(Me$_4$[14]4,11-dieneN$_4$)](ClO$_4$)$_2$ (M = Cu, Zn, Hg), *trans*-[CoLX$_2$]$^{n+}$ [X = Cl, Br, NO$_2$, NCS, N$_3$, NH$_3$, CH$_3$COO, Cl(NO$_2$)], *cis*-[CoL(CO$_3$)]ClO$_4$·2H$_2$O, and *trans*-[RhLCl$_2$]ClO$_4$·H$_2$O have also been synthesized, and their IR and d–d spectra reported.

The reaction of ethyl vinyl ketone with ethylenediamine monohydrogen-perchlorate in methanol at −5–0°C produces **82**, Et$_2$[14]dieneN$_4$, as the dihydrogenperchlorate salt.[290] Reaction of the salt with copper(II) or

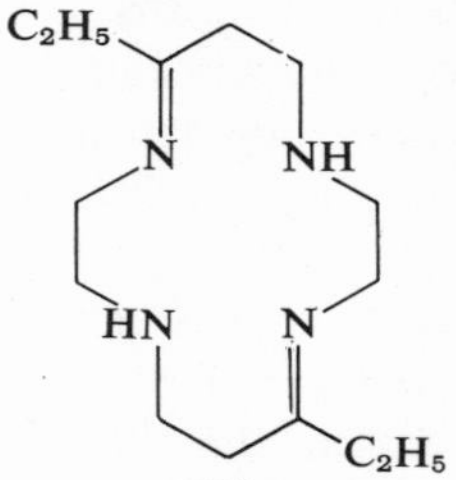

82

nickel(II) carbonate produces the complexes M(Et$_2$[14]dieneN$_4$)(ClO$_4$)$_2$ (M = Cu, Ni). A hygroscopic zinc(II) complex was also reported.

Reduction of [Ni(Me$_2$[14]dieneN$_4$)](ClO$_4$)$_2$ with nickel–aluminum alloy followed by treatment with sodium cyanide gives two tetramines,[280] **83A** and **83B**, one with *C-meso* stereochemistry, the other with *C-rac* stereochemistry; this assignment has been confirmed by X-ray crystallography on a cobalt(III) complex. Thermodynamic and kinetic studies of complexation

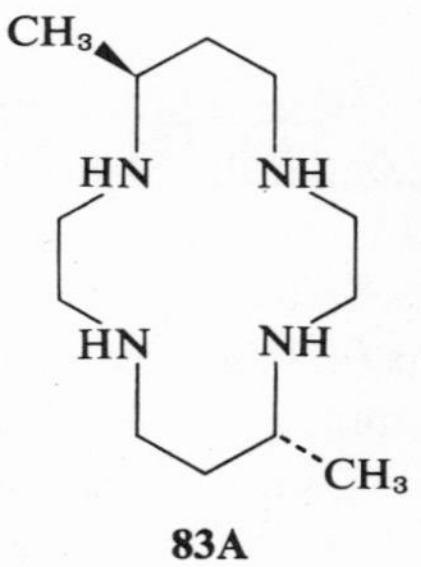

83A 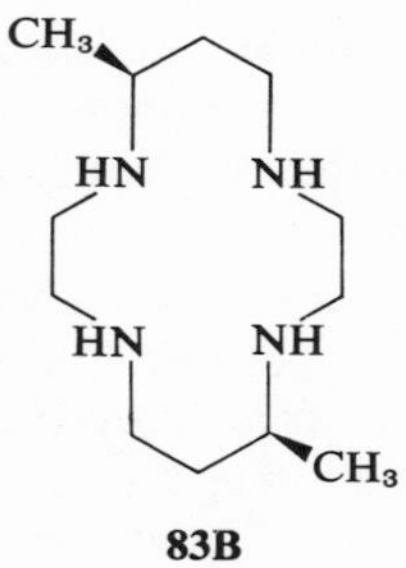   **83B**

between the *C-meso*-stereoisomer and copper(II) and nickel(II) ions have been reported,[291,292] and the complexes [ML](ClO$_4$)$_2$ (M = Cu, Ni, Zn), [PbL](NO$_3$)$_2$·1½H$_2$O, [NiLX$_2$]$^{n+}$ (X = Cl, NCS, n = 0, X = DMSO, n = 2), *trans*-[CoLCl$_2$]Cl·2H$_2$O, *trans*-[CoL(N$_3$)$_2$](N$_3$), *trans*-[CoLBr$_2$]Br·H$_2$O, *trans*-[CoL(NO$_2$)$_2$]ClO$_4$, and *cis*-[RhLCl$_2$]Cl (L = *C-ms*-Me$_2$[14]aneN$_4$) have been isolated.[293] Dei and Mani[168,169] synthesized the complexes

$VLCl_2$ and $CrLX_2$ (X = Cl, Br). *N*-methylation of the ligand L with formic-acid–formaldehyde gives the *NN'N"N'''*-tetramethyl derivative L'; the complexes $[ML'](ClO_4)_2 \cdot H_2O$ (M = Cu, Zn) have been isolated by reaction of the ligand with copper(II) and zinc(II) salts.[293]

One of the early demonstrations of the "template effect" involved the *in situ* syntheses of macrocyclic ligands from 2,6-diacetylpyridine and polyamines in the presence of metal ions.[294] Karn and Busch[295] showed that reaction between 2,6-diacetylpyridine and 3,3'-diaminodipropylamine in the presence of nickel(II) chloride in ethanol–water at 65°C gives rise to complexes of the tetradentate macrocyclic ligand **84**, $Me_2pyo[14]trieneN_4$. The dihydrate of the perchlorate salt, $Ni(Me_2pyo[14]trieneN_4)(ClO_4)_2$ was isolated

84

from the original reaction medium after addition of a concentrated aqueous solution of sodium perchlorate. The complexes $Ni(Me_2pyo[14]trieneN_4)XY$ (X = Y = Cl; X = Y = NCS; X = Y = ClO_4; X = Y = Br; X = Br, Y = ClO_4, X = Br, Y = BF_4) were also prepared; the bromo species were hydrated, and the importance of the hydrogen-bonded water molecule was discussed. It has been suggested[6] that the presence of the pyridine group in 2,6-diacetylpyridine probably causes initial tridentate chelation of the dicarbonyl compound, with activation of the —C=O groups and ready reaction with the amine. The structure of $[Ni(Me_2pyo[14]trieneN_4)Br]Br \cdot H_2O$ has been reported[296]; the nickel(II) ion is in a distorted square-pyramidal environment, and the molecules are connected by N–H $\cdots$ Br–Ni linkages. The water molecule is held between the coordinated and free bromide ions. Rich and Stucky[297] also reported the synthesis of nickel(II) and copper(II) complexes of $Me_2pyo[14]trieneN_4$; these species were isolated as their tetrachlorozincate(II) salts. Three series of cobalt(II) complexes of **84** have been prepared[298] by substitution of the appropriate cobalt(II) salt in the general procedure described earlier,[295] or by metathetical reactions on products formed by the general procedure. The compounds $Co(Me_2pyo[14]trieneN_4)X_2 \cdot nH_2O$ (X = Cl, Br, I, NO_3, NCS, ClO_4, $n = 0$, $\frac{1}{2}$, or 1), $Co(Me_2pyo[14]trieneN_4)A(ClO_4)_2$ (A = NH_3 or pyridine), and $Co(Me_2pyo[14]trieneN_4)BrX$ [X = ClO_4, PF_6, or $B(C_6H_5)_4$] all contain low-spin cobalt(II) and are five-coordinate. The zinc(II) complex $Zn(Me_2pyo[14]trieneN_4)I_2$ was also prepared and characterized. The copper(II) complex

Cu(Me$_2$pyo[14]trieneN$_4$)(ClO$_4$)$_2$·H$_2$O was obtained[299] by an *in situ* reaction similar to that reported earlier for the nickel(II) complex[295] and the species [Cu(Me$_2$pyo[14]trieneN$_4$)X]ClO$_4$·nH$_2$O (X = Cl, Br, I, n = 1; X = NCS, n = 0) were isolated by treatment of the complex with an excess of lithium salt in methanol. The manganese(II) species Mn(Me$_2$pyo[14]trieneN$_4$)X$_2$ (X = NCS, Cl·PF$_6$) have also been prepared.[300] Long and Busch[301] prepared cobalt(III) complexes of Me$_2$pyo[14]trieneN$_4$ by *in situ* synthesis of the ligand in the presence of a cobalt(II) salt followed by air oxidation of the metal ion, metathetical replacement of Br$^-$ or H$_2$O on cobalt(III), and by nitric oxide oxidation of Co(Me$_2$pyo[14]trieneN$_4$)X$_2$·H$_2$O. Four general types of complexes were prepared: [Co(Me$_2$pyo[14]trieneN$_4$)X$_2$]X·nH$_2$O (X = Cl, Br, I, NO$_3$, ClO$_4$, n = 1, 1, 0, $\frac{5}{2}$, 3, respectively), [Co(Me$_2$pyo[14]-trieneN$_4$)X$_2$]ClO$_4$ (X = NO$_2$, N$_3$, NCS, CN), [Co(Me$_2$pyo[14]trieneN$_4$)-(bident)]X$_m$·nH$_2$O (bident = oxalate, acetylacetonate, or ethylenediamine; X = ClO$_4$, NO$_3$, Br, m = 1, 2, or 3, and n = 0, 1), and [Co(Me$_2$pyo[14]-trieneN$_4$)X(NO)]B(C$_6$H$_5$)$_4$ (X = I, Br, NCS). All complexes are low-spin, six-coordinate complexes of cobalt(III).

Reaction of Co(Me$_2$pyo[14]trieneN$_4$)Br$_3$ with AgClO$_4$ in acetonitrile leads to the formation of orange crystals of [Co(Me$_2$pyo[14]trieneN$_4$)-(CH$_3$CN)$_2$](ClO$_4$)$_3$; this reaction has been shown to be general for a variety of cobalt(III) complexes containing macrocyclic ligands; the yields of the diacetonitrile adducts are generally 70–80%.[302] Prior to the isolation of the above "simple" cobalt(III) complexes, organic derivatives containing **84** and cobalt(III) had been prepared and characterized. Ochiai and Busch[303] prepared the complexes [RCo(Me$_2$pyo[14]trieneN$_4$)Br]PF$_6$·nH$_2$O (R = CH$_3$, n = 0; R = C$_2$H$_5$, n = $\frac{1}{2}$; R = CH$_2$ = CH, n = $\frac{1}{2}$) by reduction of Co(Me$_2$pyo[14]trieneN$_4$)Br$_2$·H$_2$O with sodium borohydride in the presence of the alkyl halide, or its analog under N$_2$ in methanol. This method of preparation is similar to that used to prepare other alkyl–cobalt(III) complexes containing nonmacrocyclic ligands. The procedure was extended[304] to the synthesis of complexes with R = n-C$_4$H$_9$, n = 1; R = ClCH$_2$, n = $\frac{1}{2}$; R = C$_6$H$_5$CH$_2$, n = 1; and R = NCCH$_2$CH$_3$, n = $\frac{1}{2}$. The formation reaction of the allyl derivative, using allyl bromide, proceeds in a similar manner to the other cases, but the substance is so reactive that no solid product could be isolated. The complexes *trans*-[R$_2$Co(Me$_2$pyo[14]trieneN$_4$)]-B(C$_6$H$_5$)$_4$ (R = CH$_3$, C$_6$H$_5$CH$_2$) can be prepared by reduction of the mono-alkyl species with NaBH$_4$ in methanol or sodium amalgam in acetonitrile followed by oxidative alkylation at the cobalt atom.[305] It is proposed that the reaction proceeds via a neutral alkyl cobalt(I) compound. Similar reactions involving alkyl–cobalt(III) complexes of the ligand Me$_4$[14]tetraeneN$_4$, **9**, were observed and the cobalt(I) species CH$_3$Co(Me$_4$[14]tetraeneN$_4$) was isolated.

The catalytic reduction of the azo–methine linkages in **84** was first demonstrated by Karn and Busch.[295,306] Since the reduction generates two asymmetric carbon atoms, the ligand **85**, $Me_2pyo[14]eneN_4$, exists as *meso* and racemic forms corresponding to the diastereoisomer with the methyl groups on the same side or opposite sides of the plane containing the metal

85A, R = H
85B, R = $-CH_2CH_2N(CH_3)_2$

85

ion and donor atoms. Catalytic hydrogenation of $Ni(Me_2pyo[14]trieneN_4)$-$(ClO_4)_2$ yields a yellow (α) form and a red (β) form of the species $Ni(Me_2pyo[14]eneN_4)(ClO_4)_2$; these may be separated by fractional crystallization from water. The α form corresponds to the complex containing the racemic form of the ligand, the β-diastereoisomer contains the *meso* form of the ligand. The *meso*(β) form is more abundant than the racemic(α) form ($\sim 10:1$). Treatment of either form of $Ni(Me_2[14]pyaneN_4)(ClO_4)_2$ with an excess of cyanide ion displaces the nickel(II) ion as $[Ni(CN)_4]^{2-}$ and enables the *rac* and *meso* forms of $Me_2[14]pyaneN_4$ to be isolated. By using $Ni(ms-Me_2pyo[14]eneN_4)(ClO_4)_2$ as starting material, metathetical reactions enabled the species $Ni(ms-Me_2pyo[14]eneN_4)X_2 \cdot nH_2O$ (X = Cl, Br, I, OOCH, NCS, $n = 0$; X = NO_2, $n = 1$, $X_2 = N_3 \cdot ClO_4$, $n = 1$) to be obtained; treatment with ethylenediamine, ammonia, or sodium oxalate produces the complexes $Ni(ms-Me_2pyo[14]eneN_4)L(ClO_4)_2$ (L = en, $2NH_3$) and $[Ni(ms-Me_2pyo[14]eneN_4)]ox(ClO_4)_2$; the oxalato species is assumed to be dimeric with the macrocyclic ligand folded such that the bidentate ligand occupies adjacent octahedral sites of the nickel(II) ion. The structure of the α form of $Ni(Me_2pyo[14]eneN_4)(ClO_4)_2$ has been determined[307]; the methyl groups were found to be positioned on opposite sides of the plane, confirming that this species is the racemic diastereoisomer. It was also concluded from the location of the hydrogen atoms that hydrogen adds *trans* across both double bonds of $[Ni(Me_2pyo[14]trieneN_4)]$ in the hydrogenation process.[308] Each of the three tetrahedral secondary nitrogen atoms of the coordinated reduced ligand **85** can have either of two configurations. Thus, when *ms*-$Me_2pyo[14]eneN_4$ coordinates in a plane there is the possibility of six configurational isomers. However, steric considerations indicate that only two of these are probable, viz., one with all three secondary amine hydrogen atoms on the same side of the plane (α-*trans* form), and the other with the secondary amine hydrogen atom opposite the pyridine nitrogen located on the opposite side of the plane to the other two hydrogen atoms (β-*trans* form). Ochiai and Busch isolated these isomers for Ni(ms-Me$_2$pyo[14]eneN$_4$)(ClO$_4$)$_2$ and studied their interconversion (see Chapter 6). These authors[309] also prepared and characterized cobalt(III) complexes of ms-Me$_2$pyo[14]eneN$_4$ with the general formula [Co(ms-Me$_2$pyo[14]eneN$_4$)XY](ClO$_4$)$_n$ (X = Y = Cl, Br, OH, N$_3$, NCS, CN, NO$_2$, n = 1; X = Cl, Y = I, N$_3$, n = 1; X = Y = H$_2$O, n = 3; XY = carbonate, oxalate, acetylacetonate ions, n = 1, 1, 2; XY = ethylenediamine, n = 3). The occurrence of geometrical (*cis–trans*) isomers, corresponding to folded or planar coordination of the ligand, and configurational isomers due to the asymmetric tetrahedral nitrogen atoms, was observed, and some reactions of the individual species were reported (see Chapter 6). The copper(II) complex [Cu(ms-Me$_2$pyo[14]eneN$_4$)]-(ClO$_4$)$_2$ has been obtained by reaction of the ligand monohydrate with Cu(ClO$_4$)$_2 \cdot$6H$_2$O in ethanol.[299] Complexes of general formula [Cu(ms-Me$_2$pyo[14]eneN$_4$)X]ClO$_4$ (X = Cl, Br, I) were obtained from the perchlorate complex by treatment with the appropriate lithium salt in methanol; it is proposed that these species contain five-coordinated copper(II). Iron(II) complexes of ms-Me$_2$pyo[14]eneN$_4$ have been synthesized by the reaction of an iron(II) halide with the ligand monohydrate in ethanol under N$_2$[310]; the species [Fe(ms-Me$_2$pyo[14]eneN$_4$)X]X (X = Cl, Br, I) were isolated, and from these [Fe(ms-Me$_2$pyo[14]eneN$_4$)X]PF$_6$ (X = Cl, Br, I) were obtained by treatment with LiPF$_6$. It is concluded from their physical properties that these complexes contain high-spin, five-coordinated iron(II). The species [Fe(ms-Me$_2$pyo[14]eneN$_4$)X$_2$] (X = N$_3$, NCS) are six-coordinate, and [Fe(ms-Me$_2$pyo[14]eneN$_4$)OAc]PF$_6$ is suggested to contain six-coordinated iron(II) with the ligand in a folded conformation to enable the acetate ion to function as a bidentate ligand occupying *cis* positions. Oxidation of the iron(II) complexes is achieved in acidic solution by bubbling air through the solution; the high-spin six-coordinate iron(III) complexes [Fe(ms-Me$_2$pyo-[14]eneN$_4$)X$_2$]Y (X = Cl, Y = ClO$_4$, BF$_4$; X = Br, Y = ClO$_4$, BF$_4$) are obtained.

A complete series of nickel(II) complexes containing macrocyclic ligands based on **84** with differing degrees of unsaturation have been obtained and their electrochemistry studied[311] (see Chapter 6).

Prince *et al.*[312] and Keypour and Stotter[313] have extended the range of macrocyclic ligands obtained by *in situ* reactions involving 2,6-diacetyl-pyridine and polyamines to include macrocycles of general formula **86** (**86A** is identical to **84**). Complexes with **86A** (**84**, Me$_2$pyo[14]trieneN$_4$), (ZnL)$_2 \cdot$OH$\cdot$H$_2$O$\cdot$(ClO$_4$)$_3$, ZnLBr$\cdot$ClO$_4$, ZnLBr$_2$, and NiLCl(ClO$_4$), with **86B**, NiL(ClO$_4$)$_2 \cdot n$H$_2$O (n = 0, 1, 2), CuL(ClO$_4$)$_2 \cdot$H$_2$O, and ZnLBrClO$_4$, with **86C**, CuL(ClO$_4$)$_2$, with **86D**, CuLClClO$_4 \cdot$H$_2$O and ZnLBrClO$_4$, and with **86E**, ZnLBrClO$_4$, were prepared by *in situ* reactions in either aqueous ethanol or aqueous dimethylformamide. It is interesting to note that if the amine H$_2$N(CH$_2$)$_3$NH(CH$_2$)$_2$NH$_2$ (3,2-triamine) is used with 2,6-diacetyl-

Ligand	R_1	R_2	x	y
86A	CH_3	H	3	3
86B	CH_3	H	3	4
86C	CH_3	H	2	4
86D	CH_3	CH_3	3	3
86E	H	H	3	3

86

pyridine (DAP) and copper(II) perchlorate, the complex $CuDAP(3,2$-triamine$)(ClO_4)_2$ is obtained; no macrocyclic ligand is formed even upon prolonged refluxing of the reaction mixture. It is, however, assumed that the formation of macrocyclic ligands by reaction of DAP with other triamines proceeds via a similar ternary complex. Restrictions of ring size and the strength of complexation of the metal by the triamine at the pH of the reaction are considered to be important criteria governing the *in situ* formation of macrocyclic ligands of this general type. The crystal structure of $[CuLCl]NO_3 \cdot 2H_2O$ (L = **86D**) has demonstrated a distorted square-pyramidal coordination environment about the copper(II) ion; this complex was obtained[314] from the corresponding chloro perchlorate[312] derivative by column replacement of the perchlorate ion followed by addition of sodium nitrate and evaporation and crystallization.

Recently, the cobalt(II) complex of **86D**, $[CoL](ClO_4)_2 \cdot H_2O$, was prepared[315]; treatment of this complex with chlorine results in oxidation to cobalt(III); the complex *trans*-$[CoLCl_2]ClO_4$ can be isolated. The corresponding *trans*-$[CoLBr_2]ClO_4$ complex is prepared by a similar procedure using a dilute methanolic bromine solution. If the chloro or bromo complex is suspended in methanol and warmed, a color change is observed; from the solution, crystals of the cobalt(III) complexes of **86F** ($R_1 = CH_3$, $R_2 = CH_2$, $x = y = 3$), $[CoLX]ClO_4$ (X = Cl, Br), can be obtained. The crystal structure of the chloro derivative shows that the *N*-methyl carbon atom has lost a proton and is σ bonded to the central cobalt(III) ion forming a three-membered ring.

Lotz and Kaden[316] recently reported that the condensation of N,N-bis-(3-aminopropyl)-N',N'-dimethylethylenediamine with 2,6-diacetylpyridine in the presence of nickel(II) ions and subsequent reduction of the imine groups yields nickel(II) complexes of **85B** (R = —$CH_2CH_2N(CH_3)_2$). Of the 14 possible isomers, three components were isolated by chromatography on an ion-exchange column. The major component was isolated as a triperchlorate $NiLH(ClO_4)_3$, in which the ligand **85B** is protonated at one of the amino groups. Treatment of this complex with 2,6-lutidine in acetone results in $NiL(ClO_4)_2$ (L = **85B**); the deprotonation is accompanied by a color change from yellow to blue. From additional studies in aqueous solution, it is concluded that at pH < 6.3, where the dimethylamino group is protonated, the nickel(II) ion is surrounded by the four nitrogen atoms of the macrocycle in a square-planar geometry, whereas at pH > 6.3, after deprotonation occurs, a potential pentadentate ligand is produced that induces six-coordination for the nickel(II) ion.

Voegtle *et al.*[317] have described the preparation and characterization of the copper(II) complex of **87**.

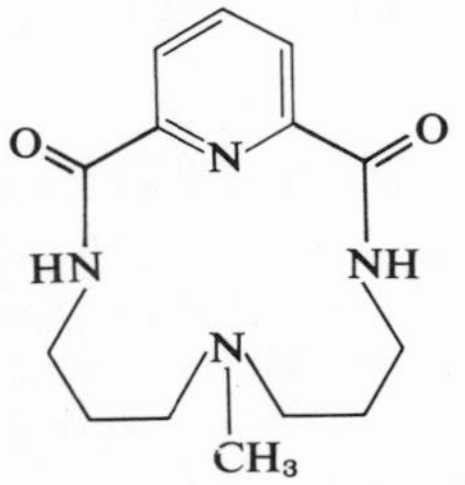

87

A macrocyclic ligand related to those described above but containing three nitrogen and one phosphorus donor atom **88**, Me$_2$pyo[14]trieneN$_3$P, has been synthesized by an *in situ* reaction involving *bis*-(3-aminopropyl)phenylphosphine, 2,6-diacetylpyridine, and NiX$_2 \cdot 6$H$_2$O (X = Br, I), followed by the addition of NH$_4$PF$_6$.[318] The diamagnetic, five-coordinate complexes [Ni(Me$_2$pyo[14]trieneN$_3$P)X]PF$_6$ (X = Br, I) are obtained. When X = Cl in

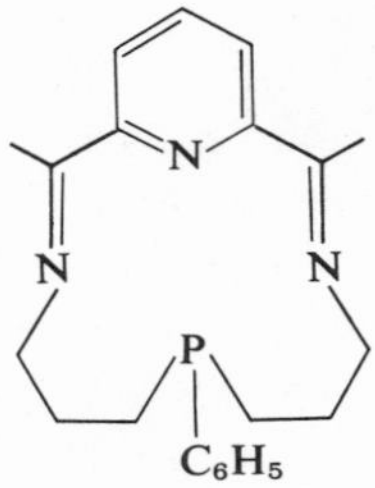

88

the above reaction, the product is the four-coordinate complex [Ni(Me$_2$pyo-[14]trieneN$_3$P)](PF$_6$)$_2$. The five-coordinate complex [Ni(Me$_2$pyo[14]triene-N$_3$P)Cl]PF$_6$ can be obtained by using an ethanolic solution of NiCl(PF$_6$) as the original source of nickel(II). Reduction of the imine groups in the macrocycle **88** has been achieved with both sodium borohydride and hydrogen over a platinum oxide catalyst; the yellow crystalline compound [Ni(Me$_2$pyo-[14]eneN$_3$P)](PF$_6$)$_2$ is obtained from the borohydride reduction of the nickel(II) complex. The reduced ligand is assigned the *meso* configuration **89** from the PMR spectrum of this compound, and may be removed from the nickel(II) ion by refluxing the compound with a large excess of cyanide ion. The square-planar complex containing the reduced ligand adds an iodide ion in ethanol to give the diamagnetic, five-coordinate species [Ni(Me$_2$pyo-[14]eneN$_3$P)I]PF$_6$.

89

Condensation reactions involving amines and carbonyl compounds have been extensively used for the formation of tetradentate macrocyclic ligands containing four nitrogen donors. These reactions have involved coordinated amines reacting with carbonyl compounds and vice versa, but one of the early *in situ* macrocyclic ligand formation reactions employed *o*-aminobenzaldehyde, in which the amine and carbonyl groups are functional groups on the same molecule. The self-condensation of *o*-aminobenzaldehyde in the absence of metal ions was first investigated by Seidel and Dick[319] and then reexamined by McGeachin,[72] who showed that in acidified aqueous solution several cyclic products, including **12** and **13**, are obtained. However, in the presence of nickel(II) ions in ethanol, the self-condensation results in the formation of complexes containing the tridentate and tetradentate macrocyclic ligands **14**, bzo$_3$[12]hexaeneN$_3$, (TRI), and **15**, bzo$_4$[16]octaeneN$_4$, (TAAB), respectively.[73,81,82] Selective crystallization of the perchlorate derivatives results in separation of the complexes containing **14** and **15**. Complexes with bzo$_3$[12]hexaeneN$_3$ are discussed in Section 2, which deals

14   **15**

with tridentate ligands. A series of nickel(II) complexes containing **15**, $Ni(bzo_4[15]octaeneN_4)X_2$ (X = BF_4, $B(C_6H_5)_4$, I, Cl, Br, NO_3, NCS), can be obtained from the perchlorate derivative (X = ClO_4) by metathesis.[82] The chloride and bromide species each contain a molecule of water and have room temperature magnetic moments that are intermediate between those expected for high-spin nickel(II), as demonstrated by the species with X = I, NO_3, NCS and low-spin nickel(II), with X = ClO_4, BF_4, $B(C_6H_5)_4$. An equilibrium between triplet and singlet states has been assumed to explain the non-Curie–Weiss temperature dependence of the magnetic susceptibility for the chloride and bromide salts.[320] The crystal structures of $Ni(bzo_4[16]-octaeneN_4)I_2 \cdot H_2O$ [high-spin nickel(II)] and $Ni(bzo_4[16]octaeneN_4)(BF_4)_2$ [low-spin nickel(II)], the macrocyclic ligand coordinates in a planar manner about the nickel(II) ion, have been reported.[321] When the self-condensation of *o*-aminobenzaldehyde is carried out in the presence of copper(II) nitrate, the complex $Cu(bzo_4[16]octaeneN_4)(NO_3)_2$, or $Cu(TAAB)(NO_3)_2$, is obtained.[81,82] No copper(II) species containing the tridentate condensate **14** have been obtained from the self-condensation reaction or by rearrangement of **12** in the presence of copper(II) salts.[76] These observations may be rationalized by the stereochemical requirements of the copper(II) ion; the Jahn–Teller distorted ion will be expected to exert certain restrictions on metal-ion-controlled self-condensation and rearrangement reactions compared with a nondistorted species such as nickel(II). These restrictions presumably eliminate the possibility of formation of the tridentate macrocyclic ligand by making the axial sites unavailable for coordination of intermediates in the macrocycle formation reactions.

Seidel[322] reported that the reaction between *o*-aminobenzaldehyde and zinc(II) chloride under anhydrous conditions in ether produced a compound with formula $C_7H_5N \cdot O \cdot 5ZnCl_2$. This reaction was repeated[82] and the product formulated as $[Zn(bzo_4[16]octaeneN_4)] \cdot [ZnCl_4]$. Busch[6] has reported using this product as an intermediate for the preparation of other metal complexes of $bzo_4[16]octaeneN_4$ by metal exchange.

A mixture of two cobalt(III) complexes containing the species $[Co(bzo_3[12]hexaeneN_3)_2]^{3+}$ and $[Co(bzo_4[16]octaeneN_4X_2]^+$ results from air oxidation of the material obtained from the self-condensation of *o*-aminobenzaldehyde in the presence of cobalt(II) salts.[323] The species are separated through their bromide derivatives since $[Co(bzo_4[16]octaeneN_4)-Br_2]Br$ is relatively insoluble in acidic methanol, whereas the *bis*-tridentate complex is much more soluble. (The latter species is discussed in Section 2.) By treating $[Co(bzo_4[16]octaeneN_4)Br_2]Br$ with the appropriate salt, the complexes $[Co(bzo_4[16]octaeneN_4)X_2]Y$ (X = Y = Cl, NCS, NO_3, N_3; X = NO_2, Y = NO_3; X_2 = $NO_3 \cdot ClO_4$, Y = ClO_4) are obtained; the macrocyclic ligand coordinates in a plane around the cobalt(III) ion with the anions in axial positions.

The complex [Fe(bzo$_4$[16]octaeneN$_4$)]FeCl$_4$ is obtained from the self-condensation of *o*-aminobenzaldehyde in an acidic ethanol solution of iron(II) chloride or by displacement of zinc(II) ions from [Zn(bzo$_4$[16]-octaeneN$_4$)]ZnCl$_4$ using iron(II) chloride.[5,111,324] Both reactions must be carried out under N$_2$. The *bis*-acetonitrile adduct [Fe(bzo$_4$[16]octaeneN$_4$)-(CH$_3$CN)$_2$]FeCl$_4$ can also be obtained; treatment of this complex with an excess of LiClO$_4$ produces dark-red crystals of [Fe(bzo$_4$[16]octaeneN$_4$)-(CH$_3$CN)$_2$](ClO$_4$)$_2$.* Iron(III) complexes of bzo$_4$[16]octaeneN$_4$ which have been obtained include Fe(bzo$_4$[16]octaeneN$_4$)FX$_2 \cdot$2H$_2$O (X = NO$_3$, ClO$_4$) and the dimeric oxo-bridged species ([Fe(bzo$_4$[16]octaeneN$_4$)]$_2$O)X$_4 \cdot$4H$_2$O (X = NO$_3$, ClO$_4$).[5,324] The oxo bridge in these species is very difficult to break; extended treatment with HF gives rise to the fluoro complexes above. The Mössbauer spectra of some of these species have been reported.[123]

Recently Potter and Taylor[326] reinvestigated the self-condensation of *o*-aminobenzaldehyde in the presence of manganese(II) ions first studied by Eichhorn and Latif.[327] With manganese(II) nitrate and chloride, the *bis*-anhydrotrimer 12 was obtained, which did not contain any manganese; with manganese(II) iodide and (Et$_4$N)$_2$MnCl$_4$ an uncharacterized red, metal-free precipitate was obtained. These findings are similar to those of Eichhorn and Latif, who also reported that no metal-containing species were obtained from the reaction of manganese(II) nitrate and *o*-aminobenzaldehyde. The nonincorporation of manganese(II) into either the tridentate or tetradentate macrocyclic ligands 14 or 15 is surprising in view of the fact that the ions VO^{2+}, Fe(II), Co(II), Ni(II), Cu(II), and Zn(II) are all capable of controlling the self-condensation of *o*-aminobenzaldehyde and give rise to complexes containing macrocyclic ligands and the "controlling" metal ion. The larger size of the manganese(II) ion has been suggested as a factor in the nonincorporation of this ion. The size and stereochemistry of the controlling metal ion are obviously important factors in the self-condensation reactions. When uranyl salts UO$_2$X$_2$ (X = Cl, Br), in which the axial sites are blocked by oxygen atoms, are used in the reaction with *o*-aminobenzaldehyde, no tridentate macrocyclic species are obtained; the uranyl complex isolated from the reaction medium contains a noncyclic tetrameric condensate which, when treated with water, liberates the *tris*-anhydrotetramer 13.[328] The mass spectrum of the uranyl complex shows a peak at 412 amu corresponding to 15, indicating that under the spectrometer conditions, the noncyclic ligand eliminates water to give the macrocycle. It is concluded from these experiments that blocking the axial sites by the oxygen atoms in the uranyl ion forces the condensation reaction to take place in the plane; however, the

---

* The iron(II) complexes Fe(bzo$_4$[16]octaeneN$_4$)(NCS)$_2$, [Fe(bzo$_4$[16]octaeneN$_4$)L$_2$]-(PF$_6$)$_2$ (L = methylimidazole, pyridine, CH$_3$CN, C$_6$H$_5$CH$_2$NC), and [Fe(bzo$_4$[16]-octaeneN$_4$)LX](PF$_6$)$_2$ (X = CO) have recently been prepared by using [Fe(bzo$_4$[16]-octaeneN$_4$)]FeCl$_4$ as intermediate.[324,325]

effective in-plane size of the uranyl ion does not allow complete ring closure to form **15**, bzo$_4$[16]octaeneN$_4$, to take place. In the case of copper(II), the condensation also takes place in the plane, the tetradentate macrocyclic ligand being produced because of the smaller in-plane size compared with that of the uranyl ion. Further experiments in this area should produce more definite conclusions concerning the mechanism of the metal-ion-controlled self-condensation of *o*-aminobenzaldehyde.

All the complexes of bzo$_4$[16]octaeneN$_4$ show remarkable stability toward concentrated acids; however, they are very reactive toward nucleophiles. Taylor *et al.*[329] have found that reaction of [Ni(bzo$_4$[16]octaeneN$_4$)]$^{2+}$ with alkoxide ions results in the incorporation of two alkoxyl functions into the complex to give neutral complexes of general formula **90** (R = CH$_3$, C$_2$H$_5$). A corresponding dimeric iron(III) complex (R = CH$_3$) containing an oxo bridge has also been reported.[123,324] The nucleophilic addition reactions

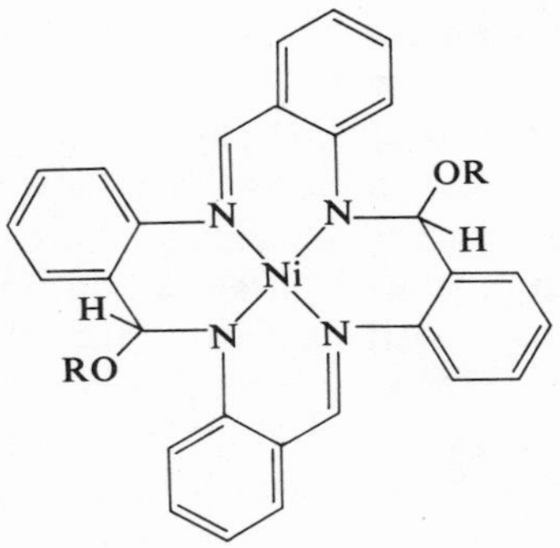

90

to nickel(II) and copper(II) complexes of bzo$_4$[16]octaeneN$_4$ were extended to include diol anions, amines, and diamines.[330,331] Polarographic studies of the nickel, copper, and cobalt complexes of bzo$_4$[16]octaeneN$_4$ have been reported by Busch *et al.*[5,332] and it has been shown that nickel(II) and copper(II) complexes of this ligand may be chemically reduced by hydrogen at relatively low pressures.[333] Chapter 6 should be consulted for discussions of the above reactions of bzo$_4$[16]octaeneN$_4$ complexes.

Recently, the self-condensation of *o*-aminobenzaldehyde in the presence of strong or mineral acids has been reinvestigated[334]; the red diacid salt produced has been assigned structure **90A** (X = BF$_4$, ClO$_4$, Br, Cl, HSO$_4$, CF$_3$SO$_3$).

Reactions of **90A** (X = BF$_4$) with metal acetates lead to the formation of complexes of bzo$_4$[16]octaeneN$_4$ with Ni(II), Co(II), Cu(II), and Fe(II). With **90A** (X = ClO$_4$) and rhodium(III) chloride, the complex[Rh(bzo$_4$-[16]octaeneN$_4$)(CH$_3$CN)$_2$](ClO$_4$)$_3$·H$_2$O is obtained. Partial hydrogenation of

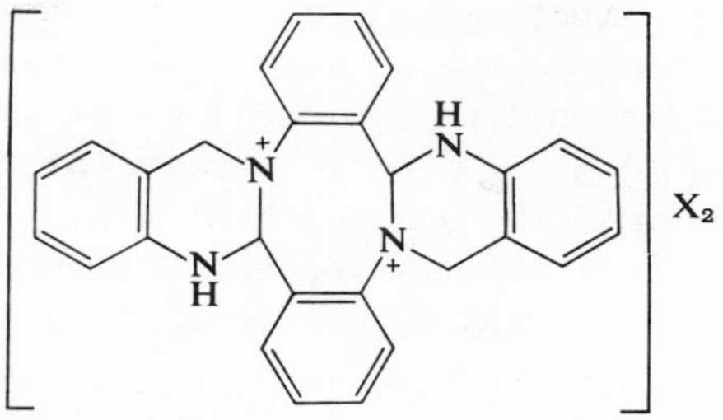

90A

90A (X = BF$_4$) with NaBH$_4$ produces the ligand **90B**; metal complexes with this ligand have been prepared, but no details are available.[334]

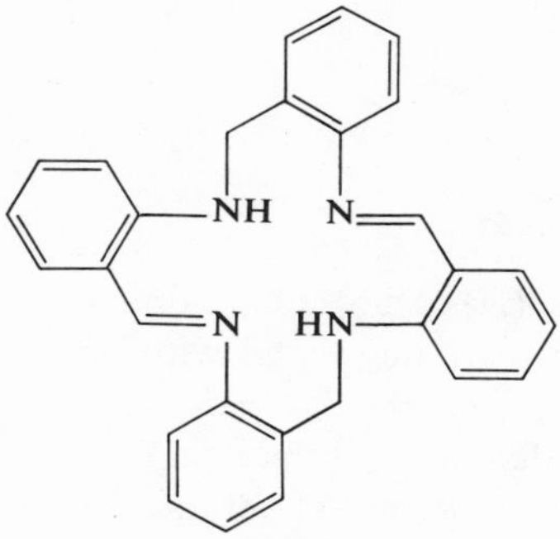

90B

Reaction of **90B** with LiAlH$_4$ or H$_2$/PtO$_2$ results in reduction of the imine bands; the macrocycle **90C** may be isolated.[334] A nickel(II) complex of **90C** has been reported[334]; it is identical to the species obtained by the hydrogen reduction of a nickel(II) complex containing bzo$_4$[16]octaeneN$_4$.[333]

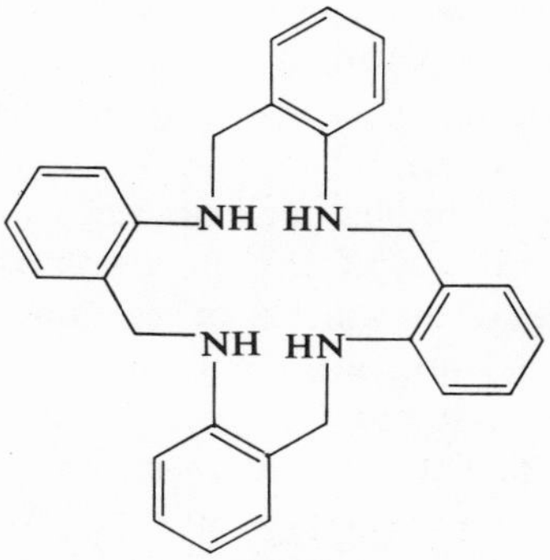

90C

Reactions of hydrazines and hydrazones with carbonyl compounds have only recently been employed for *in situ* syntheses of tetradentate macrocyclic ligands. Kerwin and Melson[41] reported that when the nickel(II) complex **91** ($R_1 = R_2 = C_6H_5$, $R_3 = R_4 = CH_3$)[333] is heated with anhydrous ethylene-diamine, the neutral nickel(II) complex **92** ($R_1 = R_2 = C_6H_5$, $R_3 = R_4 = CH_3$) is obtained. A macrocyclic complex was also obtained when 1,2-propanediamine was used under similar conditions, but with 1,3-propane-

91   **92**

diamine, a neutral nickel(II) complex containing a noncyclic ligand in which only one end of the amine had reacted with one of the coordinated —CO groups of **91** ($R_1 = R_2 = C_6H_5$, $R_1 = R_2 = CH_3$) was isolated from the reaction medium. All attempts to induce ring closure of this compound were unsuccessful. Structural assignments for the complexes containing macro-cyclic ligands have been made.[336] The related nickel(II) complex **92** ($R_1 = R_2 = R_3 = R_4 = CH_3$) is formed by reaction of ethylenediamine with the corresponding species **91**; however, when **91** [$R_1 = H$, $R_2 = C(CH_3)_3$, $R_3 = R_4 = CH_3$] is heated with anhydrous ethylenediamine under forcing conditions, no macrocyclic complex is obtained.[42,337] It was concluded from these results that the reaction of **92** with amines proceeds via attack at the carbon of the coordinated CO group, followed by elimination of a water molecule. Reactions of **91** ($R_1 = R_2 = C_6H_5$, $R_3 = R_4 = CH_3$) with ethylenediamine and 1,3-propanediamine have been followed spectrophoto-metrically in tetrahydrofuran/ethanol solution.[338,339] It was determined that the formation of the macrocyclic ligand is a two-step process with the rates of the two steps being comparable; the two steps correspond to successive reactions of the two coordinated CO groups with one molecule of ethylene-diamine. Detailed studies of the kinetics of reaction of **91** ($R_1 = R_2 = C_6H_5$, $R_3 = R_4 = CH_3$) with 1,3-propanediamine have shown that the reaction is second order in amine, first order in complex and in hydroxide ion, and dependent on the nature of the solvent. A mechanism in which two amines initially coordinate to the nickel(II) ion followed by base-catalyzed elimina-tion of a water molecule has been proposed.[339] The formation of the macrocycle by reaction with ethylenediamine follows a similar mechanism; the second ring closure step also involves a base-catalyzed water-elimination reaction. These results are similar to those obtained for the ring closure of "Jäger" complexes with amines (see p. 52).[211] The information concerning the factors that control the formation of macrocyclic complexes from **91** has led to the development of new and efficient syntheses. For example, the species **92** ($R_1 = R_2 = C_6H_5$; $R_3 = CH_3$, $R_4 = C_2H_5$, $R_3 = R_4 = C_2H_5$, $R_3 = CH_3$, $R_4 = n\text{-}C_3H_7$, $R_3 = CH_3$, $R_4 = n\text{-}C_4H_9$, $R_3 = CH_3$, $R_4 = C_6H_5$) have been obtained by reaction of the corresponding complex **91** with an excess of ethylenediamine and hydroxide ion in a tetrahydrofuran/ethanol solvent mixture. The procedure is easier to operate and gives higher yields of the desired product than the procedure involving the anhydrous amine. Reactions of **91** ($R_1 = R_2 = C_6H_5$, $R_3 = R_4 = CH_3$) with diamines of varying carbon chain length have shown that in some cases the nature of the solvent is important in determining whether a macrocyclic or nonmacrocyclic product is obtained.[339,340] Recently, Curtis[341] investigated reactions of the nickel(II) and copper(II) complexes **93**, prepared from the dialdehyde and hydrazine in the presence of the metal ion, with ketones R^1R^2CO ($R^1 =$ alkyl, $R^2 =$ H) or alkyl. The compounds **94** were obtained; a crystal structure of the copper(II) derivative has been completed.

93   **94**

Goedken and Peng found that the metal-ion-controlled condensation of butane-2,3-dione dihydrazone with formaldehyde produced complexes **95** (M = Ni(II), Fe(II), Co(II), or Cu(II)) containing a neutral macrocyclic ligand.[342–345] This macrocycle has secondary nitrogen atoms that are capable of further condensation with formaldehyde to yield a tricyclic ligand complex with two cyclic ethers fused to the macrocyclic ring, or with another molecule of dihydrazone and formaldehyde, to yield quadricyclic ligand complexes.[342] The cobalt(II) complex of **95** reacts with strong bases to produce a cobalt(I) complex that coordinates carbon monoxide,[344] and the reaction of the *bis*-acetonitrile adduct of **95** (M = Fe(II)) with carbon monoxide under basic conditions in the presence of CH_3NHNH_2 gives a complex containing carbon monoxide and methyl groups in *trans* positions about an anionic form of the macrocyclic ligand.[346] Treatment of the complex 95 (M = Co(II)) with organic hydrazines, oxygen, and alkoxide anions results in a complex series of reactions yielding molecular organo–cobalt(III) complexes.[347] Many of the above reactions serve as models for naturally occurring processes and are discussed in more detail in Chapters 6 and 10.

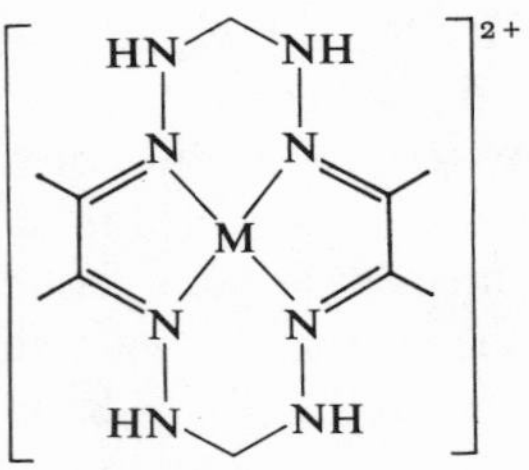

95

The product obtained from the reaction of butane-2,3-dione, hydrazine, and formaldehyde with nickel(II) perchlorate and molecular oxygen is a neutral nickel(II) complex, 96, containing a dianionic macrocyclic ligand related to that in 95 but coordinated in a different manner.[342] Two isomers

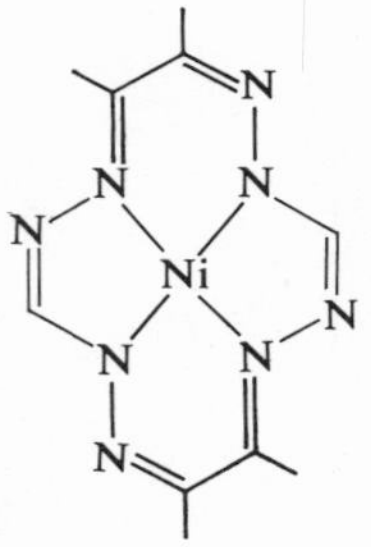

96

of the nickel(II) complex 96 have been obtained; 97 is produced by oxidation of 95 with molecular oxygen in acetonitrile solution made basic with hydrazine, and 98 is prepared in the same manner as 96 except that hydrazine must be the last reactant added. The complex 96 is the most stable isomer; both 97 and 98 undergo thermal and photochemical isomerization to 96. The structure of 96 has been reported.[343]

The reaction of the cobalt(II) complex analogous to 96 with an organic hydrazine followed by the addition of 1 atm of oxygen produces a series of organo–cobalt(III) complexes containing the same dianionic ligand L that is found in 96; molecular nitrogen is evolved during the reaction.[258] The five-coordinate species CoLR (R = CH_3, C_2H_5, n-C_3H_7, $CH(CH_3)_2$, $CH(CH_3)$-(C_2H_5), C_6H_5, $CH_2C_6H_5$, $COCH_3$, $COOC_2H_5$) have been isolated and

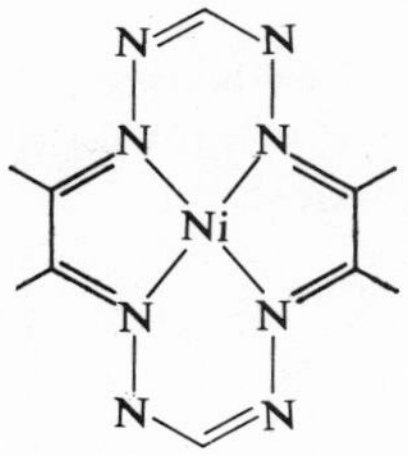

97 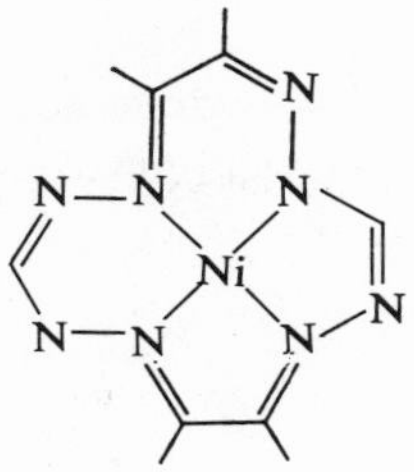   **98**

characterized by NMR and electronic spectroscopy. Yields of these species are almost quantitative based on the original cobalt(II) complex.

It is interesting to note that neither zinc(II) nor manganese(II) ions yield cyclization products under similar reaction conditions to those employed for the synthesis of the nickel(II), iron(II), and cobalt(II) complexes described above. It has been previously observed that these ions do not perform efficiently as templates for the *in situ* synthesis of tetradentate macrocyclic ligands due to their size and stereochemical restrictions.

Goedken *et al.*[348] reported the *in situ* synthesis of the complex ion **99**, containing a neutral, fully conjugated macrocyclic ligand $Me_4pyo_2[14]$-hexaene$N_4(N_2)$ from the reaction of 2,6-diacetylpyridine and hydrazine in

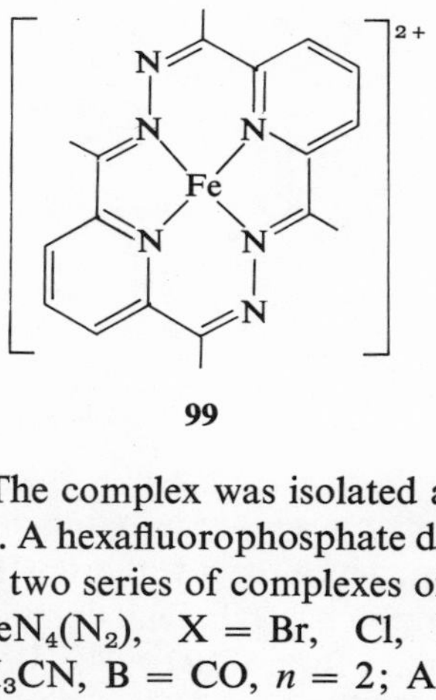

99

the presence of iron(II). The complex was isolated as the perchlorate salt of the *bis*-acetonitrile adduct. A hexafluorophosphate derivative can be prepared by a similar reaction, and two series of complexes of general formula $FeLX_2$ ($L = Me_4pyo_2[14]$hexaene$N_4(N_2)$, $X = Br$, Cl, NCS, N_3, NO_2) and $[FeLAB](PF_6)_n$ ($A = CH_3CN$, $B = CO$, $n = 2$; $A = Cl$, $B = CO$, $n = 1$; $AB = 2,2'$-bipyridine,$1,10$-phenanthroline, $n = 2$) have been obtained. It is assumed that in the bipyridine and phenanthroline complexes, the normally planar macrocyclic ligand is forced into a folded conformation. The crystal structure of $[Fe(Me_4pyo_2[14]$hexaene$N_4(N_2))(CH_3CN)_2](ClO_4)_2]$ has confirmed the mode of coordination of the planar macrocyclic ligand as depicted in **99**. A similar cobalt(II) complex has been obtained as the *bis*-hexafluoro-phosphate salt.

Reactions of 1,2-dihydrazones of cyclic α-diketones with orthoesters at elevated temperatures yield macrocycles to which the structure **100** (R = H, CH_3, C_2H_5, C_6H_5, $n = 0, 1, 2$; R = 4-Br-C_6H_4, $n = 2$) was assigned.[349]

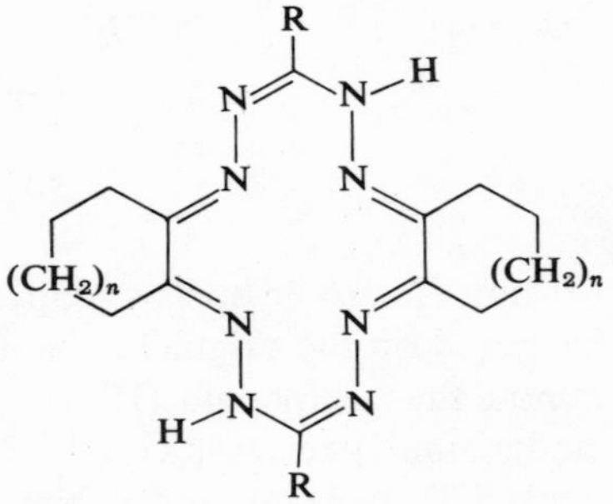

100

Later,[350] additional examples of these macrocycles (R = C_3H_7, C_4H_9, 4-CH_3-C_6H_4, OC_2H_5, $n = 0, 1, 2$; R = 2-naphthyl, $n = 1, 2$ and R = 4-Br-C_6H_4, $n = 1$) were reported and assigned the isomeric structure **101**. Baldwin

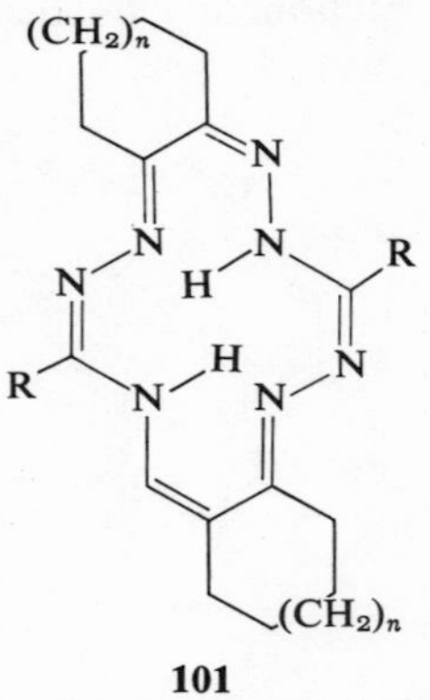

101

et al.[351] found that similar macrocycles containing a variety of R and R' groups could be prepared, which react with iron(II), cobalt(II), and nickel(II) salts in basic solution to give neutral metal complexes containing the macro-cycles in their dianionic form, **102**. The nickel(II) complexes **102A** (R = CH_3, C_2H_5) and **102C** (R = C_2H_5, n-C_4H_9) have been isolated and characterized by spectroscopic techniques; these studies have confirmed the mode of coordination of the dianionic ligand as shown in **102**. The iron(II) complex **102A** [R = $(CH_2)_7CH_3$] reacts with molecular oxygen in toluene/ pyridine solution at $-78°$ to give a red–brown solution with the absorption of 0.5 mol of O_2; this uptake is irreversible.[352] However, the iron(II) complex

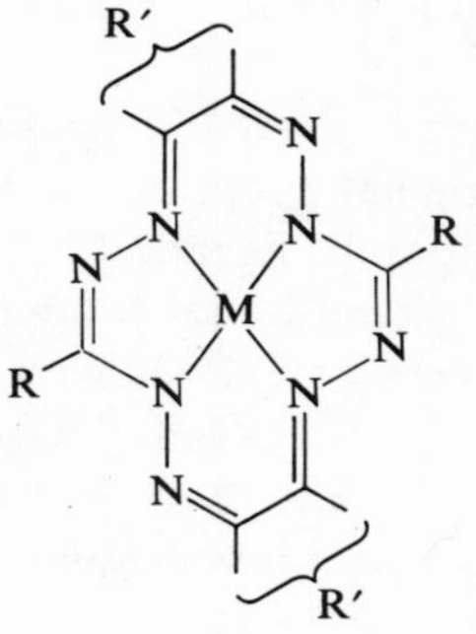

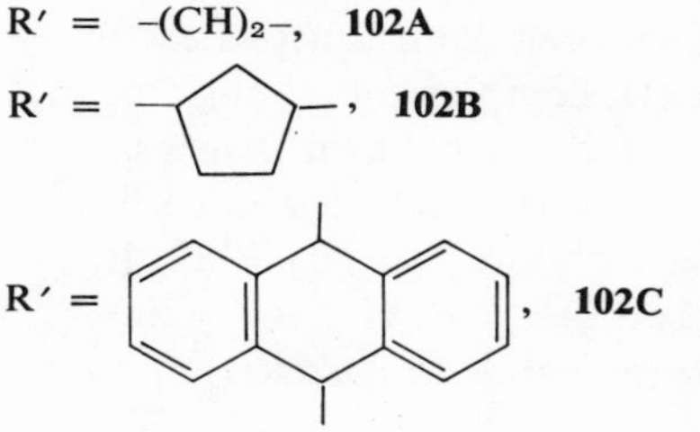

R′ = -(CH)₂-,   **102A**

R′ = (cyclopentane),   **102B**

R′ = (anthracene),   **102C**

102

102C [R = $(CH_2)_7CH_3$] reversibly binds dioxygen at $-78°$ in THF/dimethoxyethane/pyridine solution with the uptake of 1 mol of O_2. The crystal and molecular structures of the latter iron(II) complex has been reported,[353] and significant features of the structure have been related to its ability to reversibly bind dioxygen (see Chapter 10).

Ammoniacal solutions containing copper(II) ions and oxalyldihydrazide, $(CONHNH_2)_2$, react with an excess of acetaldehyde to form a green–brown solution; treatment of this solution with molecular oxygen produces a purple color.[354] It has been suggested that the copper complex produced (and the corresponding species obtained by using acetone) may be a dioxygen-containing species, and thus the system(s) may serve as models for studying the oxygen-carrying blue copper proteins. Clark *et al.*[355,356] repeated the earlier experiments; reaction of a hot aqueous solution of oxalyldihydrazide (or the *N*-phenyl derivative) with an ammoniacal solution of copper(II) perchlorate followed, after cooling, by the addition of excess acetaldehyde in the presence of air leads to a purple (or green) product. Both products contain anionic complex ions containing macrocyclic ligands resulting from ring closure reactions with acetaldehyde: **103** (purple) and **104** (green). A mechanism for the formation of these unusual species is suggested.[356]

Keyes *et al.* have recently suggested that **103** is actually a copper(III)

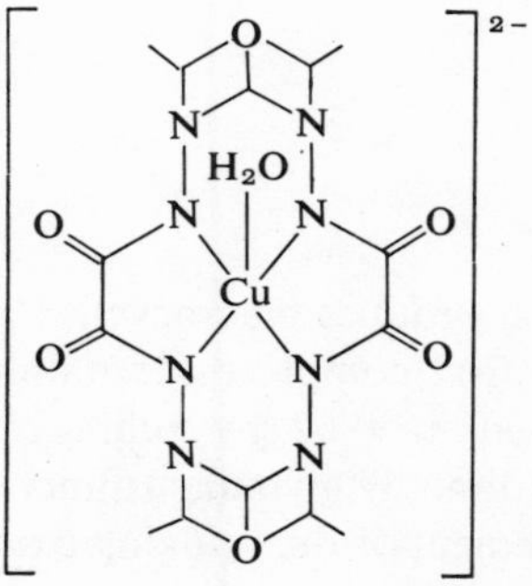

103 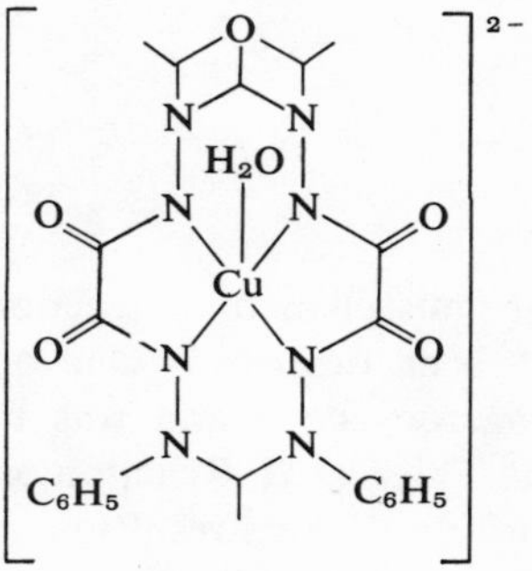   **104**

complex and that at neutral pH, the axial ligand is a hydroxyl rather than a water molecule.[357]

Reaction of the nickel(II) perchlorate complex of 6,6'-dihydrazino-2,2'-bipyridyl with refluxing aqueous acetone was originally reported to give rise to a nickel(II) complex containing a neutral macrocyclic ligand.[358] Recently, however, it has been shown that a macrocyclic ligand is not produced and that reaction with aldehydes and ketones produces complexes with ligands of general structure 105 (R_1 = H, R_2 = CH_3, C_2H_5, n-C_3H_7; R_1 = CH_3, R_2 = CH_3, C_2H_5, n-C_3H_7; R_1 = Et, R_2 = Et).[359] The mode of coordination of these ligands to nickel(II) and zinc(II) has been investigated.[360]

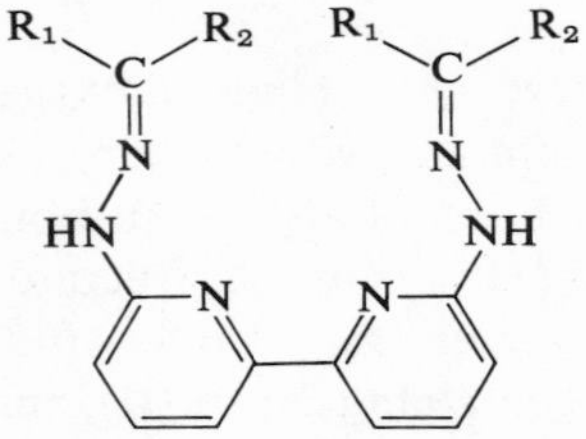

105

Russian workers[361,362] have reported the synthesis of the macrocyclic compounds 106 (R = i-C_3H_7, n-C_4H_9, n-hexyl, n-octyl); neutral nickel(II) and copper(II) complexes of the dianion derived from 106 have also been prepared.

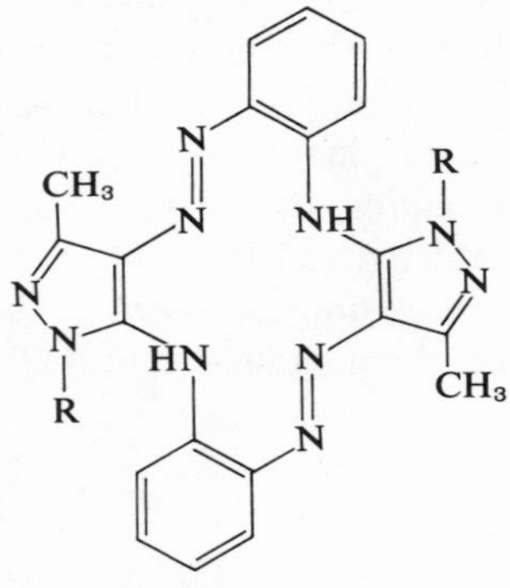

106

Several "miscellaneous" reactions that produce macrocyclic ligand complexes have been described. One of the first complexes containing a new synthetic macrocyclic ligand was reported in 1962 by Schrauzer[363] and Umland and Thierig.[364] Reaction of *bis*-(dimethylglyoximato)nickel(II) with boron trifluoride etherate results in replacement of the bridging oxime hydrogen atom by a BF_2 unit, and the formation of the complex **107** (R = F). Other complexes similar to **107** with R = Cl, alkyl, or aryl have been prepared in addition to Pd(II), Cu(II), and Fe(II) complexes analogous to **107**. The macrocyclic complex **108** (M = Ni(II)) has been prepared [6]; the cobalt(II) complex has been shown to form a 1:1 adduct with dioxygen, and be oxidized to a cobalt(III) species in protic media. [365]

107   108

Ogawa *et al.* [366,367] reported that condensation between 2,9-diamino- and 2,9-dichloro-1,10-phenanthroline in nitrobenzene in the presence of K_2CO_3 as acid acceptor gives the macrocyclic product **109**, which reacts with cobalt(II), nickel(II), and copper(II) ions in basic solution to give neutral complexes in which the ligand coordinates as a dianion.

109

Cyclic phosphorus–nitrogen compounds are known to behave as good ligands; however, the complexing ability of large ring compounds has not been extensively investigated. Marsh *et al.* [368] found that dodeca-(dimethylamido)cyclohexaphosphonitrile, $N_6P_6(NMe_2)_{12}$, reacts with $CuCl_2$ and CuCl either separately or with an equimolar mixture of the two to form a bright orange complex $[N_6P_6(NMe_2)_{12}CuCl]^+[CuCl_2]^-$ in which the 12-membered macrocyclic ligand is coordinated to a copper(II) ion in a planar manner by four nitrogen donor atoms, **110**. A coordinated chloride ion results in the copper(II) ion being five coordinated.

Several macrocycles closely related to the phthalocyanines (see Chapter 7) but with variations within the ring have been described. For example,

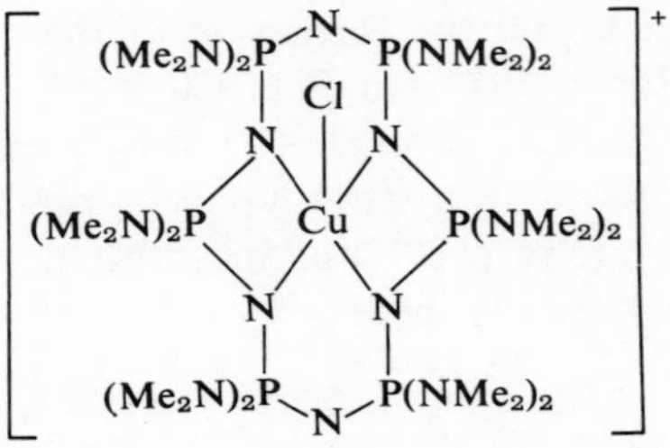

110

macrocycles of general formula **111** have been prepared and several metal complexes obtained from the dianionic form of these ligands, either by direct reaction with metal salts or by *in situ* synthesis of the ligand with the metal ion serving as a template.[255,369–372]

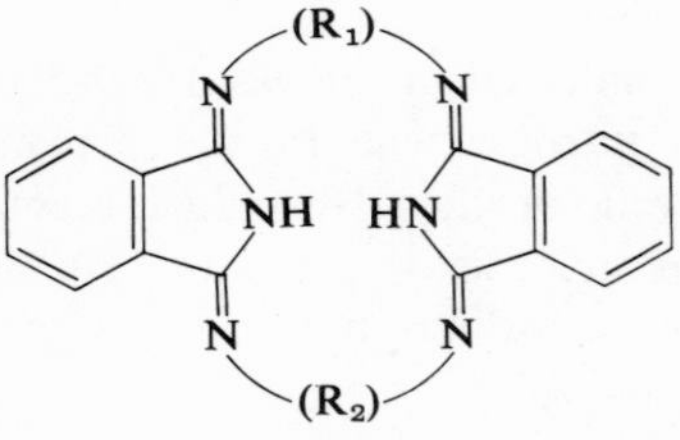

111

111A, R$_1$ = R$_2$ =

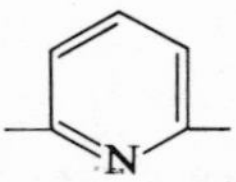

111B R$_1$ =

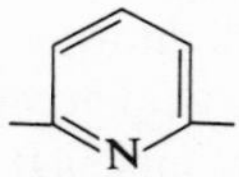

R$_2$ =

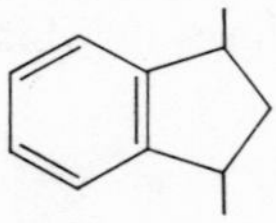

111C, R$_1$ = R$_2$ =

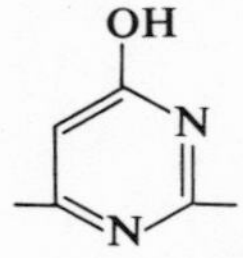

111D, R$_1$ = R$_2$ =

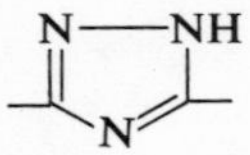

3.2. N$_2$O$_2$ Donor Atoms

When compared with the large number of tetradentate macrocyclic ligands containing four nitrogen donor atoms that have been prepared, there are very few known macrocycles with two nitrogen and two oxygen donor atoms. In general, an ether oxygen atom does not form a strong bond on coordination to a transition-metal ion, although the recent development of cyclic polyethers as ligands has demonstrated the ability of these compounds to coordinate strongly with alkali and alkaline earth metal ions (see Chapter 9).

The incorporation of donor atoms that are capable of strong coordination to transition-metal ions, e.g., nitrogen atoms, into a macrocycle that contains ether linkages should help to overcome the limitations for coordination induced by the ether oxygens. This section will deal with macrocyclic ligands with an N_2O_2 donor set joined together by groups other than simple aliphatic linkages; the latter group of macrocycles will be discussed in Chapter 9.

Kluiber and Sasso[373] reported that *bis*-(*N*-2-bromoethylsalicylaldimi-nato)nickel(II) reacts with sodium iodide in acetone to form a red, paramagnetic nickel(II) complex containing a neutral macrocyclic ligand, **112**, bzo$_2$[14]tetraene-4,11-N$_2$-1,8-O$_2$ (R = H) with an N_2O_2 donor set in which like donor atoms are arranged in *trans* positions across the ring. The crystal

112

and molecular structure of [Ni(bzo$_2$[14]tetraene-4,11-N$_2$-1,8-O$_2$)I$_2$] has confirmed the ligand structure (the nickel atom is located at a center of symmetry) and shown that it coordinates in a planar manner about the nickel(II) ion; the iodides are coordinated above and below the plane.[374] Treatment of this complex with water liberates the free ligand, demonstrating the much weaker coordinating ability of the N_2O_2 ligand compared with similar ligands with an N_4 donor set.

It has been reported[375] that nickel(II) complexes with various substituents on the ligand **112** have been synthesized. When a methoxy group is introduced into **112** (R = OCH$_3$) and the nickel(II) complex is allowed to crystallize from acetone solution at room temperature, a green compound is obtained. If the ligand and nickel(II) ions react together in boiling acetone, two green crystalline modifications can be isolated from solution; hexagonal and rhombic crystals. These modifications have identical elemental analyses. The crystal and molecular structure of the rhombic form of [Ni((MeObzo)$_2$-[14]tetraene-4,11-N$_2$-1,8-O$_2$)I$_2$] has been determined[375]; the ligand structure

112 (R = OCH$_3$) is confirmed; however, unlike the ligand with R = H, this macrocyclic ligand coordinates to the nickel(II) ion in a folded conformation. The coordination environment of the six-coordinated nickel(II) ion consists of *cis* iodides, *cis* nitrogens and *trans* oxygens, whereas that of the previously described compound[374] consists of iodides, nitrogens, and oxygens all *trans* to one another.

Armstrong and Lindoy[376,377] prepared macrocyclic ligands of general formula **113** from the dialdehydes **114** both by direct synthesis and by *in situ* procedures in the presence of nickel(II) salts. Reaction of the macrocyclic

113   **114**

ligands bzo$_2$[15]tetraene-8,12-N$_2$-1,4-O$_2$, **113** (R = –(CH$_2$)$_3$–, *n* = 2) and bzo$_2$[16]tetraene-9,13-N$_2$-1,5-O$_2$, **113** (R = –(CH$_2$)$_3$–, *n* = 3) with various nickel(II) salts in butanol yields complexes of the type [Ni(ligand)X$_2$] where X = Cl, Br, I, NCS. *In situ* condensation of the dialdehydes and diamines in butanol also leads to the above complexes and the species [Ni(ligand)X$_2$], where X = Br, I, NCS, ligand = bzo$_2$[14]tetraene-8,11-N$_2$-1,4-O$_2$, **113**, R = –(CH$_2$)$_2$–, *n* = 2; bzo$_2$[15]tetraene-9,12-N$_2$-1,5-O$_2$, **113**, R = –(CH$_2$)$_2$–, *n* = 3; Mebzo$_2$[14]tetraene-8,11-N$_2$-1,4-O$_2$, **113**, R = –CH$_2$CH(CH$_3$)–, *n* = 2; Mebzo$_2$[15]tetraene-9,12-N$_2$-1,5-O$_2$, **113**, R = –CH$_2$CH(CH$_3$)–, *n* = 3; bzo$_3$[14]pentaene-8,11-N$_2$-1,4-O$_2$, **113**, R = *o*-C$_6$H$_4$, *n* = 2; bzo$_3$-[15]pentaene-9,12-N$_2$-1,5-O$_2$, **113**, R = *o*-C$_6$H$_4$, *n* = 3. The complex [Ni(bzo$_2$[14]tetraene-8,11-N$_2$-1,4-O$_2$)I$_2$] contains a macrocyclic ligand that is isomeric with **112** (R = H, bzo$_2$[14]tetraene-4,11-N$_2$-1,8-O$_2$), the two ligands differing in location of the like donor atoms and imine linkages. Armstrong and Lindoy isolated the complex Ni(bzo$_2$[14]tetraene-4,11-N$_2$-1,8-O$_2$)(NCS)$_2$ directly from the free ligand **112** (R = H) and concluded that like the iodo derivative[33,274] the complex has a tetragonal geometry with the macrocycle coordinated in a planar manner. The electronic spectra of the nickel(II) complexes derived from **113** are similar to the nickel(II) complexes containing **112** (R = H); thus they are assumed to have similar tetragonal structures.

The addition of three molar equivalents of 1,3-diaminopropane (1,3-pn) to a suspension of $Ni(bzo_2[15]tetraene-8,12-N_2-1,4-O_2)X_2$ or $Ni(bzo_2[16]tetraene-9,13-N_2-1,5-O_2)X_2$ $(X = Br, I)$ in dry methanol results in the precipitation of $Ni(1,3-pn)_3X_2$, and the corresponding macrocycle may be isolated from the filtrate. Attempts to remove the macrocyclic ligand from $Ni(bzo_2[14]tetraene-8,11-N_2-1,4-O_2)X_2$ or $Ni(Mebzo_2[14]tetraene-8,11-N_2-1,4-O_2)X_2$ by using ethylenediamine or 1,2-propanediamine were not successful; however, when 1,3-propanediamine was used in the above procedure, the amine-exchanged macrocycle $bzo_2[15]tetraene-8,12-N_2-1,4-O_2$ was isolated in each case, suggesting that macrocyclic ring opening occurs in the displacement reactions. Recently, complexes of some of the ligands of general formula **113** with copper(II), zinc(II), and cadmium(II) have been prepared using both direct and *in situ* synthetic procedures.[378] The species $[CuL](ClO_4)_2$, $[CuLX]ClO_4$ $(X = Cl, NCS, Br)$, $[CuLBr]BF_4$ $(L = bzo_2[15]tetraene-8,12-N_2-1,4-O_2)$, $[CuL](ClO_4)_2$, $[CuLX]ClO_4$ $(X = Cl, NCS, L = bzo_2[16]tetraene-9,13-N_2-1,5-O_2)$, $[ZnLCl_2]$ $(L = bzo_2[14]tetraene-8,11-N_2-1,4-O_2, Mebzo_2[14]tetraene-8,11-N_2-1,4-O_2, bzo_2[15]tetraene-9,12-N_2-1,5-O_2)$, $[CdL_2](ClO_4)_2 \cdot nH_2O$ $(L = bzo_2[15]tetraene-8,12-N_2-1,4-O_2,$ $n = 1)$, and $Mebzo_2[14]tetraene-8,11-N_2-1,4-O_2$ $(n = \frac{1}{2})$ were isolated and characterized. The copper-halo perchlorate complexes were assigned square-pyramidal geometry and the zinc species octahedral (presumably tetragonal) geometry. The large ionic radius of the cadmium(II) ion relative to the macrocyclic ligand hole size probably leads to the formation of a 1:2 metal-ion–ligand ratio in the complexes.

Recently, cobalt(II) complexes of **113** $(R = (CH_2)_3, n = 2)$ and **113A** $(R = H, CH_3)$ have been reported.[379] The complexes are prepared by reaction between the preformed macrocyclic ligand and a slight excess of cobalt salt; both five- and six-coordinate complexes were characterized.

113A

Reduction of the diimines **113** (R = –(CH$_2$)$_2$–, –(CH$_2$)$_3$–, n = 2, 3) and the 3,13-dichloro (R = –(CH$_2$)$_3$–, n = 2) derivative with sodium borohydride produces the macrocycles **113B**; no complexes were reported in this paper.

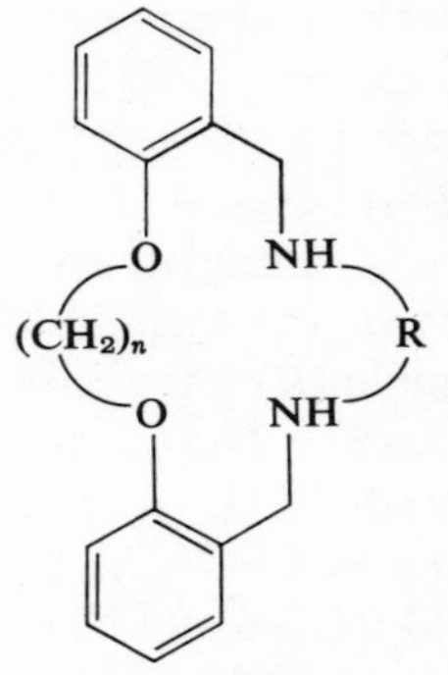

113B

The ligand **115** has been synthesized by the reaction of heptane-2,4,6-trione with ethylenediamine in ethanol.[381] Reaction of **115** with copper(II)

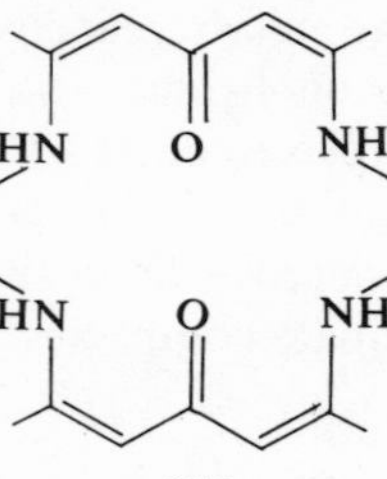

115

or nickel(II) acetate gives the neutral complex **116** (M = Cu, Ni) in which only one metal ion is incorporated into the ligand by coordination with two nitrogen and two oxygen donor atoms.[382] Vigato *et al.*[383] prepared nickel(II)

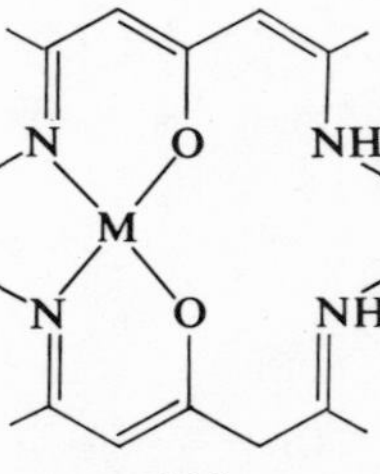

116

and cobalt(II) complexes **116** (M = Ni, Co), but reaction of **115** with copper(II) and UO_2^{2+} ions lead to partial hydrolysis and opening of the macrocyclic ring to give complexes of **117**, the copper(II) ions being coordinated by an N_2O_2 donor set and the uranyl ions by an O_4 donor set and a solvent molecule. Reactions of 1,2-propanediamine and 1,3-propanediamine

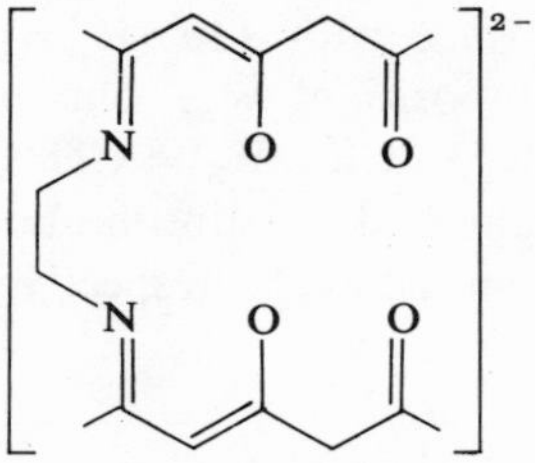

117

with heptane-2,4,6-trione in ethanol give the corresponding macrocyclic ligands **115** with $-CH_2CH(CH_3)-$ and $-(CH_2)_3-$ linkages replacing $-(CH_2)_2-$. The former macrocycle reacts with copper(II) and nickel(II) to give complexes analogous to **116** but no complexes were isolated by similar reactions employing the latter macrocycle.[382]*

Several macrocyclic ligands similar to **115** have been found to accommodate two metal ions; these are discussed in Section 6, "Binucleating Ligands."

3.3. N_2S_2 Donor Atoms

The first deliberate attempt to produce a new macrocyclic ligand system involved the formation of a nickel(II) complex containing an N_2S_2 donor atom set. Thompson and Busch[32,385] reasoned that alkylating agents with two functional groups should react with the two sulfur atoms coordinated at *cis* positions in 2,3-pentanedione-*bis*-(mercaptoethylimino)nickel(II),[386,387] **118** ($R_1 = R_2 = CH_3$), to form a new chelate ring with the carbon chain of

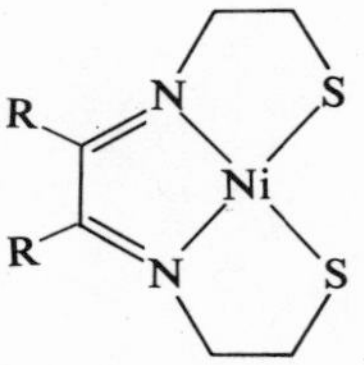

118

---

* Recently the complexation of ligands similar to **115** with Cu(II), Ni(II), and Co(II) has been discussed; in all cases only mononuclear complexes are formed.[384]

the alkylating agent linking the two sulfur atoms. Thus the reaction of α,α'-dibromo-*o*-xylene with a suspension of **118** ($R_1 = R_2 = CH_3$) in dimethylformamide results in the formation of **119** ($R_1 = R_2 = CH_3$, $X = Br$) in which the new macrocyclic ligand is coordinated in a plane about the nickel(II) ion, with bromide ions coordinated in axial positions. Similar reactions take place when $R_1 = CH_3$, $R_2 = C_2H_5$, C_5H_{11}. When 1,2-dibromoethane is used as a reactant for ring closure with **118** ($R_1 = CH_3$, $R_2 = CH_3$, C_2H_5), products of uncertain composition were obtained, although macrocyclic complexes related to **119** ($R_1 = CH_3$, $R_2 = CH_3$, C_2H_5) have been obtained by using 1,3-dibromopropane and 1,4-dibromobutane, which produced new chelate rings of six and seven members, respectively.

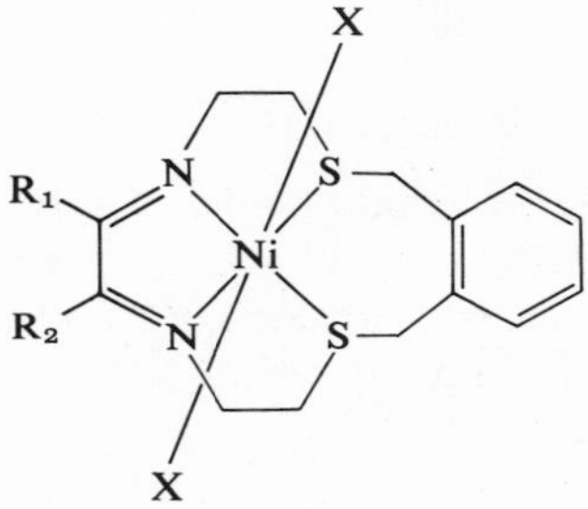

119

A series of nickel(II) complexes related to **119** ($R_1 = R_2 = CH_3$, $X = ClO_4$, I, N_3, NCS, Cl) has been obtained from the bromo derivative by treatment with a salt of the appropriate anion; the magnetic properties of these complexes have been discussed.[388] Reactions of an aromatic analog of **118** with α,α'-dibromo-*o*-xylene have led to the isolation of the macrocyclic complexes **120** ($R = CH_3$, $X = Br$, I, NCS; $R = \frac{1}{2}C_4H_8$, $X = Br$, NCS).[389]

Attempts to obtain a macrocyclic complex derived from *bis*-(*N,N*-diethyldithiocarbamate)nickel(II) by reaction with α,α'-dibromo-*o*-xylene

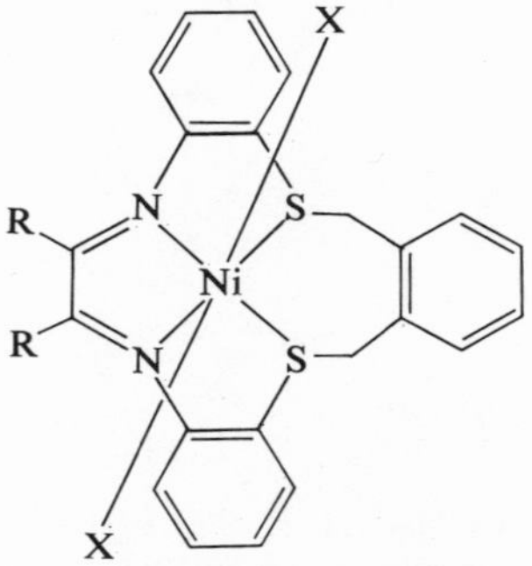

120

were not successful[390]; instead, ligand alkylation resulted in the formation of the cation **121**, isolated as the species [cation] [$Ni_5Cl_8Br_3$] and [cation]$_2$-[$NiBr_4$]. Egen and Krause[391] showed that ring closure of **118** ($R_1 = R_2 = $

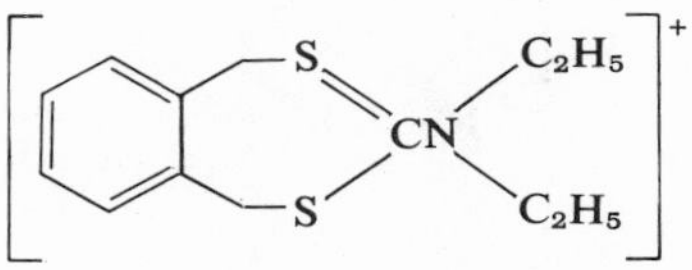

121

CH_3) takes place with S_2Cl_2 to give the maroon macrocyclic complex **122**, which contains a polysulfide chelate ring; a green hydrogen chloride adduct can also be obtained.

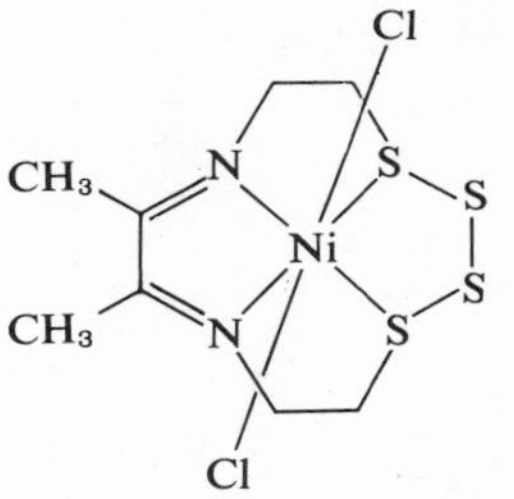

122

The phrase *kinetic template reaction* was introduced by Thompson and Busch to describe reaction in which a metal ion holds reactive groups in the correct orientation to facilitate a stereochemically selective multistep reaction. The kinetics of formation of **119** ($R_1 = R_2 = CH_3$, X = Br) was studied by Blinn and Busch.[33] This study showed that the ring closure reaction with α,α'-dibromo-*o*-xylene proceeds via a slow first step involving attack at one coordinated mercaptide followed by a second, rapid ring closure step, in contrast to the two slow consecutive steps observed for the reaction with benzyl bromide. After the first mercaptide group has reacted with α,α'-dibromo-*o*-xylene, the second mercaptide and halide are oriented such that the cyclization reaction proceeds rapidly. This control of the steric course of the reaction by the metal ion is the basis for the kinetic template effect.

The *in situ* formation of other macrocyclic ligands containing an N_2S_2 donor set has been reported. Urbach and Busch[392] prepared nickel(II) complexes $NiLX_2$ (X = Br, NCS) and $NiLX_2 \cdot H_2O$ (X = Cl, I, ClO_4, L = **123**, $Me_3bzo[15]dieneN_2S_2$) by reaction of the nickel(II) complex of 1,10-diamino-3,8-dithia-5,6-benzodecene[32] with acetone. A condensation reaction involving two acetone molecules similar to those reported earlier by Curtis takes place. Alcock and Tasker[393] reported that when a tetrahydrofuran solution

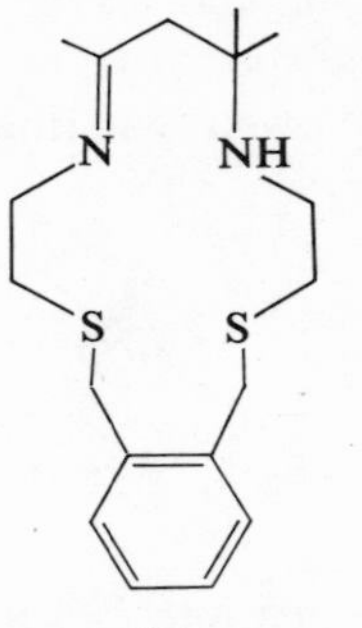

123

of the dihydrazine **124** is treated with formaldehyde and nickel(II) salts, a red–purple diamagnetic complex **125** is obtained. When cyclohexanone is employed instead of formaldehyde in the above reaction, a nickel(II) complex

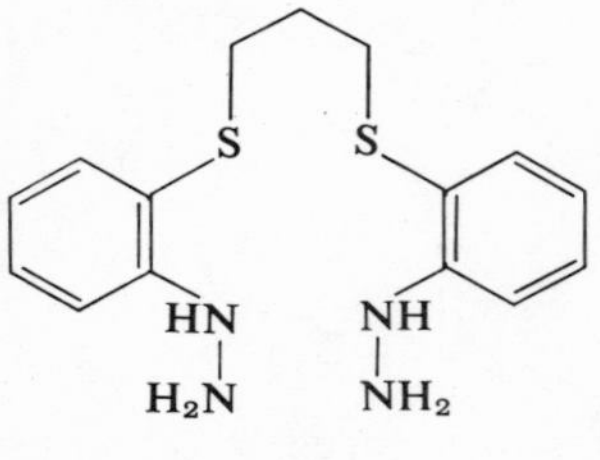

124

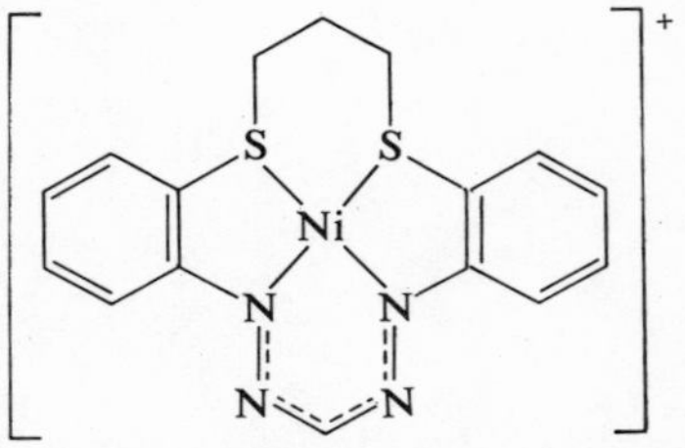

125

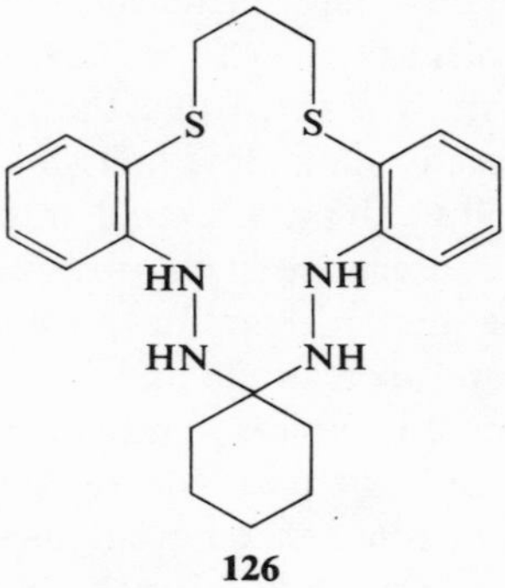

126

containing the neutral ligand **126** is obtained. Lindoy and Busch[394] described *in situ* reactions between *bis*-(2-formylphenyl)dithiabutane and ethylene-diamine or *o*-phenylenediamine in the presence of nickel(II) perchlorate; complexes containing the ligands **127** (R = $-(CH_2)_2-$, $o-C_6H_4$) are produced. When R = $-(CH_2)_2-$, the complexes $NiLX_2$ (X = ClO_4, NO_3, NCS)

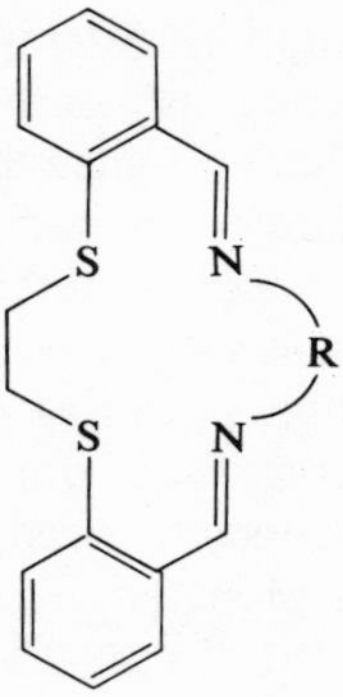

127

are obtained; when R = *o*–C_6H_4, $NiLX_2$ (X = ClO_4, Br, NCS) can be isolated.

Recently, Hay *et al.*[395] reported the formation of the macrocyclic complex **129** by reaction of the trimeric species **128** with α,α'-dibromo-*o*-xylene in methanol. It is suggested that a trimer $\rightleftharpoons$ monomer equilibrium is

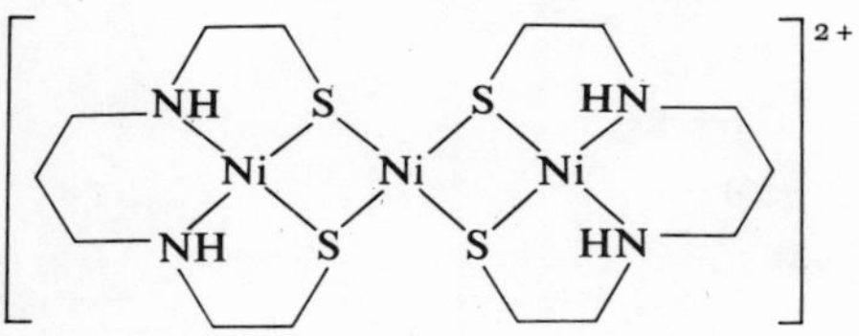

128

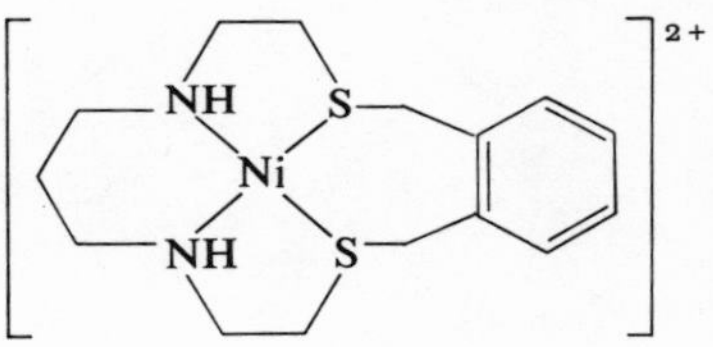

129

established in solution, with the monomeric complex reacting with the dibromide to give the macrocyclic complex. Hay and co-workers[396] also reported the synthesis of a nickel(II) complex containing the ligand **130** by an *in situ* reaction involving 1,5-*bis*-(2-aminophenylthio)pentane, acetone, and nickel(II) acetate in methanol; the complex was isolated as the dibromo derivative. The *bis*-iodo, *bis*-isothiocyanato, and bromo-perchlorato species can be obtained from the dibromo complex by treatment with the appropriate

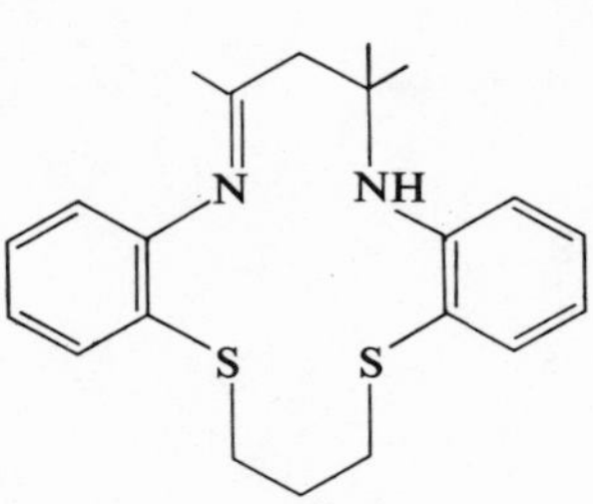

130

alkali metal salt in methanol solution. Attempts were made to prepare the 13-membered macrocyclic ring analogous to **130** by the reaction of acetone with nickel(II) acetate and 1,4-*bis*-(2-amino-phenylthio)butane. The nickel(II) complex of the reactant dithiodiamine immediately separated on mixing the reagents; no condensation product was obtained.

All the macrocyclic complexes discussed so far in this section have contained ligands with a *cis*-N_2S_2 donor atom system; there are very few examples of macrocycles containing a *trans*-N_2S_2 donor set. A 12-membered macrocycle of this latter type has been synthesized by conventional techniques[397] but no complexes have been prepared. Kolesnikov and Borodkin[398] reported the preparation of **131** from 2,5-diamino-1,3,4-thiadiazole and phthalonitrile; copper(II), nickel(II), cobalt(II), zinc(II), and lead(II) complexes containing the ligand in its dianionic form have been isolated.

131

3.4. S_4 Donor Atoms

Conventional syntheses of macrocyclic thioether compounds have been known for several years,[399,400] although few transition-metal complexes containing these compounds as ligands have been synthesized. Rosen and Busch[25,401] prepared the macrocyclic thioethers **132**, [14]aneS$_4$, and **133**, bzo[15]eneS$_4$, from the dianion of 1,4,8,11-tetrathiaundecane and 1,3-

132   **133**

dibromopropane or α,α'-dibromo-*o*-xylene, respectively. Yields of these macrocycles were not improved when nickel(II) ions were used in an analogous "template" reaction, although carrying out the condensation reaction at higher dilution does result in a significant improvement in yield.[26] The macrocycle **134** is also obtained from the reaction of 1,4,8,11-tetrathiaunde-

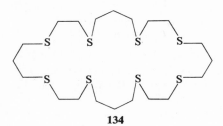

134

cane and 1,3-dibromopropane; this compound is capable of incorporating two metal ions, each with an S_4 donor atom set.[402] (See Section 6 on bi-nucleating ligands.) The nickel(II) complex Ni([14]aneS$_4$)(BF$_4$)$_2$ was obtained from nickel(II) tetrafluoroborate in nitromethane/acetic anhydride, and the ligand **132**, [14]aneS$_4$; the derivatives Ni([14]aneS$_4$)X$_2$ (X = NCS, Cl, Br, I, ClO$_4$) are obtained from the fluoroborate complex by metathesis. The complex Ni(bzo[15]eneS$_4$)(BF$_4$)$_2$ was prepared by a similar procedure to that outlined above for Ni([14]aneS$_4$)(BF$_4$)$_2$. The fluoroborate derivatives of both nickel(II) complexes and (Ni[14]aneS$_4$)(ClO$_4$)$_2$ are square-planar, whereas the other species are assigned tetragonally distorted octahedral structures. The square-planar structure of Ni([14]aneS$_4$)(BF$_4$)$_2$ has been confirmed and the EPR spectrum of its ^{63}Cu-doped crystals was reported.[403] Rosen and Busch[70] prepared the 12- and 13-membered macrocycles **135**, [12]aneS$_4$, and **136**,

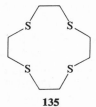

135

[13]aneS$_4$, by procedures similar to those reported earlier for the 14-membered species. The complexes Ni$_2$([12]aneS$_4$)$_3$(BF$_4$)$_4$ and Ni$_2$([13]aneS$_4$)$_3$-(BF$_4$)$_4$ have been prepared from these ligands; they both contain octahedral nickel(II) ions and the structure **137** has been proposed for the cations. The complexes react with water, which displaces the macrocyclic ligands. It was noted that when the macrocycle has four sulfur donor atoms, a 13-membered ring is unable to coordinate in a planar manner to nickel(II), although when

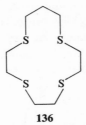

136

a macrocycle has four nitrogen donor atoms, a 13-membered ring is able to coordinate in a square plane around a nickel(II) ion. Glick *et al.*[404] reported the crystal structure of the copper(II) complex, $Cu([14]aneS_4)(ClO_4)_2$; the

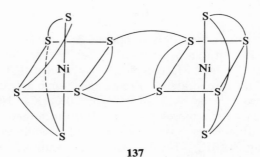

137

macrocyclic ligand coordinates in a plane with two *trans* perchlorato groups completing the tetragonal coordination environment around the copper(II) ion. The corresponding copper(I) complex, $Cu([14]aneS_4)ClO_4$, can be prepared from the copper(II) complex by reduction at 0.45 V at a platinum-working electrode and slowly crystallizing from a methanol–water mixture under a N_2 atmosphere.[405] A distorted tetrahedral environment around the copper(I) ion is achieved by coordination of three sulfur atoms from one ligand molecule and a fourth coordinated sulfur from an adjacent ligand molecule; the result is a stepwise chain polymer. A study of the kinetics and thermodynamics of complexation of the copper(II) ion by $[n]aneS_4$ ligands ($n = 12, 13, 14, 15, 16, 18$) has been reported although no complexes with these ligands were isolated.[406] Cobalt(II) and (III) complexes with $[14]aneS_4$ have been prepared.[26] Three general methods have been used for the preparation of the cobalt(III) compounds. The first involves the preparation and isolation of the cobalt(II) complex $[Co([14]aneS_4)](BF_4)_2$ by reaction of $[Co(CH_3CN)_6](BF_4)_2$ with $[14]aneS_4$, followed by reaction of the complex with coordinating anions and air oxidation; the species $[Co([14]aneS_4)X_2]BF_4$ ($X = Cl, Br$) were prepared by this method. The second method involves the reaction of a cobalt(II) salt, $[14]aneS_4$, coordinating and/or noncoordinating anions and air oxidation all in a one-step procedure; the complexes

[Co([14]aneS$_4$)X$_2$]BF$_4$ (X = Cl, NO$_2$) and [Co([14]aneS$_2$)I$_2$]B(C$_6$H$_5$)$_4$ were obtained by this method. The third synthetic procedure involves metathesis reactions between a cobalt(III) complex and suitable anions; reactions of the dichloro complex with an excess of NaNO$_2$, LiNCS, and oxalic acid have lead to isolation of [Co([14]aneS$_4$)X$_2$]$^+$ species (X = NO$_2$, NCS, $\frac{1}{2}$C$_2$O$_4$). The macrocyclic ligand coordinates to the cobalt(III) ion in a folded conformation in the species *cis*-[Co([14]aneS$_4$)X$_2$]$^+$ (X = Cl, Br, NCS, NO$_2$, $\frac{1}{2}$C$_2$O$_4$), but in a planar conformation in *trans*-[Co([14]aneS$_4$)I$_2$]B(C$_6$H$_5$)$_4$. The 15-membered macrocycle bzo[15]eneN$_4$ also readily forms cobalt(III) complexes by the first method outlined above; [Co(bzo[15]eneS$_4$)](ClO$_4$)$_2$ reacts with lithium chloride or bromide in nitromethane to give *trans*-[Co(bzo[15]eneS$_4$)X$_2$]ClO$_4$ (X = Cl, Br); however, reaction with lithium iodide leads to precipitation of the cobalt(II) complex [Co(bzo[15]eneS$_4$)I$_2$]. The dichloro complex is unstable in hot water, decomposing to the hydrated cobalt salt and free ligand. The rhodium(III) complexes [Rh([14]aneS$_4$)X$_2$]X (X = Cl, Br, I), [Rh([14]aneS$_4$)Cl$_2$]Y (Y = BF$_4$ and B(C$_6$H$_5$)$_4$), [Rh([14]ane-S$_4$)(NO$_2$)$_2$]B(C$_6$H$_5$)$_4$, and [Rh(bzo[15]eneS$_4$)Cl$_2$]Cl have been prepared by conventional synthetic techniques involving rhodium(III) halides, the ligand, and lithium salts. The macrocyclic ligands are assumed to coordinate to rhodium(III) in a folded conformation, giving rise to *cis*-[Rh(ligand)X$_2$]$^+$ species.

The ligand [14]aneS$_4$ has been shown to coordinate in an "inside-out" conformation **138** in the compound (NbCl$_5$)$_2$([14]aneS$_4$), obtained by the reaction of NbCl$_5$ with the ligand in benzene.[407,408] This is unusual since the

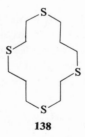

138

ligand has been shown to coordinate as structure **135** in the nickel(II) and copper(II) complexes.[403,404] It was assumed that the large niobium(V) ion was instrumental in the conformation reorganization of the ligand on coordination, although recently DeSimone and Glick[409] reported the crystal structures of two forms of [14]aneS$_4$. In all molecules (the β modification contains two distinctive centrosymmetric molecules), the macrocycle has an exodentate conformation with lone pairs on the sulfur atoms pointing out of the macrocyclic cavity. Thus the conformational reorganization may take place on coordination with small metal ions such as copper(II) or nickel(II) which are able to occupy the macrocyclic cavity. It was also suggested that "exodentate" coordination may be found when ligands coordinated to a metal ion are not readily displaced by the macrocycle, or when the metal ion is too large to fit into the macrocyclic cavity.

Schrauzer *et al.*[410] reported the formation of the nickel(II) complex **140** by reaction of **139** with α,α'-dibromo-*o*-xylene in toluene. The new macro-

139   **140**

cyclic ligand does not coordinate strongly; it is readily displaced from the nickel(II) ion on dissolution in methanol.

3.5. P_4 and P_2S_2 Donor Atoms

The macrocyclic ligand P-Ph$_4$bzo[15]eneP$_4$, **140A**, was recently prepared,[441] the final ring closure step involved a template reaction between α,α'-dibromo-*o*-xylene and the nickel(II) chloride complex of a tetradentate phosphine. The ligand can be displaced from the nickel(II) complex by treatment with aqueous NaCN. The complexes Ni(P-Ph$_4$bzo[15]eneP$_4$)X$_2$

140A

(X = BF$_4$, SCN) are prepared by metathesis from the chloride; all complexes are 2:1 electrolytes in water; a square-planar structure is suggested for the complex ion.

The macrocyclic tetraphosphine P-Ph$_4$bzo$_2$[14]dieneP$_4$, **140B**, has been prepared under conditions of high dilution,[71] and the complex Ni(P-Ph$_4$bzo[14]dieneP$_4$)ClBF$_4$ isolated. The ligand **140C**, P-Ph$_2$bzo$_2$[14]dieneP$_2$S$_2$, was also described by these authors[71]; nickel(II) and cobalt(II)

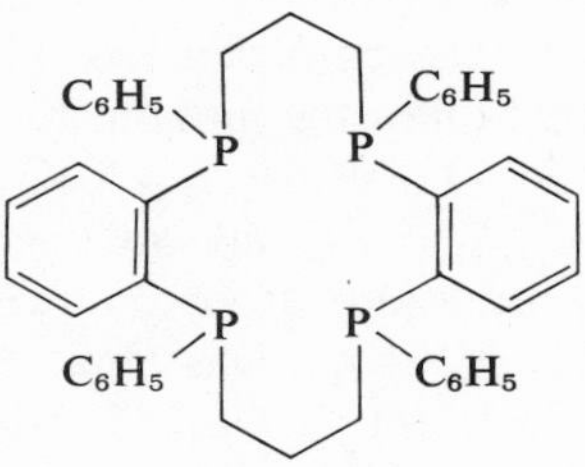

140B

complexes were obtained with this ligand although no experimental conditions were reported.

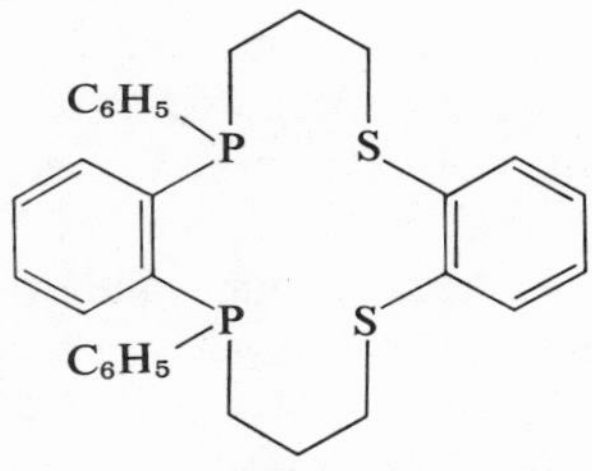

140C

4. Pentadentate Ligands

One of the early utilizations of a metal ion for an *in situ* preparation of a macrocyclic ligand involved the condensation of 2,6-diacetylpyridine with 1,4,7,10-tetraazadecane in the presence of iron(II)[294]; the pentadentate macrocyclic ligand **141A** ($m = 2, n = 2$), Me$_2$pyo[15]trieneN$_5$, was obtained

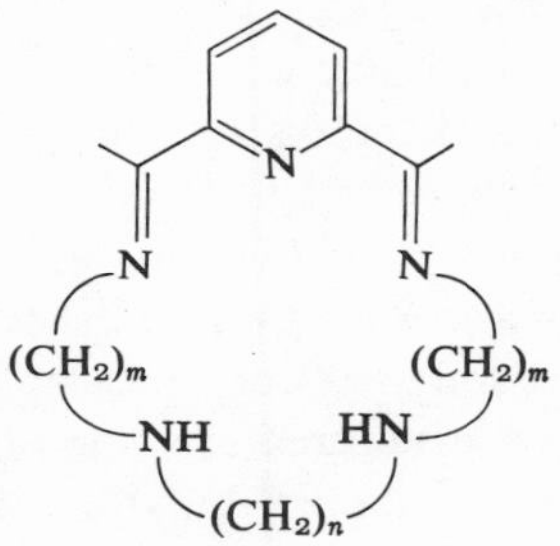

141A, $m = 2$, $n = 2$
141B, $m = 2$, $n = 3$
141C, $m = 3$, $n = 2$

141

as an iron(III) complex originally formulated as [Fe(Me$_2$pyo[15]trieneN$_5$)OH]-
(ClO$_4$)$_2$. The reaction has been reinvestigated [412,413] and the above complex
characterized as the oxygen-bridged dimer [(ClO$_4$(Me$_2$pyo[15]trieneN$_5$)Fe)$_2$-
O](ClO$_4$)$_2$·H$_2$O, with magnetic exchange occurring across the Fe–O–Fe
linkage. [412–414] Treatment of this complex with acidified solutions of sodium
salts, NaX, leads to the isolation of the monomeric species [Fe(Me$_2$pyo[15]-
trieneN$_5$)X$_2$]Y (X = Cl, Br, I, NCS, Y = ClO$_4$); the complexes with
X = Y = NCS and X = Cl, Y = BF$_4$ are obtained by the same procedure
that was used for the formation of the dimeric species. The dimer
(Fe(Me$_2$pyo[15]trieneN$_5$)SCN)$_2$(ClO$_4$)$_2$ was obtained from [Fe(Me$_2$pyo[15]-
trieneN$_5$)(NCS)$_2$]ClO$_4$ by treatment with a 20% ethylenediamine solution in
methanol. The structures of the species [(H$_2$O(Me$_2$pyo[15]trieneN$_5$)Fe)$_2$O]-
(ClO$_4$)$_4$ and [Fe(Me$_2$pyo[15]trieneN$_5$)(NCS)$_2$]ClO$_4$ have confirmed seven-
coordination about the iron(III) ions. [415] In each case, the metal ion and the
five nitrogen atoms of the ligand are coplanar; in the dimeric species axial
positions are occupied by a water molecule and the oxygen bridge atom,
whereas in the monomeric species isothiocyanato groups occupy these
positions. Alexander *et al.*[461] prepared the manganese(II) complex
Mn(Me$_2$pyo[15]trieneN$_5$)Cl$_2$·6H$_2$O by reaction of 2,6-diacetylpyridine with
1,4,7,10-tetraazadecane in the presence of MnCl$_2$·4H$_2$O in methanol.
Dehydration of the complex hexahydrate yields the monohydrate and the
anhydrous species; treatment of the last compound with perchlorate ions
produces the complex Mn(Me$_2$pyo[15]trieneN$_5$)(ClO$_4$)$_2$. The EPR spectra of
the anhydrous and hexaaquo chloro species have been reported. [417]

 Nelson and co-workers have extended the condensation reactions of
2,6-diacetylpyridine with amines to include 1,4,8,11-tetraazaundecane. In the
presence of iron(II) salts, iron(III) complexes the 16-membered potentially
pentadentate ligand **141B** ($m = 2$, $n = 3$), Me$_2$[16]pydieneN$_5$, have been
obtained. The species [Fe(Me$_2$pyo[16]trieneN$_5$)X$_2$]Y (X = Cl, Y = ClO$_4$,
PF$_6$, B(C$_6$H$_5$)$_4$, FeCl$_4$; X = Br, Y = ClO$_4$, FeBr$_4$; X = NCS, Y = ClO$_4$;
X = N$_3$, Y = ClO$_4$) have been isolated, several by metathetical reactions. [418]
The crystal and molecular structure of the *bis*-(isothiocyanato) perchlorate
complex has been determined and compared with the analogous complex
containing the 15-membered macrocyclic ligand; both species contain
iron(III) in a pentagonal-bipyramidal environment, with the macrocyclic
ligand in the equatorial plane and the isothiocyanato groups in axial positions.

 The iron(II) complexes Fe(macrocycle)Y$_2$·xH$_2$O (macrocycle = **141A**,
141B Y = Cl, Br, NCS, CN, $x = 0$, 1) can be obtained from the iron(III)
complexes by reduction with dithionite in aqueous solution in the presence of
excess Y$^-$. The compounds are stable for long periods in the solid state and
in solution in the absence of oxygen. The structures of some of these complexes have been reported [419,420]; the pentagonal-bipyramidal structure
found for the iron(III) compounds is retained in these iron(II) species.

The use of the magnesium(II) ion as a template in the *in situ* syntheses of **141A** and **141B** was demonstrated by Nelson and co-workers.[421] A mixture of 2,6-diacetylpyridine, the linear tetramine, and $MgCl_2 \cdot 6H_2O$ in a 1:1:1 molar ratio was heated at 65°C for 12 hr. Evaporation of the solvent and extraction of the residue with *n*-butanol, followed by recrystallization yielded crystals of $Mg(Me_2pyo[15]trieneN_5)Cl_2 \cdot 6H_2O$ or $Mg(Me_2pyo[16]trieneN_5)$-$Cl_2 \cdot 2H_2O$ in 40–60% yield. The importance of the magnesium(II) ions was demonstrated by carrying out the condensation reaction in the absence of Mg^{2+}, or in the presence of the Lewis acid $AlCl_3$; no macrocyclic complexes could be obtained from the products of these reactions. The structure of $Mg(Me_2pyo[15]trieneN_5)Cl_2 \cdot 6H_2O$ is similar to that found for the analogous iron(II) and iron(III) species with the coordination geometry of the magnesium(II) ion being pentagonal bipyramidal; it is isomorphous with the previously reported[416] manganese(II) complex $Mn(Me_2pyo[15]trieneN_5)$-$Cl_2 \cdot 6H_2O$. In view of the significant role played by magnesium(II) ions in various reactions of biological importance, the demonstration of its ability to serve as a template to produce macrocyclic complexes is an important observation.

Manganese(II) and zinc(II) complexes with **141A**, **141B**, and **141C** have been prepared by metal template reactions; the species $Mn(Me_2pyo[n]$-$trieneN_5)X_2$ ($n = 15$, 16, 17, $X = NCS$; $n = 17$, $X_2 = ClPF_6$) and $Zn(Me_2pyo[n]trieneN_5)X_2$ ($n = 15$, 16, 17, $X_2 = NCS \cdot ClO_4$, $n = 15$, $X = NCS$) were isolated by Dabrowiak *et al.*[300] (See also references 416 and 417.) Oxidation of the Mn(II) complexes with $NOPF_6$ leads to stable manganese(III) complexes that are proposed to exhibit pentagonal-bipyramidal coordination geometry. Solution infrared and NMR studies indicate that in solution the zinc(II) complexes display a pentagonal coordination geometry. Drew *et al.*[422] have also prepared a series of manganese(II) complexes of **141A**, **141B**, and **141C**. These authors also concluded that the manganese(II) complexes have pentagonal-bipyramidal structures with the macrocycle in the plane and the axial positions occupied by anions or water molecules. Considerable distortion is noted for the largest macrocyclic ligand, **141C**, in the complex $[Mn(Me_2pyo[17]trieneN_5)(NCS)_2]$; one of the axial ligands in this complex may be readily displaced to give the proposed pentagonal-pyramidal complex $[Mn(Me_2pyo[17]trieneN_5)NCS]ClO_4$. Dabrowiak *et al.*[300] also concluded, from IR studies, that the ligand **141C** was folded in $[Mn(Me_2pyo[17]trieneN_5)Cl]PF_6$; an "octahedral" coordination environment was proposed for the manganese(II) ion in the solid state. The reaction between $[Mn(Me_2pyo[16]trieneN_5)Cl(H_2O)]Cl \cdot H_2O$ and 1,10-phenanthroline (phen) has been investigated[423]; the product, $Mn(Me_2pyo[16]trieneN_5)$-$(ClO_4)_2 \cdot 2H_2O \cdot phen \cdot \frac{1}{2}EtOH$, shows no evidence for coordination of phenanthroline. It is concluded that in this complex, two pentagonal-bi-pyramidal cations, $[Mn(Me_2pyo[16]trieneN_5)(H_2O)_2]^{2+}$, "sandwich" a phenanthroline molecule to form a lamellar structure held together mainly by hydrogen bonds, but also by π-donor–π-acceptor and dispersion forces.

Nelson *et al.* have recently reported the preparation and crystal and molecular structures of some Cd(II), Hg(II), and Ag(I) complexes with the ligands **141A**, **141B**, and **141C**;[424] both five- and six-coordinated structures have been confirmed. In the dimeric complex, [Ag(Me$_2$pyo[16]trieneN$_5$)]$_2$-(ClO$_4$)$_2$, two almost planar [Ag(Me$_2$pyo[16]trieneN$_5$)]$^+$ cations are associated such that each metal ion has a pentagonal-pyramidal coordination environment.[425]

An *in situ* synthesis of **141D** using zinc(II) ions has been reported; the diaquo zinc complex cation containing **141D** is seven coordinate, with the equatorial N$_5$ donor set of the macrocycle almost planar and the water molecules occupying axial positions.[426] It is concluded that the rigidity of the macrocycle results in the greater degree of planarity than in previous examples of pentadentate macrocyclic ligands.

Reduction of the ligand in Mn(Me$_2$pyo[15]trieneN$_5$)Cl$_2$ with a nickel–aluminum alloy under basic aqueous conditions followed by the addition of cold concentrated nitric acid leads to the isolation of the reduced macrocycle **142**, Me$_2$pyo[15]eneN$_5$, as the tetrahydrogennitrate salt.[427]

141D   **142**

Several metal complexes have been prepared from this ligand by conventional techniques. These include the nickel(II) complexes [Ni(Me$_2$pyo-[15]eneN$_5$)X]$^{n+}$ (X = H$_2$O, CH$_3$CN, NH$_3$, n = 2; X = NCS, Cl, Br, I), the cobalt(III) complexes [Co(Me$_2$pyo[15]eneN$_5$)X]$^{2+}$ (X = Cl, Br, I, NCS, N$_3$, NO$_2$), the iron(III) complexes [Fe(Me$_2$pyo[15]eneN$_5$)X$_2$]$^+$ (X = Cl, Br, NO$_3$, NCS, N$_3$, $\frac{1}{2}$C$_2$O$_4$), and the copper(II) complex [Cu(Me$_2$pyo[15]-eneN$_5$)]$^{2+}$. In the nickel(II) and cobalt(III) complexes, the macrocycle is folded and one additional unidentate ligand is included to form six-coordinate complexes with approximate octahedral geometry; the iron(III) complexes are seven-coordinate with one bidentate or two unidentate ligands in the coordination sphere in addition to the ligand; the copper(II) complex is five-coordinate, with a suggested tetragonal structure.

Reaction of 2,6-diacetylpyridine with 3,6-dioxaoctane-1,8-diamine in the presence of a salt of Mg(II), Mn(II), Fe(III), Fe(II), Zn(II), or Ca(II) gives complexes of the 15-membered pentadentate ligand **142A** (R = Me).[428,429]

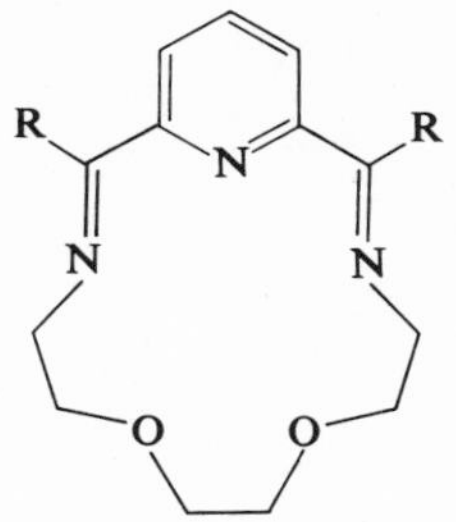

142A

If 2,6-diformylpyridine is used in place of 2,6-diacetylpyridine in the presence of Mg(II), a complex containing **142A** (R = H) is obtained. The calcium(II) ion is not effective in promoting the synthesis of either macrocyclic ligand. It is concluded that all the complexes contain the metal ion in a pentagonal-bipyramidal coordination environment.

Zinc(II) and cadmium(II) ions have been shown to induce rearrangement of 2,6-*bis*-(2-methyl-2-benzothiazolinyl)pyridine, the condensation product obtained from 2,6-diacetylpyridine and 2-aminobenzenethiol, under basic conditions to produce complexes of the deprotonated tautomeric Schiff base **143**.[430,431] These complexes are five-coordinate with the ligand in a

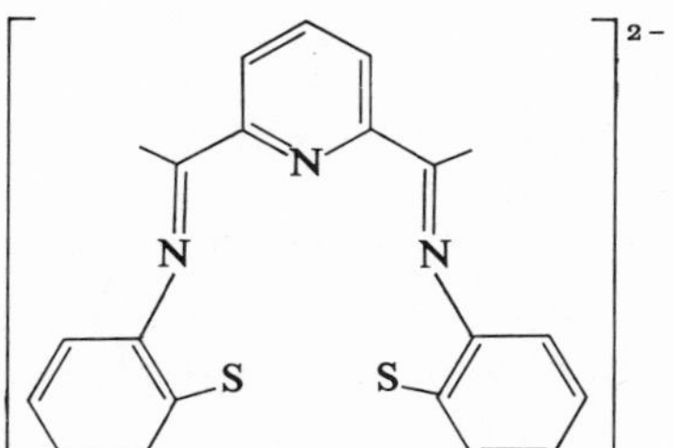

143

helical configuration.* The coordinated thiol groups in these complexes react with alkyl halides; with α,α'-dibromo-*o*-xylene or 1,4-diiodobutane ring closure takes place and zinc(II) complexes of the neutral ligands **144** (R = *o*-C_6H_4, $-(CH_2)_2-$) are obtained. It is suggested that the complexes are seven-coordinate with the macrocyclic ligand coordinating in a distorted pentagonal plane and the halide ions occupying axial positions. A similar

---

* Recently, a methylthallium(III) complex containing **143** has been reported.[432]

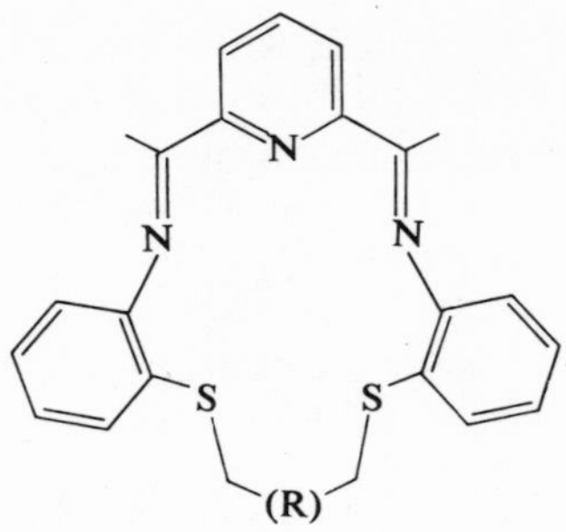

144

geometry has been found for complexes of **141** discussed earlier. Silver(I) and cadmium(II) ions have been shown to bring about the condensation of 2,6-diacetylpyridine and some ditertiaryphosphinodiamines; complexes containing the ligands **144A** ($m = 3, n = 2; m = n = 3$) have been obtained.[433]

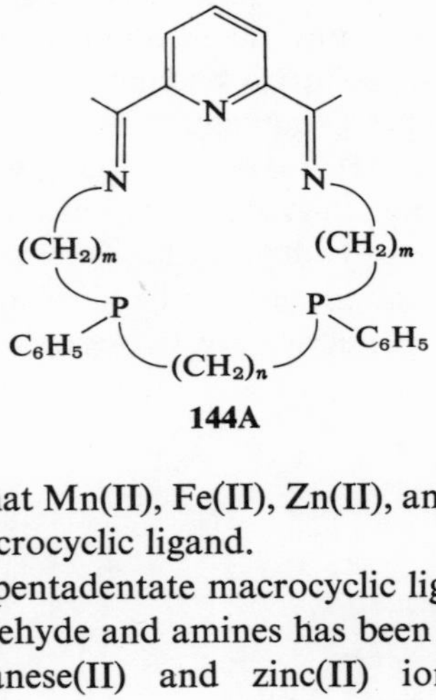

144A

It is interesting to note that Mn(II), Fe(II), Zn(II), and Hg(II) do not promote the formation of the macrocyclic ligand.

A series of neutral pentadentate macrocyclic ligands, **145**, derived from pyridine-2,6-dicarboxaldehyde and amines has been prepared by *in situ* reactions involving manganese(II) and zinc(II) ions.[434] The complexes

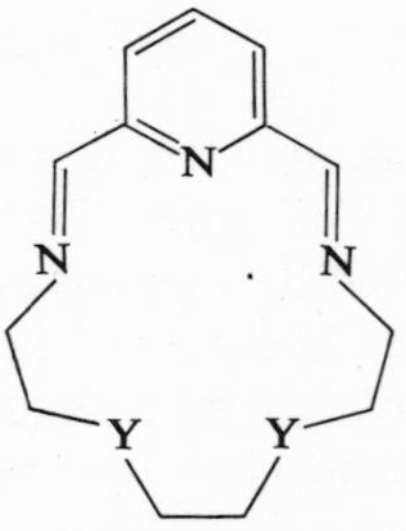

145

M(ligand)(ClO$_4$)$_2$ (M = Mn, ligand = **145**, Y = NH, O, S; M = Zn, ligand = **145**, Y = NH, O, S, have been isolated. The structure of the manganese(II) complex with **145** (Y = NH) has been determined; the metal ion is seven-coordinated, with coordinated perchlorate ions in axial positions above and below the pentagonal plane of the macrocyclic ligand.*

Neutral uranyl complexes of the macrocyclic ligands **146** (R = –(CH$_2$)$_2$–, –(CH$_2$)$_3$–, –CH$_2$·CH(CH$_3$)–, o-C$_6$H$_4$) have been prepared; they coordinate as dianionic pentadentate ligands with three nitrogen and two oxygen donor atoms.[436] Pentagonal-bipyramidal coordination is suggested, a common

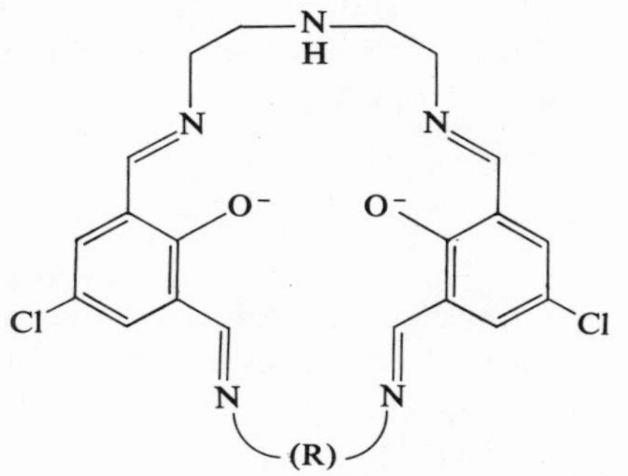

146

stereochemistry for the UO$_2^{2+}$ ion; the "super phthalocyanine" has been shown to coordinate in a plane about the uranium atom.[437]

The iron(II) complex [FeL(CH$_3$CN)](ClO$_4$)$_2$, containing the macrocyclic ligand L, **147**, has been prepared by an *in situ* reaction from the dialdehyde and ethylenediamine in the presence of iron(II) perchlorate in boiling aceto-

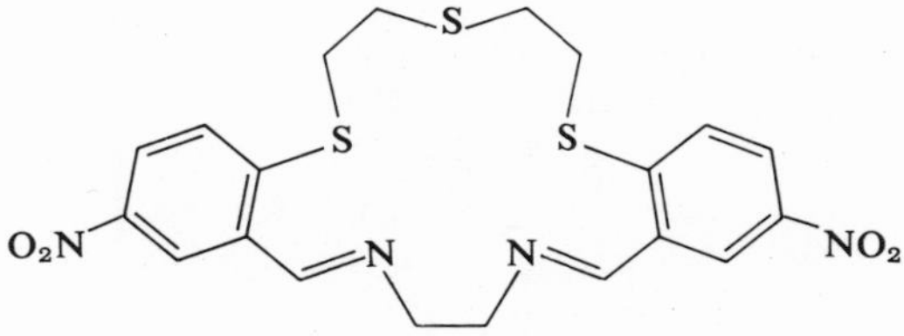

147

nitrile.[438] Recently, the copper(II) complex of **147A** has been isolated and the crystal structure reported.[439] The complex is prepared by reaction of the ligand with a copper(II) salt, the ligand having been synthesized earlier in the presence of zinc(II) acetate.

---

* Cook and Fenton[435] investigated condensation reactions between the dialdehyde and some polyfunctional diamines in the absence and presence of metal ions, Mn(II), Mg(II); macrocyclic ligands were obtained in all cases.

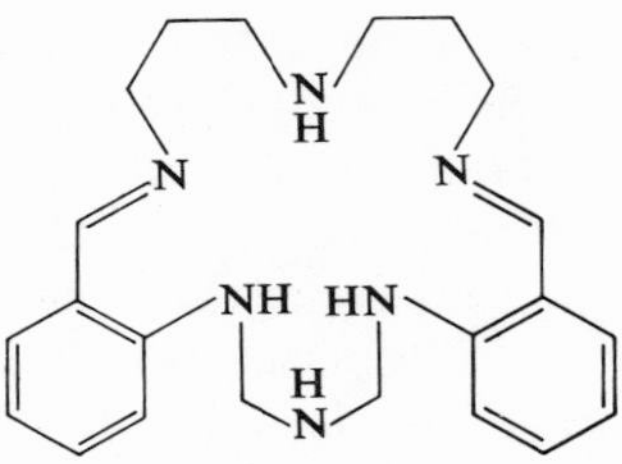

147A

Nickel(II) and copper(II) complexes of novel "basketlike" pentadentate ligands have been prepared by reaction of Ni(bzo$_4$[16]octaeneN$_4$)(BF$_4$)$_2$ and Cu(bzo$_4$[16]octaeneN$_4$)(NO$_3$)$_2$ with *bis*-(2-hydroxyethyl)methylamine and *bis*-(2-hydroxyethyl)sulfide[330] (see Chapter 6).

5. Sexadentate Ligands

In situ condensation reactions between 2,6-diacetylpyridine and poly-amines have been shown to produce a series of tetra- penta-, and sexadentate ligands.[294] When 1,4,7,10,13-pentaazatridecane (tetraethylenepentamine) is employed in the presence of iron(II) chloride, bronze crystals of a complex formulated as Fe(L-H)I$_2$ (L = **148**) are obtained. It is suggested, on the basis

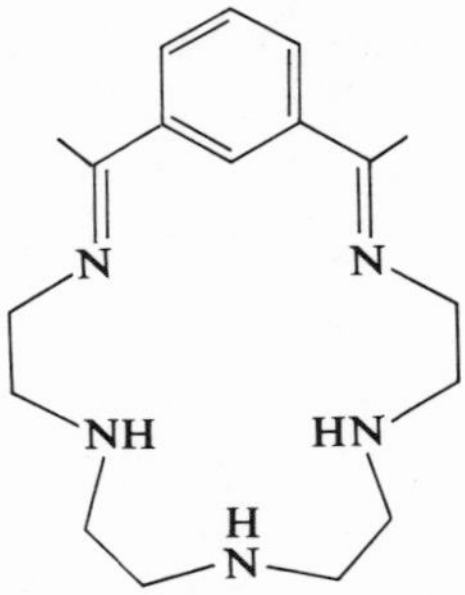

148

of color changes, that formation of an iron(II) complex with imine-type nitrogen atoms takes place after addition of the amine. Aerial oxidation of the acidified solution occurs on stirring and the product is isolated after removal of some iron(III) oxide. In view of the more recent work completed on complexes of the analogous pentadentate ligand (see Section 4) it is possible that the above complex has an oxygen-bridged dimeric structure.

    The lead(II) ion has been shown to function as a template in the condensation reactions between 2,6-diacetylpyridine or 2,6-diformylpyridine and 1,11-diamino-3,6,9-trioxaundecane; eight-coordinate complexes of the type [PbL(SCN)$_2$] (L = **148A**, R = H, CH$_3$) have been isolated.[440] The crystal

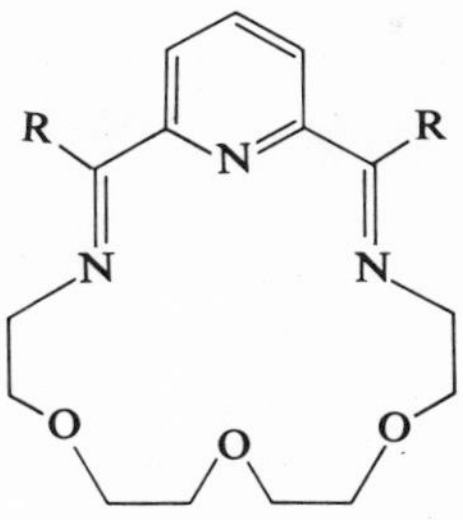

148A

and molecular structures of the complex with R = CH$_3$ have been determined; the lead ion is located within the cavity of the macrocycle with the thiocyanate ions coordinated above and below the macrocycle plane. Thiocyanates of calcium(II), strontium(II), and barium(II) are also reported to perform as templates for the above condensation reactions. Fenton *et al.* have also demonstrated the ability of Ca(II), Sr(II), and Ba(II) to function as templates in the condensation of 2,5-diformylfuran with some amines; complexes containing the ligands **148B** have been prepared.[441] The structure of

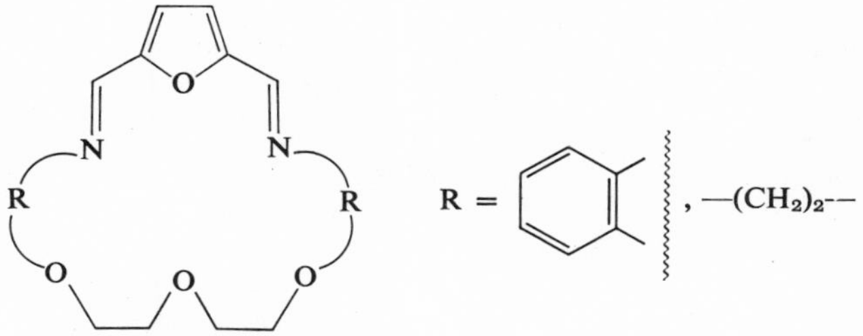

148B

the complex SrL(NCS)$_2 \cdot$H$_2$O (L = **148B**, R = –(CH$_2$)$_2$–) has been completed; the Sr(II) ion is nine-coordinated by the ligand, two isothiocyanates, and a water molecule.

The macrocycles **149** (X = NH, S, O) have been prepared by high-dilution ring closure reactions.[442,443] Complexes of the type [ML](picrate)$_2$ (L = **149**, X = NH, S, M = Ni, Co) have been prepared by treatment of the metal(II) picrate with the macrocyclic ligand in ethanol. The ligand **149** (X = NH) can coordinate almost without strain as a sexadentate ligand in two possible octahedral configurations **150** or **151**. However, the ligand **149** (X = S) can coordinate almost without strain in only one way, **150**, because

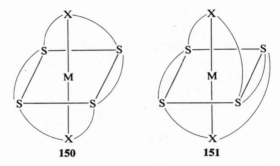

149

of the bond angles and large size of the coordinated sulfur donor atoms. The macrocycle **149** (X = O) does not form complexes with nickel(II), cobalt(II), iron(II), or copper(II) picrates, perchlorates, or chlorides. It is suggested[444]

150   **151**

that this macrocycle would coordinate as **151** to reduce strain in the ligand, and that in view of the lack of complexes formed with this macrocycle configuration **150** is preferable to **151**. Lindoy and Busch[445,446] prepared the macrocyclic ligand **152**, initially by an *in situ* reaction between 1,2-*bis*-

152

(2-aminophenylthio)ethane and 1,4-*bis*-(2-formylphenyl)-1,4-dithiabutane in the presence of a metal salt. The complexes [M(ligand)]X$_2$ (M = Ni, X = ClO$_4$, I; M = Co, X = ClO$_4$) were isolated. Treatment of the nickel(II) perchlorate complex of this ligand with hot DMF displaces the ligand, which has been characterized by a variety of techniques. The nickel(II) complex can be reprepared by reaction of the ligand with nickel(II) perchlorate in acetone solution. A molecular model indicates that **152**, which has the same sequence of donor atoms as **149** (X = NH), can only coordinate about an octahedral metal ion in the configuration **151** due to the planarity of the S–N–S units. No crystal structures have been reported for any of the complexes containing the above sexadentate ligands to confirm or refute the proposals made on the basis of molecular models.

Complexes containing macrocyclic ligands similar to **152**, viz., **153**, have been obtained by *in situ* reactions of the dialdehyde **33** (which was also used for the preparation of some tetradentate macrocyclic ligands[214] with the appropriate diamine.[447,448] The complexes M(ligand)(ClO$_4$)$_2 \cdot n$CH$_3$OH have

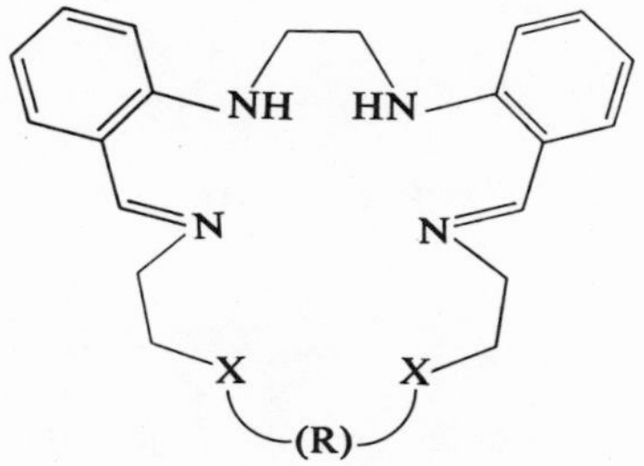

153

been isolated (ligand = **153**, X = S, R = –(CH$_2$)$_2$–, M = Zn, Ni, Co, Fe, $n = 1$; X = O, R = –(CH$_2$)$_2$–, M = Ni, $n = 1$; X = NH, R = –(CH$_2$)$_2$–, M = Zn, Fe, $n = 1$; X = NH, R = –(CH$_2$)$_3$–, M = Ni, Fe, $n = 0$). All species are assigned octahedral stereochemistries; a crystal structure of the complex [Fe(ligand)](ClO$_4$)$_2 \cdot$CH$_3$OH (ligand = **153**, X = S, R = –(CH$_2$)$_2$–) has established the mode of coordination as depicted in **154**. This is a similar configuration to that suggested by Lindoy and Busch, **151**, for complexes of **152**.[446] When acetone solutions of the complexes are treated with pyridine, the macrocyclic ligands are liberated.

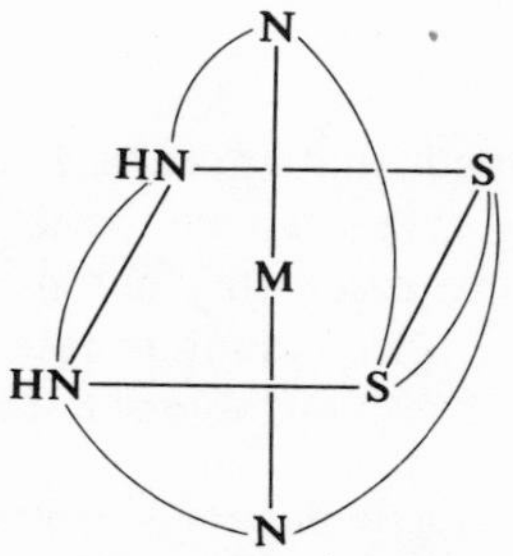

154

The reaction of [24]aneS$_6$, the analog of **143** (X = S with $-(CH_2)_3-$ linkages) with NbCl$_5$ has been reported to produce the complexes (NbCl$_5$)$_3$· ([24]aneS$_6$) and (NbCl$_5$)$_6$·([24]aneS$_8$).[408] It is possible that the macrocyclic ligand coordinates in an exodentate conformation in a similar manner to the tetradentate ligand [14]aneS$_4$,[409] although no structures have been reported to confirm this possibility.

The potentially sexadentate macrocyclic ligand **155**, [14]aneN$_4$O$_2$, has been obtained[449] as one of the condensation products from the reaction of formaldehyde with [Co(en)$_2$gly]Cl$_2$ (en = ethylenediamine, gly = glycinate ion); it is isolated as the cobalt(III) complex cation [Co(α-HOCH$_2$-serine)-([14]aneN$_4$O$_2$)]$^{2+}$. The structure of this cation has been determined; the

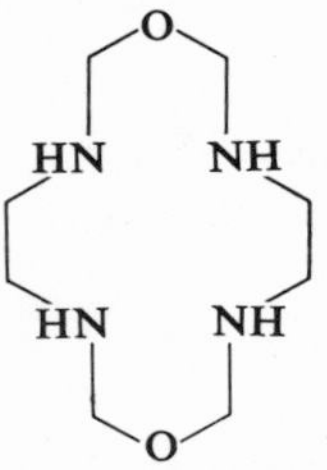

155

macrocyclic ligand is coordinated to four adjacent octahedral sites of the cobalt(III) ion by the four nitrogen donor atoms, with the amino acid occupying the other two adjacent sites. Thus the macrocycle coordinates in a similar manner to its parent compound, [14]aneN$_4$(cyclam), when other bidentate ligands are present (see Section 3). The complex [Co([14]aneN$_4$O$_2$)-C$_2$O$_4$]Cl·3H$_2$O has also been characterized.

Busch and co-workers have reported the formation of complexes containing novel sexadentate *cis-* and *trans*-hexaene ligands by the electrophilic attack of acetonitrile at the γ-carbon atoms of a coordinated tetraene ligand[243-246] (see p. 65 and Chapter 6).

Complexes containing sexadentate polyethers will not be described here; they are discussed in detail in Chapter 9.

6. Binucleating Ligands

Several macrocyclic ligands that are capable of incorporating more than one metal ion into the ring have been synthesized, most of them by *in situ* processes.* However, in some cases only one metal ion is coordinated and little tendency to incorporate a second ion is noted. For example, the ligands **115** have been shown to react with metal ions to give complexes **116**, in which

---

* A review article dealing with transition-metal complexes of binucleating ligands has recently been reported.[450]

the metal ion is four-coordinated by a *cis*-N_2O_2 donor atom set[284–383] (see Section 3.2).

Pilkington and Robson[451] reported that reaction of 2-hydroxy-5-methylisophthalaldehyde, **156**, with 1,3-diaminopropane in the presence of

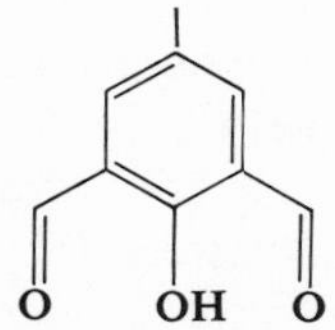

O   OH   O

156

various metal salts results in the formation of complexes, with two metal ions contained in the dianionic ligand **157** ($m = n = 3$). Species with two man-

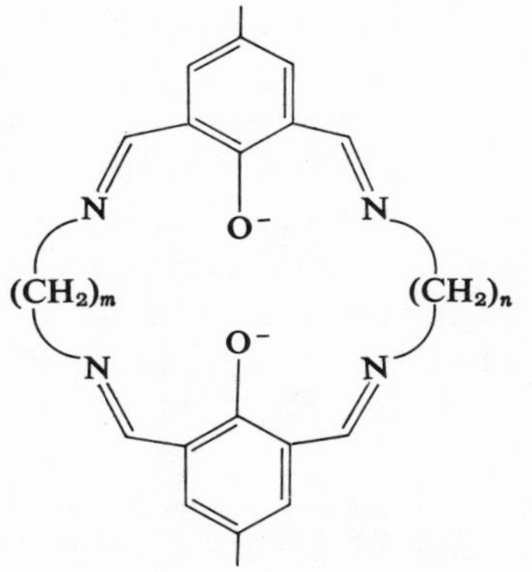

(CH₂)ₘ → $(CH_2)_m$

157

ganese(II), iron(II), cobalt(II), nickel(II), copper(II), and zinc(II) ions were isolated. Magnetic properties suggest antiferromagnetic interactions for the manganese through copper species. On the basis of magnetic and spectral evidence, the complexes were assigned binuclear structures in which each metal ion is in a distorted square-pyramidal environment consisting of a planar N_2O_2 donor set and an apical halide ion. These conclusions were confirmed by the determination of the structures of Co_2(ligand)$Br_2 \cdot$ CH_3OH[452] and Cu_2(ligand)$Cl_2 \cdot 6H_2O$[453]; the coordinated halides are on opposite sides of the macrocyclic ligand plane in both complexes. The complex (ligand $H_2)_2Ni(ClO_4)_2 \cdot 2H_2O$ was also prepared[451]; it is suggested that the nickel(II) ion is coordinated by one N_2O_2 donor set with the ligand in a neutral form, the hydrogen atoms being hydrogen bonded to the opposite (noncoordinated) nitrogen atoms of the macrocyclic ring. Oxidation of Co_2(ligand)$Br_2 \cdot CH_3OH$ with bromine has led to a variety of products depending upon the conditions of the oxidation.[454] Crystals of (ligand)Co_2-$Br_5 \cdot 2CH_3OH$ have been shown to consist of the cations [(ligand)Co(II)Co(III)-$Br_2(CH_3OH)_2]^+$, in which both metal ions are octahedrally coordinated; the cobalt(III) ion is bonded in the apical positions by two bromide ions and the cobalt(II) ion by two methanol molecules, and tribromide ions. In (ligand)$Co_2Br_3 \cdot H_2O$, the two isomeric forms of (ligand)$Co_2Br_3 \cdot 2H_2O$, and (ligand)$Co_2Br_4 \cdot CH_3OH$, the compounds are assigned binuclear Co(II)–Co(III) structures: [(ligand)Co(II)Co(III)$Br_2(H_2O)]^+Br^-$, [(ligand)Co(II)-Co(III)$Br_2(H_2O)_2]^+Br^-$ in two isomeric forms, and {[(ligand)Co(II)Co(III)-$Br_2(CH_3OH)]^+\}_2Br^-(Br_3^-)$, respectively. The crystal structures of the two isomers of (ligand)$Co_2Br_3 \cdot 2H_2O$ have been reported.[455] In the A isomer, two bromide ions are bonded to the cobalt(III) ion in apical positions relative to the N_2O_2 donor atom plane, with the two water molecules occupying similar positions about the cobalt(II) ion. In the B isomer, however, one water molecule and one bromide ion are bonded to each of the cobalt(III) and cobalt(II) ions. The complex (ligand)$Co_2Br_8 \cdot 4CH_3OH$ is diamagnetic and contains a binuclear Co(III)–Co(III) structure; it is unstable in the solid, giving off bromine and regenerating the paramagnetic Co(II)–Co(III) binuclear unit. Hoskins *et al.*[454] also reported the synthesis of the metal-free macrocycle, isolated in a protonated form, [ligand $H_4]^{2+}(Br_3^-)_2$, by carrying out the condensation reaction between 2-hydroxy-5-methylisophthalaldehyde and 1,3-propanediamine in ethanol containing concentrated hydrobromic acid. The orange solid that is deposited on cooling consists of the macrocyclic ligand salt and the hydrobromide of 1,3-propanediamine. If the reaction is carried out in methanol, [ligand $H_4]^{2+}(Br_3^-)_2$ is obtained from the solution in an analytically pure form.

    Okawa and Kida[456,457] reported that the condensation of two molecules of **156** with one molecule of ethylenediamine or 1,3-propanediamine in ethanol produces the Schiff base **158** ($m = 2, 3$). Copper(II) and nickel(II)

158

complexes of these ligands in their dianionic form can be prepared by reaction of the ligand in ethanol with an aqueous solution of the metal(II) acetate. The complexes can also be obtained by *in situ* reactions between 156, the amine, and the metal salt in stoichiometric quantities in ethanol. In these complexes, the "open side" of the ligand may be closed by reaction with one molecule of a diamine. By reaction with ethylenediamine or 1,3-propane-diamine, complexes containing 157 ($m = n = 2$; $m = 2, n = 3$; $m = n = 3$) and one metal ion, copper(II) or nickel(II), have been obtained. Reaction of these complexes with an equimolar quantity of a metal(II) salt in methanol leads to incorporation of a second metal ion into the macrocyclic ligand. For the ligands 157 ($m = n = 2$; $m = 2, n = 3$), species with two copper(II) or two nickel(II) ions have been isolated; when $m = n = 3$ only complexes with two copper(II) ions have been prepared, the chloride species is identical to the compound described earlier by Pilkington and Robson.[451] No attempt was reported to obtain binuclear complexes containing two different metal ions.*

Okawa *et al.*[459,460] have also prepared the macrocyclic complexes 159 and 160 ($m = 2, 3$), which incorporate two metal ions; with 160 (M = Ni, $n = 2$) both dia- and paramagnetic nickel(II) are present; a strong anti-ferromagnetic exchange interaction is evident in the copper(II) complexes.

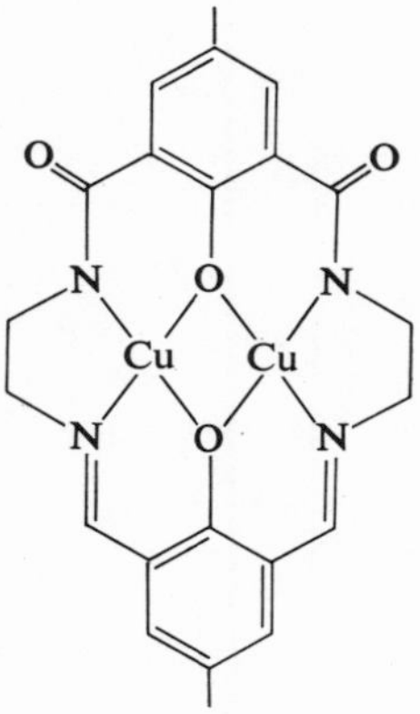

159 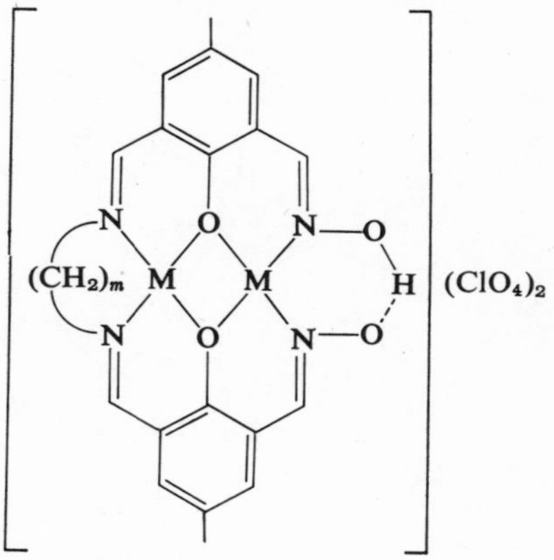　　　　　160

The binuclear copper(II) complex 161 was prepared both by direct reaction of hydrated copper(II) nitrate with the free ligand in ethanol and by an *in situ* procedure involving 2,6-diacetylpyridine and *o*-phenylenediamine in

---

* Gagné *et al.*[458] also synthesized the binuclear copper(II) complex with 157 ($m = n = 3$). Electrolysis of green solutions of this complex results in the formation of dark brown solutions from which a Cu(II)–Cu(I) complex can be isolated. Saturation of the electrolysis solution with carbon monoxide followed by addition of diethyl ether leads to the precipitation of a CO adduct of the Cu(II)–Cu(I) complex.

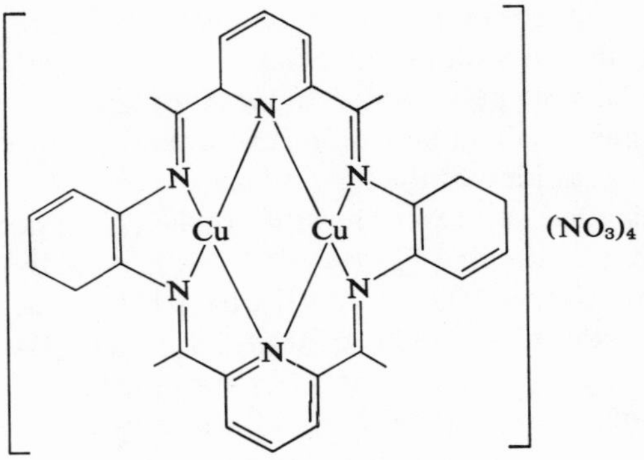

161

the presence of copper(II) ions.[461] Studies of the magnetic properties of this compound indicate a weak copper(II)–copper(II) interaction.

Rosen[462] synthesized the nickel(II) complex ion, **162**, as a fluoroborate derivative from the reaction of 1,4-dihydrazinophthalazine with 2,2-dimethoxypropane in the presence of $Ni(BF_4)_2 \cdot 6H_2O$ in methanol containing a few drops of concentrated HBF_4. The 2,2-dimethoxypropane is used as a dehydrating agent and as a source of acetone for the acid-catalyzed condensation reaction with the hydrazine. The fluoroborate and the thiocyanate,

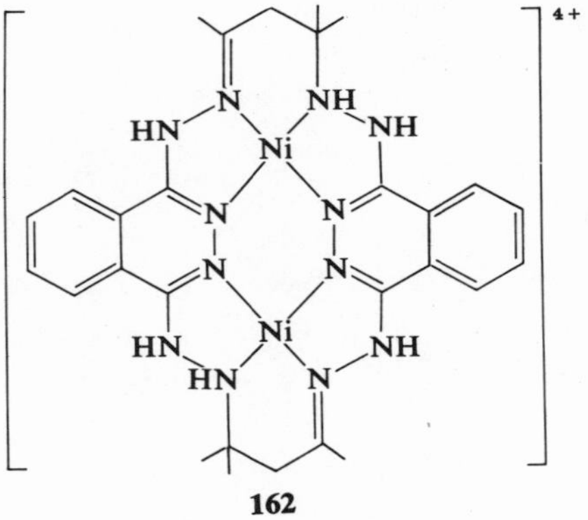

162

which may be prepared by metathesis from the fluoroborate, are paramagnetic with magnetic moments slightly below the usual values for six-coordinate nickel(II) ions. This lowering of the magnetic moment is attributed to a weak antiferromagnetic interaction between the nickel(II) ions. The six-coordination of the fluoroborate derivative is explained by stacking the macrocyclic complexes in a staggered configuration such that uncoordinated hydrazine nitrogen atoms of adjacent molecules occupy the fifth and sixth positions of nickel(II) ions within the macrocyclic ligand.

Some macrocyclic ligands are capable of incorporating two metal ions such that interaction between the metal centers is not possible due to their large separation. The ligand **41** reacts with copper(II) and nickel(II) acetates to incorporate two metal ions with N_4 donor atom sets separated by approximately 800 pm.[220] The octadentate thioether, **134**, which is obtained with [14]aneS$_4$, **132**, from the condensation of 1,4,8,11-tetrathioundecane with 1,3-dibromopropane, reacts with an excess of the hexaacetic acid derivative of nickel(II) tetrafluoroborate in nitromethane to give the complex [Ni$_2$(ligand)]-(BF$_4$)$_2 \cdot$2MeNO$_2$; the nitromethane can be removed by heating the complex at 110°C *in vacuo*.[402] Addition of an excess of potassium thiocyanate to the complex in nitromethane produces a blue solution from which the species [Ni$_2$(ligand)(NCS)$_4$]$\cdot$2H$_2$O was isolated. The tetrafluoroborate species is diamagnetic with the nickel(II) ions in a square-planar environment, whereas the thiocyanate derivative is a high-spin, nonelectrolyte in nitromethane with an electronic spectrum characteristic of six-coordinate tetragonal nickel(II). Palladium(II) and platinum(II) complexes of **134**, with formulas Pd$_4$(ligand)Cl$_8$ and Pt$_4$(ligand)Cl$_8$, respectively, have been reported; it is suggested that the ligand acts in a bidentate manner toward each of the four metal ions. Thus the ligand may coordinate with the heavier metal ions in an "exodentate" manner similar to that shown for [14]aneS$_4$ in (NbCl$_5$)$_2$[14]aneS$_4$.[407,408]

7. Clathrochelates

In the report[385] describing the deliberate synthesis of new macrocyclic compounds and an early review article[1] Busch suggested that the template effect, operative in *in situ* syntheses, may be used for the synthesis of cagelike molecules, which would completely enclose or encapsulate a metal ion. The first report of such a reaction to produce a "clathrochelate" was published in 1968.[463] In this report, Boston and Rose described the reaction between potassium *tris*-(dimethylglyoximato)cobaltate(III) and boron trifluoride etherate in ether; the orange–red complex [CoL]BF$_4$, **163** (L = [(DMG)$_3$(BF)$_2$]$^{2-}$),

163

is precipitated from solution by the addition of more ether, and may be recrystallized from an acetonitrile–ether mixture. In this complex ion, the three dimethylglyoxime ions are "capped" at both ends by BF groups. Treatment of the complex with iodide ion in acetonitrile–acetone solution leads to reduction of the cobalt from (III) to (II) while leaving the clathrochelate cage intact[464]; the neutral complex CoL crystallizes from solution within 1–2 min. The crystal structures of these two clathrochelate species have been reported.[465] The ligand structure was confirmed; the CoN_6 coordination polyhedron in [CoL]BF_4 is slightly distorted from D_3 symmetry and is midway between that of a trigonal prism and a trigonal antiprism. In CoL, the polyhedron has D_3 symmetry and is only slightly distorted from trigonal prismatic. Boston and Rose[464] also reported the formation of clathrochelates from K_3[Co(DMG)$_3$] by reaction with $SnCl_4$ and $SiCl_4$; the complexes [(n-C_3H_7)$_4$N][Co(DMG)$_3$(SnCl$_3$)$_2$] and [Co(DMG)$_3$(SiO)$_2$]HPF$_6$ were isolated. In the former species the tin(IV) is six-coordinate with three chlorine atoms and three oxygen atoms from the oxime groups bound to the tin; the latter species is formally a derivative of [OSi(OH)$_3$]$^-$ in which the three Si–OH linkages have been replaced by three oxime ester groups. The tin derivative is stable in solution, behaving as a 1:1 electrolyte in acetone, although the silicon derivative readily forms gels during normal precipitation and crystallization procedure.

Rapid, one-step syntheses of neutral clathrochelates containing iron(II) of general formula **164**, [Fe(DMG)$_3$(BX)$_2$] (X = F, OH, OCH$_3$, OC$_2$H$_5$, O-i-C$_3$H$_7$, O-n-C$_4$H$_9$) have been reported by using iron(II) salts, dimethylglyoxime, base, and either BF$_3$ or boric acid in n-butanol. When boric acid is used, the X groups are hydroxy or alkoxy groups derived from the solvent (H$_2$O, CH$_3$OH, C$_2$H$_5$OH, i-C$_3$H$_7$OH or n-C$_4$H$_9$OH).[466,467] The base should be added last and the reaction mixture kept slightly acidic. The cagelike

164

nature of **164** (X = F) has been established by X-ray diffraction; the stereochemistry about the iron(II) is intermediate between a trigonal prism and a trigonal antiprism, the twist angle being 16.5°. Attempts to prepare an iron(III) clathrochelate complex by oxidation of this compound were not successful; oxidants such as Br$_2$, I$_2$, O$_2$, H$_2$O$_2$, Ce(IV), or Cu(II) lead to decomposition of the complex or no reaction. Significant yields of the iron(II) species are also obtained if iron(III) salts are used in place of the iron(II) chloride normally employed; under the acidic conditions of the reaction medium the iron(II) complex must be thermodynamically favored.

Holm and co-workers developed a synthetic procedure for a series of six-coordinate metal(II) complexes containing a clathrochelate that encapsulates and imposes trigonal prismatic or near trigonal prismatic stereochemistry on the metal ion.[468,469] The syntheses were based on two requirements: (a) the presence in the final structure of planar, conjugated five-membered chelate rings attached to two anchoring atoms or groups centered on the

C_3 axis by linkages which prevent substantial relative motion of any three donor atoms and (b) the formation, in the last synthetic step, of the macrocyclic cage around the metal ion. The compound **165**, P(Hpox)$_3$, was synthesized by conventional synthetic procedures and then treated with cobalt(II), nickel(II), copper(II), and zinc(II) salts in acetonitrile to give the complex ions [M(P(Hpox)$_3$-H)]$^+$ (M = Co, Ni, Cu, Zn) isolated as perchlorate salts. Treatment of these complexes as fluoroborate salts with boron

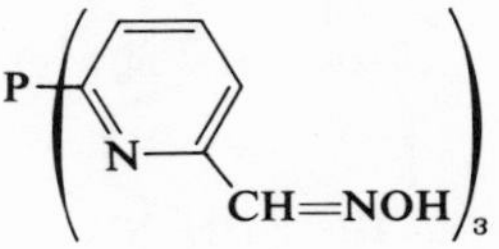

165

trifluoride etherate in acetonitrile at room temperature or reaction of **165**, the hydrated metal fluoroborate, and sodium tetrafluoroborate in equimolar ratio in acetonitrile produces the complexes [M(PccBF)]$^+$ (M = Fe, Co, Ni, Zn) containing the monoanionic ligand **166**. Both procedures failed to pro-

166

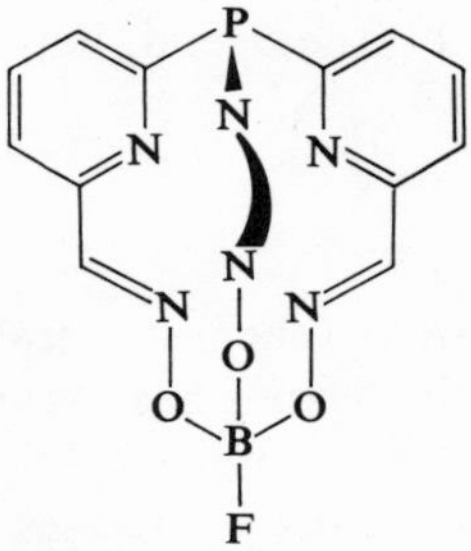

duce clathrochelates with manganese(II) and copper(II) ions. The complexes isolated were characterized by a variety of physical techniques. The structure of the nickel(II) complex has been reported,[470] showing the stereochemistry about the nickel(II) ion to be almost trigonal prismatic; the structures of the cobalt(II) and zinc(II) complexes are similar.[471,472] The iron(II) complex, however, is not isomorphous with the nickel(II), cobalt(II), and zinc(II) complexes; an X-ray study of this complex has shown the coordination geometry about the iron to be distorted by approximately 21.5° from trigonal prismatic.[473] It is suggested that the conformation of the iron(II) complex is a compromise between a trigonal prismatic geometry imposed by the ligand and an octahedral geometry preferred for the iron(II) ion due to the high crystal field stabilization energy associated with a low-spin d^6 ion. A comparison of the properties of a series of complexes, including the above species, with trigonal prismatic, trigonal antiprismatic, and intermediate stereo-chemistry has been made and the factors controlling the stereochemistry adopted have been discussed.[472]

Recently a new clathrochelate, 167, has been synthesized by the reaction of $[Co(en)_3]^{3+}$ (en = ethylenediamine) with formaldehyde and ammonia; a 95% yield of the cobalt(III) complex was reported.[474] A crystal structure

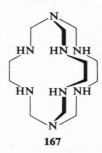

167

determination has demonstrated that synthesis occurs with retention of the chirality of the $[Co(en)_3]^{3+}$ ion and shows the sexadentate nature of the ligand with the *tris*-(methylene) amino cap at both ends of the parent ion. Oxidation and reduction of the metal ion within the cavity has been demonstrated.

8. Conclusions

It is apparent that the area of chemistry dealing with the synthesis of complexes containing macrocyclic ligands has expanded dramatically during the past few years. However, the examples described in this chapter are still somewhat limited in scope; most complexes that have been synthesized contain tetradentate nitrogen donor ligands, and the range of metal ions for which macrocyclic complexes have been reported is small, most of these ions being from the heavier first-row transition elements in their usual oxidation states. During the coming years it is anticipated that synthetic procedures will be developed for macrocyclic complexes containing a variety of different metal ions and donor atoms and with unique stereochemistries. Accumulation of knowledge concerning the mechanisms of macrocyclic complex formation should assist in the development of new and improved syntheses for these new complexes.

References

1. D. H. Busch, *Rec. Chem. Prog.* **25**, 107 (1964).
2. D. H. Busch, *Helv. Chim. Acta* **50**, 174 (1967).
3. N. F. Curtis, *Coord. Chem. Rev.* **3**, 3 (1968).

4. D. H. Busch, *Science* **171**, 241 (1971).
5. D. H. Busch, K. Farmery, V. Goedken, V. Katovic, A. C. Melnyk, C. R. Sperati, and N. Tokel, *Adv. Chem. Ser.* **100**, 44 (1971).
6. L. F. Lindoy and D. H. Busch, *Preparative Inorganic Reactions* (W. Jolly, ed.), Vol. 6, p. 1, Interscience, New York (1971).
7. J. J. Christensen, J. O. Hill, and R. M. Izatt, *Science* **174**, 459 (1971).
8. C. J. Pedersen and H. K. Frensdorff, *Angew. Chem. Int. Ed. Engl.* **11**, 16 (1972).
9. D. St. C. Black and A. J. Hartshorn, *Coord. Chem. Rev.* **9**, 219 (1972–1973).
10. J. M. Lehn, *Struct. Bonding (Berlin)* **16**, 1 (1973).
11. J. J. Christensen, D. J. Eatough, and R. M. Izatt, *Chem. Rev.* **74**, 351 (1974).
12. L. F. Lindoy, *Chem. Soc. Rev.* **5**, 421 (1975).
13. G. R. Newkome, J. D. Sauer, J. M. Roper, and D. C. Hager, *Chem. Rev.* **77**, 513 (1977).
14. R. H. Holm and T. J. Truex, *J. Am. Chem. Soc.* **93**, 285 (1971).
15. T. J. Truex and R. H. Holm, *J. Am. Chem. Soc.* **94**, 4529 (1972).
16. S. C. Tang, G. N. Weinstein, and R. H. Holm, *J. Am. Chem. Soc.* **95**, 613 (1973).
17. S. C. Tang, S. Koch, G. N. Weinstein, R. W. Lane, and R. H. Holm, *Inorg. Chem.* **12**, 2589 (1973).
18. S. Koch, R. H. Holm, and R. B. Frankel, *J. Am. Chem. Soc.* **97**, 6714 (1975).
19. C. L. Honeybourne, *Inorg. Nucl. Chem. Lett.* **11**, 191 (1975).
20. B. Bosnich, C. K. Poon, and M. L. Tobe, *Inorg. Chem.* **4**, 1102 (1965).
21. J. Van Alphen, *Rec. Trav. Chim. Pays-Bas* **55**, 835 (1936).
22. J. E. Richman and T. J. Atkins, *J. Am. Chem. Soc.* **96**, 2268 (1974).
23. E. K. Barefield, *Inorg. Chem.* **11**, 2273 (1972).
24. E. K. Barefield, F. Wagner, A. W. Herlinger, and A. R. Dahl, *Inorg. Synth.* **16**, 220 (1976).
25. W. Rosen and D. H. Busch, *J. Am. Chem. Soc.* **91**, 4694 (1969).
26. K. Travis and D. H. Busch, *Inorg. Chem.* **13**, 2591 (1974).
27. N. F. Curtis and R. W. Hay, *Chem. Commun.* **1966**, 524 (1966).
28. N. Sadasvian and J. F. Endicott, *J. Am. Chem. Soc.* **88**, 5468 (1976).
29. J. L. Love and H. K. J. Powell, *Chem. Commun.* **1968**, 39 (1968).
30. A. M. Tait and D. H. Busch, *Inorg. Nucl. Chem. Lett.* **8**, 491 (1972).
31. E. K. Barefield, F. Wagner and K. D. Hodges, *Inorg. Chem.* **15**, 1370 (1976).
32. M. C. Thompson and D. H. Busch, *J. Am. Chem. Soc.* **86**, 3651 (1964).
33. E. L. Blinn and D. H. Busch, *Inorg. Chem.* **7**, 820 (1968).
34. E. J. Jäger, *Z. Chem.* **4**, 437 (1964).
35. L. Wolf and E. G. Jäger, *Z. Anorg. Allgem. Chem.* **346**, 76 (1966).
36. E. G. Jäger, *Z. Chem.* **8**, 30 (1968).
37. E. G. Jäger, *Z. Chem.* **8**, 392 (1968).
38. E. G. Jäger, *Z. Chem.* **8**, 470 (1968).
39. E. G. Jäger, *Z. Anorg. Allgem. Chem.* **364**, 177 (1969).
40. P. Bamfield, *J. Chem. Soc. A* **1969**, 2021 (1969).
41. C. M. Kerwin and G. A. Melson, *Inorg. Chem.* **12**, 2410 (1973).
42. D. B. Bonfoey and G. A. Melson, *Inorg. Chem.* **14**, 309 (1975).
43. N. F. Curtis, *J. Chem. Soc.* **1960**, 4409 (1960).
44. D. A. House and N. F. Curtis, *Chem. Ind. (London)* **42**, 1708 (1961).
45. N. F. Curtis, Y. M. Curtis, and H. K. J. Powell, *J. Chem. Soc. A* **1966**, 1015 (1966).
46. D. A. Baldwin, R. M. Pfeiffer, D. W. Reichgott, and N. J. Rose, *J. Am. Chem. Soc.* **95**, 5152 (1973).
47. S. C. Jackals, K. Farmery, E. K. Barefield, N. J. Rose, and D. H. Busch, *Inorg. Chem.* **11**, 2893 (1972).

48. J. G. Martin, R. M. C. Wei, and S. C. Cummings, *Inorg. Chem.* **11**, 475 (1972).
49. J. G. Martin and S. C. Cummings, *Inorg. Chem.* **12**, 1477 (1972).
50. F. Wagner, M. T. Mocella, M. J. D'Aniello, A. H. J. Wang, and E. K. Barefield, *J. Am. Chem. Soc.* **96**, 2625 (1974).
51. E. K. Barefield and F. Wagner, *Inorg. Chem.* **12**, 2435 (1973).
52. C. J. Hipp and D. H. Busch, *J. Chem. Soc. Chem. Commun.* **1972**, 737 (1972).
53. C. J. Hipp and D. H. Busch, *Inorg. Chem.* **12**, 894 (1972).
54. D. C. Olson and J. Vasilevskis, *Inorg. Chem.* **8**, 1611 (1969).
55. D. C. Olson and J. Vasilevskis, *Inorg. Chem.* **10**, 463 (1971).
56. D. C. Olson and J. Vasilevskis, *Inorg. Chem.* **11**, 980 (1972).
57. D. P. Rillema, J. F. Endicott, and E. Papaconstantinou, *Inorg. Chem.* **10**, 1739 (1971).
58. J. C. Dabrowiak, F. V. Lovecchio, V. L. Goedken, and D. H. Busch, *J. Am. Chem. Soc.* **94**, 5502 (1972).
59. V. L. Goedken and D. H. Busch, *J. Am. Chem. Soc.* **94**, 7355 (1972).
60. F. V. Lovecchio, E. S. Gore, and D. H. Busch, *J. Am. Chem. Soc.* **96**, 3109 (1974).
61. H. Koyama and T. Yoshino, *Bull. Chem. Soc. Japan* **45**, 481 (1972).
62. M. S. Okamoto and E. K. Barefield, *Inorg. Chim. Acta* **17**, 91 (1976).
63. R. Yang and L. J. Zompa, *Inorg. Chem.* **15**, 1499 (1976).
64. M. Nonoyama, *Inorg. Chim. Acta* **20**, 53 (1976).
65. M. Nonoyama, *J. Inorg. Nucl. Chem.* **39**, 550 (1977).
66. S. F. Mason and R. D. Peacock, *Inorg. Chim. Acta* **19**, 75 (1976).
67. M. DeRonde, D. Driscoll, R. Yang, and L. J. Zompa, *Inorg. Nucl. Chem. Lett.* **11**, 521 (1975).
68. M. Kodama and E. Kimura, *J. Chem. Soc. Dalton Trans.* **1974**, 1473 (1974).
69. L. Fabbrizzi and L. J. Zompa, *Inorg. Nucl. Chem. Lett.* **13**, 287 (1977).
70. W. Rosen and D. H. Busch, *Inorg. Chem.* **9**, 262 (1970).
71. E. P. Kyba, C. W. Hudson, M. J. McPhaul, and A. M. John, *J. Am. Chem. Soc.* **99**, 8053 (1977).
72. S. G. McGeachin, *Can. J. Chem.* **44**, 2323 (1966).
73. G. A. Melson and D. H. Busch, *J. Am. Chem. Soc.* **87**, 1706 (1965).
74. E. B. Fleischer and E. Klem, *Inorg. Chem.* **4**, 637 (1965).
75. L. T. Taylor and D. H. Busch, *J. Am. Chem. Soc.* **89**, 5372 (1967).
76. L. T. Taylor, S. C. Vergez, and D. H. Busch, *J. Am. Chem. Soc.* **88**, 3170 (1966).
77. L. T. Taylor and D. H. Busch, *Inorg. Chem.* **8**, 1366 (1969).
78. S. C. Cummings and D. H. Busch, *J. Am. Chem. Soc.* **92**, 1924 (1970).
79. R. M. Wing and R. Eiss, *J. Am. Chem. Soc.* **92**, 1929 (1970).
80. G. Hawley and E. L. Blinn, *Inorg. Chem.* **14**, 2865 (1975).
81. G. A. Melson and D. H. Busch, *Proc. Chem. Soc.* **1963**, 223 (1963).
82. G. A. Melson and D. H. Busch, *J. Am. Chem. Soc.* **86**, 4834 (1964).
83. J. A. Elvidge and J. H. Golden, *J. Chem. Soc.* **1957**, 700 (1957).
84. R. P. Smirnov and D. B. Berezin, *Zh. Obshch. Khim.* **37**, 789 (1967).
85. V. N. Klynev and F. P. Snegireva, *Izv. Vyssh. Ucheb. Zaved. Khim. Khim. Tekhnol.* **12**, 1236 (1969) [*Chem. Abst.* **72**, 96171s].
86. J. P. Collman and P. W. Schneider, *Inorg. Chem.* **5**, 1380 (1966).
87. G. A. Kalligeros and E. L. Blinn, *Inorg. Chem.* **11**, 1145 (1972).
88. J.-M. Lehn, *Struct. Bonding (Berlin)* **16**, 1 (1973).
89. L. Y. Martin, L. J. Dehayes, L. J. Zompa, and D. H. Busch, *J. Am. Chem. Soc.* **96**, 4046 (1974).
90. E. K. Barefield and D. H. Busch, *Inorg. Chem.* **10**, 1216 (1971).
91. E. Hoyer, *Naturwissenschaften* **46**, 14 (1959).

92. M. M. Blight and N. F. Curtis, *J. Chem. Soc.* **1962**, 1204 (1962).
93. M. M. Blight and N. F. Curtis, *J. Chem. Soc.* **1962**, 3016 (1962).
94. D. A. House and N. F. Curtis, *J. Am. Chem. Soc.* **86**, 223 (1964).
95. D. A. House and N. F. Curtis, *J. Am. Chem. Soc.* **84**, 3248 (1962).
96. N. F. Curtis and T. N. Milestone, *Aust. J. Chem.* **28**, 275 (1975).
97. D. A. House and N. F. Curtis, *J. Am. Chem. Soc.* **86**, 1331 (1964).
98. N. F. Curtis, *J. Chem. Soc.* **1963**, 4109 (1963).
99. Y. M. Curtis and N. F. Curtis, *Aust. J. Chem.* **19**, 609 (1966).
100. R. Yungst and G. Stucky, *Inorg. Chem.* **13**, 2404 (1974).
101. N. F. Curtis and Y. M. Curtis, *J. Chem. Soc.* (A) **1966**, 1653 (1966).
102. Y. M. Curtis and N. F. Curtis, *Aust. J. Chem.* **18**, 1933 (1965).
103. M. F. Bailey and I. E. Maxwell, *Chem. Commun.* **1966**, 908 (1966).
104. R. R. Ryan, B. T. Kilbourn and J. D. Dunitz, *Chem. Commun.* **1966**, 910 (1966).
105. B. T. Kilbourn, R. R. Ryan, and J. D. Dunitz, *J. Chem. Soc.* (A) **1969**, 2407 (1969).
106. L. G. Warner, N. J. Rose, and D. H. Busch, *J. Am. Chem. Soc.* **89**, 703 (1967).
107. L. G. Warner, N. J. Rose, and D. H. Busch, *J. Am. Chem. Soc.* **90**, 6938 (1968).
108. F. Hanic and D. Miklos, *J. Cryst. Mol. Struct.* **2**, 115 (1972).
109. N. F. Curtis and D. A. House, *J. Chem. Soc. A* **1967**, 537 (1967).
110. T. E. MacDermott and D. H. Busch, *J. Am. Chem. Soc.* **89**, 5780 (1967).
111. D. St. C. Black and H. Greenland, *Aust. J. Chem.* **25**, 1315 (1972).
112. N. F. Curtis, D. A. Swann, T. N. Waters, and I. E. Maxwell, *J. Am. Chem. Soc.* **91**, 4588 (1969).
113. K. Hideg and D. Lloyd, *Chem. Commun.* **1970**, 929 (1970).
114. R. A. Kolinski and B. Korybut-Daszkiewicz, *Inorg. Chim. Acta* **14**, 237 (1975).
115. J. L. Love and H. K. J. Powell, *Inorg. Nucl. Chem. Lett.* **3**, 113 (1967).
116. P. R. Ireland and W. T. Robinson, *J. Chem. Soc. A* **1970**, 663 (1970).
117. D. A. Swann, T. N. Waters, and N. F. Curtis, *J. Chem. Soc. Dalton Trans.* **1972**, 1115 (1972).
118. T. Ito and D. H. Busch, *Inorg. Chem.* **13**, 1770 (1974).
119. M. V. Artenenko, E. S. Sereda, P. A. Suprunenko, and G. I. Kal'naya, *Zh. Neorg. Khim.* **19**, 60 (1974).
120. V. L. Goedken, P. H. Merrell, and D. H. Busch, *J. Am. Chem. Soc.* **94**, 3397 (1972).
121. P. H. Merrell, V. L. Goedken, D. H. Busch, and J. A. Stone, *J. Am. Chem. Soc.* **92**, 7590 (1970).
122. V. L. Goedken, J. Molin-Case, and G. G. Christoph, *Inorg. Chem.* **12**, 2894 (1973).
123. J. C. Dabrowiak, P. H. Merrell, J. A. Stone, and D. H. Busch, *J. Am. Chem. Soc.* **95**, 6613 (1973).
124. A. M. Tait and D. H. Busch, *Inorg. Chem.* **15**, 197 (1976).
125. J. Vasilevskis and D. C. Olson, *Inorg. Chem.* **10**, 1228 (1971).
126. A. M. Tait, M. Z. Hoffman, and E. Hayon, *J. Am. Chem. Soc.* **98**, 86 (1976).
127. N. Sadasvian, J. A. Kernohan, and J. F. Endicott, *Inorg. Chem.* **6**, 770 (1967).
128. G. A. Lawrance, *Inorg. Nucl. Chem. Lett.* **12**, 85 (1976).
129. W. K. Lee and C. K. Poon, *J. Chem. Soc. Dalton Trans.* **1974**, 2423 (1974).
130. C-K. Poon, C-L. Wong, and P. W. Mak, *J. Chem. Soc. Dalton Trans.* **1977**, 1931 (1977).
131. C. K. Poon, *Coord. Chem. Rev.* **10**, 1 (1973).
132. P. W. Mak and C. K. Poon, *Inorg. Chem.* **15**, 1949 (1976).
133. M. P. Liteplo and J. F. Endicott, *Inorg. Chem.* **10**, 1420 (1971).
134. E. K. Barefield and M. T. Mocella, *Inorg. Chem.* **12**, 2829 (1973).
135. D. Wester, R. C. Edwards, and D. H. Busch, *Inorg. Chem.* **16**, 1055 (1977).
136. N. F. Curtis, *J. Chem. Soc. Dalton Trans.* **1972**, 1357 (1972).

137. N. F. Curtis, *J. Chem. Soc. Dalton Trans.* **1973**, 863 (1973).
138. D. F. Cook and N. F. Curtis, *J. Chem. Soc. Dalton Trans.* **1973**, 1076 (1973).
139. N. F. Curtis, *J. Chem. Soc. Dalton Trans.* **1974**, 347 (1974).
140. N. F. Curtis, *Aust. J. Chem.* **27**, 71 (1974).
141. J. W. L. Martin and N. F. Curtis, *J. Chem. Soc. Dalton Trans.* **1975**, 88 (1975).
142. T. E. MacDermott, B. E. Sewell, and D. H. Busch, *J. Am. Chem. Soc.* **89**, 5784 (1967).
143. N. F. Curtis, *J. Chem. Soc.* **1964**, 2644 (1964).
144. N. F. Curtis, *J. Chem. Soc.* **1965**, 924 (1965).
145. N. F. Curtis, *J. Chem. Soc. C* **1967**, 1979 (1967).
146. R. A. Bauer, W. R. Robinson, and D. W. Margerum, *J. Chem. Soc. Chem. Commun.* **1973**, 289 (1973).
147. N. F. Curtis and Y. M. Curtis, *Inorg. Chem.* **4**, 804 (1965).
148. N. F. Curtis and Y. M. Curtis, *Aust. J. Chem.* **19**, 1423 (1965).
149. N. F. Curtis, *J. Chem. Soc. A* **1968**, 1579 (1968).
150. N. F. Curtis, *J. Chem. Soc. A* **1968**, 1584 (1968).
151. N. F. Curtis, D. A. Swann, and T. N. Waters, *J. Chem. Soc. Dalton Trans.* **1973**, 1408 (1973).
152. P. O. Whimp, M. F. Bailey, and N. F. Curtis, *J. Chem. Soc. A* **1970**, 1956 (1970).
153. N. F. Curtis, D. A. Swann, and T. N. Waters, *J. Chem. Soc. Dalton Trans.* **1973**, 1963 (1973).
154. P. S. Bryan and J. C. Dabrowiak, *Inorg. Chem.* **14**, 299 (1975).
155. L. G. Warner and D. H. Busch, *J. Am. Chem. Soc.* **91**, 4092 (1969).
156. E. K. Barefield and D. H. Busch, *Inorg. Chem.* **10**, 1216 (1971).
157. J. C. Dabrowiak, P. H. Merrell, and D. H. Busch, *Inorg. Chem.* **11**, 1979 (1972).
158. P. O. Whimp and N. F. Curtis, *J. Chem. Soc. A* **1966**, 867 (1966).
159. J. C. Dabrowiak and P. S. Bryan, *Inorg. Nucl. Chem. Lett.* **12**, 485 (1976).
160. W. K. Chau, W. K. Lee, and C. K. Poon, *J. Chem. Soc. Dalton Trans.* **1974**, 2419 (1974).
161. P. O. Whimp and N. F. Curtis, *J. Chem. Soc. A* **1966**, 1827 (1966).
162. N. A. P. Kane-Maguire, J. F. Endicott, and D. P. Rillema, *Inorg. Chim. Acta*, **6**, 443 (1972).
163. P. O. Whimp and N. F. Curtis, *J. Chem. Soc. A* **1968**, 188 (1968).
164. N. F. Curtis and D. F. Cook, *J. Chem. Soc. Dalton Trans.* **1972**, 691 (1972).
165. P. S. Bryan and J. C. Dabrowiak, *Inorg. Chem.* **14**, 296 (1975).
166. M. O. Kestner and A. L. Allred, *J. Am. Chem. Soc.* **94**, 7189 (1972).
167. K. B. Mertes, *Inorg. Chem.* **17**, 49 (1978).
168. A. Dei and F. Mani, *Inorg. Chim. Acta* **19**, L39 (1976).
169. A. Dei and F. Mani, *Inorg. Chem.* **15**, 2574 (1976).
170. N. F. Curtis and G. W. Reader, *J. Chem. Soc. A* **1971**, 1771 (1971).
171. J. Waters and K. Whittle, *J. Inorg. Nucl. Chem.* **34**, 155 (1972).
172. N. F. Curtis and G. W. Reader, *J. Chem. Soc. Dalton Trans.* **1972**, 1453 (1972).
173. R. Buxtorf, W. Steinmann, and T. A. Kaden, *Chimia* **28**, 15 (1974).
174. H. Stetter and K. H. Mayer, *Chem. Ber.* **94**, 1410 (1961).
175. B. Bosnich, M. L. Tobe, and G. A. Webb, *Inorg. Chem.* **4**, 1109 (1965).
176. B. Bosnich, R. Mason, P. J. Pauling, G. B. Robertson, and M. L. Tobe, *Chem. Commun.* **1965**, 97 (1965).
177. L. Hertli and T. A. Kaden, *Chimia* **29**, 304 (1975).
178. B. Bosnich, C. K. Poon, and M. L. Tobe, *Inorg. Chem.* **5**, 1514 (1966).
179. C. K. Poon and K. S. Mok, *Inorg. Chem.* **10**, 225 (1971).
180. C. K. Poon and H. W. Tong, *J. Chem. Soc. Dalton Trans.* **1974**, 1 (1974).

181. C. K. Poon and M. L. Tobe, *J. Chem. Soc. A* **1968**, 1549 (1968).
182. C. K. Liu and C. K. Poon, *J. Chem. Soc. Dalton Trans.* **1972**, 216 (1972).
183. K. S. Mok, C. K. Poon, and H. W. Tong, *J. Chem. Soc. Dalton Trans.* **1972**, 1701 (1972).
184. C. K. Poon and C. L. Wong, *J. Chem. Soc. Dalton Trans.* **1976**, 966 (1976).
185. J. Cragel, Jr., and B. E. Douglas, *Inorg. Chim. Acta* **10**, 33 (1974).
186. C. K. Poon, *Inorg. Chim. Acta* **5**, 322 (1971).
187. C. K. Poon and M. L. Tobe, *Inorg. Chem.* **7**, 2398 (1968).
188. T. F. Lai and C. K. Poon, *Inorg. Chem.* **15**, 1562 (1976).
189. E. J. Bounsall and S. P. Koprich, *Can. J. Chem.* **48**, 1481 (1970).
190. J. Ferguson and M. L. Tobe, *Inorg. Chim. Acta* **4**, 109 (1970).
191. P. K. Chan and C. K. Poon, *J. Chem. Soc. Dalton Trans.* **1976**, 858 (1976).
192. D. E. Fenton, C. Nave, and M. R. Truter, *J. Chem. Soc. Chem. Commun.* **1972**, 1303 (1972).
193. D. D. Watkins, Jr., D. P. Riley, J. A. Stone, and D. H. Busch, *Inorg. Chem.* **15**, 387 (1976).
194. L. Y. Martin, C. R. Sperati, and D. H. Busch, *J. Am. Chem. Soc.* **99**, 2968 (1977).
195. Y. Hung, L. Y. Martin, S. C. Jackels, A. M. Tait, and D. H. Busch, *J. Am. Chem. Soc.* **99**, 4029 (1977).
196. Y. Hung and D. H. Busch, *J. Am. Chem. Soc.* **99**, 4977 (1977).
197. A. Desideri, J. B. Raynor, and C.-K. Poon, *J. Chem. Soc. Dalton Trans.* **1977**, 2051 (1977).
198. P. S. Bryan and J. M. Calvert, *Inorg. Nucl. Chem. Lett.* **13**, 615 (1977).
199. M. J. D'Aniello, Jr., M. T. Mocella, F. Wagner, E. K. Barefield, and I. C. Paul, *J. Am. Chem. Soc.* **97**, 192 (1975).
200. F. Wagner and E. K. Barefield, *Inorg. Chem.* **15**, 408 (1976).
201. R. Buxtorf and T. A. Kaden, *Helv. Chim. Acta* **57**, 1035 (1974).
202. K. D. Hodges, R. G. Wollmann, E. K. Barefield, and D. N. Hendrickson, *Inorg. Chem.* **16**, 2746 (1977).
203. Y. Iitaka, M. Shina, and E. Kimura, *Inorg. Chem.* **13**, 2886 (1974).
204. R. Smierciak, J. Parsariello, and E. L. Blinn, *Inorg. Chem.* **16**, 2646 (1977).
205. L. Fabbrizzi, *Inorg. Chem.* **16**, 2667 (1977).
206. A. Anichini, L. Fabbrizzi, P. Paoletti, and R. M. Clay, *Inorg. Chim. Acta* **24**, L21 (1977).
207. E. L. Blinn, unpublished results.
208. M. C. Styka, R. C. Smierciak, E. L. Blinn, R. E. DeSimione, and J. V. Passariello, *Inorg. Chem.* **17**, 82 (1978).
209. F. Hanic, M. Handlovic, and O. Lindgren, *Coll. Czech. Chem. Commun.* **37**, 2119 (1972).
210. L. A. Funke and G. A. Melson, *Inorg. Chem.* **14**, 306 (1975).
211. L. A. Funke and G. A. Melson, unpublished results.
212. L. A. Funke, Ph.D. Thesis, Michigan State University, 1976.
213. M. Green and P. A. Tasker, *Chem. Commun.* **1968**, 518 (1968).
214. M. Green, J. Smith, and P. A. Tasker, *Inorg. Chim. Acta* **5**, 17 (1971).
215. M. Green and P. A. Tasker, *Inorg. Chim. Acta* **5**, 65 (1971).
216. D. St. C. Black, C. H. Bos Vanderzalm, and A. J. Hartshorn, *Aust. J. Chem.* **30**, 225 (1977).
217. J. Ellis, L. Gallagher, and M. Green, *Inorg. Nucl. Chem. Lett.* **9**, 185 (1973).
218. D. Losman, L. M. Engelhardt, and M. Green, *Inorg. Nucl. Chem. Lett.* **9**, 791 (1973).
219. E. M. Maslen, L. M. Engelhardt, and A. H. White, *J. Chem. Soc. Dalton Trans.* **1974**, 1799 (1974).

220. E. B. Fleischer, L. Sklar, A. Kendall-Torry, P. A. Tasker, and F. B. Taylor, *Inorg. Nucl. Chem. Lett.* **9**, 1061 (1973).
221. E. Uhlemann and M. Plath, *Z. Chem.* **9**, 234 (1969).
222. D. St. C. Black and M. J. Lane, *Aust. J. Chem.* **23**, 2039 (1970).
223. D. St. C. Black and P. W. Kortt, *Aust. J. Chem.* **25**, 281 (1972).
224. D. St. C. Black and A. J. Hartshorn, *J. Chem. Soc. Chem. Commun.* **1972**, 706 (1972).
225. D. St. C. Black, C. H. Bos Vanderzalm, and A. J. Hartshorn, *Inorg. Nucl. Chem. Lett.* **12**, 657 (1976).
226. S. C. Cummings and R. E. Sievers, *J. Am. Chem. Soc.* **92**, 215 (1970).
227. S. C. Cummings and R. E. Sievers, *Inorg. Chem.* **9**, 1131 (1970).
228. S. C. Cummings and R. E. Sievers, *Inorg. Chem.* **11**, 1483 (1972).
229. M. F. Richardson and R. E. Sievers, *J. Am. Chem. Soc.* **94**, 4134 (1972).
230. N. F. Curtis and N. B. Milestone, *Aust. J. Chem.* **27**, 1167 (1974).
231. W. H. Elfring, Jr. and N. J. Rose, *Inorg. Chem.* **14**, 2759 (1975).
232. M. S. Holtman and S. C. Cummings, *Inorg. Chem.* **15**, 660 (1976).
233. G. W. Roberts, S. C. Cummings, and J. A. Cunningham, *Inorg. Chem.* **15**, 2503 (1976).
234. B. Durham, T. J. Anderson, J. A. Switzer, J. F. Endicott, and M. D. Glick, *Inorg. Chem.* **16**, 271 (1977).
235. C. K. Poon and C. L. Wong, *Inorg. Chem.* **15**, 1573 (1976); *J. Chem. Soc. Dalton Trans.* **1977**, 523 (1977).
236. E. S. Gore and D. H. Busch, *Inorg. Chem.* **12**, 1 (1973).
237. D. St. C. Black and M. J. Lane, *Aust. J. Chem.* **23**, 2027 (1970).
238. D. A. Baldwin and N. J. Rose, *Abstracts of the 157th National Meeting of the American Chemical Society Minneapolis, Minnesota (1969)*, No. Inor 20.
239. K. Farmery and D. H. Busch, *Inorg. Chem.* **11**, 2901 (1972).
240. M. D. Glick, W. G. Schmonsees, and J. F. Endicott, *J. Am. Chem. Soc.* **96**, 5661 (1974).
241. W. A. Welsh, G. J. Reynolds, and P. M. Henry, *Inorg. Chem.* **16**, 2558 (1977).
242. D. J. Olszanski and G. A. Melson, *Inorg. Chim. Acta* **23**, L4 (1977).
243. D. P. Riley, J. A. Stone, and D. H. Busch, *J. Am. Chem. Soc.* **98**, 1752 (1976).
244. K. Bowman, D. P. Riley, D. H. Busch, and P. W. R. Corfield, *J. Am. Chem. Soc.* **97**, 5036 (1975).
245. K. B. Mertes, P. W. R. Corfield, and D. H. Busch, *Inorg. Chem.* **16**, 3226 (1977).
246. D. P. Riley, J. A. Stone, and D. H. Busch, *J. Am. Chem. Soc.* **99**, 767 (1977).
247. S. Koch, S. C. Tang, R. H. Holm, and R. B. Frankel, *J. Am. Chem. Soc.* **97**, 914 (1975).
248. M. Millar and R. H. Holm, *J. Chem. Soc. Chem. Commun.* **1975**, 169 (1975).
249. S. C. Tang and R. H. Holm, *J. Am. Chem. Soc.* **97**, 3359 (1975).
250. M. Millar and R. H. Holm, *J. Am. Chem. Soc.* **97**, 6052 (1975).
251. F. A. L'Eplattenier and A. Pugin, *Helv. Chim. Acta* **58**, 2283 (1975).
252. H. Hiller, P. Dimroth, and H. Pfitzner, *Liebigs Ann. Chem.* **717**, 137 (1968).
253. M. C. Weiss, G. Gordon, and V. L. Goedken, *Inorg. Chem.* **16**, 305 (1977).
254. P. Chave and C. L. Honeybourne, *Chem. Commun.* **1969**, 279 (1969).
255. C. L. Honeybourne and P. Burchill, *Inorg. Nucl. Chem. Lett.* **10**, 715 (1974).
256. V. L. Goedken, J. Molin-Case, and Y-A. Wang, *J. Chem. Soc. Chem. Commun.* **1973**, 337 (1973).
257. V. L. Goedken and Y-A. Park, *J. Chem. Soc. Chem. Commun.* **1975**, 214 (1975).
258. V. L. Goedken, S-M. Peng, and Y-A. Park, *J. Am. Chem. Soc.* **96**, 284 (1974).
259. M. C. Weiss and V. L. Goedken, *J. Am. Chem. Soc.* **98**, 3389 (1976).

260. M. C. Weiss and V. L. Goedken, *J. Chem. Soc. Chem. Commun.* **1976**, 531 (1976).
261. D. R. Neves and J. C. Dabrowiak, *Inorg. Chem.* **15**, 129 (1976).
262. F. C. McElroy, J. C. Dabrowiak, and D. J. Macero, *Inorg. Chem.* **16**, 947 (1977).
263. L. G. Bell and J. C. Dabrowiak, *J. Chem. Soc. Chem. Commun.* **1975**, 512 (1975).
264. V. L. Goedken, J. J. Pluth, S-M. Peng, and B. Bursten, *J. Am. Chem. Soc.* **98**, 8014 (1976).
265. M. C. Weiss, B. Bursten, S-M. Peng, and V. L. Goedken, *J. Am. Chem. Soc.* **98**, 8021 (1976).
266. F. A. L'Eplattenier and A. Pugin, *Helv. Chim Acta* **58**, 917 (1975).
267. D. P. Fisher, F. C. McElroy, D. C. Macero, and J. C. Dabrowiak, *Inorg. Nucl. Chem. Lett.* **12**, 435 (1976).
268. D. P. Fisher, V. Piermattie, and J. C. Dabrowiak, *J. Am. Chem. Soc.* **99**, 2811 (1977).
269. J. D. Goddard, *Inorg. Nucl. Chem. Lett.* **13**, 555 (1977).
270. K. Hideg and D. Lloyd, *J. Chem. Soc. C* **1971**, 3441 (1971).
271. O. H. Hankovsky, K. Hideg, D. Lloyd, and H. McNab, *J. Chem. Soc. Chem. Commun.* **1974**, 378 (1974).
272. D. F. Cook, N. F. Curtis, and R. W. Hay, *J. Chem. Soc. Dalton Trans.* **1973**, 1160 (1973).
273. D. F. Cook, *Inorg. Nucl. Chem. Lett.* **12**, 103 (1976).
274. N. F. Curtis, *J. Chem. Soc. Dalton Trans.* **1973**, 1212 (1973).
275. R. W. Hay and P. M. Gidney, *J. Chem. Soc. Dalton Trans.* **1976**, 974 (1976).
276. G. Ferguson, P. Roberts, D. Lloyd, and K. Hideg, *J. Chem. Soc. Chem. Commun.* **1977**, 149 (1977).
277. P. W. Caulkett, D. Greatbanks, R. W. Turner, and J. A. J. Jarvis, *J. Chem. Soc. Chem. Commun.* **1977**, 150 (1977).
278. J. F. Myers and N. J. Rose, *Inorg. Chem.* **12**, 1238 (1973).
279. J. F. Myers and C. H. L. Kennard, *J. Chem. Soc. Chem. Commun.* **1972**, 77 (1972).
280. R. A. Kolinski and B. Korybut-Daszkiewicz, *Bull. Acad. Pol. Sci. Ser. Sci. Chim.* **17**, 13 (1969).
281. J. Mrozinski, R. A. Koliniski, and B. Jezowski-Trzebiatowska, *Bull. Acad. Pol. Sci. Ser. Sci. Chim.* **22**, 999 (1974).
282. J. Krajeiwski, Z. Urbanczyk-Lipkowska, and P. Gluzinski, *Bull. Acad. Poln. Sci. Ser. Sci. Chim.* **22**, 955 (1974).
283. A. I. Gusiev, J. W. Krajewski, and Z, Urbanczyk, *Bull. Acad. Poln. Sci. Ser. Sci. Chim.* **22**, 387 (1974).
284. R. W. Hay and B. Jeragh, *J. Chem. Soc. Dalton Trans.* **1977**, 1261 (1977).
285. R. W. Hay and G. A. Lawrance, *J. Chem. Soc. Dalton Trans.* **1975**, 1466 (1975).
286. R. J. Restivo, J. Horney, and G. Ferguson, *J. Chem. Soc. Dalton Trans.* **1976**, 514 (1976).
287. R. W. Hay and G. A. Lawrance, *J. Chem. Soc. Dalton Trans.* **1976**, 1086 (1976).
288. C. G. Barraclough and G. A. Lawrance, *Inorg. Nucl. Chem. Lett.* **12**, 133 (1976).
289. R. W. Hay, D. P. Piplani, and B. Jeragh, *J. Chem. Soc. Dalton Trans.* **1977**, 1951 (1977).
290. R. W. Hay and D. P. Piplani, *J. Inorg. Nucl. Chem.* **38**, 1403 (1976).
291. F. P. Hinz and D. W. Margerum, *Inorg. Chem.* **13**, 2941 (1974).
292. C-T. Lin, D. B. Rorabacher, G. R. Cayley, and D. W. Margerum, *Inorg. Chem.* **14**, 919 (1975).
293. R. W. Hay and D. P. Piplani, *J. Chem. Soc. Dalton Trans.* **1977**, 1956 (1977).
294. J. D. Curry and D. H. Busch, *J. Am. Chem. Soc.* **86**, 592 (1964).
295. J. L. Karn and D. H. Busch, *Nature* **211**, 160 (1966).
296. E. B. Fleischer and S. W. Hawkinson, *Inorg. Chem.* **7**, 2312 (1968).

297. R. L. Rich and G. L. Stucky, *Inorg. Nucl. Chem. Lett.* **1**, 61 (1965).
298. K. M. Long and D. H. Busch, *Inorg. Chem.* **9**, 505 (1970).
299. L. F. Lindoy, N. E. Tokel, L. B. Anderson, and D. H. Busch, *J. Coord. Chem.* **1**, 7 (1971).
300. J. C. Dabrowiak, L. A. Nafie, P. S. Pryan, and A. T. Torkelson, *Inorg. Chem.* **16**, 540 (1977).
301. K. M. Long and D. H. Busch, *J. Coord. Chem.* **4**, 113 (1974).
302. A. M. Tait, F. V. Lovecchio, and D. H. Busch, *Inorg. Chem.* **16**, 2206 (1977).
303. E. Ochiai and D. H. Busch, *Chem. Commun.* **1968**, 905 (1968).
304. E. Ochiai, K. M. Long, C. R. Sperati, and D. H. Busch, *J. Am. Chem. Soc.* **91**, 3201 (1969).
305: K. Farmery and D. H. Busch, *J. Chem. Soc. Chem. Commun.* **1970**, 1091 (1970).
306. J. L. Karn and D. H. Busch, *Inorg. Chem.* **8**, 1149 (1969).
307. E. Dewar and E. Fleischer, *Nature* **222**, 372 (1969).
308. E. Ochiai and D. H. Busch, *Inorg. Chem.* **8**, 1798 (1969).
309. E. Ochiai and D. H. Busch, *Inorg. Chem.* **8**, 1474 (1969).
310. D. P. Riley, P. H. Merrell, J. A. Stone, and D. H. Busch, *Inorg. Chem.* **14**, 490 (1975).
311. E. K. Barefield, F. V. Lovecchio, N. E. Tokel, E. Ochiai, and D. H. Busch, *Inorg. Chem.* **11**, 283 (1972).
312. R. H. Prince, D. A. Stotter, and P. R. Woolley, *Inorg. Chim. Acta* **9**, 51 (1974).
313. H. Keypour and D. A. Stotter, *Inorg. Chim. Acta* **19**, L48 (1976).
314. M. R. Caira, L. R. Nassimbeni, and P. R. Woolley, *Acta Cryst.* **B31**, 1334 (1975).
315. C-K. Poon, W-K. Wan, and S. S. T. Liao, *J. Chem. Soc. Dalton Trans.* **1977**, 1247 (1977).
316. T. J. Lotz and T. A. Kaden, *J. Chem. Soc. Chem. Commun.* **1977**, 15 (1977).
317. F. Voegtle, E. Weber, W. Wehner, R. Naetscher, and J. Gruetze, *Chem. Ztg.* **98**, 562 (1974).
318. J. Riker-Nappier and D. W. Meek, *J. Chem. Soc. Chem. Commun.* **1972**, 442 (1972).
319. F. Seidel and W. Dick, *Berichte* **B60**, 2018 (1927).
320. G. A. Melson and D. H. Busch, *J. Am. Chem. Soc.* **86**, 4830 (1964).
321. J. W. Hawkinson and E. B. Fleischer, *Inorg. Chem.* **8**, 2402 (1969).
322. F. Seidel, *Berichte* **B59**, 1894 (1926).
323. S. C. Cummings and D. H. Busch, *Inorg. Chem.* **10**, 1220 (1971).
324. V. Katovic, S. G. Vergez, and D. H. Busch, *Inorg. Chem.* **16**, 1716 (1977).
325. I. W. Pang and D. V. Stynes, *Inorg. Chem.* **16**, 2192 (1977).
326. W. C. Potter and L. T. Taylor, *Inorg. Chem.* **15**, 1329 (1976).
327. G. L. Eichhorn and R. A. Latif, *J. Am. Chem. Soc.* **76**, 5180 (1954).
328. C. P. Lau and G. A. Melson, unpublished results.
329. L. T. Taylor, F. L. Urbach, and D. H. Busch, *J. Am. Chem. Soc.* **91**, 1072 (1969).
330. V. Katovic, L. T. Taylor, and D. H. Busch, *J. Am. Chem. Soc.* **91**, 2122 (1969).
331. V. Katovic, L. T. Taylor, and D. H. Busch, *Inorg. Chem.* **10**, 458 (1971).
332. N. E. Tokel, V. Katovic, K. Farmery, L. B. Anderson, and D. H. Busch, *J. Am. Chem. Soc.* **92**, 400 (1970).
333. V. Katovic, L. T. Taylor, F. L. Urbach, W. H. White, and D. H. Busch, *Inorg. Chem.* **11**, 479 (1972).
334. J. S. Skuratowicz, I. L. Madden, and D. H. Busch, *Inorg. Chem.* **16**, 1721 (1977).
335. C. M. Kerwin and G. A. Melson, *Inorg. Chem.* **11**, 726 (1972).
336. G. A. Melson, *Inorg. Chem.* **13**, 994 (1974).
337. G. A. Melson and D. B. Bonfoey, *Inorg. Nucl. Chem. Lett.* **9**, 875 (1973).
338. K. Nafisi-Movaghar and G. A. Melson, *Inorg. Chem.* **14**, 2015 (1975).

339. G. M. Shalhoub, Ph.D. Thesis, Michigan State University, 1976.
340. G. M. Shalhoub and G. A. Melson, unpublished results.
341. N. F. Curtis, unpublished results.
342. V. L. Goedken and S. M. Peng, *J. Chem. Soc. Chem. Commun.* **1973**, 62 (1973).
343. V. L. Goedken and S. M. Peng, *J. Am. Chem. Soc.* **95**, 5773 (1973).
344. V. L. Goedken and S. M. Peng, *J. Chem. Soc. Chem. Commun.* **1974**, 914 (1974).
345. S-M. Peng, G. C. Gordon, and V. L. Goedken, *Inorg. Chem.* **17**, 119 (1978).
346. V. L. Goedken and S. M. Peng, *J. Am. Chem. Soc.* **96**, 7826 (1974).
347. V. L. Goedken and S. M. Peng, *J. Chem. Soc. Chem. Commun.* **1975**, 258 (1975).
348. V. L. Goedken, Y-A. Park, S. M. Peng, and J. M. Norris, *J. Am. Chem. Soc.* **96**, 7693 (1974).
349. H. Neunhoeffer and L. Motitschke, *Tetrahedron Lett.* **9**, 655 (1970).
350. H. Neunhoeffer, J. Stastney, and L. Motitschke, *Tetrahedron Lett.* **20**, 1601 (1971).
351. J. E. Baldwin, R. H. Holm, R. W. Harper, J. Huff, S. Koch, and T. J. Truex, *Inorg. Nucl. Chem. Lett.* **8**, 393 (1972).
352. J. E. Baldwin and J. Huff, *J. Am. Chem. Soc.* **95**, 5757 (1973).
353. R. G. Little, J. A. Ibers, and J. E. Baldwin, *J. Am. Chem. Soc.* **97**, 7050 (1975).
354. J. F. Boas, J. R. Pilbrow, G. J. Troup, C. Moore, and T. D. Smith, *J. Chem. Soc. A* **1969**, 965 (1969).
355. G. R. Clark, B. W. Skelton, and T. N. Waters, *J. Chem. Soc. Chem. Commun.* **1972**, 1163 (1972).
356. G. R. Clark, B. W. Skelton, and T. N. Waters, *J. Chem. Soc. Dalton Trans.* **1976**, 1528 (1976).
357. W. E. Keyes, J. B. R. Dunn, and T. M. Loehr, *J. Am. Chem. Soc.* **99**, 4527 (1977).
358. J. Lewis and K. P. Wainwright, *J. Chem. Soc. Chem. Commun.* **1974**, 169 (1974).
359. J. Lewis and K. P. Wainwright, *J. Chem. Soc. Dalton Trans.* **1977**, 734 (1977).
360. J. Lewis and K. P. Wainwright, *J. Chem. Soc. Dalton Trans.* **1977**, 739 (1977).
361. V. M. Dziomko, L. G. Fedosyuk, and K. A. Dunaevskaya, *Zh. Obshch. Khim.* **45**, 2488 (1975).
362. V. M. Dziomko, L. G. Fedosyuk, K. A. Dunaevskaya, and Yu. S. Ryabokobylko, *Koord. Khim.* **2**, 39 (1976).
363. G. N. Schrauzer, *Chem. Ber.* **95**, 1438 (1962).
364. F. Umland and D. Thierig, *Angew. Chem.* **74**, 388 (1962).
365. M. Green and G. Tauzher, *Trans. Met. Chem.* **1**, 1 (1975/76).
366. S. Ogawa, T. Yamaguchi, and N. Gotoh, *J. Chem. Soc. Chem. Commun.* **1972**, 577 (1972).
367. S. Ogawa, T. Yamaguchi, and N. Gotoh, *J. Chem. Soc. Perkin Trans.* **1974**, 976 (1974).
368. W. C. Marsh, N. L. Paddock, C. J. Stewart, and J. Trotter, *Chem. Commun.* **1970**, 1190 (1970).
369. J. A. Elvidge and R. P. Linstead, *J. Chem. Soc.* **1952**, 5008 (1952).
370. P. Bamfield and P. A. Mack, *J. Chem. Soc. C* **1968**, 1961 (1968).
371. P. Bamfield and D. G. Wilkinson, *J. Chem. Soc. C* **1968**, 2409 (1968).
372. R. P. Smirnov, V. A. Gnedina, and V. F. Borodkin, *Khim. Geterotskil. Soedin.*, 1102 (1969) [*Chem. Abst.* **73**, 10243c].
373. R. W. Kluiber and G. Sasso, *Inorg. Chim. Acta* **4**, 226 (1970).
374. D. L. Johnston and W. D. Horrocks, Jr., *Inorg. Chem.* **10**, 687 (1971).
375. R. A. Lalancette, D. J. Macchia, and W. F. Furey, *Inorg. Chem.* **15**, 548 (1976).
376. L. G. Armstrong and L. F. Lindoy, *Inorg. Nucl. Chem. Lett.* **10**, 349 (1974).
377. L. G. Armstrong and L. F. Lindoy, *Inorg. Chem.* **14**, 1322 (1975).

378. L. F. Lindoy, H. C. Lip, L. F. Power, and J. H. Rea, *Inorg. Chem.* **15**, 1724 (1976).
379. L. G. Armstrong, L. F. Lindoy, M. McPartlin, G. M. Mockler, and P. A. Tasker. *Inorg. Chem.* **16**, 1665 (1977).
380. P. G. Grimsley, L. F. Lindoy, H. C. Lip, R. J. Smith, and J. T. Baker, *Aust. J. Chem,* **30**, 2095 (1977).
381. T. Yano, T. Ushijima, M. Sasaki, H. Kobayashi, and K. Ueno, *Bull. Chem. Soc. Japan* **45**, 2452 (1972).
382. D. E. Fenton and S. E. Gayda, *J. Chem. Soc. Chem. Commun.* **1974**, 960 (1974).
383. P. A. Vigato, M. Vidali, U. Casellato, R. Graziani, and F. Benetollo, *Inorg. Nucl. Chem. Lett.* **11**, 595 (1975).
384. D. E. Fenton and S. E. Gayda, *J. Chem. Soc. Dalton, Trans.* **1977**, 2095 (1977).
385. *Chem. Eng. News* **Sep. 17**, 57 (1962).
386. M. C. Thompson and D. H. Busch, *J. Am. Chem. Soc.* **84**, 1762 (1962).
387. M. C. Thompson and D. H. Busch, *J. Am. Chem. Soc.* **86**, 3651 (1964).
388. G. R. Brubaker and D. H. Busch, *Inorg. Chem.* **5**, 2114 (1966).
389. M. S. Elder, G. M. Prinz, P. Thornton, and D. H. Busch, *Inorg. Chem.* **7**, 2426 (1968).
390. J. C. Previdi and R. A. Krause, *Inorg. Chem.* **15**, 462 (1976).
391. N. B. Egen and R. A. Krause, *J. Inorg. Nucl. Chem.* **31**, 127 (1969).
392. F. L. Urbach and D. H. Busch, *Inorg. Chem.* **12**, 408 (1973).
393. N. W. Alcock and P. A. Tasker, *J. Chem. Soc. Chem. Commun.* **1972**, 1239 (1972).
394. L. F. Lindoy and D. H. Busch, *Inorg. Nucl. Chem. Lett.* **5**, 525 (1969).
395. R. W. Hay, A. L. Galyer, and G. A. Lawrance, *J. Chem. Soc. Dalton Trans.* **1976**, 939 (1976).
396. R. W. Hay, G. A. Lawrance, and U. R. Stone, *J. Chem. Soc. Dalton Trans.* **1976**, 942 (1976).
397. F. Vogtle and L. Schunder, *Chem. Ber.* **102**, 2677 (1969).
398. N. A. Kolesnikov and V. F. Borodkin, *Izv. Vyssh. Ucheb. Zaved. Khim. Khim. Teknol* **15**, 880 (1972) [Chem. Abst. **77**, 108911b].
399. N. B. Tucker and E. E. Reid, *J. Am. Chem. Soc.* **55**, 775 (1933).
400. L. A. Ochrymowycz, C. P. Mak, and J. D. Michna, *J. Org. Chem.* **39**, 2079 (1974).
401. W. Rosen and D. H. Busch, *Chem. Commun.* **1969**, 148 (1969).
402. K. Travis and D. H. Busch, *Chem. Commun.* **1970**, 1041 (1970).
403. P. H. Davis, L. K. White, and R. L. Belford, *Inorg. Chem.* **14**, 1753 (1975).
404. M. D. Glick, D. P. Gavel, L. L. Diaddario, and D. B. Rorabacher, *Inorg. Chem.* **15**, 1190 (1976).
405. E. R. Dockal, L. L. Diaddario, M. D. Glick, and D. B. Rorabacher, *J. Am. Chem. Soc.* **99**, 4530 (1977).
406. T. E. Jones, L. I. Zimmer, L. L. Diaddario, D. B. Rorabacher, and L. A. Ochrymowycz, *J. Am. Chem. Soc.* **97**, 7163 (1975).
407. R. E. DeSimone and M. D. Glick, *J. Am. Chem. Soc.* **97**, 942 (1975).
408. R. E. DeSimone and T. M. Tighe, *J. Inorg. Nucl. Chem.* **38**, 1623 (1976).
409. R. E. DeSimone and M. D. Glick, *J. Am. Chem. Soc.* **98**, 762 (1976).
410. G. N. Schrauzer, R. K. Y. Ho, and R. P. Murillo, *J. Am. Chem. Soc.* **92**, 3508 (1970).
411. T. A. DelDonno and W. Rosen, *J. Am. Chem. Soc.* **99**, 8051 (1977).
412. S. M. Nelson, P. Bryan, and D. H. Busch, *Chem. Commun.* **1966**, 641 (1966).
413. S. M. Nelson and D. H. Busch, *Inorg. Chem.* **8**, 1859 (1969).
414. W. M. Reiff, C. J. Long, and W. A. Baker, *J. Am. Chem. Soc.* **90**, 6347 (1968).
415. E. Fleischer and S. Hawkinson, *J. Am. Chem. Soc.* **89**, 720 (1967).

416. M. D. Alexander, A. Van Heuvelen, and H. G. Hamilton, Jr., *Inorg. Nucl. Chem. Lett.* **6**, 445 (1970).
417. A. Van Heuvelen, M. D. Lundeen, H. G. Hamilton, and M. D. Alexander, *J. Chem. Phys.* **50**, 489 (1969).
418. M. G. B. Drew, A. H. bin Othman, P. D. A. McIlroy, and S. M. Nelson, *J. Chem. Soc. Dalton Trans.* **1975**, 2507 (1975).
419. M. G. B. Drew, A. H. bin Othman, W. E. Hill, P. D. A. McIlroy, and S. M. Nelson, *Inorg. Chim. Acta* **12**, L25 (1975).
420. M. G. B. Drew, A. H. bin Othman, P. McIlroy, and S. M. Nelson, *Acta Crystallogr.* **32B**, 1029 (1976).
421. M. G. B. Drew, A. H. bin Othman, S. G. McFall, and S. M. Nelson, *J. Chem. Soc. Chem. Commun.* **1975**, 818 (1975).
422. M. G. B. Drew, A. H. bin Othman, S. G. McFall, P. D. A. McIlroy, and S. M. Nelson, *J. Chem. Soc. Dalton Trans.* **1977**, 438 (1977).
423. M. G. B. Drew, A. H. bin Othman, S. G. McFall, and S. M. Nelson, *J. Chem. Soc. Chem. Commun.* **1977**, 558 (1977).
424. S. M. Nelson, S. G. McFall, M. G. B. Drew, A. H. bin Othman, and N. B. Mason, *J. Chem. Soc. Chem. Commun.* **1977**, 167 (1977).
425. S. M. Nelson, S. G. McFall, M. G. B. Drew, and A. H. bin Othman, *J. Chem. Soc. Chem. Commun.* **1977**, 370 (1977).
426. Z. P. Haque, D. C. Liles, M. McPartlin, and P. A. Tasker, *Inorg. Chim. Acta* **23**, L21 (1977).
427. M. C. Rakowski, M. Rycheck, and D. H. Busch, *Inorg. Chem.* **14**, 1194 (1975).
428. D. H. Cook, D. E. Fenton, M. G. B. Drew, S. G. McFall, and S. M. Nelson, *J. Chem. Soc. Dalton Trans.* **1977**, 446 (1977).
429. M. G. B. Drew, A. H. bin Othman, S. G. McFall, P. D. A. McIlroy, and S. M. Nelson, *J. Chem. Soc. Dalton Trans.* **1977**, 1173 (1977).
430. L. F. Lindoy, D. H. Busch, and V. L. Goedken, *J. Chem. Soc. Chem. Commun.* **1972**, 683 (1972).
431. L. F. Lindoy and D. H. Busch, *Inorg. Chem.* **13**, 2494 (1974).
432. K. Hendrick, R. W. Matthews, and P. A. Tasker, *Inorg. Chim. Acta* **25**, L31 (1977).
433. J. deO. Cabral, M. F. Cabral, M. G. B. Drew, S. M. Nelson, and A. Rodgers, *Inorg. Chim. Acta* **25**, L77 (1977).
434. N. W. Alcock, D. C. Liles, M. McPartlin, and P. A. Tasker, *J. Chem. Soc. Chem. Commun.* **1974**, 727 (1974).
435. D. H. Cook and D. E. Fenton, *Inorg. Chim. Acta* **25**, L95 (1977).
436. U. Casellato, M. Vivaldi, and P. A. Vigato, *Inorg. Nucl. Chem. Lett.* **10**, 437 (1974).
437. V. W. Day, T. J. Marks, and W. A. Wachter, *J. Am. Chem. Soc.* **97**, 4519 (1975).
438. D. St. C. Black and I. A. McLean, *Inorg. Nucl. Chem. Lett.* **6**, 675 (1970).
439. C. Griggs, M. Hasan, K. F. Henrick, R. W. Matthews, and P. A. Tasker, *Inorg. Chim. Acta* **25**, L29 (1977).
440. D. E. Fenton, D. H. Cook, and I. W. Nowell, *J. Chem. Soc. Chem. Commun.* **1977**, 274 (1977).
441. D. E. Fenton, D. H. Cook, I. W. Nowell, and P. E. Walker, *J. Chem. Soc. Chem. Commun.* **1977**, 623 (1977).
442. D. St. C. Black and I. A. McLean, *Chem. Commun.* **1968**, 1004 (1968).
443. D. St. C. Black and I. A. McLean, *Tetrahedron Lett.* **1969**, 3961 (1969).
444. D. St. C. Black and I. A. McLean, *Aust. J. Chem.* **24**, 1401 (1971).
445. L. F. Lindoy and D. H. Busch, *Chem. Commun.* **1968**, 1589 (1968).
446. L. F. Lindoy and D. H. Busch, *J. Am. Chem. Soc.* **91**, 4690 (1969).
447. E. B. Fleischer and P. A. Tasker, *Inorg. Nucl. Chem. Lett.* **6**, 349 (1970).

448. P. A. Tasker and E. B. Fleischer, *J. Am. Chem. Soc.* **92**, 7072 (1970).
449. R. J. Gene, M. R. Snow, J. Springborg, A. J. Herlt, A. M. Sargeson, and D. Taylor, *J. Chem. Soc. Chem. Commun.* **1976**, 285 (1976).
450. U. Casellato, M. Vidali, and P. A. Vigato, *Coord. Chem. Rev.* **23**, 31 (1977).
451. N. H. Pilkington and R. Robson, *Aust. J. Chem.* **23**, 2225 (1970).
452. B. F. Hoskins and G. A. Williams, *Aust. J. Chem.* **28**, 2607 (1975).
453. B. F. Hoskins, N. J. McLeod, and H. A. Schaap, *Aust. J. Chem.* **29**, 515 (1976).
454. B. F. Hoskins, R. Robson, and G. A. Williams, *Inorg. Chim. Acta* **16**, 121 (1976).
455. B. F. Hoskins and G. A. Williams, *Aust. J. Chem.* **28**, 2593 (1975).
456. H. Okawa and S. Kida, *Inorg. Nucl. Chem. Lett.* **7**, 751 (1971).
457. H. Okawa and S. Kida, *Bull. Chem. Soc. Japan* **45**, 1759 (1972).
458. R. R. Gagné, C. A. Koval, and T. J. Smith, *J. Am. Chem. Soc.* **99**, 8367 (1977).
459. H. Okawa, M. Honda, and S. Kida, *Chem. Lett.* **1**, 1027 (1972).
460. H. Okawa, T. Tokii, Y. Muto, and S. Kida, *Bull. Chem. Soc. Jpn.* **46**, 2464 (1973).
461. R. W. Stotz and R. C. Stoufer, *J. Chem. Soc. Chem. Commun.* **1970**, 1682 (1970).
462. W. Rosen, *Inorg. Chem.* **10**, 1832 (1971).
463. D. R. Boston and N. J. Rose, *J. Am. Chem. Soc.* **90**, 6859 (1968).
464. D. R. Boston and N. J. Rose, *J. Am. Chem. Soc.* **95**, 4163 (1973).
465. G. A. Zakrzewski, C. A. Ghilardi, and E. C. Lingafelter, *J. Am. Chem. Soc.* **93**, 4411 (1971).
466. S. C. Jackels, S. S. Dierdorf, N. J. Rose, and J. Zektzer, *J. Chem. Soc. Chem. Commun.* **1972**, 1291 (1972).
467. S. C. Jackels and N. J. Rose, *Inorg. Chem.* **12**, 1232 (1973).
468. J. E. Parks, B. E. Wagner, and R. H. Holm, *J. Am. Chem. Soc.* **92**, 3500 (1970).
469. J. E. Parks, B. E. Wagner, and R. H. Holm, *Inorg. Chem.* **10**, 2472 (1971).
470. M. R. Churchill and A. H. Reis, Jr., *J. Chem. Soc. Chem. Commun.* **1970**, 879 (1970).
471. M. R. Churchill and A. H. Reis, Jr., *J. Chem. Soc. Dalton Trans.* **1973**, 1570 (1973).
472. E. Larsen, G. N. LaMar, B. E. Wagner, J. E. Parks, and R. H. Holm, *Inorg. Chem.* **11**, 2652 (1972).
473. M. R. Churchill and A. H. Reis, Jr., *J. Chem. Soc. Chem. Commun.* **1971**, 1307 (1971).
474. I. I. Creaser, J. MacB. Harrowfield, A. J. Herlt, A. M. Sargeson, J. Springborg, R. J. Gene, and M. R. Snow, *J. Am. Chem. Soc.* **99**, 3181 (1977).

Thermodynamics and Kinetics of Cation–Macrocycle Interaction

John D. Lamb, Reed M. Izatt, James J. Christensen, and Delbert J. Eatough

1. Introduction

The intense interest recently generated in the so-called macrocyclic ligands finds roots in the extraordinary stability of their cation complexes and their peculiar ability to selectively bind certain cations in preference to others. This chapter deals specifically with the thermodynamic and kinetic quantities associated with the interactions of cations with these ligands in solution. The thermodynamics of these reactions has received considerably more attention than has their kinetics and, consequently, the bulk of the chapter will deal with the former. Of particular interest are those factors that are responsible for and that influence the remarkable binding strength and selectivity of the wide variety of ligands studied thus far. In this chapter we will summarize work concerning these factors which has been published through 1976. One of these factors, the macrocyclic effect, has aroused particular interest and will be analyzed in detail.

Our present knowledge of macrocyclic ligand–cation interactions developed from three approaches during the past two decades. These approaches together with corresponding major review articles or textbooks describing them follow: (i) *Antibiotic macrocycles.* Naturally occurring macrocycles were shown to be capable of actively transporting metal ions across membranes, beginning with the antibiotic valinomycin in 1964.[1] Since that time an extensive literature has arisen, which describes the complexation and carrier

John D. Lamb, Reed M. Izatt, James J. Christensen, and Delbert J. Eatough · Departments of Chemistry and Chemical Engineering and the Thermochemical Institute, Brigham Young University, Provo, Utah 84602.

abilities of many such naturally occurring ligands, including enniatin B, nonactin, and the other macrotetrolides, monensin, gramicidin and many analogs of these compounds.[2,3] In addition, large molecules like cyclodextrin have been shown to complex neutral organic molecules.[4] (ii) *Porphyrin analogs.* Tetradentate macrocyclic ligands with nitrogen donors originated simultaneously in two different laboratories in the early 1960s[5] and have undergone intense investigation due to their similarity to the naturally occurring corrin and porphyrin rings found in metalloenzymes and elsewhere. Numerous metal complexes have been reported for the wide variety of these ligands that have been synthesized.[6] Most studies have involved the crystalline complexes, and consequently little is known of their solution chemistry in terms of stability or kinetics of complexation. (iii) *Macrocyclic polyethers.* The cyclic polyethers, which in many ways resemble the antibiotic ligands, were first reported in 1967 by Pedersen,[7] who noted their unusual affinity for alkali metal ions and their cation selectivity characteristics. These crown compounds, as Pedersen labeled them, and their thia and aza derivatives have since enjoyed considerable interest in terms of their complexation properties in solution not only with the alkali cations but also with other univalent and bivalent metal ions and with simple and substituted protonated amines.[8–14] Certain chiral derivatives of the crowns are able to distinguish between enantiomers of optically active species, making them of particular biological interest.[15–18] Such selective binding of organic cations by these compounds has been treated under the title "host–guest chemistry" by Cram and co-workers. Other crown compounds have been incorporated onto polymer chains for application in cation separations,[19,20] and their use in liquid membranes has been investigated.[21] A logical extension of crown ether chemistry is found in the cryptand compounds or macrobicyclic ligands of Lehn and co-workers.[22] These compounds differ from the others in that three bridges, rather than two as in the case of crown ethers, meet at common atoms to form a central cavity into which cations may be incorporated.

Examples of the three types of macrocyclic ligands are given in Table 13, which is a general table of thermodynamic values. The structures of many macrocycles also appear in this table, to which repeated reference will be made throughout the chapter. Additional reviews covering a broad range of the chemistry of these compounds are also available.[23,24]

2. Parameters Determining Cation Selectivity and Complex Stability

Probably the most intriguing characteristic of macrocyclic compounds is their ability to selectively bind certain cations in preference to others which may be present in solution. Indeed, much of the current work in this field involves seeking out new ligands which will be specific for particular cations.

Various characteristics of the ligand, cation, and reaction medium are considered to be responsible for selectivity, and these may be rationalized in terms of a Born–Haber cycle, to which reference has been made previously.[25] Selectivity patterns have been well characterized particularly for the macrocyclic compounds of the crown and cryptand classes, and these have been discussed in detail.[8,9,16,23,24,26] In the following sections we shall examine the parameters that influence the selectivity and binding properties of macrocycles for which thermodynamic data are available.

2.1. Relative Sizes of Cation and Ligand Cavity

2.1.1. Cyclic Polyethers

From the inception of crown ether investigation it was recognized that these ligands complex most strongly those metal ions whose ionic crystal radius best matches the radius of the cavity formed by the ring upon complexation.[7] Although relatively few thermodynamic data are available, this same observation has been made for two classes of macrocycles, which shall be referred to as the cyclic tetraamine[27,28] and tetrathia[29] ligands. Examples of these two ligand types are compounds **47–49** and **61–65** in Table 13. In considering size relationships, it is reasoned that bond energies between ligand and cation will be greatest when all donor groups can fully participate. If the macrocyclic ring is too large, the metal ion will "fall through" the cavity (in this simplistic picture we consider the ligand inflexible) or if the ring is too small the metal ion can only saddle up to it, not fit inside. Figure 1 illustrates how the stability constants in the case of 18-crown-6 (Table 13, **12**) vary with the ratio of cation to cavity diameters.

One serious question which the size relationship approach raises is: how does one determine the "size" of the cavity in the ligand? Originally different types of molecular models were used to find a range of values for the cavity diameter. As demonstrated in Table 1, the difference is rather large between the Corey–Pauling–Kolton (CPK) models and the Fisher–Hirschfelder–Taylor (FHT) models used. Since that time X-ray crystallographic studies have made it possible to accurately determine the positions of ring atoms in the complexed and/or uncomplexed form of the ligand. These interatomic distances may be adjusted by subtracting the van der Waals radii of the donor atoms to give what is probably a much better estimation of the correct ring cavity size.[30] It may be noted from the table that the CPK models give ring sizes most consistent with the X-ray results. By comparison to this determination of cavity dimension, the small variation between values for cation radii determined by different methods is essentially insignificant. Metal cation radii have been taken from Shannon and Prewitt[31] for oxides of coordination number 6 to produce the plot in Figure 1.

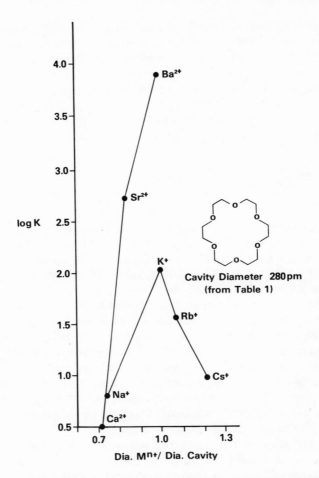

FIGURE 1. Selectivity of 18-crown-6: log K values for reaction of 18-crown-6 with metal cations in H_2O vs. ratio of cation diameter to 18-crown-6 cavity diameter. Value for Ca^{2+} reported <0.5.[9]

As is evident from Figure 1, the maximum stability for complexes of 18-crown-6 with both the alkali and alkaline earth cations occurs at a metal ion to cavity diameter ratio of unity. It is fortuitous that in each of these series there is a metal ion of exactly the correct size to fit the ring (K^+ and Ba^{2+}). Other cations for which data are known, namely, Ag^+, Tl^+, Hg^{2+}, and Pb^{2+}, also seem to follow this size rule, showing greater stability for complexes with cations of diameter more closely matching that of the ligand cavity.[32] Table 2 illustrates that in the case of 18-crown-6, increased stability of the complexes of K^+ and Ba^{2+} over those of other ions in the series is

TABLE 1. Cavity Diameters for Various Cyclic Polyethers

Molecule	Cavity diameter (pm)		
	CPK models[a]	FHT models[a]	From X-ray structure[b]
12-Crown-4	120	150	
Benzo-15-crown-5	170	220	172 from Na$^+$ complex
			184 from K$^+$ complex
18-Crown-6	260	320	274–286 from K$^+$ complex
			267–284 from Rb$^+$ complex
			273–285 from Cs$^+$ complex

[a] From reference 7.
[b] From reference 30.

TABLE 2. Log K, ΔH, and $T\Delta S$ for the Formation of 18-Crown-6 Complexes with Metal Ions in H_2O at 25°C[25,32]

Ion	log K	ΔH (kcal mol^{-1})	$T\Delta S$ (kcal mol^{-1})
Na$^+$	0.8	−2.25	−1.16
K$^+$	2.03	−6.21	−3.40
Rb$^+$	1.56	−3.82	−1.7
Cs$^+$	0.99	−3.97	−2.6
Sr^{2+}	2.72	−3.61	0.1
Ba^{2+}	3.87	−7.58	−2.3

largely due to the enthalpy term. This undoubtedly corresponds to the greater electrostatic bond energy for those ions that better fit the ligand cavity.

For cyclic polyethers smaller than 18-crown-6, little work has been reported except in the case of 15-crown-5 and its derivatives (Table 13, **9–11**). As demonstrated in Figure 2a, almost no cation selectivity is seen for 15-crown-5 in H_2O, which is consistent with its cavity being too small even for the Na$^+$ ion. For benzo-15-crown-5 and cyclohexo-15-crown-5, trends are difficult to distinguish because of the formation of 2:1 complexes. For crowns larger than 18-crown-6, complexes of variable stoichiometry also complicate the data, as seen in Figure 2b. These large crown ethers, like the small crowns, are not as selective as 18-crown-6. Although it is true in every case for the larger crowns (except dibenzo-27-crown-9) that Na$^+$ is much less strongly complexed than any other ion, an expected trend of successively stronger binding of larger ions by larger ligands is not observed. Where stability peaks occur, as with dibenzo-27-crown-9 and dibenzo-30-crown-10, the strongest complexes are formed with K$^+$ or Rb$^+$, which are considerably smaller than

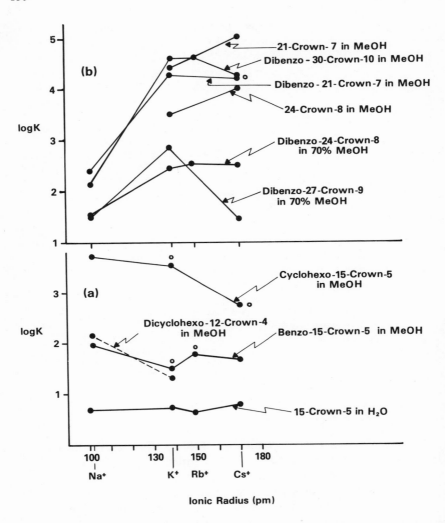

FIGURE 2. Selectivity of cyclic polyethers of various sizes: log K values (from Table 13) for reaction of several crown ethers with alkali metal ions vs. cation radius. (a) Crowns smaller than 18-crown-6. (b) Crowns larger than 18-crown-6. Data points labeled ○ indicate 2:1 complex formed.

the cavities of either of these two rings assuming a planar configuration for the macrocycles. These results are not unexpected when it is considered that even the cavity of 21-crown-7 is more than spacious enough to accommodate the large Cs$^+$ ion. Thus for the crowns larger than 18-crown-6 which are too large for any of the cations in the alkali or alkaline earth series, cavity size ceases to be of major importance in determining selectivity and other factors become increasingly significant. Since, on the other hand, 15-crown-5 is too small to comfortably accommodate even Na^+, it may be concluded that for the cyclic polyethers correlation of selectivity with cavity size is restricted essentially to 18-crown-6 and possibly 21-crown-7 and their substituted analogs. It is fair to say that the data support the view that in general smaller

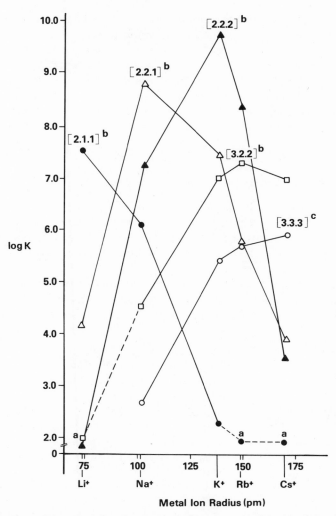

FIGURE 3. Selectivity of cryptands: log K values for reaction of several cryptands with alkali metal cations vs. cation radius.[33] Data points: *a*—value reported <2.0; *b*—in 95% MeOH; *c*—in MeOH.

macrocycles bind smaller cations better than larger ones and vice versa. However, the severe limitations to this generalization must not be overlooked.

2.1.2. Macrobicyclic Ligands

Much more striking correspondence of cavity size to complex stability is found in the selectivity of the cryptand or macrobicyclic ligands of Lehn and co-workers. Figure 3 illustrates how in proceeding through a series of these ligands of ever increasing size, each of the alkali metal ions is preferentially bound in its turn.[33] Table 3 demonstrates that the cation which is located at each selectivity peak in Figure 3 has an ionic radius very close to the ligand cavity radius (from CPK models) at least for cryptands [2.1.1], [2.2.1], and [2.2.2] (Table 13, **37**, **38**, and **40**). Here again, selectivity becomes less pronounced for the larger macrobicycles, most likely owing to ligand flexibility. The shape of the selectivity curves is primarily enthalpic in origin, as in the case of the cyclic polyethers.[26]

2.2. Arrangement of Ligand Binding Sites

2.2.1. Synthetic Macrocycles with Metal Cations

It would be overly simplistic to think of macrocycle–cation binding as the insertion of a metal ion into a preformed rigid cavity or the fastening of an organic cation onto a fixed template. Even the small 15-crown-5 ring is fairly flexible with respect to the orientation of its donor groups in space, while larger ligands are free to fold in such a way as to leave no real cavity at all. Indeed, the hydrophobic carbon backbone together with the mutual repulsion of the basic donor groups militate against the assumption of a "ready" cavity configuration in polar solvents. X-ray crystallographic work

TABLE 3. Cryptands and Their Preferred Alkali
Metal Ions: Cavity and Ionic Radii

Cryptand	Cavity radius[a] (pm)	Ion of maximum log K	Ionic radius[b] (pm)
[2.1.1]	80	Li^+	74
[2.2.1]	110	Na^+	102
[2.2.2]	140	K^+	138
[3.2.2]	180	Rb^+	149
[3.3.3]	240	Cs^+	170

[a] From CPK models.[33]
[b] From reference 31.

has thrown considerable light on the structures of these compounds, although Grell *et al.*[3] correctly noted the hazards of assuming that structures in the crystal can be taken to represent structures of species in solution. Such structural studies show that uncomplexed cyclam in the diprotonated form (Figure 4, structure 1) exists in the same conformation as in its Cu(II) or Ni(II) complexes (Figure 4, structure 2). This is likely due to the intramolecular H bonding found in the free ligand.[34] The tetrathia equivalent of cyclam, on the other hand, has all donor groups pointed outward in the free ligand (Figure 4, structure 3) while its Cu(II) and Ni(II) complexes are similar to those of cyclam with donor atoms facing into the ring (Figure 4, structure 4).[35–37]

Cryptand ligands can assume any of three conformations according to the configuration of the two nitrogen bridgeheads—namely, endo–endo,

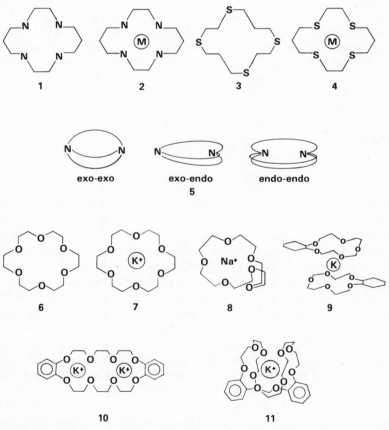

FIGURE 4. Simplified structures of several macrocycles and their cation complexes.

exo–endo, or exo–exo (Figure 4, structure 5). X-ray studies show the free [2.2.2] cryptand to exist in the endo–endo conformation in the crystal. In all the known cryptate complexes, the configuration is also endo–endo with the metal ion inside the cavity.[38] Uncomplexed 18-crown-6 exists in an open cavity conformation (Figure 4, structure 6) with its oxygen atoms in a plane. The cavity is slightly more oblong than its similar K^+ complex (Figure 4, structure 7).[39] The ligand folds somewhat around the smaller Na^+ ion (Figure 4, structure 8).[40]

The effect of ligand flexibility is well illustrated by the structures of the complexes of various crown ethers with K^+. With the small benzo-15-crown-5, a 2:1 complex is formed having the structure given in Figure 4, structure 9. With dibenzo-24-crown-8, a 1:2 complex forms (Figure 4, structure 10) and with dibenzo-30-crown-10 the ligand wraps itself completely around the ion (Figure 4, structure 11).[41] This ability of the larger, more flexible crown ethers to accommodate more than one metal ion or to arrange themselves so as to conform to cation size very much reduces their selectivity.

Macrobicyclic ligands are more rigid than cyclic polyethers over a broader range of cavity sizes, accounting for their wider ranges of selectivity. Above the size of the [2.2.2] cryptand, however, flexibility again takes its toll (see Figure 3). For this reason, Lehn *et al.*[33] conclude that rigid ligands display "peak selectivity," while flexible ligands exhibit "plateau selectivity." It must be kept in mind, however, that ligands that are not flexible but that are too small to accommodate the metal ion may form 2:1 complexes which likewise greatly alter selectivity patterns.

2.2.2. Naturally Occurring Macrocycles with Metal Cations

Studies of the interaction of naturally occurring macrocycles with metal cations have been extensively reviewed.[2,3] Unlike the synthetic macrocycles, antibiotics like valinomycin (Table 13, **1**) exhibit some peak selectivity despite their very large size. Valinomycin has been shown to fold upon itself by means of six intramolecular H bonds to provide a nearly octahedral coordination shell into which K^+ or Rb^+ can fit well. The coordinating carbonyl oxygens are held in complexing position relatively well even in the case of the free ligand in nonpolar solvents. This "closed" conformation of the ligand has been found in the solid K^+ complex, and there is good spectral evidence that the same structure holds in solution. With smaller cations the ligand is unable to contract further; consequently stability is greatly reduced because the cation is unable to hold the ligand tight to itself. Flexibility is therefore a crucial factor influencing the binding of cations to this ligand.

The macrotetrolides like nonactin (Table 13, **5**) are also K^+ specific, though certainly to a lesser extent. Unlike valinomycin, they form simple "wrapped" structures much like that of dibenzo-30-crown-10 cited earlier, being 8-coordinate via four carbonyl and four ether oxygens. The differing ability of these two types of complexing groups to approach different-sized metal ions may be critical to understanding the macrotetrolide selectivity.

2.2.3. Binding of Organic Cations

Ammonium ion and protonated organic amine cations complex with crown ethers probably by way of hydrogen bonds to the ether oxygen atoms. The smaller 15-crown-5 offers a more favorable arrangement of oxygen donors than does the larger 18-crown-6, which probably explains the greater affinity of the former for NH_4^+.[32,42] Likewise, it has been postulated that the greater affinity of nonactin for NH_4^+ over that of valinomycin is probably due to the availability of a tetrahedral arrangement of oxygen donors in the former compared to an octahedral arrangement in the latter.[43] Cram and co-workers have also noted the importance of donor arrangement in complexing *t*-butylammonium ions with cyclic polyethers and their derivatives in chloroform. They find the complexes of rigid cyclic "hosts" to be more stable than those of their flexible open-chain counterparts by factors of several powers of 10.[16]

2.3. Type and Charge of Cation

2.3.1. Alkali and Alkaline Earth Cations

The binding of alkali and alkaline earth metal ions to macrocyclic ligands may be considered to be electrostatic in nature. Ligand basic groups tend to distribute themselves as evenly as possible over what is essentially a sphere of positive charge. Wide variation in coordination number and geometry is thus possible. For example, K^+ is 10-coordinate in its *bis*-benzo-15-crown-5 complex, 6-coordinate planar in its 18-crown-6 complex, with possible additional solvation, 8-coordinate in its dibenzo-24-crown-8 complex, presumably 9-coordinate in its dibenzo-27-crown-9 complex, 10 coordinate in its dibenzo-30-crown-10 complex, 6-coordinate octahedral in its valinomycin complex, and 8-coordinate cubic in its nonactin complex.[30] For these two groups of cations there are no real stereochemical requirements; all that is necessary is that the ligand provide an electronically basic environment to replace all or part of the cation's solvation shell.

The size of alkali and alkaline earth metal ions is primarily responsible for their complexing characteristics. Smaller ions like Li^+ are so strongly solvated that considerably more energy must be expended in the desolvation step than for larger ions like Cs^+. Often this desolvation energy is not replaced in the complexation step because of ligand–solvent characteristics (see Section 2.7). On the other hand, the larger cations are unable to attract and organize the ligand as well as smaller ones. These two effects cause the ΔH values and thus the stabilities of the complexes of cations at the two extremes of the size spectrum to be smaller than those of intermediate size. Thus, selectivity peaks for K^+ and Ba^{2+} cations which are of "intermediate" size, are generally higher than those for larger or smaller cations (e.g., see Figure 3). By the same token, large dipositive ions often have higher stability constants than monopositive ions of similar size, whereas the opposite is true when comparing small cations of differing charge. For example, in Figure 1 is illustrated the fact that while 18-crown-6 prefers Na^+ over similarly sized Ca^{2+}, among larger cations it prefers Ba^{2+} over K^+. These effects resulting from cation size are made manifest in both the enthalpy and entropy of the complexation reaction as described by Izatt *et al.*[8,32]

Of particular interest in the application of alkaline earth selectivity is the recent reporting of the successful separation of ^{40}Ca and ^{44}Ca using dicyclohexo-18-crown-6.[44] This separation was made by means of a multistage exchange system between the cyclic polyether dissolved in chloroform or methylene chloride and concentrated aqueous $CaCl_2$.

2.3.2. Other Metal Cations

Metal ions found in other parts of the Periodic Table have not been as widely studied as have the alkali and alkaline earth cations. Some have affinities for donors other than oxygen and are more prone to classical ligand interactions than those previously mentioned. For instance, Ag^+ and Hg^{2+} show a preference for sulfur donors.[45] Tl^+ and Pb^{2+} are quite similar to the alkali and alkaline earth cations in their macrocycle binding properties and tend to form complexes even more stable than those of other metal ions of similar size.[32] Ni^{2+} has square planar stereochemical requirements, which make the cyclic tetraamines ideal ligands for this cation. The study of Ni^{2+} and Cu^{2+} has been limited almost exclusively to reactions with these ligands and their tetrathia analogs (Table 13).

The reactions of lanthanide and actinide metal cations with macrocycles have received little attention. Complexes have been prepared in the solid for all of the lanthanide nitrates with benzo-15-crown-5 from acetone solution and for some of these with dibenzo-18-crown-6 from acetonitrile.[46,47] However, controversy exists over whether stoichiometric complexes of the latter ligand are formed with the higher atomic weight lanthanides. The crystal structure of *cis-syn-cis*-dicyclohexo-18-crown-6 with lanthanum(III) nitrate has been determined, showing the metal ion to be bound approximately in the plane of the ligand by the six ether oxygens and in addition by three bidentate nitrate anions.[48]

Care must be taken in assuming real complexes from the synthesis of solids before confirmation by X-ray crystallography as illustrated by an original report of a complex between 18-crown-6 and UO_2^+ which was later revised because cocrystallization rather than complex formation was responsible.[49–51]

2.3.3. Organic Cations

Considerable work has been dedicated to synthesizing new macrocycles with the particular goal of selectively binding organic cations of differing structures and/or chiralities. Much that has been accomplished in this field may be attributed to Cram and co-workers.[16–18] These authors define the enantiomer distribution coefficient (EDC) of a particular host as the ratio of the distribution coefficient between a water phase containing a racemic cation mixture and a chloroform phase containing the host macrocycle of the cation enantiomer that forms the most stable complex to the distribution coefficient of the enantiomer that forms the least stable complex. One of the highest EDC values reported was 12, found for the interaction of the chiral crown shown in Figure 5 with racemic phenylglycine methyl ester. Such ligands have also been shown to complex metal ions in chloroform.

In a study of the binding of various amines with 18-crown-6 in methanol, Izatt *et al.*[11] report that formation constants decrease in the order NH_4^+, $RNH_3^+ > R_2NH_2^+ > R_3NH^+$. In the case of the organic amines, this stability order is identical to the earlier observed permeability sequence for protonated organic amines in glyceryl dioleate bilayers doped with valinomycin, nonactin, or gramicidin.[52] The log K trend is primarily of enthalpic origin, the $T\Delta S$ term being essentially constant for the systems studied (Table 13, **12**). These results are explained in terms of the number of hydrogen atoms on the protonated organic amine available for H bonding to the crown oxygens. Cation steric factors are also important, just as in the case of the host–guest interactions investigated by Cram.

2.4. Type of Donor Atom

Substitution of nitrogen for an ether oxygen in a crown ether ring reduces the affinity of the ligand for the alkali and alkaline earth metal ions while

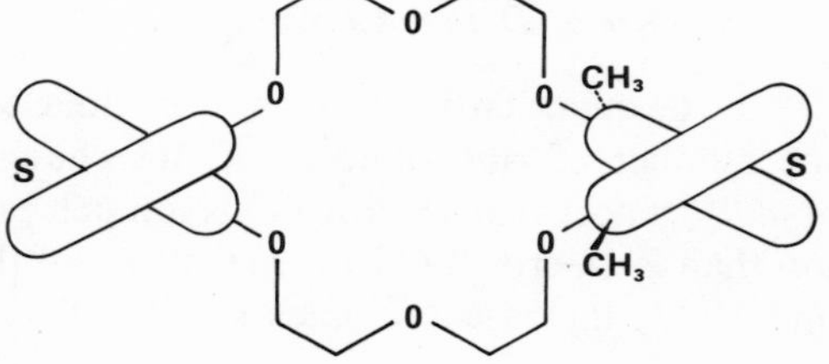

FIGURE 5. A chiral crown ether used in the separation of enantiomers of optically active protonated amines. Each elongated oval represents a naphtho group.[16]

producing little change in or even enhancing the stability of the complexes of post-transition-metal ions of similar sizes, e.g., Ag^+.[53] This effect is illustrated by the stabilities of complexes formed by this metal ion with 18-crown-6 and its aza- and diaza-substituted analogs listed in Table 13. Substitution of sulfur for oxygen produces even more dramatic results—the stabilities of Ag^+ and Hg^{2+} complexes are enhanced; those of the other metal ions studied, including Tl^+ and Pb^{2+}, reduced.[45] Were they to be turned into the ring, sulfur donor atoms would serve to decrease the size of the cavity, thus changing its cavity size characteristics. However, X-ray studies show that even for crown-4, crown-5, and crown-6 rings, free ligands direct their sulfur atoms away from the cavity.[54] It has been shown recently that the mercury atom in the $HgCl_2$ complex with 1,4-dithia-18-crown-6 binds only the two sulfur atoms and lies outside the ring.[55] Thus, binding of sulfur-containing crowns to Ag^+ and Hg^{2+} may not be of the inclusion type, which could account for the high incidence of 2:1 (ligand to metal) complexes in these systems. However, no structures of complexes of sulfur-substituted crown ethers with simple metal ions not covalently bound to other ligands have been reported.

Substitution of sulfur for the four nitrogens of the tetramine ligands like cyclam greatly reduces their affinity for Cu^{2+} (compare Table 13, **49** and **63**). In both the tetramine and tetrathia cases, the metal ion has been shown to occupy the ligand cavity.[37]

Various donor effects have been investigated in the binding of *t*-butyl-ammonium ion to host molecules.[16] Substitution of one oxygen in 18-crown-6 by a pyridine nitrogen almost doubles the ligand's binding power to this cation. On the other hand, two or three pyridines evenly distributed around the ring reduce complex stability considerably. Aromatic furan oxygens provide very poor binding, as would be expected on the basis of their low basicity.

Cyclic polyether-esters have been synthesized and their complexing characteristics studied by Izatt *et al*.[56] In addition to ether oxygens, these compounds provide carbonyl oxygen atoms, which may be available for cation binding (Table 13, **20**, **21**). One of these compounds (**20**) shows the same unusual preference for K^+ over Ba^{2+} exhibited by valinomycin, which employs exclusively carbonyl oxygen atoms as its binding groups.

2.5. Number of Donor Atoms

Little quantitative work has been done to investigate the result of varying the number of ring donors without changing the size of the ring. Cram *et al*.[16] report that 18-crown-5 is a much poorer host for *t*-butylammonium ion than is 18-crown-6. In the reactions of thia-substituted crowns with Ag^+ and Hg^{2+}, increased complex stability is associated with increased number of sulfur atoms in the ring (see Table 13). The [2.2.2] cryptand produces complexes of far greater stability than its analog [2.2.C$_8$] (Table 13, **39**), in which one of the bridges is a simple carbon chain containing no ether oxygens.[33] Although ΔH data are not yet available for these systems, it is expected that these differences in stability result from larger heats of formation due to an increased number of sites of interaction, at least in the latter two cases. This reasoning would explain in part the greater stability of cryptate complexes over crown complexes, which has been shown to be primarily enthalpic in origin.[26] One of the most striking features of the [2.2.C$_8$] ligand is its marked selectivity for K$^+$ over Ba^{2+}, which corresponds to the unusual stability sequence of valinomycin noted earlier. This unusual selectivity sequence is attributed to the shielding of the complex from the solvent by the hydrophobic groups, which disallows the recouping of energy lost in desolvating the cation. Desolvation energy expended is greater in the dipositive case, resulting in destabilization of the complexes of the $+2$ ions.

2.6. Substitution on the Macrocyclic Ring

2.6.1. Benzo and Cyclohexo Derivatives

Dietrich and co-workers[57] have shown that addition of benzene rings to one or two bridges of the [2.2.2] cryptand alters the selectivity of this ligand considerably. The presence of one benzo substituent causes a rise in the stability of the Na$^+$ complex with a concurrent drop in the stabilities of the complexes of both K$^+$ and Ba^{2+} in 95% methanol. Addition of another benzene substituent into a second bridge causes the stability of the Ba^{2+} complex to drop slightly below that of K$^+$ (compare Table 13, **40**, **41**, and **42**). The same effect is observed in comparing 18-crown-6 to dibenzo-18-crown-6. In methanol, the formation constant of the Ba^{2+} complex of 18-crown-6 is larger than that of the K$^+$ complex by a factor of 10. Dibenzo-18-crown-6, on the other hand, displays the opposite preference, binding K$^+$ better than Ba^{2+} in methanol by approximately the same amount (compare Table 13, **12** and **19**). These results may be explained in terms of ligand bulkiness leading to the isolation of the cation from the solvent, which results in essentially the same effect as in the case of the [2.2.C$_8$] ligand noted in Section 2.5. It may also be theorized that the aromatic ring lends rigidity to the ligand and withdraws electrons from the basic oxygen donors, thus decreasing the strength of the metal–ligand interaction.

Substitution onto crown ethers of cyclohexo rather than benzo groups has a less dramatic effect on the stability of complexes or on cation selectivity. The ligand containing the aliphatic substituent is the more flexible of the two and thus more closely echoes the binding properties of the unsubstituted ring.[32] Notable variation between the cation binding stabilities in water of the two isomers of dicyclohexo-18-crown-6 (Table 13, **14** and **15**) is likely due to differences in solvation of their complexes resulting from different positioning of the bulky substituents.

2.6.2. Substituted Benzo Derivatives

Schori and Jagur-Grodzinski[58] have shown that the dinitro derivative of dibenzo-18-crown-6 (Table 4) has markedly reduced affinity for Na^+ as compared to the unsubstituted ligand in DMF. This is consistent with the electron-withdrawing character of the $-NO_2$ substituent, which may delocalize electrons from the oxygen donors. Conversely, substitution by $-NH_2$

TABLE 4. Formation Constants of Substituted Benzo-Crowns with Na^+

Crown	X	log K
(in DMF at 30°C[58])	H NH_2 NO_2	2.69 2.76 1.99
(in acetone at 25°C[59])	NH_2 CH_3 H Br COOH $COOCH_3$ CHO NO_2	3.91 3.60 3.54 3.3 3.21 3.09 3.05 2.65
	CH_3 H CHO $CONHC_4H_9$ NO_2	5.09 4.72 4.59 4.51 4.67

TABLE 5. Stability Constants of K^+-Macrotetrolide
Complexes in Acetonitrile[61]

Name	Structure	log K
Nonactin	$R_1 = R_2 = R_3 = R_4 = CH_3$	4.43
Monactin	$R_1 = R_2 = R_3 = CH_3, R_4 = C_2H_5$	4.78
Dinactin	$R_1 = R_2 = CH_3, R_3 = R_4 = C_2H_5$	5.24
Trinactin	$R_1 = CH_3, R_2 = R_3 = R_4 = C_2H_5$	5.44

results in a slight increase in complex stability consistent with the electron-donating character of this substituent. In a detailed study of this aromatic substituent effect, Ungaro et al.[59] find that 4′ substitution in benzo-15-crown-5 alters the stability constant of the Na^+ complex in acetone by a factor of 18, comparing the amino- to the nitro-substituted species (see Table 4). A Hammett correlation was found for the Na^+ complexes with the substituted benzo-15-crown-5 series, but not for the Na^+ or K^+ complexes with the benzo-18-crown-6 series. Selectivity for K^+ over Na^+ was much less for the $-NO_2$-substituted benzo-18-crown-6 than in the case of the corresponding $-CH_3$-substituted ligand. A similar observation was made by Pannell et al.[10] who studied the effect of aromatic substitution on the selectivity of dibenzo-18-crown-6 in methylene chloride. They found that the normal selectivity of this crown for K^+ over Na^+ was actually reversed with the addition of electron-withdrawing substituents of sufficient strength. From these results it appears that aromatic substituents can play a major role in determining both the strength of complexation and the selectivity toward cations of benzo-crowns due to their contribution to or detraction from donor basicity. This inductive effect described for aromatic substituents has also been invoked to explain the enhancement of complex stability in acetonitrile by nonaromatic substituents such as the series of macrotetrolides shown in Table 5.[61] Moreover, the number and position of methyl substituents has been reported as significant in determining the rate of complexation in water of substituted cyclic tetraamines with Cu^{2+} and Ni^{2+}.[62]

2.6.3. Multiple Crowns

Bourgoin et al.[63] have reported the interaction of a series of *bis*-benzo-15-crown-5 compounds with K^+ and NH_4^+ picrates in THF. The two crown moieties are connected by chains of varying length and structure. They find that in extending the connecting chain from two to eight methylene groups, complex stability with K^+ passes through a maximum at that position where the complexed crown moieties are parallel, the cation being sandwiched between. Oxygens in the connecting chain enhance complex stability probably by adding flexibility to the dimer.

2.6.4. Chiral Barriers

The addition of substituents to serve as chiral barriers is essential to the design of host molecules capable of enantiomeric differentiation. Simple chiral crowns have been synthesized by tetra substitution of 18-crown-6.[15,64] More elaborate structures, generally involving binaphthyl units, have also been prepared by Cram and co-workers.[16] In addition to providing a chiral environment, arms added to the host may serve to allow manipulation of the hydrophobic–lipophilic balance of the host or provide counterions for ionic guests. Complete resolution of amino acid esters by such optically active host compounds has been achieved.[16]

2.7. Solvent

In the complexation process, macrocycles must compete with solvent molecules for the cations in solution. As a result, variation of the solvent can produce significant changes in the apparent binding properties of these ligands. Specifically, solvents of low dielectric constant and solvating power lead to greater complex stability than those that tend to strongly solvate the cations. In addition, selectivity for certain cations over others may be altered according to the nature of the reaction medium.

2.7.1. Effect of Solvent on Stability

Frensdorff noted that stability constants for the reaction of cyclic polyethers with metal cations were three to four decades higher in methanol than in water.[53] The same effect is observed in the complexation reactions of cryptands, although the degree to which stability is enhanced for methanol over water covers a wider range.[33] The binding constants for K^+ with valinomycin and enniatin B in various methanol/water mixtures shown in Figure 6 demonstrates that the same trend is observed for the antibiotic macrocycles.[3] Similar solvent effects have been observed in comparing reactions in various other media.[59,65–67]

Izatt et al.[68,69] have studied the reaction of 18-crown-6 with cations in water/methanol solvents of varying composition. Their results, illustrated in

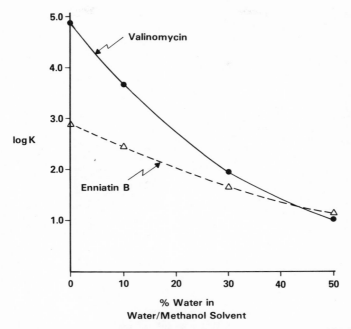

FIGURE 6. Variation of log K for formation of K^+ complexes of valinomycin and enniatin B with solvent composition.[3]

Figure 7, support the findings of Kauffman *et al.*[26] that the enhancement of stability in methanol over water is primarily of enthalpic origin. Indeed, both studies report that the difference between reaction entropies found in water and those found in methanol actually oppose this enhancement. The enthalpic stabilization is explained by the expenditure of less energy in the cation desolvation step in the solvent of lower dielectric constant.

Dielectric constant is not the only solvent parameter that influences the stability of macrocyclic complexes. Matsuura *et al.*[70] report that log K for the formation of alkali-metal-ion–dibenzo-18-crown-6 complexes increases in three solvents in the order log K in dimethylsulfoxide $<$ log K in dimethyl-formamide $<$ log K in propylene carbonate. However, dielectric constants of these solvents increase in a different order: DMF ($\varepsilon = 36.7$) $<$ DMSO ($\varepsilon = 46.7$) $<$ PC ($\varepsilon = 64.4$). They conclude that the donor number of the solvent is also an important factor in influencing complex stability. Their results tend to reinforce those of Schori and Jagur-Grodzinski,[58] who found that the difference in complexing power of dibenzo-18-crown-6 in dimeth-oxyethane and dimethylformamide is smaller than would be expected simply on the basis of polarity and must be attributed to the bidentate character of dimethoxyethane.

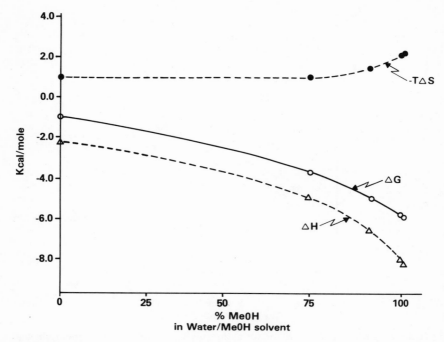

FIGURE 7. Variation with solvent composition of ΔG, ΔH, and $-T\Delta S$ for reaction of Na$^+$ with 18-crown-6.[69]

2.7.2. Effect of Solvent on Selectivity

Agostino et al.[71] have studied the reactions of the alkali metal ions with dicyclohexo-18-crown-6 in three alcohols of successively larger carbon chain lengths: methanol, ethanol, and n-propanol. Their results, shown in Figure 8, demonstrate that while the ligand is consistently selective for K$^+$ over Na$^+$ and Cs$^+$, selectivity between Na$^+$ and Cs$^+$ reverses over the range of solvents studied. Changes in selectivity patterns have also been observed for the complexation of the alkali metal ions with dibenzo-18-crown-6 among the solvents methanol, dimethylformamide, and dimethylsulfoxide.[67] This same ligand prefers Ba^{2+} over K$^+$ in water, but in methanol shows just the opposite preference (Table 13, **19**).

Cation selectivity by the macrobicyclic ligands is also affected by the nature of the solvent as illustrated in Figure 9. The selectivity curve is steeper in methanol especially on the side of the curve corresponding to the larger cations. These results correlate with those of Arnett and Moriarity,[65] who studied complexation of alkali cations with dicyclohexo-18-crown-6 in various solvents. They reported that the stabilities of the complexes of large cations are less affected by solvent than those of smaller ones. The greatest selectivity

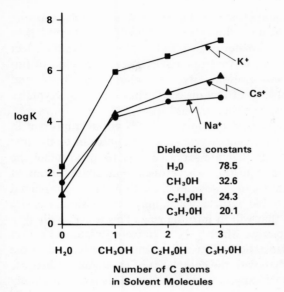

Dielectric constants

H_2O	78.5
CH_3OH	32.6
C_2H_5OH	24.3
C_3H_7OH	20.1

FIGURE 8. Log K values for reaction of dicyclohexo-18-crown-6 with Na^+, K^+, and Cs^+ in solvents of varying dielectric constant.[69]

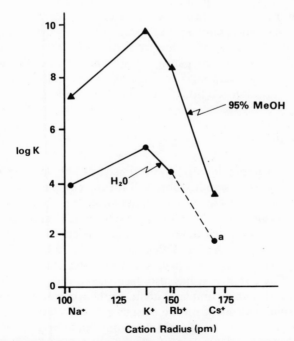

FIGURE 9. Selectivity of [2.2.2] cryptand for alkali cations in H_2O and in 95% MeOH. Data point labeled *a* indicates value reported < 2.[33]

for large over small cations should therefore be observed in strongly solvating solvents, which will have greater affinity for the small cations and thus destabilize their complexes to a greater degree. In less polar media, the effect of the interaction of cation and solvent upon selectivity is eclipsed in importance by ligand size considerations. By the same token, Lehn and Sauvage[33] conclude that solvents of low dielectric constant favor cryptates of monovalent ions over those of bivalent ions.

The solvation of the macrocycle–cation complex is not as important in determining the selectivity of the macrocycle as is the solvation of the free cation. In complexes of ligands that are large enough to encapsulate the cation the solvent is not able to distinguish between complexed cations of identical charge but different sizes. In these cases, the cation may be considered to be embedded in a hydrophobic particle, which may be very similar in size and shape for many cations that will fit into the ring. This is especially true of the large antibiotic complexes, which very effectively isolate the cation from the solvent.[72] This "isosteric" effect will be more pronounced in the case of cryptate complexes than with the complexes of crown ethers because the latter leave the cation more exposed to the solvent. This factor probably contributes to the higher selectivity of the cryptate complexes. Danesi et al.[73] point out that different alkali cation complexes of dibenzo-18-crown-6 are not isosteric in nitrobenzene/toluene unless water is present. The water presumably occupies the cation binding sites above and below the plane of the crown ring, effectively isolating the cation from the nitrobenzene–toluene solvent. These authors' study of ion pairing in solvents of low dielectric constant underscores the importance of considering cation–anion association when studying complexation reactions in this type of medium.

3. Macrocyclic Effect

The term "macrocyclic effect" was coined by Cabbiness and Margerum[74] to describe the greater stability observed for the complexes of cyclic ligands over those of open-chain ligands of similar structure. These authors reported that the enhancement of stability of cyclic tetramine ligands over their linear counterparts was about ten times larger than the chelate effect observed for Cu^{2+} with monodentate and multidentate amine complexes. This increase in complexing power of cyclic ligands over their linear counterparts appears to be common. For instance, it has also been shown that the cyclic polyethers form much more stable complexes than do their corresponding open-chain analogs. Despite agreement on the existence of the macrocyclic effect, controversy has arisen over its specific thermodynamic origin. While it is well established that the chelate effect, of which the macrocyclic effect is at least in concept an extrapolation, is of entropic origin,[75] no agreement has been reached as to whether the macrocyclic effect is a result of more favorable enthalpy or entropy terms in the cyclic ligand reactions. Various thermodynamic and kinetic approaches have been proposed to explain this effect, and these will be presented and discussed in the following sections.

3.1. Tetramines

In reporting the macrocyclic effect as observed for the cyclic tetramines, Cabbiness and Margerum[74] rejected the possibility of explaining its origins in terms of changes of translational entropy. Instead they proposed that ligand solvation and configuration were the important factors to be considered. In a later publication Hinz and Margerum[76] underscored this original hypothesis with a detailed study of the thermodynamic properties of the Ni^{2+} complexes in water of the tetramine ligands shown in Table 6. The magnitude of the

TABLE 6. Thermodynamics of Formation of Ni^{2+} Complexes with Tetramine Ligands at 25°C ($\mu = 0.1$) in Water[76]

Ligand	log K	ΔH (kcal mol^{-1})	$T\Delta S$ (kcal mol^{-1})
(2,3,2-tet)	15.3	−16.8	4.1
(cyclam)	22.2	−31	0.6
(Me$_2$ cyclam)	21.9	−28	2.4

macrocyclic effect is illustrated by comparing the thermodynamic formation constant of the complex of the open-chain ligand to that of its very similar cyclic counterparts. The average $\Delta \log K$ between cyclic and noncyclic ligands is $6.75 \log K$ units. Table 6 illustrates that enthalpy contributions predominate in lending enhanced stability to the cyclic complexes. In fact, entropy changes are in opposition to the macrocyclic effect in this system.

Hinz and Margerum[76] explain their observations in terms of ligand solvation. Release of solvent molecules from the metal ion and the ligand result in positive ΔH and ΔS changes. Assuming that desolvation of the Ni^{2+} ion is the same in the reaction with the cyclic and the noncyclic ligands only desolvation of the ligands is considered for comparison. The cyclic ligand, being much more compact, is less solvated than the noncyclic, and consequently less energy is expended in its desolvation step. The authors point out that using $\Delta H^0 = -7.3 \text{ kcal mol}^{-1}$ for the bond energy of a single H-bond, the data indicate that the cyclic ligand is solvated with two fewer water molecules than the open-chain ligand. In addition, more entropy is expended to wrap the open-chain ligand around the metal ion than to simply insert the ion into the preformed cyclic ring. It is proposed that while this latter entropy term is an important contributing factor toward the enhancement of stability of macrocyclic ligand complexes, it is very much outweighed in H-bonding solvents by the enthalpy contribution due to ligand solvation. In media where H bonding with the solvent is weaker or less available, the configurational entropy contribution should become more dominant. This solvation enthalpy stabilization approach to explaining the macrocyclic effect is reiterated by Dei and Gori[77] in a study of the enthalpies of reaction of these same ligands with Cu^{2+} in water.

In a 1973 communication Paoletti *et al.*,[78] who had previously reviewed the thermodynamics of Cu^{2+}-polyamine complexes,[79] reported the preliminary results of a study of Cu^{2+}-tetramine complexes. They proposed that the macrocyclic effect results from a combination of enthalpy and entropy factors. After a more extensive study and reassessment of the results, however, it was concluded that only entropy contributions were important.[28] Their results, combined with those of Kodama and Kimura[80,81] are listed in Table 7. As is readily apparent, ΔH actually has a destabilizing effect in going toward the cyclic ligand in this series, while $T\Delta S$ is responsible for increased stability of the cyclic ligand. These ligands differ from those used by Hinz and Margerum in that all branches connecting donor groups are of equal size. Consideration was also given to the reaction of Cu^{2+} with larger tetramine ligands and their straight-chain counterparts. Arguments were presented, citing specifically an empirically derived relationship between ΔH for formation of Cu^{2+} complexes and the frequency of the maximum of their electronic absorption spectra in aqueous solution described earlier,[78] which lead to the conclusion that entropy must be of primary importance also in this case. In the same study, comparison between the hydration energies of

TABLE 7. Thermodynamics of Formation of Complexes of Cu^{2+}
with Tetramine Ligands in Water[28]

Ligand	$\log K$	ΔH (kcal mol^{-1})	$T \Delta S$ (kcal mol^{-1})
(en)	19.7	−25.2	1.7
(2,2,2-tet)	20.1	−21.6	5.8
(cyclen)	24.8	−18.3	15.3

cyclic tetramines and ethylene diamine illustrates that these do not differ appreciably, implying perhaps that although fewer waters are bound to the cyclic ligand, these water molecules may each be forming two hydrogen bonds instead of one. These results raise questions concerning the ligand solvation enthalpy explanation proposed by Hinz and Margerum.[76]

Results from a polarographic study of the Cu^{2+} complex of cyclen (Table 7) by Kodama and Kimura[80,81] were used by Fabrizzi et al.[28] in reporting the above comparison. Kodama and Kimura concluded from their results that the enhanced stability of the metal complex of the cyclic ligand arose entirely from the favorable changes in ΔS, and that the apparent contrast between their results with Cu(cyclen)$^{2+}$ and those of Hinz and Margerum with Ni(cyclam)$^{2+}$ could not be fully explained by differences between the metal ions and rings. It is clear, however, that the positioning of the metal ion with respect to the ligand is of critical importance to the nature of any complex formed. Space filling molecular models show that Ni^{2+} or Cu^{2+} can fit into the cavity of the larger cyclam-type ligands, and this observation is confirmed by X-ray crystallographic studies.[34,82,83] However, molecular models demonstrate that the cavity of the smaller cyclen is too small to accommodate these cations. Thus, the question arises as to whether these two studies measure the same "macrocyclic effect." Further crystallographic and thermodynamic work should be useful in resolving this dilemma.

3.2. Cyclic Polyethers

In his original work describing the thermodynamics of cyclic polyether complexation reactions, Frensdorff noted the remarkable increase in stability of the complexes of cyclic polyethers over their linear counterparts.[53] Specifically, in comparing the complexes of Na^+ and K^+ with pentaglyme and 18-crown-6 in methanol he noted a 10^3–10^4 enhancement of the stability constant in the cyclic ligand complexes. He hypothesized that lower stability in the open-chain ligand results from its inability to completely envelop the cation because of electrostatic repulsion between the terminal oxygens and the unfavorable entropy change involved in wrapping the ligand around the cation. The enthalpy changes for these and other reactions in methanol have since been determined[69,84] and help clarify the nature of the macrocyclic effect. The above observations are summarized in Table 8.

There are distinct advantages in studying the macrocyclic effect using crown ethers as opposed to tetraamine ligands: (i) the ligands are uncharged at neutral pH and their complexing ability is not pH dependent, (ii) among the metal ions which form complexes with these ligands are the alkali and alkaline earth cations, which can be considered to be simple charged spheres, unlike transition-metal ions, which have specific stereochemical preferences, (iii) the reaction kinetics are rapid so that equilibrium measurements are readily obtained.

Comparison has been made between the thermodynamic properties of the complexes of pentaglyme and dibenzo-18-crown-6.[84] However, care must be taken in comparing stabilities when more than one parameter (in this case both cyclization and substitution) is varied. Indeed, even the presence of minor ring substituents has given rise to considerable changes in selectivity (see Section 2.6).

Examination of the data in Table 8 reveals no reproducible trends in ΔH or ΔS among the cations studied to explain the macrocyclic effect. In fact, although in every case stability of the cyclic ligand complex far surpasses that of the linear ligand, in no two cases do the results agree upon the origin of this effect. The sodium complex seems very much stabilized by the entropy term, with little enthalpy contribution. It was just such an observation comparing the stabilities of Pb (18-crown-6)$^{2+}$ to those of Pb(tetraglyme)$^{2+}$ and Pb(tetraglyme)$_2^{2+}$ that led Kodama and Kimura[85] to conclude that the macrocyclic effect in that case was accounted for entirely by the favorable entropy contribution. It should be noted, however, that tetraglyme has one fewer donor oxygen atom than 18-crown-6. Stabilization of the potassium complex in Table 8, on the other hand, is due totally to the enthalpy contribution. Finally, the barium complex is stabilized by both enthalpy and entropy, although the enthalpy term predominates. Complex structure may be important to the understanding of these results. While the crown ligand is wrapped around the Na^+ ion (Section 2.2.1 and Figure 4, structure 8) in the

TABLE 8. Thermodynamics of Formation of Metal Complexes of Cyclic and Noncyclic Polyethers at 25°C in 99 wt.% Methanol[69]

Ligand	Cation	log K	ΔH (kcal mol^{-1})	$T\Delta S$ (kcal mol^{-1})
(structure)	K^+	2.05	-6.37	-3.57
	Ba^{2+}	3.96	-6.71	-1.31
(structure)	Na^{+a}	1.0^a	-9.14^a	-7.7^a
	K^+	2.27	-8.16	-5.06
	Ba^{2+}	2.51	-5.64	-2.22
(structure)	Na^+	4.33	-8.11	-2.20
	K^+	6.05	-13.21	-4.96
	Ba^{2+}	7.0	-10.38	-0.83

^{a} From Ref. 84.

case of K^+ (Section 2.2.1 and Figure 4, structure 7) and probably Ba^{2+} (same radius as K^+),[31] the ligand lies essentially flat with the metal ion fitting snugly in its cavity. Unfortunately, no structural data are available for the open-chain ligand complexes, making comparisons between complexes difficult.

The results with cyclic polyethers and their noncyclic analogs show that the source of the macrocyclic effect is not simply defined and that different systems may be responding to different stabilizing factors. This may be the origin of some of the seemingly contradictory results obtained with the various tetramine complexes of Cu^{2+} and Ni^{2+}.

3.3. Solvation Effects

In order to better understand the roles of solvent and hydrogen bonding in the macrocyclic effect, Izatt *et al.*[69] have studied the reaction of two open-chain polyethers of different hydrogen bonding capacities and of 18-crown-6 with K^+ and Ba^{2+} in 90 and 99 wt. % methanol solvents. The results are listed in Table 9. If, as postulated by Hinz and Margerum,[76] the differences in solvation of cyclic ligands as opposed to open-chain ligands is largely

TABLE 9. Equilibrium Constants for Reaction of K^+ and Ba^{2+} with Cyclic and Noncyclic Polyethers in 90% and 99% Methanol at 25°C[69]

Ligand	Cation	log K	
		90% MeOH	99% MeOH
	K^+	1.91	2.05
	Ba^{2+}	3.42	3.96
	K^+	1.96	2.27
	Ba^{2+}	2.34	2.51
	K^+	5.35	6.05
	Ba^{2+}	6.56	7.0

responsible for the macrocyclic effect, then comparison of the stabilities of complexes of these two types of ligand made in solvents of varying solvating strengths should give differing results. Reactions in solvents of lower ligand binding power should show a smaller macrocyclic effect than solvents of higher polarity or hydrogen bonding capability. The logical extrapolation of the solvation enthalpy explanation would be that in the gas phase, where solvation has no effect, there should be minimal macrocyclic effect observed. Hydrogen bonds between the solvent and ligand should involve less energy in 99% than in 90% methanol owing to the lower dielectric constant of the 99% solution. Thus, if desolvation of the open-chain ligand is responsible for robbing its complex of enthalpic stability, the difference in stabilities of the open-chain ligands and 18-crown-6 should be smaller in 99% methanol. Just the opposite is shown to be the case in Table 9. While the difference in log K between the K^+-glyme and K^+-crown complexes in 90% methanol is approximately 3.4, in 99% methanol it is approximately 3.8. The same effect is seen in all the systems studied—the macrocyclic effect is greater in the solvent of lower dielectric constant.

3.4. Mixed Donor Groups

Frensdorff[53] first noted that when nitrogen or sulfur is substituted in the polyether ring for oxygen, the stabilizing influence of cyclization of the ligand disappears, i.e., the macrocyclic effect as defined by Cabbiness and Margerum[74] is not present. Specifically, he compared the binding of Ag^+ with 1,10-diaza-18-crown-6 (log K = 7.8) to that with

$$H_2NCH_2CH_2OCH_2CH_2OCH_2CH_2NH_2$$

(log K = 7.9) and the binding of Ag^+ with aza-18-crown-6 (log K = 3.3) to that with $CH_3OCH_2CH_2NH_2$ (log K = 3.2). Anderegg[86] studied the binding in water of Cd^{2+} and Hg^{2+} with the former two ligands. As in the reaction of Ag^+ with these ligands, no macrocyclic effect is observed. However, in all these cases the cyclic and noncyclic ligands are strictly comparable.

Izatt et al.[45] have compared the stabilities of several linear S-containing ligands to their cyclic analogs, and representative results appear in Table 10. The stabilities of the 2:1 (ligand to metal) Ag^+ and Hg^{2+} complexes of the linear ligand differ only slightly from those of the cyclic ligand. Indeed, for both Hg^{2+} and Ag^+ the 2:1 complex with the linear ligand is more stable than that of the crown. In addition, the ΔH values for the 1:1 reaction are also similar. In Section 2.4 we have already noted that it is doubtful that the metal ion is situated in the ring cavity during complexation with sulfur-substituted crown ethers. If only outwardly turned sulfur donors are participating in complexation with the cyclic ligand, no "macrocyclic effect" is to be expected. For example, for the $PdCl_2$ complex of 1,10-dithia-18-crown-6, Metz et al.[87] reports that, in the crystal, only the ligand sulfur

TABLE 10. Thermodynamics of Formation of Ag^+ and Hg^{2+} Complexes of Thia-Substituted Polyethers in H_2O at $25°C$[45]

Ligand	Metal ion	ΔH^a (kcal mol^{-1})	log K^b	ΔH^b (kcal mol^{-1})	$T\Delta S^b$ (kcal mol^{-1})
	Ag^+	-14.13	3.06	-3.68	0.68
	Hg^{2+}	-14.02	3.22	-7.09	-3.64
	Ag^+	-16.57	2.70	-1.00	2.80
	Hg^{2+}	-16.10	2.91	-5.00	-1.10

^{a} For reaction $M + L \rightarrow ML$.
^{b} For reaction $ML + L \rightarrow ML_2$.

atoms participate in coordination with Pd. The Pd atom lies outside the ring, which is bent like a bow. This coordination is reminiscent of that found in the $HgCl_2$-1,4-dithia-18-crown-6 crystal (Section 2.4). The absence of a macrocyclic effect in these thia-substituted ligands is probably attributable to the fact that only part of the ring is participating in coordination.

3.5. Multiple Juxtapositional Fixedness

Busch and co-workers[5] have attributed the enhanced stabilities exhibited in the macrocyclic effect to what they term "multiple juxtapositional fixedness," based on the fact that the stability of macrocyclic complexes is due to their very slow rates of decomposition. Data of Cabbiness and Margerum[88] (Table 11) illustrate dramatically that the decomposition rate of the Cu^{2+} cyclic tetramine complex is much slower than that of its linear counterpart. Indeed, the low dissociation rate of the complex is sufficient to more than overcome the fact that the formation rate is less favorable for the cyclic ligand. These results conform to the hypothesis that the straight-chain ligand can undergo successive S_N1 replacement steps of the nitrogen donors by solvent molecules beginning at one end of the ligand. In acidic media the dissociated groups are quickly protonated and rendered unusable for com-

TABLE 11. Kinetic Data for Cu^{2+} Complexes at 25°C

Ligand	Solvent	k_f $(M^{-1}s^{-1})$	k_d (s^{-1})
	H_2O[88]	8.9×10^4	4.1
	H_2O[88]	5.8×10^{-2}	3.6×10^{-7}
	80% MeOH[29]	4.1×10^5	3.0×10^4
	80% MeOH[29]	2.8×10^4	9

plexation. The cyclic ligand, however, cannot be displaced by such a simple mechanism because a ring has no end. In this case a mechanism resembling a bimolecular substitution is proposed, requiring the distortion of the ligand such that one coordination bond is significantly weakened and finally broken. Such a mechanism accounts for enhanced stability of cyclic ligands even in cases where the metal ion does not fit into the cavity. The thermodynamic consequence of this mechanistic constraint on dissociation is invariably the higher stability of cyclic ligands.

A study by Jones and co-workers[29] supports this kinetic approach to understanding the macrocyclic effect. Their studies involved the tetrathia analogs of the tetraamine ligands. Data for the last two ligands in Table 11 illustrate that the slow dissociation step of the cyclic tetrathia complex is responsible for its extra stability over the complex of the linear ligand. X-ray crystallographic investigation demonstrates that the Cu^{2+} ion fits into the ring cavity in this case.[89] These ligands are free of protonation considerations and should not be significantly solvated. The authors conclude that configurational effects in the dissociation step are responsible for the stability of the cyclic complex and that these effects should be manifest primarily in the entropy term. In addition they suggest that solvation effects must be important only in the dissociation step, and thus only for the complexed or partially complexed species and not the free ligands.

3.6. Cryptate Effect

An extension of the macrocyclic effect has been observed for the cryptate complexes of Lehn and co-workers.[33] They describe the enhancement of complex stability observed by the addition of another connecting bridge onto the macrocyclic ring to form macrobicyclic ligands or cryptands. The increases in complex stability over monocyclic ligands are very large and have received the name "cryptate effect" or "macrobicyclic cryptate effect."[33] This effect is illustrated in Table 12. By simply joining the dangling arm of ligand 2 to the opposite side of the molecule to form ligand 3 an increase in stability constant for the K^+ complex of $\sim 10^5$ is observed. This enhancement of stability observed for the cryptate effect is even larger than the increase ($\sim 10^4$) observed in closure of the 18-crown-6 ring to produce the macrocyclic effect (see Table 9). Thus, creation of a suitably enclosed cavity increases the stability of the resulting metal complex more than does provision of a ring alone.

From results of a calorimetric study of the reaction of certain cryptands with K^+ to form cryptate complexes, Kauffman et al.[26] conclude that the cryptate effect is of enthalpic origin. This conclusion was based on the comparison of the thermodynamic properties of the K^+ complex of [2.2.2] to that of dicyclohexo-18-crown-6 (see Table 13, **14**, **15**, and **40**). It should be noted that these macrocyclic compounds are not strictly comparable. Their conclusion that the cryptate effect is a result of enthalpic stabilization agrees well with the data reported by Anderegg[86] for the Ba^{2+} and Sr^{2+} complexes of monocyclic compound **34** for bicyclic compound **40** of Table 13. However, corresponding data for two of the post-transition-metal ions are contradictory, i.e., Hg^{2+} seems to show no cryptate effect and the Cd^{2+}-bicyclic complex is entropy stabilized. Further study to elucidate the thermodynamic origin of the cryptate effect using comparable compounds appears warranted.

3.7. Summary

The thermodynamic data reported to date indicate that macrocycles fall into three categories with respect to macrocyclic and cryptate effects: (i) macrocycles that contain only one kind of donor atom (O, N, or S) show a

TABLE 12. Equilibrium Constants for the Reaction (in 95 Vol. % Methanol at 25°C) of Several Metal Ions with Macrocyclic Ligands and a Macrobicyclic Ligand of Similar Structure[33]

Ligand	log K with metal ion				
	Na^+	K^+	Ca^{2+}	Sr^{2+}	Ba^{2+}
1	3.26	4.38	4.4	6.1	6.7
2	3.35	4.80			
3	7.21	9.75	7.60	11.5	12

macrocyclic effect, (ii) macrocycles that contain two kinds of donor atoms (O, and S, O and N) do not show a macrocyclic effect, and (iii) macrocycles containing two kinds of donor atoms (O and N) do show a cryptate effect. As to whether the macrocyclic and cryptate effects have enthalpic or entropic origins, the available data are far from conclusive. Systems have been studied in which the macrocyclic effect is a result of either enthalpic stabilization or entropic stabilization, or a combination of these. Too few data are available to establish the thermodynamic origin of the cryptate effect. An understanding of the thermodynamic origin of the macrocyclic and cryptate effects will require study of numerous carefully selected comparable systems. Compounds should be included that contain different types of donor atoms as well as those with the same donor atom, and studies should be made using a variety of solvents.

The kinetic approach to explaining the macrocyclic effect in terms of multiple juxtapositional fixedness is in good agreement with what few data are available. However, its logical thermodynamic outgrowth, that of entropic stabilization of the cyclic complexes, has not been observed consistently.

4. Table of Thermodynamic Data

In Table 13 (pages 180–210), thermodynamic data for the reaction of cations with macrocyclic ligands in solution are summarized. The data reported include log K, ΔH, and ΔS values for reactions of the following type:

$$ML_{i-1}^{n+} + L \rightleftharpoons ML_i^{n+} \qquad (i = 1 \text{ or } 2)$$

The value of i for each reaction is listed. Reactions are listed for macrocycles in the following order: (i) naturally occurring macrocycles, (ii) cyclic polyethers, (iii) nitrogen-substituted cyclic polyethers, (iv) cryptands, (v) cyclic tetraamines, (vi) sulfur-substituted cyclic polyethers, (vii) macrocyclic ligands with all sulfur donors. This collection should make possible the rapid perusal of thermodynamic results obtained for macrocyclic ligands and provide a quick reference for those interested in studying their selectivity in greater detail than can be discussed in the body of the chapter. In order to conserve space an attempt has been made to be critical in selecting data to be included in the table. In most cases where different researchers have studied the same system, values are reported for the most direct means of measurement (e.g., direct reaction as opposed to substitution or competition reactions) and the most common reaction medium. Data are listed only for reactions at 25°C or, where these are not available, at the temperature closest to 25°C. Many more ligands have been synthesized than are listed here. For some, approximate stability constants with certain cations have been reported. These are not cited. Nor are the numerous reports of complexes of macrocyclic ligands with various cations in the solid phase for which solution thermodynamic data are not available.

5. Kinetics

Complexation kinetics of the cyclic polyethers and cryptands have received little attention by comparison to the thermodynamics of their cation interactions. The naturally occurring antibiotic macrocycles, on the other hand, have been better characterized in terms of their reaction kinetics.[2] Kinetic studies have already been cited with respect to the macrocyclic effect observed for the tetradentate nitrogen and sulfur ligand A comprehensive review of the kinetics of cation complexation of the macrocycles has recently been published.[96] Only a brief summary of the kinetic results found in the literature will be presented here.

5.1. Antibiotic Macrocycles

As a first approximation, reaction of monovalent cations with valinomycin in methanol may be represented by[3,96]

$$M^+_{\text{solv}} + V'_{\text{solv}} \underset{k_{21}}{\overset{k_{12}}{\rightleftharpoons}} M^+_{\text{solv}}V' \underset{k_{32}}{\overset{k_{23}}{\rightleftharpoons}} MV^+$$

where M^+ represents the cation, "solv" the solvent, and V valinomycin. In the first step, a diffusion-controlled bimolecular collision between one of the open forms of valinomycin (V') and the solvated cation is followed by subsequent replacement of some solvent molecules around the cation by ester carbonyl oxygens of the ligand. This process continues as long as the original conformation of the ligand will allow the rapid formation of an intermediate complex $M^+_{\text{solv}}V'$. In the second step, the ligand folds itself around the cation, displacing the remaining solvent molecules. The latter step is rate determining.

The mechanism described is unquestionably an oversimplified description of what is happening on the molecular level. Evidence has been presented that in methanol valinomycin exists in no fewer than five conformations in rapid equilibrium. However, a more complicated mechanism is not required to account for the data obtained. Some representative valinomycin kinetic data are illustrated in Figure 10, which shows how both formation and dissociation rates of the cation complex contribute to maximum stability for Rb$^+$, although the latter predominate in importance.

5.2. Cyclic Polyethers

The kinetics of reaction of Na$^+$ with dibenzo-18-crown-6 in DMF have been studied by ^{23}Na NMR by Shchori *et al.*[97] The authors concluded that the dynamic equilibrium involves interaction of solvated Na$^+$ with uncoordinated ligand rather than a bimolecular exchange equilibrium wherein one Na$^+$ is replaced by another. Their rate constant $K_f \sim 6 \times 10^7 M^{-1} \text{s}^{-1}$

TABLE 13. Log K, ΔH, and ΔS Values for the Interaction of Macrocycles with Cations in Solution

Ligand	Ion	i	log K		ΔH^b		ΔS^c	Medium^a	Ref.
			Value	Method^a	Value	Method^a			
Valinomycin[2]									
1	Li$^+$	1	<0.7	Spec				MeOH	2, 3
	Na$^+$	1	0.67	Spec				MeOH	2, 3
	K$^+$	1	1.08	EMF				MeOH	2, 3
		1	4.79	NMR				MeOH	90
		1	4.90	Spec				MeOH	2, 3
		1	4.43	Cond				MeOH	2
		1			-4.54	Cal		MeOH	91
		1	6.30	Con				EtOH	2, 3
		1			-8.9		-2.16	EtOH	2
	Rb$^+$	1	5.12	NMR				MeOH	90
		1	5.26	Spec				MeOH	2, 3
		1	6.46	Con				EtOH	2
	Cs$^+$	1	4.32	NMR				MeOH	90
		1	4.41	Spec				MeOH	2
		1	5.81	Con				EtOH	2
	Tl$^+$	1	4.36	NMR				MeOH	90
		1	3.73	Spec				MeOH	2
	Ag$^+$	1	3.90	Spec				MeOH	2
	Mg^{2+}	1	<0.7	Spec				MeOH	2
	Ca^{2+}	1	2.70	Spec				MeOH	2
	Sr^{2+}	1	2.23	Spec				MeOH	2
	Ba^{2+}	1	3.34	Spec				MeOH	2
	NH$_4^+$	1	1.67	Spec				MeOH	2
Enniatin B[2]									
2	Li$^+$	1	1.28	Spec				MeOH	3
	Na$^+$	1	2.41	Spec				MeOH	3

		log K	Method	ΔH	Method	ΔS	Solvent	Ref.
K^+	1	3.41	ORD				EtOH	3
	1	2.92	Spec				MeOH	3
	1	3.81	ORD				EtOH	3
Rb^+	1	2.74	Spec				MeOH	3
Cs^+	1	2.34	Spec				MeOH	3
Mg^{2+}	1	1.2					MeOH	2
Ca^{2+}	1	3.0					MeOH	2
Sr^{2+}	1	2.65					MeOH	2
Ba^{2+}	1	2.93					MeOH	2
Mn^{2+}	1	0.6					MeOH	2
NH_4^+	1	1.92	Spec				MeOH	3
Cyclo-(Pro-Gly)[3] **3**								
Li^+	1	2.26	CD				80% MeOH, 20°C	92
Na^+	1	0.34	CD				H_2O, 20°C	92
	1	2.04	CD				80% MeOH, 20°C	92
K^+	1	1.46	CD				80% MeOH, 20°C	92
Mg^{2+}	2	5.00	CD				Acnit, 20°C	92
		2.81	CD				Acnit, 20°C	92
Ca^{2+}	1	2.11	CD				H_2O, 20°C	92
	1	5.04	CD				Acnit, 20°C	92
	1	3.15	CD				80% MeOH, 20°C	92
Ba^{2+}	1	2.62	CD				80% MeOH, 20°C	92
Cyclodextrin[4] **4**								
Anilinium ClO_4^-	1	1.5	Cal	−12.3	Cal	−35	H_2O	4
$HClO_4$	1	1.6	Cal	−7.5	Cal	−17	H_2O	4
$NaClO_4$	1	1.3	Cal	−9.7	Cal	−23	H_2O	4
Pyridine	1	2.2	Cal	−2.5	Cal	2	H_2O	4
2-Amino-benzoic acid	1	5.0	Cal	−0.3	Cal	21	H_2O	4

continued overleaf

TABLE 13 (continued)

Ligand	Ion	i	log K		ΔH^b		ΔS^c	Medium^d	Ref.
			Value	Method^a	Value	Method^a			
Nonactin$^{(2)}$ 5	Na$^+$	1	3.97	Polg				Acnit μ, 22°C	66
		1	2.71	Cal	-2.65	Cal	3.49	MeOH	93
		1	3.27	Cal	-6.55	Cal	-7.02	EtOH	93
	K$^+$	1	4.43	Polg				Acnit μ, 22°C	66
		1	4.49	Cal	-10.4	Cal	-14.41	MeOH	93
		1	5.26	Cal	-12.5	Cal	-17.8	EtOH	93
	Rb$^+$	1	3.87	Polg				Acnit μ, 22°C	66
		1	3.81	NMR				MeOH	90
	Cs$^+$	1	2.59	Polg				Acnit μ, 22°C	66
		1	3.23	NMR				MeOH	90
	Tl$^+$	1	4.15	NMR				MeOH	90
	Ba^{2+}	1	1.61	Vap				MeOH, 30°C	93
		1	2.30	Vap				EtOH, 30°C	93
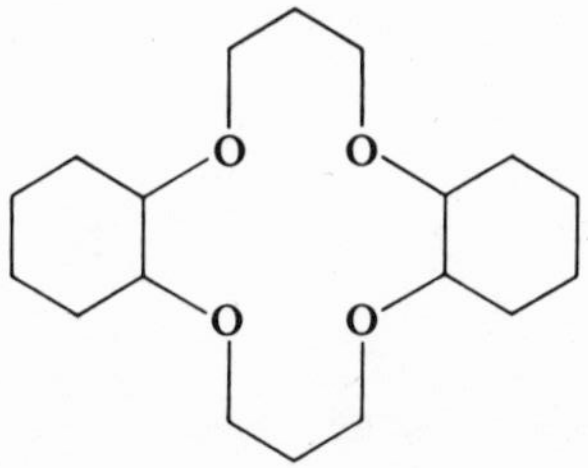	Na$^+$	1	2.18	Pot				MeOH μ	9
	K$^+$	1	1.30	Pot				MeOH μ	9

6

Structure	Cation		log K	method	ΔH	method	$T\Delta S$	Solvent	μ	Ref
7	Na^+F^-	1	2.18	Spec				THF		9
8	Na^+	1	1.41	Pot				MeOH	μ	9
9	Na^+	1	0.70	Cal	−1.50	Cal	−1.8	H_2O		32
	K^+	1	0.74	Cal	−4.1	Cal	−10.4	H_2O		32
	Rb^+	1	0.62	Cal	−1.90	Cal	−3.5	H_2O		32
	Cs^+	1	0.8	Cal	−1.3	Cal	−0.7	H_2O		32
	Ag^+	1	0.94	Cal	−3.23	Cal	−6.5	H_2O		32
	Tl^+	1	1.23	Cal	−4.01	Cal	−7.8	H_2O		32
	Sr^{2+}	1	1.95	Cal	−0.9	Cal	6	H_2O		32
	Ba^{2+}	1	1.71	Cal	−1.14	Cal	4	H_2O		32
	Hg^{2+}	1	1.68	Cal	−3.60	Cal	−4.0	H_2O		32
	Pb^{2+}	1	1.85	Cal	−3.26	Cal	−2.5	H_2O		32
	NH_4^+	1	1.71	Cal	−0.24	Cal	7.0	H_2O		32

continued overleaf

TABLE 13 (continued)

Ligand	Ion	i	log K Value	log K Method[a]	ΔH[b] Value	ΔH[b] Method[a]	ΔS[c]	Medium[d]	Ref.
10	Li⁺	1	<1.0	Pot				H_2O μ	9
	Na⁺	1	<0.3	Pot				H_2O μ	9
	K⁺	1	3.71	Pot				MeOH μ	9
		1	0.6	Pot				H_2O μ	9
		1	3.58	Pot				MeOH μ	9
		2	1.88	Pot				MeOH μ	9
	Cs⁺	1	2.78	Pot				MeOH μ	9
		2	1.91	Pot				MeOH μ	9
11	Na⁺	1	0.72	Cal	−1.77	Cal	−2.6	20% MeOH	68
		1	1.17	Cal	−2.63	Cal	−3.5	40% MeOH	68
		1	1.64	Cal	−3.78	Cal	−5.2	60% MeOH	68
		1	1.99	Cal	−3.82	Cal	−3.7	70% MeOH	68
		1	2.26	Cal	−8.32	Cal	−17.6	80% MeOH	68
		1	3.54	Con				Ace μ	59
	K⁺	1	1.20	Cal	−1.8	Cal	−0.5	20% MeOH	68
		1	1.92	Cal	−2.51	Cal	0.4	40% MeOH	68
		1	1.5	Cal				70% MeOH	68
		2	4.15	Cal	−13.9	Cal	−27.6	70% MeOH	68
		1	2.2	Cal				80% MeOH	68
		2	4.80	Cal	−15.50	Cal	−30.0	80% MeOH	68
		1	0.97	Pot				50% THF	9
	Rb⁺	1	0.38	Cal	−2.33	Cal	−6.1	H_2O	68
		1	1.8	Cal				70% MeOH	68
		2	3.77	Cal	−12.0	Cal	−23.5	70% MeOH	68
		1	0.46	Pot				50% THF	9

Cation	n	log K	Method	ΔG		ΔH		Solvent	Ref
Cs^+	1	1.70	Cal	−2.43	Cal	−0.4		70% MeOH	68
Ca^{2+}	1			0	Cal			70% MeOH	68
Sr^{2+}	1			0	Cal			70% MeOH	68
Ba^{2+}	1			0	Cal			70% MeOH	68
Co^{2+}	1			0	Cal			70% MeOH	68
Cu^{2+}	1			0	Cal			70% MeOH	68
Zn^{2+}	1			0	Cal			70% MeOH	68
Cd^{2+}	1			0	Cal			70% MeOH	68
Pb^{2+}	1	2.04	Cal	−5.11	Cal	−7.8		70% MeOH	68
NH_4^+	1			0	Cal			70% MeOH	68
Na^+	1	0.8	Cal	−2.25	Cal	−3.9		H_2O	32
	1	2.76	Cal	−4.89	Cal	−3.8		70% MeOH	68
	1	3.66	Cal	−6.64	Cal	−5.5		90% MeOH	69
	1	4.36	Cal	−8.36	Cal	−8.1		MeOH	69
K^+	1	2.03	Cal	−6.21	Cal	−11.4		H_2O	32
	1	4.33	Cal	−9.68	Cal	−12.7		70% MeOH	68
	1	5.35	Cal	−11.77	Cal	−15.0		90% MeOH	69
	1	6.05	Cal	−13.41	Cal	−17.3		MeOH	69
Rb^+	1	1.56	Cal	−3.82	Cal	−5.7		H_2O	32
	1	3.46	Cal	−9.27	Cal	−15.3		70% MeOH	68
Cs^+	1	0.99	Cal	−3.97	Cal	−8.1		H_2O	32
	1	2.84	Cal	−8.09	Cal	−14.1		70% MeOH	68
	1	4.62	Pot					MeOH μ	9
	2	1.3	Pot					MeOH μ	9
Ag^+	1	1.50	Cal	−2.17	Cal	−0.4		H_2O	32
Tl^+	1	2.27	Cal	−4.44	Cal	−4.5		H_2O	32
Ca^{2+}	1	<0.5	Cal					H_2O	9
Sr^{2+}	1	2.51	Cal	−4.27	Cal	−2.8		70% MeOH	68
	1	2.72	Cal	−3.61	Cal	0.3		H_2O	32
	1	5.0	Cal	−7.49	Cal	−2.5		70% MeOH	68

continued overleaf

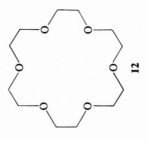

12

TABLE 13 *(Continued)*

Ligand	Ion	i	log K Value	log K Method[a]	ΔH[b] Value	ΔH[b] Method[a]	ΔS[c]	Medium[d]	Ref.
	Ba^{2+}	1	3.87	Cal	-7.58	Cal	-7.9	H_2O	32
		1	6.0	Cal	-10.66	Cal	-8.5	70% MeOH	68
		1	6.56	Cal	-10.33	Cal	0.8	90% MeOH	69
		1	7.0	Cal	-10.23	Cal	0.7	MeOH	69
	Hg^{2+}	1	2.42	Cal	-4.69	Cal	-4.7	H_2O	32
	Pb^{2+}	1	4.27	Cal	-5.16	Cal	2.2	H_2O	32
		1	6.5	Cal	-9.19	Cal	-1.1	70% MeOH	68
	NH_4^+	1	1.23	Cal	-2.34	Cal	-2.2	H_2O	32
	$CH_3NH_3^+$	1	4.27	Cal	-9.27	Cal	-11.6	MeOH	11
	$C_2H_5NH_3^+$	1	4.25	Cal	-10.71	Cal	-14.3	MeOH	11
	$C_3H_7NH_3^+$	1	3.99	Cal	-10.65	Cal	-17.51	MeOH	11
	$C_3H_6NH_3^+$	1	3.98	Cal	-9.99	Cal	-15.33	MeOH	11
	$CH_2{=}CH$	1	3.90	Cal	-9.85	Cal	-15.2	MeOH	11
	$^+H_3N{-}CH_2$ $CH{\equiv}C{-}$	1	4.02	Cal	-10.52	Cal	-17.0	MeOH	11
	$^+H_3N{-}CH_2$	1	4.13	Cal	-10.39	Cal	-16.0	MeOH	11
	$NH_2NH_3^+$	1	4.21	Cal	-10.43	Cal	-15.7	MeOH	11
	$HONH_3^+$	1	3.99	Cal	-9.01	Cal	-12.0	MeOH	11
	$(CH_3)_3CNH_3^+$	1	2.90	Cal	-7.76	Cal	-12.7	MeOH	11
	$(CH_3)_2NH_2^+$	1	1.76	Cal	-6.73	Cal	-14.5	MeOH	11
	$(C_2H_5)_2NH_2^+$	1	0.85	Cal	-5.40	Cal	-14.3	MeOH	11
	$C_4H_8NH_2^+$	1	1.98	Cal	-7.36	Cal	-15.6	MeOH	11

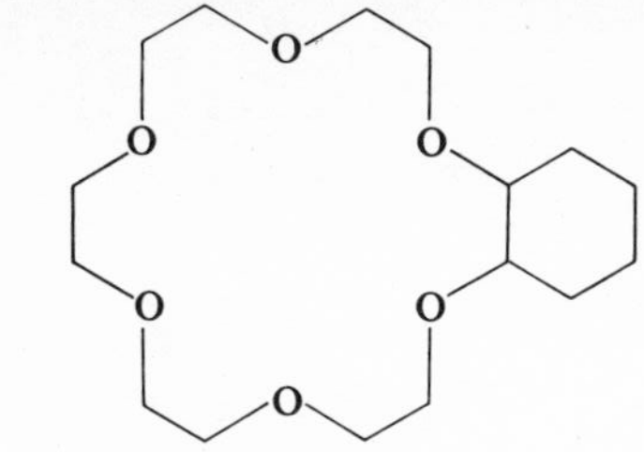

13

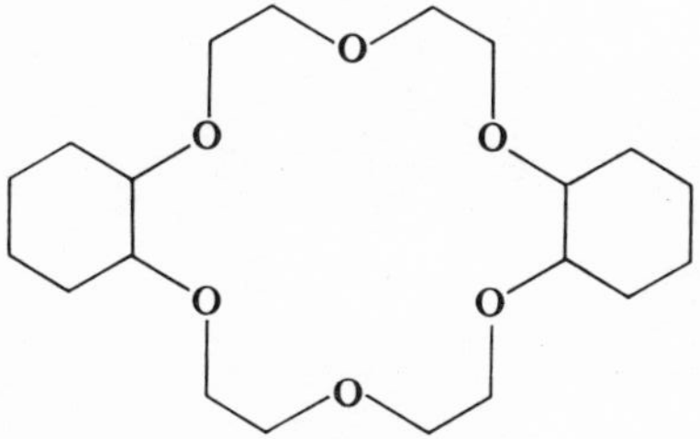

Cation	n	log K				Solvent		Ref
Li$^+$	1	<0.7	Pot			H$_2$O	μ	9
Na$^+$	1	0.8	Pot			H$_2$O	μ	9
K$^+$	1	4.09	Pot			MeOH	μ	9
	1	1.90	Pot			H$_2$O	μ	9
	1	5.89	Pot			MeOH	μ	9
Cs$^+$	1	0.8	Pot			H$_2$O	μ	9
	1	4.30	Pot			MeOH	μ	9
Ag$^+$	2	1.52	Pot			MeOH	μ	9
	1	1.7–1.9	Pot			H$_2$O	μ	9
NH$_4^+$	1	1.1	Pot			H$_2$O	μ	9

14

Cation	n	log K		ΔH		value	Solvent		Ref
Li$^+$	1	0.6	Pot				H$_2$O	μ	9
Na$^+$	1	4.08	Pot				MeOH	μ	9
K$^+$	1	1.21	Cal	0.16	Cal	6.1	H$_2$O		32
	1	2.02	Cal	−3.88	Cal	−3.8	H$_2$O		9
	1	6.01	Pot				MeOH	μ	9
Rb$^+$	1	1.52	Cal	−3.33	Cal	−4.2	H$_2$O		9
Cs$^+$	1	0.96	Cal	−2.41	Cal	−3.7	H$_2$O		9
	1	4.61	Pot				MeOH	μ	9
	2	0.59	Pot				MeOH	μ	9
Ag$^+$	1	2.36	Cal	0.07	Cal	11.0	H$_2$O		32
Tl$^+$	1	2.44	Cal	−3.62	Cal	−1.0	H$_2$O		32
Hg$_2^{2+}$	1	1.93	Cal	−2.16	Cal	1.6	H$_2$O		32
	2	0.6–1.8	Cal	6.1	Cal	2.6	H$_2$O		32
Sr^{2+}	1	3.24	Cal	−3.68	Cal	2.5	H$_2$O		32
Ba^{2+}	1	3.57	Cal	−4.92	Cal	−0.2	H$_2$O		32
Hg^{2+}	1	2.75	Cal	−0.71	Cal	10.2	H$_2$O		32
Pb^{2+}	1	4.95	Cal	−5.58	Cal	3.9	H$_2$O		32
NH$_4^+$	1	1.33	Cal	−2.16	Cal	−1.2	H$_2$O		32
CH$_3$NH$_3^+$	1	0.82	Cal	−0.77	Cal	1.2	H$_2$O		32
	1	0.69	Cal	−1.57	Cal	−2.1	H$_2$O		32

continued overleaf

TABLE 13 (continued)

Ligand	Ion	i	log K		ΔH^b		ΔS^c	Medium^d	Ref.
			Value	Method^a	Value	Method^a			
15	Na$^+$	1	0.69	Cal	-1.57	Cal	-2.1	H$_2$O	32
		1	1.7	Cal	-2.5	Cal	-0.6	DMSO	9
		1	~4.5	Cal	~ -8.6	Cal	~ -8.3	EtOH	9
		1			-5.6	Cal	-1.9	MeOH	9
		1						MeOH μ	9
	K$^+$	1	3.68	Pot					9
		1	1.63	Cal	-5.07	Cal	-12.2	H$_2$O	9
		1	2.7	Cal	-7.7	Cal	-13.5	DMSO	9
		1			-10.5	Cal	-10.5	MeOH	9
		1						MeOH μ	9
	Rb$^+$	1	5.38	Pot				H$_2$O	9
		1	0.87	Cal	-3.97	Cal	-9.3	MeOH μ	9
	Cs$^+$	1	3.49	Pot				H$_2$O μ	9
		2	0.9	Pot					9
	Ag$^+$	1	1.59	Cal	-2.09	Cal	0.3	H$_2$O	32
	Tl$^+$	1	1.83	Cal	-4.29	Cal	-6.0	H$_2$O	32
	Hg$_2^{2+}$	1	1.57	Cal	-4.34	Cal	-7.4	H$_2$O	32
		2	1.1	Cal	-5.7	Cal	-14	H$_2$O	32
	Sr^{2+}	1	2.64	Cal	-3.16	Cal	1.5	H$_2$O	9
	Ba^{2+}	1	3.27	Cal	-6.20	Cal	-5.8	H$_2$O	9
	Hg^{2+}	1	2.60	Cal	-2.55	Cal	3.3	H$_2$O	32
	Pb^{2+}	1	4.43	Cal	-4.21	Cal	6.2	H$_2$O	32
	NH$_4^+$	1	0.80	Cal	-3.41	Cal	-7.8	H$_2$O	9
	CH$_3$NH$_3^+$	1	0.66	Cal	-0.90	Cal	0	H$_2$O	32

Structure	Cation					
16	Li^+	1	<0.6	Con	H_2O μ	9
	Na^+	1	1.42	Con	H_2O μ	9
	K^+	1	2.08	Con	H_2O μ	9
	Cs^+	1	0.9	Con	H_2O μ	9
	Rb^+	1	1.53	Con	H_2O μ	9
	NH_4^+	1	1.28	Con	H_2O μ	9
17	Na^+	1	3.59	Pot	MeOH	64
	K^+	1	4.48	Pot	MeOH	64
	Rb^+	1	4.66	Pot	MeOH	64
	$t\text{-}butNH_3^+$	1	<1.5	NMR	$CDCl_3$	64
	$PhCH_2NH_3^+$	1	6.18	NMR	$CDCl_3$	64
18	Na^+	1	4.72	Con	Ace μ	59
	K^+	1	5.10	Con	Ace μ	59

continued overleaf

TABLE 13 (continued)

Ligand	Ion	i	log K Value	log K Method[a]	ΔH[b] Value	ΔH[b] Method[a]	ΔS[c]	Medium[d]	Ref.
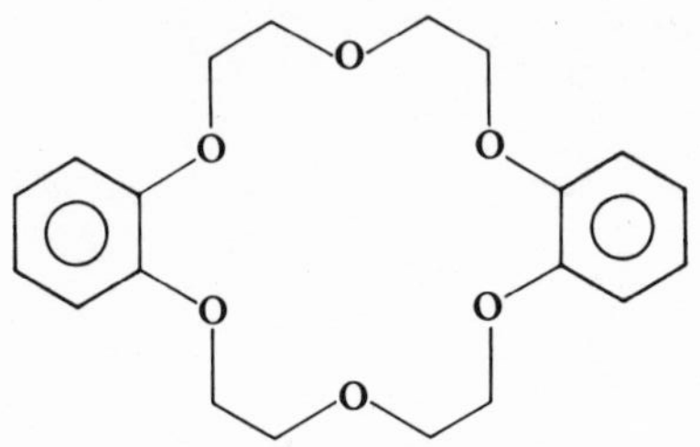	Li⁺	1	<0	Spec				H₂O	94
	Na⁺	1	3.0	Con				DMF	70
		1	3.26	Con				PC	70
		1	1.16	Spec				H₂O	94
		1	4.5	Cal	−7.48	Cal	−4.49	MeOH	84
		1	2.69	Con	−6	T	−7	DMF, 30°C	9
		1	3.66	Con	−3.9	T	3.9	DME, 30°C	58
		1	5.04	Con				Acnit μ	9
		1	3.30	Con				DMSO	70
		1	3.34	Con				DMF	70
		1	3.88	Con				PC	70
	K⁺	1	1.67	Spec				H₂O	94
		1	5.00	Pot				MeOH μ	9
		1	5.1	Cal	−9.58	Cal	−8.77	MeOH	84
		1	4.83	Con				Acnit μ	9
		1	1.87	Pot				50% THF	9
		1	2.5	Cal	−5.5	Cal	−6.9	DMSO	9
		1	3.42	Con				DMSO	70
		1	3.56	Con				DMF	70
		1	5.08	Con				PC	70
	Rb⁺	1	1.08	Spec				H₂O	94
		1	1.35	Pot				50% THF	9
		1	3.37	Con				DMSO	70
		1	3.54	Con				DMF	70
		1	3.76	Con				PC	70

19

Cation		log K					Solvent	Ref
Cs^+	1	0.83	Spec				H_2O	94
	1	3.55	Pot				MeOH μ	9
	2	2.92	Pot				MeOH μ	9
	1	3.30	Con				DMSO	70
	1	3.48	Con				DMF	70
	1	3.55	Con				PC	70
Ag^+	1	1.41	Spec				H_2O	94
Tl^+	1	1.50	Spec				H_2O	94
Sr^{2+}	1	1.0	Spec				H_2O	94
Ba^{2+}	1	~1.95	Spec				H_2O	94
	1	4.28	Cal	−5.06	Cal	2.65	MeOH	84
$BaCl^+$	1	~2.15	Spec				H_2O	94
Pb^{2+}	1	1.89	Spec				H_2O	94
La^{3+}	1	<0	Spec				H_2O	94
NH_4^+	1	~0.30	Spec				H_2O	94
t-butNH_3^+	1	4.11	NMR				Chlor	16
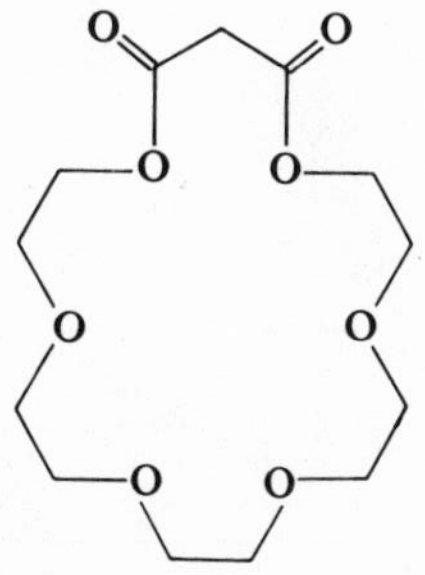 Na^+	1	1.8	Cal	−1.1	Cal	4.7	MeOH	56
K^+	1	2.55	Cal	−7.91	Cal	−14.9	MeOH	56
Ba^{2+}	1	1.41	Cal	−4.88	Cal	−10	MeOH	56

20

continued overleaf

TABLE 13 (continued)

Ligand	Ion	i	log K Value	log K Method[a]	ΔH[b] Value	ΔH Method[a]	ΔS[c]	Medium[d]	Ref.
21	Na⁺	1	2.5	Cal	−2.27	Cal	3.7	MeOH	56
	K⁺	1	2.79	Cal	−5.87	Cal	−6.9	MeOH	56
	Ba²⁺	1	3.1	Cal	−0.46	Cal	12.7	MeOH	56
22	K⁺	1	4.41	Pot				MeOH μ	9
	Cs⁺	1	5.02	Pot				MeOH μ	9

Cation	n	log K	Method	Solvent	Ref
Cs⁺	1	1.9	Pot	H₂O μ	9

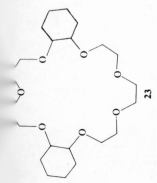

23

Cation	n	log K	Method	Solvent	Ref
Na⁺	1	2.40	Pot	MeOH μ	9
K⁺	1	4.30	Pot	MeOH μ	9
Cs⁺	1	4.20	Pot	MeOH μ	9
	2	1.9	Pot	MeOH μ	9

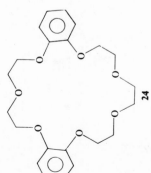

24

Cation	n	log K	Method	Solvent	Ref
K⁺	1	3.48	Pot	MeOH μ	9
Cs⁺	1	4.15	Pot	MeOH μ	9

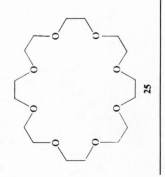

25

continued overleaf

TABLE 13 (continued)

Ligand	Ion	i	log K Value	log K Method[a]	ΔH[b] Value	ΔH[b] Method[a]	ΔS[c]	Medium[d]	Ref.
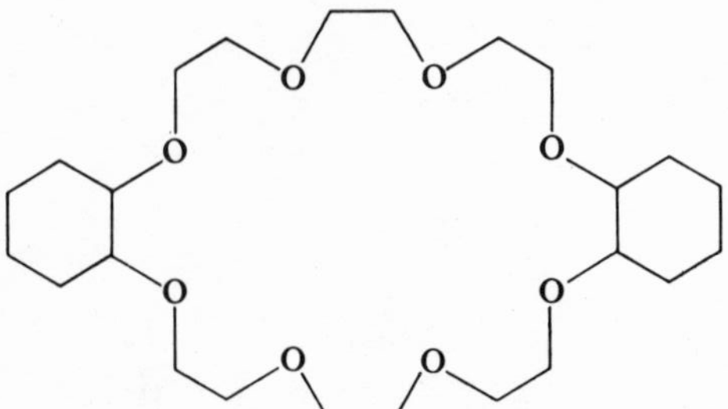	Cs+	1	1.9	Pot				MeOH μ	9
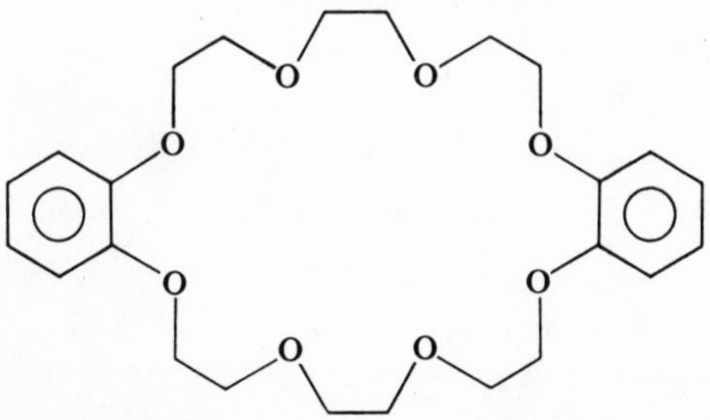	Na+	1	1.54	Cal	−7.75	Cal	−18.9	70% MeOH	68
	K+	1	2.42	Cal	−8.54	Cal	−17.6	70% MeOH	68
		1	3.49	Pot				MeOH	9
	Rb+	1	2.55	Cal	−8.72	Cal	−17.6	70% MeOH	68
	Cs+	1	2.48	Cal	−8.93	Cal	−18.6	70% MeOH	68
		1	3.78	Pot				MeOH	9

26

27

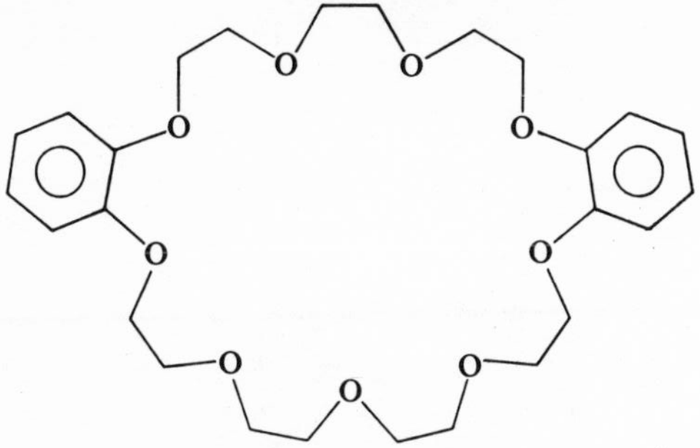

Cation	n	log K	method	value	method	value	solvent	ref
Na^+	1	1.50	Cal	-11.74	Cal	-32.5	70% MeOH	68
K^+	1	2.86	Cal	-9.50	Cal	-18.8	70% MeOH	68
Cs^+	1	1.42	Cal	-6.14	Cal	-14.1	70% MeOH	68

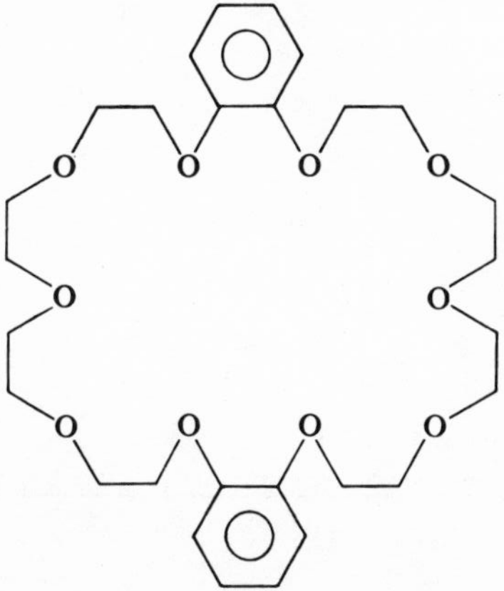

28

Cation	n	log K	method	value	method	value	solvent	ref
Na^+	1	2.114	Spec	-4	T	-3.74	MeOH μ	9
K^+	1	4.568	Spec	-11.5	T	-17.7	MeOH μ	9
	1	1.35	Pot				50% THF	9
Rb^+	1	4.643	Spec	-12.7	T	-21.4	MeOH μ	9
	1	1.56	Pot				50% THF	9
Cs^+	1	4.230	Spec	-11.2	T	-18.2	MeOH μ	9
Tl^+	1	4.505	Spec	-11	T	-16.3	MeOH μ	9
NH_4^+	1	2.431	Spec	-5.5	T	-7.32	MeOH μ	9

29

continued overleaf

TABLE 13 (continued)

Ligand	Ion	i	log K		ΔH[b]		ΔS[c]	Medium[d]	Ref.
			Value	Method[a]	Value	Method[a]			
30	K⁺	1	3.90	Pot				MeOH μ	9
31	K⁺	1	3.90	Pot				MeOH μ	9
	Ag⁺	1	3.3	Pot				MeOH μ	9
32	K⁺	1	3.20	Pot				MeOH μ	9

Cation	n	log K		ΔH		TΔS	Solvent	Ref.
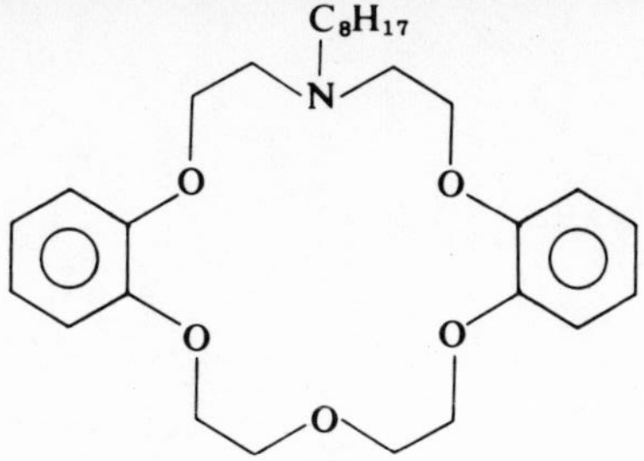 K⁺		4.10	Pot				MeOH μ	9
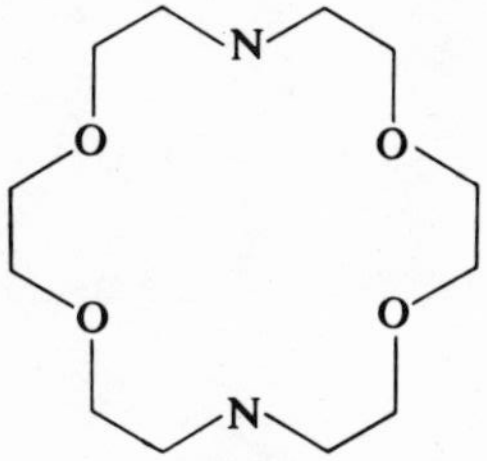 K⁺	1	2.04	pH				MeOH μ	86
Ag⁺	1	7.8	pH	−9.15	Cal	5.0	H₂O	86
Sr²⁺	1	2.56	pH	−2.6	Cal	3.1	H₂O	86
Ba²⁺	1	2.97	pH	−3.0	Cal	3.6	H₂O	86
Cd²⁺	1	5.25	pH	−0.7	Cal	21.6	H₂O	86
Hg²⁺	1	17.85	pH	−17.15	Cal	24.3	H₂O	86
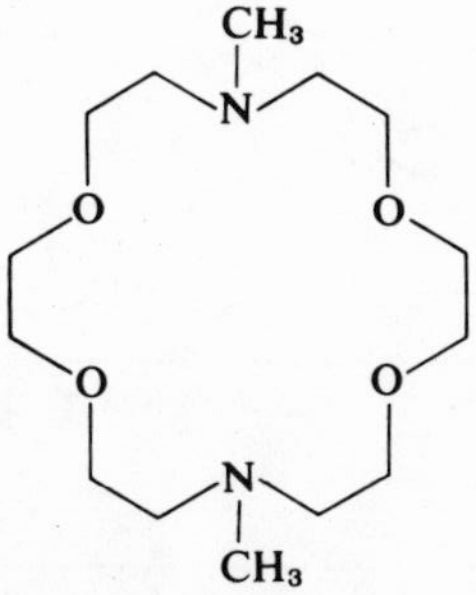 Na⁺	1	3.26	Pot				95% MeOH μ	33
K⁺	1	4.38	Pot				95% MeOH μ	33
Rb⁺	1	4.3	Pot				MeOH μ	33
Ca²⁺	1	4.4	Pot				95% MeOH μ	33
Sr²⁺	1	6.1	Pot				95% MeOH μ	33
Ba²⁺	1	6.7	Pot				95% MeOH μ	33

33

34

35

continued overleaf

TABLE 13 (continued)

Ligand	Ion	i	log K		ΔH[b]		ΔS[c]	Medium[d]	Ref.
			Value	Method[a]	Value	Method[a]			
36	K$^+$	1	1.63	Pot				MeOH μ	9
37	Li$^+$	1	5.5	Pot	−5.1	Cal	8.05	H$_2$O μ	26, 33
		1	7.58	Pot				95% MeOH μ	33
	Na$^+$	1	3.2	Pot	−5.4	Cal	−3.0	H$_2$O μ	26, 33
		1	6.08	Pot				95% MeOH μ	33
	K$^+$	1	<2.0	Pot				H$_2$O μ	33
		1	2.26	Pot				95% MeOH μ	33
	Rb$^+$	1	<2.0	Pot				H$_2$O μ	33
		1	<2.0	Pot				95% MeOH μ	33
	Cs$^+$	1	<2.0	Pot				H$_2$O μ	33
		1	<2.0	Pot				95% MeOH μ	33
	Mg^{2+}	1	2.5	Pot				H$_2$O μ	33
		1	4.0	Pot				95% MeOH μ	33
	Ca^{2+}	1	2.50	Pot	−0.1	Cal	11.07	H$_2$O μ	26, 33
		1	4.34	Pot				95% MeOH μ	33
	Sr^{2+}	1	<2.0	Pot				H$_2$O μ	33
		1	2.90	Pot				95% MeOH μ	33
	Ba^{2+}	1	<2.0	Pot				H$_2$O μ	33
		1	<2.0	Pot				95% MeOH μ	33

38

Cation	n	log K	Method	ΔH		TΔS	Solvent	Ref
Li^+	1	2.50	Pot	0.0	Cal	11.4	H_2O μ	26, 33
	1	4.18	Pot	−5.35	Cal	6.77	95% MeOH μ	33
Na^+	1	5.40	Pot				H_2O μ	26, 33
	1	8.84	Pot	−6.8	Cal	−4.7	95% MeOH μ	33
K^+	1	3.95	Pot				H_2O μ	26, 33
	1	7.45	Pot				95% MeOH μ	33
Rb^+	1	2.55	Pot	−5.4	Cal	−6.5	H_2O μ	26, 33
	1	5.80	Pot				95% MeOH μ	33
Cs^+	1	<2.0					H_2O μ	33
	1	3.90	Pot				95% MeOH μ	33
Mg^{2+}	1	<2.0					H_2O μ	33
	1	<2.0					95% MeOH μ	33
Ca^{2+}	1	6.95	Pot	−2.9	Cal	22.1	H_2O μ	26, 33
	1	9.61	Pot				95% MeOH	33
Sr^{2+}	1	7.35	Pot	−6.1	Cal	13.1	H_2O μ	26, 33
	1	10.65	Pot				95% MeOH μ	33
Ba^{2+}	1	6.30	Pot	−6.3	Cal	7.71	H_2O μ	26, 33
	1	9.70	Pot				95% MeOH μ	33

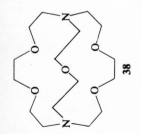

39

Cation	n	log K	Method	Solvent	Ref
Li^+	1	<2.0	Pot	MeOH μ	33
Na^+	1	3.5	Pot	MeOH μ	33
K^+	1	5.2	Pot	MeOH μ	33
Rb^+	1	3.4	Pot	MeOH μ	33
Cs^+	1	2.7	Pot	MeOH μ	33
Ba^{2+}	1	<2.0	Pot	95% MeOH	57

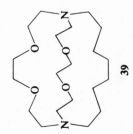

— continued overleaf

TABLE 13 (continued)

Ligand	Ion	i	log K Value	log K Method[a]	ΔH[b] Value	ΔH[b] Method[a]	ΔS[c]	Medium[d]	Ref.
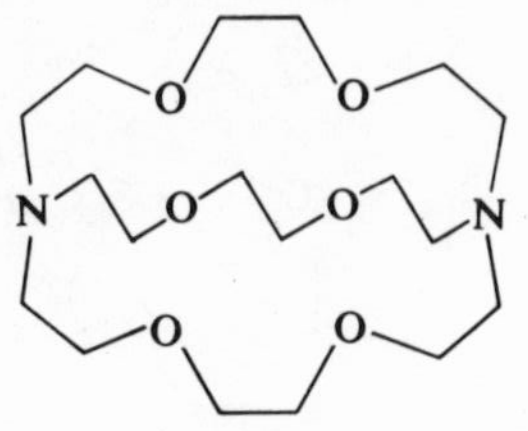	Li^+	1	<2.0	Pot				H_2O μ	33
	Na^+	1	1.8	Pot			-7	95% MeOH μ	33
		1	3.90	Pot	-7.4	Cal	-6	H_2O μ	26,33
		1	4.11	pH	-7.4	Cal	~-2.62	H_2O μ	86
	K^+	1	7.21	Pot	-10.6	Cal	-14.1	95% MeOH	26,33
		1	5.40	Pot	-11.4	Cal	-11.5	H_2O μ	26,33
		1	5.58	pH	-11.0	Cal	-19.2	H_2O μ	86
	Rb^+	1	9.75	Pot	-19.0	Cal	-19.8	95% MeOH	26,33
		1	4.35	Pot	-11.8	Cal	-20.9	H_2O μ	26,33
		1	4.06	pH	-11.8	Cal	-27.4	H_2O μ	86
	Cs^+	1	8.40	Pot	-19.6	Cal	-23	95% MeOH μ	26,33
		1	<2.0	Pot				H_2O μ	33
		1	3.54	Pot	-11.9	Cal	1	95% MeOH μ	26,33
	Ag^+	1	9.6	pH	-12.8	Cal	-14.8	H_2O μ	86
	Tl^+	1	5.5	pH	-13.2	Cal		H_2O μ	86
	Mg^{2+}	1	<2.0	Pot				95% MeOH μ	33
	Ca^{2+}	1	<2.0	Pot			19.5	H_2O μ	33
		1	4.40	Pot	-0.2	Cal	20.2	H_2O μ	26,33
		1	4.57	pH	-0.2	Cal	12.6	H_2O μ	86
		1	7.60	Pot	-6.6	Cal	2.01	95% MeOH	26,33
	Sr^{2+}	1	8.00	Pot	-10.3	Cal	2.2	H_2O μ	26,33
		1	8.26	pH	-10.6	Cal	5.23	H_2O μ	86
		1	11.5	Pot	-14.1	Cal	-4.02	95% MeOH	26,33
	Ba^{2+}	1	9.50	Pot	-14.1	Cal	-3.7	H_2O μ	26,33
		1	9.7	pH	-14.3	Cal	-12.8	H_2O μ	86
		1	12	Pot	-20.1	Cal	33.0	95% MeOH	26,33
	Cd^{2+}	1	6.8	pH	0.5	Cal	29.8	H_2O μ	86
	Hg^{2+}	1	18.2	pH	-15.95	Cal	10.1	H_2O μ	86
	Pb^{2+}	1	12.36	pH	-13.8	Cal		H_2O μ	86

40

Compound	Cation	n	log K	ΔH	ΔS	Method	Solvent	Ref.
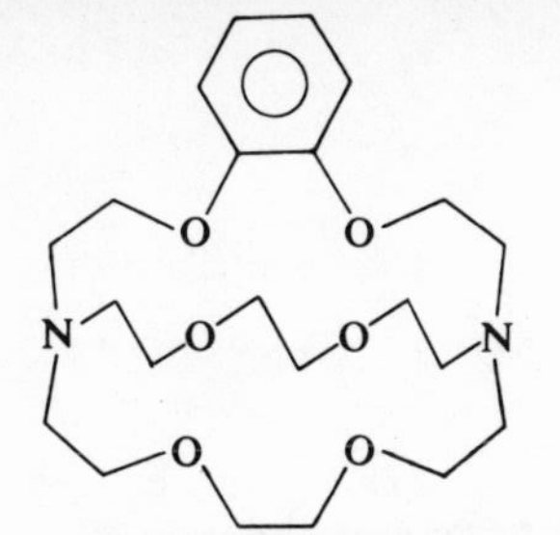 **41**	Na$^+$	1	7.4			Pot	95% MeOH	57
	K$^+$	1	9.05			Pot	95% MeOH	57
	Ba^{2+}	1	11.05			Pot	95% MeOH	57
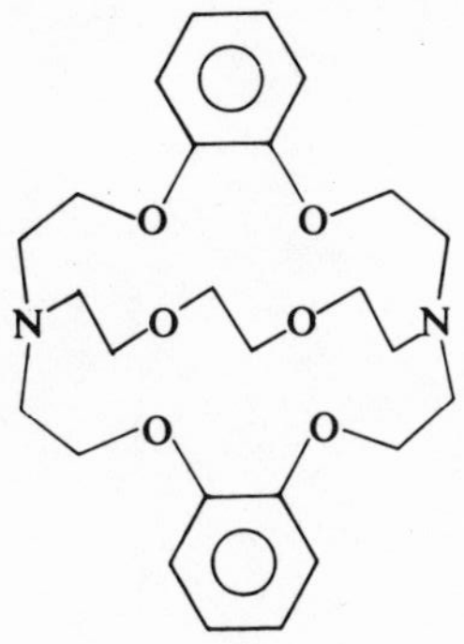 **42**	Na$^+$	1	7.3			Pot	95.5% MeOH	57
	K$^+$	1	8.6			Pot	95.5% MeOH	57
	Ba^{2+}	1	8.5			Pot	95.5% MeOH	57
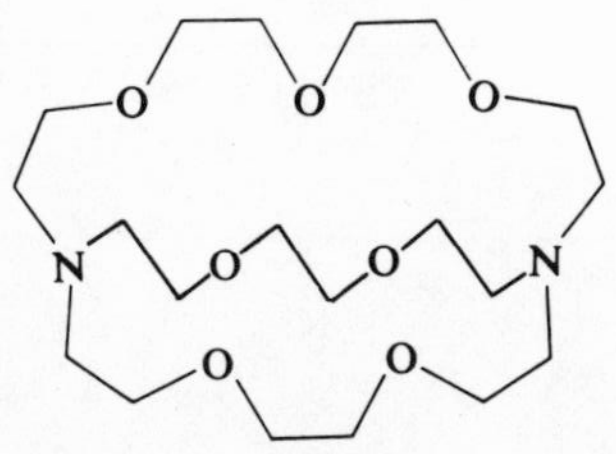 **43**	Li$^+$	1	<2.0			Pot	H$_2$O μ	33
		1	<2.0			Pot	95% MeOH μ	33
	Na$^+$	1	1.65			Pot	H$_2$O μ	33
		1	4.57			Pot	95% MeOH μ	26, 33
	K$^+$	1	2.2	−3.0	0	Cal	H$_2$O μ	33
		1	7.0			Pot	95% MeOH μ	26, 33
	Rb$^+$	1	2.05	−4.2	−4.7	Cal	H$_2$O μ	33
		1	7.30			Pot	95% MeOH μ	33

41

42

43

continued overleaf

TABLE 13 (continued)

Ligand	Ion	i	log K Value	log K Method[a]	ΔH[b] Value	ΔH[b] Method[a]	ΔS[c]	Medium[a]	Ref.
	Cs^+	1	2.0	Pot	−5.4	Cal	−9.9	H_2O μ	26, 33
			7.0	Pot				95% MeOH μ	33
	Mg^{2+}	1	<2.0	Pot				H_2O μ	33
			<2.0	Pot				95% MeOH μ	33
	Ca^{2+}	1	2.0	Pot	0.16	Cal	9.6	H_2O μ	26, 33
			4.74	Pot				95% MeOH μ	33
	Sr^{2+}	1	3.40	Pot	−3.3	Cal	4.4	H_2O μ	26, 33
			7.06	Pot				95% MeOH μ	33
	Ba^{2+}	1	6.00	Pot	−6.2	Cal	6.7	H_2O μ	26, 33
			10.40	Pot				95% MeOH μ	33
	Li^+	1	<2.0	Pot				H_2O μ	33
	Na^+	1	<2.0	Pot				H_2O μ	33
			3.2	Pot				MeOH μ	33
	K^+	1	<2.0	Pot				H_2O μ	33
			6.0	Pot				MeOH μ	33
	Rb^+	1	<0.7	Pot				H_2O μ	33
			6.15	Pot				MeOH μ	33
	Cs^+	1	<2.0	Pot				H_2O μ	33
			>6.0	Pot				MeOH μ	33
	Mg^{2+}	1	<2.0	Pot				H_2O μ	33
	Ca^{2+}	1	2.0	Pot				H_2O μ	33
	Sr^{2+}	1	2.0	Pot				H_2O μ	33
	Ba^{2+}	1	3.65	Pot				H_2O μ	33

44

Structure	Cation	n	log K	Method	Solvent	Ref.
45	Li⁺	1	<2.0	Pot	H_2O μ	33
	Na⁺	1	<2.0	Pot	H_2O μ	33
	K⁺	1	2.7	Pot	MeOH μ	33
		1	<2.0	Pot	H_2O μ	33
	Rb⁺	1	5.4	Pot	MeOH μ	33
		1	<0.5	Pot	H_2O μ	33
	Cs⁺	1	5.7	Pot	MeOH μ	33
		1	<2.0	Pot	H_2O μ	33
		1	5.9	Pot	MeOH μ	33
	Mg²⁺	1	<2.0	Pot	H_2O μ	33
	Ca²⁺	1	<2.0	Pot	H_2O μ	33
	Sr²⁺	1	<2.0	Pot	H_2O μ	33
46	Na⁺	1	4.5	Pot	MeOH	9
	K⁺	1	5.8	Pot	MeOH	9
	Rb⁺	1	6.2	Pot	MeOH	9
	Cs⁺	1	>6.0	Pot	MeOH	9
	Ag⁺	1	6.0	Pot	H_2O	9
		1	>9.5	Pot	MeOH	9
	Ca²⁺	2	>6.0	Pot	MeOH	9
		1	6.53	Pot	H_2O	95
	Sr²⁺	1	6.97	Pot	H_2O	95
	Ba²⁺	1	8.0	Pot	H_2O	95

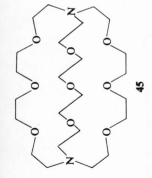

45

46

continued overleaf

TABLE 13 (continued)

Ligand	Ion	i	log K		ΔH^b		ΔS^c	Medium[a]	Ref.
			Value	Method[a]	Value	Method[a]			
47	Cu²⁺	1	24.8	Polg	−18.3	T	51.4	H₂O	81
48	Ni²⁺	1	22.2	Spec	−31.0	T	−2	H₂O μ	81
	H⁺	1	11.6	Pot				H₂O	9
	H⁺	1	10.7	Pot				H₂O	9
	H⁺	1	2.7	Pot				H₂O	9
	H⁺	1	2.3	Pot				H₂O	9
	Ni²⁺	1	21.9	Spec	−28	T	8	H₂O	76
	Cu²⁺	1	20	Spec				H₂O (blue complex)	9
	Cu²⁺	1	28	Spec				H₂O (red complex)	9

Cation	n	log K		ΔH		ΔS	Solvent	Ref.
Ag⁺	1			0	Cal		70% MeOH	45
Hg²⁺	2			−13.82	Cal		70% MeOH	45
Hg²⁺	1	5.61	Cal	−12.43	Cal		70% MeOH	45
Hg²⁺	2			−14.95		−24.5	70% MeOH	45
Ag⁺	1	2.71	Cal	−10.23	Cal	−21.9	H₂O	45
Ag⁺	1			−10.39	Cal		70% MeOH	45
Ag⁺	2			−10.54	Cal	−18.5	70% MeOH	45
Hg²⁺	1	3.68	Cal	−12.00	Cal		70% MeOH	45
Hg²⁺	2			−12.44	Cal		70% MeOH	45
Pb²⁺	1	0.94	Cal	−5.86	Cal	−15.4	H₂O	45
Ag⁺	1	4.26	Cal	−16.78	Cal	−20.1	70% MeOH	45
Ag⁺	2			−14.94	Cal		70% MeOH	45
Hg²⁺	1			−16.69	Cal		70% MeOH	45
Hg²⁺	2			−14.94	Cal		70% MeOH	45

50

51

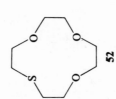

52

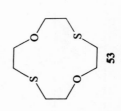

53

continued overleaf

TABLE 13 (continued)

Ligand	Ion	i	log K Value	Method[a]	ΔH[b] Value	Method[a]	ΔS[c]	Medium[d]	Ref.
54	Ag^+	1	4.97	Cal	−9.37	Cal	−8.7	H_2O	45
		2	2.45	Cal	−3.47	Cal	−0.4	H_2O	45
	Tl^+	1	0.80	Cal	−7.69	Cal	−22.1	H_2O	45
	Hg^{2+}	2			−16.87	Cal		H_2O	45
	Pb^{2+}	1	1.65	Cal	−5.14	Cal	−9.7	H_2O	45
55	Ag^+	1	5.56	Cal	−12.12	Cal	−15.2	H_2O	45
		2	3.31	Cal	−5.57	Cal	−3.5	H_2O	45
	Tl^+	1	<0.6	Cal				H_2O	45
	Hg^{2+}	1			−11.32	Cal		H_2O	45
		2	5.06	Cal	−7.79	Cal	−3.0	H_2O	45
	Pb^{2+}	1	1.21	Cal	−5.65	Cal	−13.4	H_2O	45
	Ag^+	1			−16.57	Cal		H_2O	45
		2	2.70	Cal	−1.00	Cal	9.0	H_2O	45
	Tl^+	1	<0.2	Cal				H_2O	45
	Hg^{2+}	1			−16.10	Cal		H_2O	45
		2	2.91	Cal	−5.00	Cal	−3.5	H_2O	45
	Pb^{2+}	1	1.62	Cal	−7.60	Cal	−18.1	H_2O	45

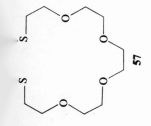

57

Cation	n	log K	Method	ΔH	Method	TΔS	Solvent	Ref
Hg²⁺	1	5.06	Cal		Cal	−19.7	H₂O	45
Pb²⁺	2	2.63	Cal	−11.32	Cal		H₂O	45
	1			−7.79	Cal	−3.0	H₂O	45
				−8.83	Cal	−17.6	H₂O	45

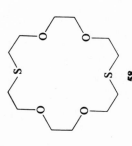

58

Cation	n	log K	Method	ΔH	Method	TΔS	Solvent	Ref
K⁺	1	1.15	Pot				MeOH	9
Ag⁺	1	4.34	Pot				H₂O	9
Tl⁺	1	0.93	Cal	−16.86	Cal		H₂O	45
Hg²⁺	1			−11.00	Cal	−32.6	H₂O	45
Pb²⁺	1	3.13	Cal	−17.67	Cal		H₂O	45
				−21.17	Cal	−56.7	H₂O	45

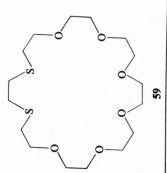

59

Cation	n	log K	Method	ΔH	Method	TΔS	Solvent	Ref
Ag⁺	1	4.47	Cal	−14.31	Cal	−27.5	H₂O	45
	2	4.98	Cal	−2.09	Cal	15.8	H₂O	45
Tl⁺	1			0	Cal		H₂O	45
Hg²⁺	1			−13.72	Cal		H₂O	45
Pb²⁺	1			0	Cal		H₂O	45

continued overleaf

TABLE 13 (continued)

Ligand	Ion	i	log K		ΔH^b		ΔS^c	Medium^d	Ref.
			Value	Method^a	Value	Method^a			
	Ag$^+$	1			-13.83	Cal		H$_2$O	45
	Tl$^+$	1			0	Cal		H$_2$O	45
	Hg^{2+}	1			-13.53	Cal		H$_2$O	45
60	Pb^{2+}	1			0	Cal		H$_2$O	45
61	Cu^{2+}	1	2.38	Spec				80% MeOH μ	89
	Cu^{2+}	1	2.43	Spec				80% MeOH μ	89

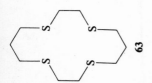

63	Cu^{2+}	1	3.49	Spec	80% MeOH μ	89
64	Cu^{2+}	1	2.36	Spec	80% MeOH μ	89
65	Cu^{2+}	1	0.95	Spec	80% MeOH μ	89

continued overleaf

TABLE 13 (continued)

Ligand	Ion	i	log K		ΔH^b		ΔS^c	Medium^a	Ref.
			Value	Method^a	Value	Method^a			
66	Cu^{2+}	1	2.99	Spec				80% MeOH μ	89

^{a} Abbreviations: NMR—nuclear magnetic resonance; Spec—spectroscopic techniques; EMF—EMF techniques; Cond—conductometric techniques; Cal—calorimetry; Polg—polarographic techniques; ORD—optical rotatory dispersion; CD—circular dichroism; T—temperature dependence of log k; VAP—vapor pressure osmometry; pH—pH titration.

^{b} ΔH in kcal mol^{-1}. A zero appearing in the ΔH column indicates that no significant amount of heat was produced in the calorimetric experiment, signifying that either $\Delta H = 0$ or no significant amount of reaction took place.

^{c} ΔS in cal (K mole)$^{-1}$

^{d} (i) Abbreviations: MeOH—methanol; EtOH—ethanol; Acnit—acetonitrile; THF—tetrahydrofuran; Chlor—chloroform; DMSO—dimethylsulfoxide; DMF—N,N-dimethylformamide; PC—propylene carbonate; DME—dimethoxyethane; Ace—acetone. (ii) %: In cases where "Medium" is expressed as percent of one component, the other component is H_2O unless otherwise indicated. For information on whether wt. % or vol. %, see original article. (iii) μ: The letter μ appearing in the "Medium" column signifies adjustment of ionic strength by addition of a salt to the reaction medium. (iv) Temperature: The temperature of reaction is 25°C unless otherwise indicated.

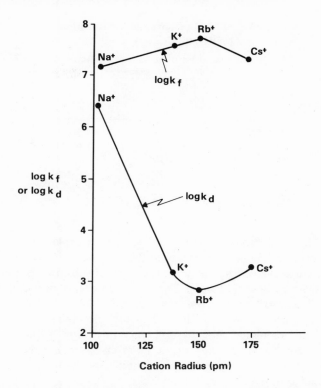

FIGURE 10. Variation of rate constants of formation (k_f) and dissociation (k_d) for alkali metal complexes of valinomycin with cation radius.[3]

is very close to that reported for the interaction of Na^+ with monactin in MeOH.

Chock[98] studied the interaction of monovalent cations with dibenzo-30-crown-10 in methanol. His results are listed in Table 14. The least complicated reaction scheme that could be devised was

$$\text{(fast)} \quad \begin{matrix} \text{crown}_1 \\ \Updownarrow \\ \text{crown}_2 \end{matrix} \quad + \text{ M}^+ \underset{k_d}{\overset{k_f}{\rightleftharpoons}} \text{ M-crown}^+$$

which involves reaction of the metal ion with crown molecules rapidly inter-converting between the different conformations. Interestingly, this sequence of steps is just the opposite of those chosen for valinomycin complexation. The specific rates in both directions are considerably faster than those for valinomycin, and curves similar to those in Figure 10 would be much flatter for this ligand, which is much less selective. The dissimilarity between the

TABLE 14. Rate Constants for Reaction
in MeOH at 25°C, $\mu = 0.15^{(98)}$:

$$M^+ + \text{Dibenzo-30-crown-10} \underset{k_d}{\overset{k_f}{\rightleftharpoons}} M\text{-crown}^+$$

Cation	k_f $(M^{-1}s^{-1})$	k_d (s^{-1})
Na^+	$>1.6 \times 10^7$	$>1.3 \times 10^5$
K^+	6×10^8	1.6×10^4
Rb^+	8×10^8	1.8×10^4
Cs^+	8×10^8	4.7×10^4
Tl^+	8×10^8	2.5×10^4
NH_4^+	$>3 \times 10^7$	$>1.1 \times 10^5$

data for NH_4^+ and those for the metal cations is explained in the more stringent conformational requirements of the organic cation.

Kinetic studies have been made on the effect of substituents and solvents on crown ethers by Schori *et al.*[58] The activation energy of decomplexation of Na^+ by dibenzo-18-crown-6 was the same (52.7 kJ mol^{-1}) in all solvents studied, but considerably less than that for dicyclohexo-18-crown-6 in methanol. It is postulated that the difference in energy is that required to effect a conformational rearrangement, more energy being required in the case of the aromatic substituent. Thus, flexibility of the ligand is seen as an important factor in the decomplexation step.

The kinetics of cation complexation by 15-crown-5 and 18-crown-6 in methanol have been studied by Eyring and co-workers.[99,100] These results are shown in Table 15. They conclude that the high selectivity of 18-crown-6 for K^+ over other alkali cations arises from the slowness of the dissociation

TABLE 15. Rate Constants for Reaction
at 25°C in H_2O, $\mu = 0.3^{(99,100)}$:

$$M^+ + \text{18-Crown-6} \underset{k_d}{\overset{k_f}{\rightleftharpoons}} M\text{-crown}^+$$

Cation	k_f $(M^{-1}s^{-1})$	k_d (s^{-1})
Na^+	$\sim 2.2 \times 10^8$	3.4×10^7
K^+	$\sim 4.3 \times 10^8$	3.7×10^6
Rb^+	$\sim 4.4 \times 10^8$	1.2×10^7
Cs^+	$\sim 4.3 \times 10^8$	4.4×10^7
Ag^+	$\sim 11.3 \times 10^8$	3.6×10^7
Tl^+	9.9×10^8	5.3×10^6
NH_4^+	$\sim 5.6 \times 10^8$	4.4×10^7

step. This same observation is made for the reactions of dibenzo-18-crown-6 with K^+ and Rb^+ by Shporer and Luz[101] explaining the preference of this ligand for K^+. These results seem to bear out the "best fit" explanation for cation selectivity.

5.3. Macrobicyclic Ligands

From kinetic studies with cryptate complexes, Lehn et al.[22] concluded that, as in the case of the cyclic polyethers, a simple complexation–decomplexation exchange mechanism was to be preferred over the bimolecular exchange process. Representative data are listed in Table 16. It is concluded that the greater stability of macrobicyclic complexes over crown ether complexes is due to the slower rate of decomplexation of the former. Indeed, comparing data in Tables 15 and 16, a factor of 10^6 separates the slower macrobicycles from 18-crown-6. They also conclude that the stability sequence of cryptates follows the trend in dissociation rates, which is in the opposite direction to rates of exchange of H_2O molecules in the cation hydration shell. Work by Loyola et al.[102] supports this view, showing that binding of Ca^{2+} to [2.1.1], [2.2.1], and [2.2.2] increases mainly with decreasing decomplexation rates. It is noted that association rates are considerably lower than diffusion-controlled rates. The enhanced associative rate for Tl^+ with [2.2.2] (and for Ag^+ with 18-crown-6) may reflect the importance of the role that metal ion orbitals may be playing in macrocycle binding. The increased stability of dipositive cations is attributed to their slowness of decomplexation when compared to monovalent cations.

TABLE 16. Rate Constants for Reaction in Water

$$M^{n+} + L \underset{k_d}{\overset{k_f}{\rightleftharpoons}} ML^{n+}$$

Complex	Temperature (°C)	k_f $(M^{-1}\,s^{-1})$	k_d (s^{-1})	Method	Reference
[2.1.1] $Li^+ClO_4^-$	25	0.98×10^3	4.9×10^{-3}	^7Li NMR	103
[2.1.1] $Ca^{2+}(Cl^-)_2$	25	1.6×10^2	1.6×10^2	SF[a]	102
[2.2.1] $Ca^{2+}(Cl^-)_2$	25	1.2×10^4	1.9×10^{-3}	SF[a]	102
[2.2.2] Na^+Cl^-	3	$2 \times 10^{5\,b}$	27	^1H NMR	22
[2.2.2] K^+Cl^-	36	$7.5 \times 10^{6\,b}$	38	^1H NMR	22
[2.2.2] Rb^+Cl^-	9	$7.5 \times 10^{5\,b}$	38	^1H NMR	22
[2.2.2] Tl^+Cl^-	40	$2.5 \times 10^{8\,b}$	60	^1H NMR	22
[2.2.2] $Ca^{2+}(Cl^-)_2$	25	6.6×10^3	0.26	SF[a]	102
[2.2.2] $Sr^{2+}(Br^-)_2$	25	$6 \times 10^{3\,b}$	10^{-4}	Pot	22
[2.2.2] $Ba^{2+}(Cl^-)_2$	25	$3 \times 10^{4\,b}$	10^{-5}	Pot	22

[a] Stopped flow.
[b] These values are approximate—calculated from data at two different temperatures.

Comparing reaction of [2.1.1] and [2.2.1] with Li$^+$ in pyridine, it was found that the rate of decomplexation for the latter larger ligand is 10^4 times as rapid as for the smaller.[103] This effect may be attributed to the loose fit of Li$^+$ in the larger ligand and to the greater flexibility of this macrocycle.

ACKNOWLEDGMENT

This work was supported in part by the National Science Foundation (CHE 76-10991) and the National Institutes of Health (GM18811).

References

1. B. C. Pressman, in: *Inorganic Biochemistry* (G. I. Eichhorn, ed.), Vol. 1, p. 203, Elsevier, Amsterdam (1973).
2. Y. A. Ovchinnikov, V. T. Ivanov, and A. M. Shkrob, *Membrane-Active Complexones*, Elsevier, Amsterdam (1974).
3. E. Grell, T. Funck, and F. Eggers, in: *Membranes* (G. Eisenman, ed.), Vol. 3, p. 1, Marcel Dekker, New York (1975).
4. E. A. Lewis and L. D. Hansen, *J. Chem. Soc. Perkin Trans. 2* **1973**, 2081 (1973).
5. D. H. Busch, K. Farmery, V. Goedken, V. Katovic, A. C. Melnyk, C. R. Sperati, and N. Tokel, *Advan. Chem. Ser.* **100**, 44 (1971).
6. L. F. Lindoy, *Chem. Soc. Rev.* **4**, 421 (1975).
7. C. J. Pedersen, *J. Am. Chem. Soc.* **89**, 7017 (1967).
8. R. M. Izatt, D. J. Eatough, and J. J. Christensen, *Struct. Bonding* **16**, 161 (1973).
9. J. J. Christensen, D. J. Eatough, and R. M. Izatt, *Chem. Rev.* **74**, 351 (1974).
10. D. Midgley, *Chem. Soc. Rev.* **4**, 549 (1975).
11. R. M. Izatt, N. E. Izatt, B. E. Rossiter, J. J. Christensen, and B. L. Haymore, *Science* **199**, 994 (1978).
12. C. J. Pedersen, in: *Synthetic Multidentate Macrocyclic Compounds* (R. M. Izatt and J. J. Christensen, eds.), Academic Press, New York (1978).
13. G. W. Gokel and H. D. Durst, *Synthesis* 168 (March 1976).
14. A. C. Knipe, *J. Chem. Ed.* **53**, 618 (1976).
15. J. M. Girodeau, J. M. Lehn, and J. P. Sauvage, *Angew. Chem. Int. Ed. Engl.* **14**, 764 (1975).
16. D. J. Cram, R. C. Helgeson, L. R. Sousa, J. M. Timko, M. Newcomb, P. Moreau, F. DeJong, G. W. Gokel, D. H. Hoffman, L. A. Domeier, S. C. Peacock, K. Madan, and L. Kaplan, *Pure Appl. Chem.* **43**, 327 (1975).
17. D. J. Cram, in: *Applications of Biochemical Systems in Organic Chemistry*, Part 2 (J. B. Jones, ed.), John Wiley and Sons, New York (1976).
18. D. J. Cram and J. M. Cram, *Science* **183**, 803 (1974).
19. S. C. Shah, S. Kopolow, and J. Smid, *J. Polym. Sci.* **14**, 2023 (1976).
20. E. Blasius, W. Adrian, K.-P. Janzen, and G. Glautke, *J. Chromatogr.* **96**, 89 (1974).
21. E. L. Cussler and D. F. Evans, *Sep. Purif. Methods* **3**, 399 (1974).
22. J. M. Lehn, *Struct. Bonding* **16**, 1 (1973).
23. J. J. Christensen, J. O. Hill, and R. M. Izatt, *Science* **174**, 459 (1971).
24. C. Kappenstein, *Bull. Soc. Chim. Fr.*, 89 (1974).
25. R. M. Izatt, L. D. Hansen, D. J. Eatough, J. S. Bradshaw, and J. J. Christensen, in: *Metal–Ligand Interactions in Organic Chemistry and Biochemistry*, Part 1 (B. Pullman and N. Goldblum, eds.), D. Reidel, Dordrecht, Holland (1977).
26. E. Kauffmann, J. M. Lehn, and J. P. Sauvage, *Helv. Chim. Acta* **59**, 1099 (1976).

27. L. Y. Martin, L. J. DeHayes, L. J. Zompa, and D. H. Busch, *J. Am. Chem. Soc.* 96, 4046 (1974).
28. L. Fabbrizzi, P. Paoletti, and A. B. P. Lever, *Inorg. Chem.* 15, 1502 (1976).
29. T. E. Jones, L. L. Zimmer, L. L. Diaddario, D. B. Rorabacher, and L. A. Ochrymowycz, *J. Am. Chem. Soc.* 97, 7163 (1975).
30. N. K. Dalley, in: *Synthetic Multidentate Macrocyclic Compounds* (R. M. Izatt and J. J. Christensen, eds.), Academic Press, New York (1978).
31. R. D. Shannon and C. T. Prewitt, *Acta Cryst.* B25, 925 (1969).
32. R. M. Izatt, R. E. Terry, B. L. Haymore, L. D. Hansen, N. K. Dalley, A. G. Avondet, and J. J. Christensen, *J. Am. Chem. Soc.* 98, 7620 (1976).
33. J.-M. Lehn and J. P. Sauvage, *J. Am. Chem. Soc.* 97, 6700 (1975).
34. C. Nave and M. R. Truter, *J. Chem. Soc. Dalton Trans.* 1974, 2351 (1974).
35. P. H. Davis, L. K. White, and R. L. Belford, *Inorg. Chem.* 14, 1753 (1975).
36. R. E. DeSimone and M. D. Glick, *J. Am. Chem. Soc.* 97, 942 (1975).
37. M. D. Glick, D. P. Gavel, L. L. Diaddario, and D. B. Rorabacher, *Inorg. Chem.* 15, 1190 (1976).
38. B. Metz, D. Moras, and R. Weiss, *J. Chem. Soc. Perkin Trans.* 2, 1976, 423 (1976).
39. J. D. Dunitz and P. Seiler, *Acta Cryst.* B30, 2739 (1974).
40. M. Dobler, J. D. Dunitz, and P. Seiler, *Acta Cryst.* B30, 2741 (1974).
41. M. R. Truter, *Struct. Bonding* 16, 71 (1973).
42. I. Goldberg, *Acta Cryst.* B31, 2592 (1975).
43. G. Eisenman and S. J. Krasne, *MTP Int. Rev. Sci. Biochem. Ser.* 2, 27 (1975).
44. B. E. Jepson and R. DeWitt, *J. Inorg. Nucl. Chem.* 38, 1175 (1976).
45. R. M. Izatt, R. E. Terry, L. D. Hansen, A. G. Avondet, J. S. Bradshaw, N. K. Dalley, T. E. Jensen, B. L. Haymore, and J. J. Christensen, *Inorg. Chim. Acta* 30, 1 (1978).
46. A. Cassol, A. Seminaro, and G. DePaoli, *Inorg. Nucl. Chem. Lett.* 9, 1163 (1973).
47. R. B. King and P. R. Heckley, *J. Am. Chem. Soc.* 96, 3118 (1974).
48. M. E. Harman, F. A. Hart, M. B. Hursthouse, G. P. Moss, and P. R. Raithby, *J. Chem. Soc. Chem. Commun.* 1976, 396 (1976).
49. R. M. Costes, G. Folcher, N. Keller, P. Plurien, and P. Rigny, *Inorg. Nucl. Chem. Lett.* 11, 469 (1975).
50. R. M. Costes, G. Folcher, P. Plurien, and P. Rigny, *Inorg. Nucl. Chem. Lett.* 11, 13 (1976).
51. G. Bombieri, G. DePaoli, A. Cassol, and A. Immirzi, *Inorg. Chim. Acta* 18, L23 (1976).
52. J. H. Moreno and J. M. Diamond, in: *Membranes* (G. Eisenman, ed.), Vol. 3, p. 383, Marcel Dekker, New York (1975).
53. H. K. Frensdorff, *J. Am. Chem. Soc.* 93, 600 (1971).
54. N. K. Dalley, J. S. Smith, S. B. Larson, K. L. Matheson, J. J. Christensen, and R. M. Izatt, *J. Chem. Soc. Chem. Commun.* 1975, 84 (1975).
55. N. K. Dalley and S. B. Larson, 33rd Northwest Regional Meeting, American Chemical Society, Seattle, Washington, June, 1978, paper no. 138.
56. R. M. Izatt, J. D. Lamb, G. E. Maas, R. E. Asay, J. S. Bradshaw, and J. J. Christensen, *J. Am. Chem. Soc.* 99, 2365 (1977).
57. B. Dietrich, J.-M. Lehn, and J. P. Sauvage, *J. Chem. Soc. Chem. Commun.* 1973, 15 (1973).
58. E. Shchori and J. Jagur-Grodzinski, *Isr. J. Chem.* 11, 243 (1973).
59. R. Ungaro, B. El Haj, and J. Smid, *J. Am. Chem. Soc.* 98, 5198 (1976).
60. K. H. Pannell, W. Yee, G. S. Lewandos, and D. C. Hambrick, *J. Am. Chem. Soc.* 99, 1457 (1977).

61. K. Angelis, *Anal. Lett.* **8**, 895 (1975).
62. W. Steinmann and T. A. Kaden, *Helv. Chim. Acta* **58**, 1358 (1975).
63. M. Bourgoin, K. H. Wong, J. Y. Hui, and J. Smid, *J. Am. Chem. Soc.* **97**, 3462 (1975).
64. W. D. Curtis, D. A. Laidler, J. F. Stoddart, and G. H. Jones, *J. Chem. Soc. Chem. Commun.* **1975**, 833 (1975).
65. E. M. Arnett and T. C. Moriarity, *J. Am. Chem. Soc.* **93**, 4908 (1971).
66. T. H. Ryan, J. Koryta, A. Hofmanova-Matejkova, and M. Brezina, *Anal. Lett.* **7**, 335 (1974).
67. J. J. Dechter and J. I. Zink, *J. Am. Chem. Soc.* **98**, 845 (1976).
68. R. M. Izatt, R. E. Terry, D. P. Nelson, Y. Chan, D. J. Eatough, J. S. Bradshaw, L. D. Hansen, and J. J. Christensen, *J. Am. Chem. Soc.* **98**, 7626 (1976).
69. J. D. Lamb, Synthetic Macrocyclic Ligands: Stability, Selectivity and Transport of Their Cation Complexes, Ph.D. Dissertation, Brigham Young University, Provo, Utah (1978).
70. N. Matsuura, K. Umemoto, Y. Takeda, and A. Sasaki, *Bull. Chem. Soc. Jpn.* **49**, 1246 (1976).
71. A. Agostiano, M. Caselli, and M. Della Monica, *J. Electroanal. Chem.* **74**, 95 (1976).
72. G. Eisenman, S. Ciani, and G. Szabo, *J. Membrane Biol.* **1**, 294 (1969).
73. P. R. Danesi, R. Chiarizia, C. Fabiani, and C. Domenichini, *J. Inorg. Nucl. Chem.* **38**, 1226 (1976).
74. D. K. Cabbiness and D. W. Margerum, *J. Am. Chem. Soc.* **91**, 6540 (1969).
75. D. Munro, *Chem. Br.* **13**, 100 (1977).
76. F. P. Hinz and D. W. Margerum, *Inorg. Chem.* **13**, 2941 (1974).
77. A. Dei and R. Gori, *Inorg. Chim. Acta* **14**, 157 (1975).
78. P. Paoletti, L. Fabbrizzi, and R. Barbucci, *Inorg. Chem.* **12**, 1961 (1973).
79. E. K. Barefield, F. V. Lovecchio, N. E. Tokel, E. Ochiai, and D. H. Busch, *Inorg. Chem.* **11**, 283 (1972).
80. M. Kodama and E. Kimura, *J. Chem. Soc. Chem. Commun.* **1975**, 326 (1975).
81. M. Kodama and E. Kimura, *J. Chem. Soc. Dalton Trans.* **1976**, 116 (1976).
82. B. Bosnich, M. L. Tobe, and G. A. Webb, *Inorg. Chem.* **4**, 1109 (1965).
83. N. F. Curtis, D. A. Swann, and T. N. Waters, *J. Chem. Soc. Dalton Trans.* **1973**, 1408 (1973).
84. P. U. Früh and W. Simon, *Protides of the Biological Fluids—20th Colloquium* (H. Peeters, ed.), Pergamon Press, New York (1973).
85. M. Kodama and E. Kimura, *Bull. Chem. Soc. Jpn.* **49**, 2465 (1976).
86. G. Anderegg, *Helv. Chim. Acta* **58**, 1218 (1975).
87. B. Metz, D. Moras, and R. Weiss, *J. Inorg. Nucl. Chem.* **36**, 785 (1974).
88. D. K. Cabbiness and D. W. Margerum, *J. Am. Chem. Soc.* **92**, 2151 (1970).
89. T. E. Jones, D. B. Rorabacher, and L. A. Ochrymowycz, *J. Am. Chem. Soc.* **97**, 7485 (1975).
90. G. Cornelius, W. Gärtner, and D. H. Haynes, *Biochemistry* **13**, 3052 (1974).
91. P. U. Früh, J. T. Clerc, and W. Simon, *Helv. Chim. Acta* **54**, 1445 (1971).
92. V. Madison, M. Atreyi, C. M. Deber, and E. R. Blout, *J. Am. Chem. Soc.* **96**, 6725 (1974).
93. Ch. U. Züst, P. U. Früh, and W. Simon, *Helv. Chim. Acta* **56**, 495 (1973).
94. E. Shchori, N. Nae, and J. Jagur-Grodzinski, *J. Chem. Soc. Dalton Trans.* **1975**, 2381 (1975).
95. J. M. Lehn and M. E. Stubbs, *J. Am. Chem. Soc.* **96**, 4011 (1974).
96. G. W. Liesegang and E. M. Eyring, in: *Synthetic Multidentate Macrocyclic Compounds* (R. M. Izatt and J. J. Christensen, eds.), Academic Press, New York (1978).

97. E. Shchori, J. Jagur-Grodzinski, Z. Luz, and M. Shporer, *J. Am. Chem. Soc.* **93**, 7133 (1971).
98. P. B. Chock, *Proc. Natl. Acad. Sci. USA* **69**, 1939 (1972).
99. G. W. Liesegang, M. M. Farrow, N. Purdie, and E. M. Eyring, *J. Am. Chem. Soc.* **98**, 6905 (1976).
100. G. W. Liesegang, M. M. Farrow, F. A. Vazquez, N. Purdie, and E. M. Eyring, *J. Am. Chem. Soc.* **99**, 3240 (1977).
101. M. Shporer and Z. Luz, *J. Am. Chem. Soc.* **97**, 665 (1975).
102. V. M. Loyola, R. G. Wilkins, and R. Pizer, *J. Am. Chem. Soc.* **97**, 7382 (1975).
103. Y. M. Cahen, J. L. Dye, and A. I. Popov, *J. Phys. Chem.* **79**, 1292 (1975).

4

Structural Aspects

Neil F. Curtis

1. Introduction

1.1. Scope and Organization

This chapter is concerned with structural aspects of metal-ion compounds formed by macrocyclic ligands. Compounds of the porphyrin and phthalocyanine macrocycles, which are discussed in Chapters 7 and 8, are excluded. Compounds of the oxygen donor macrocycles, the cyclic polyethers and polycyclic "cryptates," which are discussed in Chapter 9, are also excluded, except for the two reported structural studies of transition-element derivatives. Only macrocycles for which X-ray structural studies have been reported are included, and the listing of reported structures is hopefully comprehensive through 1976.*

The majority of the macrocyclic ligands included have nitrogen donor atoms, usually four, and the most common macrocycle ring size is 14. Most structures reported have central metal ions from the latter half of the first transition series, with singlet ground-state nickel(II) the most common metal ion.

The chapter is divided into a number of sections which each deal with macrocycles of a particular type. The classes, which are to some extent arbitrary, are as follows.

Class 1. Cyclic amines (polyaza saturated macrocycles) with secondary or tertiary amine donor atoms (Section 2).

---

* Structures reported after this date, to late 1978, and some inadvertently omitted, have been listed in an Addendum at the end of the chapter, but are not included in the general discussion.

Neil F. Curtis · Department of Chemistry, Victoria University of Wellington, Wellington, New Zealand.

Class 2. Cyclic imines (polyaza unsaturated macrocycles) with imine nitrogen, or imine plus amine donor atoms (Section 3).

Class 3. Cyclic amine/imines with 2,6-fused pyridine rings, with pyridine nitrogen plus amine, or plus amine and imine donor atoms (Section 4).

Class 4. Tetraazamacrocycles with 2-imino(or 2-amido)-benzaldimine chelate rings, with imine, or imine plus deprotonated secondary amine donor atoms (Section 5).

Class 5. Dibenzo[b, i]-1,4,8,11-tetraazacyclotetradec-2,4,6,9,11,13-hexanato(2-) [dibenzo-tetraazacyclotetradecinato(2-)] compounds (Section 6).

Class 6. Cyclic hydrazines and hydrazones (macrocycles with adjacent nitrogen atoms) (Section 7).

Class 7. Tetraoxa- and tetrathiamacrocycles (cyclic tetraethers and tetrathia-ethers) (Section 8).

Class 8. Macrocycles with more than one type of heteroatom (Section 9).

Class 9. Binucleating macrocycles (macrocycles which coordinate two metal ions *within* the macrocycle) (Section 10).

Class 10. Cyclohexaphosphazenes (four nitrogen donor atoms in a N_6P_6 heterocycle) (Section 11).

Class 11. Clathrochelates (tricyclomacrocycles which encapsulate metal ions) (Section 12).

1.2. Order of Tabulation

Within each class, macrocycles are arranged in a numbered sequence determined by the following rules, in decreasing order of precedence.

1. Increasing ring size.
2. Increasing unsaturation.
3. Increasing ring carbon atom substitution.
4. Increasing donor atom substitution (i.e., secondary before tertiary amines).

For each macrocycle, compounds are listed in the following order.

1. Free ligand, protonated ligand.
2. Coordination compounds in order of atomic number of central atom.
3. For any central atom, in order of increasing oxidation state.
4. For any central atom ion, in order of increasing spin multiplicity.

Each macrocyclic ligand is given a serial number, determined by the class of the macrocycle (first digit), and the sequence within the class (e.g., **2.13**). The individual compounds of macrocycles, for which structures have been reported, are given alphabetical sequence indication, e.g., (**2.13c**), used in the structure listing, tabulation of dimensions, molecular structure figures, and in the general discussion. The additional structures given in the Addendum are listed according to the same sequence rules. If a structure of a macrocyclic ligand has already appeared in the main body of the text, the same code number is used; otherwise, the number sequence for the section is continued where it left off.

Reported structures are tabulated, with the structural formula of the ligand (and in a proportion of cases with a representation of the molecular structure), and with description of significant aspects of the structure. Features of the structures of each of the classifications are discussed at the beginning of each section, and structural parameters are tabulated, where appropriate. Since the latter are intended mainly for comparative purposes, bond lengths have been rounded off to the nearest picometer, and angles to the nearest degree, with average values for *chemically* equivalent dimensions quoted, unless there is a large difference in the values. Minor variations in the dimensions which arise from packing effects, etc. are thus ignored, and generally the standard deviation is appreciably less than the last digit (where uncertainty extends to the last digit quoted, standard deviations are included).

For each published structure the reliability index R, together with the number of reflections used for the refinement, is given and any problems that hindered refinement, such as disorder, are mentioned. Structures based on photographic data are indicated by the symbol P. The nature of the publication is indicated by symbols:

***: Full paper, with publication (or deposition) of structure factors and full structural data, and with some discussion and interpretation of significant features of the structure.

**: A paper reporting a structural study, often in preliminary "note" form, but including all significant dimensions. Atom site parameters are published or available.

*: A paper reporting a structural study, and including some significant dimensions.

No asterisk: A structure reported with scant details, or reported as having been completed, but not published.

The donor atoms of each macrocycle are numbered on the structural formula, as for the systematic name used for the macrocycle, and these numbers are used in the discussion and tabulated dimensions. These numbers in general are *not* the same as the crystallographic atom numbering in the original paper.

2. Class 1: Cyclic Amines—Saturated Polyaza Macrocycles

2.1. Introduction

This section describes structures of complexes formed by saturated cyclic amines. Structures of cyclic amines with fewer than three nitrogen atoms are not considered and no structures have been reported for amines with more than four nitrogen atoms. For cyclic triamines, there is one preliminary report of a structural study of a [9]aneN$_3$ compound, **1.1**. The remainder of the structures are of cyclic tetramine compounds, and a part from two [12]aneN$_4$, **1.2**, and one [13]aneN$_4$, **1.3**, structures, all have [14]aneN$_4$ macrocycles. The majority of the structures have the macrocycle in planar coordination (square-planar, or *trans*-octahedral arrangements), but several structures with the macrocycles in folded coordination (*cis*-octahedral or trigonal bipyramidal arrangements) have been reported. Apart from studies of one cyclic tetra-tertiary amine, **1.8**, the macrocycles are all cyclic tetrasecondary amines.

Key dimensions of cyclic amine complexes are listed in Table 1 (planar coordination) and Table 2 (folded coordination).

2.2. Configurations and Conformations of Coordination Cyclic Tetramines

A coordinated cyclic tetraamine can, in principle, occur in several configurations arising from the chiral coordinated nitrogen centers present. For [14]aneN$_4$ macrocycles, with alternating five- and six-membered chelate rings, these configurations have been designated I–V (Figure 1) by Bosnich *et al.*[1] and this terminology will be used. An equivalent set of configurations occurs for [13]aneN$_4$ and [15]aneN$_4$ macrocycles, with two configurations derived from II, but for [12]aneN$_4$ and [16]aneN$_4$ macrocycles with all five-membered, or all six-membered chelate rings, configurations III and IV become equivalent. For the [14]aneN$_4$ system the relative energies of the

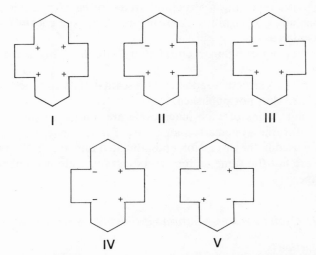

FIGURE 1. Configurations of coordinated 1,4,8,11-tetraazacyclotetradecane. The + indicates that the NH group is above the plane of the (flattened) macrocycle; the − indicates that it is below.

TABLE 1. Dimensions for Compounds of Cyclic Tetramines in Planar Coordination

Compound[a]	Subst.[b]	Ion[c]	Spin[d]	M–N[e]	Config.[f]	MN_4[g]	C.No.[h]	i	j	k (Chelate rings[i,j,k])
							[13]aneN₄ Compounds			
1.3a	Me₃	Ni(II)	0	186(2)	III	±2	4	(1–4)	90°	17, −50
								(4–7)	85°	−68, −82
								(7–10)	90°	−48, 38
								(10–13)	95°	102, 79(−57, 182), 101[l]
							[14]aneN₄ Compounds			
1.4a		—		206[n]	III	i		(5)°		{−25}, 26, −20, {34}[m]
								(6)°		{34}, 25–27, 25, {−25}
1.4b		2H⁺		205[n]	III	i		(5)°		−43, 32
								(6)°		81, 47, 84
1.4d		Co(II)	½	198	III	i	4[p]	(5)	86°	
								(6)	94°	
1.4f		Ni(II)	1	206	III	i	6[q]	(5)	86°	Gauche
								(6)	94°	Chair
1.5a	Me₂	Co(III)		201(1)	III	i	6[r]	(5)		33, −46
1.6a	Me₂Ph₂	Cu(II)		206	III	[2], −3, 2	4[s]	(4–8)		91(92), 37, 77(32)
								(11–1)		−94(−106), −56, −93(−79)
1.7b	Me₆	Ni(II)	0	190	V	[4], 4, −6	4	(5)	88°	−58, 12
								(6)	93°	92(95, 232), 45, −44(−57)
1.7c	Me₆	Ni(II)	0	193	I	±6	4	(5)	88°	40, −25
								(4–8)	88°	−89(−62, −234), −49, −75(−35)
								(11–1)	96°	−129(−126, −250), −136, −129(−144)

continued overleaf

TABLE 1 (continued)

Compound[a]	Subst.[b]	Ion[c]	Spin[d]	M–N[e]	Config.[f]	MN$_4$[g]	C.No.[h]	Chelate rings[i,j,e]		
								i	j	k
1.7d	Me$_6$	Ni(II)	0	193	III	i	4	(5)	86°	−58, 13
								(4–8)	94°	97(42, 247), 127, 10(14)
								(11–1)	94°	−72(−25, −221), −32, −83(−82)
1.8a	N-Me$_4$	Ni(II)	1	216	III	?	6^t	(5)	87°	−43, 39^l
								(6)	93°	106, 81, 104
1.8b	N-Me$_4$	Ni(II)	1	210	I	[33]	5^u	(5)	84°	41, −18^l
								(4–8)	92°	109, 97, 109
								(11–1)	94°	62, 18, 62
Additional Entries for [14]aneN$_4$ Compounds[v]										
1.4i		Cu(II)		202(4)		i	6^w	(5)	86°	37, −35^m
								(6)	94°	84, 48, 84
1.5b	Me$_2$	Ni(II)	0	193	III	i	4	(5)	86°	−44, 24^m
								(6)	94°	80, 31, 66
1.11a	Me$_4$	Ni(II)	0	194^x	III	i	4	(5)	87°	−36, 35^m
								(6)	93°	75, 49, 88
1.12a	Me$_2$Et$_2$	Ni(II)	I(?)^t	199	III	i	6(?)^y	(5)	86°	−34, 40^m
								(6)	94°	75, 37, 81
1.7j	Me$_6$	Cu(II)		204	III	0^z	6^{aa}	(5)	86°	^{m}
								(6)	94°	
1.7h	Me$_6$	Ag(II)		217	III	i	6^{bb}	(5)	84°	^{m}
								(6)	96°	

[a] Compounds are numbered as in the listing of structures.
[b] Substituents on the macrocycle.
[c] Central metal ion and oxidation state.

d Spin quantum number S of central ion, where significant.

e Mean M–N distance, in picometers.

f Configuration of nitrogen centers, as represented in Figure 1.

g Characteristics of MN_4 plane. i indicates centrosymmetric structure. A number in square brackets indicates the displacement of the metal from the N_4 plane. Following numbers indicate displacement of the nitrogen atoms from the MN_4 plane, in picometers. This displacement is generally tetrahedral in character; ±2 indicates displacement of +2 for N(1), N(8); −2 for N(4), N(11); 2, −3 indicates displacement of +2 for N(1), N(8), and −3 for N(4), N(11).

h Coordination number. Where additional ligands are present, M-donor distances are shown in a footnote.

i Chelate rings indicated by numbers of included nitrogen atoms, e.g., (1–4). The numbering is as in the formulas and is generally different from the crystallographic numbering in the original publication. Where pairs of chelate rings are identical, or similar, in dimensions, the symbols (5) and (6) indicate five- and six-membered chelate rings, respectively. For similar chelate rings, average values of dimensions are listed.

j Chelate (NMN) angle.

k Displacement from MN_2 chelate plane, in picometers; ring atoms in numerical sequence, with values for substituent atoms in parentheses. Displacement from N_4 coordination plane indicated by inclusion of nitrogen donor atom values in [ ]. Deviation from mean ring plane indicated by inclusion of donor atom values in { }.

l Values calculated from published atom positions.

m Inclusion of torsion (dihedral) angle data in publication.

n Mean distance "center"-N.

o Nominal chelate rings, including center.

p Weak interaction with *trans* perchlorates, Co–O, 241 pm.

q *Trans* chlorides; Ni–Cl, 249 pm.

r *Trans* azides; Co–N, 195 pm.

s Weak interaction with *trans* nitrate; Cu–O, 250(2) pm.

t Dinuclear structure with *trans* bridging and terminal azide; Ni–N(terminal); 211(1); Ni–N(bridging), 215(1) pm. Angles N–N–Ni, 138° (terminal), 142° (bridging).

u Square-pyramidal structure with axial azide; Ni–N, 195(1) pm, angle N–N–Ni, 136°.

v See the Addendum at the end of this chapter.

w Weak interaction with *trans* perchlorates; Cu–O, 257(4) pm.

x Ni–N(1), 192 pm, Ni–N(2), adjacent methyl, 197 pm.

y See Addendum. Interaction with *trans* perchlorates; Ni–O, 276 pm.

z CuN_4 coplanar within experimental error.

aa Weak interaction with *trans* perchlorates; Cu–O, 268 pm.

bb Weak interaction with *trans* nitrates; Ag–O, 280 pm.

TABLE 2. Dimensions for Compounds of Cyclic Tetramines in Folded Coordination^a

Compound^b	Subst.^{c}	Ion^d	NMN angles^e	M–N^{f,g}	Chelate rings^{h,i,j} h	i	j
			[12]aneN$_4$ Compounds				
1.2a		Co(III)	(1,7) 164°	197^k	(1–4)	85°	−32, 39
			(4,10) 94°	195	(7–10)	84°	−31, −77
			[14]aneN$_4$ Compounds				
1.4e		Co(III)	(1,8) 173°	200	(5)	87°	25, −45
			(4,11) 96°	^{l}	(6)	90°	Chair
1.7e	Me$_6$	Ni(II)	(1,8) 175°	215(1)^m	(5)	85°	−27, 48
			(4,11) 103°	210(1)	(6)	92°	75(38, 225), 45, 90(89)
1.7f	Me$_6$	Ni(II)	(1,8) 172°	211	(1–4)	85°	57, −20
			(4,11) 92°	^{n}	(4–8)	89°	96(51, 220), 151, 43(97)
					(8–11)	85°	65, −3
					(11–14)	92°	92(52, 236), 58, 91(94)
1.7g	Me$_6$	Cu(II)	(4,11)^o 115°	200(2)^o	(5)		Gauche
				206(2)^{pq}	(6)		Chair

^{a} All with nitrogen configuration V, in *cis*-octahedral coordination, except for **1.6f**, which has a trigonal-bipyramidal arrangement.
^{b} Compounds are numbered as in the listing of structures.
^{c} Substituents on the macrocycle.
^{d} Central metal ion and oxidation state. The Ni(II) compounds have $S = 1$.
^{e} For example, (1,7) indicates the angle N(1)–M–N(7).
^{f} Mean M–N distance, in picometers. Where the distances to the "fold-line" and "equatorial" nitrogen atoms differ significantly they are quoted separately, fold-line nitrogen values first (atom numbers as for adjacent NMN angles).
^{g} Other ligands and M–donor-atom distances shown in footnotes.
^{h} Chelate ring identification.
^{i} Chelate (NMN) angle.
^{j} Deviations from NMN plane. For g, h, i, see footnotes to Table 1.
^{k} *Cis* nitro groups; Co–N, 193 pm; angle NCoN, 85°.
^{l} Diaminoethane as additional ligand; Co–N, 200 pm; angle NCoN, 84°.
^{m} Chelate acetate as additional ligand; Ni–O, 211(1) pm; angle ONiO, 62°.
^{n} Chelate acetylacetonate as additional ligand; Ni–O, 205(1) pm; angle ONiO, 87°.
^{o} Axial nitrogens of trigonal bipyramid.
^{p} Equatorial nitrogens of trigonal bipyramid.
^{q} With chloride as bridging group for dimer, in an equatorial site; Cu–Cl, 250 pm; angles in equatorial plane, ClCuN, 124°, 122°.

various configurations in planar and in folded coordination have been discussed.[1] In planar coordination, a major factor determining the relative energies of the configurations appears to be the conformations possible for the individual chelate rings making up the macrocyclic complex. The conformations of bidentate five- and six-membered chelate rings are related to those of cyclopentane and cyclohexane, respectively, and have been extensively investigated,[2] and the same conformations are possible when the chelate ring is incorporated into a linear, or cyclic, polyamine. For the five-membered (diaminoethane) chelate ring, a range of gauche conformations from symmetric (carbon atoms displaced equally on opposite sides of the MN_2 plane) to highly asymmetric (both carbon atoms on the same side of the MN_2 plane) have comparable strain energy, but strain energy rapidly increases if the chelate ring is forced into an eclipsed conformation (both carbon atoms displaced by about the same amount on the same side of the MN_2 plane, i.e., torsion angle about the central C–C bond appreciably less than 60°). Similarly, for the six-membered (1,3-diaminoethane) chelate ring, a range of chair conformations has comparable energies, but strain energy is greater for the twist and greater again for the boat families of conformations.

The conformations which can be adopted by each chelate ring for a coordinated cyclic polyamine are restricted by the configuration of the coordinated nitrogen centers. For planar coordination of a [14]aneN_4 macrocycle the optimum gauche (five-membered) and chair (six-membered) conformations of all chelate rings are possible only for configuration III, and this configuration is present for the complexes of [14]aneN_4 in planar coordination, **1.4d,g,h**, and also for the free ligand, **1.4a**, and a salt, **1.4b**. For folded coordination (one "fold-line" NMN angle $\sim 180°$, the other $\sim 90°$ for *cis*-octahedral or $\sim 120°$ for trigonal-bipyramidal arrangements), the macrocycle is highly strained unless the fold-line NH groups are on the same side of the molecular plane. Optimum chelate ring conformations for folded [14]aneN_4 macrocycles are possible only for configuration V (which has the other *trans* pair of NH groups on the opposite side of the molecular plane), and this configuration is present for all such structures reported, **1.4e,f**, and **1.7g**.

For [12]aneN_4 macrocycles, configuration V would apparently permit all chelate rings to adopt optimum conformations for both planar and folded coordination. However, for folded coordination, this configuration would result in severe intraligand interactions, and configuration II is present for the structures reported, **1.2a,b** (and for the structure of a compound of a related [12]aneO_4 macrocycle, **7.1a**), with two of the chelate rings in highly asymmetric gauche conformations. The conformation of segments of the macrocycle for these compounds closely resembles those of β-1,9-diamino-3,6-diazaoctane(trien) complexes.[3]

For the [13]aneN_4 macrocycle, no configuration permits all chelate rings to adopt optimum conformations. For compound **2.3a** the [13]aneN_4 macrocycle is in planar coordination to $S = 0$ Ni(II) with configuration III, with the "central" five-membered chelate ring in an eclipsed conformation.

Details of the conformations of the coordinated macrocyclic amines are shown in Tables 1 and 2 as displacement of the ring atoms from the MN_2 chelate planes. Conformations of chelate rings can also be represented by torsional angles, which in addition provide an indication of the localization of torsional strain, and papers which include torsional angle data are indicated in Tables 1 and 2. However, the total strain present arises from the combination of torsional, bond angle, and bond length strain, and consideration of only one of these factors can be misleading. The presence of bond angle strain is indicated by distortions of the coordination environment (e.g., from coplanarity of the MN_4 set for "planar" compounds) and of bond angles from their strain-free values throughout the molecule (e.g., at the nitrogen atoms, and in particular, at the central carbon atom of the six-membered chelate rings where values up to 121° have been observed). Small variations in bond lengths throughout the molecule are also indicative of varying bond length strain.

2.3. Metal-Ion–Nitrogen Distances

Coordination compounds with macrocyclic ligands can be considered from different viewpoints which give some insight into the structural factors involved. When a metal ion is coordinated within a macrocycle, the donor atoms are constrained to occupy the coordination sites about the metal ion, with the lone-pair electrons oriented toward the metal ion. Constraints on the system are greatest when the metal and donor atoms are coplanar. The noncoordinated cyclic amine can in principle adopt a variety of conformations, which in most cases bear little relationship to the *endo* conformation required for planar coordination. However, the ligand can generally adopt a relatively strain-free *endo* conformation with the donor atoms approximately coplanar, and strain is introduced if the donor atoms are required to move inwards or outwards from these preferred sites to accommodate a metal ion. The center–nitrogen distances for the least strained conformation with approximately coplanar nitrogen atoms have been calculated for the cyclic amines [12]aneN_4 to [16]aneN_4. The values ([12]aneN_4, 183 pm; [13]aneN_4, 192 pm; [14]aneN_4, 207 pm; [15]aneN_4, 222 pm; and [16]aneN_4, 238 pm) indicate that the optimum M–N distance increases by 10–15 pm per ring member.[4]

For cyclam, [14]aneN_4, structures of the free amine **1.4a** and a diprotonated salt **1.4b** have been reported, and these have *endo* conformations similar to those present for planar coordination compounds of this macrocycle, with center–N distances of 206 and 205 pm, respectively, in agreement with the calculated value. Because metal-ion–donor distances are relatively inflexible, it would be expected that for the coordinated macrocycles the M–N distance would vary much less with ring size. However, compressive strain (for larger-than-optimum ions) or tensive strain (for smaller-than-optimum ions) leads to noticeable increases, or decreases, respectively, in the ligand field experienced by the ion. Spectral data support the view that the Co(III) ion is approximately the optimum size for the [14]aneN_4 macrocycle, and the $S = 1$ Ni(II) ion for the [15]aneN_4 macrocycle.[4] Unfortunately, structural data are sparse, but it can be noted that for the $S = 1$ Ni(II) complex of [14]aneN_4, **1.4g**, the Ni–N distance is 206 pm, at the bottom of the 206–216 pm range observed for complexes of this ion with diaminoethane, and less than the 212-pm value observed for [14]aneN_4 macrocycles in folded coordination where macrocyclic constraints are relaxed (**1.7e,f**).

For folded coordination, no special macrocycle constraints are present and the M–N-donor bond lengths are similar to those observed for noncyclic ligands with similar donor atoms, e.g., the mean Co(III)–N values of 196 and 200 pm, respectively, observed for [12]aneN$_4$ and [14]aneN$_4$ compounds, **1.2a** and **1.4e**.

For three $S = 0$ Ni(II) square-planar compounds of a Me$_6$[14]aneN$_4$, the mean Ni–N distance is 192 pm, similar to the values reported[5,6] for complexes with noncyclic amines. The Ni–N distance for a [13]aneN$_4$ complex of this ion is marginally less at 186(2) pm, possibly indicating some macrocyclic constriction.

The tetraazamacrocyclic complexes can also be considered as formally derived from complexes of the appropriate linear tetraamines by bridging the terminal amino group by a –(CH$_2$)$_2$– or –(CH$_2$)$_3$– bridge, with formation of a fourth (ring closing) chelate ring. Some indication of the strain likely to be introduced by the ring closure can be obtained by comparing the "open" NMN angle with the usual NMN angle for the appropriate chelate ring, about 85° for the five-membered, and about 92° for the six-membered chelate. [Strictly the chelate "bite" (N–N distance) should be considered, but if attention is restricted to the similarly sized M(II) and M(III) ions of the latter half of the first transition series, for which most structural information about macrocyclic complexes is available, comparison of the chelate angles is adequate.] Triethylenetetraamine in planar coordination shows terminal NMN angles close to 100°.[5,6] For planar coordination, ring closure by a –(CH$_2$)$_3$– bridge to give a [13]aneN$_4$ complex will introduce less strain than closure by a –(CH$_2$)$_2$– bridge to give a [12]aneN$_4$ complex. No planar complexes of [12]aneN$_4$ macrocycles have been reported. The 12,12-dimethyl derivative of [13]aneN$_4$ has been reported in both planar and folded coordination,[7] but the only structure reported, **1.3a**, has the amine in planar coordination to $S = 0$ Ni(II). The conformation of the trien moiety in this compound is very similar to that found for [Ni(trien)](ClO$_4$)$_2$,[5] and the short Ni–N distance may be a consequence of the "macrocyclic constrictive effect."

The linear tetraamine 1,5,8,12-tetraazadodecane (3,2,3-tet) forms cobalt-(III) complexes with the amine in planar coordination, with terminal NCoN angles of 89.5°.[8] Ring closure by a –(CH$_2$)$_2$– or a –(CH$_2$)$_3$– bridge will introduce a small amount of strain in opposite senses. This is in agreement with the conclusions of Busch *et al.* that the optimum cyclic amine ring size for Co(III) is between [14] and [15].[4]

2.4. Substituents on the Macrocycle

Orientations of substituents are represented in Table 2 in terms of displacements from the MN$_2$ plane of the chelate ring.

2.4.1. Carbon Substituents

Substituents on the carbon atoms of coordinated 1,2- or 1,3-diamines, in their usual gauche or chair conformations can, by analogy with substituted cyclopentanes and cyclohexanes, be described as equatorial or axial. For isolated chelates the equatorial orientation has minimum intraligand strain energy, and is normally adopted.[9] For a linear polydentate ligand with substituents, the optimum (i.e., minimum total-strain energy) arrangement is determined by the balance of the minimum-energy configuration/conformation of the unsubstituted ligand and the preferred equatorial orientation of as many substituents as possible, and the same principles apply for cyclic ligands.[10,11]

The introduction of a single substituent on a ring carbon atom generally introduces a chiral center. Most macrocyclic amines for which structures have been reported have substituents in symmetrical pairs, leading to *rac* or *meso* configurations of the *ligand*. Often *meso* ligand configurations are compatible with configuration III of the nitrogen centers and planar coordination, while *rac* ligand configurations are compatible with configuration V and folded coordination, leading to distinct stereochemical behavior of the isomers of the macrocycle.

The most comprehensive set of structures illustrating this effect is provided by the amine 5,5,7,12,12,14-hexamethyl-1,4,8,11-tetraazacyclotetradecane, Me_6[14]aneN_4, **1.7**.[10] The *C-rac* isomer (tet *b*) forms an extensive set of octahedral complexes with a chelate occupying the two additional *cis* coordination sites, as for [Ni(*rac*-Me_6[14]aneN_4)(OAc)]ClO_4, **1.7e**. This has nitrogen configuration V, as for [14]aneN_4 in folded coordination, **1.4e**, with two equatorial and one axial methyl substituents on each six-membered chelate ring. Treatment of such nickel(II) compounds with acid leads to removal of the chelate by protonation, while preserving the nitrogen configuration (since acid inhibits nitrogen inversion), giving the metastable square-planar complex **1.7b**, with retention of configuration V.[10] For this configuration the NH groups of the six-membered chelate rings are *trans*, giving rise to a strained twist conformation, which still places the substituents in equatorial (two) and axial (one) orientations (see Figures 2 and 3).

When this cation is allowed to isomerize (neutral or basic solution) the configuration of the isolated salt depends upon the anion present. For this amine in planar coordination no one nitrogen configuration can have both optimum macrocycle conformation and optimum substituent orientation. The perchlorate crystallizes with configuration (III), optimum for the unsubstituted macrocycle (and also for the *C-meso* amine), but for the *C-rac* amine with the usual conformation, one six-membered chelate ring would be 1,3-diaxially substituted, which is highly strained. Strain is minimized by adoption of a

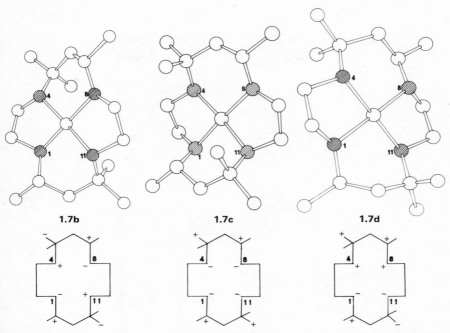

FIGURE 2. Structures of three square-planar isomeric (*C-rac*-5,5,7,12,12,14-hexamethyl-1,4,8,11-tetraazacyclotetradecane)nickel(II) cations, **1.7b**, **1.7c**, and **1.7d**, with nitrogen configurations V, I, and III, respectively. Overall configurations are shown on the formulas below each structure, with symbolism as for Figure 1, but including orientation of axial methyl substituents. Note the twist conformation of the six-membered chelate rings of **1.7b** and the twist boat conformation of chelate ring N(4)–N(8) of **1.7d**. Reproduced from reference 31 with the permission of the Chemical Society.

twist-boat conformation which orients the single methyl group equatorially. With tetrachlorozincate as the anion, the stable form has configuration I, **1.7c**, with all NH groups on the same side of the molecular plane. Early conformational studies[10,11] suggested that for this configuration the five-membered chelate rings, with *cis* NH groups, would be forced into strained eclipsed conformations. However, the structure shows that the strain is spread over the macrocycle, the six-membered chelate rings adopting two different chair conformations, one more strongly puckered than the other, while the five-membered chelate rings adopt asymmetric gauche conformations. A similar macrocycle conformation is present for the N-Me$_4$[14]aneN$_4$ complex **1.8b**, which also has configuration I. The energies of the cation in configurations I and II are apparently sufficiently close so that the differences in lattice

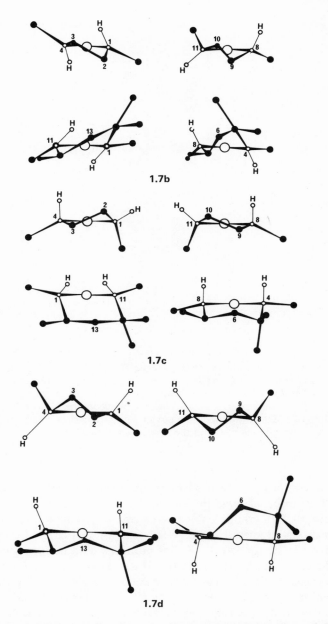

FIGURE 3. Conformations of the five- and six-membered chelate rings of the cations of three isomeric (*C-rac*-5,5,7,12,12,14-hexamethyl-1,4,8,11-tetraazacyclotetradecanes)-nickel(II) cations **1.7b**, **1.7c**, and **1.7d**, with nitrogen configurations V, I, and III, respectively. Carbon atoms are projected on a plane through the nitrogen atoms and normal to the N–Ni–N plane of each chelate ring. Atom numbers are as in Figure 3. Reproduced from reference 31 with the permission of the Chemical Society.

energies arising from changing the anion can render one form or the other stable.[31]

The *C-meso* isomer of this amine, $Me_6[14]aneN_4$, usually occurs in planar coordination, but one structure has been reported for a compound with this amine in folded coordination, **1.7f**.[12] This has nitrogen configuration V, which, for the usual conformation of the macrocycle, would again lead to 1,3-diaxial substituents on one chair conformation six-membered chelate ring. A distortion is present which swivels one of these methyl groups to an equatorial site with a twist boat conformation of the chelate ring, analogous to that found for **1.7d** (Figure 4 and Table 2).

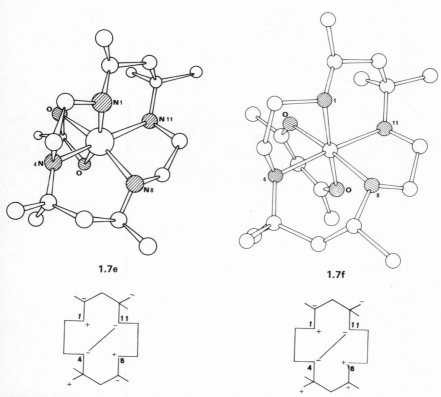

1.7e

1.7f

FIGURE 4. Structures of the cations of octahedral nickel(II) compounds with *C-rac*, **1.7e**, and *C-meso*, **1.7f**, 5,5,7,12,12,14-hexamethyl-1,4,8,11-tetraazacyclotetradecane with folded macrocycle and with chelate acetate and acetylacetonate, respectively. Overall configurations and fold-line are shown on the formulas below each structure, with symbolism as for Figure 2. Note the difference in conformation of chelate rings N(4)–N(8) for the two structures. Reproduced from references 10 and 31, respectively, by permission of the Chemical Society.

2.4.2. Nitrogen Substituents

For coordinated cyclic tertiary amines, isomerization of the nitrogen configuration is not possible without breaking the metal–nitrogen bond. Reactions of nickel(II) with tetra-N-methyl[14]aneN$_4$, **1.8**, result in a relatively unstable cation with configuration I, presumably determined by the mechanism of the metal-ion insertion reaction.[13] This square-pyramidal cation, [Ni(N-Me$_4$[14]aneN$_4$)N$_3$]$^+$, **1.8b**, has the azide ion coordinated on the same side of the macrocycle as the four methyl substituents, with the nickel(II) ion 33 pm from the N$_4$ plane (Figure 5). Forcing the nickel ion down into the N$_4$ plane to give normal planar coordination would introduce severe interaction between the methyl groups.[13] The macrocycle conformation present is similar to that found for the Me$_6$[14]aneN$_4$ compound, **1.7c**, also with configuration I, with the five-membered chelate rings in asymmetric gauche conformation and the six-membered chelate rings in chair conformations with one ring more puckered than the other. N-methylation of preformed [Ni([14]aneN$_4$)]$^{2+}$ yields the tetra-N-methylated cation in the optimum configuration III, **1.8a**[14] (Figure 5). This cation is very resistant to removal of the nickel, being unaffected by reagents such as cyanide which are effective for the other cyclic amines. The cation in configuration I in contrast is very labile. The difference in reactivity is attributed to the exposed nature of the metal ion in configuration I,[14] but complexes of other cyclic amines which fold readily, such as [Ni(C-rac-[14]aneN$_4$)]$^{2+}$, similarly expose the metal ion but are nonlabile. The inability of the tertiary nitrogen centers to change configuration provides a more likely explanation.

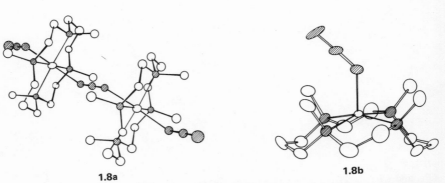

1.8a   **1.8b**

FIGURE 5. Structures of two azido compounds of (1,4,8,11-tetramethyl-1,4,8,11-tetraazacyclotetradecane)nickel(II), **1.8a** and **1.8b**, with nitrogen configurations III, and I, respectively. Reproduced from references 14 and 34 by permission of the copyright holder, The American Chemical Society.

2.5. Chelate Angles

For a bidentate chelate, the angle subtended at the central atom is a compromise between the optimum value of the bonding angle at the central atom (usually 90°), the metal–donor-atom distance, and the preferred "bite" of the chelate. For M(II) and M(III) ions of the latter half of the first transition series (with which this review is largely concerned), with diamine chelates, common values are $\sim 86°$ for five-membered and $\sim 92°$ for six-membered rings, and appreciable deviation from these values is indicative of a strained conformation. Linking two or more five(or six)-membered chelate rings for a linear polyamine introduces cumulative strain, which is greatest for coplanar arrangement of the metal ion and donor atoms. Polyamines with alternating five- and six-membered chelate rings with alternate chelate angles greater than, and less than 90°, minimize this effect. For cyclic polyamines in planar coordination the constraints are greater, as the chelate angles must sum to 360°. Fourteen-membered cyclic tetramines, with two five- and two six-membered chelate rings, preferably alternating, will have least cumulative chelate angle strain and are by far the most common. Chelate angles for these compounds are similar to those of 1,2-diaminoethane and 1,3-diaminopropane compounds, respectively, indicating that little chelate angle strain is introduced by cyclization. For the [13]aneN$_4$ compound, **1.3a**, with three linked five-membered chelate rings, two have chelate angles of 90°, indicative of strain. No structures of saturated [15] or [16] amine macrocycle compounds have been reported, but it can be noted that the [16]dieneN$_4$ macrocycle of **2.17a** has a saturated six-membered chelate ring with chelate angle of 87°, in the chair conformation. Cumulative strain of the chelate angles may be relieved by relaxing the coplanarity requirement of the metal-ion–donor set, and many of the structures of cyclic tetramines show distortion of the MN$_4$ set of a tetrahedral character. This effect is much greater for the [15]-N$_5$ and [16]-N$_5$ macrocycles of Section 3, which have abnormally small chelate angles, with appreciable distortion from coplanarity. For these macrocycles, chair conformation six-membered chelate rings have chelate angles ranging from 98° (**3.3a**) to 75° (**3.6b**).

Chelate angle constraints are less for folded coordination of the macrocycle, and the similarity of the chelate angles for planar and folded [14]aneN$_4$ compounds provides further indication of the absence of appreciable strain for this ring system in planar coordination.

2.6. Listing of Structures of Compounds of Cyclic Amines

See introduction for rules determining order of compounds. Key dimensions are listed in Tables 1 and 2.

[9]aneN₃ Compounds

1.1 2-methyl-1,4,7-triazacyclononane. Me[9]aneN₃, $C_7H_{17}N_3$

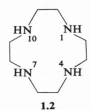

1.1

1.1a [Co(L)₂]I₃

Structural study mentioned in reference 15. Facial coordination of two macrocycles to Co(III). The chelate rings all have gauche conformations.

[12]aneN₄ Compounds

1.2 1,4,7,10-tetraazacyclotetradecane. [12]aneN₄, cyclen, $C_8H_{20}N_4$

1.2

1.2a *cis*-[Co(L)(NO₂)₂]Cl. $R = 0.026$ for 1146 refl.***[3]

Folded coordination [configuration II with N(1)–N(7) fold-line], with *cis*-nitro groups completing an octahedral arrangement about Co(III). An approximate mirror plane passes through Co, N(4), and N(10).

1.2b *cis*-[Co(L)(CO₃)]ClO₄·H₂O. $R = 0.053$ for 881 refl.***[16]

Folded coordination to Co(III) with chelate carbonate completing an octahedral arrangement. The configuration is the same as, and the conformation similar to, that for **1.2a.**

[13]aneN₄ Compounds

1.3 12,12-dimethyl-1,5,8,11-tetraazacyclotridecane. Me₂[13]aneN₄, $C_{11}H_{26}N_4$

1.3

Nickel(II) complex formed by reduction of 1-ene complex,[7] which is formed by reaction of 1,8-diamino-3,6-diazaoctane with 2-methylpropanal in the presence of Ni(II).[17]

1.3a   [Ni(L)](ClO$_4$)$_2$. $R = 0.147$ for 1733 refl. (P)***[18]

Planar coordination to $S = 0$ Ni(II), with configuration III, and with approximate mirror symmetry through C(12), Ni, and the midpoint of C(5)–C(6). The "central" five-membered chelate ring N(4)–N(7) is eclipsed. Disorder of the perchlorate ions hindered refinement.

[14]aneN$_4$ Compounds

1.4   1,4,8,11-tetraazacyclotetradecane
[14]aneN$_4$, cyclam, C$_{10}$H$_{24}$N$_4$

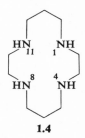

1.4

1.4a   [L]. $R = 0.047$ for 1827 refl. [19]

The unit cell contains two independent molecules, each with an inversion center, and with very similar dimensions. The molecules have very similar *endo* conformations, which closely resemble the conformations found for the Co(II) and Ni(II) complexes, **1.4d,e**, except that the N(4)–H and N(11)–H bonds lie approximately in the macrocycle plane. The *endo* conformation is stabilized by hydrogen bonding N(4)–H $\cdots$ N(8) and N(11)–H $\cdots$ N(1) across the 1,3-diamine links (cf. the tetrathia analog, **7.2a,b**, and Me$_2$Ph$_2$[14]aneN$_4$ **1.6a**).

1.4b   [H$_2$(L)](ClO$_4$)$_2$. $R = 0.049$ for 1009 refl.***[20]

The diprotonated macrocycle is centrosymmetric, and in an *endo* conformation similar to that found for the amine **1.4a**, and for planar complexes **1.4d,e,g,h**, except that the nominal six-membered chelate rings including the center of symmetry have boat rather than chair conformations.

1.4c   [Li(L)](ClO$_4$) and [Li$_2$(L)](ClO$_4$)$_2$[21]

Very preliminary report. The nitrogen atoms of [Li(L)](ClO$_4$)$_2$ are crystallographically equivalent.

1.4d   [Co(L)(ClO$_4$)$_2$]. $R = 0.047$ for 1827 refl.***[22]

Planar coordination to $S = \frac{1}{2}$ Co(II) with perchlorate oxygen atoms

1.4e

FIGURE 6. Structure of cation of **1.4e** diaminoethane-(1,4,7,10-tetraazacyclotetradecane)cobalt(III), with the macrocycle in folded coordination with nitrogen configuration II. Reproduced from reference 3 by permission of the copyright holder, The American Chemical Society.

completing a tetragonally elongated octahedral arrangement. The centrosymmetric molecule has nitrogen configuration III.

   1.4e $[Co(L)(en)]Cl_3$, $3H_2O$. $R = 0.052$ for 2608 refl.***[23]

   Folded coordination to Co(III), with diaminoethane completing an octahedral arrangement. The cation has configuration V with approximate twofold symmetry. (See Figure 6.)

   1.4f *cis*-$[Co(L)Cl_2]Cl$[24]

   Folded coordination to Co(III) with *cis* chlorides completing an octahedral arrangement, and with nitrogen configuration V.

   1.4g *trans*-$[Ni(L)Cl_2]$. $R = 0.047$*[25]

   Planar coordination to $S = 1$ Ni(II), with *trans* chlorides completing an octahedral arrangement. The centrosymmetric molecule has configuration III.

   1.4h $[Cu(L)](ClO_4)_2$. $R = 0.056$ for 1545 refl.***[26]

   Planar coordination to Cu(II) in configuration III, with *trans* perchlorates weakly interacting.

   1.5 5,12-dimethyl-1,4,8,11-tetraazacyclotetradecane
             5,12-*meso* = isomer a, $Me_2[14]aneN_4$, $C_{12}H_{28}N_4$

1.5

Ligand formed by reduction of imine groups of 4,11-diene, **2.5**.

1.5a   *trans*-[Co(*meso*-L)(N$_3$)$_2$]N$_3$. $R = 0.073$ for 1291 refl.***[27]

Planar coordination to Co(III) with *trans* azide ions completing an octahedral arrangement. The nitrogen configuration is III, with the methyl groups equatorially oriented.

1.6   5,12-dimethyl-8,14-diphenyl-1,4,8,11-tetraazacyclotetradecane

       Me$_2$Ph$_2$[14]aneN$_4$, C$_{24}$H$_{32}$N$_4$

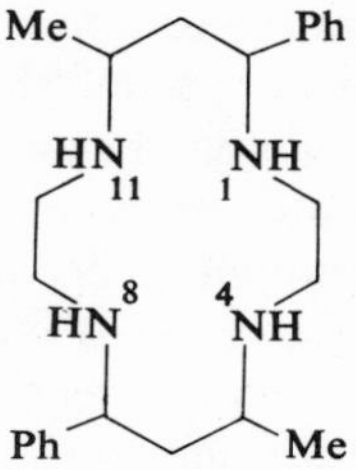

1.6

Ligand formed by reduction of imine groups of 4,11-diene, **2.9**.

1.6a   (L). $R = 0.081$ for 1011 refl.*[28]

The macrocycle is close to centrosymmetric, with the N(1) and N(8) hydrogen atoms directed toward the center of the macrocycle (cf. **1.4a**).

1.6b   [Cu(*meso*-L)(ONO$_2$)$_2$]. $R = 0.071$ for 740 refl.*[29]

The isomer of the amine present has the 5,12- and 8,14-carbon centers in *meso* configurations. The macrocycle is in planar coordination to Cu(II), with nitrate ions weakly coordinated in the axial sites. The configuration of the nitrogen centers is III, with the substituents all equatorially oriented.

1.7   5,5,7,12,12,14-hexamethyl-1,4,8,11-tetraazacyclotetradecane

       Me$_6$[14]aneN$_4$, 7,14-*rac* = tet *b*, 7,14-*meso* = tet *a*, C$_{16}$H$_{36}$N$_4$

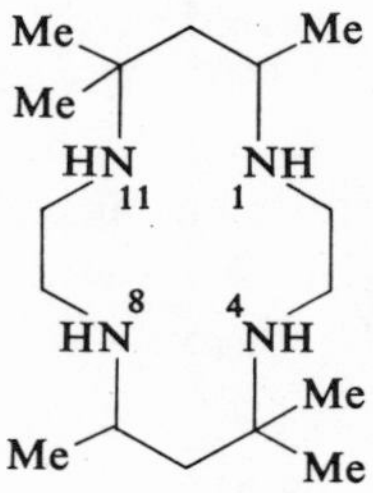

1.7

Ligand formed by reduction of imine groups of the 4,11-diene, **2.10**.

1.7a   [Mn(*meso*-L)Cl$_2$]Cl·3H$_2$O[30]

1.7b   [Ni(*rac*-L)](ClO$_4$)$_2$, α isomer. $R = 0.071$ for 1536 refl.***[31]

(See Figure 2.) Planar coordination to $S = 0$ Ni(II), with nitrogen configuration V, (α-*rac*), and with the nickel ion lying on an approximate twofold axis. The six-membered chelate rings have twist conformations.

1.7c [Ni(*rac*-L)]ZnCl$_4 \cdot$H$_2$O, β isomer. $R = 0.039$ for 1698 refl.***[31]

(See Figure 2.) Planar coordination to $S = 0$ Ni(II), with nitrogen configuration I (β-*rac*). (See discussion on page 230.)

1.7d [Ni(*rac*-L)](ClO$_4$)$_2$, γ isomer. $R = 0.075$ for 1198 refl. P***[31]

(See Figure 2.) Planar coordination to $S = 0$ Ni(II), with nitrogen configuration III (γ-*rac*). The six-membered chelate ring N(8)–N(11) has the normal chair conformation, but the chelate ring N(4)–N(8) adopts a twist boat conformation to reduce intersubstituent interaction. (See discussion on page 230 and Figure 3.) The compound crystallizes in a space group that requires the nickel ion to lie on a center of symmetry although the cation is inherently chiral. The overall "shape" determined largely by the peripheral methyl substituents is sufficiently similar for the two six-membered chelate rings to permit adoption of a disordered structure.

1.7e [Ni(*rac*-L)(CH$_3$COO)]ClO$_4$. $R = 0.12$ for 2753 refl. (P)***[11]

(See Figure 4.) Folded coordination, with N(4)–N(11) fold-line, to $S = 1$ Ni(II), nitrogen configuration V, with chelate acetate completing an octahedral arrangement.

1.7f [Ni(*meso*-L)(acac)]ClO$_4$. $R = 0.062$ for 1591 refl.***[31]

(See Figure 4.) Folded coordination, with N(4)–N(11) fold-line, to $S = 1$ Ni(II), nitrogen configuration V, with planar chelate acetylacetonate completing an octahedral arrangement. The six-membered chelate rings have dissimilar conformations, as for **1.7d** (see discussion on page 233).

1.7g [{Cu(*rac*-L)}$_2$Cl](ClO$_4$)$_3$. $R = 0.115$ for 1039 refl.*[32]

Folded coordination to copper(II), with nitrogen configuration V. Bridging chloride completes a trigonal-pyramidal arrangement about the copper ion, with the N(1)–N(7) fold-line as axis of the bipyramid. The two halves of the dimer are related by a twofold axis through the bridging chloride ion.

1.7h [Ag(*meso*-L)(ONO$_2$)$_2$].[33] $R = 0.033$ for 2268 refl.***[33]

Planar coordination to Ag(II), with *trans* in nitrate oxygen atoms completing a tetragonally distorted arrangement.

1.8 1,4,8,11-tetramethyl-1,4,8,11-tetraazacyclotetradecane
     N-Me$_4$[14]aneN$_4$, C$_{14}$H$_{32}$N$_4$

1.8

1.8a [{Ni(L)}$_2$(N$_3$)$_3$]. $R = 0.11$ for 2638 refl.**[14]

(See Figure 5.) Planar coordination to $S = 1$ Ni(II), with one bridging and one terminal azide ion in the axial sites completing an octahedral arrangement. The two halves of the dimer are related by a center of symmetry at the central nitrogen of the bridging azide ion. The nitrogen configuration is III with the [Ni(L)]$^{2+}$ moiety approximately centrosymmetric. Large thermal parameters indicate considerable disorder of the perchlorate ion, and of the carbon atoms of the five-membered chelate rings. The cation [N(L)]$^{2+}$ was formed by N-methylation of the [Ni([14]aneN$_4$)]$^{2+}$ cation.[14] The magnetic behavior of the dimer has been reported.[34]

1.8b [Ni(L)(N$_3$)]I$_4$. $R = 0.085$ for 1469 refl.**[35]

(See Figure 5.) Planar coordination to $S = 1$ Ni(II) with an azide ion completing a square-pyramidal arrangement. The nitrogen configuration is I, with the four methyl groups lying on the same side of the molecular plane as the azide. The azide ion, Ni, C(4), and C(9) lie on a mirror plane. The cation [Ni(L)]$^{2+}$ was formed by reaction of nickel(II) with the preformed ligand.[13] (See discussion on page 234.)

3. Class 2: Cyclic Imines and Cyclic Amine–Imines (Unsaturated Polyaza Macrocycles with all Nitrogen Atoms Coordinated)

3.1. Discussion of Structures

The structures reported in this section are characterized by the presence of imino nitrogen donor atoms. In many cases the macrocycles also have coordinated secondary amino nitrogen atoms. Except for one compound with a delocalized "diiminato" six-membered chelate ring, **2.2a**, and one compound with a delocalized 15-π-electron system, **2.15a**, the double bonds of the imine groups are effectively localized, and the stereochemistry is related to that of the saturated analogs, modified by the stereochemical characteristics of the imine group: planarity, 120° bond angles, and shorter C=N bond lengths. The mean value for C=N for macrocycles with isolated imine groups of this section is 127 pm, the smallest value 116(3) pm for **2.4a**, and the largest value 131(1) pm for **2.10aa**.

Many of the structures reported are for symmetrically substituted diamine–diimine "diene" tetraazamacrocycles, which can occur with two nitrogen configurations, with the N–H groups on the same, or on opposite sides of the molecular plane. For macrocycles with *trans* amino groups (e.g., [14]4,11-dieneN$_4$, **2.10**), NH groups on the same side of the molecular plane correspond to the *N-rac* configuration, while for the macrocycles with *cis* amino groups (e.g., [14]4,14-dieneN$_4$, **2.11**), NH groups on the same side correspond to the *N-meso* configuration. For the *trans*-Me$_6$[14]4,11-dieneN$_4$ macrocycle, **2.10**, coordinated to $S = 0$ Ni(II)[55] or to Co(III),[56] the *N-rac* and *N-meso* configurations are of comparable stability. For the *cis*-Me$_6$[14]-

4,14-diene macrocycle, **2.11**, the $S = 0$ Ni(II) complex has been prepared only for the *N-rac* configuration, **2.11b**, which models suggest is least strained. However, the *trans*-dicyano Co(III) compound, **2.11a**, which has the *N-meso* macrocycle configuration, has both the five-membered chelate rings in eclipsed conformations. Apparently in this case the unfavorable macrocycle conformation is compensated for by lower interaction energy with the cyanide ligands. For the Me$_8$[14]4,11-diene macrocycle, **2.12**, coordinated to $S = 0$ Ni(II), the *C-rac* isomer has been prepared in only the *N-rac* configuration, **2.12a**, and the *C-meso* isomer in only the *N-meso* configuration. Similar relationships between carbon and nitrogen chiralities may be present for other macrocycles, but structural reports often do not mention the relative stabilities of possible isomeric forms.

The three reported structures of [14]4,7-dieneN$_4$ macrocycles (**2.5a**, **2.5Aa,b**) have the *N-meso* configuration, which permits the saturated six-membered chelate ring to adopt the chair conformation.

The presence of the imino group reduces the flexibility of the macrocycles compared with their saturated analogs, and in particular prevents "folding" of the macrocycle at the imino donor groups. Compounds with folded macrocycles are formed for [14]4,11-dieneN$_4$ macrocycles, with the fold-line through the amine groups, and in the *N-rac* configuration [i.e., NH groups on the same side of the molecular planes (**2.10a,f**)], but not for [14]4,14-dieneN$_4$ macrocycles.

The presence of imino groups also restricts the conformational freedom of the macrocycle, and it is notable that compounds **2.10d,e** which have the same [Ni(*N-rac*-Me$_6$[14]4,11-diene)]$^{2+}$ cation, but different anions, and which crystallize in different space groups, have very similar conformations of the macrocycle. The conformation of a [Ni(*N-rac*-Me$_8$[14]4,11-dieneN$_4$)]$^{2+}$ compound, **2.12a**, is generally similar to that of the Me$_6$ analog, **2.10b,c**, except that the macrocycle is somewhat more puckered. For the pairs of compounds **2.10b,c**, **2.13a,b**, and **2.5Aa,b**, Me$_6$[14]4,11-dieneN$_4$, Me$_4$[14]tetraeneN$_4$, and Me$_2$[14]dieneN$_4$-one macrocycles are coordinated to cobalt(II) and to cobalt(III), respectively, with different axial ligands, and again the conformations of each macrocycle are similar for the two oxidation states.

3.2. Conformation of Macrocycles

The conformation of the macrocycles of these compounds can be considered in terms of the conformations of the component chelate rings, represented in Figure 7.

3.2.1. Six-Membered Amine–Imine Chelate Rings

A number of structures have been reported for tetraaza "diene" macrocycle compounds, which have two secondary amine and two imine donor

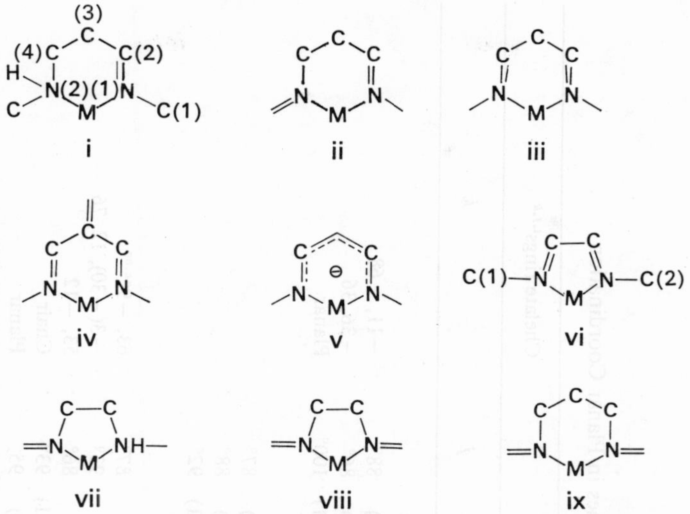

FIGURE 7. Chelate rings present for cyclic imine and cyclic amine–imine compounds.

atoms. Complexes of these macrocycles are formed by reaction of metal diamine complexes with ketones or with β-unsaturated carbonyl compounds, and diprotonated salts of the macrocycles are formed by reaction of the monoprotonated diamines with these reagents. The complexes have the chelate ring (i) of Figure 7, which is present for the bidentate chelate 2-methyl-2-amino-4-iminopentane, for which structures of a number of compounds have been reported.[36] Chelate ring (i) is related to cyclohexene, for which conformations derived from the chair and boat conformations of cyclohexane are possible, the "chair" (variously called "half-chair" and "sofa") having minimum strain energy. All structures of complexes with chelate ring (i) have the imine group MN(1)C(1)C(2)C(3) more or less planar, usually with this imine plane twisted with respect to the MN_2 chelate plane, and with the chelate ring in a "half-chair" conformation. The two equivalent half-chair conformations of cyclohexene become nonequivalent for the chelate ring because of the longer M–N bonds and $\sim 92°$ NMN bond angle. Two conformations, A and B, can be distinguished for the complexes by displacements from the MN_2 plane (Tables 3 and 4, Figure 8), atom C(4) being displaced further on the same side of this plane than C(3) for A, and less than, or on the opposite side of this plane for B. They can also be distinguished by displacements with respect to the imine plane, atoms C(3) and C(4) being displaced on opposite sides of this plane for A, and on the same side for B (Table 5). Interconversion of conformations A and B results in interchange of the axial–equatorial substituent sites for atoms C(3), C(4), and N(2). The conformation adopted depends upon interactions with the linking chelate rings, and for

TABLE 3. Dimensions for Compounds of Cyclic Imines in Planar Coordination

Compound[a]	Subst.[b]	Ion[c]	Spin[d]	Config.[f]	M–N[e,l,m] N	M–N[e,l,m] Length	MN_4[g]	C.N.[h]	i	j	Chelate rings[i,j,k] k
							[13]dienato(1−)N_4 Compounds				
2.2a	Me_2	Ni(II)	0	rac	(1,10)	183(1)[l]	±11	4	(1–4)	88°	−11, −68
					(4,7)	188(1)[m]			(4–8)	86°	−36, 36
									(10–1)	100°	Planar
							[13]10,13-dieneN_4 Compounds				
2.3a	Me_2	Ni(II)	0		(1,10)	189[l]	[30] ± 4	5[a]	(1–4)	87°	
					(4,7)	190[m]			(4–7)	88°	
									(10–1)	92°	
							[14]4,11-dieneN_4 Compounds				
2.4a	Me_2	Co(III)	0	rac	(1,8)	201(1)[l]	[3]0, −2	6[o]	(5)	87°	63, −9[aa,bb]
					(4,11)	198(1)[m]			(6)	93°	−3(−30), 12, 76
2.5a	Me_4	Co(II)	½	meso	(4,8)	192[l]	[15]	5[p]	(5)	86°	53, −12
					(11,1)	199[m]			(11–1)	93°	Chair
2.5Aa	Me_2	Co(II)	½(?)	meso	(4,8)	192(1)[l]	[6]	6[q]	(5)	87°	Planar
					(11,1)	197(1)[m]			(4–8)	94°	−26, 41[aa,bb]
									(11–1)	93°	24(24), 60(123), 24(20)
											86, 48, 82

2.5Ab	Me$_2$O Co(III)		meso	(4,8) 192[l] / (11,1) 197[m]	[4]		(5) / (4-8) / (11-1)	86° / 94° / 94°	-25, 43[aa,bb] / 23(16), 71(151), 29(31) / 79, 40, 78
2.6a	Me$_4$ Ni(II)	0	rac	(4,11) 188(3)[l] / (1,8) 193(3)[m]			(5) / (5)	87(?)° / 93(?)°	46, -36[bb] / -8(-66), 4, 82(228)
2.6b	Me$_4$ Ni(II)	0	rac	(4,11) 184(4)[l] / (1,8) 195(4)[m]	[4]2, -4	4	(1-4) / (8-11) / (4-8) / (11-1)	90 / 88° / 89° / 91°	13, -47 / 58, -43 / -5(-43), 24, 90(106) / 7(2), 41, 100(244)
2.7a	Me$_4$ Ni(II)	0	meso	(4,11) 193[l] / (1,8) 193[m]	–	4			
2.8a	Me$_4$ Ni(II)	0	meso	(4,11) 191[l] / (1,8) 194[m]	–	4	(5) / (6)	86° / 94°	61, -2(148)[b] / 14(39), 8, -72 / -57, 12[b] / -9(-4), -29, 51(16, 199)
2.10b	Me$_6$ Co(II)	½	meso	(4,11) 194[l] / (1,8) 197[m]	–	6[s]	(5) / (6)	85° / 95°	
2.10c	Me$_6$ Co(III)		meso	(4,11) 192[l] / (1,8) 199[m]	–	6[t]	(5) / (6)	86° / 94°	
2.10d	Me$_6$ Ni(II)	0	rac	(4,11) 186(1)[l] / (1,8) 190(1)[m]	[5]3, -6	4	(6) / (5)	87° / 93°	32, -34[b,c] / 12(-10), 45, 96(99, 240) / [4]20, -46[-6] / [-6]12(-10), 47, 94(103, 218)[4]
2.10e	Me$_6$ Ni(II)	0	rac	(4,11) 188[l] / (1,8) 192[m]	[2]4, -6	4	(6) / (5)	87° / 93°	

[14]4,14-diene N$_4$ Compounds

2.11a	Me$_6$ Co(III)		meso	(1,4) 194(2)[l] / (8,11) 198(2)[m]	[4]3, -5	6[u]	(1-4) / (8-11) / (6)	86(1)° / 81(1)° / 97(1)°	-57, -17 / -93, -98 / 36(54), 78, 15(30, 164)

continued overleaf

TABLE 3 (continued)

Compound[a]	Subst.[b]	Ion[c]	Spin[d]	Config.[f]	M–N[e,l,m] N	M–N Length	M–N MN_4[g]	C–N[h]	i	j	Chelate rings[i,j,k] k
2.11b	Me_6	Ni(II)	0	*rac*	(1,4)	189[l]	±7	4	(1–4)	86°	−33, 33[aa,bb]
					(8,11)	193[m]			(8–11)	88°	38, −38
									(6)	93°	2(33), −22, −86(−227, −88)
2.12a	Me_8	Ni(II)	0	*rac*	(4,11)	189(1)[l]	[3]5, −7	4	(5)	88°	[4],16, −59(−209)[−7]
					(1,8)	193(1)[m]			(6)	92°	[7]−7(15), −56, −106(−115, −240)[−4]
[14]1,3,8,10-tetraeneN₄ Compounds											
2.13a	Me_4	Co(II)	½			190(1)[l]	'	6[v]	(5)	83°	1(28), −9(−27)[aa,bb]
									(6)	97°	−3, −71, 9
2.13b	Me_4	Ni(II)	1			201[l]		6			
[14]1,3,7,11-tetraeneN₄ Compounds											
2.14a	Me_6	Ni(II)	0			189(1)[l]	±18	4[w]	(1–4)[w]	84°	17, −17
									(8–11)[w]	86°	39, −39
									(6)	96°	51(33, 202), −22, −19(−54)
[14]hexaenato(1−)N₄ Compounds											
2.15a	Me_4	Ni(II)	0			187	[10]	x	(5)	86°	Close to planar[y]
									(6)	94°	Close to planar

[15]4,7,12,15-tetraeneN₄ Compounds

2.16a	Ni(II)	0		(1,4)	186ᶦ	?	4	(1-4) 83°	Eclipsed
				(8,12)	190ᶦ			(4-8) 91°	Boat
								(8-12) 95°	Boat

[16]1,12-dieneN₄ Compounds

2.17a	Ni(II)	1	meso	(1,13)	210ᶠ	±7	6ᶻ	(1-5) 94°	[7]−42(−47), −97, 1[7]
				(5,9)	214ᵐ			(5-9) 87°	[−7]89, 67, 99[7]
								(9-13) 92°	[7]13, 108, 38(18)[−7]
								(13-1) 87°	[−7]−86, −11, 75[7]

a-e, g-k See footnotes to Table 1 for symbolism.

f rac = N-rac, meso = N-meso for coordinated diamine macrocycles.

l Bonds to imine nitrogen.

m Bonds to secondary amine nitrogen.

n Axially coordinated iodide in square-pyramidal arrangement; Ni-I, 296 pm.

o Trans thiocyanates; Co-N, 189 pm.

p Oxygen of water completes square-pyramidal arrangement; Co-O, 228 pm; Co displaced 55 pm toward O. Fluoride of hexafluorophosphate weakly coordinated in trans-octahedral site; Co-F, 256 pm.

q Trans water, Co-O, 228(1), 239(1) pm.

r Trans chloride, Co-Cl, 223 pm.

s Tetragonally coordinated water; Co-O, 248 pm.

t Trans ammonias; Co-N, 195 pm.

u Trans cyanides; Co-C, 192(3) pm.

v Tetragonally coordinated water; Co-O, 229 pm.

w Note numbering difference between 2.11 and 2.14.

x Nickel-nickel bonded dimer; Ni-Ni, 306 pm. (See Table 6 for further data.)

y Macrocycle close to planar (maximum displacement, C(14), 4 pm) with methyl substituents on parallel eclipsed macrocycles folded apart.

z Trans-coordinated water; Ni-O, 222 pm and chloride; Ni-Cl, 243 pm.

aa Original publication includes torsional angle data.

bb Some, or all, of deviations from chelate planes calculated from atom site data in original publication.

cc Reference 60 includes comparison of torsion angles for 2.10d, 2.11a, and 2.11b.

TABLE 4. Dimensions for Compounds with Cyclic Imines in Folded Coordination

Compound[a]	Subst.[b]	Ion[c]	Spin[d]	NMN angles[n]	M–N[e,l,m]	i	j	k (Chelate rings[i,j,k])
				[12]eneN₃ Compounds				
2.1a[o]	Me₃	Ni(II)	1	[p]	(1) 205(1)[l]	(1–5)	90°	−30(−13), −86, −113(−146, −238)
					(5) 206(1)[m]	(5–9)	101°	30, −43, 35
					(9) 202(1)[m]	(9–1)	90°	113, 100, 120
				[14]4,11-dieneN₄ Compounds				
2.10a[q]		Fe(II)	2	(1,8) 160°	(4,11) 213(1)[l]	(8–11)	81°	135, 0
				(4,11) 118°	(8,1) 217(1)[m]	(4–8)	88°	166(−59), 37, 7(8, −11e)
2.10f[r]		Cu(II)		(1,8) 174°	(4,11) 203[l]	(5)	84°	−38, 31
				(4,11) 135°	(8,1) 206[m]	(6)	94°	−31(−49), −56, 31(5, 176)

$a–e, l–k$ See footnotes to Tables 1 and 2 for symbolism.

l Metal–imine-nitrogen distance (picometers).

m Metal–secondary-amine distance (picometers).

n For example, (1,8) indicates the angle M(1)–M–N(8).

o Square-pyramidal arrangement with nitrogen atoms of two thiocyanate ions, N_a, N_b, $N(1)$, and $N(9)$ in the basal plane; $Ni–N_a$, 199(1), $Ni–N_b$, 210(1) pm.

p Angles in coordination sphere: $N(1)–N_a$, 93°; $N(1)–N_b$, 174°; $N(5)–N_a$, 154°; $N(5)–N_b$, 87°; $N(9)–N_a$, 105°; $N(9)–N_b$, 95°; $N_a–N_b$, 88°.

q Distorted square-pyramidal arrangement with Fe–Cl axial; Fe–Cl, 231 pm.

r Trigonal-bipyramidal arrangement, with disordered bridging cyanide, Cu–cyanide donor atom, 213-pm mean value.

TABLE 5. Conformations of Six-Membered Amine–Imine Chelate Rings

Compound	Metal	Configuration	Deviation from imine plane[a]		Conformation[b]
			C	N	
		[14]4,11-dieneN$_4$ Compounds			
2.6a	Ni(II)	*N-rac-C-meso*	44	−41	A
2.4a	Co(III)	*N-rac*	61	−14	A
2.6b	Ni(II)	*N-rac-C-meso*	52	−37	A
2.10b	Co(II)	*N-meso*	86	26	B
2.10f	Cu(II)	*N-rac*(folded)	96	48	B
		[14]4,14-dieneN$_4$ Compounds			
2.11a	Co(II)	*N-meso*	105	61	B
2.11b	Ni(II)	*N-rac*	59	−22	A
		[14]1,3,7,10-tetraeneN$_4$ Compounds			
2.14a	Ni(II)		12	73	B

[a] Deviation (picometers) of atoms C(3) and N(2) from plane of atoms M, N(1), C(1), C(2) of chelate ring (i) of Figure 7.
[b] As represented in Figure 8.

planar coordination of the [14]4,11-dieneN$_4$ and [14]4,14-dieneN$_4$ macrocycles, the *N-rac* configuration is present with conformation A, and the *N-meso* configuration is present with conformation B for all structures reported. The [14]4,11-dieneN$_4$ compounds in folded coordination, with *N-rac* configuration, have conformation B.

Compounds with chelate ring (i) often show large deviations from strain-free tetrahedral bond angles at the amine nitrogen atom, and particularly at the central carbon atom, where values of ∼119° are common, and values of 123° and 124° are shown for **2.4a**.

The related six-membered chelate ring (ii), with two imine donor atoms, is present for the tetraimine compound **2.14a** with a conformation similar to that of the related diimine compound **2.11b**.

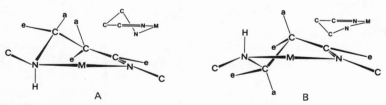

FIGURE 8. Conformations of type A and B of six-membered chelate rings with amine and imine nitrogen donor atoms.

3.2.2. Dienato Six-Membered Chelate Ring

The conjugated, delocalized dienato chelate ring (v) of Figure 7, present for the [13]dienatoN$_4$(1$-$) macrocycle compound **2.2a**, is planar and symmetrical, as is usual for "diiminato" derivatives of acetylacetone. For this compound the chelate angle has the unusually large value of 100°, which may be indicative of strain associated with the planar coordination of a [13]N$_4$ macrocycle with three linked five-membered chelate rings.

3.2.3. Diimine Six-Membered Chelate Rings

The diimine chelate ring (iii) is expected to be least strained in a planar or flattened boat conformation. This chelate ring is present for the [14]dieneN$_4$ macrocycle compound **2.5a**, where it is close to planar, and for the clathrochelate macrocycle compound **2.18a**, where it is constrained to adopt a boat conformation by the clathrochelate bridgehead structure.

The conjugated diimine chelate ring (iv) is expected to be least strained in a planar, or flattened boat conformation, and this chelate ring has a boat conformation for the [13]dieneN$_4$-dinuclear macrocycle of **2.3a**, for the [15]tetraeneN$_4$ macrocycle of **2.16a**, and for the [14]dieneN$_4$-one macrocycle of **2.5Aa,b**. For **2.3a** the $S = 0$ Ni(II) ion is displaced 40 pm from the N$_4$ plane, toward an apically coordinated iodide ion, probably indicating that this macrocycle is smaller than optimum for the ion (cf. **1.3a** and **2.2a**). For the [14]dieneN$_4$-one macrocycle, **2.5A**, the chelate ring (iv) has a boat conformation for the Co(II) compound, **2.5Aa**, and a slightly twisted boat conformation for the Co(III) compound, **2.5Ab**.

The CCC bond angle at the central methylene group of the diimine chelate ring of **2.5a** is 128(2)°, indicating considerable strain from the tetrahedral value, and providing a possible explanation for the facile reaction of compounds of this type to yield derivatives with macrocycles such as **2.3** and **2.5A** which have trigonal carbon at this point.

3.2.4. Diimine Five-Membered Chelate Rings

The diimine chelate ring (vi) is present for the [14]tetraeneN$_4$ macrocycles **2.13** and **2.14**. The isolated diimine chelate ring (vi) is planar for dimethylglyoxime derivatives, for example, but when linked within linear or cyclic polydentate systems the planar imino group interacts with the linking chelate rings. The displacement of atoms C(1) and C(2) on the same side of the MN$_2$ plane causes adoption of an "envelope" conformation related to the eclipsed conformation of diaminoethane, while displacement of these atoms on the opposite sides of this plane causes adoption of a "pseudogauche" conformation, as observed for the N(1)–N(4) chelate ring of compound **2.14a**. For the

[14]tetraeneN$_4$ compound, **2.13a**, where the linking chelate rings are relatively flexible saturated six-membered rings, the five-membered diimine chelate ring (vi) is essentially planar.

The bond lengths within the α-diimine five-membered chelate rings for the compounds **2.13a** and **2.14a** indicates no significant delocalization of the double-bond character (cf. compounds of Section 6 which have this chelate ring, some with appreciable delocalization).

3.2.5. Saturated Chelate Rings with Imine Donor Atoms

Possible conformations of saturated chelate rings with imine donor atoms, (vii)–(ix), are constrained by the planarity of the imine group. The five-membered chelate ring (vii), with one imine and one amine donor group, adopts a gauche conformation for a number of structures reported for [14]4,11-dieneN$_4$ macrocycle compounds. The five-membered chelate ring (viii), with two imine donor groups, is present for the [14]4,14-dieneN$_4$ macrocycle, **2.11**. For the Ni(II) compound **2.11b** the macrocycle has the *N-rac* configuration (twofold symmetry) with a gauche conformation of the chelate ring, while for the Co(III) compound **2.11a** with the *N-meso* configuration (approximate mirror symmetry), an eclipsed conformation is present. The [16]1,12-dieneN$_4$ compound **2.17a** has one diamine and one diimine saturated six-membered chelate ring, both with 87° chelate angles. The diamine chelate ring has a chair conformation (*N-meso* configuration), while the diimine chelate ring (ix) adopts a twist conformation, even though this requires some distortions of the remainder of the molecule. A twist conformation is also present for chelate ring (ix) in the clathrochelate compound **2.18a**, although in this case the *cis*-octahedral coordination geometry of the clathrochelate ligand favors the twist conformation. The chelate ring (ix) is also present for the [15]tetraeneN$_4$ compound **2.16a**, where it has a boat conformation.

3.3. Substituents on the Macrocycle

3.3.1. Substituents on Six-Membered Chelate Rings

Substituents on imine carbon atoms are confined to equatorial orientations by the planarity of the imino group. Substituents on C(3) or C(4) of the six-membered amine–imine chelate ring (i) may be equatorially or axially oriented. Structures of compounds with $S = 0$ Ni(II) of two tetramethyl-substituted macrocycles that have single methyl substituents on C(3), **2.7a**, and C(4), **2.6a**, respectively, have been reported. The compounds **2.7a** with C(2) and C(3) methyl groups, and **2.6a** with C(2) and C(4) methyl groups both have *N-rac-C-rac* configurations, with the non-imine methyl group axially oriented. In both cases the compound could adopt a configuration with the same macrocycle conformation, but with these substituents equatorially oriented, suggesting that the axial orientation is preferred. For **2.6b**, which has the *C-meso* configuration, the *N-rac* configuration is again present, with one methyl substituent axially and the other equatorially oriented. The C(4) phenyl-substituted macrocycle of **2.9a** is present with the *C-meso-N-meso* configuration, with the phenyl groups equatorially oriented. A number of macrocycles with the C(2), C(4), C(4) trimethyl-substituted chelate ring have been reported for both the *N-rac* and *N-meso* configuration, and these have the *gem* dimethyl groups in axial–equatorial orientations. For the tetraimine compound, **2.14a**, the conformation of the macrocycle in the solid is similar to that of the analogous diimine compound, **2.11a**, although somewhat more puckered (i.e., the macrocycle has a chiral conformation, without chiral centers), but the *gem* methyl groups become equivalent in solution on the NMR time scale, by flexing of the macrocycle.[37]

3.3.2. Substituents on Five-Membered Chelate Rings

Substituents on imine carbon atoms [chelate ring (vi) of Figure 7] are confined to equatorial orientation by the planar imino group for compound **2.13a**.

Substituents on saturated carbon atoms of chelate rings with imine donor atoms, (vii) and (viii), can adopt equatorial or axial orientations, as for the saturated macrocycles. Because any substituent on the imine carbon atom is confined to an equatorial orientation, substituents on the adjacent carbon atoms of the five-membered chelate ring (vii) adopt axial orientation to minimize the intersubstituent interaction, as for **2.8a** and **2.12a**. The interaction is sufficient in the case of **2.12** to restrict observed configurations to *C-rac-N-rac* and *C-meso-N-meso*.

3.4. Metal Ion–Nitrogen Distances

The compounds in this section have metal ion to imine nitrogen, and in most cases also metal ion to secondary amine nitrogen bonds (Tables 3 and 4). The metal–amine-nitrogen bond lengths are similar to those observed for the analogous, completely saturated macrocycles of Section 1. For macrocycles with both amine and imine nitrogen donor atoms, the M–N distance is generally shorter for the imine nitrogen, the mean difference for the reported structures being 4 pm. For $S = 0$ Ni(II) compounds, for which most structures have been reported, the Ni–N distances span the ranges 184–193 pm for imine and 190–196 pm for amine nitrogen. A similar difference in the M–N distances is shown for compounds of a bidentate amine–imine chelate.[36] One Co(III) compound, **2.4a**, has M–N longer for imine nitrogen, 201(1) pm, than for amine nitrogen, 198(1) pm, but large thermal motion of the imine carbon atom may indicate some disorder in the vicinity which could account for the abnormally long Co–N(imine) distance. Structures have been reported for $S = \frac{1}{2}$ Co(II) and Co(III) compounds of three macrocycles, **2.5Aa,b**, **2.10b,c**, and **2.13a,b**. In each case the Co–N distances are very similar for the two oxidation states, although the distances to the axial ligands are much greater for the Co(II) compounds.

The [13]dienatoN$_4$(1 –) compound, **2.2a**, has $S = 0$ Ni–N distances of 183(1) pm (delocalized "imine") and 188(1) pm (amine), which are marginally shorter than the values observed for the [14]N$_4$ macrocycles and may indicate a macrocycle constrictive effect. The [13]dieneN$_4$ compound, **2.3a**, has the $S = 0$ Ni(II) displaced by 40 pm from the N$_4$ plane, which may also be a reaction to a smaller-than-optimum macrocycle size. The [15]tetraeneN$_4$ compound, **2.16a**, has $S = 0$ Ni–N (imine) distances of 186 and 190 pm, similar to the values for the [14] macrocycles, but the unusual macrocycle conformation with eclipsed five-membered and boat conformation six-membered chelate rings may be adopted to reduce the Ni–N distances. The only [16]dieneN$_4$ structure reported, **2.17a**, has $S = 1$ Ni–N distances of 210 pm (imine) and 214 pm (amine). The latter distance is within the range of Ni–N distances for noncyclic amines, and can be compared with the 206-pm value for the [14]aneN$_4$ compound, **1.4g**.

3.5. Listing of Reported Structures of Cyclic Imine and Cyclic Amine–Imine Compounds

See introduction for rules determining order of compounds. Key dimensions are listed in Tables 3 and 4.

[12]eneN$_3$ Compounds

2.1 2,4,4-trimethyl-1,5,9-triazacyclododec-1-ene
Me$_3$[12]eneN$_3$, C$_{12}$H$_{25}$N$_3$

2.1

Complexes formed by reaction of 1,5,9-triazanonane (dipropylenetriamine) complexes with acetone.[38]

2.1a [Ni(L)(NCS)$_2$]. $R = 0.068$ for 1126 refl.***[38]

Facial coordination to $S = 1$ Ni(II) with *iso*-thiocyanato groups completing a distorted square-pyramidal arrangement, with N(1), N(5), and the two thiocyanate nitrogen atoms in the basal plane. The nitrogen configuration is pseudo-*rac*, i.e., with both NH groups on the same side of the (flattened) macrocycle. One of the methyl substituents on C(4) is situated 290 pm from the nickel ion in the "vacant" octahedral site. The saturated chelate rings have chair conformations, very flattened in the case of N(5)–N(9).

[13]*dienato*N₄(1-) Compounds

2.2 11,13-dimethyl-1,4,7,10-tetraazacyclotridec-10,12-dienato(1-)

Me₂[13]dienato(1-)N₄, AT⁻, $[C_{11}H_{21}N_4]^-$

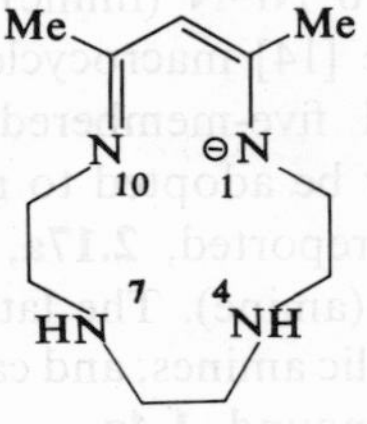

2.2

Complex formed by reaction of 1,4,7,10-tetraazaundecane (triethylenetetramine), with 2,4-pentanedione, in the presence of metal ions.[39]

2.2a [Ni(L)]ClO₄. $R = 0.072$ for 659 refl.***[40]

Planar coordination to $S = 0$ Ni(II). A twofold axis passes through C(12), Ni, and the midpoint of the C(5)–C(6) bond (*rac* configuration). The bond lengths within the six-membered chelate ring indicate the presence of a delocalized dienyl (2,4-pentane diiminato) chelate ring [C⋯C, 140(1) pm; C⋯N, 134(2) pm; angle at central carbon, 126(2)°], with this chelate ring close to planar. Compare with **2.5a**, where protonation at the central atom produces the localized diimine.

2.3 *bis*-(11,13-dimethyl-1,4,7,10-tetraazacyclotetradec-10,12-diene-11-enyl), $C_{22}H_{44}N_8$

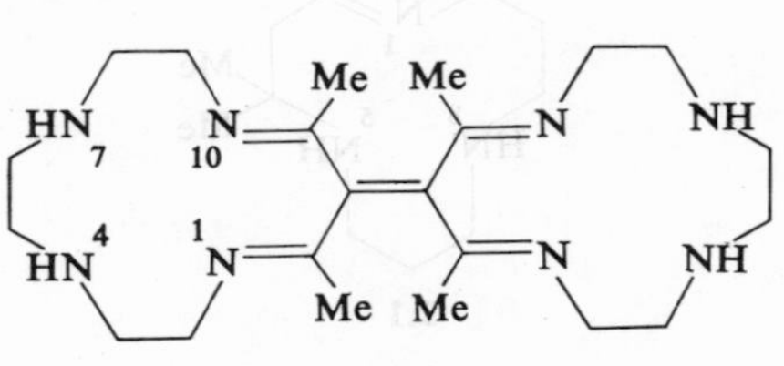

2.3

Dinuclear complex formed slowly from [Ni(2.2)]ClO₄ under strongly acidic conditions.[41]

2.3a [Ni₂(L)I₂]I₂·2H₂O. $R = 0.0819$ for 4808 refl.**[41]

Planar coordination of each N_4 donor set of the ligand to $S = 0$ Ni(II) with iodide completing a square-pyramidal arrangement, with the nickel ion displaced by 30 pm from the N_4 plane toward the iodide ion. The halves of the dimer are related by a center of symmetry at the midpoint of the linking double bond. The close methyl–methyl contacts (330 pm) force distortions from planarity which localize the double bonds. The bonding about each imine group, and about the central double bond, is essentially planar, but the central carbon atom of the six-membered chelate ring is displaced by 55 pm from the plane defined by the two C=N bonds in that ring. (See also **2.5Aa,b** and **2.16a**.)

[14]dieneN₄ Compounds

2.4   5,12-dimethyl-1,4,8,11-tetraazacyclotetradec-4,11-diene
Me₂[14]dieneN₄, C₁₂H₂₄N₄

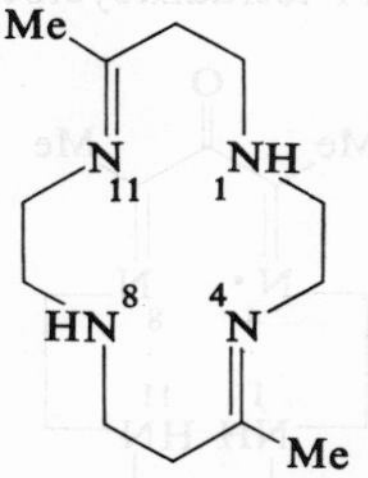

2.4

Ligand formed by reaction of diaminoethane monohydroperchlorate with prop-1-ene-3-one (methylvinyl ketone).[42]

2.4a   *trans*-[Co(L)(NCS)₂]ClO₄, H₂O. $R = 0.075$ for 1049 refl.***[43]

Planar coordination to cobalt(III), with *trans-iso*-thiocyanato groups completing an octahedral arrangement. The nitrogen configuration is racemic, with an approximate twofold axis through the cobalt and thiocyanate ions.

2.5   5,7-dimethyl-1,4,8,11-tetraazacyclotetradec-4,7-diene
Me₂[14]dieneN₄, C₁₂H₂₄N₄

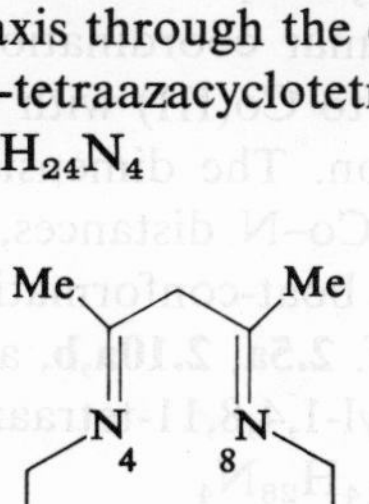

2.5

Complex formed by reaction of 1,4,8,11-tetraazaundecane with 2,4-pentanedione in the presence of metal ions.[44]

2.5a   *trans*-[Co(L)(OH$_2$)(PF$_6$)]PF$_6$. R = 0.0953 for 3064 refl.***[45]

Planar coordination to $S = \frac{1}{2}$ Co(II) with the oxygen of water completing a square-pyramidal arrangement. Fluorine of hexafluorophosphate interacts weakly in the *trans*-"octahedral" site. The nitrogen configuration is *meso*, with both NH groups on the same side of the molecular plane as the coordinated water molecule, and possibly hydrogen bonded to it. The molecule possesses an approximate mirror plane through the cobalt, central carbon atom of the six-membered chelate rings, oxygen of the water, and linear F–P–F of the hexafluorophosphate. The bond lengths for the dimethyl-substituted six-membered chelate ring indicate the presence of the isolated diimine (C–C, 150 pm; C=N, 128 pm; cf. **2.3a**). The bond angle at the central methylene group of the chelate ring N(4)–N(8) is 128(2)°, indicative of considerable strain (cf. **2.2a** and **2.5Aa,b**).

2.5A   5,7-dimethyl-1,4,8,11-tetraazacyclotetradec-4,7-diene-6-one,

2.5A

C$_{12}$H$_{22}$N$_4$O

Complexes formed by oxygenation of Co(II) complex of the diene **2.5**.[46]

2.5Aa   *trans*-[Co(L)(H$_2$O)$_2$](ClO$_4$)$_2$. R = 0.062 for 1835 refl.***[46]

2.5Ab   *trans*-[Co(L)(Cl)$_2$]ClO$_4$. R = 0.038 for 1894 refl.***[46]

The macrocycle is in planar coordination to $S = \frac{1}{2}$ Co(II) with *trans* water ligands for **2.5Aa**, and to Co(III) with *trans* chloride for **2.5Ab**, both with the *N-meso* configuration. The dimensions and conformation of the M(L) moiety, including the Co–N distances, are very similar for the two compounds, except that the boat-conformation chelate ring N(4)–N(8) is slightly twisted for **2.5Aab** (cf. **2.5a**, **2.10a,b**, and **2.13a,b**).

2.6   5,7,12,14-tetramethyl-1,4,8,11-tetraazacyclotetradec-5,11-diene
        Me$_4$[14]dieneN$_4$, C$_{14}$H$_{28}$N$_4$

Ligand formed by reaction of diaminoethane monohydroperchlorate and pent-2-ene-4-one.[42]

2.6a   [Ni(7,14-*rac*-L)](ClO$_4$)$_2$. R = 0.106 for 523 refl.**[47] Minor yellow product VIIBα described in Ref. 50b.

Planar coordination to $S = 0$ Ni(II). The carbon, C(7), C(14), and

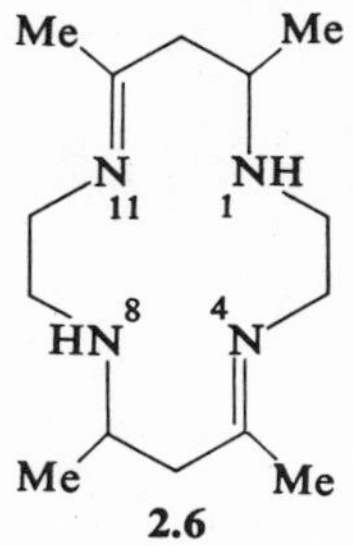

2.6

nitrogen, N(1), N(8), chiral centers are both racemic, with the chiral CH and NH groups on opposite sides of the molecular plane. The cation has twofold symmetry, with the methyl substituents on C(7) and C(14) axially oriented.

2.6b Ni(7,14-*meso*-L)](ClO$_4$)$_2$. $R = 0.1437$ for 775 refl. (P)**[48]

Major orange product VIIIAα described in Ref. 50b.

Planar coordination to $S = 0$ Ni(II). The nitrogen centers N(8), N(11) have the *rac* configuration, and the carbon centers C(7), C(14) have the *meso* configuration with one methyl substituent C(7) equatorially, and one C(14) axially oriented.

2.7   5,6,12,13-tetramethyl-1,4,8,11-tetraazacyclotetradec-4,11-diene
       Me$_4$[14]dieneN$_4$, C$_{14}$H$_{28}$N$_4$, 5,6,12,13-Me$_4$[14]dieneN$_4$

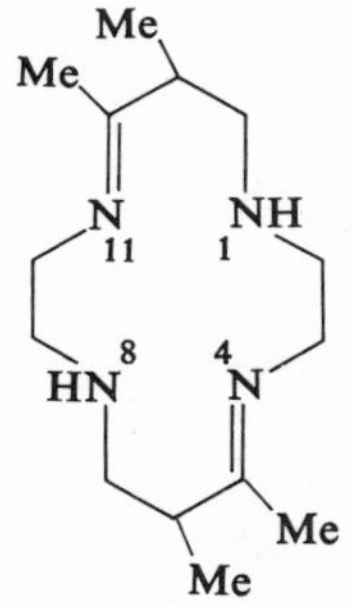

2.7

Ligand formed by reaction of 1,2-diaminoethane monohydroperchlorate and 2-methyl-prop-1-ene-3-one.[42]

2.7a [Ni(6,13-*meso*-L)](ClO$_4$)$_2$. $R = 0.039$.[49]

Planar coordination to $S = 0$ Ni(II). The cation is centrosymmetric (both chiral nitrogen and carbon centers *meso*), with adjacent chiral NH and CH groups on the opposite sides of the molecular plane. The methyl substituents on C(6) and C(13) are axially oriented.

2.8   3,5,10,12-tetramethyl-1,4,8,11-tetraazacyclotetradeca-4,11-diene
       Me$_4$[14]dieneN$_4$, C$_{14}$H$_{28}$N$_4$, 3,5,10,12-Me$_4$[14]dieneN$_4$

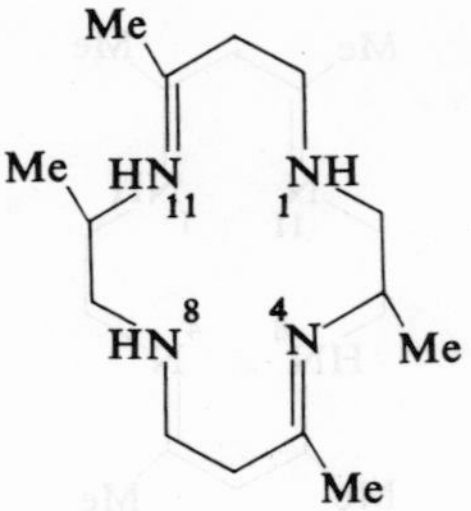

2.8

Ligand formed by reaction of methylvinyl ketone with monohydroperchlorate of 1,2-diaminopropane.[50a)]

    2.8a [Ni(3,10-*meso*-L)](ClO$_4$)$_2$. $R = 0.042$ for 1520 refl.***[51)]

    Planar coordination to $S = 0$ Ni(II). The cation is centrosymmetric, with the nitrogen and carbon configurations both *meso*. The methyl substituents on the five-membered chelate rings are axially oriented. (cf. **2.12a**.)

    2.9 5,12-dimethyl-7,14-diphenyl-1,4,8,11-tetraazacyclotetradec-4,11-diene, Me$_2$Ph$_2$[14]dieneN$_4$, C$_{24}$H$_{28}$N$_4$

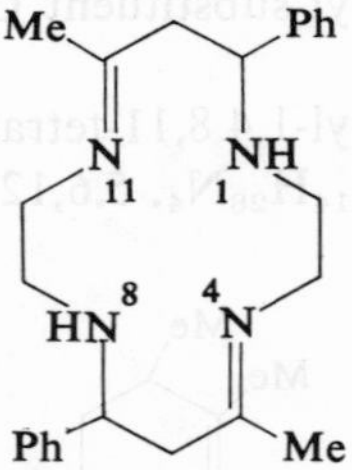

2.9

Ligand formed by reaction of benzylidene acetone and diaminoethane.[52)]

    2.9a [Cu(7,14-*meso*-L)(NO$_3$)$_2$]. $R = 0.055$ for 854 refl.*[53)]

    Planar coordination to copper(II), with *trans* weakly coordinated unidentate nitrate ions. The configuration is 1,8-*meso*-7,14-*meso*, with the adjacent chiral CH and NH groups on opposite sides of the molecular plane.

    2.10 5,7,7,12,14,14-hexamethyl-1,4,8,11-tetraazacyclotetradec-4,11-diene, Me$_6$[14]4-11-dieneN$_4$, C$_{16}$H$_{32}$N$_4$

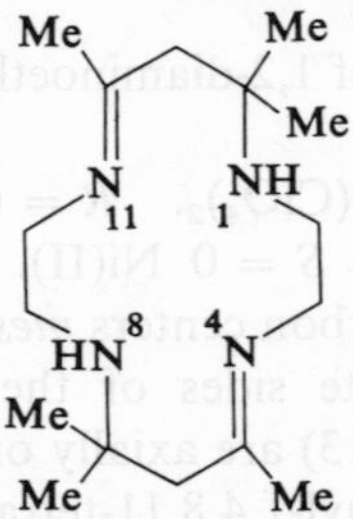

2.10

Ligand formed by reaction of monoprotonated diaminoethane with acetone.[54] Metal-ion complexes formed by reaction of diaminoethane complexes with acetone.[55,56]

2.10aa *trans*-[Cr(L)(NO$_2$)(NO)]PF$_6$. $R = 0.061$ for 1853 refl.***[57]

Planar coordination of the macrocycle to $S = \frac{1}{2}$ Cr(I), with nitric oxide and a nitro group coordinated axially, with ONCoN lying along a twofold axis (i.e., *N-rac*). The chromium atom is displaced by 22 pm out of the equatorial plane [which has a slight (± 1) tetrahedral distortion] toward the nitric oxide. The Cr–N distances are amine, 204; imine, 202; nitric oxide, 168; nitro, 220 pm.

2.10a [Fe(L)Cl]I. $R = 0.079$ for 1192 refl.***[58]

Folded coordination of the macrocycle to $S = 2$ Fe(II), with chloride at the apex of a distorted square-pyramidal arrangement, with a twofold axis through Fe–Cl. The nitrogen configuration is racemic, with NH groups on the same side of molecular plane as the chloride. Alternatively, the coordination sphere may be regarded as distorted trigonal-bipyramidal with 118°, 121°, 121° equatorial angles and with a bent (160°) N(1)FeN(8) axis. The iodide ions are disordered.

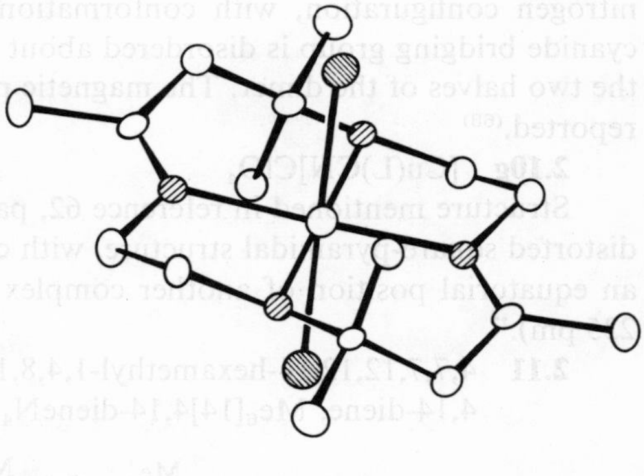

2.10b

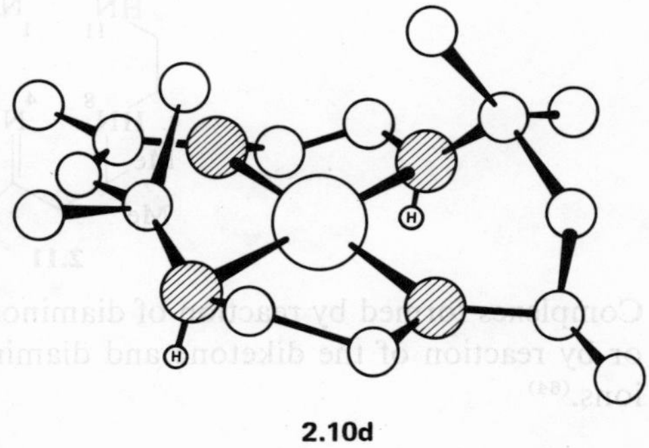

2.10d

FIGURE 9. Structures of compounds of 5,7,7,12,14,14-hexamethyl-1,4,8,11-tetraazacyclotetradec-4,11-diene with cobalt(III), **2.10b**, and nickel(II), **2.10d**, with the *N-meso* and *N-rac* configurations, respectively. Reproduced from reference 22, **2.10b**, by permission of the copyright holder, The American Chemical Society, and from reference 60, **2.10d**, by permission of the Chemical Society.

2.10b   *trans*-[Co(L)(H$_2$O)$_2$](BF$_4$)$_2$. $R = 0.068$ for 1442 refl.***[22]
(See Figure 9.)
   2.10c   *trans*-[Co(L)(NH$_3$)$_2$](ClO$_4$)$_3$. $R = 0.066$ for 1533 refl.*[59]

The macrocycle is in planar coordination to $S = \frac{1}{2}$ Co(II) with *trans* water (**2.10b**), and to Co(III) with *trans* ammonia (**2.10c**) completing octahedral arrangements. The compounds are centrosymmetric (*meso* nitrogen configurations). The dimensions within the Co(L) moiety are very similar for the two compounds, but the Co–axial-donor distances are much longer for Co(II) than for Co(III) (cf. **2.5Aa,b** and **2.13a,b**).
   2.10d   [Ni(L)](ClO$_4$)$_2$. $R = 0.12$ for 2048 refl. (P)***[60]
(See Figure 9.)
   2.10e   [Ni(L)](NCS)$_2$, H$_2$O. $R = 0.071$ for 3536 refl.***[61]

Planar coordination of the macrocycle to $S = 0$ Ni(II). Both structures have the racemic nitrogen configuration, with an approximate twofold axis through the nickel, with very similar conformations of the macrocycle for the two compounds.
   2.10f   [{Cu(L)}$_2$CN](ClO$_4$)$_3$. $R = 0.071$ for 2138 refl.***[62]

Folded coordination to copper(II), with bridging cyanide completing a trigonal-bipyramidal arrangement [N(1)CuN(8) axial, 174°]. Racemic nitrogen configuration, with conformation similar to that for **2.10a**. The cyanide bridging group is disordered about an inversion center which relates the two halves of the dimer. The magnetic properties of the dimer have been reported.[63]
   2.10g   [Cu(L)CN]ClO$_4$

Structure mentioned in reference 62, page 2407. Described as "having a distorted square-pyramidal structure, with cyanide bridging from an axial to an equatorial position of another complex (Cu–CN = 200 pm, Cu–NC = 235 pm)."
   2.11   4,7,7,12,12,14-hexamethyl-1,4,8,11-tetraazacyclotetradec-4,14-diene, Me$_6$[14]4,14-dieneN$_4$, C$_{16}$H$_{32}$N$_4$

2.11

Complexes formed by reaction of diaminoethane complexes with acetone,[55] or by reaction of the diketone and diaminoethane in the presence of metal ions.[64]

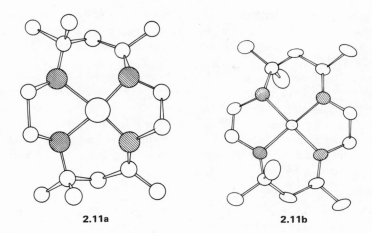

2.11a   **2.11b**

FIGURE 10. Structures of compounds of 5,7,7,12,12,14-hexamethyl-1,4,8,11-tetraaza-cyclotetradec-4,14-diene with cobalt(III), **2.11a** omitting the *trans* cyanide ligands, and with nickel(II), **2.11b**, with the *N-meso* and *N-rac* configurations, respectively. Reproduced from references 66 and 67 by permission of The Chemical Society.

2.11a   *trans*-[Co(L)(CN)$_2$]ClO$_4$.   R = 0.99 for 1063 refl. (P)***[65]

(See Figure 10.) Planar coordination to cobalt(III) with *trans* cyano groups completing an octahedral arrangement. The nitrogen configuration is *meso* with an approximate mirror plane through the cobalt and cyanide ions, and the midpoints of the five-membered chelate rings, which are in eclipsed conformations, giving a saddle shape to the macrocycle.

2.11b   [Ni(L)](ClO$_4$)$_2$.   R = 0.062 for ～1000 refl.***[66]

(See Figure 10.) Planar coordination to S = 0 Ni(II). The nitrogen configuration is racemic, with a twofold axis through the nickel and the midpoints of the five-membered chelate rings (cf. the tetraimine analog, **2.14a**).

2.12   3,5,7,7,10,12,14,14,14-octamethyl-1,4,8,11-tetraazacyclotetradec-4,11-diene, Me$_8$[14]dieneN$_4$, C$_{18}$H$_{36}$N$_4$

Me—C=N—C—Me
       | 11 |
Me   N   NH   Me
       | |
      HN   8   4   N—Me
Me   N
Me   Me

2.12

Complexes of 3,10-*rac* and 3,10-*meso* isomers formed by reaction of 1,2-diaminopropane complexes with acetone.[67] 3,10-*meso* isomer of ligand formed by reaction of monoprotonated 1,2-diaminopropane with acetone.[68]

2.12a [Ni(3,10-*rac*-L)](ClO$_4$)$_2$. $R = 0.0767$ for 2211 refl.***[69]

Planar coordination to $S = 0$ Ni(II). The 3,10-carbon and 1,8-nitrogen chiral centers are both racemic, with the CH and NH groups on opposite sides of the molecular plane. The methyl substituents on the five-membered chelate rings adopt axial orientations, on the side of the molecular plane opposite to the axial component of the *gem* dimethyl groups (cf. **2.8a**). The macrocycle of **2.12a** has a conformation similar to, but more puckered than that of the hexamethyl analog, **2.10d,e**.

[14]tetraeneN$_4$ Compounds

2.13   2,3,9,10-tetramethyl-1,4,8,11-tetraazacyclotetradec-1,3,8,10-tetraene, TIM, Me$_4$[14]tetraeneN$_4$, C$_{14}$H$_{28}$N$_4$

2.13

Complexes formed by reaction of 1,3-diaminopropane and 2,3-butanedione in the presence of metal ions.[70]

2.13a   *trans*-[Co(L)(OH$_2$)$_2$](ClO$_4$)$_2$. $R = 0.077$ for 496 refl.***[22,71]

2.13b   *trans*-[Co(L)(NH$_3$)$_2$](ClO$_4$)$_3$. $R = 0.067$ for 1595 refl.[71]

Planar coordination to $S = \frac{1}{2}$ cobalt(II) with axially coordinated water, **2.13a**, and to cobalt(III) with axially coordinated ammonia, **2.13b**. Both cations are centrosymmetric, and the metal–macrocycle moiety has essentially the same dimensions in the two structures. The five-membered chelate rings are essentially planar, as are the six-membered chelate rings, apart from the central carbon atoms. The methyl substituents on the five-membered chelate rings are displaced in opposite directions out of the molecular plane. The cobalt–axial-donor bond length is much longer for the Co(II) compound (cf. **2.5Aa,b** and **2.10b,c**).

2.13c   *trans*-[Ni(L)(C$_3$H$_4$N$_2$)$_2$](PF$_6$)$_2$[72]

Planar coordination to $S = 1$ Ni(II) with *trans*-imidazole ligands completing octahedral coordination.

2.14   5,5,7,12,14,14-hexamethyl-1,4,8,11-tetraazacyclotetradec-1,3,7,11-tetraene, *cis*-[14]tetraene, Me$_6$[14]tetraeneN$_4$, C$_{16}$H$_{28}$N$_4$

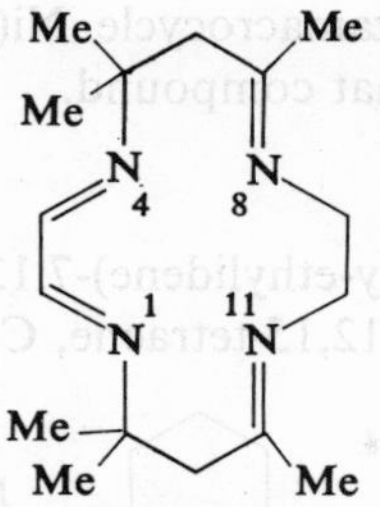

2.14

Complex formed by oxidative dehydrogenation of [Ni(Me$_6$[14]diene)]$^{2+}$.[73]

2.14a [Ni(L)](ClO$_4$)$_2$. $R = 0.012$ for 1229 refl. (P)***[74]

Planar coordination to $S = 0$ Ni(II). A twofold axis passes through the nickel and the midpoints of the two five-membered chelate rings. The compound is close to isostructural with the 4,14-diene analog (**2.11b**).

[14]hexaenato(1−)N$_4$ Compounds

2.15 5,7,12,14-tetramethyl-1,4,8,11-tetraazacyclotetradec-1,5,7,9,11,13-hexaenato(1−), Me$_4$[14]hexaenato(1−)N$_4$, [C$_{14}$H$_{18}$N$_4$]$^-$

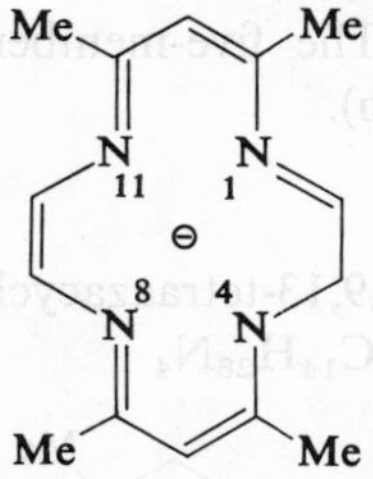

2.15

Metal-ion compounds formed by hydride abstraction from the 4,7,12,14-tetraene complex with the triphenylmethyl cation.[75]

2.15a [{Ni(L)}$_2$](CF$_3$SO$_3$)$_2$. $R = 0.049$ for 1978 refl.****[76]

(See Figure 21.) The $S = 0$ Ni(II) ion is in planar coordination. The cation is a centrosymmetrical confacial dimer with Ni–Ni distance of 306 pm. The macrocycles of the dimer are essentially planar and are eclipsed, with mean separation of the N$_4$ planes of 320 pm, with the Ni atoms displaced by 10 pm from the N$_4$ plane towards each other. The unsaturation in the macrocycle is essentially delocalized, interatomic distances for the six-membered and five-membered chelate rings being C–C, 140 pm; C–N, 135 pm; and C–C, 138(2) pm; C–N, 137(1) pm, respectively, close to the values for benzene and pyridine. The structure and bonding of this compound are very similar to that of an octaazamacrocycle Ni(II) compound **6.8b**, and are discussed in Section 6 with that compound.

[15]tetraeneN₄ Compounds

2.16 6,14-*bis*-(1-methoxy-ethylidene)-7,13-dimethyl-1,4,8,12-tetraaza-cyclopentadec-4,7,12,15-tetraene, $C_{19}H_{30}N_4O_2$

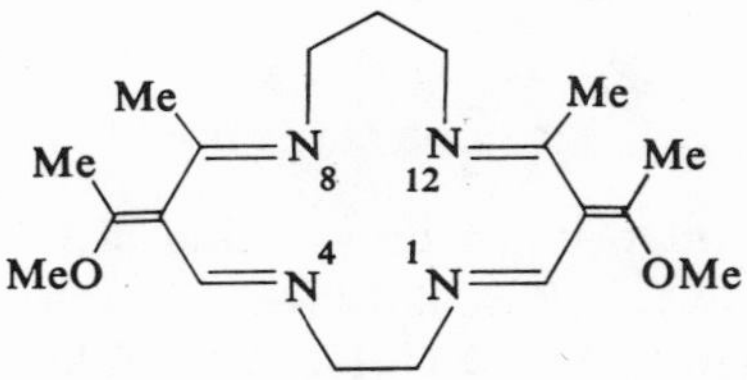

2.16

2.16a [Ni(L)₂](ClO₄)₂. $R = 0.046$ for 2228 refl.*[77]

Planar coordination to $S = 0$ Ni(II), with a mirror plane through the nickel, C(10), and the midpoint of C(2)–C(3). The N(8)–N(12) chelate ring adopts a boat conformation (Ni and C segments tilted up by 35° and 56°, respectively, from the plane defined by the two C–N bonds). The chelate rings N(4)–N(8) and N(1)–N(12) also adopt a boat conformation, with the C=N bonds approximately parallel, and with the Ni and C segments tilted up by 33° and 25°, respectively. The five-membered chelate ring N(1)–N(4) is eclipsed (cf. **2.3a** and **2.5Aa,b**).

[16]dieneN₄ Compounds

2.17 2,12-dimethyl-1,5,9,13-tetraazacyclohexadec-1,12-diene
Me₂[16]dieneN₄, $C_{14}H_{28}N_4$

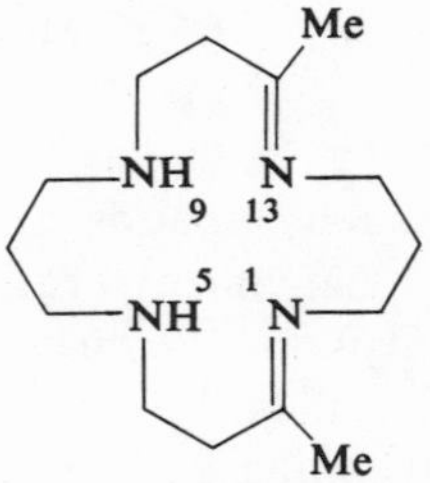

2.17

Complex formed by reaction of prop-1-ene-3-one with 1,3-diaminopropane in the presence of nickel(II).[78]

2.17a *trans*-[Ni(L)(OH₂)Cl]Cl. $R = 0.048$ for 1465 refl.*[79]

Planar coordination to $S = 1$ Ni(II) with oxygen of water and a chloride ion completing a *trans*-octahedral arrangement. The nitrogen configuration is *meso*, with the NH groups on the same side of the molecular plane as the

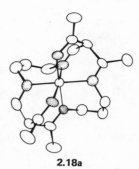

     2.18a

water molecule. The chelate ring conformations are N(1)–N(5) and N(9)–
N(13), half chair; N(5)–N(9), chair; N(13)–N(1), twist (cf. **2.18a**).

    2.18   *cis*-[3,11-*bis*-(1-iminoethyl)-2,12-dimethyl-1,5,9,13-tetraazacyclo-
hexadec-1,4,9,12-tetraene, $C_{18}H_{30}N_6$

2.18

Complex formed by reaction of complex of 1,4,8,12-tetraimine macrocycle
2.11 with methyl cyanide.[80]

    2.18a   [Fe(L)](PF$_6$)$_2$. $R = 0.064$ for 2787 refl.***[80]

    (See Figure 11.) The macrocycle is coordinated as a sexadentate ligand
to $S = 0$ Fe(II), with the pendant imine groups derived from the methyl
cyanide in *cis* sites. The three imine groups from each bridgehead are parallel,
and the donor atoms define an equilateral triangle, the triangles from the
two bridgeheads being twisted relative to each other by 52°, i.e., with a
trigonal twist angle ϕ of 8°. The two saturated six-membered chelate rings
have twist conformations. (Structure discussed further in Section 11.)

4. Class 3: Macrocycles Including a 2,6-Pyridyl Group

4.1. Discussion of Structures

    The metal-ion macrocycle compounds in this section were all prepared
by reaction of 2,6-diacetyl pyridine with linear tri- or tetramines in the

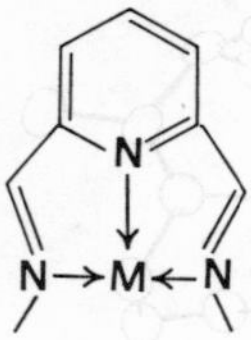

FIGURE 12. Chelate rings associated with 2,6-pyridyl groups of macrocycles of Section 4.

presence of metal ions.[81–83] The compounds so formed have one pyridine nitrogen donor atom linked to two imine nitrogen donor atoms by five-membered chelate rings (Figure 12) plus one or two amine donor atoms. For $[Ni(Me_2pyo[14]eneN_4)]^{2+}$, **3.1a**, the two imine groups of $[Ni(Me_2pyo[14]-trieneN_4)]^{2+}$, **3.2a**, have been reduced, introducing two chiral carbon and two chiral nitrogen centers.[84] Macrocycles with the 2,6 pyridyl group are also included elsewhere (**6.10**, **6.12**, **8.7A**) and in the Addendum. (The properties of complexes of these macrocycles are included in a more general review of macrocycles with fused pyridyl rings.)[85]

Key dimensions for the compounds are listed in Table 6.

4.1.1. Pyo[14]N₄ Macrocycles

Three structures have been reported for pyo[14]N_4 compounds of this type, two with $S = 0$ Ni(II), and one with Cu(II) (Figure 13). All have the macrocycle in planar coordination, with some "tetrahedral" distortion of the four donor atoms from planarity, which causes a tilting of the pyridine ring with respect to the MN_4 plane. All show some displacement of the metal ion from the N_4 plane toward a coordinated anion which completes a square-pyramidal arrangement. Square-pyramidal coordination is unusual for $S = 0$ Ni(II), and its general presence suggests that this is a favored geometry for these macrocycles.

The $[Ni(Me_2pyo[14]trieneN_4)]^{2+}$ cation of **3.2a** shows approximate mirror symmetry [through the pyridine nitrogen N(17), N(7), and the nickel ion], with the five-membered chelate rings approximately planar, and with the six-membered chelate rings in mirror-related chair conformations. For the copper(II) compound of the *N*-methyl macrocycle **3.3a**, the six-membered chelate rings have chair conformations, but with an approximate twofold relationship.

The $[Ni(Me_2pyo[14]eneN_4)]^{2+}$ cation (with imine groups reduced), **3.1a**, has the *C-rac-N-rac* configuration. The five-membered chelate rings remain approximately planar, with one methyl group pseudoequatorially and the other pseudoaxially oriented. The six-membered chelate rings have chair conformations, with a very approximate twofold relationship. The chelate ring N(3)–N(7) has the NH groups *trans*, i.e., on opposite sides of the flattened macrocycle, but adopts a flattened chair conformation, although with an unusually large chelate angle of 102° [deviations from good N(3), C(4), C(6)

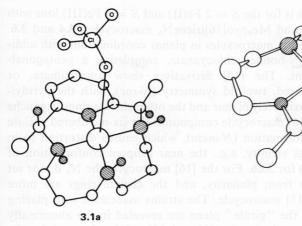

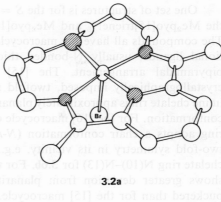

3.1a

3.2a

FIGURE 13. Structures of cations of compounds with pyo[14]N₄ macrocycles, **3.1a**, **3.2a**, and **3.3a**. Reproduced from references 85, 86, and 89 by permission of the copyright holders, the publishers of *Nature*, The American Chemical Society, and the publishers of *Acta Crystallographica*, respectively.

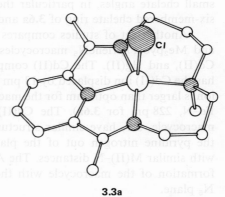

3.3a

plane: Ni, −40 pm; C(5), 74 pm]. The chelate ring N(7)–N(11) has the NH groups *cis*, and adopts a highly buckled chair conformation with a smaller-than-usual chelate angle of 89° [deviation from good N(7), C(8), C(10), N(11) plane: Ni, 115 pm; C(9), −61 pm]. The pyridine ring is tilted with respect to the N₄ plane, as for the diimine compounds, and also twisted with respect to the N(1)N(3)N(11) plane. These asymmetries arise because of the inherent asymmetry of the configuration present, and the coordination of a perchlorate ion on one side of the molecular plane.

4.1.2. *Pyo[15]N₅, pyo[16]N₅, and pyo[17]N₅ Macrocycles*

Studies have been reported for compounds of four pentadentate pyo-N₅ macrocycles, with "nonpyridyl" chelate ring systems of linked 5,5,5-, 5,6,5-, and 6,5,6-membered chelate rings.

One set of structures is for the $S = 2$ Fe(II) and $S = \frac{5}{2}$ Fe(III) ions with the Me$_2$pyo[15]trieneN$_5$ and Me$_2$pyo[16]trieneN$_5$ macrocycles, **3.4** and **3.6**. The compounds all have the macrocycles in planar coordination, with additional ligands, usually N-bonded thiocyanate, completing a pentagonal-bipyramidal arrangement. The [15] derivatives show approximate, or crystallographically imposed, twofold symmetry (*N-rac*), with the pyridyl-imine chelate rings approximately planar and the other chelate rings in gauche conformation. For the [16] macrocycle compounds the six-membered chelate ring adopts a chair conformation (*N-meso*), which causes distortion from two-fold symmetry in its vicinity, e.g., the near eclipsed conformation of chelate ring N(10)–N(13) for **3.6b**. For the [16] macrocycle the N$_5$ donor set shows greater deviation from planarity, and the chelate rings are more puckered than for the [15] macrocycle. The strains associated with placing the five donor atoms in the "girdle" plane are revealed in the abnormally small chelate angles, in particular the 77° and 75° values observed for the six-membered chelate ring of **3.6a** and **3.6b**.

Another set of studies compares structures of the Me$_2$pyo[16]trieneN$_5$ and Me$_2$pyo[17]trieneN$_5$ macrocycles **3.6** and **3.7** with the d^{10} ions Ag(I), Cd(II), and Hg(II). The Cd(II) compound with the [16] macrocycle, **3.6d**, has the Cd(II) ion displaced by 13 pm from the N$_5$ plane, suggesting that this ion is larger than optimum for the macrocycle (Cd–N, 240 pm compared with Fe–N, 228 pm for **3.6a**). The Cd(II) and Hg(II) compounds of the [17] macrocycle, **3.7**, have similar structures, with the macrocycle folded, with the pyridine nitrogen out of the plane of the other nitrogen atoms, and with similar M(II)–N distances. The Ag(I) compound in contrast has a conformation of the macrocycle with the pyridine nitrogen close to the best N$_5$ plane.

4.1.3. Metal-Ion–Nitrogen Bond Lengths

For the M(II) transition-metal ions the M–N distances for all the macrocycles increase in the order pyridine < imine < amine nitrogen. For the Fe(III) compounds of the [15]-N$_5$ and [16]-N$_5$ macrocycles the sequence is pyridine < amine < imine. For the different types of nitrogen donor atoms, the Fe–N distances are: Fe(II) $\sim$ Fe(III) for imine, Fe(II) > Fe(III) for amine, and Fe(II) < Fe(III) for pyridine. The Fe–N distances for the N$_5$ macrocycles are abnormally long, and show only marginal differences between the Fe(II) and Fe(III) compounds, and between the [15] and [16] macrocycles, suggesting that the distances are determined largely by the strain arising from placement of the five donor atoms in the plane. The Fe–axial-ligand distances for these compounds are substantially greater for Fe(II) than for Fe(III), as for the Co(II)–Co(III) pairs of compounds **2.10b,c**, **2.5Aa,b**, and **2.13a,b**.

4.2. Listing of Reported Structures of Compounds of Macrocycles Including a 2,6-Pyridyl Group

See introduction for rules determining order of compounds. Key dimensions for the compounds for which structural details have been published are listed in Table 6.

pyo[14]eneN₄ Compounds

3.1 2,12-dimethyl-3,7,11,17-tetraazabicyclo[11.3.1]septadeca-1(7),13,15-triene, Me$_2$pyo[14]eneN$_4$, CRH, C$_{15}$H$_{26}$N$_4$

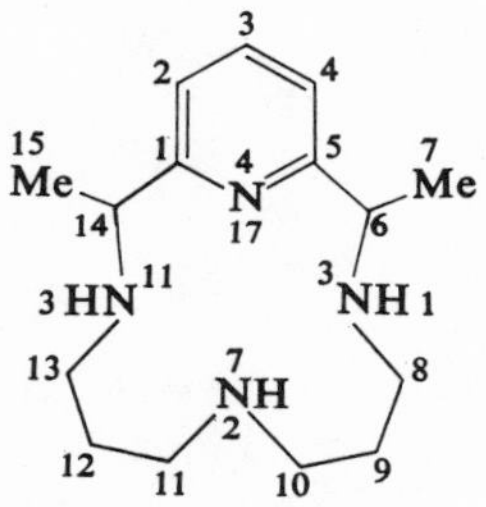

3.1

Complex formed by reduction of imine groups of the diimine compound **3.a**.[84]

3.1a [Ni(2,12-*rac*-L)](ClO$_4$)$_2$. $R = 0.13$ for 1331 refl.**[86]

(See Figure 13.) Planar coordination to $S = 0$ Ni(II). The crystallographic atom numbering which was omitted from the original paper is shown in the formula. One perchlorate ion is weakly coordinated to give a square-pyramidal arrangement. (See discussion on page 266.)

Pyo[14]trieneN₄ Compounds

3.2 2,12-dimethyl-3,7,11,17-tetraazabicyclo[11.3.1]septadec-1(17),2,11-13,15-pentaene, Me$_2$pyo[14]trieneN$_4$, CR, C$_{15}$H$_{22}$N$_4$

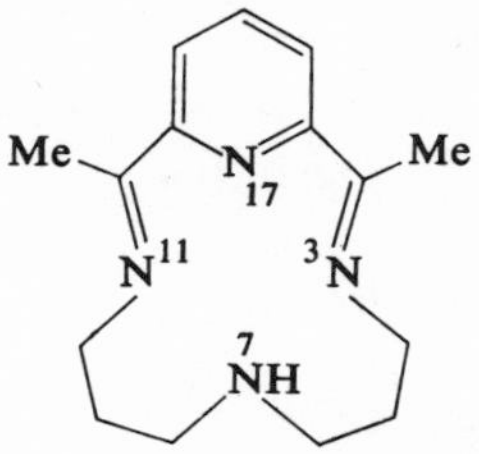

3.2

Complex formed by reaction of 2,6-diacetylpyridine and 1,5,9-triazaundecane in the presence of metal ions.[83]

3.2a [Ni(L)Br]Br·H$_2$O. $R = 0.066$ for 1074 refl.***[87]

(See Figure 13.) Planar coordination to $S = 0$ Ni(II) with bromide coordinated at the apex of a square-pyramidal arrangement.

3.3   2,7,12-trimethyl-3,7,11,17-tetraazabicyclo[11.3.1]septadec-
1(17),2,11,13,15-pentaene, Me$_2$(N–Me)[15]trieneN$_4$, C$_{16}$H$_{24}$N$_4$

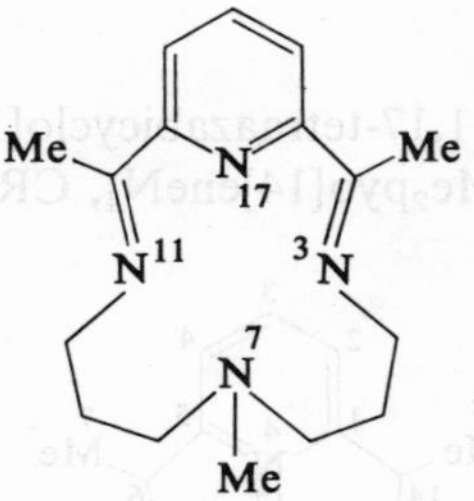

3.3

Complex formed by reaction of 2,6-diacetyl pyridine with 5-methyl-1,5,9-triazaundecane in the presence of metal ions.[88]

3.3a   [Cu(L)Cl]Cl. $R = 0.108$ for 2190 refl.***[89]

(See Figure 13.) Planar coordination to copper(II) with a chloride ion weakly coordinated to complete a square-pyramidal arrangement. The unit cell contains two independent cations with very similar dimensions.

Pyo[15]*trieneN$_5$ Compounds*

3.4   2,13-dimethyl-3,6,9,12,18-pentaazabicyclo[12.3.1]octadec-
1(18),2,12,14,16-pentaene, Me$_2$pyo[15]trieneN$_5$, C$_{15}$H$_{23}$N$_5$

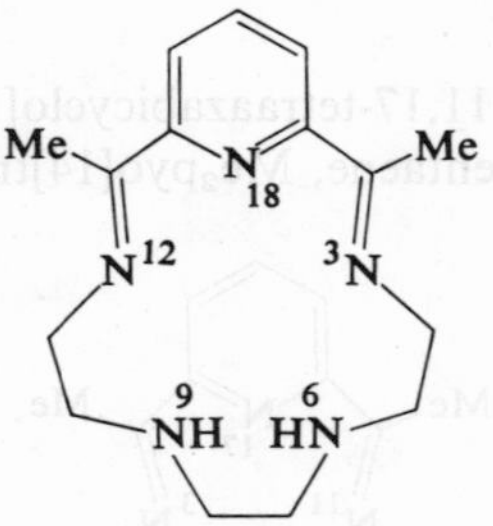

3.4

Complex formed by reaction of 2,6-diacetyl pyridine with 1,4,7,10-tetra-azaundecane in the presence of metal ions.

3.4a   [Mg(L)(H$_2$O)$_2$]Cl$_2$. $R = 0.082$ for 1782 refl.[90]

Planar coordination of the macrocycle, with axially coordinated water completing a pentagonal-bipyramidal arrangement. The maximum deviation of any nitrogen atom from the MgN_5 plane is 18 pm. Mg–N ranges from 224–231 pm, and Mg–O is 210 pm. The Mn(II) analog is isomorphous.[90]

3.4aa $[Mn(L)(H_2O)_2](ClO_4)_2 \cdot 1,10$-phenanthroline $\cdot \frac{1}{2}$ethanol.

$R = 0.10$ for 950 refl.*[91]

The macrocycle is in planar coordination to $S = \frac{5}{2}$ Mn(II), with maximum deviation of any N atom from the MnN_5 plane of 3 pm. Two water molecules complete a pentagonal-bipyramidal arrangement about the manganese atom [Mn–N, 234(3)–241(3) pm; Mn–O, 222(2) pm]. The phenanthroline molecules are sandwiched between adjacent macrocycle planes in a lamella structure, the angle between the macrocycle and phenanthroline planes being 2.7°.

3.4b *trans*-[Fe(L)(NCS)$_2$]. $R = 0.058$ for 1057 refl.***[92]
(See Figure 14.)

3.4c *trans*-[Fe(L)(OH$_2$)$_2$]Cl $\cdot$ ClO$_4$. $R = 0.073$ for 1285 refl.*[93]

Planar coordination of macrocycle to $S = 2$ Fe(II), with nitrogens of thiocyanate, **3.4b**, or oxygens of water, **3.4c**, coordinated axially to complete pentagonal-bipyramidal arrangements. The cation of **3.4c** has imposed twofold symmetry, while the molecule of **3.4b** has approximate twofold symmetry [i.e., NH groups at N(6) and N(9) on opposite sides of the molecular plane, *rac* configuration].

3.4d *trans*-[Fe(L)(NCS)$_2$]ClO$_4$. $R = 0.090$ for 1531 refl.***[94]

$R = 016$ for 1278 refl.*[95]

Planar coordination to $S = \frac{5}{2}$ Fe(III), with nitrogens of thiocyanate completing a pentagonal-bipyramidal arrangement, similar to that of the

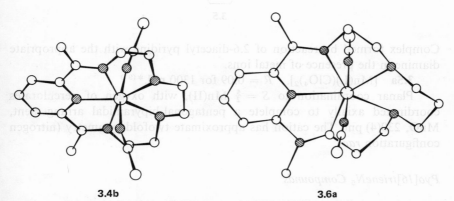

3.4b   **3.6a**

FIGURE 14. Structures of *trans*-dithiocyanato-iron(II) compounds with Me$_2$pyo[15]dieneN$_5$ and Me$_2$pyo[16]dieneN$_5$ macrocycles, **3.4b** and **3.6a**. Reproduced from reference 93 by permission of The Chemical Society.

Fe(II) compound **3.4b**. The cation has approximate twofold symmetry, *N-rac* configuration. The perchlorate ion shows rotational disorder.

3.4e   *trans*-[{Fe(L)(OH)}$_2$O](ClO$_4$)$_4$[(95)]

Planar coordination to $S = \frac{5}{2}$ Fe(III) with water and bridging oxo groups completing a pentagonal-bipyramidal arrangement about each ion. The Fe–O–Fe bridge is linear; Fe–O, 180 pm; Fe–OH$_2$, 215 pm. "Extensive disorder" of the perchlorate atoms, together with limited data, precluded effective refinement.

3.4f   [Zn(L)(NCS)$_2$], $\frac{1}{2}$(C$_2$H$_4$Cl$_2$). $R = 0.084$ for 4500 refl.[(96)]

Planar coordination to Zn(II) with the nitrogens of two thiocyanate ions completing a distorted pentagonal-bipyramidal arrangement.

Pyo[15]-pentaeneN$_5$ Compounds

3.5   dibenzo[d,j]-2,13-dimethyl-3,6,9,12,18-pentaazabicyclo[12.3.1]-octadec-1(18),2,5,11,13,15,17-septaene, Me$_2$bzo$_2$pyo[15]-pentaeneN$_5$, C$_{23}$H$_{23}$N$_5$

3.5

Complex formed by reaction of 2,6-diacetyl pyridine with the appropriate diamine in the presence of metal ions.

3.5a   [Mn(L)(ClO$_4$)$_2$]. $R = 0.09$ for 1300 refl.*[(97)]

Planar coordination to $S = \frac{5}{2}$ Mn(II), with oxygen of perchlorates coordinated axially to complete a pentagonal-bipyramidal arrangement, MnO, 230(4) pm. The cation has approximate twofold symmetry (nitrogen configuration *rac*).

Pyo[16]trieneN$_5$ Compounds

3.6   2,14-dimethyl-3,6,10,13,19-pentaazabicyclo[13.3.1]nonadec-1(19),2,13,15,17-pentaene, Me$_2$pyo[16]trieneN$_5$, C$_{16}$H$_{25}$N$_5$

Complex formed by reaction of 1,4,8,11-tetraazaundecane with 2,6-diacetyl-pyridine in the presence of metal ions.

3.6

3.6a *trans*-[Fe(L)(NCS)$_2$]. $R = 0.069$ for 1131 refl.*** [92]
3.6b *trans*-[Fe(L)(NCS)$_2$]ClO$_4$. $R = 0.077$ for 1473 refl.*** [94]
(See Figure 14.) Planar coordination to $S = 2$ Fe(II), **3.7a**, or $S = \frac{5}{2}$ Fe(III), **3.6b**, with *trans-iso*-thiocyanato groups completing a pentagonal-bipyramidal arrangement. The nitrogen configuration is *meso* [i.e., NH groups at N(6) and N(10) on the same side of the molecular plane]. The cations show elements of twofold symmetry, although the distortions are much greater than for **3.4b,d** especially near the six-membered chelate ring.
3.6c [{Ag(L)}$_2$](ClO$_4$)$_2$. $R = 0.066$ for 2066 refl.* [98]
The silver(I) ion is coordinated to the five nitrogen atoms of one macrocycle, and to one imine nitrogen atom of a second macrocycle to give a dimeric cation with C_2 symmetry, with each metal ion in a pentagonal-pyramidal arrangement. The silver and bridging imine nitrogen atoms lie 59 and 88 pm above the N$_5$ plane, respectively. The Ag–N distances are: pyridine, 243 pm; nonbridging imine, 243 pm; bridging imine, 257 pm (in-plane), 275 pm (axial) amine, 240, 253 pm, respectively. There are no significant differences in dimensions about the bridging and nonbridging imine nitrogen atoms, the bridging imine group remaining close to planar, with C=N of 129(2) pm.
3.6d [Cd(L)Br]$_n$[CdBr$_4$]$_{2n}$. $R = 0.092$ for 1732 refl.* [99]
The macrocycle is close to planar (maximum deviation of any nitrogen atom from the N$_5$ plane is 5 pm), with the Cd(II) ion displaced by 13 pm from this plane. The Cd–N distances range from 236(2) to 243(3) pm. Two bromide ions are coordinated apically (Cd–Br, 278 and 304 pm; CdBrCd, 143°) to produce a chain polymeric structure.

Pyo[17]*triene*N$_5$ *Compounds*

3.7 2,15-dimethyl-3,7,10,14,20-pentaazabicyclo[14.3.1]-1(20),2,14,16,18-pentaene, Me$_2$pyo[17]trieneN$_5$, C$_{17}$H$_{27}$N$_5$
Complexes formed by reaction of 1,5,8,12-tetraazadodecane with 2,6-diacetylpyridine in the presence of metal ions.

TABLE 6. Dimensions for Compounds with 2,6-Pyridyl Groups

Compound[a]	Ion[b]	Spin[c]	MN_x plane[a]	M–L[e]		Chelate rings[f,g,h]	
					f	g	h
Me₂pyo[14]eneN₄ Compounds							
3.1a	Ni(II)	0	[6], 18, −15, 18, −15	(3,11) 189(?)	(3-7)	102(?)°	−53, 41, −49*
				(7) 186(?)	(7-11)	89(?)°	122, 135, 108
				(17) 180(?)	(11-17)	86(?)°	−6(−157), −8
				$OClO_3^-$ 277(?)	(17-3)	84(?)°	−16, −43(34)
Me₂pyo[14]trieneN₄ Compounds							
3.2a	Ni(II)	0	[14], 0, 12, −6, 14	(3,11) 191(1)	(3-7)	97°	45, 6, 70*
				(7) 193(2)	(11-17)	83°	5(12), −1
				(17) 181(1)			
				Br^- 279			
3.3a	Cu(II)	0	[13], −6, −11, 6, −14	(3,11) 200(2)	(3-7)	98°	76, 45, 91*
				(7) 201(2)	(11-17)	80°	−4(−18), 3
				(17) 191(2)			
				Cl^- 250(1)			
Me₂pyo[15]trieneN₅ Compounds							
3.4b	Fe(II)	2	[−2], 19, −25, 25, −15, −2	(3,12) 225(1)	(3-6)	72°	−9, −69
				(6,9) 228(1)	(6-9)	75°	−37, 37
				(18) 216(1)	(9-12)	71°	69, 10
				OH_2 212(1)	(12-18)	72°	[8], −4(−7), 2[0]
					(18-3)	71°	[0], 5, 6(2), [8]
3.4c	Fe(II)	2	[0], −8, 13, −13, 8, 0	(3,12) 226	(3-6)	73°	7, 68
				(6,9) 226	(6-9)	75°	37, −37
				(18) 222	(18-3)	69°	(planar)
				OH_2 222			
3.4d	Fe(III)	5/2	[0], 8, −11, 10, −5, −2	(3,12) 224(1)	(3-6)	72°	[−5,] −7(−16), −9, [−2]
				(6,9) 222(1)	(6-9)	77°	−34, 38
				(18) 220(1)	(9-12)	72°	67, 6
				NCS^- 200(1)	(12-18)	70°	[−2], 2, 4(9), [8]
					(13-3)	70°	

	Metal (b)	Spin (c)	Ring-atom deviations (h)	Donor atoms (e)	M–donor distance (d)	Chelate ring (f)	NMN angle (g)	Deviations from N_x plane (i)
$Me_2bzo_2pyo[15]pentaeneN_5$ Compounds								
3.5a	Mn(II)	$5/2$		(3,12)	226(4)	(3–6)	72°	24, −44
				(6,9)	230(1)	(6–9)	78°	
				(18)	222(1)	(12–18)	71°	
$Me_2pyo[16]trieneN_5$ Compounds								
3.6a	Fe(II)	2	[−2], 19, −25, 25, −15, −2	(3,13)	227(1)	(3–6)	74°	−105, −86, −101
				(6,10)	235(1)	(6–10)	77°	−17, 51
				(19)	223(1)	(10–13)	73°	[−15], −31(−56), −19, [−2]
				NCS^-	213(1)	(13–19)	69°	[−2], 10, 30(62), [19]
						(19–3)	70°	[0], 5, 6(2), [8]
3.6b	Fe(III)	$5/2$	[−1], 22, −29, 28, −15, −5	(3,13)	223(1)	(3–6)	71°	36, −39
				(6,10)	228(1)	(6–10)	76°	−114, −111, −119
				(19)	225(1)	(10–13)	75°	51, −12
				NCS^-	201(1)	(13–19)	74°	[−15], −44(78), −28, [−5]
						(19–3)	70°	[−5], 16, 36(81), [22]
							68°	
$Me_2pyo[17]trieneN_5$ Compounds								
3.7a	Mn(II)	$5/2$	[14], 31, −16, −9, 26, −46	(3,14)	237	(3–7)	79°	70, 33, 77
				(7,10)	241	(7–10)	74°	24, −46
				(20)	231	(10–14)	77°	−71, −1, 71
				NCS^-	217(1)	(14–20)	67°	[−12], 0(20), −2, [−4]‡
					229(1)	(20–3)	68°	[−4], −3, −2(14), [−10]‡

a Compounds are numbered as in the listing of structures.

b Central metal ion and oxidation state.

c Spin quantum numbers of central ion, where significant.

d Characteristics of MX_n plane. Displacement of the metal ion from the N_x plane. Following numbers show displacement of ring donor atoms from the N_x plane in numerical sequence.

e Metal–donor-atom distances. Average values for chemically equivalent donor atoms shown where the difference is nonsignificant.

f Chelate rings indicated by ring numbers of donor atoms.

g Chelate (NMN) angles.

h Deviations of ring atoms from MN_2 chelate plane, in numerical sequence, values for substituents in parentheses. Deviations from N_x plane indicated by values for donor atoms in square brackets. Where values for chemically equivalent chelate rings are similar, average values are shown. Values calculated from data in original publication indicated by asterisk.

i Deviations from plane of pyridine ring and linked five-membered chelate rings. (Deviation of Mn atom from this plane, −77 pm.)

3.7

3.7a [Mn(L)(NCS)$_2$]. $R = 0.058$ for 2023 refl.*** [100]
3.7b [Ag(L)](ClO$_4$)$_2$. $R = 0.099$ for 1709 refl.* [99]
3.7c [Cd(L)Br]Br, H$_2$O. $R = 0.052$ for 2728 refl.* [99]
3.7d [Hg(L)Br]$_2$[Hg$_2$Br$_6$]. $R = 0.064$ for 1320 refl.* [99]

For **3.7a,b,c** the macrocycle is coordinated in a conformation which has the metal and four nonpyridine nitrogen atoms approximately coplanar, and with the plane of the pyridine ring and linked five-membered chelate rings folded out of this plane, and with the pyridine nitrogen atom appreciably displaced from the plane of the other four nitrogen atoms (Dihedral angle, displacement: **3.7a**—42°, 92 pm; **3.7b**—42°, 131 pm; **3.7c**—49°, 138 pm.) The conformation of the macrocycle in **3.7d** approximates to a C$_2$ distortion, with the pyridine nitrogen closest to the AgN$_5$ plane. For the $S = \frac{5}{2}$ Mn(II) compound, **3.7a**, two thiocyanate nitrogen atoms are coordinated to complete a pentagonal-bipyramidal arrangement, which is distorted toward a capped trigonal prism. The nitrogen configuration is *rac*, and one six-membered chelate ring adopts a chair conformation, the other a twist conformation. The Mn–N distance and Mn–N–C angles for the two thiocyanate ions are unequal (Table 6), and the more weakly bonded thiocyanate ion on the more crowded side of the macrocycle is readily displaced to form six-coordinate derivatives. [100] The Cd(II) and Hg(II) compounds **3.7b,c** have similar structures, with one bromide ion completing a six-coordinate arrangement. For the Ag(I) compound, **3.7b**, the asymmetric unit contains two independent molecules with similar geometries, one of the molecules having the six-membered chelate rings disordered.

5. Class 4: Tetraazamacrocycles with 2-Imino(or 2-amido)-benzaldimine Chelate Rings

5.1. Discussion of Structures

The macrocyclic ligands of this section are characterized by the presence of the 2-iminobenzaldimine chelate ring [(a) of Figure 15], or the 2-amido-benzaldimine chelate ring [(b) of Figure 15]. These chelate rings, with two

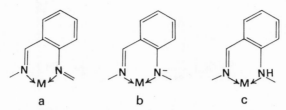

a   b   c

FIGURE 15. (a) 2-iminobenzaldimine chelate ring of **4.1a** and **4.2a,b**, formed by cyclic tri- and tetramerization, respectively, of 2-aminobenzaldehyde. (b) 2-Amidobenzaldimine chelate ring of **4.3a** and **4.4a**. (c) 2-Aminobenzaldimine chelate ring of **8.9a**.

trigonal nitrogen donor atoms, are closely related to the N–O chelate ring of salicylaldimine compounds (including the macrocycles **8.2**, **8.3**, and **8.5A**), for which many structural studies have been reported, and details of conformation, etc., reviewed.[101] These chelate rings are usually close to planar, but with the benzene ring slightly twisted with respect to the MN_2 plane.

Complexes of the macrocycles **4.1** and **4.2**, which have only the 2-iminobenzaldimine chelate ring, (a), are formed by cyclic trimerization,[102] or tetramerization,[103] respectively, of 2-aminobenzaldehyde in the presence of metal ions. The macrocycles are relatively rigid, and the tetraazamacrocycle has been reported only in planar coordination. These macrocycles are formally [12] and [16] annulenes, respectively, but the structures reveal a pattern of alternating single- and double-bond lengths around the macrocycle, and hence little delocalization of the double-bond character, presumably because of the stability of the aromatic system.

The macrocycles **4.3** and **4.4** have two 2-amidobenzaldimine chelate rings (b), the deprotonated form of the 2-aminobenzaldimine chelate ring (c), which is present for the $[20]N_4S_2$ macrocycle **8.9**. See also **4.5**, **4.6**, and **4.7** in the Addendum.

5.2. Listing of Structures of Tetraazamacrocycles with 1-Imino(or 1-amido)-2-aldiminobenzene Chelate Rings (o-Iminobenzaldimine and o-Amidobenzaldimine Derivatives)

4.1   tribenzo[b,f,j][1,5,9]triazacyclododec-2,4,6,8,10,12-hexaene, tribenzo[b,f,j]triazacyclododecine
        TRI
        $Bzo_3[12]hexaeneN_3$, $C_{21}H_{15}N_3$
4.1a $[Co(L)_2]I_3$. $R = 0.043$ for 3293 refl.***[104]

Two ligand molecules are coordinated facially to a cobalt(III) ion. The three nitrogen atoms of each ligand define parallel planes 236 pm apart, with the nitrogen atom triangles of the two ligands rotated by 8° from the octahedral configuration toward the trigonal prismatic configuration. The two

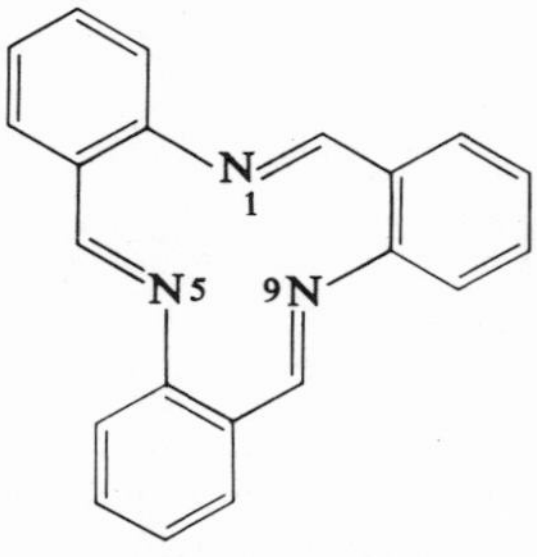

4.1

propeller-shaped macrocycles have the same chirality (*rac* configuration), with close interaction between pairs of benzene rings on the two ligand molecules. The Co(III) ion lies 118(2) pm from the N_3 plane, with a mean Co–N distance of 192 pm, which is normal for Co(III) salicylaldimine compounds.

4.1b [Ni(L)(NO$_3$)H$_2$O]NO$_3$. $R = 0.088$ for 1487 refl.*** (105)

The macrocycle is coordinated facially, with two water oxygens and one nitrogen oxygen forming the other facial set of ligands which complete an octahedral arrangement about $S = 1$ Ni(II). The macrocycle is propeller shaped, and hence chiral. The Ni(II) ion lies 127(2) pm from the N_3 plane, with a mean Ni–N distance of 203(1) pm, which is normal for $S = 1$ Ni(II) salicylaldimine compounds. The Ni–O distances are 211(1), 209(1) pm to water and 208(1) pm to nitrate.

Bzo$_4$[16]octaeneN$_4$ Compounds

4.2   tetrabenzo[b,f,j,n]-1,5,9,13-tetraazacyclohexadec-2,4,6,8,10,12,14, 16-octaene, tetrabenzo[b,f,j,n][1,5,9,13]tetraazacyclohexadecine Bzo$_4$[16]octaeneN$_4$, TAAB, C$_{28}$H$_{20}$N$_4$

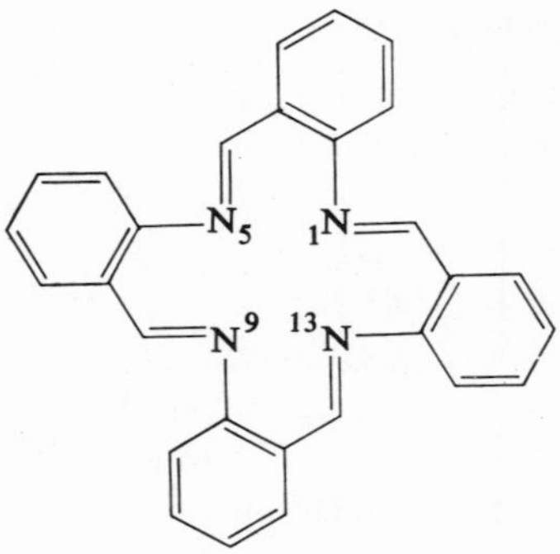

4.2

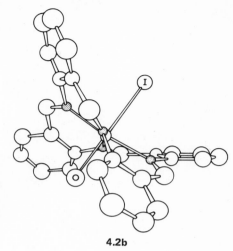

FIGURE 16. Structure of nickel(II) compound of cyclic tetramer of 2-amino-benzaldehyde, with *trans* coordinated water and iodide, **4.2b**. Reproduced from reference 104 by permission of the copyright holder, The American Chemical Society.

4.2b

4.2a [Ni(L)](BF$_4$)$_2$. R = 0.141 for 1709 refl.***[106]

4.2b *trans*-[Ni(L)I(H$_2$O)]I. R = 0.109 for 1368 refl.***[106]

(See Figure 16.) Planar coordination to $S = 0$ Ni(II) for **4.2a** and to $S = 1$ Ni(II) with *trans*-coordinated water and iodide for **4.2b**. The Ni–N distances of 190(2) pm for **4.2a** and 203(3) pm for **4.2b** are similar to those of other compounds with noncyclic salicylaldimines coordinated to $S = 0$ Ni(II) or to $S = 1$ Ni(II), respectively. These values indicate that the macrocycle has adjusted to optimize the metal-ion–nitrogen distance, since these ions are probably smaller than optimum for the [16] macrocycle. The appreciable tetrahedral distortions of the NiN$_4$ group ($\pm$24 pm for **4.2a** and $\pm$20 pm for **4.2b**), which are probably a consequence of this contraction of the N$_4$ donor set, lead to a saddle shape for the macrocycle (approximate S_4 symmetry), compared to the propeller shape of the triaza macrocycle of **4.1a,b**. For **4.2b** the distances to the axial ligands are Ni–O, 220(3) pm; Ni–I, 290(1) pm. For **4.2a** there is one close Ni–F contact [270(2?) pm] with a BF$_4{}^-$ ion.

For both **4.2a** and **4.2b** refinement was hindered by a rotational disorder.

Bzo$_2$[14]tetraenato(2–)N$_4$ Compounds

4.3 dibenzo[e,k]-1,4,8,11-tetraazacyclotetradec-4,6,12,14-tetraenato(2–), bzo$_2$[14]tetraenato(2–)N$_4$, [C$_{18}$H$_{18}$N$_4$]$^{2-}$
Complex formed in low yield during preparation of **4.4a**.[107]

4.3a [Ni(L)]. R = 0.058 for 1272 refl.***[107]

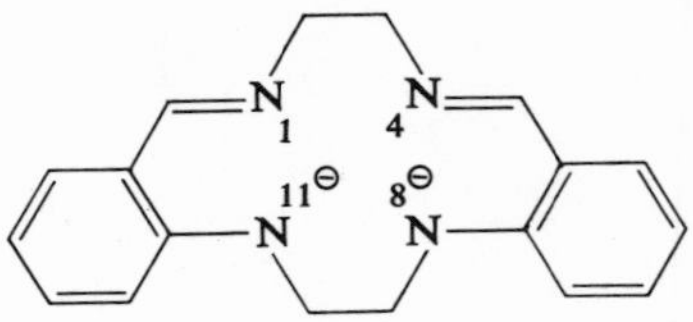

4.3

4.4 dibenzo[e,k]pyrazino[h]-1,4,8,11-tetraazacyclotetradec-4,6,12,14-tetraenato(2−),[$C_{20}H_{18}N_6$]$^{2-}$

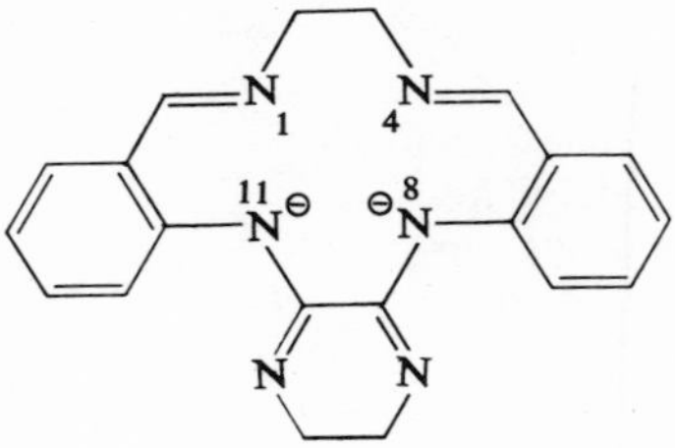

4.4

Ligand formed by reaction of diaminoethane and the appropriate dialdehyde.[107]

4.4a [Ni(L)]. $R = 0.048$ for 995 refl.***[107]

Planar coordination to $S = 0$ Ni(II) for **4.3a** and **4.4a**.

The chelate ring (b) of Figure 15 is present for the [14] macrocycles **4.3** and **4.4**, with five-membered linking chelate rings. The MN_4 group shows a slight (± 5 pm) tetrahedral distortion. The Ni–N distances to the imine nitrogen atom N(1) and N(4) and the trigonal anionic nitrogen atom N(8) and N(11) are similar at 186 pm. (The M—N= and M–O$^-$– distances are generally similar for salicylaldimine complexes.) The chelate ring N(1)–N(4) is gauche for **4.3a** [(85°; [−5], −26, 32, [5], symbolism as for Table 1), and asymmetric gauche for **4.4a** (86°; [−4], −1, 50,[5]). The chelate ring N(8)–N(11) is flattened gauche for **4.1a** (88°; [−5], −3, 17, [4]), and eclipsed for **4.1b** (87°; [−4], 48, 66, [5]). The pyrazine ring is tilted by 31° from the MN_4 plane. Appreciable shortening of the N(8)–C(9) and N(11)–C(10) bond lengths indicates significant extension of the pyrazine delocalization. See also **4.5a** in Addendum.

6. Class 5: Dibenzo[b,i]-1,4,8,11-tetraazacyclotetradec-2,4,6,9,11-hexaenato(2−) Compounds

6.1. Discussion of Structures

The compounds described in this section are characterized by the presence of the dianionic macrocycle dibenzo[b,i]-1,4,8,11-tetraazacyclotetradec-2,4,6,9,11,13-hexaenato(2−) (Figure 17). The macrocycle is the

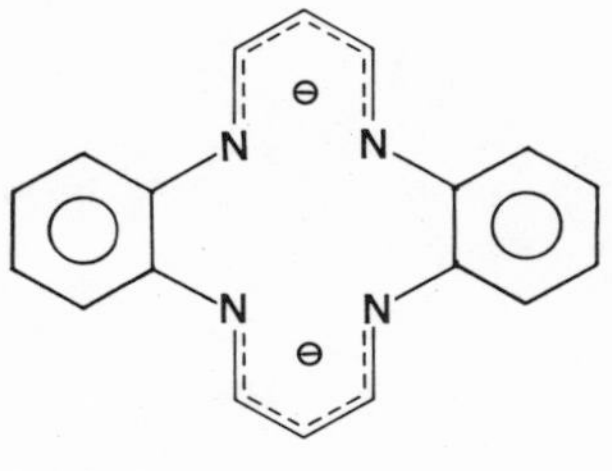

FIGURE 17. Dibenzo[b,i]-1,4,8,11-tetraazacyclotet-
radec-2,4,6,9,11-hexaenato(2−) macrocycle characteristic of compounds of Section 6.

dianion obtained by 1,8 deprotonation of the 1,8-dihydro[14]annulene. The anion, with 16 π-electrons, has nominal antiaromatic character, unlike the porphyrins and phthalocyanins which satisfy the $(4n + 2)$ π-electron rule and have nominal aromatic character, with extensive double-bond delocalization.

For all the structures of coordination compounds of this tetraazacyclo-tetradecinato macrocycle, the delocalization is complete for the six-membered "diiminato" chelate ring, which is essentially planar and with equal C–N and C–C bond lengths. There is little indication of the extension of the delocalization into the "1,2-diaminobenzene" five-membered chelate rings, as the N(1)–C(2)-type bond lengths are typically about 142 pm. This can be compared with the 146 pm observed for a 1,2-diaminobenzene chelate[108] and 132 pm observed for an o-benzoquinone diimine chelate,[109] bearing in mind that the change in bonding arrangement at the nitrogen would be expected to cause shortening of the C–N bond length compared with the diaminobenzene compound [cf. the distances N(1)–C(2) of 142(1) pm and C(3)–N(4) of 148(1) pm for **2.2a**].

The structure of the diprotonated, tetramethyl-substituted ligand **5.3a**, which is isomorphous with its square-planar $S = 1$ iron(II) derivative **5.3b**, shows similar bond lengths around the macrocycle to those of the metal-ion derivatives. This indicates some delocalization for the diiminato moiety for the protonated ligand, which is considered to arise from a planar bond configuration at the protonated nitrogen, with the lone pair occupying a p orbital, and interacting with the other $p\pi$ electrons in the system. (This is in addition to tautomerism involving interchange of the N–H and N---H bonds.)

The unsubstituted macrocyclic ligand for the Ni(II) complex **5.3a** is close to planar, but for the methyl-substituted derivatives, interactions between the methyl groups and the adjacent aromatic rings cause distortions from planarity, arising from twisting about the C–N bonds of the diamino-benzene chelate rings; the macrocycles adopting saddle-shaped conformations, with the aromatic rings and the diiminato chelate rings tilted on opposite sides of the coordination plane (Figure 18). The twisting of the macrocycle is not responsible for the double-bond localization, as the pattern of bond

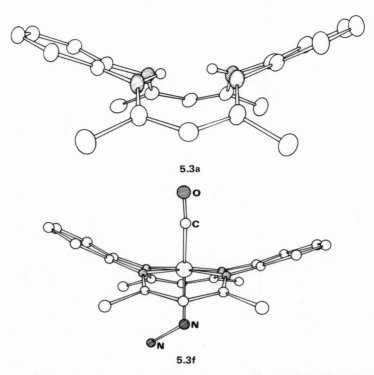

5.3a

5.3f

FIGURE 18. Structures of dibenzo[b,i]-1,4,8,11-tetraazacyclotetradec-2,4,6,9,11-hex-aene, **5.3a**, and the Fe(II) compound of the deprotonated macrocycle with *trans* carbon monoxide and hydrazine **5.3f**. Reproduced from references 110 and 112 by permission of the copyright holder, The American Chemical Society.

lengths, and in particular the length of the C–N bonds of the "1,2-diamino-benzene" chelate ring, are similar for the planar unsubstituted, and the twisted methyl-substituted compounds. The absence of appreciable bond delocalization around the macrocycle is attributed to the inherent stability of the aromatic benzene rings, and the delocalized "diiminato" chelate rings.

One structural difference observed between the diprotonated ligand **5.3a** and the isomorphous iron(II) compound **5.3c** concerns the N–N distance (chelate bite) of the 1,2-diaminobenzene moiety, which decreases from 271 pm for the protonated macrocycle to $\sim$257 pm for the Fe(II) compound, corresponding to a decrease in the chelate angle from 90° to 84°. This change is attributed to the need to redirect the lone pairs of the nitrogen atoms towards the metal ion [cf. the 276-pm N–N distance and 82° NNiN angle for a simple 1,2-diaminobenzene chelate of $S = 1$ Ni(II)].[108]

A consequence of the saddle shape of the tetramethyl-substituted macro-cycle **5.3** is that the lone-pair electrons on the nitrogen atoms are directed out of the N_4 plane, on the same side as the benzene rings are tilted. The coordination compounds of this macrocycle all have the metal ion displaced from the N_4 plane.

For the diprotonated macrocycle **5.3a** the distance from the "center of the N_4 set" to the nitrogen atoms is 190 pm. It has been suggested that the optimum center–nitrogen distance will decrease to 185–187 pm for the dianionic macrocycle because of delocalization effects, and that the macrocycle will be unable to adjust to accommodate appreciably larger ions.[110] The Ni–N distances for the square-planar $S = 0$ Ni(II) compounds of the unsubstituted and dimethyl macrocycles are 187 and 185 pm, respectively, and the NiN_4 set is coplanar for **5.1a**, or nearly so for **5.2a**. The Fe–N distance for the square-planar $S = 1$ Fe(II) compound of the tetramethyl macrocycle **5.3b** is also near the optimum at 190 pm, but the Fe(II) ion is 11 pm from the N_4 plane. Square-pyramidal compounds formed by $S = 0$ Fe(II), $S = \frac{1}{2}$ Fe(III), and $S = 1$ Co(III) all have M–N distances close to 190 pm, and have the metal ion about 25 pm from the N_4 plane. Square-pyramidal compounds formed by the larger $S = \frac{5}{2}$ ions Fe(III) and Mn(II) have M–N distance unusually short at 200 and 212 pm, respectively (usual M–N distances for these ions are about 215 and 225 pm, respectively), and have displacements of the metal ion from the N_4 plane increasing with ion size at 60 and 73 pm, respectively.

Two structures have been reported for octahedral $S = 0$ Fe(II) derivatives, with pyridine, **5.3e**, and hydrazine, **5.3f**, coordinated *trans* to carbon monoxide. The carbon monoxide is coordinated on the same side of the molecular plane as the benzene rings are tilted for the five-coordinate carbon monoxide derivative **5.3d** and the hydrazine derivative, but on the opposite side for the pyridine compound.

Correlations have been drawn between the displacement of the metal ion from the N_4 plane, the tilting of the benzene rings, the dihedral angle at the C–N bond of the "diaminobenzene" chelate rings, and the extent of bond delocalization within the "diiminato" chelate rings.[110–112] The occurrence of the $S = 1$ square-planar Fe(II) compound **5.3c** provides an example of the stabilization of this geometry and spin state for Fe(II) by azamacrocycles, other examples including a porphyrin,[113] a phthalocyanin,[114] and an octaaza[14]annulene, **6.9a**.[115] Compound **5.3i** with the unusual $S = 1$ Co(III) ion in square-pyramidal coordination is another example of the ability of these macrocycles to stabilize unusual spin states.

6.2. Listing of Structures of $Bzo_2[14]hexaenato(2-)N_4$ Compounds

See introduction for rules determining order of compounds. Key dimensions are listed in Table 7.

TABLE 7. Dimensions for Bzo$_2$[14]hexaenatoN$_4$ Compounds

Compound	Ion[a]	Spin[b]	Ligands[c]	MN$_4$[d]	M–N[e]	M–X[f]	Tilt acac.[g]	Tilt bz.[h]
5.1a	Ni(II)	0		0	187			
5.2a	Ni(II)	0		2	185		21°, 17°	25°, 12°
5.3a	H$_2$				190		37°	25°
5.3c	Fe(II)	1		11	192		22°	21°, 26°
5.3d	Fe(II)	0	CO	29	193	169	24°	23°
5.3e	Fe(II)	0	CO	5[i]	194(2)	170	20°	21°
			C$_5$H$_5$N			209		
5.3f	Fe(II)	0	CO	11[j]	194	175	22°	20°, 26°
			NH$_2$NH$_2$			212		
5.3b	Mn(II)	$\frac{5}{2}$	N(Et)$_3$	73	212	225	35°	21°
5.3g	Fe(III)	$\frac{5}{2}$	Cl	60	200	225	33°	22°
5.3i	Co(III)	1	I	23	190	256	21°	22°
5.3h	Fe(III)	$\frac{1}{2}$	C$_6$H$_5$	23	191	193		

[a] Metal ion and oxidation state.
[b] Spin quantum number S.
[c] Additional ligands present.
[d] Displacement of M from MN$_4$ plane (picometers).
[e] Mean M–N distance (picometers).
[f] M–donor-atom distance for additional ligand (picometers).
[g] Dihedral angle between N$_4$ plane and plane of "diiminato" chelate rings.
[h] Dihedral angle between N$_4$ plane and plane of benzene rings.
[i] Fe displaced toward pyridine.
[j] Fe displaced toward CO.

5.1   dibenzo[b,i]-1,4,8,11-tetraazacyclotetradec-2,4,6,9,11,13-hexaenato(2−)
[C$_{18}$H$_{14}$N$_4$]$^{2-}$

5.1

5.1a   [Ni(L)]. $R = 0.041$ for 3112 refl.**[116]
Planar coordination to $S = 0$ Ni(II). The unit cell contains two independent molecules with virtually identical dimensions, each with the nickel(II) ion at a center of symmetry. The whole molecule is close to planar.

5.2   5,14-dimethyl-dibenzo[b,i]-1,4,8,11-tetraazacyclotetradec-2,4,6,9,11,13-hexaenato(2−)
[C$_{20}$H$_{18}$N$_4$]$^{2-}$

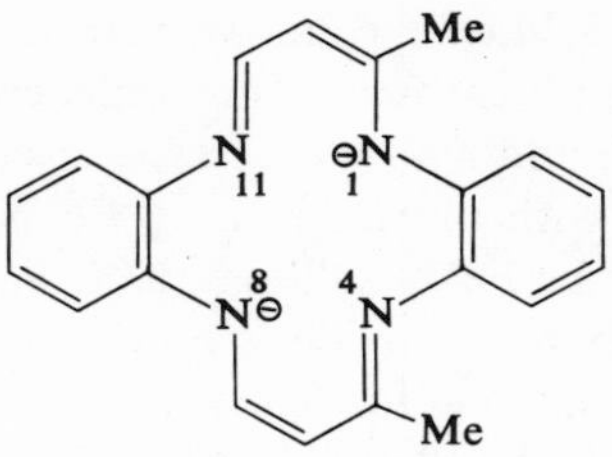

5.2

Ligand formed by condensation of *NN'-o*-phenylene-*bis*-1-amino-1-buten-3-one) with *o*-phenylenediamine.

5.2a [Ni(L)]. $R = 0.094$ for 1140 refl.***(117)

Planar coordination to $S = 0$ Ni(II). The molecule has a saddle-shaped distortion from planarity, i.e., with the five-membered chelate rings with benzo substituents tilted to one side of the molecular plane, and the six-membered diiminato chelate ring tilted to the other side.

5.3 5,7,12,14-tetramethyl-dibenzo[b,i]-1,4,8,11-tetraazacyclotetradec-2,4,6,9,11,13-hexaenato(2−)

$[C_{22}H_{22}N_4]^{2-(118)}$

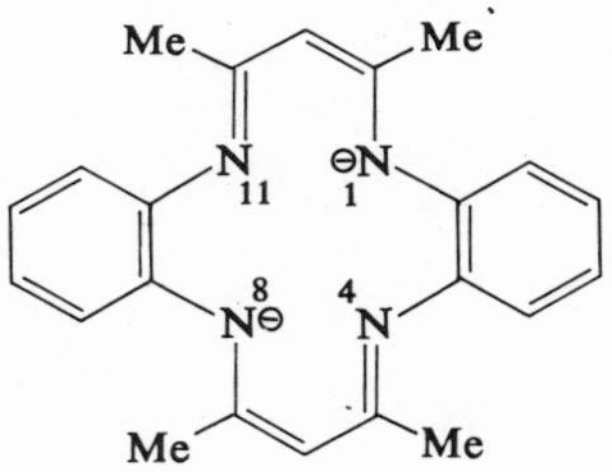

5.3

5.3a [H$_2$(L)]. $R = 0.071$ for 4503 refl.***(110)

(See Figure 18.) The diprotonated ligand is isomorphous with the Fe(II) complex **5.3c**, and has a generally similar structure to that of its compounds.

5.3b [Mn(L){N(C$_2$H$_5$)$_3$}]. $R = 0.056$ for 3883 refl.***(111)

Planar coordination to $S = \frac{5}{2}$ Mn(II), with triethylamine completing a square-pyramidal arrangement. Decomposition of the crystals hindered data collection.

5.3c [Fe(L)]. $R = 0.057$ for 3553 refl.***(110)

Square-planar coordination to $S = 1$ Fe(II).

5.3d [Fe(L)(CO)], $\frac{1}{2}$(C$_7$H$_8$). $R = 0.049$ for 3488 refl.***(112)

The toluene solvate molecule is disordered.

5.3e [Fe(L)(CO)(C$_5$H$_5$N)]. $R = 0.042$ for 3696 refl.***(112)

5.3f [Fe(L)(CO)(NH$_2$NH$_2$)]. $R = 0.0719$ for 2802 refl.***(112)

(See Figure 18.) For **5.3d,e,f** the macrocycle is in planar coordination to $S = 0$ Fe(II). In each case carbon monoxide is coordinated axially, while for **5.3d** and **5.3e**, pyridine and hydrazine, respectively, are coordinated *trans* to the carbon monoxide with unusually long Fe–N distances. **5.3d**: Fe–C, 169 pm; **5.3e**: Fe–C, 173 pm; Fe–N, 209 pm. **5.3f**: Fe–C, 175; Fe–N, 212 pm. For **5.3d** and **5.3f** the carbon monoxide is on the same side of the ligand as the benzene rings are tilted, while for **5.3e** it is on the opposite side.

5.3g   [Fe(L)Cl]·CH$_3$CN. $R = 0.064$ for 5907 refl.***[111]

Planar coordination to $S = \frac{5}{2}$ Fe(III) with chloride completing a square-pyramidal arrangement. The methyl cyanide is not coordinated.

5.3h   [Fe(L)(C$_6$H$_5$)]. $R = 0.066$ for 4206 refl.***[119]

Planar coordination to $S = \frac{1}{2}$ Fe(III), with a σ-bonded phenyl group completing a square-pyramidal arrangement.

5.3i   [Co(L)I]·CHCl$_3$. $R = 0.052$ for 2929 refl.***[111,120]

Planar coordination to $S = 1$ Co(III), with iodide completing a square-pyramidal arrangement. The Co–I bond lies along a twofold axis.

5.4   13(β-vinyl)-5,7,12,14-tetramethyl-dibenzo[b,i][1,4,8,11]-tetraazacyclotetradec-2,4,6,9,11,14-sexaenato(2−) [C$_{24}$H$_{23}$N$_4$]$^{2-}$

5.4

Complex formed by reaction of cobalt(II) complex of **4.3** with acetylene.[111]

5.4a   [Co(L)(C$_6$H$_5$N)][121]

The four nitrogen atoms of the macrocycle, the vinyl carbonium carbon, and the pyridine nitrogen complete octahedral coordination about cobalt(III).

7. Class 6: Cyclic Hydrazines and Hydrazones

7.1. Discussion of Structures

The compounds in this section are characterized by the presence of adjacent nitrogen atoms in the macrocycle, and are formed by condensation reactions of hydrazones, or hydrazines, with carbonyl compounds. All except **6.11** and **6.12** have a 1,2,4,5,8,9,11,12-octaazacyclotetradecane macrocycle, with varying unsaturation, substitution, and/or deprotonation. All have four nitrogen atoms of the macrocycle coordinated in a planar arrangement. Isomers can arise from the set of nitrogen atoms coordinated, as exemplified by **6.8**, for which the nickel(II) complex has been isolated in three isomeric forms, with nitrogen atoms 1,5,8,12-coordinated, D_{2h} symmetry (**6.8b**); atoms 2,5,9,12-coordinated, C_{2h} symmetry (**6.8c**); and a further form with atoms 1,5,9,12-coordinated for which no structure has been reported.[122]

Many of the structures reported are of compounds derived from the tetraimine macrocycle formed by reaction of 2,3-butanedione (diacetyl), hydrazine, and formaldehyde, in the presence of metal ions (**6.3–6.9**). These macrocyclic complexes are much more reactive than their tetraaza analogs, and in particular suffer facile oxidative dehydrogenation, often leading to migration of the double bonds.

Many of the compounds can be formally represented with conjugated double bonds, and the extent of delocalization of these bonds is of some interest. Several structures have been reported for compounds in which the oxidative dehydrogenation has proceeded to the octaaza [14] annulene **6.8**, **6.9**. In one such case, **6.8b**, equivalence of bond lengths around the ring indicates that delocalization is essentially complete, while in other cases a pattern of alternating bond lengths is observed, although the actual bond lengths indicate that some delocalization is generally present for the conjugated systems (see individual compounds for discussion). Bond lengths around the macrocycles are listed in Table 8.

The compounds have chelate rings with noncoordinated nitrogen atoms of three skeletal types: A, B, and C of Figure 19.

The tetraaza six-membered chelate ring A is present with bonding arrangements A1–A5 for reported structures. For the more highly unsaturated macrocycles several valence isomers can be drawn, and the formulas represent the most important contributors to the structure as shown by the observed bond lengths.

The saturated chelate ring A occurs with deprotonated "amide" donor atoms, A1, for **6.1a** and **6.2a**, and with imine donor atoms, A2, for **6.3a,b** and **6.11a**. For **6.6a** and **6.4a** one six-membered chelate ring is of type A2, and in the latter case the other is the deprotonated variant A3. For **6.2a** and **6.3a,b** and for one of the chelate rings of **6.1a** the ring is fused to a chair conformation oxadiazine ring. The chelate rings A1–A3 have trigonal nitrogen donor atoms, and usually adopt a conformation with the four nitrogen atoms close to planar, and with the central carbon atoms displaced from this plane. The metal ion lies close to the plane, or is displaced on the same side as the carbon to produce a boat conformation as for **6.1a**, **6.3b**, and **6.4a**. For the [15]diene-$N_4(N_2)$ compound **6.11a** the chelate ring A2 adopts a twist conformation, and the overall conformation of the macrocycle resembles that of the related [14]dieneN_4 compound **2.14a**.

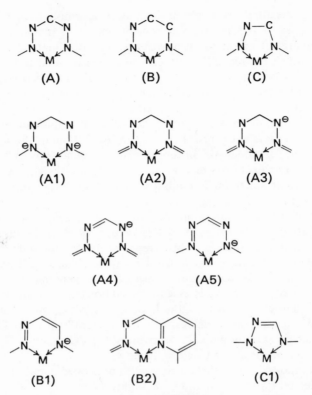

FIGURE 19. Chelate rings present for macrocycles of Section 7.

The "enyl" chelate ring A4 is present for **6.6a**, **6.7a**, and **6.8** (configuration I, as for **6.8a,c**. For **6.7a** and **6.8a**. equivalent C–N distances indicate delocalization over the N⋯C⋯N segment, and the absence of appreciable shortening of the N–N bonds indicates that little delocalization extends to the conjugated imine groups. The NCN bond angle has the unusually large value of $\sim 140°$ for these compounds. In the case of **6.8b** extensive delocalization of the double-bond character is present (see below) and this NCN bond angle is 135°.

The "dienyl" chelate ring A4, present for compounds **6.5a** and **6.7a** [chelate ring N(1)–N(5)] is planar, with equivalent N–N and N–C bond distances indicating delocalization over the chelate ring, as for the related "diiminato" chelate ring (v) of Figure 7, which is present for macrocycles **2.2** and **5.1–5.4**. The central NCN bond angle for these compounds is about 130°.

The six-membered triaza chelate ring B1, together with the triaza five-membered chelate ring C1 is present for compounds **6.8c** and **6.9a**. The observed pattern of alternating bond lengths indicates little delocalization of the double-bond character. The triaza chelate ring B2 is present for **6.10a**, and the bond lengths again indicate little delocalization of the double-bond character.

Several of the compounds have α-diimine five-membered chelate rings. For **6.3a,b**, as for **2.14a**, the C–N and C–C distances of about 129 and 150 pm indicate little delocalization of the double-bond character. For **6.4a**, **6.8a**, and particularly **6.6a**, shortening of the C–C and lengthening of the C–N distances indicates delocalization of the double-bond character, which is essentially complete for **6.8b** and **12.15a**.

7.1.1. Metal-Ion–Nitrogen Distances

The metal-ion–nitrogen distances for the compounds in Section 6 are listed in Table 8. These distances are generally shorter than those observed for metal-ion–imine-nitrogen distances (Table 3). The very short Fe(II)–N distance for **6.9a** and Co(I)–N distance for **6.6a** have been attributed to a macrocyclic constrictive effect arising from the small-optimum center–nitrogen distance for these highly unsaturated macrocycles, and to special π bonding characteristics of the α-diimine system. Short M–N distances are observed with other metal ions and for other macrocycles of this type without the α diimine group, so it appears that the effect is a more general one, and that the effective radius of the ligating "hydrazone" nitrogen atom in these compounds is less than that of imine nitrogen. Short M–N distances are also observed for compounds with delocalized diiminato chelate rings (v, Figure 7), e.g., **2.2a**, **5.1–5.3**. For the pyridyl macrocycles of Section 3, the M–N distance is generally appreciably shorter for pyridine nitrogen than for secondary amine, or imine donor atoms (Table 6). However, for the $Me_4pyo_2[14]hexaeneN_4(N_2)$ compound, **6.10a**, the Fe–N distances for pyridine and hydrazone nitrogen atoms are similar. The $bzo_2[14]$-tetraeneN$_2$S$_2$(N$_2$) compounds, **8.4a**, **8.5a**, with a chelate ring of type A4, also have short [184(1) pm] Ni–N distances.

For the [15]dieneN$_4$(N$_2$) Cu(II) compound, **6.11a**, the mean Cu–N distance of 200 pm is ~10 pm longer than the value for the [14]N$_4$(N$_4$) compounds, **6.1a**, **6.2a**, **6.3a**, and comparable with the 203(4)-pm value observed for the [14]aneN$_4$ compounds, **1.4h**.

The appreciable difference in Ni–N distances for **6.7a** for the donor atoms associated with chelate rings of types A4 and A5 (Figure 19) can be noted.

7.1.2. Bond Angles

The chelate angles for these compounds are unremarkable, averaging about 85° for the five-membered, and about 95° for the six-membered chelate rings. The chelate angle at the twist-boat conformation tetraaza chelate ring of **6.11a** is, as often observed for this conformation, smaller at 89°.

7.2. *Listing of Structures of Cyclic Hydrazine and Hydrazone Compounds*

See introduction for rules determining order of compounds. Key dimensions are listed in Table 8. See also **6.13**, **6.14**, and **6.15** in the Addendum.

[14]aneato(4−)N₄(N₄) Compounds

6.1 7,14,16,17-tetramethyl-6,8-diphenyl[15]oxa-[1,2,5,6,8,9,12,13]-octaazatricyclo[11.3.1]sept-3,4,10,11-tetraonato(4−) $[C_{24}H_{26}N_8O_5]^{4-}$

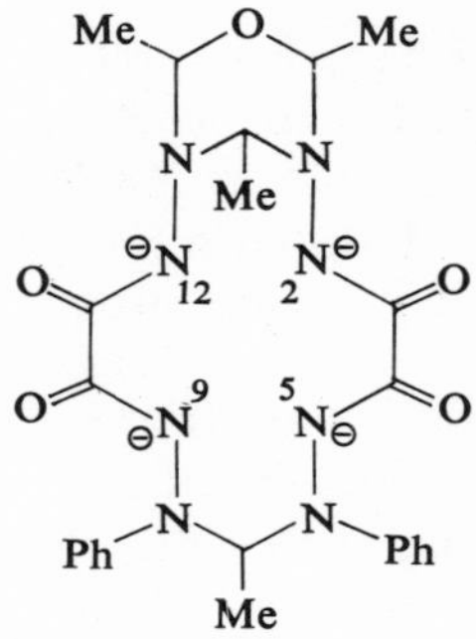

6.1

Intensely colored copper complex formed by oxygenation of ammoniacal solutions of copper(II), oxaldihydrazone, and acetaldehyde.

6.1a $(NH_4)_2[Cu(L)(H_2O)]$, $4H_2O$ or $(NH_4)_2[Cu(L)(OH)]$, $4H_2O$. $R = 0.077$ for 1184 refl.***[123]

6.2 7,9,16,18,19,20-septamethyl[8,17]dioxa-[1,2,5,6,10,11,14,15]-octaazatricyclo[13.3.1⁶,¹⁰]eicosa-3,4,12,13-tetraonato(4−) $[C_{16}H_{34}N_{10}O_6]^{4-}$

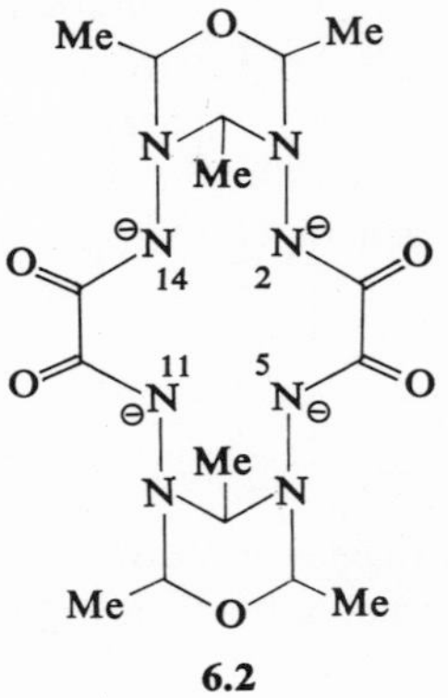

6.2

TABLE 8. Dimensions for Compounds of Cyclic Hydrazines and Hydrazones

Compound	Metal ion[a]	Distances[b]				
		a	b	c	d	e
6.1a	Cu(II)/(III)[c]	189(1)	140(2)	147(2)	134(2)	155(2)
6.2a	Cu(II)/(III)[c]	187(1)	141(2)	147(2)	133(2)	151(2)
6.3a	Co(III)	190	141	149	128	150
6.3b	Cu(II)[c]	194	142	148	128	150
6.4a	Fe(II) $S = 0$	190	139	145	131	145
6.5a	Ni(II) $S = 0$	181	129	134	148	154
6.6a	Co(I) $S = 0$	187	e	e	133	143
6.7a	Ni(II) $S = 0$	{179[f] / 187[g]}	129 / 138	134 / 132	150 / 129	150
6.8a	Co(III)	187	138	133	131	146
6.8b	Ni(II) $S = 0^f$	183	134	134	136	140
2.15a	Ni(II) $S = 0^f$	187	135	140	137(1)	138(2)

		a	a′	f	g	h	i	j	k	l
6.8c	Ni(II) $S = 0$	178	182	135	131	146	132	141	131	136
6.9a	Fe(II) $S = 1$	185	183	138	131	146	131	141	131	139
6.10a	Fe(II) $S = 0$	189[j]	190	137	130	148	135[k]	134[i]	147	133

		a	a′	m	n	o	p	q	r
6.11a	Cu(II)	201	198	152	150	150	125	144	147

[a] Coordination geometry *trans*-octahedral [Co(1) and Co(III)], or square planar (other metal ions), unless otherwise specified.
[b] Bonds shown on formulas below (picometers). Mean distances are quoted where there are small differences for chemically equivalent bonds.

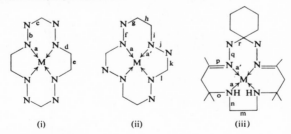

(i)   (ii)   (iii)

[c] Square-pyramidal coordination.
[d] *Trans*-octahedral coordination.
[e] Distances uncertain because of disordered structure.
[f] Distances pertaining to the nonconjugated chelate ring N(1)–N(5), adjacent to the hydrazine substituent.
[g] Distances pertaining to chelate ring N(8)–N(12).
[h] Bonds included in fused pyridyl ring.
[i] Pyridine nitrogens N(19), N(20).
[j] Hydrazone nitrogens N(3), N(12).
[k] Bonds included in fused pyridyl ring.

Compound formed similarly to **6.1a**, with monophenyl derivative of oxaldihydrazone.

6.2a $(NH_4)_2[Cu(L)(H_2O)]$, $\frac{3}{4}NH_4ClO_4$ or $(NH_4)_2[Cu(L)(OH)]$,
$\frac{3}{4}NH_4ClO_4$. $R = 0.088$ for 1836 refl.***[123]

(See Figure 20.) Both compounds have the macrocycle in planar co-ordination to a copper atom (by nitrogen atoms 2,5,9,12 for **6.1a** and 2,5,11,14 for **6.2a**). The original report proposed copper(II), with oxygen of a water molecule completing a square-pyramidal arrangement (Cu–O, 274 pm), but a later report, referring specifically to compound **6.1a**, suggests that the compounds have copper(III), with axially coordinated hydroxide.[124] The copper atom is displaced from the N_4 plane toward the water/hydroxide (17 pm for **6.1a** and 15 pm for **6.2a**). The anion of **6.1a** has mirror symmetry, through Cu, C(7), O(15), and C(17), and hence the N_4 group is planar, but for **6.2a** this group shows a small tetrahedral distortion (± 4 pm). The C–O distances of the oxalyl moieties (~ 124 pm), indicate that these groups have a keto rather than an enol structure, while the C–C and N–N distances indicate single bonds. The C–N distances [133(2) pm] are short for single bonds but similar to those found for amides.

The oxalyl moieties (five-membered chelate rings) are approximately coplanar, but these planes are tilted with respect to the coordination plane both on the same side as the water ligand. The tetraaza six-membered chelate rings have boat conformations, both tilted to the side opposite to the five-membered chelate rings, giving the macrocycle a saddle-shaped conformation. The methyl substituent on the six-membered chelate ring is axially oriented,

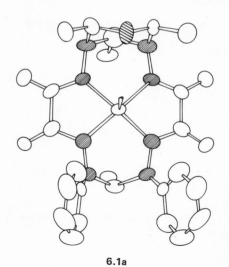

6.1a

FIGURE 20. Structure of cation of copper-(II)/(III) compound of the [14]ane(4 –)-$N_4(N_4)$ macrocycle, **6.1a**, with coordinated (water molecule)/(hydroxide ion) omitted. Reproduced from reference 123 by permission of The Chemical Society.

while the phenyl substituents for **6.1a** are equatorially oriented. The oxadiazine ring fused to the six-membered chelate ring has a chair conformation, with the methyl substituents equatorially oriented. The oxygen atoms of these rings thus both lie on the same side of the copper atom as the coordinated water molecule.

For **6.2a** the ammonium and perchlorate ions are distributed over a number of equivalent sites, and for both compounds there is an extensive hydrogen-bonding network.

[14]tetraeneN$_4$(N$_4$O$_2$) Compounds

6.3 3,4,12,13-tetramethyltricyclo[18.1.1^{6,10}][8,17]dioxa-1,2,5,6,10,11, 14,15-octaazacyclotetradec-2,4,11,13-tetraene (C$_4$H$_{24}$N$_8$O$_2$)

6.3

Complexes formed by acid-catalyzed reaction of complexes of **6.7** with formaldehyde.[125]

6.3a *trans*-[Co(L)(NCCH$_3$)$_2$](ClO$_4$)$_3$. $R = 0.148$ for 4231 refl.***[125]

6.3b [Cu(L)(H$_2$O)](ClO$_4$)$_2$. $R = 0.072$ for 2129 refl.***[125]

Planar coordination (by nitrogen atoms 2,5,11,14) to cobalt(III) or copper(II), respectively, with *trans* methyl cyanide molecules completing octahedral coordination about cobalt(III) [Co–N, 189(1) pm], and with a water molecule completing square-pyramidal coordination about copper(II) [Cu–O, 226 pm], with the copper displaced by 25 pm from the N$_4$ plane toward the oxygen. The copper compound has a twofold axis through the water oxygen–copper bond, and the cobalt compound has approximate C_{2v} symmetry with the axis through the methyl cyanide–cobalt bonds (i.e., both compounds have the isomer of the ligand with both oxygen atoms on the same side of the "molecular plane"); in the case of **6.3b** the oxygen atoms are on the opposite side of the plane to the water. The conformation of the macrocycle is similar to that present for **6.1a**.

[14]tetraenato(1 −)N₄(N₄) Compounds

6.4 6,7,13,14-tetramethyl[1,2,4,5,8,9,11,12]octaazacyclotetradec-
      5,7,12,14-tetraenato(1 −)
      $[C_{10}H_{19}N_8]^-$

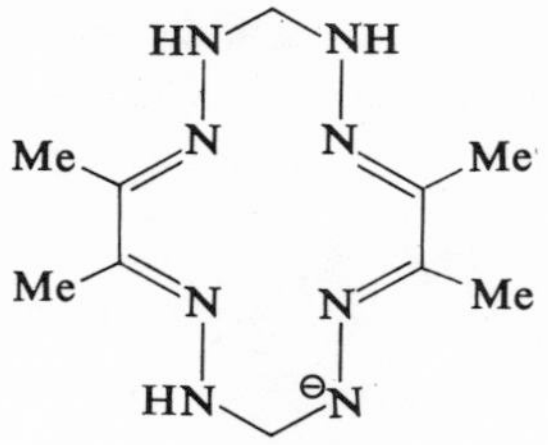

6.4

Complex **6.4a** formed from tetraimine complex by reaction with methyl
hydrazine, potassium *tert*-butoxide, and carbon monoxide.[126]
6.4a [Fe(L)(CH₃)(CO)]. $R = 0.042$ for 872 refl.*[126]
Planar coordination (by nitrogen atoms 1,5,8,12) to $S = 0$ Fe(II), with
trans σ-bonded methyl and carbon monoxide, completing an octahedral
arrangement. The CH₃–Fe–C–O axis lies at the intersection of two perpendicular mirror planes, with the deprotonated nitrogen site disordered over
the four equivalent positions. The Fe–CH₃ bond at 208 pm is longer than
expected, attributed to the *trans* effect of the CO group (Fe–CO, 177 pm).
The iron atom is displaced by 19 pm from the coordination plane toward the
CO group. The macrocycle is saddle shaped, with boat conformation
tetraaza six-membered chelate rings tilted on the same side of the molecular
plane as the CO, while the planar diimine five-membered chelate rings are
tilted in the opposite direction (both carbon atoms 27 pm from the N₄ plane).

[14]tetraenato(2 −)N₄(N₄) Compounds

6.5 6,7,13,14-tetramethyl[1,2,4,5,8,9,11,12]octaazacyclotetradec-
      1,3,8,10-tetraenato(2 −)
      $[C_{10}H_{18}N_6]^{2-}$

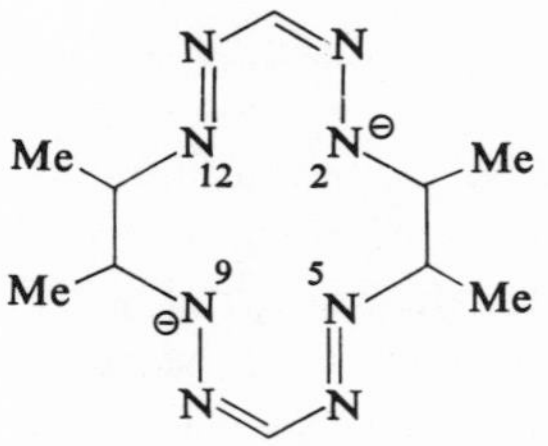

6.5

Complex formed by reduction of [Ni(**6.8**)] with BH_4^-, followed by oxygenation.[127]

6.5a [Ni(L)]. $R = 0.035$ for 1394 refl.***[127]

Planar coordination (by nitrogen atoms 1,5,8,12) to $S = 0$ Ni(II), with the nickel ion at a center of symmetry. The six-membered chelate rings are close to planar (with a small "boat-shaped" distortion), and the symmetrical bond lengths (N–N, 129 pm; C–N, 134 pm) indicate delocalization analogous to that of 2,4-pentanediiminato complexes. The five-membered chelate rings have symmetrical gauche conformations, with the methyl substituents all axially oriented (deviations from mean plane of macrocycle $\pm$ (C, 25 pm; C(methyl), 172 pm).

[14]pentaenato(1−)$N_4(N_4)$ Compounds

6.6 6,7,12,14-tetramethyl[1,2,4,5,8,9,11,12]octaazacyclotetradec-1,5,8,12,13-pentaenato(1−)
$[C_{10}H_{17}N_8]^{-1}$

6.6

Compound **6.6a** formed from cobalt(II) tetraimine complex by reaction with strong base plus carbon monoxide.[128]

6.6a [Co(L)(CO)]. $R = 0.050$ for 1782 refl.*[128]

Planar coordination (by nitrogen atoms 1,5,8,12) to $S = 0$ Co(I), with carbon monoxide completing a square-pyramidal arrangement, and with the cobalt displaced 40 pm from the N_4 plane toward the CO. The Co–N distance, 187 pm, is abnormally short. The cobalt atom and carbon monoxide lie on a crystallographic mirror plane which passes through the midpoints of the five-membered chelate rings. The two dissimilar six-membered chelate rings are randomly oriented in a disordered structure, which limited refinement and the validity of dimensions, particularly those involving the central carbon atoms of these chelate rings. The α-diimine five-membered chelate rings show appreciable delocalization (C–N, 133 pm; C–C, 142 pm).

[14]pentaenato(2−)$N_4(N_4)$ Compounds

6.7 6,7,13,14-tetramethyl-6,13-dihydrazine[1,2,4,5,8,9,11,12]octa-azacyclotetradec-1,3,7,9,12-pentaenato(2−)
$[C_{10}H_{20}N_{12}]^{2-}$

6.7

Complex formed by oxidation of the Ni(II) tetraimine complex in the presence of hydrazine.[129]

6.7a [Ni(L)](ClO$_4$)$_2$. $R = 0.059$ for 2129 refl.***[129]

Planar coordination (by nitrogen atoms 1,5,8,12) to $S = 0$ Ni(II). The chiral centers 6 and 14 have the *meso* configuration, i.e., with the methyl groups on the same side of the molecular plane. An approximate mirror plane passes through the nickel ion and the central carbon atoms of the six-membered chelate rings. The six-membered chelate ring N(1)–N(5) is of the symmetrical "dienyl" (A5) type, and the chelate ring N(8)–N(12) is of the symmetrical "enyl" (A4) type.

[14]hexaenato(2−)N$_4$(N$_4$) Compounds

6.8 6,7,13,14-tetramethyl[1,2,4,5,8,9,11,12]octaazacyclotetradec-2,4,7,9,13,14-hexaenato(2−)
[C$_{10}$H$_{14}$N$_8$]$^{2-}$

6.8 (I) D_{2h} **6.8 (II)** C_{2h}

Complexes formed by oxidative dehydrogenation of the tetraimine complexes formed by condensation of 2,3-butanedione, hydrazine and formaldehyde in the presence of metal ions.[130]

6.8a [Co(L)(CH$_3$)(NH$_2$NHCH$_3$)].*[130]

Planar coordination (by nitrogen atoms 1,5,8 and 12, isomer I) to cobalt(III) with *trans* methyl and methylhydrazine ligands completing an octahedral arrangement (Co–C, 200 pm; Co–N, 208 pm). The macrocycle is planar, with a three-atom "enyl" $(N \cdots C \cdots N)^-$ delocalized system in the six-membered chelate rings and an essentially isolated α-diimine system in the five-membered chelate rings. The compound shows an anomalous ^{1}H NMR spectrum attributed to thermal population of a low-lying paramagnetic state.

6.8b [{Ni(L)}$_2$].　$R = 0.104$ for 2077 refl.***[131]

(See Figure 21.)

6.8c　[Ni(L)].　$R = 0.0221$ for 961 refl.***[122, 131]

Planar coordination to $S = 0$ Ni(II). For **6.8b** the macrocycle is co-ordinated by nitrogen atoms 1,5,8,12 (D_{2h} symmetry) and the molecule is a confacial dimer, with nickel–nickel bond length of 279 pm and with eclipsed macrocycles. Inability to grow sufficiently large crystals for the data collection contributed to the high R value. For **6.8c** the macrocycle is coordinated by nitrogen atoms 2,5,9,12 (C_{2h} symmetry) in a square-planar, monomeric arrangement.

In both cases the macrocycle is essentially planar, although for the dimer the nickel atoms are displaced by 10 pm from the N_4 plane toward each other.

For the compound **6.8b**, with a confacial dimeric structure, the macro-cycles are eclipsed and close to parallel, with mean interplanar separation of 300 pm. Repulsion between methyl groups of the two macrocycles causes small distortion from planarity. It is suggested[131] that 2 of the 16 π electrons of each octaaza annulene are in π^* orbitals (antiaromatic system) which are split for the dimer sufficiently so that population of the lower member of the pair stabilizes the eclipsed dimer, effectively by formation of two π bonds. This makes the bonding pattern of the macrocycle closer to a 14 π aromatic system, accounting for the bond delocalization. For the compound of **6.8a**,

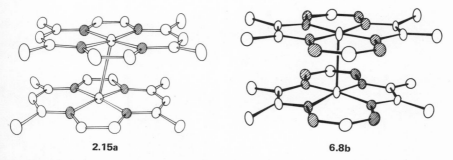

2.15a　　　　　　　　　　　　**6.8b**

FIGURE 21. Structures of two confacial dimeric compounds of nickel(II) with Me$_4$[14]-hexaenato(1−)N$_4$(N$_4$), **2.15a**, and Me$_4$[14]hexaenato(2−)N$_4$(N$_4$), **6.8b**, macrocycles. Reproduced from references 76 and 131 by permission of the copyright holder, The American Chemical Society.

cobalt(III) does not have the ability of nickel(II) to form metal–metal bonded dimers, and the macrocycle retains its antiaromatic character with alternating bond lengths, although these lengths are modified compared with isolated single and double bonds, indicating some delocalization. For the compound **6.8c**, with C_{2h} symmetry (2,5,9,12-bonded), it is suggested that the negative charge on the macrocycle resides largely on nondonor atoms, unlike the D_{2h} arrangement, where it is largely on the donor atoms, and that this difference is sufficient to favor formation of the Ni–Ni bond in one case and not the other.

The nickel–nickel bond is considered to arise from a d_{z^2} σ bond plus a $d_{x^2-y^2}$ δ bond. The compound is diamagnetic, but does show an isotropic ESR signal, attributed to spin density delocalized over the π-bonding system of the ligand.

For both the monomer and dimer the molecules are stacked in the crystal, with intermolecular nickel–nickel distances of 486 pm for **6.8c** and 380 pm for **6.8b**, respectively.

Compound **2.15a** with a Me$_4$[14]hexaenato(1 −)N$_4$ macrocycle has a very similar structure to **6.8b**, with a 306-pm nickel–nickel bond and 319-pm separation between the N$_4$ planes of the eclipsed macrocycles. This macrocycle formally has 15 π electrons, one less than for **6.8**, and the intermacrocycle interaction consists of one weak π bond, instead of two as for **6.8b**.[76]

6.9   3,10-di(n-octyl)di-(9,10-dihydroanthryl)[f.m.][1,2,4,5,8,9,11,12]-
octaazacyclotetradec-2,5,7,9,12,14-hexaenato(2 −)
[C$_{50}$H$_{54}$N$_8$]$^{2-}$ [132]

6.9a   [Fe(L)]. $R = 0.085$ for 2651 refl.*** [133]

6.9

Planar coordination to $S = 1$ Fe(II). (See also **5.3c.**) The molecule is centrosymmetric, with nitrogen atoms 2,5,9, and 12 coordinated to produce a complex analogous to that present for **6.8c**. The bond lengths around the macrocycle indicate a pattern of alternating double and single bonds, very similar to those found for **6.8c**. The macrocycle is essentially planar, except that the six-membered chelate rings are tilted by 3° with respect to the co-ordination plane. This distortion, which is not present for the nickel compound **6.8c**, is attributed to the larger-than-optimum size of the iron atom for this macrocycle, which also shows up in unusually short Fe–N distances, attributed to the resultant macrocycle constrictive effect.

The dihydroanthryl groups form a "hydrophobic pocket" about the iron atom 456 pm deep (from the hydrogen atom positions) which is relevant to the functioning of this compound as a reversible oxygen carrier.

$Me_4pyo_2[14]hexaeneN_4(N_2)$ Compounds

6.10 2,5,11,14-tetramethyl-3,4,12,13,19,20-hexaazatricyclo-[13.3.1.1^{6,10}]-eicosa-1(19),2,4,6,8,10(20),11,13,15,17-decaene
$C_{18}H_{18}N_6$

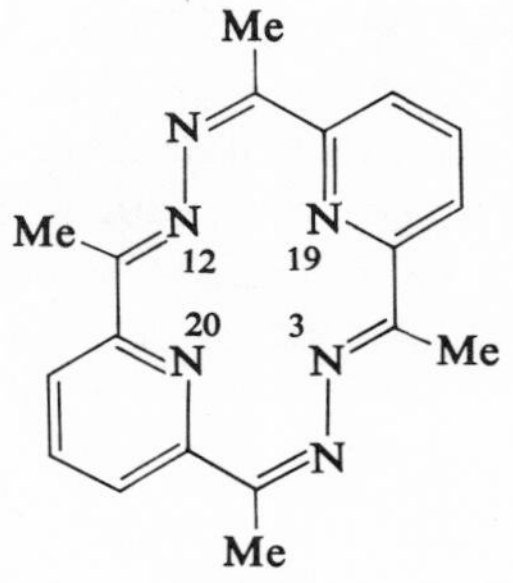

6.10

Complex formed by reaction of 2,6-diacetyl pyridine and the substituted hydrazine in the presence of metal ions.[134]

6.10a [Fe(L)(NCCH$_3$)$_2$](ClO$_4$)$_2$. $R = 0.089$ for 2058 refl.***[134]

Planar coordination by nitrogen atoms 3,20,12, and 19 to $S = 0$ Fe(II), with *trans* methyl cyanide nitrogen atoms completing a centrosymmetric octahedral arrangement. The structure shows disorder, associated with rotation by 180° about the N(1)FeN(8) axis, which limited refinement.

The macrocycle is close to planar, with an alternating pattern of bond lengths (although the C=N bonds are slightly longer than normal for isolated imine groups, which may indicate some delocalization).

[15]dieneN₄(N₂) Compounds

6.11 9,11,11,16,16,18-hexamethyl-7,8,12,15,19-hexaazaspiro[5.15]-
eicosa-8,18-diene, 6,8,8,13,13,15-hexamethyl-1,2,4,5,9,12-
hexaazacyclopentdec-5,14-diene-3-spirocyclohexane
$C_{20}H_{35}N_6$

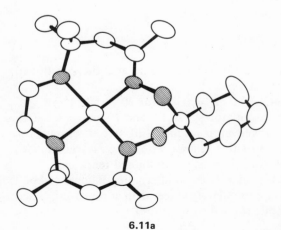

6.11

Complex formed by reaction of 4,4,9,9-tetramethyl-5,8-diazadodec-2,11-di-
hydrazone complex with cyclohexanone.[135]

6.11a   [Cu(L)](ClO₄)₂[135]

(See Figure 22.) Planar coordination (by nitrogen atoms 1,5,9,12) to
copper(II). The chiral nitrogen centers N(9) and N(12) are racemic, with a
highly approximate twofold axis through C(3), Cu, and the midpoint of
C(10)–C(11). The tetraaza six-membered chelate ring adopts a twist conformation, while the cyclohexane ring adopts a chair conformation. The
five-membered chelate ring is gauche and the six-membered "imine" chelate
rings have conformations similar to those found for the compounds of

6.11a

FIGURE 22. Structure of cation
of copper(II) compound with
Me₄C₆H₁₀[15]dieneN₄(N₂) macrocycle, **6.11a**.

Section 3. There is appreciable tetrahedral distortion of the CuN_4 "plane," with *trans* NCuN angles of 163°.

$Me_3pyo_3[15]pentaene\ N_5(N_2)$ *Compounds*

6.12 2,17-dimethyl-3,4,15,16,22,23,24-septa-azatetracyclo[16.3.1.1^{5,9}-.1^{10,14}]tricosa-1(22),2,5,7,9(23),10,12,14(24),16,18,20-undecane, (8,15-dihydro-2,6-dimethyltripyrido-[c,d:i,j:l,m][1,4,7,8,10,13,15]-heptaazapentadecen-N^1,N^{2b},N^7,N^{8b},N^{12}) C_{19},H_{17},N_7

6.12

Complexes formed by condensation of 2,6-diacetylpyridine and 5,5-dihydrazino-2,2-bipyridyl in the presence of metal ions.[136]

6.12a [Zn(L)(H$_2$O)$_2$](NO$_3$)$_2$. $R = 0.0097$ for 1403 refl.*[136]

Planar coordination of the macrocycle by the three pyridine (22,23,24) and two imine nitrogen atoms (3,16), with two water molecules coordinated axially to give a pentagonal-bipyramidal arrangement. The macrocycle is close to planar, with no donor atom deviating by more than 1 pm from the N_5 donor atom plane. The rigidity of the macrocycle results in the Zn–N distances to pyridine nitrogen atoms (23,24) being appreciably shorter at 204 pm than to the pyridine nitrogen atom 22 at 228 pm or the imine nitrogen atoms 3 and 16 at 230 pm. The mean Zn–O distance is 220 pm (cf. compounds of **3.4** and **3.5**, and **6.13** in the Addendum).

8. Class 7: Cyclic Tetraethers, and Tetrathiaethers (*Tetraoxo- and Tetrathiamacrocycles*)

8.1. Discussion of Structures

This section describes structures of compounds formed by cyclic polyethers and polysulfides with transition-metal ions.

The polyethers are poor ligands for transition-metal ions, and reported complexes are few and not particularly stable. The only structure reported, **8.1a**, is for [12]aneO$_4$, the tetraoxo analog of [12]aneN$_4$, cyclen **1.1**, coordinated to copper(II). As for [12]aneN$_4$, the macrocycle is in folded coordination in a configuration analogous to II (Figure 1), the pyramidally bonded oxygen atoms acting as chiral centers.

Structures have been reported for two isomeric forms of the free ligand [14]aneS$_4$, the sulfur analog of [14]aneN$_4$, cyclam **1.3**, with a total of three independent molecules. All have a similar rectangular "exodentate" conformation, with the sulfur atom lone pairs directed outward (Figure 23).

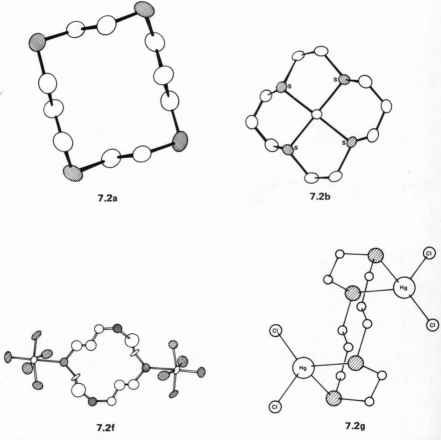

7.2a   **7.2b**

7.2f   **7.2g**

FIGURE 23. Structures of 1,4,8,11-tetrathiacyclotetradecane (α isomer), **7.2a**, the cation of the nickel(II) compound with endodentate coordination **7.2b**, and compounds with niobium(IV), **7.2f**, and mercury(II), **7.2g** with exodentate coordination of the macrocycle. Reproduced from references 144, **7.2e**, 141, **7.2a**, and 141, **7.2b**, by permission of the copyright holder, The American Chemical Society, and from reference 137, **7.2f**, by permission of The Chemical Society.

Exodentate conformations are also present for compound **7.2f**, where the macrocycle is in unidentate coordination to two Nb(IV) ions, and for **7.2g**, where the macrocycle is in bidentate coordination to two Hg(II) ions, forming five-membered chelate rings (Figure 23). The conformation of this macrocycle can be compared to the "endodentate" conformation of the tetraaza analog **1.3a**, which is stabilized by hydrogen bonding across the N–(CH$_2$)$_3$–N portions of the ring, and the exodentate conformation of cyclotetradecane.[138]

The Ni(II) and Cu(II) complexes of [14]aneS$_4$, **7.2c** and **7.2d**, have very similar conformations of the macrocycle, closely resembling that present for the tetraaza analogs **1.3d** and **1.3f**. One structural distinction between the sulfur and nitrogen compounds is the set of chelate angles, which are close to 90° for the S$_4$ compounds (cf. 86° and 94° for the N$_4$ compounds), leading to nearly perfect square planar arrangements about the metal ions. The chelate rings of the S$_4$ compounds are more puckered than their N$_4$ analogs (Table 1), and this is also reflected in larger torsional angles for the S$_4$ compounds.[139] The pyramidally bonded sulfur atoms are chiral centers, and the centrosymmetric complex cations of **7.2b** and **7.2c** have configurations analogous to III of [14]aneN$_4$ (Figure 1) as found for **1.3c** and **1.3d**. The difference between the Ni–S and Cu–S distance is 12 pm, similar to the difference between the $S = 0$ Ni–N and Cu–N distance observed for [14]aneN$_4$ compounds (Section 2).

8.2. Listing of Structures of Tetraoxa- and Tetrathiamacrocycles

[12]aneO$_4$ Compounds

7.1 1,4,7,10-tetraoxacyclododecane, [12]aneO$_4$
C$_8$H$_{16}$O$_4$
7.1a [Cu(L)Cl$_2$]. $R = 0.034$ for 1038 refl.*** [140]

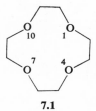

7.1

The macrocycle is in folded coordination to copper(II), which has a tetragonally elongated coordination sphere, with oxygen atoms O(1) and O(7) of the macrocycle (Cu–O, 212 pm), and two chloride ions (Cu–Cl, 222 pm) in the equatorial plane, which is an approximate mirror plane. The remaining two oxygen atoms are coordinated tetragonally (Cu–O, 234 and 240 pm). Angles: O(1)CuO(7), 135°; O(14)CuO(10), 84°.

[14]aneS$_4$ Compounds

7.2   1,4,8,11-tetrathiacyclotetradecane
[14]aneS$_4$, C$_{10}$H$_{20}$S$_4$

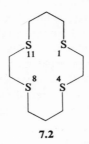

7.2

7.2a   [L], α isomer. $R = 0.027$ for 1098 refl.***[141]
(See Figure 23.)
7.2b   [L], β isomer. $R = 0.055$ for 1988 refl.***[141]
The α form contains one, and the β form two independent molecules.
All have exodentate conformations, with approximately rectangular molecules. The α molecule and one of the β molecules (β_1) have very similar conformations, while the β_2 molecule has a substantially different conformation, although disorder in the S–C–C–S portion of the molecules hindered effective refinement.
7.2c   [Ni(L)](BF$_4$)$_2$. $R = 0.050$ for 1871 refl.***[142]
(See Figure 23.)
7.2d   [Cu(L)](ClO$_4$)$_2$. $R = 0.032$ for 968 refl.***[139]
Planar coordination to $S = 0$ Ni(II) (Ni–S, 218 pm), or to copper(II) (Cu–S, 230 pm), in a nearly perfect square-planar arrangement, with the metal ion at a center of symmetry. The perchlorate ions are weakly coordinated in the tetragonal sites for the copper compound (Cu–O, 265 pm), the chelate angles are all 90°; chelate ring conformation gauche or chair, respectively [deviations from MS$_2$ plane; **7.2c**, (5) 41, -44; (6) -145; -134, -148; **7.2d**, (5), 30, -34; (6) -152, -137, -148 pm]. Torsional angles for **7.2c**, **7.2d**, and the [14]aneN$_4$ complexes with Co(II), **1.4d**, and Cu(II) **1.4h** have been compared.[139]
7.2e   [Cu(L)]ClO$_4$. $R = 0.042$ for 985 refl.*[143]
The compound has a zig-zag chain polymeric structure, with the Cu(I) ion coordinated to three sulfur atoms of one macrocycle and one sulfur atom of a second macrocycle, with an approximately tetrahedral arrangement about the copper atom. The mean Cu–S distance of 232 pm is close to that for the Cu(II) compound **7.2d**. The *trans* SCuS angle is 130° and the chelate SCuS angles are 89° (five membered) and 106° (six membered), with the SCuS angles involving the bridging sulfur in the range 107°–113°.

7.2f [Nb$_2$(L)Cl$_{10}$]C$_6$H$_6$. $R = 0.109$ for 1062 refl.*[144]

(See Figure 23.) The macrocycle is in exodentate unidentate coordination to two niobium atoms. The conformation of the macrocycle is similar to that found for the free ligand molecule β_2, **7.2b**, with similar disorder. An approximately octahedral arrangement about each Nb(IV) ion is completed by five chloride ions [Nb–S, 271(1); Nb–Cl, 230(1) pm].

7.2g [Hg$_2$(L)Cl$_4$]. $R = 0.038$ for 1675 refl.*[137]

(See Figure 23.) The macrocycle is in exodentate–bidentate coordination to two Hg(II) ions forming five-membered chelate rings (SHgS, 83°). An approximately tetrahedral arrangement about each Hg(II) ion is completed by two chloride ions (Hg–S, 258, 270 pm; Hg–Cl, 242 pm). (Cf. **7.2h** in the Addendum.)

9. Class 8: Macrocycles with More Than One Type of Heteroatom

9.1. Discussion of Structures

The compounds in this section have metal ions coordinated to macrocycles with heteroatoms of more than one element. The stereochemistry is as diverse as the macrocycles present, and significant features are discussed individually. The pair of compounds **8.2a** and **8.3a** with closely related macrocycles in planar and folded coordination to nickel(II), and the compounds **8.5a,b,c** with the same macrocycle coordinated to $S = 1$ Ni(II), Pd(II), and Ag(I), respectively, provide interesting comparisons.

9.2. Listing of Structures of Compounds

See introduction for rules determining order of compounds. Key dimensions are given for the individual compounds.

[14]aneN$_4$(O$_2$) Compounds

8.1 1,8-dioxa-3,6,10,13-tetraazacyclotetradecane
C$_8$H$_{20}$H$_4$O$_2$

8.1

Complex formed by reaction of coordinated diaminoethane with formaldehyde.[145]

8.1a   cis-$[Co(L)(C_4H_8NO_4)]$. $R = 0.040$ for 2430 refl.*[145]

Folded coordination by the four nitrogen atoms to cobalt(III) with chelated α-hydroxymethylserine completing an octahedral arrangement. The configuration of the nitrogen centers is V (as for folded [14]aneN$_4$ compounds), and the conformation of the macrocycle is generally similar to that of [14]aneN$_4$ macrocycles in folded coordination (mean Co–N, 198 pm).

Bzo$_2$[14]tetraeneN$_2$O$_2$ Compounds

8.2   dibenzo[f,m]-1,8-dioxa-4,11-diazacyclotetradec-1,3,10-tetraene, 7,18-dioxa-10,21-diazatricyclo[8.4.0^{12,17}]dicosa-1,3,5,10,12,14-16,21-octaene
$C_{18}H_{18}N_2O_2$

8.3   3,14-methoxy-7,8-dioxa-10,21-diazatricyclo[8.4.0^{12,17}]dicosa-1,3,5,10,12,14,16,21-octaene
$C_{20}H_{22}N_2O_4$ (dimethoxy derivative of **8.2**)

8.2 R = H
8.3 R = OMe

Complexes formed by reaction of bis-[N-2-bromoethylsalicylaldiminato)-nickel(II), or its p-methoxy analog, with sodium iodide in acetone.[146]

8.2a   trans-$[Ni(L)I_2]$. $R = 0.047$ for 943 refl.***[147]
(See Figure 24.)
8.3a   cis-$[Ni(L)I_2]$. $R = 0.038$ for 2686 refl.***[148]
(See Figure 24.) Both compounds have $S = 1$ Ni(II) with a $N_2O_2I_2$ coordination environment. For red **8.2a** the macrocycle is in planar coordination, with *trans* coordinated iodide ions in a centrosymmetric arrangement. For green **8.3a** the macrocycle is in folded coordination (N–Ni–N = 94°, O–Ni–O = 157°) and with *cis* coordinated iodide ions (I–Ni–I = 92°). The bond lengths show some unusual features which cannot be entirely steric in origin. Mean values for **8.2a** followed by those for **8.3a**: Ni–N, 192(1), 205; Ni–O, 207(1), 222; Ni–I, 288(1), 273 pm.

These can be compared with values for noncyclic salicylaldimine chelates of Ni–N, 205 pm; Ni–O, 201 pm; the Ni–O value of 207 pm for **8.6a**, and the Ni–I value for the [16]N$_4$ macrocycle compound **4.2b** of 290 pm. For **8.2a**

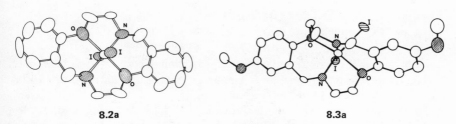

8.2a   8.3a

FIGURE 24. Structures of diiodo compounds of two bzo$_2$[14]tetraeneN$_2$O$_2$ macrocycles with nickel(II), with *trans* **8.2a** and *cis* **8.3a** configurations. Reproduced from references 147 and 148 by permission of the copyright holder, The American Chemical Society.

the Ni–N distance appears to be abnormally short, with a value more typical of $S = 0$ Ni(II), while for **8.3a** the Ni–O distances appear to be abnormally long, and the Ni–I distances appear to be abnormally short (cf. **8.5Aa**).

Bzo$_2$[14]tetraenato(1 −)N$_2$S$_2$(N$_2$) Compounds

8.4 Dibenzo[f,m]-1,5-dithia-8,9,11,12-tetraazacyclotetradeca-6,8,10,13-tetraenato(1 −)
[C$_{16}$H$_{15}$S$_2$N$_4$]$^-$

8.5 10-(3-hydroxypropyl)-dibenzo[f,m]-1,5-dithia-8,9,11,12-tetraaza-cyclotetradeca-6,8,10,13-tetraenato(1 −)
[C$_{19}$H$_{21}$S$_2$N$_4$]$^-$

8.4 R = H
8.5 R = (CH$_2$)$_3$OH

Ni(II) complexes prepared by oxygenation of the linear dihydrazine plus Ni(II) in appropriate solvents. The compound **8.5a** is formed as a minor product in tetrahydrafuran.[149a,150]

8.4a [Ni(L)]ClO$_4$, CH$_3$OH. $R = 0.011$ for 1096 refl.*[149a]
8.4b [Ni(L)]ClO$_4$. $R = 0.077$ for 2299 refl.***[149b]
8.5a [Ni(L)]ClO$_4$. $R = 0.067$ for 2180 refl.***[150]

For all compounds the macrocycle is in planar coordination to $S = 0$ Ni(II). The macrocycles are close to planar, except for the chair conformation dithia chelate ring. The tetraazachelate ring [type A5, Figure 19] is planar and symmetrical, and the bond lengths indicate delocalization analogous to that of diiminato compounds. Significant dimensions for the two compounds **8.4b** and **8.5a** are: distances Ni–S, 217, 216 pm; Ni–N, 184(1), 183(1) pm; N–N, 128(2), 129(2) pm; N(9)–C(10), 134(2), 135(2) pm; angles SNiS, 88°, 88°; SNiN, 89°, 90°; NNiN, 94°, 93°; NCN, 132(1)°, 127(1)°. (Compare $S = 0$ Ni–S for [14]aneS$_4$ of 218 pm.) Compound **8.4a** was initially described as a "monohydrate",[149a] later as a "mono-methanol solvate."[149b] Refinement of **8.4a** was hindered by the triclinic space group (cf. compounds of Section 6).

Bzo$_2$[15]tetraeneN$_2$O$_2$ Compounds

8.5A Dibenzo[e,n]1,4-dioxa-8,12-diazacyclopentadeca-5,7,12,14-
tetraene, 2,5-dioxo-13,17-diazatricyclo[17.4.0^{6,11}]tricosa-
1(19),6(11),7,9,12,17,20,22-octaene
C$_{19}$H$_{20}$N$_2$O$_2$

8.5A

8.5Aa *cis*-[Co(L)(NCS)$_2$]. $R = 0.058$ for 1208 refl.***[151]

The macrocycle is in folded coordination to $S = \frac{3}{2}$ cobalt(II), with the nitrogens of two thiocyanate ions completing a very distorted octahedral arrangement. The folding is unusual in that each of the three pairs of similar donor atoms is *cis*. The bond angles between cobalt and adjacent donor atoms of the macrocycle are all small [O(1)–O(4), 76; O(4)–N(8), 81; N(8)–N(11), 85; N(11)–O(1), 81°], while these angles for *trans* donor atoms of the macrocycle are large [O(1)–N(8), 96; O(4)–N(11), 151°]. The Co–O distances to the ether donor atoms [233(1) pm *trans* to NCS$^-$ and 217(1) pm *trans* to N(8) are long, as for **8.3a**. The mean Co–N distances 212(1), macrocycle, and 202(1), thiocyanate are more normal. An isomeric, presumably *trans*-thiocyanato, compound was prepared (cf. **4.3a**, **4.5a**, **8.2a**, and **8.3a**).

[15]aneN₂OS₂ Compounds

8.6 1-oxa-7,10-dithia-4,13-diazacyclopentadecane C₁₀H₂₂N₂OS₂

(Chemical structure diagram)

8.6

8.6a [Ni(L)(H₂O)](NO₃)₂. $R = 0.052$ for 3956 refl.***(152)

(See Figure 25.) The five donor atoms of the macrocycle and a water molecule are coordinated to $S = 1$ Ni(II), in the octahedral coordination configuration shown, with *meso* configuration. [Mean bond lengths: Ni–O, 207 pm; Ni–N, 206 pm; Ni–S, 242 pm; Ni–OH₂, 209 pm. Chelate ring angles, and deviation from metal–donor-atom planes (pm) O(1)–N(4), 82°, −70, −7; N(4)–S(7), 84°, 74, 26, S(7)–S(10), 87°, 11, −46; S(10)–N(13), 99P, 28, 78; N(13)–O(1), 82°, 32–39. Angles O(1)–Ni–S(10), 167°; N(4)–Ni–N(13), 164°; S(7)–Ni–OH₂, 172°.]

8.6b [Pd(L)](NO₃)₂. $R = 0.047$ for 3580 refl.***(153)

(See Figure 25.) The Pd(II) ion is in square-pyramidal coordination to all five hetero atoms of the macrocycle, with the two sulfur atoms (Pd–S, 226 pm) and the two nitrogen atoms (Pd–N, 208 pm) in the basal plane, and the oxygen weakly coordinated apically (Pd–O, 278 pm). The Pd(II) ion is displaced by 23 pm from the N₂S₂ plane toward the oxygen atom. The chelate angles are SPdS, 84°; SPdN, 88°; NPdN, 100°, with the SPdS chelate ring

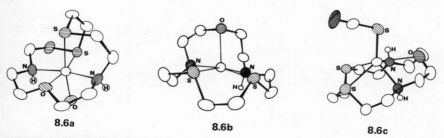

8.6a   8.6b   8.6c

FIGURE 25. Structures of compounds of [15]aneN₂OS₂ macrocycle with nickel(II), **8.6a**, palladium(II), **8.6b**, and silver(I), **8.6c**. Reproduced from references 152, 153, and 154, respectively, by permission of the publishers of *Acta Crystallographica.*

eclipsed (deviations from SPdS plane: C(8), −109 pm; C(9), −151 pm), and with one SPdN chelate ring gauche, and the other highly asymmetrically gauche (deviations from SPdN planes: C(5), 25 pm; C(6), −64 pm; C(11), −25 pm; C(12), −74 pm).

8.6c [Ag(L)(SCN)]. $R = 0.040$ for 3273 refl.***[154]

(See Figure 25.)

The silver(I) ion ion is coordinated within the cavity of the macrocycle to the two nitrogen atoms (Ag–N, 256 and 245 pm), the two sulfur atoms (Ag–S, 271 and 301 pm) of the macrocycle, and the sulfur atom of the thiocyanate ion (Ag–S, 253 pm). Interaction of the oxygen atom of the macrocycle with the silver ion is weak (Ag–O, 288 pm), and the silver ion can be considered to be in square-pyramidal coordination, with two nitrogen and two sulfur atoms of the macrocycle in the "plane" and with the thiocyanate sulfur axially coordinated. The N_4S_2 plane shows a small "tetrahedral" distortion, and the silver ion is displaced by 106 pm from the plane toward the apical sulfur atom. Chelate angles are: NAgS, 74°; SAgS, 73°; NAgN, 102°. The structure has two crystallographically distinct molecules but there are no significant differences between them (cf. **8.7a**).

8.7   1-oxa-7,10-diaza-4,13-dithiacyclopentadecane

$C_{10}H_{22}N_2OS_2$

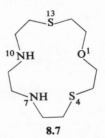

8.7

8.7a [Ag(L)SCN]. $R = 0.037$ for 1200 refl.*[155a,b]

The silver(I) ion is in distorted square-pyramidal coordination to the nitrogen atoms (Ag–N, 253(1) and 245(1) pm), and the sulfur atoms (Ag–S, 268 and 286 pm) of the macrocycle, and with the sulfur atom of the thiocyanate ion coordinated apically (Ag–S, 254 pm). The Ag–O distance of 372(1) pm indicates a very weak interaction. Chelate angles are: NAgN, 74°; SAgN, 76°; SAgS, 110° (cf. **8.6c**).

Pyo[15]trieneN₃O₂ Compounds

8.7A   2,13-dimethyl-6,9-dioxa-3,12,18-triazabicyclo[12.13.1]octadeca-
1(18),2,12,14,16-pentaene
$C_{15}H_{21}N_3O_2$

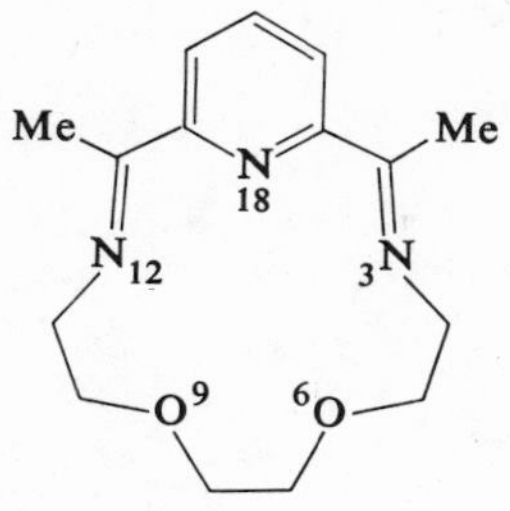

8.7A

Complexes formed by reaction of 2,6-diacetyl pyridine and 3,6-dioxaoctane-1,8-diamine in the presence of metal ions.[155c]

8.7Aa [Mn(L)(NCS)$_2$]. $R = 0.075$ for 1427 refl.[155c]

The $S = \frac{5}{2}$ Mn(II) ion and the five donor atoms of the macrocycle are almost coplanar (maximum deviation of donor atoms from N$_3$O$_2$ plane 8 pm), with thiocyanate nitrogen coordinated axially to give a pentagonal bipyramidal arrangement. As for the compounds **3.4b** and **3.4d** of the nitrogen analog with Fe(II) and Fe(III), there is a twofold distortion of the macrocycle. The mean Mn–donor-atom distances are: pyridine, 221(1) pm; imine, 225(1) pm; ether 229(1) pm, thiocyanate 227 pm. The mean chelate angles are N$_{18}$–N$_3$, 72°; N$_3$–O$_6$, 73° (deviation from MnNO plane: −3, 64 pm); O(6)–O(9), 73° (deviation 32, −41 pm).

[18]aneN$_2$O$_4$ Compounds

8.8 1,4,10,13-tetraoxa-7,16-diazacyclooctadecane
C$_{12}$H$_{26}$H$_2$O$_4$

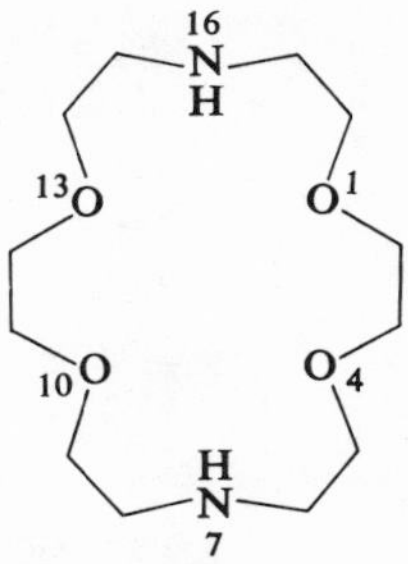

8.8

8.8a [Cu(L)Cl$_2$]. $R = 0.044$ for 1569 refl.***[156a,b]
(See Figure 26.)
8.8b [Cu(L)Br$_2$].[156c]

The copper(II) ion of **8.8a** is in tetragonally elongated octahedral coordination with two chloride ions (Cu–Cl, 231 pm) and the two nitrogen

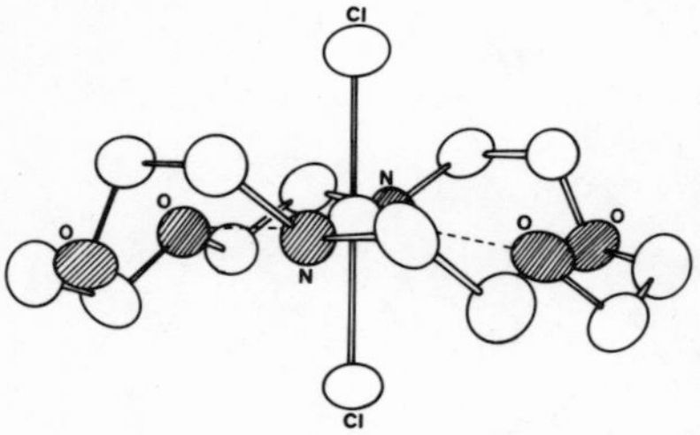

8.8a

FIGURE 26. Structure of dichloro copper-(II) compound with [18]aneN$_2$O$_4$ macrocycle, **8.8a**. The two chlorides and two nitrogens are strongly coordinated and the two oxygens with "dotted" Cu–O bonds are weakly coordinated. Reproduced from reference 156 by permission of the publishers of *Acta Crystallographica*.

atoms of the macrocycle (Cu–N, 203 pm) in the equatorial plane, and with two of the oxygen atoms O(7) and O(16) weakly coordinated in the axial sites (Cu–O, 273 pm), to produce alternating five- and eight-membered chelate rings. The oxygen atoms O(1) and O(10) are not coordinated (Cu–O, 365 pm). The coordinated nitrogens have a *rac* configuration. The macrocycle has two-fold symmetry, the axis being approximately collinear with the Cl–Cu–Cl axis. Compound **8.8b** has a similar structure.

Bzo$_4$[20]sexaeneN$_4$S$_2$ Compounds

8.9   1,4-dithia-7,11,14,18-tetraazatetrabenzo[e,i,o,s]cyclodicosa-5,7,9,15,17,19-sexaene
      C$_{30}$H$_{28}$N$_4$S$_2$

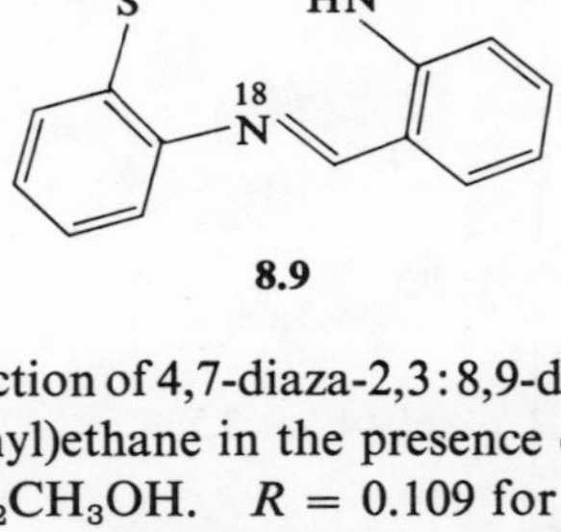

8.9

Complexes formed by reaction of 4,7-diaza-2,3:8,9-dibenzo-decane-1,10-dione and 1,2-*bis*-(2-aminophenyl)ethane in the presence of metal ions.[157]

    8.9a   [Fe(L)](ClO$_4$)$_2$CH$_3$OH. $R = 0.109$ for 1311 refl.***[157]

(See Figure 27.) The macrocycle is coordinated to $S = 0$ Fe(II) by all six donor atoms in an octahedral configuration which has approximate two-fold symmetry. One perchlorate ion is disordered. Mean bond lengths are

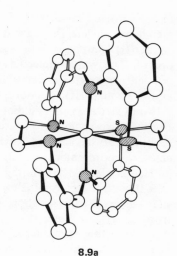

FIGURE 27. Structure of cation of Fe(II) compound with sexadentate [20]N₄S₂ macrocycle, **8.9a**. Reproduced from reference 157 by permission of the copyright holder, The American Chemical Society.

8.9a

Fe–N, 198(2) pm (distances to amine and imine nitrogens not significantly different), Fe–S, 224 pm. Deviations from Fe–donor-atom planes (pm) are S(1)–S(4), 39, −25; S(4)–N(7), −25, −27; N(7)–N(11), 43, 92, 77; N(11)–N(14), 47, 102, 47; N(18)–S(1), −37, 18. The "benzo" S–N five-membered chelate rings are eclipsed, and the 2-aminobenzaldimine N–N six-membered chelate rings (c of Figure 15) have boat conformations.

10. Class 9: Binucleating Macrocycles

10.1. Discussion of Structures

This section deals with structures of macrocycles capable of coordinating two metal ions within the macrocycle. A number of macrocycles of this type are known, but structures have been reported for only one example, **9.1**, which has four nitrogen and two oxygen donor atoms.

Structures of binuclear complexes, where the ligand consists of two linked tetraazamacrocycles, each with a coordinated metal ion, **2.3a**, and of two metal–macrocycle moieties linked by metal–metal bonds, **2.15a** and **6.8b**, are reported in other sections.

The compounds have the same general structure, with the four nitrogen and two oxygen atoms approximately coplanar, with one metal ion in each N_2O_2 donor set. For **9.1a** two $S = \frac{3}{2}$ Co(II) ions are present, with a bromide ion completing square-pyramidal coordination about each metal ion. The molecule is centrosymmetric, with one bromide on each side of the molecular plane. The methanol is weakly coordinated in half of the "octahedral" sites in a disordered manner. The central methylene groups of the saturated six-membered chelate rings are disordered over two sites. The structure of the copper(II) compound **9.1e** is generally similar to that of the cobalt(II) compound. The water molecules are not coordinated, but form a hydrogen-bonded network. The compounds **9.1b,c,d** all contain one $S = \frac{3}{2}$ cobalt(II) and one cobalt(III) ion coordinated within each macrocycle. The compounds **9.1b** and **9.1c** are isomeric, both cobalt ions being in octahedral coordination. For **9.1b** two bromide ions are coordinated to the cobalt(III) and two water molecules to the cobalt(II), while for **9.1c** one bromide and one water molecule are coordinated to each cobalt ion, one of each on each side of the molecular plane. The compound **9.1d** has a structure generally similar to that of **9.1b** with the unusual $[Br_3]^-$ anion present. Assignment of the cobalt(III) sites was made from the shorter bond lengths, particularly to the axial ligands, from one cobalt atom.

For **9.1a,e** the macrocycle is essentially planar, with the metal ions displaced from the N_2O_4 donor sets, on opposite sides of the macrocycle. For **9.1b** the macrocycle is bent, with a dihedral angle between the N_2O_4 donor sets of 21.6°, and with the benzene rings and central methylene groups all tilted to the same side of the molecular plane. Compound **9.1c** has an intermediate geometry, with a dihedral angle between the two N_2O_2 planes of 0.6°, and with the methylene groups displaced on opposite sides of the molecular plane, toward the adjacent water ligands.

10.2. Listing of Structures of Binucleating Macrocycles

Key dimensions are listed in Table 9.

9.1   11,23-dimethyl-3,7,15,19-tetraazatricyclo[19.3.1.1⁹,¹³]hexacosa-2,7,9,11,13(26),14,19,21(25),22,24-decaene-25,26-diolate(2−) $[C_{24}H_{26}N_4]^{2-}$

9.1

Ligand formed by reaction of 2-hydroxy-5-methyl-*iso*-phthalaldehyde and 1,3-diaminopropane.

TABLE 9. Dimensions of Compounds of the Binucleating Macrocycle **9.1**

Compound	Ion	M–M[a]	Plane[b]	M–N[c]	M–O[c]	M–X[d]	M–O[e]
9.1a	Co(II)	316	30[f]	204	208	256[g]	250[h]
9.1b	Co(II)	313	7	189(2), 211(2)	201(1), 213(2)	224[g]	
	Co(III)		0	194(2), 200(2)	183(2), 198(1)		209(2), 221(2)
9.1c	Co(II)	313	10[f]	203	202	265[g]	217
	Co(III)		3[i]	188, 195	198	235[g]	200
9.1e	Cu(II)	313	21[f]	198	198	258[j]	

[a] Inter-metal-ion separation (picometers).
[b] Deviation of metal ion from plane of N_2O_2 donor set (picometers).
[c] Metal-ion–macrocycle-donor-atom separation (picometers).
[d] Metal-ion–halide separation (picometers).
[e] Metal-ion–axial-oxygen-donor-atom separation (picometers).
[f] Square-pyramidal geometry.
[g] Coordinated bromide.
[h] Methanol oxygen, disordered over half available sites *trans* to bromide ion.
[i] Toward bromide.
[j] Coordinated chloride.

9.1a $[Co_2(L)Br_2(CH_3OH)]$. $R = 0.049$ for 2004 refl.***[158]
9.1b $[Co_2(L)Br_2(OH)_2]Br$, isomer A. $R = 0.074$ for 1883 refl.***[159]
9.1c $[Co(L)Br_2(OH)_2)_2]Br$, isomer B. $R = 0.101$ for 2006 refl.***[159]
9.1d $[Co_2(L)Br_2(CH_3OH)_2]Br_3$. $R = 0.15$ for 1993 refl.***[160]
The data set was degraded by the rapid decomposition of the crystal by X rays.
9.1e $[Cu_2(L)Cl_2]6H_2O$. $R = 0.049$ for 1369 refl.***[161]
A space group ambiguity limits confidence in details of the structure.

11. Class 10: Cyclic Phosphazenes

11.1. Discussion of Structures

Many cyclic phosphazenes are known, and coordination compounds of these macrocycles have been described.[162] The structures of two transition-metal-ion derivatives of octamethylcyclotetraphosphazene (L) have been reported. A copper(II) compound has a copper(II) ion (as $CuCl_3^-$) coordinated externally to the ring, with N–Cu of 204 pm, and with the "opposite" nitrogen protonated to give the compound $[H(L)_4CuCl_3]$.[163] The formally related cobalt compound $[H(L)_4]_2CoCl_4$ has the discrete $CoCl_4^{2-}$ anion, with the Co(II) ion not coordinated by the phosphazene.[164] A tungsten derivative, $W(L)(CO)_4$ has the W(O) coordinated by one ring nitrogen and one dimethylamino nitrogen.[165a] A platinum compound of the methylaminophosphazine has the Pt(II) coordinated by two *trans* ring nitrogens and two chlorides, giving a square-planar arrangement with the macrocycle in a saddle-shaped conformation.[165b]

Structures have been reported for two compounds of cyclic hexaphosphazenes which have the metal ions coordinated within the macrocycle.

11.2. Listing of Structures of Cyclic Phosphazene Compounds

10.1   Dodeca(dimethylamino)cyclohexaphosphazene
$C_{12}H_{72}N_6P_6$

10.1

10.1a   (L). $R = 0.07$ for 777 refl. (P)***(166)

The molecule has $\bar{3}$ symmetry, with the N_6P_6 ring highly puckered, with the nitrogen atoms lying 39 pm, and the phosphorus atoms 56 pm, from the mean ring plane. The two NMe_2 substituents on each phosphorus atom have their P–N bonds approximately parallel to, and normal to the mean plane, with the axial substituents on adjacent phosphorus atoms lying on opposite sides of the plane.

10.1b $[Co(L)Cl][Co_2Cl_6]$. $R = 0.093$ for 2695 refl.***(167a)

10.1c $[Cu(L)Cl][CuCl_2]$. $R = 0.083$ for 1103 refl.***(167b)

The $S = \frac{3}{2}$ Co(II) and the Cu(II) ions are coordinated to four nitrogen atoms of the phosphazene ring (producing alternating three- and six-membered chelate rings), and to a chloride ion, to give a geometry which can be regarded as square pyramidal (Cl apical) or trigonal bipyramidal [N(1) and N(4) axial]. The macrocycle consists of two approximately planar portions folded along N(1)–N(3), with a dihedral angle of ~130°. The copper–chlorine bond is a twofold axis, while the cobalt–chlorine bond is an approximate twofold axis. The anions present are linear $[ClCuCl]^-$ and $[Cl_2CoCl_2CoCl_2]^{2-}$, respectively.

Regarding the structures as trigonal pyramidal, dimensions (Co, Cu) are: $M–N_{eq}$, 206(1), 211(1) pm; $M–N_{ax}$, 226(1), 203(1) pm; M–Cl, 227, 228 pm; with NMN chelate angles of 71°, 71° and 101°, 91°.

12. Class 11: Clathrochelates

12.1. Discussion of Structures

A number of "cage" ligands able to completely surround a metal ion are known, and these have been termed "clathrochelates." The tertiary-amine–polyether "cryptates" form many complexes with "A"-type elements, but few transition-metal-ion derivatives have been reported, and only one of these structurally investigated. For **11.4a** the cobalt(II) ion is coordinated within the cage to two tertiary amine and five ether oxygen donor atoms in a distorted pentagonal-bipyramidal arrangement [Co–N (mean) 222 pm, Co–O (mean) 218 pm].

Structures have been reported for one fully saturated, **11.4**, and three unsaturated types of clathrochelate ligands with six nitrogen donor atoms: **11.1**, with a symmetrical cage of oxime donor nitrogen capped with BF groups; **11.2**, with three pyridine and three oxime donor nitrogens (capped with phosphorus and BF, respectively); and **11.3**, with a symmetrical cage of six hydrazone nitrogens, capped with 1,3,5-triazacyclohexane rings (and related to the macrocyclic ligands of Section 6).

These clathrochelate ligands are related to the trifurcated hexadentate ligands (py$_3$TPN)[167c] and (py$_3$tren),[168] which have bridgehead carbon and nitrogen atoms, respectively, and to (py$_3$tach),[169] which is capped by a 1,3,5-substituted cyclohexane ring (Figure 28), and to the tetraazamacrocycle with two pendant iminoethyl groups, **2.18a**.

For each of the types of unsaturated clathrochelate described, the free ligand would be expected to adopt a conformation with the six donor nitrogen atoms defining a trigonal prism, with a cage size determined by the particular molecular geometry. The cages are able to deform with relatively little strain toward an octahedral arrangement of the donor atoms, by rotation of the two triangular arrays of nitrogen atoms relative to each other (specified by a twist angle ϕ which is 0° for a trigonal prism and 60° for an octahedron. This deformation is accompanied by a reduction in cage size. On steric grounds, therefore, it would be expected that the complex of a metal ion which is the optimum size for the free ligand would adopt a trigonal prismatic arrangement, with ϕ inversely proportional to ion size for smaller ions, and with a trigonal prismatic arrangement with some distortion of the ligand to increase the cage size for larger ions. The steric effect is complicated by an electronic effect

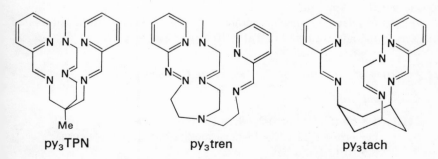

FIGURE 28. Some tripod clathrochelate ligands.

TABLE 10. Dimensions for Clathrochelate Compounds

Compound	Metal ion	Spin	M–N^a	$\phi^{\circ b}$	Reference
11.1a	Co(II)	$S = \frac{3}{2}$	197	8.6	172
11.1b	Co(III)		189	31.2	172
11.2a	Fe(II)	$S = 0$	195	21.7	173
11.2b	Co(II)	$S = \frac{3}{2}$	209	0.9	174
11.2c	Ni(II)	$S = 1$	204	1.6	175
11.2d	Zn(II)		209	1.2	176
11.3a	Fe(II)	$S = 0$	192	21	177
2.18a	Fe(II)	$S = 0$	195	8	80
py$_3$TPN	Fe(II)	$S = 0$	196	54	167
	Zn(II)		215	28	167
py$_3$tren	Fe(II)	$S = 0$	196	54	168
py$_3$tach	Fe(II)	$S = 0$	196	43	169
	Ni(II)	$S = 1$	209	32	169
	Zn(II)		215	28	169

^{a} Mean metal–nitrogen distance (picometers).
^{b} Twist angle; $\phi = 0°$ for trigonal prism and $\phi = 60°$ for octahedron.

arising from the difference in ligand field stabilization energies between octahedral and trigonal prismatic coordination for the particular electron configuration, which increases in the sequence Zn(II) < Co(II), $S = \frac{3}{2}$ < Ni(II), $S = 1$ < Fe(II), $S = 0$ < Co(III), $S = 0$. This is much the same sequence as the ionic size, and hence the two effects cannot be distinguished.

The most extensive set of data is available for **11.2** for which structures of Fe(II), Co(II), Ni(II), and Zn(II) complexes have been determined. The structure of the $S = 1$ Ni(II) complex is close to trigonal prismatic, and this ion is considered to be approximately the optimum size for this cage. The compound of the smaller Fe(II) ion has a ϕ value of 21.7°, while the complexes of the larger cobalt(II) and zinc(II) complexes have approximately trigonal prismatic coordination, but with the ligands showing distortion from C_{3v} symmetry which is attributed to an expansion of the cage to accommodate ions larger than optimum.

Structural data for these compounds and for some compounds of related trifurcated ligands are listed in Table 10.

12.2. *Listing of Structures of Clathrochelate Compounds*

N_6 *Clathrochelate Compounds*

11.1 1,8-*bis*-(fluoroboro)-2,7,9,14,15,20-hexaoxa-3,6,10,13,16,19-hexaaza-4,5,11,12,17,18-hexa-methyl-bicyclo[6,6,6]eicosa-3,5,10,12,16,18-hexaenato(2−)
[C$_{12}$H$_{18}$B$_2$F$_2$N$_6$O$_6$]$^{2-}$

11.1

The complexes are formed by condensation of dimethylglyoxime and boron trifluoride about a metal ion.[170]

11.1a [Fe(L)], $0.5(C_6H_6)$ [171]

11.1b [Co(L)]. $R = 0.054$ for 666 refl.*** [172]

11.1c [Co(L)]BF_4. $R = 0.061$ for 1318 refl.*** [172]

(See Figure 29.) In each structure the metal ion ($S = \frac{5}{2}$ Fe(II), $S = \frac{1}{2}$ Co(II), or Co(III), respectively) is coordinated within the cavity of the ligand to the six nitrogen atoms, in a configuration intermediate between octahedral and trigonal prismatic. The cobalt(II) ion lies at a site of D_3 symmetry, while the cobalt(III) ion lies at a site of C_2 symmetry, with the twofold axes passing through the cobalt ion and the midpoint of one of the five-membered chelate rings.

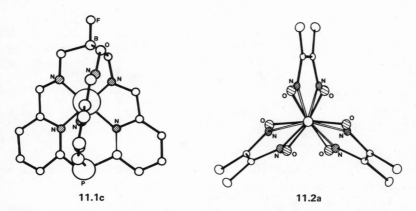

11.1c   **11.2a**

FIGURE 29. Structures of compounds of two cage clathrochelate ligands with Co(III), **11.1c**, and with Fe(II), **11.2a**, viewed down the threefold axis. Reproduced from references 172 and 176, respectively, by permission of the copyright holder, The American Chemical Society.

11.2 Fluoroboro-*tris*-(2-aldoximo-6-pyridyl)phosphinate
$[C_{18}H_2BFN_6O_3P]^-$

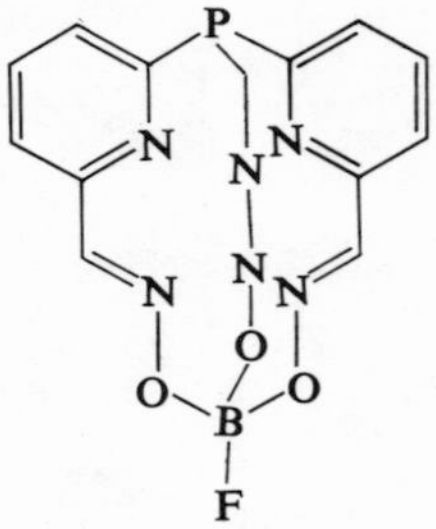

11.2

Complexes formed by reaction of the appropriate tripodal oxime complex
with boron trifluoride.

11.2a [Fe(L)]BF$_4$, CH$_2$Cl$_2$. $R = 0.075$ for 2181 refl.***[173]
(See Figure 29.)

11.2b [Co(L)]BF$_4$, CH$_3$CN. $R = 0.105$ for 1126 refl.***[174]

11.2c [Ni(L)]BF$_4$. $R = 0.0732$ for 1781 refl.***[175]

11.2d [Zn(L)]. $R = 0.082$ for 2728 refl.***[176]

In each structure the metal ion ($S = 0$ Fe(II), $S = \frac{3}{2}$ Co(II), $S = 1$
Ni(II), and Zn(II), respectively) is coordinated within the cavity of the
clathrochelate ligand. The nickel(II), cobalt(II), and zinc(II) compounds have
approximate C_{3v} symmetry.

11.3 5,6,12,13,18,19-hexamethyltetracyclo[8.8.4^{8,16}.1^{3,21}][1,3,4,7,8-
10,11,14,17,20,21]dodecaazatetracos-4,6,11,13,18,19-hexaene
$C_{18}H_{24}N_{12}$

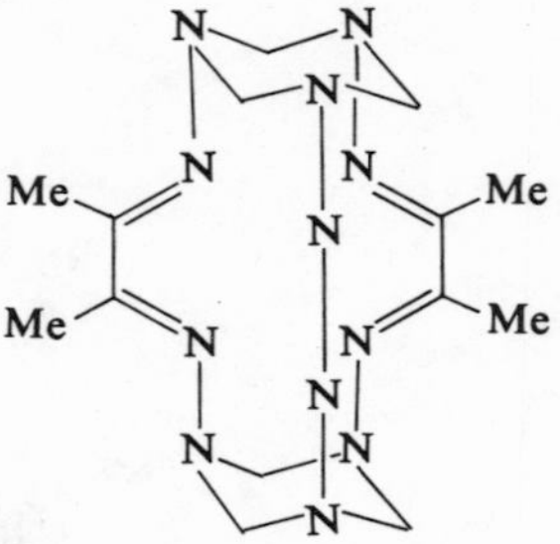

11.3

Complex formed by acid-catalyzed condensation of 2,3-butane dihydrazone
with formaldehyde in the presence of metal ions.[177]

11.3a [Fe(L)]ZnCl$_4$, H$_2$O. $R = 0.053$ for 4117 refl.[178]

The $S = 0$ Fe(II) ion is coordinated within the cavity of the clathrochelate with a coordination geometry between trigonal prismatic and
octahedral.

11.4 (S)-1,3,6,8,10,13,16,19-octaazatricyclo[6.6.6]eicosane, sepulchrate
C$_{12}$H$_{30}$N$_8$

11.4

Complex formed by condensation of *tris*-(diaminoethane)cobalt(III) form-aldehyde, and ammonia[179] (cf. **8.1**).

11.4a [Co(L)]Cl$_3$·H$_2$O. $R = 0.038$ for 3166 refl.*[179]

The cobalt(III) ion is coordinated to the six secondary amine nitrogen atoms within the cavity of the clathrochelate. The chirality of the [Co(en)$_3$]$^{3+}$ ion is retained and the cation has overall symmetry close to D_3. The mean Co–N distance is 199 pm.

N$_2$O$_5$ "Cryptate" Compounds

11.5 1,10-diaza-4,7,13,16,21,24-hexaoxobicyclo[8.8.8]hexacosane
C$_{18}$H$_{36}$N$_2$O$_6$

11.5a [Co(L)][Co(NCS)$_4$]. $R = 0.033$ for 3247 refl.*[180]

The $S = \frac{3}{2}$ Co(II) ion is within the cavity of the clathrochelate ligand and coordinated to all seven donor atoms in a distorted pentagonal-bipyramidal arrangement. Mean Co–O, 218 pm; mean Co–N, 222 pm.

N$_2$O$_4$S$_2$ "Cryptate" Compounds

11.6 1,10-diaza-4,7,13,16-tetraoxo-21,24-dithiabicyclo[8.8.8]-
hexacosane
C$_{18}$H$_{36}$N$_2$O$_4$S$_2$

11.6a Pd(L)Cl$_2$. $R = 0.040$ for 420 refl.***[181]

The palladium(II) ion is a square-planar coordination to the two sulfur atoms of the macrocycle and two chloride ions, with the palladium ion outside the cavity of the ligand. Mean Pd–S, 226 pm; chelate angle SPdS, 91°.

13. Conclusion

A survey of structures is restricted to reported studies of compounds chosen for a variety of reasons, often because other means of establishing the structure of an interesting compound were inconclusive. Thus compounds studied are often in some way unusual, or at least atypical, making com-parisons difficult, since the compounds chosen seldom form convenient series in which metal ion, ring size, substitution pattern, etc., are varied in a systematic manner.

The majority of the macrocycles for which structures have been reported have four nitrogen donor atoms, most commonly in a [14]-membered macrocycle. In consequence most compounds have metal ions from the latter part of the first transition series. Some compounds of the more flexible saturated macrocycles are in folded coordination, but for the unsaturated macrocycles planar coordination is usual. Square-planar coordination extends from Cu(II) through $S = 0$ Ni(II) and $S = \frac{1}{2}$ Co(II) to the unusual $S = 1$ Fe(II). Pentaaza pyridyl macrocycle compounds occur for pentagonal-bipyramidal arrangement for $S = \frac{5}{2}$ Mn(II), Fe(III), $S = 2$ Fe(II), and various d^{10} ions. Metal–donor-atom distances for the more flexible macrocycles tend to be similar to those of related but noncyclic ligands. Where the metal ion is smaller than optimum the ligand distorts to produce "normal" distances. Where the metal ion is larger than optimum, folded coordination, or exclusion of the metal ion from the donor atom plane is common, although in some instances smaller-than-usual metal–donor distances are indicative of a "macrocycle constrictive effect."

ACKNOWLEDGMENTS

Provision of prepublication information about structures by Professors J. E. Endicott, G. Fergusson, V. L. Goedken, J. W. Krajewski, and D. B. Rorabacher is gratefully acknowledged.

Addendum: Additional Structure Listings

We list here additional structures reported after submission of the manuscript, or previously omitted. Where a structure of a macrocycle compound is listed in the main sequence, the same code number is used, otherwise the number sequence for the section is continued, and a structural formula shown.

Addendum to Section 2

[9]aneN₃ Compounds

1.9   1,4,7-triazacyclononane
[9]aneN₃, $C_9H_{21}N_3$

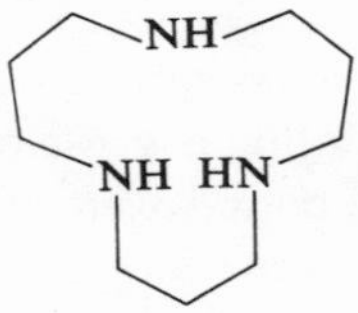

1.9

1.9a [Ni(L)$_2$](NO$_3$)Cl, H$_2$O. R = 0.034 for 1795 refl.[(182)]

Facial coordination by two macrocycles to give a trigonally distorted octahedral arrangement about S = 1 Ni(II). Mean Ni–N, 211 pm.

[12]aneN$_4$ Compounds

1.10 2R,5R,8R,11R-2,5,8,11-tetraethyl-1,4,7,10-tetraazacyclododecane
[Et$_4$12]aneN$_4$, C$_{16}$H$_{36}$N$_4$

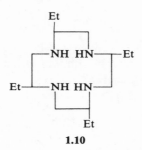

1.10

1.10a [L]$^{(183)}$

The macrocycle adopts a square conformation, like cyclododecane, with the unsubstituted carbon atoms at the corners.

[14]aneN$_4$ Compounds

1.4 1,4,8,11-tetraazacyclotetradecane
[14]aneN$_4$, C$_{10}$H$_{24}$N$_4$

1.4i [Co(L)(CH$_3$)(OH$_2$)](ClO$_4$)$_2$$^{(184)}$

Planar coordination to Co(III) with methyl and water completing a tetragonal arrangement.

1.5 5,12-dimethyl-1,4,8,11-tetraazacyclotetradecane
Me$_2$[14]aneN$_4$, C$_{12}$H$_{28}$N$_4$

1.5b [Ni(L)](ClO$_4$)$_2$. R = 0.033 for 1651 refl.***[(185)]

Planar coordination to S = 0 Ni(II). The cation is centrosymmetric, with nitrogen configuration III. The methyl substituents are axially oriented, although the compound was crystallized under conditions which permitted nitrogen isomerization, indicating that this is the thermodynamically stable form.

1.11 5,6,12,13-tetramethyl-1,4,8,11-tetraazacyclotetradecane
Me$_4$[14]aneN$_4$, C$_{14}$H$_{32}$N$_4$

1.11a [Ni(L)](ClO$_4$)$_2$. R = 0.033 for 1203 refl.***[(186)]

Planar coordination to S = 0 Ni(II), with nitrogen configuration (III), and with the methyl substituents in equatorial orientation, for a centrosymmetric structure.

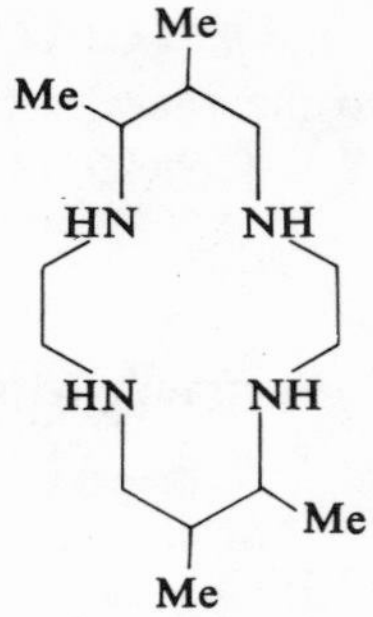

1.11

1.12   5,12-diethyl-7,14-dimethyl-1,4,8,11-tetraazacyclotetradecane
Et$_2$Me$_2$[14]aneN$_4$, C$_{16}$H$_{36}$N$_4$

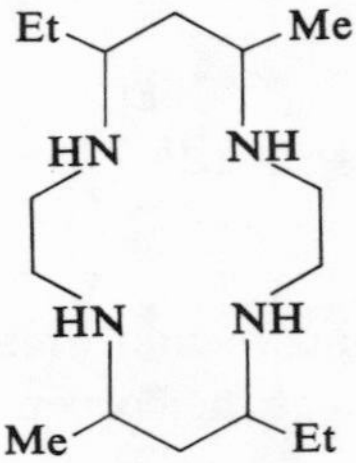

1.12

1.12a   [Ni(L)](ClO$_4$)$_2$. $R = 0.042$ for 1720 refl.*** [187]

Planar coordination to Ni(II), with a centrosymmetric arrangement,
with the substituents all equatorially oriented. The compound is paramagnetic
in nitromethane, but the spin state of the solid is not reported. With NiO to
tetragonal perchlorate oxygen atom separation of 276 pm, the assumed $S = 1$
ground state appears improbable.

1.7   5,5,7,12,12,14-hexamethyl-1,4,8,11-tetraazacyclotetradecane
          Me$_6$[14]aneN$_4$, C$_{16}$H$_{36}$N$_4$

1.7i   [(rac-L), H$_2$O]. $R = 0.063$ for 2014 refl.** [188]

The macrocycle adopts a conformation with the nitrogen atoms non-
coplanar. Hydrogen bonds N(8)–H$\cdots$N(4), N(1)–H$\cdots$N(11), N(8)$\cdots$H–O,
N(1)$\cdots$H–O, and N(11)–H$\cdots$O (separate water molecule) are present.

1.7j   [Cu(meso-L)(ClO$_4$)$_2$]. $R = 0.054$ for 2945 refl.*** [189]

Planar coordination to Cu(II), with configuration (III). Perchlorate
oxygen atoms complete a tetragonal arrangement. The molecule has approximate twofold symmetry (through the midpoints of the five-membered chelate
rings). A reference [190] to an unpublished structure of this compound is
mentioned [191] together with preliminary information about the structure of
another isomer.

[16]aneN₄ Compounds

   1.13   1,5,9,13-tetraazacyclohexadecane
          [16]aneN₄, $C_{12}H_{28}N_4$

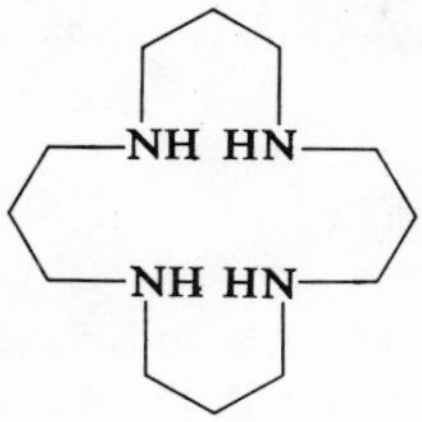

1.13

1.13a [L][(192)]

The macrocycle adopts an *endo* conformation with the four nitrogen atoms in a square (edge 293 pm), with alternative NH groups above and below the plane and with the trimethylene chains in "pseudo-chair" conformations, also alternatively above and below the plane.

Addendum to Section 3

[12]eneN₃ Compounds

   2.19   2,4,4,7-tetramethyl-1,5,9-triazacyclododeca-1-ene
          Me₃(N–Me)[12]eneN₃, $C_{13}H_{27}N_3$

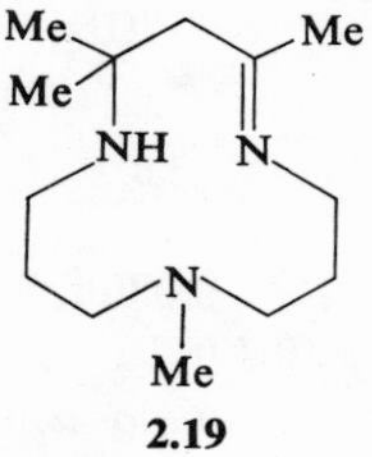

2.19

2.19a [{Cu(L)}₂CO₃](ClO₄)₂, HCON(CH₃)₂. *R* = 0.068 for 1126 refl.**[(193)]

Coordination of macrocycle to Cu(II) is in the same configuration and in a very similar conformation to that found for **2.1a**. The carbonate ion is bidentate to both Cu(II) ions, with one oxygen atom bonded to both Cu(II) ions, with Cu–O–Cu linear, to give a square-pyramidal arrangement (tertiary amine N(8) apical). The halves of the dinuclear cation are related by a twofold axis through the carbon and shared oxygen of the CO_3^{2-}. The compound is diamagnetic by antiferromagnetic spin coupling (cf. **2.1a**). $Cu-N_1$, 196 pm; $Cu-N_5$, 198 pm; $Cu-N_8$, 220 pm; Cu–O (bridging), 204 pm; Cu–O, 203 pm.

[14]tetraeneN₄ Compounds

2.13   2,3,9,10-tetramethyl-1,4,8,11-tetraazacyclotetradeca-1,3,8,10-
tetraene
$Me_4[14]tetraeneN_4$, $C_{14}H_{28}N_4$

2.13d $[Fe(L)(NCCH_3)_2](PF_6)_2^{(194)}$

2.13e $[Co(L)Cl_2]PF_6^{(194)}$

Planar coordination of the macrocycle to Fe(II), with *trans* methyl cyanide nitrogen; and to Co(III) with *trans* chloride, respectively.

Addendum to Section 4

Pyo[15]eneN₅ Compounds

3.8   2,13-Dimethyl-3,6,9,12,18-pentaazabicyclo-[12,3,1]-octadec-
1(18),14,16-triene
$Me_2pyo[15]eneN_5$, $C_{15}H_{27}N_5$

3.8

3.8a $[Co(L)Cl](ClO_4)_2$. $R = 0.086$ for 1336 refl.*[195]

3.8b $[Cu(L)](PF_6)_2$. $R = 0.078$ for 1608 refl.*[195]

Both compounds have the C-*meso* configuration. In both cases the metal, pyridine, and adjacent nitrogen atoms are approximately coplanar. For the Cu(II) compound, the other two nitrogen atoms are coordinated above and below this plane, respectively, to give a trigonal-bipyramidal arrangement. For the Co(III) compound, one of these nitrogen atoms is in the same plane, *trans* to the pyridine nitrogen, and the other, N(9), is coordinated *trans* to a chloride ion in an octahedral arrangement. Cu–N distances in numerical sequence: 201(1), 210(1), 210(1), 204(1), 190(1) pm. Co–N distances in numerical sequence: 194(1), 190(1), 191(1), 199(1), 180(1) pm. Co–Cl, 217 pm.

Pyo[15]trieneN₅ Compounds

3.4 2,13-dimethyl-3,6,9,12,18-pentaazabicyclo[12,3,1]octadec-
1(18),2,12,14,16-pentaene
$Me_2pyo[15]trieneN_5$, $C_{15}H_{23}N_5$
3.4g $[Cd(L)(ClO_4)]_n(ClO_4)_n$. $R = 0.10$ for 975 refl.*[195]
The macrocycle is in planar coordination to Cd(II) with bridging perchlorate ions forming chains of pentagonal-bipyramidal units. The ionic perchlorate is disordered (Cd–N, 231(2)–241(3) pm, Cd–O, 247(2) pm).

Pyo[16]trieneN₅ Compounds

3.6 2,14-dimethyl-3,6,10,13,19-pentaazabicyclo[13,3,1]nonadec-
1(19),2,13,15,17-pentaene
$Me_2pyo[16]trieneN_5$, $C_{16}H_{25}N_5$
3.6c $[Mn(L)(H_2O)_2](ClO_4)_2$. $((1,10\text{-phenanthroline})0.5(C_2H_5OH)$.
$R = 0.10$ for 950 refl.*[197]
Planar coordination of the macrocycle to the Mn(II), with water molecules completing a pentagonal-bipyramidal arrangement. The 1,10-phenanthroline molecules are not coordinated, but are sandwiched between adjacent macrocycle planes in a lamellar structure.
3.6e $[\{Ag(L)\}_2](ClO_4)_2$. $R = 0.066$ for 2066 refl.*[198]
Dimeric cation with C_2 symmetry, formed by axial interaction of Ag(1) ion coordinated within approximately planar macrocycle with an imine nitrogen of the second macrocycle. The Ag(I) ion and the bridging nitrogen are displaced by 59 and 88 pm from the mean plane of the other four nitrogen atoms, although the maximum deviation from the AgN_5 plane is only 45 pm. (Ag–N distances: pyridine, 243; imine 243, bridging imine 257, 275 (axial), amine 240, 243 pm).

Pyo₂pyrrole₂[16]hexaeneN₄(N₄) Compounds

3.9 Pyo₂pyrrole₂[16]hexaenato(2−)N₄(N₄)
$[C_{26}H_{14}N_8]^{2-}$

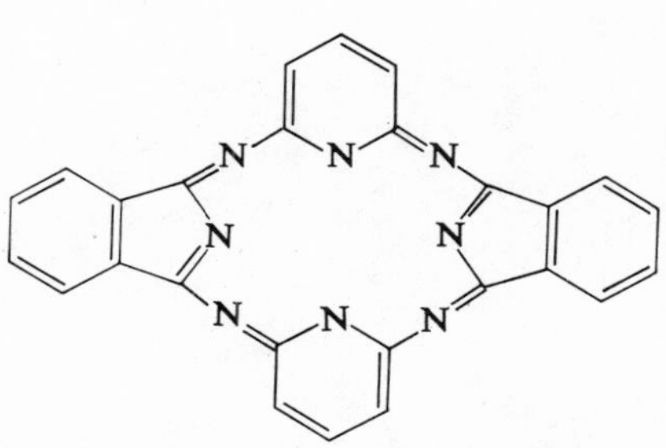

3.9

3.9a   [Ni(L)]. $R = \sim 0.19$ (P, projection)[199]
Planar coordination to $S = 0$ Ni(II).

Addendum to Section 5

Bzo$_2$[16]tetraenato(2−)N$_4$ Compounds

4.5   Dibenzo[c,i]-1,5,8,12-tetraazacyclohexadeca-1,11-dienate(2−)
bzo$_2$[16]tetraenato(2−)N$_4$, [C$_{20}$H$_{22}$N$_4$]$^{2-}$

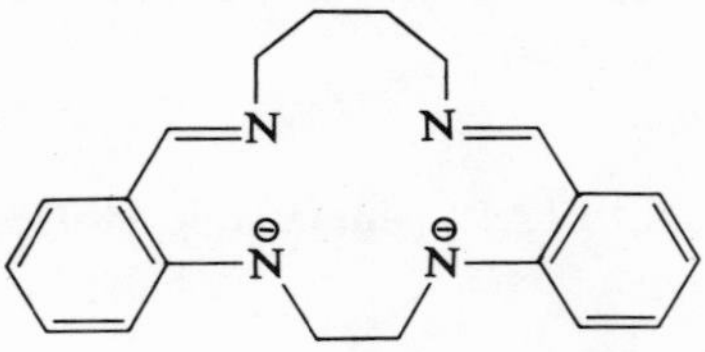

4.5

4.5a   [Cu(L)]. $R = 0.06$ for 1495 refl.*[200]
The macrocycle is in distorted planar coordination to Cu(II), with an
angle of 160° between the planes Cu,N(1),N(5) and Cu,N(8),N(12). The five-
membered chelate ring including N(5) and N(8) has a gauche conformation,
and the seven-membered chelate ring including N(1) and N(12) has a "half-
boat" conformation. Cu–N distances in numerical sequence are: 193(1),
200(1), 198(1), 194(1). Chelate angles in same sequence: 91, 97, 92, 87°.
Angles N(1) Cu N(8), 159°; N(5) Cu N(12), 163° (cf. **4.4a**).

Bzo$_4$[16]tetraenatoN$_4$ Compounds

4.6   2,10-Acetylmethyl-tetrabenzo[b,f,j,n]-1,5,9,13-tetraazacyclo-
hexadec-2,4,8,10,12,14,16-octaene, C$_{34}$H$_{30}$N$_4$O$_2$

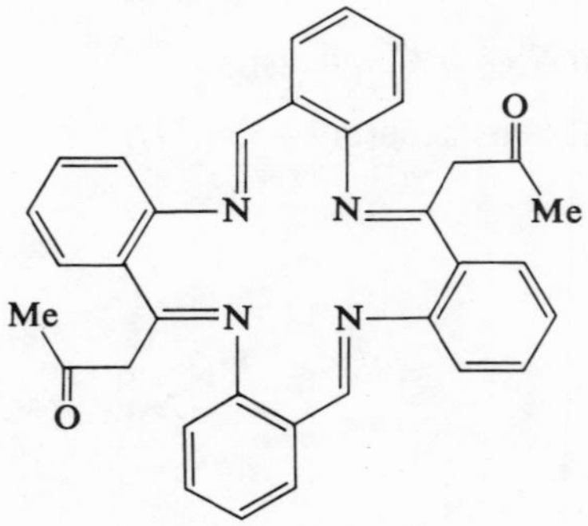

4.6

4.6a   Ni(L). $R = 0.095$*[201]
Structure similar to **4.2a**. Mean Ni–N, 190 pm.

Bzo$_2$[20]tetraeneN$_5$ Compounds

4.7 Dibenzo[c,j]-1,5,9,13,17-pentaazacycloeicosa-1,12-diene
C$_{23}$H$_3$N$_5$

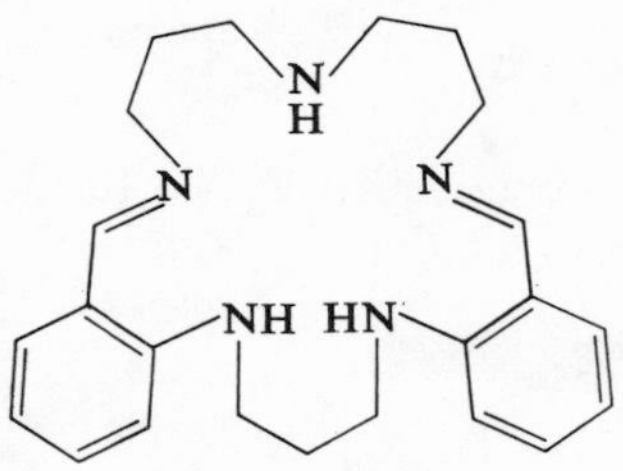

4.7

4.7a [Cu(L)](ClO$_4$)$_2$. $R = 0.0957$ for 1546 refl.* (202)

The macrocycle is in square-pyramidal coordination to Cu(II), with one secondary amino group adjacent to a benzene ring, N(5), in a site slightly distorted from apical. The carbon atoms of the six-membered chelate rings linking N(17) to N(1) and N(13), and the perchlorate oxygen atoms, are disordered. Cu–N distances in numerical sequence are 196(1), 222(1) apical, 207(1), 196(1), 206(2) pm.

Addendum to Section 7

Pyo phen[15]hexaeneN$_5$(N$_2$) Compounds

6.13 2,13-dimethyl-(1,10-phenanthrolino)[e,j]-3,4,6,9,11,12,18-septazabicyclo-octadec-1(18),2,5,7,9,12,14,16-septaene
C$_{23}$H$_{21}$N$_7$

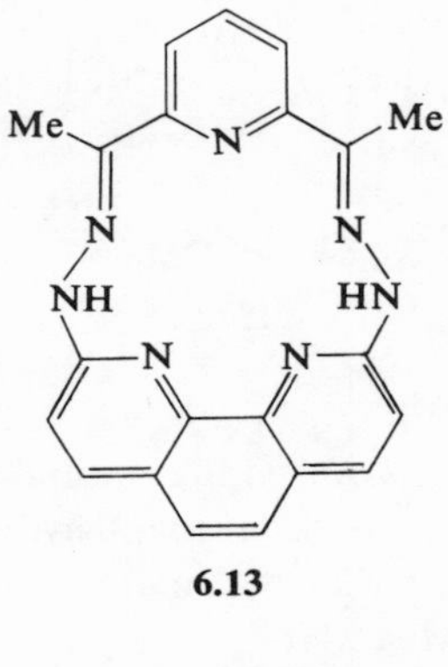

6.13

6.13a [Mn(L)Cl]BF$_4$. $R = 0.099$ for 1709 refl.** (203)

The five donor nitrogen atoms (1,3,6,9,12) are approximately coplanar (maximum deviation 9 pm) with the Mn(II) 53 pm above the plane, and with axial chloride completing a pentagonal-pyramidal arrangement (Mn–N in sequence, 217, 235, 220, 219, 231, all (1) pm, Mn–Cl, 235 pm).

[16]tetraeneN$_4$(N$_2$) Compounds

6.14   3,4,7,9,14,14,16-octamethyl-1,2,5,6,10,13-hexaazacyclo-
           octadeca-2,4,6,16-tetraene C$_{18}$H$_{34}$N$_6$

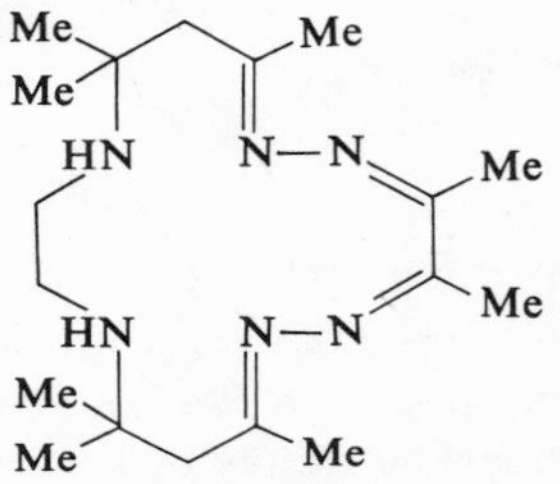

6.14

6.14a   [Ni(L)](ClO$_4$)$_2$. $R = 0.066$ for 2154 refl.[204]

The macrocycle is in planar coordination to $S = 0$ Ni(II), by N(1), N(6), N(10), and N(13), in the N-*rac* configuration. The cation has approximate C$_2$ symmetry, with the eight-membered ring in a twist conformation (cf. **6.11a**).

6.15   4-hydroxy-3,4,7,9,9,14,14,16-octamethyl-1,2,5,6,10,13-
           hexaazacyclo-octadeca-2,6,16-triene C$_{18}$H$_{36}$N$_6$O

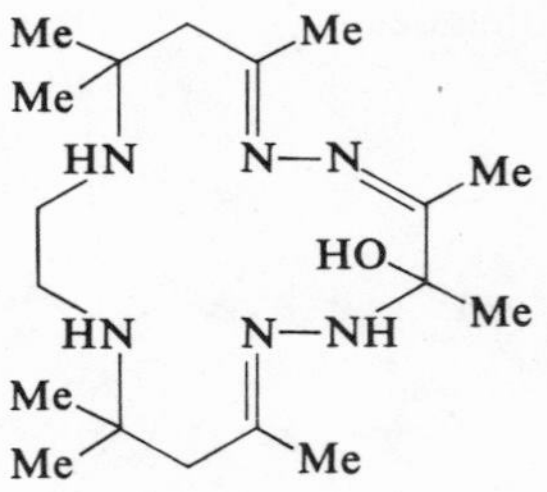

6.15

6.15a   [Ni(L)(OH$_2$)](ClO$_4$)$_2$ · 2H$_2$O. $R = 0.079$ for 3254 refl.[204]

The macrocycle, in the N-*rac* configuration, is in folded coordination by N(1), N(6), N(10), N(13), and the pendant hydroxy substituent on C(4), to $S = 1$ Ni(II), with a water mole coordinated *cis* to the hydroxy group in an octahedral arrangement. Mean Ni–N, 208 pm, Ni–O, 217 pm.

Addendum to Section 8

[12]aneO$_4$ Compounds

7.1 1,4,7,10-tetraoxacyclododecane, $C_8H_{16}O_4$
7.1 $[UO_2(L)(OH_2)_2](NO_3)_2$. $R = 0.102$ for 791 refl. (P)***[205]
The U(IV) ion is coordinated within the macrocycle. The four ether oxygen atoms [U–O, 177(12), 205(8) pm] and two water oxygen atoms [U–O, 237(8) pm] are arranged in a planar hexagon which makes an angle of 23(4)° to the linear uranyl (OUO) axis [U–O, 165(4) pm] of the centrosymmetric cation.

[14]aneS$_4$ Compounds

7.2 1,4,8,11-tetrathiacyclotetradecane, $C_{10}H_{20}S_4$
7.2h $[Hg(L)(OH_2)](ClO_4)_2$. $R = 0.116$ for 1693 refl.***[206]
The macrocycle is in endodentate coordination, with the sulfur atoms coplanar, with the Hg(II) 48 pm from this plane, and with a water molecule completing a square-pyramidal arrangement. [Hg–S, 257(5); Hg–O, 235(4) pm; SHgS angles 76(1)°, 87(1)°, and 154°] (cf. **7.2g**).

[16]aneS$_4$ Compounds

7.3 1,5,9,3-tetrathiocyclohexadecane, $C_{12}H_{24}S_4$

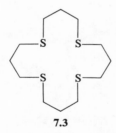

7.3

7.3h $[C_2H_5\text{—}O\text{—}Mo(L)\text{=}O \cdots Mo(L)\text{=}O](CF_3SO_3)_3$. $R = 0.085$ for 3149 refl.[207,208a]
A dinuclear structure with planar coordination of the macrocycles about each Mo(IV), with the macrocycles staggered with respect to each other. The cation has pseudo D_{4d} symmetry with linear O—Mo=O$\cdots$Mo=O axis. The mean Mo–S distance is 248(2) pm.
7.3i $[HS\text{–}Mo(L)\text{–}Mo(L)\text{–}SH](CF_3SO_3)_2 \cdot 2H_2O$. $R = 0.090$ for 2637 ref.***[208a]
7.3j $[HS\text{–}Mo(L) = O](CF_3SO_3)$. $R = 0.047$ for 2509 refl.***[208b]

[18]aneO$_6$ Compounds

7.4   1,4,7,10,14,16-hexaoxacyclo-octadecane,18-crown-6, $C_{12}H_{30}N_6$

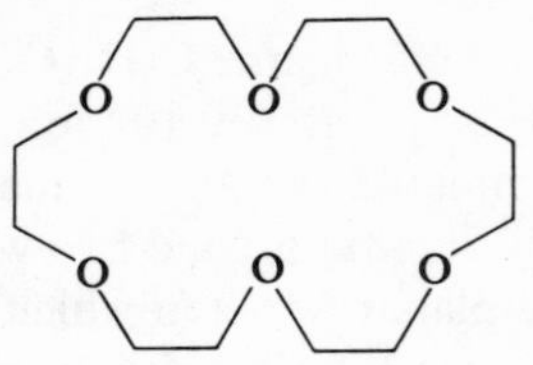

7.4

7.4a $[Mn(NO_3)(OH_2)_5]NO_3 \cdot (L) \cdot H_2O$ [209]

The Mn(II) ion is in octahedral coordination to five water oxygen atoms and one nitrate oxygen atom, with the macrocycle not coordinated.

Addendum to Section 9

[14]tetraenato(1−)N$_4$(O$_2$B) Compounds

8.10   5,6,12,13-tetramethyl-2,2-difluoro-1,3-dioxo-4,7,11,14-tetraaza-
2-boracyclotetradec-4,6,11,13-tetraenato(1−)
Cyclops, $[C_{11}H_{18}BF_2N_4O_2]^-$

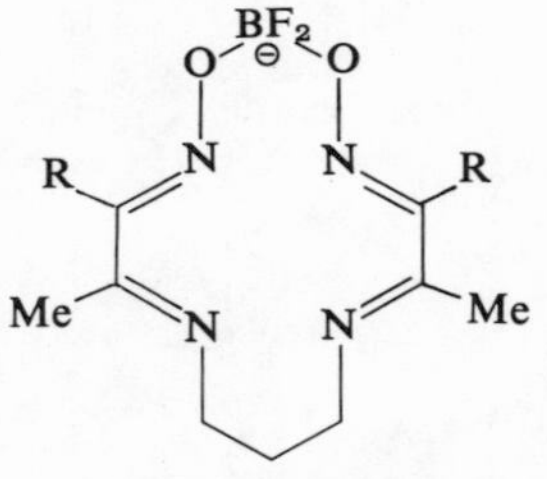

8.10 R = Me
8.10A R = Et

8.10a $[Cu(L)(NCO)]$. $R = 0.045$ for 1528 refl.***[210]

The four nitrogen atoms are approximately coplanar, with the Cu(II) 58 pm from this plane (Cu–N, 200 pm), with cyanate nitrogen coordinated in the apical site (Cu–N, 204 pm), which is unusually short for an apical interaction.

8.10b $[Cu(L)]$. $R = 0.054$ for 2882 refl.***[211a]

Distorted square-planar coordination to Cu(I).

8.10c $[Cu(L)CO]$. $R = 0.055$ for 2511 refl.***[211b]

Square-pyramidal coordination to Cu(I), with the Cu 96 pm from the N$_4$ plane, and with apically coordinated carbon monoxide, Cu–C, 178 pm.

8.10A   5,13-diethyl-6,12-dimethyl-2,2-difluoro-1,3-dioxo-4,7,11,14-
tetraaza-2-boracyclotetrac-4,6,11,13-tetraenato(-1), $[C_{13}H_{22}BF_2N_4O_2]^-$.

8.10Aa   *trans*-$[Rh(L)(Ch_3)I]$. $R = 0.031$ for 2977 refl. ***[212a]

The four nitrogen atoms of the macrocycle and the Rh(III) are close to coplanar, with methyl carbon and iodide completing an octahedral arrangement which has mirror symmetry. The boron atom lies 50 pm above the coordination plane. Distances: Rh–N, 198, Rh–C, 208; Rh–I, 281 pm.

8.10Ab *trans*[Rh(L)(SeC$_6$H$_5$)$_2$]. $R = 0.052$ for 3659 refl.***[(212b)]

[14]tetraenato(2−)N$_4$(O$_4$B$_2$) Compounds

8.11 5,6,12,13-tetramethyl-2,2,9,9-tetrafluoro-1,3,8,10-tetraoxa-4,7,11,14-tetraaza-2,9-diboracyclotetradec-3,7,10,14-tetraenato(2−), [C$_8$H$_{12}$B$_2$F$_4$N$_4$O$_4$]$^{2-}$

8.11 R = Me
8.12 R = Ph

8.11a [{Ni(L)}$_2$]. $R = 0.093$ for 964 refl.***[(213,214)]

The macrocycle is in planar coordination by the four nitrogen atoms to $S = 0$ Ni(II). Two nickel–macrocycle units are linked by a weak nickel–nickel bond (Ni–Ni, 321 pm), which lies along a C$_2$ axis, with the two macrocycle units of each dimer rotated by 92° relative to each other. The nickel and macrocycle atoms, apart from boron, are close to coplanar, with the BF$_2$ group tilted away from the other member of the dimer. The dimensions within the nickel–macrocycle unit are similar to those of the parent dimethylglyoxime compound, with Ni–N 187(1) pm and chelate angles of 84° and 97°. The nickel atoms are stacked, as for the dimethylglyoxime compound, which has Ni–N of 325 pm, but the bulky BF$_2$ groups prevent formation of the chain arrangement, and alternative Ni–Ni distances of 321 and 465 are present.

8.11b [{Ni(L)(C$_6$H$_7$N)}$_2$]. $R = 0.073$ for 3443 refl. (P)***[(214)]

This aniline adduct of the previous entry has a similar molecular structure, but with aniline nitrogen coordinated (Ni–N, 271 pm), to give an octahedral arrangement about the $S = 0$ Ni(II). The Ni–Ni distance is increased to 365 pm.

8.12 5,6,12,13-tetraphenyl-2,2,9,9-tetrafluoro-1,3,8,10-tetraoxa-4,7,11,14-tetraaza-2,9-diboracyclotetradec-3,7,10,14-tetraenato(2−), [C$_{20}$H$_{22}$B$_2$F$_2$N$_4$O$_4$]$^{2-}$

8.12a [Ni(L)(1,10-phenanthroline)]·2CH$_3$COCH$_3$. $R = 0.085$ for 2008 refl. (P)***[(215)]

The macrocycle is in folded coordination by the four nitrogen atoms to $S = 1$ Ni(II), with 1,10-phenanthroline nitrogens completing a distorted octahedral arrangement, which has twofold symmetry. The *trans* N·Ni·N angles are 106° and 152°, with chelate angles of 87° and 97°. The mean Ni–N (macrocycle) distance is 212 pm.

$Bzo_2pyo[17]pentaeneN_3S_2$ Compounds

8.13   Dibenzo[d,l]-2,15-dimethyl-6,11-dithia-3,14,20-triazabicyclo [16,3,1]tricosa-1(23),2,4,12,14,20,22-septaene $Me_2bzo_2pyo[17]pentaeneN_3(S_2)$, $C_{25}H_{25}N_3S_2$

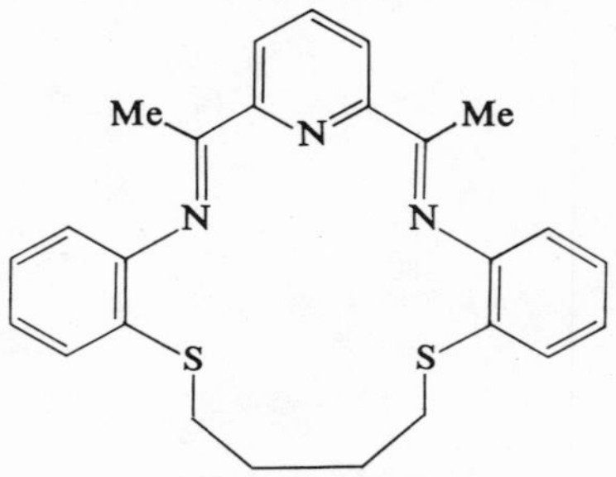

8.13

8.13a   [Zn(L)I$_2$]. $R = 0.065$ for 3262 refl.[216]

The asymmetric unit contains two molecules with similar geometries. The Zn(II) ions are in square-pyramidal coordination, with the three nitrogen atoms in basal sites, with one iodide in an equatorial site, and the other in the apical site. The basal atoms are coplanar, with Zn 57 pm above this plane. The sulfur atoms are not coordinated [Sn–N (pyridine), 210(1); Z–N (imine), 226(1); Zn–I (basal), 255; Zn–I (apical), 262 pm].

[18]aneO$_4$S$_2$ Compounds

8.14   1,4,10,13-tetraoxa-7,16-dithiacyclo-octadecane $C_{12}H_{24}O_4S_2$

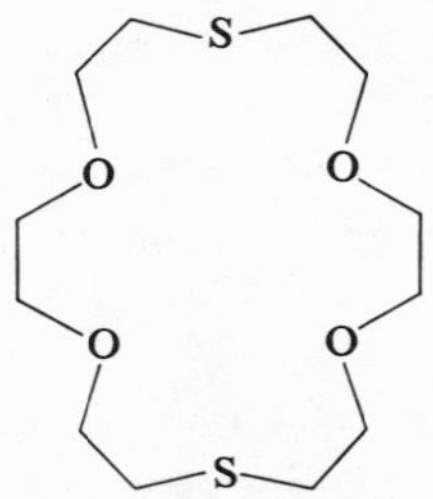

8.14

8.14a   [Pd(L)Cl$_2$]. $R = 0.036$ for 1356 refl.***[217]

The macrocycle is coordinated to Pd(II) by the two sulfur atoms (PdS,

230 pm; SPdS, 87° with two *cis* chloride ions completing a square-planar arrangement (PdCl, 231 pm; ClPdCl, 91°). The lone pairs of the oxygen atoms are directed away from the palladium ion (PdO 394–416 pm) (cf. **7.2g**, **11.6a**).

[18]aneO$_4$N$_2$ Compounds

    8.8c 1,4,10,13-tetraoxa-7,16-diazacyclo-octadecane
          C$_{12}$H$_{24}$O$_4$N$_2$
          [L]. $R = 0.044$***[218]

The eighteen-membered ring has a crown conformation with C_{2h} (approximate D_{3d}) symmetry.

    8.8d [K(L)NCS]. $R = 0.065$***[219]

The eighteen-membered ring has a crown conformation with C_i symmetry, with a potassium ion at the center of symmetry, with the K(L) units linked by NCS groups.

    8.8e [Pb(L)(SCN)$_2$]. $R = 0.047$ for 2180 refl.[220]

The macrocycle is in a crown conformation, with the four oxygen atoms and Pb(II) coplanar, and with the nitrogen atoms 59(1) pm from this plane. Sulfur-bonded thiocyanate ions complete a distorted hexagonal bipyramidal arrangement about the lead, with a twofold axis through the lead atom and the centers of the C(2)–C(3) and C(11)–C(12) bonds. The Pb–O (278 pm) and Pb–N (275 pm) distances are significantly longer than normal, indicating weak interactions of an ion–dipole nature. The Pb–S distance (289 pm) is similar to that in lead sulfide, with SPbS of 158°.

Pyr[18]trieneN$_3$O$_3$ Compounds

    8.15 3,15,21-triaza-6,9,12-trioxabicyclo[15,3,1]heneicosa-
          1(21),2,15,17,19-pentaene
          C$_{15}$H$_{21}$O$_3$N

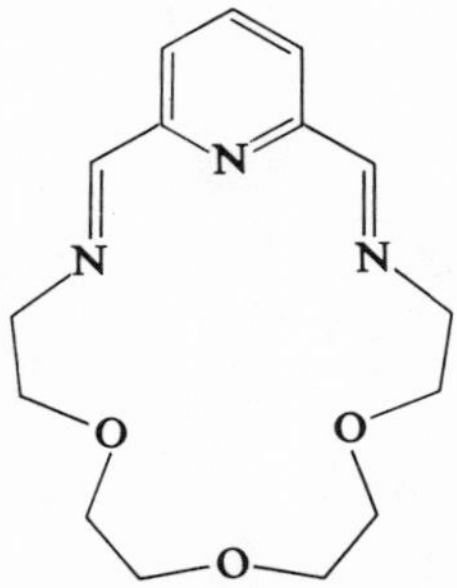

8.15

    8.15a [Ca(L)(NCS)$_2$]. $R = 0.040$ for 2084 refl.**[221]
    8.15b [Sr(L)(NCS)$_2$(H$_2$O)]. $R = 0.047$ for 1999 refl.**[221]
The metal ions are coordinated to all six donor atoms of the macrocycle.

The Ca(II) ion is located .within the mean plane of the N and O atoms, whereas the larger Sr(II) ion is 53 pm from this plane. The angle $N(21) \cdot M \cdot O(9)$ (pyridine nitrogen to central oxygen) is 146° for the Ca compound and 176° for the Sr compound. The M–N and M–O distances are similar at 264 pm for Ca and 278 pm for Sr. The coordination number of Ca is increased to 8 by *trans* coordination of two thiocyanate nitrogen atoms. The coordination number of Sr is increased to 9 by coordination of two thiocyanate nitrogen atoms on the same side of the molecular plane, and oxygen of water (Sr–O, 259 pm) on the opposite (concave) side of the molecular plane.

Addendum to Section 10

$Pyo_2[30]sexaeneN_6O_4$ Compounds

9.2 2,13,19,30-tetramethyl-6,9,23,26-tetraoxa-3,12,20,29,35,36-sexaazatricyclo[32.1.1^{14}]-hexatriacontan-1(35),2,12,14(36),15,17, 19,29,31,33-decaene
$C_{34}H_{42}N_{10}O_4$

9.2a [Pb$_2$(L)(SCN)$_4$]. $R = 0.087$ for 1286 refl.** [222]

Two Pb(II) ions are coordinated within the macrocycle; to one pyridine nitrogen, the two adjacent imine nitrogens, and two oxygen atoms, which are all approximately coplanar (maximum deviation 44 pm), with thiocyanate sulfur atoms above and below this plane completing a seven-coordinate arrangement. The two halves of the molecule are related by a twofold axis, with the two N_3O_2 coordination planes intersecting at 65°. Distances are: Pb–N, 254(3), 256(3), 247(2); Pb–O, 296(3), 288(3); Pb–S, 291(2), 300(2); Pb–Pb, 527 pm.

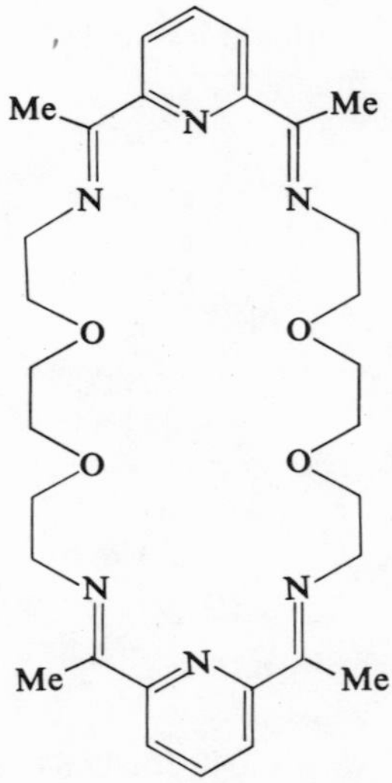

9.2

References

1. B. Bosnich, C. K. Poon, and M. L. Tobe, *Inorg. Chem.* **4**, 1102 (1965).
2. (a) A. M. Sargeson, *Trans. Metal. Chem.* **3**, 303 (1966). (b) J. K. Beattie, *Acc. Chem. Res.* **4**, 253 (1971). (c) C. J. Hawkins, *Absolute Configurations of Metal Complexes*, Wiley-Interscience, New York (1971). (d) F. A. Jurnak and K. N. Raymond, *Inorg. Chem.* **11**, 3149 (1972). (e) S. R. Niketic, K. Rasmusson, F. Woldbye, and S. Lifson, *Acta Chem. Scand.* **A30**, 497 (1976).
3. Y. Iitaka, M. Shina, and E. Kimura, *Inorg. Chem.* **13**, 2886 (1974).
4. L. Y. Martin, L. J. De Hayes, L. J. Zompa, and D. H. Busch, *J. Am. Chem. Soc.* **96**, 4046 (1974).
5. A. McPherson, M. G. Rossman, D. W. Margerum, and M. R. James, *J. Coord. Chem.* **1**, 73 (1971).
6. (a) D. J. Roger, V. H. Schievelbein, A. R. Kalyanaraman, and J. A. Bertrand, *Inorg. Chim. Acta* **6**, 307 (1972). (b) F. Hanic, F. Pavelicik, and D. Cjepesova, *Kristallografiya* **17**, 10 (1972).
7. (a) N. F. Curtis and G. W. Reader, *J. Chem. Soc. A* **1971**, 1771 (1971). (b) *J. Chem. Soc. Dalton Trans.* **1972**, 1453 (1972).
8. N. C. Payne, *Inorg. Chem.* **11**, 1376 (1972); *Inorg. Chem.* **12**, 1151 (1973).
9. J. R. Gollogly and C. J. Hawkins, *Inorg. Chem.* **8**, 1168 (1969).
10. P. O. Whimp, M. F. Bailey, and N. F. Curtis, *J. Chem. Soc. A* **1970**, 1956 (1970).
11. L. G. Warner and D. H. Busch, *J. Am. Chem. Soc.* **91**, 4092 (1969); D. H. Busch, *Co-ordination in Chemistry—Papers Presented in Honour of Professor J. C. Bailer, Jr.*, Plenum Press, New York (1969).
12. N. F. Curtis, D. A. Swann, and T. N. Waters, *J. Chem. Soc. Dalton Trans.* **1973**, 1963 (1973).
13. E. K. Barefield and F. Wagner, *Inorg. Chem.* **12**, 2435 (1973).
14. F. Wagner, M. T. Mocella, and M. J. D'Aniello, A. H. J. Wang, and E. K. Barefield, *J. Am. Chem. Soc.* **96**, 2625 (1974).
15. S. F. Mason and R. D. Peacock, *Inorg. Chim. Acta* **19**, 75 (1976) (attributed Y. Saito, personal communication).
16. J. H. Loehlin and E. B. Fleischer, *Acta Crystallogr.* **B32**, 3063 (1976).
17. D. A. House and N. F. Curtis, *J. Am. Chem. Soc.* **84**, 3248 (1962).
18. J. M. Waters and K. R. Whittle, *J. Inorg. Nucl. Chem.* **34**, 155 (1972).
19. S. M. Peng and V. L. Goedken, private communication.
20. C. Nave and M. R. Truter, *J. Chem. Soc. Dalton Trans.* **1974**, 2351 (1974).
21. D. E. Fenton, C. Nave, and M. R. Truter, *J. Chem. Soc. Chem. Commun.* **1972**, 1032 (1972).
22. J. F. Endicott, J. Lilie, J. M. Kuszaj, B. S. Ramaswamy, W. G. Schmonsees, M. G. Simic, M. D. Glick and D. P. Rillema, *J. Am. Chem. Soc.* **99**, 429 (1977).
23. T. F. Lai and C. K. Poon, *Inorg. Chem.* **15**, 1562 (1976).
24. G. B. Robertson and P. O. Whimp, personal communication.
25. B. Bosnich, R. Mason, P. J. Pauling, G. B. Robertson, and M. L. Tobe, *Chem. Commun.* **1965**, 97 (1965).
26. P. A. Tasker and L. Sklar, *Cryst. Mol. Struct.* **5**, 329 (1975).
27. R. J. Restivo, G. Ferguson, R. W. Hay, and D. P. Piplani, *J. Chem. Soc. Dalton Trans.* **1978**, 1131 (1978).
28. G. Ferguson, P. Roberts, D. Lloyd, and K. Hideg, *J. Chem. Soc. Chem. Commun.* **1977**, 149 (1977).
29. D. F. Cook, *Inorg. Nucl. Chem. Lett.* **12**, 103 (1976).
30. E. Brackett and C. Pfluger, *Proc. Amer. Cryst. Assoc., Summer Meeting*, The Pennsylvania State University, University Park, Pa. *1974*, 228 (1974).

31. N. F. Curtis, D. A. Swann, and T. N. Waters, *J. Chem. Soc. Dalton Trans.* **1973**, 1963 (1973).
32. R. A. Bauer, W. R. Robinson, and D. W. Margerum, *J. Chem. Soc. Chem. Commun.* **1973**, 289 (1973).
33. K. S. Bowman, *Inorg. Chem.* **17**, 49 (1978).
34. C. G. Pierpoint, D. N. Hendrickson, D. M. Duggan, F. Wagner, and E. K. Barefield, *Inorg. Chem.* **14**, 604 (1975).
35. M. J. D'Aniello, M. T. Mocella, F. Wagner, E. K. Barefield, and I. C. Paul, *J. Am. Chem. Soc.* **97**, 192 (1975).
36. F. Hanic and M. Serator, *Chem. Zvesti* **18**, 572 (1964); F. Hanic and D. Machajdik, *Chem. Zvesti* **23**, 3 (1969); P. Domiano, A. Musatti, and C. Pelizzi, *Cryst. Struct. Commun.* **4**, 185 (1975).
37. N. F. Curtis, *J. Chem. Soc. A* **1971**, 2834 (1971); V. L. Goedken and D. H. Busch, *Inorg. Chem.* **10**, 2679 (1971).
38. J. W. L. Martin, N. F. Curtis, and J. H. Johnston, *J. Chem. Soc. Dalton Trans.* **1978**, 68 (1978).
39. S. C. Cummings and R. E. Sievers, *Inorg. Chem.* **9**, 1131 (1970).
40. M. F. Richardson and R. E. Sievers, *J. Am. Chem. Soc.* **94**, 4134 (1972).
41. J. A. Cunningham and R. E. Sievers, *J. Am. Chem. Soc.* **95**, 7183 (1973).
42. R. A. Kolinski and D. Korybut-Daszkiewicz, *Pol. Sci.* **17**, 13 (1969); *Inorg. Chim. Acta* **14**, 237 (1975).
43. R. J. Restivo, J. Horney, and G. Ferguson, *J. Chem. Soc. Dalton Trans.* **1976**, 514 (1976).
44. J. G. Martin and S. C. Cummings, *Inorg. Chem.* **12**, 1477 (1973).
45. G. W. Roberts, S. C. Cummings, and J. A. Cunningham, *Inorg. Chem.* **15**, 2503 (1976).
46. B. Durham, T. J. Anderson, J. A. Switzer, J. F. Endicott, and M. D. Glick, *Inorg. Chem.* **16**, 271 (1977).
47. J. Krajewski, Z. Urbancyzk-Lipowska, and P. Gluzinski, *Bull. Acad. Pol. Sci. Ser. Sci.* **22**, 955 (1974).
48. A. I. Gusiev, J. W. Krajewski, and Z. Urbanczyk, *Bull. Acad. Pol. Sci. Ser. Sci. Chim.* **22**, 387 (1974).
49. G. Ferguson, R. J. Restivo, and R. W. Hay, *Acta Cryst.* **B35**, 159 (1979).
50. (a) R. A. Kolinski and Z. Kubaj, in: Proceedings of the XV International Conference on Coordination Chemistry, p. 67, Moscow, 1973. (b) R. A. Kolinski and B. Korybut-Daszkiewicz, *Inorg. Chim. Acta* **14**, 237 (1976).
51. J. Krajewski, Z. Urbanczyk-Lipowska, and P. Gluzinski, *Rocz. Chem.* **48**, 1821 (1974); *Rocz. Chem.* **51**, 2385 (1977).
52. K. Hideg and D. Lloyd, *J. Chem. Soc. C* **1971**, 3441 (1971); *Chem. Commun.* **1970**, 929 (1970).
53. G. Ferguson, personal communication.
54. (a) N. F. Curtis and R. W. Hay, *Chem. Commun.* **1966**, 524 (1966). (b) R. W. Hay, G. A. Lawrance, and N. F. Curtis, *J. Chem. Soc. Perkin Trans. I* **1977**, 591 (1977).
55. (a) N. F. Curtis, Y. M. Curtis, and H. K. J. Powell, *J. Chem. Soc. A* **1966**, 1015 (1966). (b) N. F. Curtis, *J. Chem. Soc. Dalton Trans.* **1972**, 1357 (1972); L. G. Warner, N. J. Rose, and D. H. Busch, *J. Am. Chem. Soc.* **89**, 703 (1967).
56. N. Sadasivan, J. A. Kernohan, and J. F. Endicott, *Inorg. Chem.* **6**, 770 (1967).
57. D. Wester, R. C. Edwards, and D. H. Busch, *Inorg. Chem.* **16**, 1055 (1977).
58. V. L. Goedken, J. Molin-Case, and G. G. Cristoph, *Inorg. Chem.* **12**, 2894 (1973).
59. M. D. Glick, J. M. Kusaj, and J. F. Endicott, *J. Am. Chem. Soc.* **95**, 5097 (1973).

60. M. F. Bailey and I. E. Maxwell, *J. Chem. Soc. Dalton Trans.* **1972**, 938 (1972); *Chem. Commun.* **1966**, 908 (1966).
61. F. Hanic and D. Miklos, *J. Cryst. Mol. Struct.* **2**, 115 (1972).
62. R. G. Jungst and G. D. Stucky, *Inorg. Chem.* **13**, 2404 (1974).
63. D. M. Duggan, R. G. Jungst, K. R. Mann, G. D. Stucky, and D. N. Hendrickson, *J. Am. Chem. Soc.* **96**, 3443 (1974).
64. G. R. Hedwig, J. L. Love, and H. K. K. Powell, *Aust. J. Chem.* **23**, 981 (1970).
65. P. R. Ireland and W. T. Robinson, *J. Chem. Soc. A* **1970**, 663 (1970).
66. B. T. Kilbourn, R. R. Ryan, and J. D. Dunitz, *J. Chem. Soc. A* **1969**, 2407 (1969); *Chem. Commun.* **1966**, 910 (1966).
67. (a) M. M. Blight and N. F. Curtis, *J. Chem. Soc.* **1962**, 1204 (1962). (b) N. F. Curtis, D. A. Swann, T. N. Waters, and I. E. Maxwell, *J. Am. Chem. Soc.* **91**, 4588 (1969).
68. N. F. Curtis, *J. Chem. Soc. Dalton Trans.* **1973**, 863 (1973).
69. D. A. Swann, T. N. Waters, and N. F. Curtis, *J. Chem. Soc. Dalton Trans.* **1972**, 1115 (1972).
70. S. C. Jackels, K. Farmery, E. K. Barefield, N. J. Rose, and D. H. Busch, *Inorg. Chem.* **11**, 2893 (1972).
71. M. D. Glick, W. G. Schmonsees, and J. F. Endicott, *J. Am. Chem. Soc.* **96**, 5661 (1974).
72. (a) L. P. Torre and E. C. Lingafelter, *Proc. Am. Cryst. Assoc., Winter Meeting, 1971*, 62 (1971) (b) L. P. Torre, *Diss. Abstr.* **32**, 2641-B (1971).
73. N. F. Curtis, *J. Chem. Soc. A* **1971**, 2834 (1971); V. L. Goedken and D. H. Busch, *Inorg. Chem.* **10**, 2679 (1971).
74. I. E. Maxwell and M. F. Bailey, *J. Chem. Soc. Dalton Trans.* **1972**, 935 (1972); *Chem. Commun.* **1966**, 883 (1966).
75. (a) T. J. Truex and R. H. Holm, *J. Am. Chem. Soc.* **94**, 4529 (1972). (b) M. Millar and R. H. Holm, *J. Am. Chem. Soc.* **97**, 6052 (1975). (c) *J. Chem. Soc. Chem. Commun.* **1975**, 169 (1975).
76. S. M. Peng, J. A. Ibers, M. Millar, and R. H. Holm, *J. Am. Chem. Soc.* **98**, 8037 (1976).
77. P. W. R. Corfield, J. D. Moskren, C. J. Hipp and D. H. Busch, *J. Am. Chem. Soc.* **95**, 4465 (1973).
78. J. F. Myers and N. J. Rose, *Inorg. Chem.* **12**, 1238 (1973).
79. J. F. Meyers and C. H. L. Kennard, *J. Chem. Soc. Chem. Commun.* **1972**, 77 (1972).
80. K. Bowman, D. P. Riley, D. H. Busch, and P. W. R. Corfield, *J. Am. Chem. Soc.* **97**, 5036 (1975).
81. J. D. Curry and D. H. Busch, *J. Am. Chem. Soc.* **86**, 592 (1964).
82. R. L. Rich and G. I. Stucky, *Inorg. Nucl. Chem. Lett.* **1**, 85 (1965).
83. J. Karn and D. H. Busch, *Nature* **211**, 160 (1966).
84. J. L. Karn and D. H. Busch, *Inorg. Chem.* **8**, 1169 (1969).
85. G. R. Newkome, J. D. Sauer, J. M. Roper, and D. C. Hager, *Chem. Rev.* **77**, 513 (1977).
86. R. Dewar and E. B. Fleischer, *Nature* **222**, 372 (1969).
87. E. B. Fleischer and S. W. Hawkinson, *Inorg. Chem.* **11**, 2312 (1968).
88. R. H. Prince, D. A. Stotter, and P. R. Woolley, *Inorg. Chim. Acta* **9**, 51 (1974).
89. M. R. Caira, L. R. Nassibeni, and P. R. Woolley, *Acta Crystallogr.* **B31**, 1334 (1975).
90. M. G. B. Drew, A. H. bin Othman, S. G. McFall, and S. M. Nelson, *J. Chem. Soc. Chem. Commun.* **1975**, 818 (1975).
91. M. G. B. Drew, A. H. bin Othman, S. G. McFall, and S. M. Nelson, *J. Chem. Soc. Chem. Commun.* **1977**, 558 (1977).

92. (a) M. G. B. Drew, A. H. bin Othman, and S. M. Nelson, *J. Chem. Soc. Dalton Trans.* **1976**, 1394 (1976). (b) M. G. B. Drew, A. H. bin Othman, W. E. Hill, P. McIlroy, and S. M. Nelson, *Inorg. Chim. Acta* **12**, L25 (1975).

93. M. G. B. Drew, A. H. bin Othman, P. D. A. McIlroy, and S. M. Nelson, *Acta Crystallogr.* **B32**, 1029 (1976).

94. M. G. B. Drew, A. H. bin Othman, P. D. A. McIlroy, and S. M. Nelson, *J. Chem. Soc. Dalton Trans.* **1975**, 2507 (1975).

95. E. Fleischer and S. W. Hawkinson, *J. Am. Chem. Soc.* **89**, 720 (1967).

96. M. G. B. Drew and S. M. Nelson, *Acta Crystallogr.* **A31**, S140 (1975).

97. N. W. Alcock, D. C. Liles, M. McPartlin, and P. A. Tasker, *J. Chem. Soc. Chem. Commun.* **1974**, 727 (1974).

98. S. M. Nelson, S. G. McFall, M. G. B. Drew, and A. H. bin Othman, *J. Chem. Soc. Chem. Commun.* **1977**, 370 (1977).

99. S. M. Nelson, S. G. McFall, M. G. B. Drew, A. H. bin Othman, and N. B. Mason, *J. Chem. Soc. Chem. Commun.* **1977**, 167 (1977).

100. M. G. B. Drew, A. H. bin Othman, S. G. McFall, P. D. A. McIlroy, and S. M. Nelson, *J. Chem. Soc. Dalton Trans.* **1977**, 438 (1977).

101. (a) R. H. Holm and M. J. O'Connor, *Progr. Inorg. Chem.* **14**, 241 (1971). (b) M. D. Hobday and T. D. Smith, *Coord. Chem. Rev.* **9**, 311 (1972). (c) H. S. Maslen and T. N. Waters, *Coord. Chem. Rev.* **17**, 137 (1975). (d) M. Calligaris, G. Nardin, and L. Randaccio, *Coord. Chem. Rev.* **7**, 385 (1971).
    Chem. **4**, 199 (1968).

102. G. A. Melson and D. H. Busch, *J. Am. Chem. Soc.* **87**, 1706 (1965).

103. G. A. Melson and D. H. Busch, *Proc. Chem. Soc.* **1963**, 223 (1963).

104. R. M. Wing and R. Eiss, *J. Am. Chem. Soc.* **92**, 1929 (1970).

105. E. B. Fleischer and E. Klem, *Inorg. Chem.* **4**, 637 (1965).

106. S. W. Hawkinson and E. B. Fleischer, *Inorg. Chem.* **8**, 2402 (1969).

107. E. N. Maslen, L. M. Engelhardt, and A. H. White, *J. Chem. Soc. Dalton Trans.* **1974**, 1799 (1974).

108. G. S. Hall and R. H. Soderberg, *Inorg. Chem.* **7**, 2300 (1968).

109. R. C. Elder, D. Koran, and H. B. Mark, *Inorg. Chem.* **13**, 1644 (1974).

110. V. L. Goedken, J. J. Pluth, S. M. Peng, and B. Bursten, *J. Am. Chem. Soc.* **98**, 8014 (1976).

111. M. C. Weiss, B. Bursten, S. M. Peng, and V. L. Goedken, *J. Am. Chem. Soc.* **98**, 8021 (1976).

112. V. L. Goedken, S. M. Peng, J. Molin-Norris, and Y. Park, *J. Am. Chem. Soc.* **98**, 8391 (1976).

113. J. P. Collman, J. L. Hoard, N. Kim, G. Lang, and C. A. Reed, *J. Am. Chem. Soc.* **97**, 2676 (1976).

114. J. F. Kirner, W. Dow, and W. R. Scheidt, *Inorg. Chem.* **16** (1977).

115. R. G. Little, J. A. Ibers, and J. E. Baldwin, *J. Am. Chem. Soc.* **97**, 7049 (1975).

116. M. C. Weiss, G. Gordon, and V. L. Goedken, *Inorg. Chem.* **16**, 305 (1977).

117. F. Hanic, M. Handlovic, and O. Lindgren, *Coll. Czech. Chem. Commun.* **37**, 2119 (1972).

118. E. G. Jäger, *Z. Anorg. Allgem. Chem.* **364**, 177 (1967); F. Hiller, P. Dimroth, and H. P. Fitzner, *Chem. Ber.* **137**, 717 (1968).

119. V. L. Goedken, S. M. Peng, and Y. Park, *J. Am. Chem. Soc.* **96**, 284 (1974).

120. M. C. Weiss and V. L. Goedken, *J. Chem. Soc. Chem. Commun.* **1976**, 531 (1976).

121. M. C. Weiss, G. C. Gordon, and V. L. Goedken, *J. Am. Chem. Soc.* **101**, 857 (1979).

122. V. L. Goedken and S. M. Peng, *J. Am. Chem. Soc.* **95**, 5773 (1973).

123. G. R. Clark, B. W. Skelton, and T. N. Waters, *J. Chem. Soc. Dalton Trans.* **1976**, 1528 (1976).
124. W. E. Keyes, J. B. R. Dunn, and T. M. Loehr, *J. Am. Chem. Soc.* **99**, 4527 (1977); W. E. Keyes, W. E. Swartz, and T. M. Loehr, *Inorg. Chem.* **17**, 3316 (1978).
125. S. M. Peng and V. L. Goedken, *Inorg. Chem.* **17**, 820 (1978).
126. V. L. Goedken and S. M. Peng, *J. Am. Chem. Soc.* **96**, 7826 (1974).
127. G. C. Gordon, S. M. Peng, and V. L. Goedken, *Inorg. Chem.* **17**, 3578 (1978).
128. V. L. Goedken and S. M. Peng, *J. Chem. Soc. Chem. Commun.* **1974**, 914 (1974).
129. G. C. Gordon, S. M. Peng, and V. L. Goedken, *Inorg. Chem.* **17**, 3578 (1978).
130. V. L. Goedken and S. M. Peng, *J. Chem. Soc. Chem. Commun.* **1975**, 258 (1975).
131. S. M. Peng and V. L. Goedken, *J. Am. Chem. Soc.* **98**, 8500 (1976).
132. J. E. Baldwin and J. Huff, *J. Am. Chem. Soc.* **95**, 5757 (1973).
133. R. G. Little, J. A. Ibers, and J. E. Baldwin, *J. Am. Chem. Soc.* **97**, 7049 (1975).
134. V. L. Goedken, Y. Park, S. M. Peng, and J. M. Norris, *J. Am. Chem. Soc.* **96**, 7693 (1974).
135. N. F. Curtis, J. S. de Courcy, and T. N. Waters, to be published.
136. Z. P. Hasque, D. C. Liles, M. McPartlin, and P. A. Tasker, *Inorg. Chim. Acta* **23**, L21 (1977).
137. N. W. Alcock, N. Herron, and P. Moore, *J. Chem. Soc. Chem. Commun.* **1976**, 886 (1976).
138. P. Groth, *Acta Chem. Scand.* **A30**, 155 (1976).
139. M. D. Glick, D. P. Gavel, L. L. Diaddario, and D. B. Rorabacher, *Inorg. Chem.* **15**, 1190 (1976).
140. F. P. van Remootere, F. P. Boer, and E. C. Steiner, *Acta Crystallogr.* **B31**, 1420 (1975).
141. R. E. De Simone and M. D. Glick, *J. Am. Chem. Soc.* **98**, 762 (1976).
142. P. H. Davis, K. L. White, and R. L. Bedford, *Inorg. Chem.* **14**, 1753 (1975).
143. E. R. Dockal, L. L. Diadario, M. D. Glick, and D. B. Rorabacher, *J. Am. Chem. Soc.* **99**, 4530 (1977).
144. R. E. De Simone and M. D. Glick, *J. Am. Chem. Soc.* **97**, 942 (1974).
145. R. J. Geue, M. R. Snow, J. Springborg, A. J. Herlt, A. M. Sargeson, and D. Taylor, *J. Chem. Soc. Chem. Commun.* **1976**, 285 (1976).
146. R. W. Kluiber and G. Sasso, *Inorg. Chim. Acta* **4**, 226 (1970).
147. D. L. Johnston and W. D. Horrocks, *Inorg. Chem.* **10**, 687 (1971).
148. R. A. Lalancette, D. J. Macchia, and W. F. Furey, *Inorg. Chem.* **15**, 548 (1976).
149. (a) N. W. Alcock and P. A. Tasker, *J. Chem. Soc. Chem. Commun.* **1972**, 1239 (1972). (b) P. B. Donaldson, P. A. Tasker, and N. W. Alcock, *J. Chem. Soc. Dalton Trans.* **1976**, 2262 (1976).
150. P. B. Donaldson, P. Haria, and P. A. Tasker, *J. Chem. Soc. Dalton Trans.* **1976**, 2382 (1976).
151. L. G. Armstrong, L. F. Lindoy, M. McPartlin, G. M. Mockler, and P. A. Tasker, *Inorg. Chem.* **16**, 1665 (1977).
152. P. R. Louis, B. Metz, and R. Weiss, *Acta Crystallogr.* **B30**, 774 (1974).
153. P. R. Louis, D. Pelissard, and R. Weiss, *Acta Crystallogr.* **B30**, 1889 (1974).
154. P. R. Louis, D. Pelissard, and R. Weiss, *Acta Crystallogr.* **B32**, 1480 (1976).
155. (a) P. R. Louis, F. Arnaud, R. Weiss, and M. J. Schwing-Weill, *Inorg. Nucl. Chem. Lett.* **13**, 31 (1977). (b) P. R. Louis, Y. Agnus, and R. Weiss, *Acta Crystallogr.* **B33**, 1418 (1977). (c) M. G. B. Drew, A. H. bin Othman, S. G. McFall, P. D. A. McIlroy, and S. M. Nelson, *J. Chem. Soc. Dalton Trans.* **1977**, 1173 (1977).
156. (a) P. M. Herceg and R. Weiss, *Acta Crystallogr.* **B29**, 542 (1973); (b) P. M. Herceg and R. Weiss, *Inorg. Nucl. Chem. Lett.* **6**, 435 (1970); (c) P. M. Herceg and R. Weiss, *Rev. Chim. Mineral* **10**, 509 (1973); *Chem. Abstr.* **80**, 88316t (1974).

157. P. A. Tasker and E. B. Fleischer, *J. Am. Chem. Soc.* **92**, 7072 (1970).

158. B. F. Hoskins and G. A. Williams, *Austral. J. Chem.* **28**, 2607 (1975).

159. B. F. Hoskins and G. A. Williams, *Austral. J. Chem.* **28**, 2593 (1975).

160. B. F. Hoskins, R. Robson, and G. A. Williams, *Inorg. Chim. Acta* **16**, 121 (1976).

161. B. F. Hoskins, N. J. McLeod, and H. A. Schaap, *Austral. J. Chem.* **29**, 515 (1976).

162. (a) N. L. Paddock, *Quart. Rev.* **18**, 168 (1964). (b) R. A. Shaw, B. Fitzsimmons, and B. C. Smith, *Chem. Rev.* **62**, 247 (1962).

163. J. Trotter and S. H. Whitlow, *J. Chem. Soc. A* **1970**, 455 (1970).

164. J. Trotter and S. H. Whitlow, *J. Chem. Soc. A* **1970**, 460 (1970).

165. (a) H. P. Calhoun, N. L. Paddock, and J. Trotter, *J. Chem. Soc. Dalton Trans.* **1973**, 2708 (1973). (b) R. W. Allen, J. P. O'Brien, and H. R. Allcock, *J. Am. Chem. Soc.* **99**, 3987 (1977).

166. A. J. Wagner and A. Vos, *Acta Crystallogr.* **B24**, 1423 (1968).

167. (a) W. Harrison and J. Trotter, *J. Chem. Soc. Dalton Trans.* **1973**, 61 (1973). (b) W. C. Marsh, N. L. Paddock, C. J. Stewart, and J. Trotter, *J. Chem. Soc. Chem. Commun.* **1970**, 1190 (1970); W. C. Marsh and J. Trotter, *J. Chem. Soc. A* **1971**, 1482 (1971). (c) E. B. Fleischer, A. E. Gebala, D. R. Swift, and P. A. Tasker, *Inorg. Chem.* **11**, 2775 (1972); P. B. Donaldson, P. A. Tasker, and N. W. Alcock, *J. Chem. Soc. Dalton Trans.* **1977**, 1160 (1977).

168. (a) L. J. Wilson and N. J. Rose, *J. Am. Chem. Soc.* **90**, 6041 (1968). (b) C. Meali and E. C. Lingafelter, *Chem. Commun.* **1970**, 885 (1970).

169. D. A. Durham, F. A. Hart, and D. Shaw, *J. Inorg. Nucl. Chem.* **29**, 509 (1967).

170. D. R. Boston and N. J. Rose, *J. Am. Chem. Soc.* **90**, 6859 (1968).

171. M. Dunaj-Jurco and E. C. Lingafelter, in: Proceedings of the Seminar on Crystallochemistry and Coordination of Metallorganic Compounds, Bratislava, Czechoslovakia, 1973 [*Chem. Abstr.* **80**, 137987y (1974)].

172. G. A. Zakrewski, C. A. Ghilardi, and E. C. Lingafelter, *J. Am. Chem. Soc.* **93**, 4411 (1971).

173. M. R. Churchill and A. H. Reiss, *Inorg. Chem.* **11**, 2299 (1972).

174. M. R. Churchill and A. H. Reiss, *J. Chem. Soc. Dalton Trans.* **1973**, 1570 (1973).

175. M. R. Churchill and A. H. Reiss, *Inorg. Chem.* **11**, 1811 (1972).

176. M. R. Churchill and A. H. Reiss, *Inorg. Chem.* **12**, 2280 (1973).

177. V. L. Goedken and S. M. Peng, *J. Chem. Soc. Chem. Commun.* **1973**, 62 (1973).

178. V. L. Goedken, S. M. Peng, and J. J. Pluth, private communication.

179. I. I. Creaser, J. MacB. Harrowfield, A. J. Herlt, A. M. Sargeson, J. Springborg, R. J. Geue, and M. R. Snow, *J. Am. Chem. Soc.* **99**, 3181 (1977).

180. F. Mathieu and R. Weiss, *J. Chem. Soc. Chem. Commun.* **1973**, 386 (1973).

181. P. R. Louis, J. C. Thierry, and R. Weiss, *Acta Crystallogr.* **B30**, 753 (1974).

182. L. J. Zompa and T. N. Margulis, *Inorg. Chim. Acta* **28**, L157 (1978).

183. T. Sakurai, K. Kobayashi, K. Tsuboyama, and S. Tsuboyama, *Acta Crystallogr.* **B34**, 1144 (1978).

184. J. F. Endicott, D. Halko, T. S. Roche, M. D. Glick, and W. Bulter, *Proc. Int. Conf. Coord. Chem.* **343** (1974); *Chem. Abstr.* **86**, 42726 (1977).

185. J. W. Krajewski, Z. Urbanczyk-Lipkowska, and P. Gluzinski, *Bull. Acad. Pol. des Sci. Ser. Sci. Chem.* **25**, 853 (1977).

186. J. W. Krajewski, Z. Urbanczyk-Lipkowski, and P. Gluzinski, *Pol. J. Chem.* **52**, 1513 (1978).

187. J. W. Krajewski, Z. Urbanczyk-Lipkowski, and P. Gluzinski, *Bull. Acad. Pol. des Sci. Ser. Sc. Chem.* **15**, 939 (1977).

188. J. W. Krajewski, Z. Urbanczyk-Lipowska, and P. Gluzinski, *Cryst. Struct. Comm.* **6**, 817 (1977).

189. Ei-Ichiro Ochiai, S. J. Rettig, and J. Trotter, *Can. J. Chem.* **56**, 267 (1978).
190. R. M. Clay, P. Murray-Rust, and J. Murray-Rust, private communication.
191. R. W. Hay, C. R. Clark, *J. Chem. Soc. Dalton Trans.* **1977**, 1148 (1977).
192. W. L. Smith and J. D. Ekstrait, *175th Amer. Chem. Soc. Meeting, Abstr. Inor.*, 184 (1978).
193. J. W. L. Martin, J. H. Johnston, and N. F. Curtis, *J. Chem. Soc. Dalton Trans.* **1978**, 68 (1978).
194. H. W. Smith, G. W. Svetich, and E. C. Lingafelter, *Amer. Cryst. Assoc., Summer Meeting 1973, Abstract* 174 (1973).
195. M. G. B. Drew and S. Hollis, *Inorg. Chim. Acta* **29**, L231 (1978).
196. M. G. B. Drew, S. Hollis, S. G. McFall, and S. M. Nelson, *J. Inorg. Nucl. Chem.* **40**. 1956 (1978).
197. M. G. B. Drew, A. H. bin Othman, S. G. McFall, and S. M. Nelson, *J. Chem. Soc, Chem. Commun.* **1977**, 558 (1977).
198. S. M. Nelson, S. G. McFall, M. G. B. Drew, and A. H. bin Othman, *J. Chem. Soc. Chem. Commun.* **1977**, 370 (1977).
199. J. C. Speakman, *Acta Crystallogr.* **6**, 784 (1953).
200. D. Losman, L. M. Engelhardt, and M. Green, *Inorg. Nucl. Chem. Letters* **9**, 791 (1973).
201. B. Kamenar and B. Kaitner, *Izv. Jugosl. Cent. Kristafiju. Ser. A* **78**, 379 (1976).
202. C. Griggs, M. Hasan, K. F. Henrick, R. W. Mathews, and P. A. Tasker, *Inorg. Chim. Acta* **25**, L29 (1977).
203. M. M. Bishop, J. Lewis, T. D. O'Donoghue, and P. Raithby, *J. Chem. Soc. Chem. Commun.* **1978**, 476 (1978).
204. A. R. Davis, F. W. B. Einstein, and N. F. Curtis, to be published.
205. N. Armaguan, *Acta Crystallogr.* **B33**, 2281 (1977).
206. N. W. Alcock, N. H. Herron, and P. Moore, *J. Chem. Soc. Dalton Trans.* **1978**, 394 (1978).
207. R. E. DeSimone, M. D. Glick, and J. Craigel, *174th A.C.S. Meeting, Abstract Inorg.*, 154 (1977).
208. (a) J. Cragel, V. B. Petts, M. D. Glick, and R. E. DeSimone, *Inorg. Chem.* **17**, 2885 (1978). (b) R. E. DeSimone and M. D. Glick, *Inorg. Chem.* **17**, 3574 (1978).
209. A. Knüchel, J. Kopf, J. Oehler, and G. Rudolph, *Inorg. Nucl. Chem. Lett.* **14**, 61 (1978).
210. O. P. Anderson and J. C. Marshall, *Inorg. Chem.* **17**, 1258 (1978).
211. (a) R. R. Gagné, J. L. Allison, and G. C. Lisensky, *Inorg. Chem.* **17**, 3563 (1978). (b) R. R. Gagné, J. L. Allison, R. S. Gall, and C. A. Koval, *J. Am. Chem. Soc.* **99**, 7170 (1977).
212. (a) J. P. Collman, P. A. Christian, S. Current, P. Denisevich, T. R. Halbert, E. R. Schmittou, and K. O. Hodgson, *Inorg. Chem.* **15**, 223 (1978). (b) J. P. Collman, R. K. Ruthcock, J. P. Sen, T. D. Tullius, and K. O. Hodgson, *Inorg. Chem.* **11**, 2947 (1976).
213. F. S. Stephens and R. S. Vagg, *Acta Crystallogr.* **B33**, 3159 (1977).
214. A. J. Charlson, F. S. Stephens, R. S. Vagg, and E. C. Watton, *Inorg. Chem. Acta* **25**, L51 (1977), R. S. Vagg and E. C. Watton, *Acta Crystallogr.* **B34**, 2715 (1978).
215. F. S. Stephens and R. S. Vagg, *Acta Crystallogr.* **B33**, 3165 (1977).
216. M. G. B. Drew and S. Hollis, *J. Chem. Soc. Dalton Trans.* **1978**, 511 (1978).
217. B. Metz, D. Moras, and R. Weiss, *J. Inorg. Nucl. Chem.* **36**, 785 (1974).
218. M. Hercog and R. Weiss, *Bull. Soc. Chim. Fr.* **1972**, 549 (1972).
219. D. Moras, B. Metz, M. Herces, and R. Weiss, *Bull. Soc. Chim. Fr.* **1972**, 551 (1972).
220. B. Metz and R. Weiss, *Acta Crystallogr.* **B29**, 1088 (1973).

221. D. E. Fenton, D. H. Cook,P. I. W. Norwell, and E. Walker, *J. Chem. Soc. Chem. Commun.* **1978**, 279 (1978).
222. M. G. B. Drew, A. Rodgers, M. McCann, and S. M. Nelson, *J. Chem. Soc. Chem. Commun.* **1978**, 415 (1978).

5

Ligand Field Spectra and Magnetic Properties of Synthetic Macrocyclic Complexes

F. L. Urbach

1. Introduction

Investigations of ligand field spectra, magnetic susceptibilities, and electron spin resonance provide the basis for the assignment of electronic configurations of macrocyclic complexes with transition-metal ions. This chapter reviews those studies that have provided detailed interpretations of the electronic properties of the synthetic macrocyclic complexes. Studies of the naturally occurring macrocycles, e.g., porphyrins or corrins, or the closely related phthalocyanines, are excluded, as they are treated elsewhere in this volume. Several earlier reviews [1-4] have included information about the electronic structures of the synthetic macrocyclic complexes.

Macrocyclic ligands offer several important advantages for spectroscopic and magnetic studies of transition-metal complexes. Considerable use has been made of the ability of tetradentate macrocycles to maintain a constant planar stereochemistry about a metal ion while additional noncyclic axial ligands are varied systematically. These studies of the influence of tetragonal geometries on the electronic configurations of metal ions have been greatly aided by macrocyclic ligands. [1,4-7] Macrocyclic ligands normally produce stronger ligand fields than comparable nonmacrocyclic ligands owing to a constraining or constrictive effect. [5-7] This property often leads to ground state electronic configurations which would not occur in the absence of the macrocyclic ligand. Less frequent, but of considerable importance, are those

---

F. L. Urbach · Department of Chemistry, Case Western Reserve University, Cleveland, Ohio 44106.

cases where the constraints imposed on the metal ion are sufficiently strong to produce distorted, strained, or unusual coordination geometries. Elegant examples of ligand-imposed geometries are the distorted octahedral and trigonal prismatic coordination of certain clathrochelate complexes [8-10] and the pentagonal-bipyramidal structures [11] achieved with planar pentagonal macrocycles.

The dramatically enhanced metal-ion binding constants for macrocyclic ligands (the macrocyclic effect) [12,13] provide a strong likelihood that macrocyclic complexes maintain their coordination geometry in solution. This effect simplifies solution behavior by limiting the number of complex species in equilibrium. The strong binding of macrocyclic complexes often allows a dependable correlation between solid state structures as determined by X-ray crystallography and the stereochemistry exhibited by the complex in solution. The presence of a macrocyclic ligand normally allows the stereochemistry of a metal complex to be maintained during changes in the oxidation state of the metal ion, although there are notable exceptions [14] to this behavior. This decreased lability of macrocyclic ligands often assists in the stabilization of metal ions in unusual or reactive oxidation states.

In the following sections the published reports that deal with detailed studies of the ligand field parameters of transition-metal complexes with synthetic macrocyclic ligands are reviewed. The presentation is arranged by individual metal ions, subdivided according to oxidation state and spin state, where appropriate. All of these studies have been done with metal ions in the latter half of the first transition series.

2. Nickel Complexes

2.1. Nickel(II) Macrocyclic Complexes

More complexes of synthetic macrocyclic ligands have been made with nickel(II) than with any other metal ion. This preference for nickel(II) reflects, in part, the ability of this ion to act as a metal template for ligand-forming reactions, the lack of interfering redox or hydrolysis reactions, and a relatively well understood ligand field description of the various coordination geometries of nickel(II). Earlier reviews [1,4] have discussed the ligand field spectra of nickel(II) macrocycles with an emphasis on the interpretation of the spectra of high-spin, tetragonal complexes with a planar macrocyclic ligand and two axial donors. The model for tetragonal $3d^8$ complexes is analogous to that derived [15] for cobalt(III) $(3d^6)$ complexes in D_{4h} symmetry and is illustrated in Figure 1.

Normally only the splitting of the lowest-energy octahedral band $(^3T_{2g} \leftarrow {}^3A_{2g})$ is observed with $^3B_{2g} \leftarrow {}^3B_{1g}(\nu_2)$ being indicative of the in-plane ligand field strength, Dq^{xy}:

$$\nu_2 = 10Dq^{xy}$$

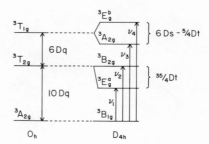

FIGURE 1. Ligand field diagram for a d^8 ion in O_h and D_{4h} symmetries.

The transition to $^3E_g{}^a(\nu_1)$ is a measure of the tetragonal splitting and the strength of the axial ligands:

$$\nu_1 = 10Dq - (35/4)Dt$$

The tetragonal splitting parameter is given by

$$Dt = (4/35)(\nu_2 - \nu_1)$$

and is positive when the axial donors are weaker than the macrocycle donor. Dq^z may be calculated for the out-of-plane ligands from

$$Dq^z = (2\nu_1 - \nu_2)/10$$

The above treatment utilizes only ν_1 and ν_2 to derive the spectrochemical parameters Dq^{xy}, Dq^z, and Dt. Recently, a very complete study[6] has been published, which utilizes a complete fit of the triplet–triplet spectrum of tetragonal nickel(II) complexes with both the weak-field[16] and strong-field[17,18] theoretical descriptions. Spectra for several new tetragonal nickel(II) complexes are reported, and some previously published spectra are reexamined and their spectrochemical parameters are revised.

For nickel(II) complexes with a strong in-plane field and no axial donors, spin-paired, square-planar complexes are formed. These complexes are characterized by magnetic moments of 0.5–0.6 B.M., attributed to temperature-independent paramagnetism (TIP), and by an absorption band at ~ 450 nm ($\varepsilon \sim 100$ liter mol^{-1} cm^{-1}). This characteristic band is usually assigned as $^1A_{2g} \leftarrow {}^1A_{1g}(D_{4h})$ or $^1B_{1g} \leftarrow {}^1A_{1g}(D_{2h})$ following Maki,[19] although several $d \rightarrow d$ transitions are anticipated for a square-planar Ni(II) chromophore and this assignment is less than satisfactory.

2.1.1. Ni[12-16]aneN₄ Complexes and Substituted Derivatives

Representative saturated N_4 macrocyclic ligands that have yielded nickel(II) complexes are illustrated in **1–14**. The smallest of these macrocycles, [12]aneN₄ **(1)**[20] and N-bzl₄[12]aneN₄ **(2)**,[21] cannot encompass[20-22] the

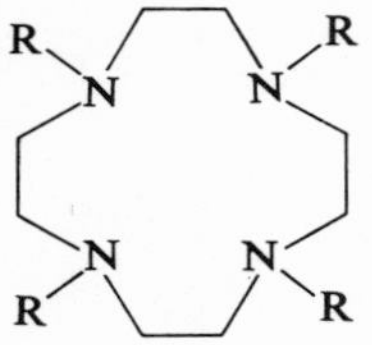

1 [12]aneN$_4$
  (R = H)
2 N-Bzl$_4$[12]aneN$_4$
  (R = C$_6$H$_5$CH$_2$)

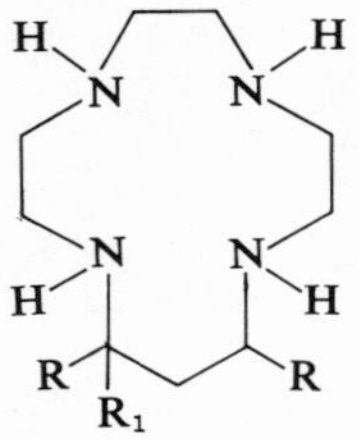

3 [13]aneN$_4$
  (R = R$_1$ = H)
4 8,10-Me$_2$[13]aneN$_4$
  (R = CH$_3$; R$_1$ = H)
5 8,8,10-Me$_3$[13]aneN$_4$
  (R = R$_1$ = CH$_3$)

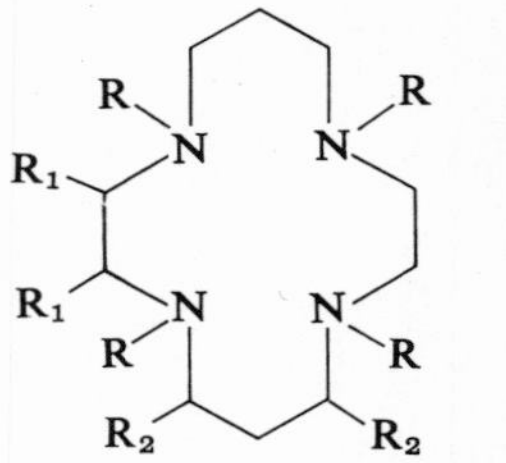

6 [14]aneN$_4$
  (R = R$_1$ = R$_2$ = H)
7 N-Me$_4$[14]aneN$_4$
  (R = CH$_3$; R$_1$=R$_2$=H)
8 2,3-Me$_2$[14]aneN$_4$
  (R = R$_2$ = H;
  R$_1$ = CH$_3$)
9 5,7-Me$_2$[14]aneN$_4$
  (R = R$_1$ = H;
  R$_2$ = CH$_3$)

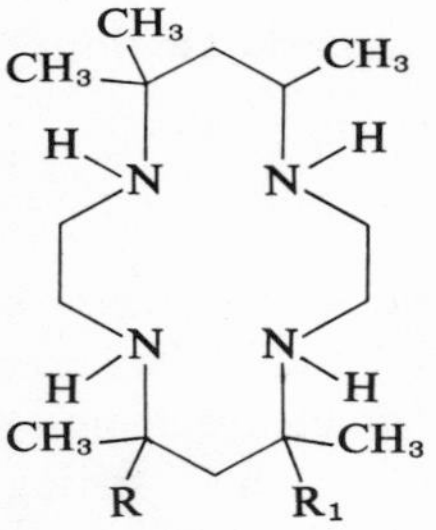

10 trans-Me$_6$[14]aneN$_4$
  (R = H; R$_1$ = CH$_3$)
11 cis-Me$_6$[14]aneN$_4$
  (R = CH$_3$; R$_1$ = H)

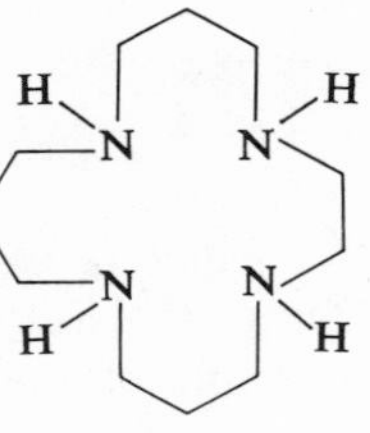

12 [15]aneN$_4$

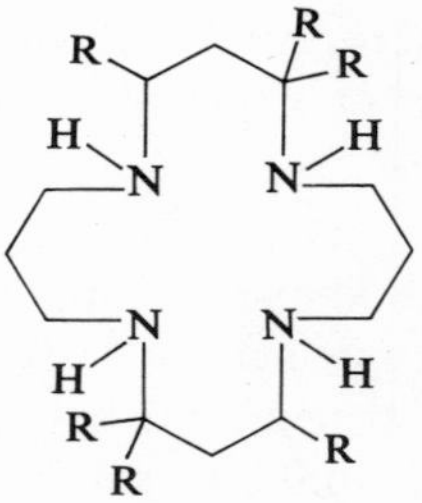

13 [16]aneN$_4$
  (R = H)
14 Me$_6$[16]aneN$_4$
  (R = CH$_3$)

metal ion in a square-planar arrangement and form only high-spin five- and six-coordinate complexes with the ligands in a *cis*-folded arrangement. The ligands [13–16]aneN$_4$ and their substituted derivatives (**3–14**) form singlet ground-state, square-planar nickel(II) complexes[3,23-30] with anions that have little tendency to coordinate, such as ClO$_4^-$, BF$_4^-$, PF$_6^-$, and I$^-$. The characteristic $d \rightarrow d$ band for these complexes (Table 1) is thought to include the manifold of transitions expected for square-planar Ni(II) since no further weak transitions are observed at $\lambda > \sim 235$ nm. The circular dichroism spectrum[27] of $(+)$[Ni((−)-5,14-*cis*-Me$_6$[14]aneN$_4$)]$^{2+}$ also exhibits a single band nearly coincident with the isotropic absorption band and reveals no additional transitions within this broad band. The energy of this band is markedly insensitive to ring size or substituent effects for the [13–16]aneN$_4$ complexes

TABLE 1. Absorption Spectra of Representative Nickel(II) Complexes with [12–16]aneN$_4$ Ligands

Complex	Ligand	$\lambda_{max}(\varepsilon)^a$	Ref.
S = 0 (*Square-planar complexes*)			
[Ni(8,10-Me$_2$[13]aneN$_4$)](PF$_6$)$_2$	**4**	428 (145)^b	23
[Ni(8,8,10-Me$_3$[13]aneN$_4$)]$^{2+}$ ^{c}	**5**	429 (119)	3
[Ni([14]aneN$_4$)](ClO$_4$)$_2$	**6**	460 (68)^b	24
[Ni([14]aneN$_4$)]I$_2$	**6**	480^d	24
[Ni(5,7-Me$_2$[14]aneN$_4$)](PF$_6$)$_2$	**9**	441 (62)^b	23
[Ni(*ms-trans*-Me$_6$[14]aneN$_4$)](ClO$_4$)$_2$	**10**	462 (79); 235 (15,600); 206 (10,500)^b	25
[Ni(*α-C-rac-trans*-Me$_6$[14]aneN$_4$)(ClO$_4$)$_2$	**10**	443 (70); 228 (12,700); 204 (9300)^b	25
[Ni(*β-C-rac-trans*-Me$_6$[14]aneN$_4$)(ClO$_4$)$_2$	**10**	450 (104); 231 (10,600); 204 (10,200)^b	25
[Ni(*C-rac-cis*-Me$_6$[14]aneN$_4$)]$^{2+}$ ^{c}	**11**	444 (111)	3
[Ni(*trans*-Me$_6$[14]aneN$_4$)](ClO$_4$)$_2$	**10**	471 (66); 233 (13,830); 200 (9270)^e	26
[Ni(*cis*-Me$_6$[14]aneN$_4$)](ClO$_4$)$_2$	**11**	471 (73); 233 (13,280); 200 (9460)^e	26
(+)-[Ni((−)-5,14-*cis*-Me$_6$[14]aneN$_4$)](ClO$_4$)$_2$	**11**	459 (94); 230 (11,300)^e; 461 (+2.02); 250 (−2.78); 224 (+6.38)^{e,f}	27
[Ni(*trans*-Me$_6$[16]aneN$_4$)]$^{2+}$ ^{c}	**14**	450 (106)	3
[Ni(1-Me[14]aneN$_4$)](ClO$_4$)$_2$		460 (58)^b	28, 29
[Ni(1,11-Me$_2$[14]aneN$_4$)](ClO$_4$)$_2$		470 (35)^b	28
[Ni(1,4-Me$_2$[14]aneN$_4$)](ClO$_4$)$_2$		465^g	29
[Ni(*N*-Me$_4$[14]aneN$_4$)](ClO$_4$)$_2$	**7**	519 (184)^g; 520^h	29, 30
S = 1 (*Five-coordinate complexes*)			
[Ni(*N*-bzl$_4$[12]aneN$_4$)Cl]Cl·0.5H$_2$O	**2**	1560sh (25); 1460 (5); 1090 (10); 715 (100); 440 (245)^i	21
[Ni(*N*-Me$_4$[14]aneN$_4$)Cl]ClO$_4$	**7**	1785 (9); 1110 (4); 869sh (11); 820 (54); 602sh (9); 524 (12); 406 (170)^g	30
[Ni(*N*-Me$_4$[14]aneN$_4$)Br]ClO$_4$	**7**	1896 (10); 1170 (4); 881sh (11); 726 (50); 619sh (13); 540 (11); 433 (172)^g	30
[Ni(*N*-Me$_4$[14]aneN$_4$)NCS]ClO$_4$	**7**	1495 (11); 1060 (3); 901sh (8); 657 (55); 573sh (14); 495sh (8); 400 (147)^g	30
[Ni(*N*-Me$_4$[14]aneN$_4$)(H$_2$O)]$^{2+}$	**7**	1130 (23); 800sh (44); 654 (29); 511 (71); 392 (104)^b; 660 (24); 520 (60); 400 (78)^j	30 28

continued overleaf

TABLE 1 (continued)

Complex	Ligand	$\lambda_{max}(\varepsilon)^a$	Ref.
S = 1 (*Six-coordinate tetragonal complexes*)			
[Ni([14]aneN$_4$)Cl$_2$]	6	670, 520^d	24
		1163, 678,^{k} 658,^{k} 515,	6
		367sh,^{k} 343^l	
[Ni(2,3-Me$_2$[14]aneN$_4$)Cl$_2$]	8	1190, 680,^{k} 671,^{k} 521,	6
		369sh, 345^l	
[Ni(5,7-Me$_2$[14]aneN$_4$)Cl$_2$]	9	1185 (3); 685 (4)	23
		526 (7); 343 (30)^m	
[Ni(*trans*-Me$_6$[14]aneN$_4$)Cl$_2$]	10	1235, 699, 543, 357^d	37
		1190, 680,^{k} 662,^{k} 530,	6
		375sh, 348^l	
[Ni(*N*-Me$_4$[14]aneN$_4$)(NCS)$_2$]	7	975 (10); 820sh (5);	30
		610 (13); 390 (23)^g	
[Ni(*ms-trans*-Me$_6$[14]aneN$_4$)(C$_2$O$_4$)]·3H$_2$O	10	1100, 676, 538, 352^d	38
[Ni([15]aneN$_4$)Cl$_2$]	12	1154, 807,^{k} 693,^{k} 568,	6
		383sh,^{k} 366^l	
[Ni([16]aneN$_4$)Cl$_2$]	13	1285, 896, 743, 635, 390^l	6
S = 1 (*Six-coordinate, cis-folded complexes*)			
[Ni([12]aneN$_4$)(NO$_3$)]NO$_3$	1	920 (28); 555 (19)	20
		355 (19)^i; 870, 545, 350^h	
[Ni([12]aneN$_4$)Cl$_2$]	1	950, 585, 375^h	20
[Ni([12]aneN$_4$)Br$_2$]·0.5C$_2$H$_5$OH	1	945, 590, 375^h	20
		945 (20); 545 (12);	
		340 (19)^i	
[Ni(*N*-bzl$_4$[12]aneN$_4$)(NO$_3$)]NO$_3$·0.5H$_2$O	2	1025 (25); 850sh (5);	21
		610 (30); 384 (70)^i	
{[Ni(8,10-Me$_2$[13]aneN$_4$]$_2$C$_2$O$_4$}(PF$_6$)$_2$	4	905 (57); 808sh (46);	23
		550 (64); 345 (138)^e	
{[Ni(5,7-Me$_2$[14]aneN$_4$]$_2$C$_2$O$_4$}(PF$_6$)$_2$	9	910 (20); 805 (13);	23
		543 (28); 348 (61)^e	
{[Ni(*rac-trans*-Me$_6$[14]aneN$_4$]$_2$C$_2$O$_4$}(ClO$_4$)$_2$	10	990, 581, 368^d	38

^{a} In nm.
^{b} H$_2$O solution.
^{c} Anion not specified.
^{d} Solid state, diffuse reflectance.
^{e} Acetonitrile solution.

^{f} Values of $\Delta\varepsilon$ in parentheses.
^{g} Nitromethane solution.
^{h} Solid state, diffuse transmittance.
^{i} Methanol solution.

^{j} H$_2$O solution, pH = 7.
^{k} Predicted by Gaussian analysis.
^{l} Solid state, diffuse transmittance at 77 K.
^{m} Chloroform solution.

except when substitution occurs on the amine donor atoms.[28–30] The characteristic band for square-planar [Ni(*N*-Me$_4$[14]aneN$_4$)](ClO$_4$)$_2$ occurs at 519 nm, a significantly lower energy than for the complexes with secondary amine donors. The weakening of the ligand field strength of [14]aneN$_4$ ligands by *N*-methylation occurs monotonically with increasing substitution as shown by Figure 2 for several series of complexes.[29] The origin of this decrease in donor strength is not obvious since gas-phase studies[31] show tertiary amines

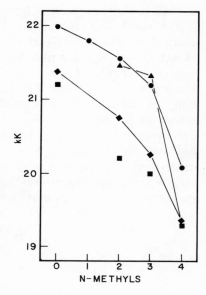

FIGURE 2. Change in absorption maximum for alkylated forms of [14]aneN$_4$ (●), 1,4-N-Me$_2$[14]aneN$_4$ (▲), *trans*-Me$_6$[14]aneN$_4$ (◆), and *cis*-Me$_6$[14]aneN$_4$ (■) as a function of the number of N-methyl groups. Data were obtained on the *bis*-perchlorate salts in nitromethane. Reprinted from reference 29 with permission. Copyright by The American Chemical Society.

to have a greater intrinsic basicity than secondary amines and steric interactions with the N-methyl groups do not appear to be significant in the case of a *trans*-diazido complex of [Ni(N-Me$_4$[14]aneN$_4$)]$^{2+}$ for which a crystal structure is available.[32]

High-spin, five-coordinate nickel(II) complexes have been reported for the tetra-N-alkylated ligands, N-bzl$_4$[12]aneN$_4$[21] and N-Me$_4$[14]aneN$_4$,[28,30,33] in donor solvents or in the solid state with certain coordinating anions. In the case of the 12-membered macrocycle, where folding about the metal ion must occur, the spectrum of the chloride derivative (Table 1) is suggested to parallel that of a known trigonal-bipyramidal complex.[34] In contrast, a square-pyramidal configuration for the complex [Ni(N-Me$_4$[14]aneN$_4$Cl]$^+$ is proposed on the basis of NMR studies of the corresponding Zn(II) derivative.[30] This structural assignment gains support from the crystal structure of [Ni(N-Me$_4$[14]aneN$_4$)N$_3$]ClO$_4$,[35] which shows the nitrogen donors to be coplanar with all four methyl groups on one side of the macrocycle.

In the presence of coordinating anions, high-spin tetragonal nickel(II) complexes may be achieved (Table 1) with [14–16]aneN$_4$ ligands and substituted derivatives. The ligand [13]aneN$_4$, although capable of forming a square-planar nickel(II) complex,[3,36] apparently folds on the metal ion in the presence of donor anions and produces *cis*-octahedral complexes[5,6] analogous to the behavior of [12]aneN$_4$.[20]

The detailed study of the tetragonal spectra of the *trans*-dihalotetramine series [Ni([14–16]aneN$_4$)X$_2$] by Busch and co-workers[5,6] describes the decrease in Dq^{xy} with increasing ring size: [14]aneN$_4$ = 1460 cm^{-1}, [15]aneN$_4$ = 1240 cm^{-1}, and [16]aneN$_4$ = 1100 cm^{-1} (averaged values). The abnormally high value for [14]aneN$_4$ is attributed to a constrictive effect of the 14-membered ring about nickel(II). The larger ligand [15]aneN$_4$ is proposed to have a more ideal fit and to give Dq values in line with other amine donors. The lowered value of Dq^{xy} for [16]aneN$_4$ results from a dilative effect. This study also showed essentially no effect of C-methyl substitution on the Dq^{xy} values for [14]aneN$_4$ ligands, even when the substitution becomes extensive. Averaged Dq^{xy} values for 2,3-Me$_2$[14]aneN$_4$,[6] 5,7-Me$_2$[14]aneN$_4$,[23] and *trans*-Me$_6$[14]ane$_4$[6] are 1450, 1470, and 1443 cm^{-1}, respectively. This result is in contrast to a prior report[1] of Dq^{xy} = 1160 cm^{-1} for *trans*-Me$_6$[14]aneN$_4$ and suggestions[23,36] that this low value be attributed to the steric crowding of axial methyl groups on the chelate rings.

15
8,10,10-Me$_3$[13]7-eneN$_4$

16
5,7,7-Me$_3$[14]4-eneN$_4$

17
Me$_6$[14]4,11-dieneN$_4$

18
Me$_6$[14]4,14-dieneN$_4$

19
Me$_2$[14]1,3-dieneN$_4$

20
Me$_4$[14]1,3,8,10-tetraeneN$_4$

21
Me$_6$[15]4,12-dieneN$_4$

22
2,12-Me$_2$[16]1,12-dieneN$_4$

23
Me$_6$[16]4,12-dieneN$_4$

2.1.2. Ni[13–16]eneN₄ and Ni[13–16](poly)eneN₄ Complexes

Some representative N_4 macrocyclic ligands containing one–four double bonds are shown in **15–23**. The preponderance of complexes formed with these ligands are square-planar, singlet ground state species[3,26,39–49] (Table 2) since, in general, they display a lessened tendency to coordinate anions or

TABLE 2. Absorption Spectra of Representative Nickel(II) Complexes with [13–16]eneN₄ and [13–16](poly)eneN₄ Ligands

Complex	Ligand	$\lambda_{max}(\varepsilon)$ [a]	Ref.
S = 0 (*Square-planar complexes*)			
[Ni(8,10,10-Me₃[13]7-eneN₄)](ClO₄)₂	**15**	437 (92); 282 (5200); 229 (13,000); 216 (14,600)[b]	39
[Ni(8,10-Me₂[13]7,10-dieneN₄)](PF₆)₂		405 (103)[c]	40
[Ni(8,10-Me₂[13]7,10-dienatoN₄)]PF₆	**28**	495 (136)[c]	40
		485 (128); 358 (1640)[b,d]	41
[Ni(8,10-Me₂-9-CHO[13]7,10-dienatoN₄)]PF₆		461 (288)[c]	40
[Ni(8-CF₃-10-Me[13]7,10-dienatoN₄)]BF₄		474 (153); 396 (1940); 370 (3950)[b]	41
α-[Ni(5,7,7-Me₃[14]4-eneN₄)]ZnCl₄	**16**	429 (72)[e]	42
β-[Ni(5,7,7-Me₃[14]4-eneN₄)]ZnCl₄	**16**	446 (86)[e]	42
[Ni(5,12-Me₂[14]4,11-dieneN₄)](ClO₄)₂		446 (89); 279 (4900); 227 (13,000)[b]	43
[Ni(5,7-Me₂[14]4,7-dieneN₄)](PF₆)₂		546 (35)[f]	44
[Ni(5,7-Me₂[14]4,7-dienatoN₄)]PF₆	**29**	546 (90)[f]	44
[Ni(*ms*-Me₆[14]4,11-dieneN₄)](ClO₄)₂	**17**	441 (89); 282 (5400); 230 (15,200); 216 (17,280)[g]	26
		435 (101); 282 (5000); 229sh, 216 (16,000)[b]	3
[Ni(*rac*-Me₆[14]4,11-dieneN₄)]²⁺ [h]	**17**	431 (93); 282 (5000); 233sh, 216 (17,000)[b]	3
[Ni(*rac*-Me₆[14]4,14-dieneN₄)]²⁺ [h]	**18**	438 (106); 275 (4080); 222 (22,700)[g]	26
		435 (108); 276 (4000); 222 (22,000)[b]	3

continued overleaf

TABLE 2 (continued)

Complex	Ligand	$\lambda_{max}(\varepsilon)$ [a]	Ref.
[Ni(rac-5,12-Me$_2$-7,14-Ph$_2$[14]4,11-dieneN$_4$)](ClO$_4$)$_2 \cdot$0.5H$_2$O		435 (116)[t], 481[j]	45
[Ni(ms-5,12-Me$_2$-7,14-Ph$_2$[14]4,11-dieneN$_4$)](ClO$_4$)$_2$		469[j]	45
[Ni(Me$_6$[14]1,4,8,11-tetraeneN$_4$)]$^{2+}$ [h]		442 (159)[b]	3
[Ni(Me$_2$pyo[14]tetraeneN$_4$)](ClO$_4$)$_2$	24	415 (1100); 390 (1450); 310sh (3380)[b]	46
[Ni(Me$_2$pyo[14]dieneN$_4$)](ClO$_4$)$_2$ [k]	26	450 (320); 365 (2240)[a]	46
[Ni(C-ms-Me$_2$pyo[14]eneN$_4$)](ClO$_4$)$_2$ (α' isomer)	27	470 (160)[c]	47
(β' isomer)		440 (80)[c]	
[Ni(9,11-Me$_2$[15]8,11-dieneN$_4$)](PF$_6$)$_2$	30	543 (10); 431 (68); 358 (135)[f]	44
[Ni(Me$_6$[15]4,12-dieneN$_4$)](ClO$_4$)$_2 \cdot$H$_2$O	21	461 (131); 282 (3300); 239 (11,200); 222 (12,600)[b]	39
[Ni(2,12-Me$_2$[16]1,12-dieneN$_4$)](PF$_6$)$_2$ [k]	22	437 (104)[c]; 449[l]	48
[Ni(Me$_6$[16]4,12-dieneN$_4$)](ClO$_4$)$_2$	23	436 (100); 277sh 230sh (15,200); 222 (16,300)[b]	49

$S = 1$ (*Tetragonal six-coordinate*)

Complex	Ligand	$\lambda_{max}(\varepsilon)$ [a]	Ref.
β-[Ni(5,7,7-Me$_3$[14]4-eneN$_4$)(NCS)$_2$]	16	971, 685sh, 510, 368[j]	42
[Ni(Me$_2$[14]1,3-dieneN$_4$)(NCS)$_2$]	19	889, 640, 568,[m] 468[m,n]	6
[Ni(N-ms-5,12-Me$_2$-7,14-Ph$_2$[14]4,11-dieneN$_4$)Cl$_2$]$\cdot$2H$_2$O		1080, 649, 505[j]	45
[Ni(N-ms-5,12-Me$_2$-7,14-Ph$_2$[14]4,11-dieneN$_4$)(NCS)$_2$]		952, 680, 505[j]	45
[Ni(ms-Me$_6$[14]4,11-dieneN$_4$(NCS)$_2$]	17	926, 800sh, 645, 488[j]	52
[Ni(Me$_6$[14]4,11-dieneN$_4$)(NCS)$_2$]	17	890, 635, 549, 475[n]	6
[Ni(ms-5,7,12,14-Et$_4$-7,14-Me$_2$[14]4,11-dieneN$_4$)(NCS)$_2$]		1030, 625, 498[j]	54
[Ni(Me$_4$[14]1,3,8,10-tetraeneN$_4$)(NCS)$_2$]	20	837, 565, 526, 435[m,n]	6
[Ni(cis-Me$_6$[14]tetraeneN$_4$)(NCS)$_2$]		909, 476, 376[j]	3
[Ni(ms-Me$_2$pyo[14]eneN$_4$)(NCS)$_2$]	27	911, 712, 591,[m] 501, 339sh, 331[n]	6
[Ni(Me$_2$pyo[14]trieneN$_4$)(NCS)$_2$]	25	885, 538,[m] 532,[m] 408[n]	6
[Ni(Me$_6$[15]4,12-dieneN$_4$)(NCS)$_2$]	21	962, 833, 555, 351[j]	39
[Ni(Me$_6$[16]4,12-dieneN$_4$)(NCS)$_2$]	23	968, 818, 606,[m] 549, 351sh[n]	6
		935, 575[j]	39

TABLE 2 *(continued)*

Complex	Ligand	$\lambda_{max}(\varepsilon)$ [a]	Ref.
= 1 *(Six-coordinate,* cis-*folded configuration)*			
α-[Ni(5,7,7-Me₃[14]4-eneN₄)(acac)]ClO₄ [o]	16	980, 568,[j] 962 (41); 568 (49)[c]	42
α-{[Ni(5,7,7-Me₃[14]4-eneN₄)]₂C₂O₄}(ClO₄)₂	16	943, 565[j]	42
[Ni(N-rac-5,12-Me₂-7,14-Ph₂[14]4,11-dieneN₄)(acac)]ClO₄·0.5H₂O		885, 526[j]	45
{[Ni(N-rac-5,12-Me₂-7,14-Ph₂[14]4,11-dieneN₄)]₂C₂O₄}(ClO₄)₂		926, 546[j]	45
[Ni(N-rac-5,12-Me₂-7,14-Ph₂[14]4,11-dieneN₄)(en)](ClO₄)₂·2H₂O [p]		943, 541[j]	45
[Ni(rac-Me₆[14]4,11-dieneN₄)(acac)]ClO₄ [o]	17	952, 562[j]	25
{[Ni(rac-Me₆[14]4,12-dieneN₄)]₂C₂O₄}(ClO₄)₂	17	926, 559[j]	25
{[Ni(2,12-Me₂[16]1,12-dieneN₄)]₂C₂O₄}(PF₆)₂	22	909 (8.5); 552 (22.5)[c]	48
[Ni(2,12-Me₂[16]1,12-dieneN₄)(acac)]PF₆ [o]	22	917 (9.5); 555 (11.7)[c]	48
{[Ni(ms-Me₂pyo[14]eneN₄)]₂C₂O₄}(ClO₄)₂	27	910, 800sh, 580, 360sh[l]	55

n nm.   [b] H₂O solution.   [c] Acetone solution.   [d] BF₄⁻ salt.   [e] Dilute HClO₄.
Methanol solution.   [g] Acetonitrile solution.   [h] Anion not specified.   [i] Methanol/HClO₄.
olid state, diffuse reflectance.
xists as equilibrium between $S = 0$ (square-planar) and $S = 1$ (six-coordinate) species in coordinating
olvents.   [l] Solid state, diffuse transmittance.   [m] Predicted by Gaussian analysis.
olid state, diffuse transmittance at 77 K.   [o] acac = acetylacetonate.   [p] en = ethylenediamine.

solvent than do the corresponding saturated ligands. Some exceptions to this behavior are noted in Table 2. One exception is the series 24–27; where the parent complex Ni(Me₂pyo[14]trieneN₄)²⁺ (25) is fully high spin, six-coordinate in aqueous solution[50,51] but the nickel(II) complexes of 24, with increasing unsaturation, or 26 and 27, with decreasing unsaturation, show less tendency to coordinate solvent in axial positions.[46,47] The ability of this series to undergo axial coordination is 24 < 25 > 26 > 27. Proton magnetic resonance provides a sensitive technique to detect the presence of triplet species by the large chemical shifts produced in ligand resonances.[48] By this technique, Ni(5,12-Me₂[14]4,11-dieneN₄)²⁺ and Ni(Me₂[16]1,12-dieneN₄)²⁺ (22) were shown[48] to exist in equilibrium between square-planar and six-coordinate species in the solvents H₂O, CH₃CN, acetone/H₂O, acetone/pyridine, and CH₃CN/H₂O, whereas Ni(Me₆[14]4,11-dieneN₄)²⁺[52] (17) and Ni(Me₆[16]4,12-dieneN₄)²⁺[53] (23) exist totally as $S = 0$ species in these solvents. It was concluded[48] that the more hindered hexamethyl species are less likely to coordinate axial donors as a result of the presence of gem-dimethyl groups blocking the axial positions. Another factor that has been proposed to account for the generally low tendency of the complexes with unsaturated macrocycles to coordinate axial donors is the strong in-plane ligand fields generated by these ligands.[1,6]

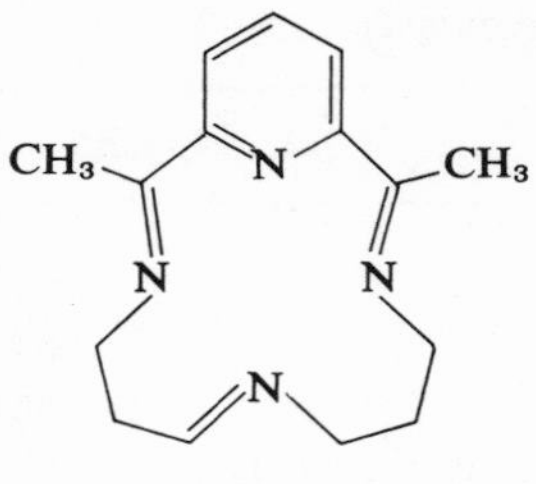

24
Me₂pyo[14]tetraeneN₄

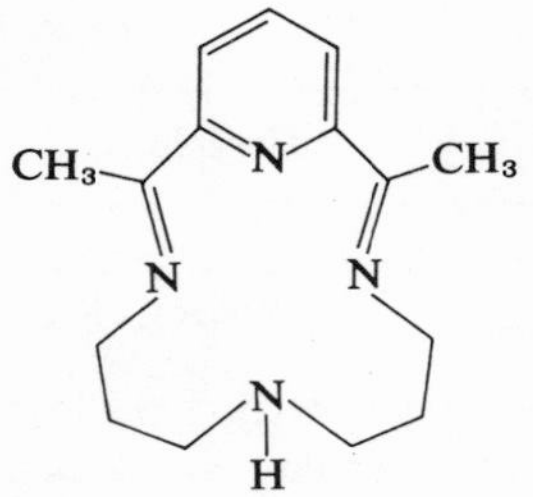

25
Me₂pyo[14]trieneN₄

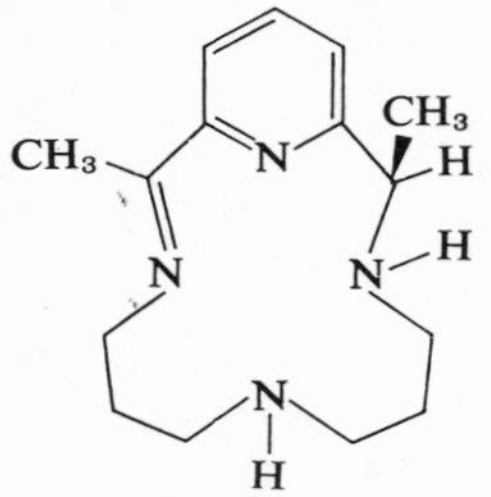

26
Me₂pyo[14]dieneN₄

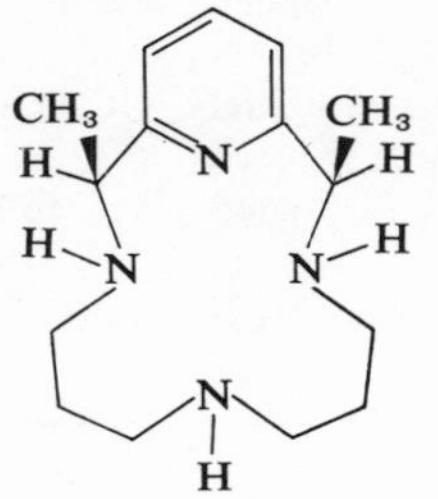

27
ms-Me₂pyo[14]eneN₄

For essentially all of the singlet ground state complexes in Table 2 the characteristic $d \rightarrow d$ band occurs over a relatively narrow energy range and systematic variations of the ligand field strengths of the macrocycles with ring size, degree of unsaturation, or number of substituents are not observed.

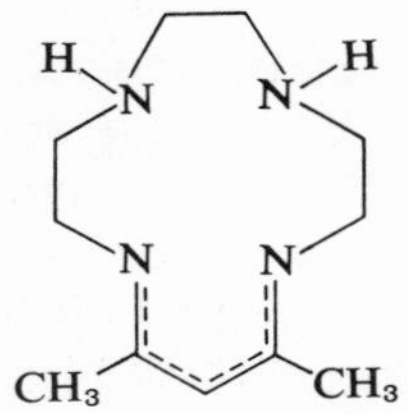

28
8,10-Me₂[13]7,10-dienatoN₄

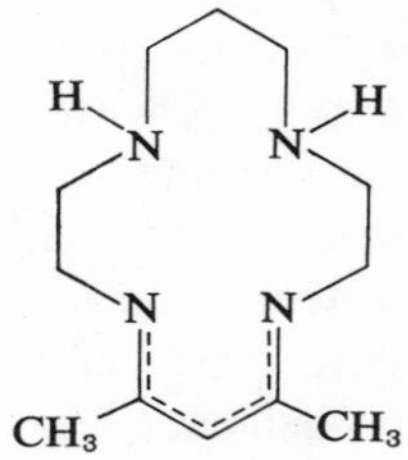

29
5,7-Me₂[14]4,7-dienatoN₄

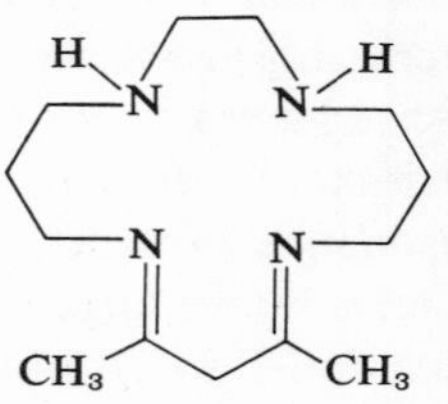

30
9,11-Me₂[15]8,11-dieneN₄

Ligands **28–30** are notably weaker[40,41,44] than other [13–15]-membered unsaturated N_4 macrocycles. No change in the ligand field strength of **29** occurs upon protonation[44] to give the neutral ligand, 5,7-Me₂[14]dieneN₄, but protonation of **28** produces a dramatic shift of the $d \rightarrow d$ band to higher energy[40] (Table 2). This increase of ligand field strength upon protonation of **28** was compared[40] to the differences in ligand field strength of 2,4-pentanedione in its neutral (976 cm⁻¹) and anionic (903 cm⁻¹) forms.

The high-spin complexes containing [14–16]eneN$_4$ or [14–16](poly)eneN$_4$ ligands (Table 2) are invariably six-coordinate with either a *trans*-axial configuration[3,6,39,42,45,52] with two monodentate donors or a *cis*-folded configuration[25,42,45,48,55] with bidentate acetylacetonate, ethylenediamine, or bridging oxalate ligands. In the case of the folded structures the ligand field strengths of the different donors are effectively averaged and their electronic spectra are essentially those expected for an octahedral geometry. The transitions at ∼950, ∼550, and ∼350 nm are assigned as $^3A_{2g} \rightarrow {}^3T_{2g}$, $^3A_{2g} \rightarrow {}^3T_{1g}(F)$, and $^3A_{2g} \rightarrow {}^3T_{1g}(P)$, respectively. For the tetragonal complexes, pseudo-octahedral spectra are also observed when the difference between the ligand field strengths of the macrocycle and the axial donors is not great.

In many cases, the additional band splittings for tetragonal nickel(II) anticipated from Figure 1 are observed and more detailed assignments are possible. The spectra for six of the *trans*-axial thiocyanato complexes in Table 2 have been treated[6] by the complete tetragonal model. Less spectral information is accessible for the complexes of unsaturated ligands since the imine functions produce metal-to-ligand and ligand-to-metal charge transfer transitions which obscure the higher energy portion of the high-spin nickel(II) transitions. In addition, the unsaturated macrocycles invariably produce stronger in-plane ligand fields, and high-spin complexes are possible only for relatively strong axial ligands. Greater reliance was placed on Gaussian resolution and computer fitting in the interpretation of the spectra of these complexes, and the Dq^{xy} values[6] were found to be Me$_2$pyo[14]trieneN$_4$ (25), 1866 cm^{-1}; Me$_4$[14]1,3,8,10-tetraeneN$_4$ (20), 1767 cm^{-1}; Me$_6$[14]4,11-dieneN$_4$ (17), 1569 cm^{-1}; Me$_2$[14]1,3-dieneN$_4$ (19), 1553 cm^{-1}; Me$_2$pyo[14]-eneN$_4$ (27), 1398 cm^{-1} and Me$_6$[16]4,12-dieneN$_4$ (23), 1223 cm^{-1}. The greater ligand field strength of the unsaturated ligands compared to the aneN$_4$ ligands arises from the stronger donor ability of imines versus secondary amines and the constrictive effect of a smaller metal-ion site produced by the unsaturated linkages in the macrocyclic ring. An increase in the extent of unsaturation invariably produces larger Dq^{xy} values with the addition of the third and fourth imines generating an abnormally large increase. This observation suggests that the ring constriction is the more important effect for producing strong ligand fields in extensively unsaturated macrocycles.

2.1.3. Nickel(II) Complexes with Highly Delocalized N$_4$ Macrocycles

The highly unsaturated tetraaza macrocycle which has received the most attention with respect to nickel(II) complexes is TAAB (31)[1,56–58] and its related derivatives TAABH$_8$ (32)[59] and TAAB(OR)$_2^{2-}$ (33).[60,61] In the absence of coordinating anions, Ni(TAAB)$^{2+}$ is a diamagnetic, square-planar complex. On the other hand, an extensive series of high-spin complexes Ni(TAAB)X$_2$, with X = NCS$^-$, N$_3^-$, Cl$^-$, NO$_3^-$, Br$^-$, and I$^-$, has been prepared and their spectra treated by the simple tetragonal crystal field

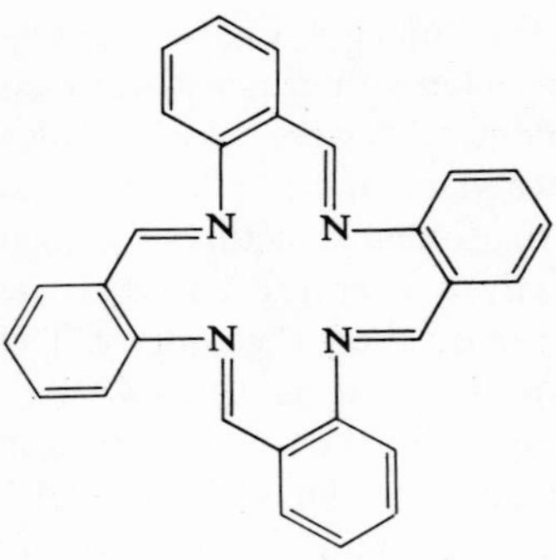

31
TAAB

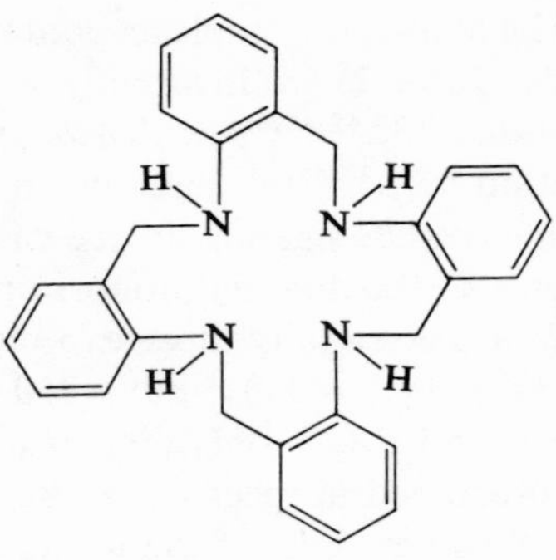

32
TAABH$_8$

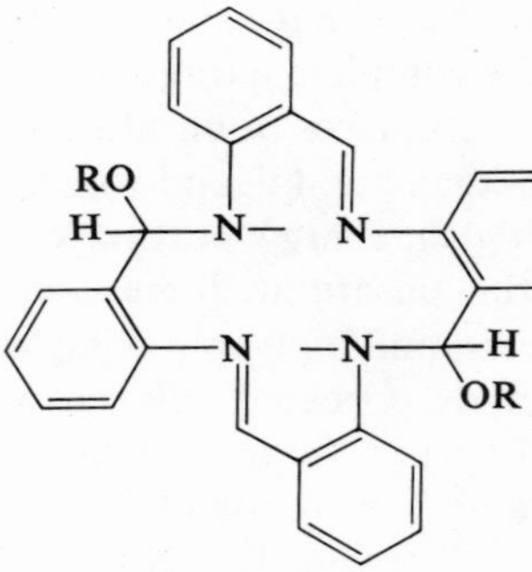

33
TAAB(OR)$_2^{2-}$

model.[1,57,62] An early value[1] of Dq^{xy} for TAAB has recently been revised to $Dq^{xy} = 1465$ cm^{-1} on the basis of a more detailed study.[6] A similar analysis of a series of tetragonal complexes of Ni(TAABH$_8$)$^{2+}$ originally yielded a value[59] of $Dq^{xy} = 1147$ cm^{-1} for this reduced ligand and this value was later revised to 1115 cm^{-1}.[6] The addition of nucleophiles across two of the azomethine linkages in Ni(TAAB)$^{2+}$ leads to complexes of the dianionic ligands, Ni(TAAB(OR)$_2$) (**33**), which always produce low-spin configurations.[60,61]

The spectra of a series of low-spin nickel(II) complexes[63,64] with the unsaturated tetraaza ligands **34–39** consist of intense absorptions throughout the visible and ultraviolet regions for which detailed transition assignments were not possible. Ligand **38** is of particular importance since it provides a simple model for the inner corrin ring of B$_{12}$-type compounds. The spectrum of the nickel(II) complex with this ligand compares favorably with spectra of several Ni(II)-corrinoid complexes.[64] Intense spectra were also observed for the low-spin Ni(II) complexes of several closely related ligands.[65–69]

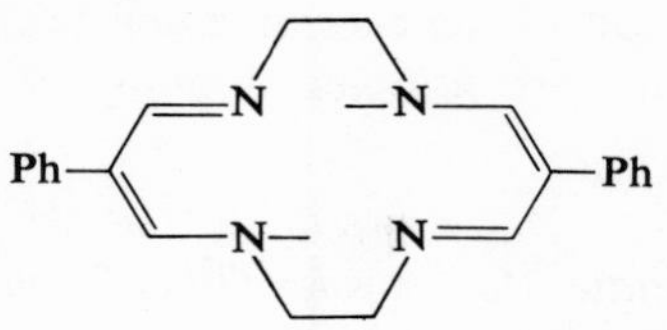

34

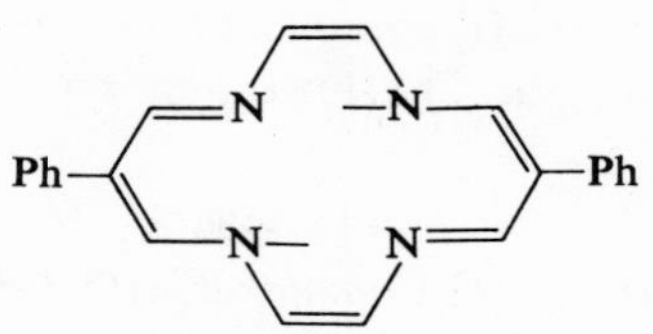

35

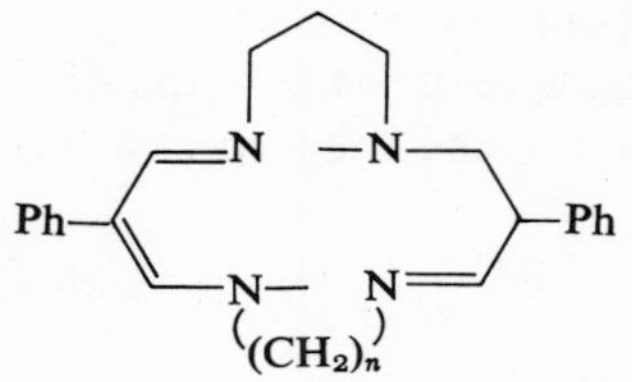

36 $n = 2$
37 $n = 3$

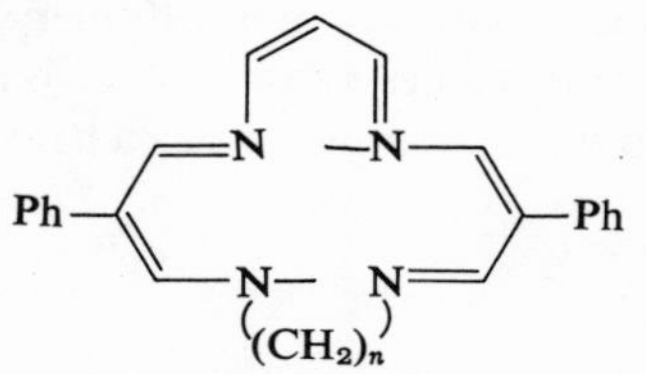

38 $n = 2$
39 $n = 3$

Low-spin nickel(II) complexes of the macrocyclic Schiff-base ligands[70] of type **40** exhibit a low-energy band in the region 550–650 nm, which is attributed to the characteristic band of square-planar nickel(II) complexes. Further resolution of this band and also of intraligand transitions is provided by the circular dichroism spectra[71] of optically active analogs of **40**. The related Ni(II) complex of **41**[72] is important in that it represents a dibenzo-corrin chromophore. The dinitro-substituted analogs (**42**)[73] of these Schiff-base macrocycles exhibit intense transitions in the visible region which mask the Ni(II) $d \rightarrow d$ band. Strong, low-energy transitions are also observed

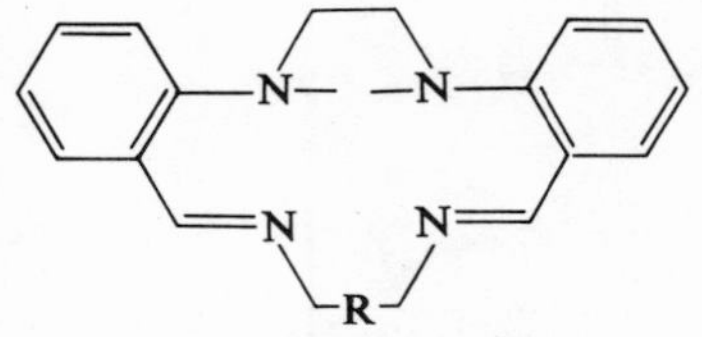

40

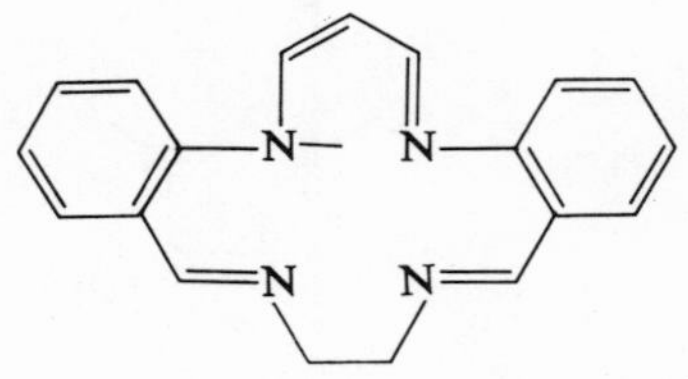

41

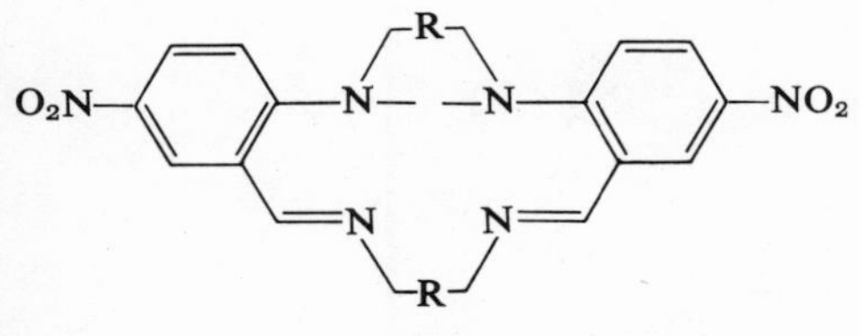

42

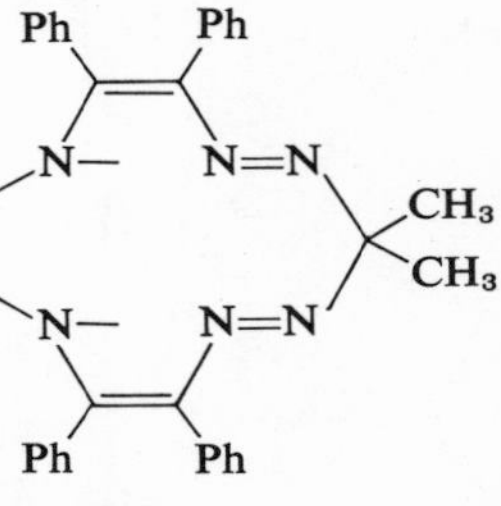

43

for the nickel(II) complex of **43**[74] as a result of the extensive delocalization of the ligand. The lowest-energy band in this complex is assigned[74] as $^1B_{1g} \leftarrow {}^1A_{1g}$.

2.1.4. Nickel(II) Complexes with Tetradentate Macrocycles Other Than N₄

Nickel(II) complexes with tetradentate macrocyclic ligands having donor atoms other than N_4 are not numerous and only a few have undergone extensive spectral studies (Table 3). Two *cis*-N_2S_2 ligands, **44**[1,75] and **45**,[76] have each provided a series of nickel(II) complexes and their spectra have been

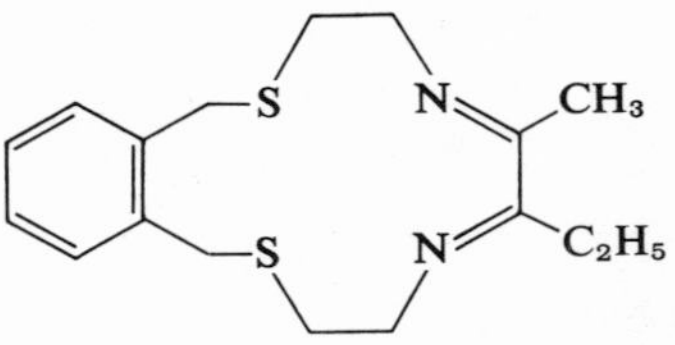

44

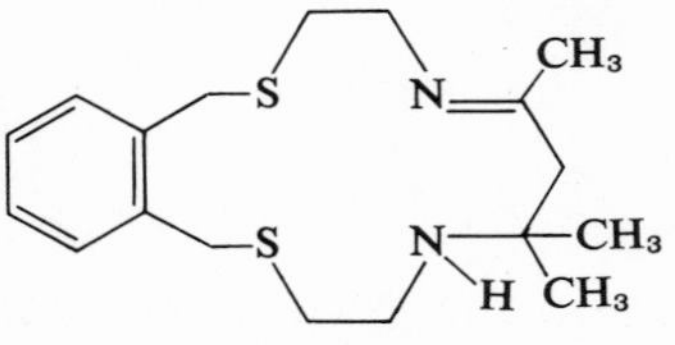

45

investigated. Both ligands produce diamagnetic, square-planar Ni(II) complexes in the absence of coordinating anions, and complexes of **45** retain this configuration in aqueous solution. Analysis of the spectra of a series of tetragonal complexes of **44** has yielded $Dq^{xy} = 1062$ cm^{-1} for this ligand.[1,75]

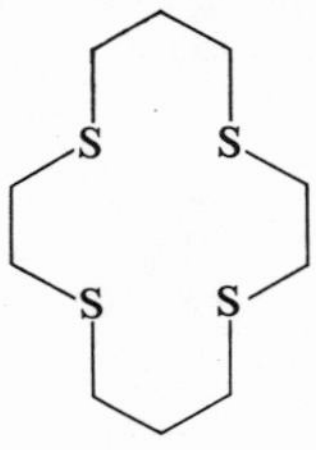

46
[14]aneS₄

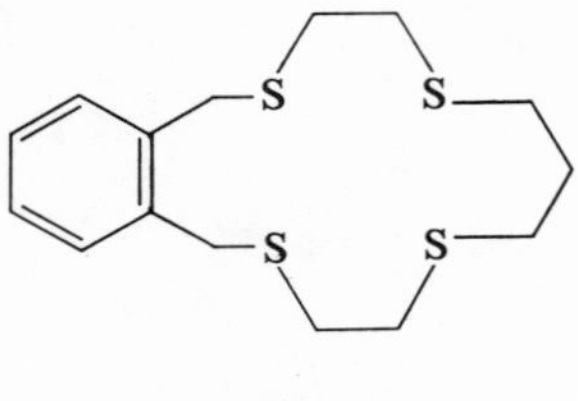

47
Bzo[15]eneS₄

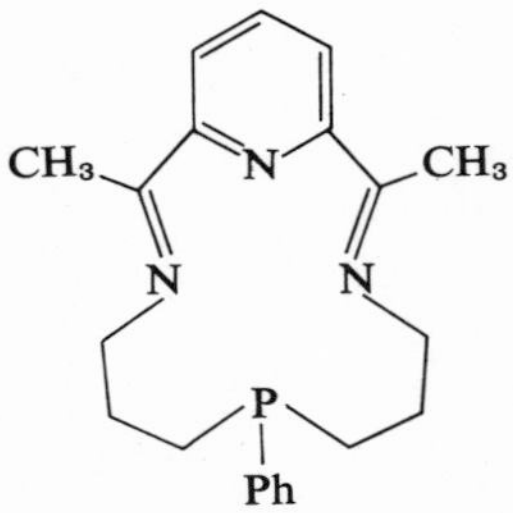

48
P-PhMe₂pyo[14]trieneN₃P

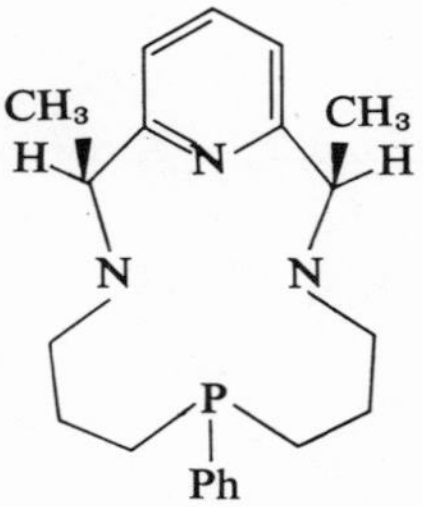

49
ms-P-PhMe₂pyo[14]eneN₃P

TABLE 3. Ligand Field Spectra of Some Nickel(II) Complexes with
N_2S_2, S_4, and N_3P Macrocyclic Ligands

Complex	Ligand	$\lambda_{max}(\varepsilon)$ [a]	Ref.
Diamagnetic complexes (S = 0)			
[Ni(2-Me-3-Etbzo[14]trieneN$_2$S$_2$)](ClO$_4$)$_2$	44	617; 500; 402[b]	75
[Ni(2-Me-3-Etbzo[14]trieneN$_2$S$_2$]I$_2$	44	637; 592; 420[b]	75
[Ni(2,4,4-Me$_3$bzo[15]dieneN$_2$S$_2$)](ClO$_4$)$_2$·H$_2$O	45	488[c]; 488(29.9)[d]	76
[Ni([14]aneS$_4$)](BF$_4$)$_2$	46	494(263); 410(9.75)sh[e]	78
		494(336)[e]	79
[Ni(Bzo[15]eneS$_4$)](BF$_4$)$_2$	47	510(273); 450(142)sh[e]	78
[Ni(Me$_2$Phpyo[14]trieneN$_3$P)X]PF$_6$ [f]	48	~525; 370[g or h]	80
Paramagnetic complexes (S = 1)			
[Ni(2-Me-3-Etbzo[14]trieneN$_2$S$_2$)Cl$_2$]	44	1080; 935; 543; 403[b]	75
[Ni(2-Me-3-Etbzo[14]trieneN$_2$S$_2$(NCS)$_2$]	44	926; 855; 543; 417[b]	75
[Ni(2,4,4-Me$_3$bzo[15]dieneN$_2$S$_2$)Cl]Cl·H$_2$O	45	1176(5.8); 893(2.7); 758(7.1); 585(15.0)[d]	76
[Ni(2,4,4-Me$_3$bzo[15]dieneN$_2$S$_2$)(NCS)$_2$]	45	945; 867; 559[c]	76
[Ni([14]aneS$_4$)Cl$_2$]	46	1080(25)sh; 940(48); 610(28)[e]	78
[Ni([14]aneS$_4$)(NCS)$_2$]	46	1010(34)sh; 915(54); 570(28)[e]	78

[a] In nm.   [b] Solid state reflectance.   [c] Solid state diffuse transmittance.
[d] Methanol solution.   [e] Nitromethane solution.   [f] X = Cl⁻, Br⁻, or I⁻.
[g] Dichloromethane solution.   [h] Acetonitrile solution.

The Cl⁻, Br⁻, and I⁻ derivatives of **45** are five-coordinate in the solid state
and in methanol solution and these spectra have been interpreted[76] on the
basis of a square-pyramidal model.[77] The SCN⁻ derivative exhibits a
pseudo-octahedral spectrum and is assigned[76] a six-coordinate geometry.

The tetrathiaether macrocycles **46**[78,79] and **47**[78] produce square-planar
nickel(II) complexes as the BF$_4$⁻ or ClO$_4$⁻ salts, and tetragonal high-spin
complexes with coordinating anions (Table 3). The calculated value of $Dq^{xy} =$
1070 cm⁻¹ for [14]aneS$_4$ places it as the weakest ligand field for symmetrical
tetradentate macrocycles.[78] The phosphorus-containing ligands **48** and **49**,
which are analogs of **25** and **27**, have been prepared via a template reaction on
nickel(II) and subsequent reduction with sodium borohydride.[80] Both ligands
form low-spin, square-planar ([NiL](PF$_6$)$_2$) and trigonal-bipyramidal, five-
coordinate ([NiL(X)]PF$_6$) complexes (L = **48** or **49**). Examples of *cis*- and
trans-N$_2$O$_2$ Schiff-base macrocycles **50–52** have been synthesized from
salicylaldehydes and have been coordinated to nickel(II).[81–83] The spectra
of an extensive series of high-spin complexes of **50** (presumed to have a
trans-axial configuration) are reported to be pseudo-octahedral but with
some additional splittings which could not be assigned.[81] The di-iodo and

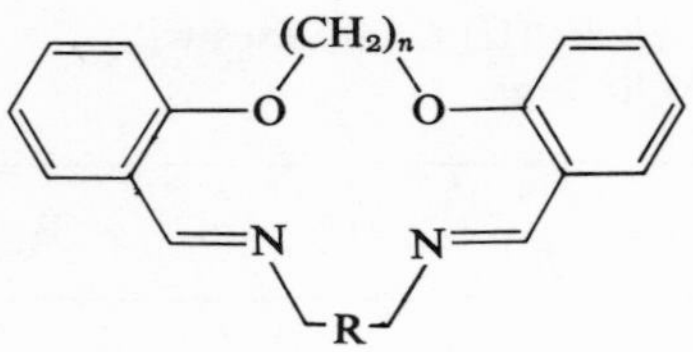

50

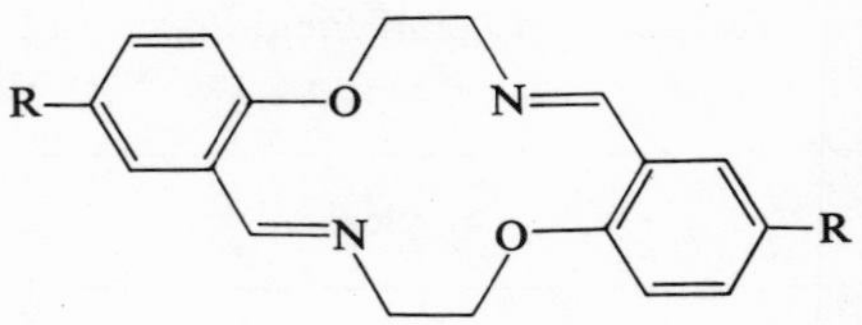

51 R = H
52 R = OCH₃

(Structures: 50, 51, 52 with R substituents)

di-thiocyanato derivatives of **51** exhibit pseudo-octahedral spectra.[82] The crystal structure of the di-iodo complex of **51** reveals a *trans*-tetragonal configuration[83] with a planar macrocycle, where the structure[84] of **52** as the iodide derivative is that of a folded macrocycle with *cis*-iodides, *cis*-imine nitrogen, and *trans*-oxygen donors.

2.1.5. Nickel(II) Complexes with Other Than Tetradentate Macrocycles

Relatively few examples of nickel(II) complexes with macrocyclic ligands providing other than tetradentate coordination have been reported. The high-spin, mono-, and *bis*-nickel(II) complexes[85–87] of the tridentate N_3 ligand derived from the trimeric condensation of *o*-aminobenzaldehyde exhibit pseudo-octahedral ligand field spectra with some evidence of a trigonal distortion. Two pentadentate macrocycles N_5[88] and N_2S_2O[89] also produce high-spin, pseudo-octahedral Ni(II) complexes. Several examples of monocyclic sexadentate macrocycles have been synthesized including N_4S_2,[90] N_4O_2,[90] and N_2S_4[91,92] donor atom arrays. The nickel(II) complexes with these ligands are invariably high spin and exhibit pseudo-octahedral spectra. Totally encapsulated nickel(II) species have been produced with the bicyclic clathrochelate ligand **53**.[8,9,93] The ligand field spectrum of the high-spin nickel(II) complex has been treated in detail[9] in terms of the degree of trigonal distortion produced by the constrained ligand.

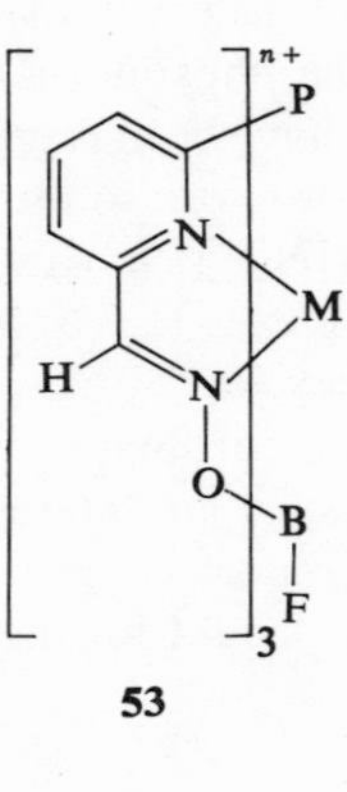

53

2.1.6. Magnetically Anomalous Mononuclear and Binuclear Nickel(II) Macrocycles

The complexes $Ni(TAAB)Br_2 \cdot H_2O$ and $Ni(TAAB)Cl_2 \cdot H_2O$ (TAAB = **31**) exhibit room temperature magnetic moments of 1.47 and 1.68 B.M., respectively, intermediate between that expected for $S = 0$ and $S = 1$ species.[94] From a study of the temperature dependence of the magnetic susceptibilities of these compounds, Melson and Busch[94] concluded that a spin equilibrium between singlet and triplet states was responsible for the anomalous magnetic behavior. An intermediate value of $\mu_B = 1.57$ for $NiLBr_2$ (L = **44**), on the other hand, has been interpreted to arise from a mixture of $S = 0$ and $S = 1$ species in the crystal.[75] In this case the magnetic susceptibility obeys the Curie–Weiss law, which rules out the possibility of a spin state equilibrium.[75] The complex $Ni(2,12-Me_2[16]1,12-dieneN_4)Br_2 \cdot 2H_2O$ (ligand **22**) also exhibits an intermediate value for μ_B, but detailed magnetic studies have not been carried out.[48]

A few examples of unusual magnetic behavior in nickel(II) complexes of binucleating macrocyclic ligands have been reported.[95–97] Binuclear Ni(II) complexes of type **54**, where $n = m = 2$ or $n = 2$, $m = 3$, are diamagnetic indicating a square-planar structure. This conclusion is supported by the observation of a single $d \rightarrow d$ band at ~ 550 nm.[96] For the complex with $n = m = 3$, and the counter ions = Cl^-, high-spin behavior ($\mu_B = 3.15$) was observed at room temperature but deviations from Curie–Weiss behavior below 120 K were interpreted as resulting from antiferromagnetic interactions.[95] From spectral evidence Pilkington and Robson[95] concluded that the two nickel(II) ions were square pyramidal with an apical coordinated chloride. Complex **55** is unique in that it contains both a diamagnetic and a paramagnetic nickel(II) ion.[97]

2.2. Macrocyclic Complexes of Nickel(I) and Nickel(III)

The initial report[98] of Ni(III) complexes with synthetic macrocyclic ligands involved the mild oxidation by nitric acid of Ni(II) complexes with *meso-* or racemic-Me_6[14]aneN_4, **10**. Products with the composition [Ni(Me_6-[14]aneN_4)(NO_3)$_2$]$ClO_4 \cdot H_2O$, [Ni(Me_6[14]aneN_4)(NO_3)$_2$]$NO_3 \cdot HNO_3$, and [Ni(Me_6[14]aneN_4)SO_4]$ClO_4 \cdot \frac{1}{3}H_2SO_4$ were isolated and characterized by their magnetic susceptibilities.[98] The dinitrato complexes exhibited magnetic moments typical for one unpaired electron and obeyed the normal Curie–Weiss relationship. Infrared spectra were consistent with monodentate coordination of the nitrate ions and the complexes were therefore described as low-spin six-coordinate d^7 species. The room temperature magnetic moment of the sulfate derivative was subnormal ($\mu_B = 1.54$) and anti-ferromagnetic coupling via sulfate bridges was invoked. Ni(III) and Ni(I) complexes of *cis-* and *trans-*Me_6[14]aneN_4 and *cis-* and *trans-*Me_6[14]dieneN_4 were prepared by controlled potential electrolyses and isolated as perchlorate salts.[26] The magnetic moments of the Ni(III) complexes ranged from 1.77 to 2.03 B.M., consistent with one unpaired electron (low-spin $3d^7$) as described above. The formally $3d^9$ Ni(I) species yielded values of $\mu_B = 1.70$–1.89. UV–visible spectra of the Ni(I) and Ni(III) species in acetonitrile were presented without interpretation.[26]

The formulation of Ni(I) or Ni(III) oxidation states from magnetic susceptibility data alone is not warranted since metal-ion stabilized anion or cation-radical ligands can give similar results. This ambiguity has been resolved in two extensive EPR studies.[36,99] In the first of these studies, a series of complexes, [Ni(2,3-Me_2[14]aneN_4)X_2]ClO_4, was prepared with X = Cl^-, Br^-, NCO^-, NO_3^-, and $\frac{1}{2}SO_4^{2-}$, by oxidation with $(NH_4)_2S_2O_8$ or concentrated HNO_3.[99] The solid complexes gave $\mu_B = 1.74$–2.05, except for the sulfate ($\mu_B = 1.54$), for which antiferromagnetic interactions were suggested. The anisotropic EPR data (Table 4) indicate a low-spin $3d^7$ Ni(III) species rather than a cation radical ligand. The superhyperfine splittings observed for the Br^- and NCO^- derivatives confirm their six-coordinate nature and all of the complexes are assigned low-spin six-coordinate Ni(III) formulations. As part of an extensive study of the electrochemical behavior of Ni(II) complexes with N_4 macrocyclic ligands, a series of 1e oxidation and reduction products

TABLE 4. Room-Temperature EPR Data for the Complexes [Ni(2,3-Me_2[14]aneN_4)X_2]ClO_4[99]

X	$g_\perp$ [a]	$g_\parallel$ [a]	g_{soln} [b]
Br	2.171	2.022	2.121, $\langle A \rangle = 57G$
Cl	2.181	2.025	2.124
NCO	2.169	2.055	2.129, $\langle A \rangle = 22G$
NO_3	2.221, 2.182[c]	2.103	2.159
$\frac{1}{2}SO_4$	2.16[d]		2.165[e]

[a] Powder.   [b] In CH_3CN unless otherwise stated.
[c] Nonaxially symmetric.
[d] Very broad ($\sim$1000 G) symmetric resonance.   [e] In H_2O.

were characterized by EPR.[36] Oxidation of the complexes with neutral ligands gave low-spin, six-coordinate Ni(III) derivatives rather than ligand cation radicals as evidenced by the anisotropic EPR parameters obtained: $g_{\parallel} \sim 2.020$; $g_{\perp} \sim 2.190$, $A_{\parallel} = 21.3\text{--}25.4$ G; $A_{\perp} = 16\text{--}21$ G. Similarly, the Ni(II) complex with the dianionic ligand, $Me_2Ac_2[14]$tetraenatoN$_4$, **56**, gave

56

a square-planar Ni(III) derivative upon oxidation. EPR parameters for this species in acetonitrile (at 77 K) were $g_{\parallel} = 2.138$, $g_{\perp} = 2.016$. The EPR spectra for the six-coordinate Ni(III) complexes with neutral macrocyclic ligands are consistent with an $(e_g)^4(b_{2g})^2(a_{1g})^1$ electronic configuration, whereas the square-planar Ni(III) species is assigned an $(e_g)^4(a_{1g})^2(b_{2g})^1$ ground state. These electronic configurations are based on theoretical considerations derived for $3d^7$ Co(II) systems.[100] Anisotropic EPR parameters and the observation of nitrogen superhyperfine splitting in the $g_{\parallel}$ bands characterize the oxidation product of a nickel(II) complex with a pentadentate macrocycle as a six-coordinate Ni(III) complex.[88]

In contrast to this metal-ion localized redox behavior, a detailed examination[101] of the EPR of the cation [Ni(Me$_4$[14]hexaenatoN$_4$)]$^+$ and selectively deuterated derivatives revealed that the oxidation state changes in the series $[M(L)]^0 \rightleftharpoons [M(L)]^+ \rightleftharpoons [M(L)]^{2+}$ are effectively confined to the delocalized ligand system. Over 80 hyperfine lines were resolved for the spin-doublet nickel cation and $\langle g \rangle = 2.001$. Ligand free radical nature is similarly ascribed to [Ni(Ph$_2$[14]hexaenatoN$_4$)]$^+$ from the observation of $g_{AV} = 2.01$ with linewidth of 9 G.[64] Ligand radical species have also been invoked in the base-prompoted reduction of nickel(III) complexes with [14]aneN$_4$ and ms-Me$_6$[14]aneN$_4$.[102]

3. Copper Complexes

3.1. Macrocyclic Copper(II) Complexes

Square-planar copper(II) complexes[25,27,42–45,104–107] with N$_4$ macrocyclic ligands normally exhibit a single, broad $d \rightarrow d$ band in the visible region (Table 5). For complexes of [14]aneN$_4$ and [14]dieneN$_4$ ligands this band occurs at approximately 510 nm with $\varepsilon \sim 125$ liter mol^{-1} cm^{-1}. Three or four $d \rightarrow d$ transitions are predicted for these complexes according to the symmetry but are not usually resolved as is typical for square-planar

TABLE 5. Absorption Spectra of Representative Copper(II) Complexes with Macrocyclic Ligands

Complex	Ligand	$\lambda_{max}(\varepsilon)^a$	Ref.
[Cu(8,10-Me$_2$[13]7,10-dieneN$_4$)](PF$_6$)$_2$		529 (175); 329 (428); 256 (5800)^b	103
[Cu(8,10-Me$_2$[13]7,10-dienatoN$_4$)]PF$_6$	28	567 (84); 441 (82); 338 (12,700); 217 (26,000)^c	103
[Cu(*ms-trans*-Me$_6$[14]aneN$_4$)](ClO$_4$)$_2$	10	515 (133); 272 (8310)^b 510 (132); 270 (5200)^b	104 25
[Cu(*rac-trans*-Me$_6$[14]aneN$_4$)](ClO$_4$)$_2$	10	510 (184); 270 (5000)^b	25
{[Cu(*ms-trans*-Me$_6$[14]aneN$_4$)]$_2$CN}(ClO$_4$)$_3$	10	833; 690^d	105
[Cu(*ms-trans*-Me$_6$[14]aneN$_4$)CN]ClO$_4 \cdot$H$_2$O	10	926; 641^d 617 (219)^c	105
(+)-[Cu((−)-5,14-*cis*-Me$_6$[14]aneN$_4$)](ClO$_4$)$_2$	11	518 (138); 271 (6800)^e; 658 (+0.023); 515 (+0.18); 299 (−4.20); 258 (+7.00)^{e,f}	27
[Cu(*N*-Me$_4$[14]aneN$_4$)](ClO$_4$)$_2$	7	627 (227); 583 (214)^b	30
[Cu(*N*-Me$_4$[14]aneN$_4$)Br]ClO$_4$	7	787 (122)^g	30
[Cu(5,7,7-Me$_3$[14]4-eneN$_4$)]ZnCl$_4$	16	518 (149)^b	42
[Cu(5,12-Me$_2$[14]4,11-dieneN$_4$)](ClO$_4$)$_2$		508 (87); 265 (5100); 244 (6300)^b	43
[Cu(5,7-Me$_2$[14]4,7-dieneN$_4$)](PF$_6$)$_2$		500 (150); 345 (2500); 230 (12,000)^c	44
[Cu(5,7-Me$_2$[14]4,7-dienatoN$_4$)]PF$_6 \cdot$H$_2$O	29	625 (100); 426 (200); 342 (20,000); 230 (25,000)^c	44
[Cu(*rac*-Me$_6$[14]4,11-dieneN$_4$)](ClO$_4$)$_2$	17	505 (126); 260 (5400)^b; 504 (105); 260 (5460); <200 (>11,000)^b	106 104
[Cu(*ms*-Me$_6$[14]4,11-dieneN$_4$)](ClO$_4$)$_2$	17	501 (122); 260 (6550); <200 (>11,000)^b	104
[Cu(Me$_8$[14]4,11-dieneN$_4$)](ClO$_4$)$_2$		509 (110); 262 (5930)^b 518 (105)^b	106 107
[Cu(Et$_4$Me$_2$[14]4,11-dieneN$_4$)](ClO$_4$)$_2$		508 (131); 263 (7000)^b	106
[Cu(*ms*-5,12-Me$_2$-7,14-Ph$_2$[14]4,11-dieneN$_4$)](ClO$_4$)$_2$		526 (130)^e	45
[Cu(Me$_6$[16]4,12-dieneN$_4$)](ClO$_4$)$_2$	23	590 (280); 286 (6100); 212 (7000)^b	49

^{a} In nm. $\quad$ ^{b} H$_2$O solution. $\quad$ ^{c} Methanol solution. $\quad$ ^{d} Solid state, diffuse reflectance.

^{e} Acetonitrile solution. $\quad$ ^{f} Values of $\Delta\varepsilon$ in parentheses. $\quad$ ^{g} Nitromethane solution.

copper(II) coordination. Increased resolution of this band envelope into two components is achieved by circular dichroism (CD) measurements[27] of an optically active [14]aneN$_4$ complex of Cu(II). The position of the $d \rightarrow d$ band is often observed to be solvent dependent, indicating axial interaction between donor solvents and the copper ion.[3,104,105,107] Spectral evidence for solvent and ionic association with [Cu(Me$_6$[14]4,11-dieneN$_4$)]$^{2+}$ has been reported[3,104,105] and approximate equilibrium constants have been evaluated[104] for association with various anions. In the case of cyanide, bridged five- and six-coordinate species have been isolated.[105] The low-energy $d \rightarrow d$ bands in these complexes are characteristic of five- and six-coordinated Cu(II)[105] and are found also in [Cu(Me$_4$[14]aneN$_4$)Br]$^+$, which is proposed to have a five-coordinate geometry.[30]

All of the [14]4,11-dieneN$_4$ complexes of copper(II) exhibit a strong band at ~ 260 nm ($\varepsilon \sim 5000$–8000). This band is also present in the spectra of Cu(II) complexes of [14]aneN$_4$ and on this basis is assigned to a charge transfer transition.[3,25,104,106] Endicott and co-workers[104] have suggested that the unusual broadness of the 260-nm band in [Cu(Me$_6$[14]4,11-dieneN$_4$)]$^{2+}$ arises from the overlap of both Cu $\rightarrow$ L and L $\rightarrow$ Cu transitions. This proposal gains support from CD measurements,[27] which resolve this band into two components. An intense band at < 200 nm ($\varepsilon > 11,000$) in the [14]4,11-dieneN$_4$ complexes disappears in the [14]aneN$_4$ analogs and is assigned as a $\pi \rightarrow \pi^*$ transition of the azomethine chromophore.[3,25,104]

The complexes [Cu(8,10-Me$_2$[13]7,10-dienatoN$_4$)]$^{+}$[103] and [Cu(5,7-Me$_2$[14]4,7-dienatoN$_4$)]$^{+}$[44] exhibit a different pattern of $d \rightarrow d$ transitions. Two bands are resolved in the ligand field spectra of these complexes and they have been tentatively assigned as the $^2A_{1g} \leftarrow {}^2B_{1g}$ and $^2E_g \leftarrow {}^2B_{1g}$ transitions of square-planar copper(II).[44,103] When these two complexes are protonated to yield Cu(II) complexes with neutral N$_4$ macrocycles, only one $d \rightarrow d$ band is observed at ~ 500–530 nm.[44,103]

Copper(II) complexes with Schiff-base macrocycles of type **40** and related analogs also display a single $d \rightarrow d$ band in the range 650–740 nm.[70] Increased resolution of this band has been achieved by circular dichroism measurements[71] of optically active derivatives of **40**. The ligand field bands of nitro-subsituted analogs **42** are obscured by intense intraligand transitions.[73] Mononuclear copper(II) complexes of binucleating Schiff-base ligands also exhibit a single $d \rightarrow d$ band in the same wavelength region.[96,108,109]

Unusual spectral properties have been reported[110] for the copper(II) complexes with the series of tetrathia macrocycles [12–16]aneS$_4$ (Table 6). The intense band at ~ 600 nm is similar to that observed for blue copper sites[111–113] in certain proteins and suggests that thioether coordination to the copper(II) ion may be present in these proteins. A distorted coordination geometry is not responsible for the intense band since similar spectra are observed for [Cu([14]aneS$_4$)]$^{2+}$, with a square-planar geometry,[114] and for

TABLE 6. Spectral Data for
Copper(II)-Tetrathiaether Macrocycles[a,b]

Compound	Ligand	$\nu_{max}(\varepsilon)^c$
$[Cu([12]aneS_4)]^{2+}$	—	675(2000), 387(6000)
$[Cu([13]aneS_4)]^{2+}$	—	625(1800), 390(6000)
$[Cu([14]aneS_4)]^{2+}$	**46**	570(1900), 390(8200)
$[Cu([15]aneS_4)]^{2+}$	—	565(1140), 414(8000)
$[Cu([16]aneS_4)]^{2+}$	—	603(800), 440(6100)

[a] Taken from reference 110.
[b] In 80% methanol, 20% water (wt./wt.) at 25°C, $\mu = 0.1\ M$ (HClO_4).
[c] In nm.

$[Cu([12]aneS_4)]^{2+}$ and $[Cu([13]aneS_4)]^{2+}$ where the ligands are too small to permit planar coordination. The high redox potentials ($E_0' \sim +0.7$ V vs. SHE in 80% methanol) for these complexes also mimic the behavior of blue copper proteins, but the EPR parameters for $[Cu([14]aneS_4)](ClO_4)_2$[115] are not in line with the unusual parameters reported[111,112] for the blue copper sites. The EPR spectrum of $[Cu([14]aneS_4)]^{2+}$ is markedly similar[115] to that observed for other CuS_4 complexes (Table 7).[116,117]

3.2. Magnetic Interactions in Binuclear Macrocyclic Copper Complexes

Mononuclear macrocyclic complexes containing Cu(II) invariably behave as magnetically dilute species and exhibit values of μ_B in the range 1.8–2.0. A variety of dicopper complexes with binucleating macrocyclic ligands, on the other hand, show various degrees of antiferromagnetic interactions. A preliminary report[118] suggests a weak Cu–Cu interaction in the dicopper com-

TABLE 7. EPR Parameters for Some Four-Coordinate
Copper(II) Thio Chelates[a]

Compound	$g_{\parallel}$	$g_{\perp}$	$|A_{\parallel}|^b$	$|A_{\perp}|^b$	$\langle|A|\rangle^b$
$[Cu([14]aneS_4)]^{2+}$ [c,d]	2.088	2.027	172	48	89
$[^{63}Cu[14]aneS_4)]^{2+}$ [d,e]	2.087	2.026	172	45	87
$[Cu(MNT)_2]^{2-}$ [f]	2.086	2.026	162	39	80
$[Cu(DTC)_2]^g$	2.084	2.020	159	36	79

[a] Taken from reference 115.   [b] In units of 10^{-4} cm^{-1}.   [c] Powder sample.
[d] In the corresponding Ni host.   [e] Single crystal.
[f] MNT = dithiomaleonitrile, SC(CN)C(CN)S^{2-}.
[g] DTC = diethyldithiocarbamate, $(C_2H_5)_2NCS_2^-$.

plex of **57**, but no detailed studies have been carried out. The dicopper complex of **54**, where $n = m = 3$, exhibits strong antiferromagnetic interaction with $\mu_B = 0.59$ at room temperature.[95] Variable-temperature magnetic susceptibility measurements confirm the antiferromagnetic behavior.[95] Similar magnetic behavior is observed for the dicopper complexes of **58**[97] ($\mu_B = 0.59$), **59**[119] ($\mu_B = 0.79$), and related complexes.

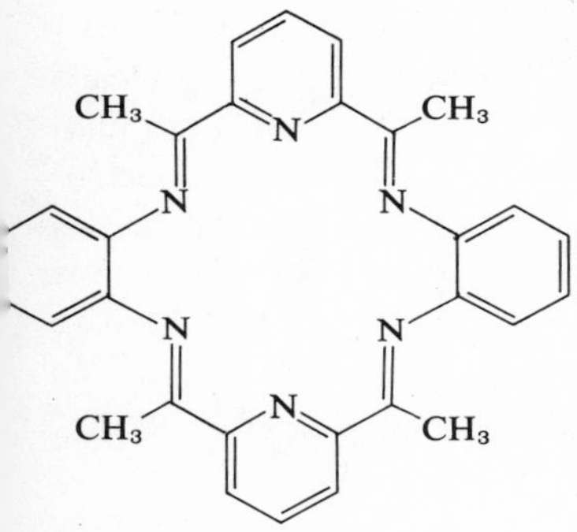

57

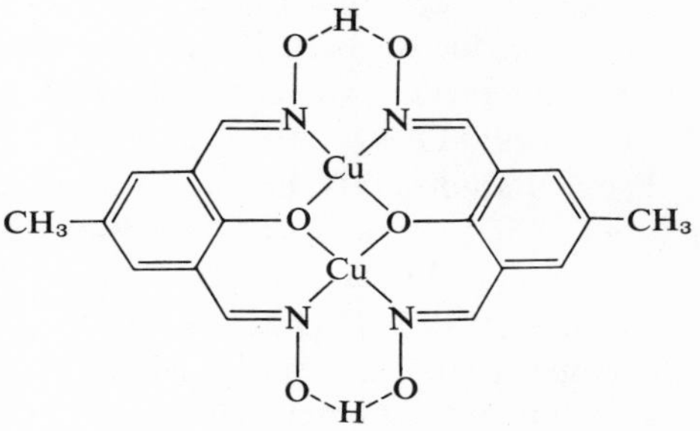

58

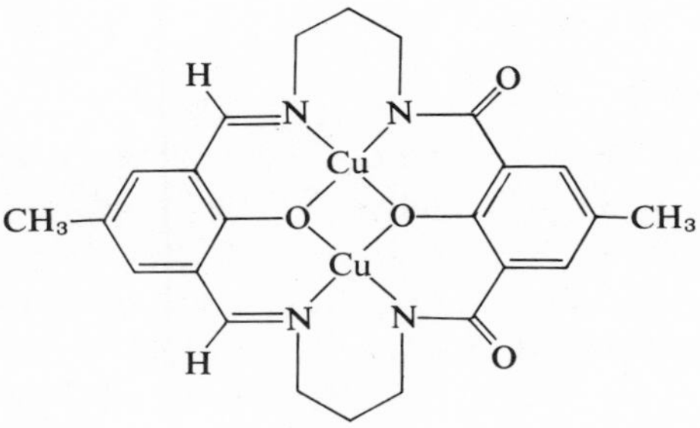

59

3.3. Macrocyclic Complexes of Copper(I) and Copper(III)

The spectra of electrochemically produced Cu(I) derivatives of *ms*-Me$_6$[14]4,11-dieneN$_4$, **17**, and *ms-trans*-Me$_6$[14]aneN$_4$, **10**, in acetonitrile solution have been reported.[120] The unsaturated *trans*-diene complex exhibits an intense Cu → L charge transfer band at 415 nm ($\varepsilon = 5700$) and an additional transition at 283 nm ($\varepsilon = 5600$). A similar spectrum for a transient *trans*-diene Cu(I) complex in aqueous solution has been obtained during a pulse radiolysis study.[121] The low-energy charge transfer band is absent in the saturated analog which exhibits bands at 560 nm (7), 241 nm (7570), and 219 nm (9670).

The complexes [Cu(ms-Me$_6$[14]4,11-dieneN$_4$)]$^{3+}$ and [Cu(ms-$trans$-Me$_6$-[14]aneN$_4$)]$^{3+}$ were prepared by controlled potential oxidation of the corresponding Cu(II) species.[120] Both Cu(III) derivatives undergo spontaneous reduction in acetonitrile solution at room temperature but are stable at -15 to $-20°$C. The spectra of the $3d^8$ Cu(III) complexes consisted of intense bands at 425 nm (15,000), 375 nm (12,000), and 275 nm (6700) for the tetraamine complex and 395 nm (14,530) and 335 nm (12,690) for the diene complex. No d–d bands are observed and the observed intense bands are attributed to charge transfer transitions.

The reduction of [Cu(TAAB)]$^{2+}$, (TAAB = **31**) with H$_2$/PtO$_2$ or Hg leads to a diamagnetic royal blue complex [Cu(TAAB)]$^+$.[59] This species is formulated as [Cu^{III}(TAAB^{2-})]$^+$ rather than [Cu^I(TAAB^0)]$^+$ on the basis of electrochemical studies[122] and the lowering of the C=N stretching frequency as being consistent with the reduction of the TAAB ligand. The UV–visible spectrum of Cu(TAAB)PF$_6$ exhibits the following bands: 690 nm (6410), 350 nm (13,600), 310 nm (18,200); Cu(TAAB)H$_g$Cl$_3 \cdot$3H$_2$O, 710 nm (5,280), 360 nm (11,200), 325 nm (14,400). These intense parity allowed bands support the argument for a delocalized TAAB^{2-} ligand.

4. Cobalt Complexes

4.1. Cobalt(II) Macrocyclic Complexes

Although a large number of macrocyclic complexes of cobalt(II) have been reported, only a few detailed studies of their ligand field spectra have appeared. One such study is the description of a series of cobalt(II) complexes[123] with Me$_2$pyo[14]trieneN$_4$ (**25**) as low-spin, five-coordinate species. The spectra of the complexes, [Co(Me$_2$pyo[14]trieneN$_4$)X]$^+$ and [Co(Me$_2$-pyo[14]trieneN$_4$)A]$^{2+}$, were interpreted using an energy level diagram[124] for trigonal bipyramidal complexes of cobalt(II). The magnetic moments for most of these complexes fall in the range $\mu_B = 1.91$–2.09, which is consistent with low-spin pentacoordinate cobalt(II) but lower than generally reported for square-planar cobalt(II).[125–127] The complexes [Co(Me$_2$pyo[14]trieneN$_4$)-Br]X, where X$^-$ = ClO$_4^-$, PF$_6^-$, or B(C$_6$H$_5$)$_4^-$, exhibit low magnetic moments ($\mu_B = 1.05$–1.20) indicative of a cobalt–cobalt interaction.[123] Preliminary reports of dimeric formulations for [Co(Me$_6$[14]4,14-dieneN$_4$)CN]$_2^{2+}$[128] (ligand = **18**) and [Co(TAAB)Br]$_2^{2+}$[129] containing cobalt–cobalt bonding are apparently erroneous. Both complexes are Co(III) derivatives, [Co(Me$_6$-[14]4,14-dieneN$_4$)(CN)$_2$]ClO$_4$[130] and [Co(TAAB)Br$_2$]Br[123] thus accounting for the low magnetic moments.

Another series of complexes,[131] [Co(Me$_6$[14]1,4,8,11-tetraeneN$_4$)(X)]-ClO$_4$ and [Co(Me$_6$[14]1,4,8,11-tetraeneN$_4$)(S)](ClO$_4$)$_2$, have also been shown to be pentacoordinate. In this case the ligand field spectra have been tentatively interpreted with an energy level diagram[132] for low-spin, square-pyramidal d^7 systems. The magnetic moments for the tetraene series are all in the range 1.97–2.13 B.M. and show no evidence of magnetic interactions between the cobalt(II) ions. The pentacoordinate nature of these complexes is verified by EPR measurements. The observed g values for [Co(Me$_6$[14]1,4, 8,11-tetraeneN$_4$)(CH$_3$CN)](ClO$_4$)$_2$ in acetonitrile solution at 77 K are $g_x = 2.323$, $g_y = 2.243$, and $g_z = 2.017$. Superhyperfine splittings ($A_N = 14.1$ G) corresponding to interaction with one nitrogen donor in an axial position are observed superimposed on the hyperfine structure ($A_z = 113$ G) of the g_z component. Similar EPR behavior is observed for [Co(Me$_6$[14]1,4,8,11-tetraeneN$_4$)](ClO$_4$)$_2$ in nitromethane solution with $A_z = 109$ G and $A_N = 16.1$ G. EPR spectra for the tetraene complexes are very similar to those reported for cobalt(II) base-coordinated cobalamin and for five-coordinate cobaloxime and cobinamide complexes.[133,134]

Green and Tasker[70] have shown that the spectra of a series of low-spin cobalt(II) complexes with **40** compare closely with spectra observed for acyclic Schiff-base analogs but have not given detailed spectral assignments. Subsequently, additional d–d bands in the near infrared region were reported[135] for the acyclic complexes together with a tentative ligand field assignment.[135,136] The magnetic moments for the square-planar Co(II) complexes of **40** lie in the range 2.15–2.51 B.M. Cobalt(II) complexes of a similar series of macrocyclic ligands have also been reported[73] without spectral interpretation. The EPR parameters for the Co(II) complex of **40** with $R = en$ are $g_{xx} = 2.713$, $g_{yy} = 1.979$, and $g_{zz} = 1.993$ with A_{xx}^{Co}, A_{yy}^{Co}, and A_{zz}^{Co} equal to 42.8, 30.4, and 22.0 G, respectively.[137] These values are quite similar to those reported for the acyclic analog,[135] which was assigned a $^2A_1$ ground state based on the theoretical model of Maki *et al*.[100] More recent studies of the acyclic analog utilizing a more sophisticated model including second-order and quartet terms have concluded that the ground state is $^2B_1$.[138,139] Circular dichroism studies[71] of optically active derivatives of **40** show that the same stereochemical effects dictate the observed CD of the macrocycles as for the acyclic Schiff-base complexes.[135] These effects are the conformation of the central chelate ring and a stereospecific pseudo-tetrahedral distortion of the donor atom array as detected by exciton splitting in the CD bands attributed to the azomethine chromophore.

Cobalt(II) complexes of Me$_6$[14]4,11-dieneN$_4$[140,141] and *trans*-Me$_6$[14]-aneN$_4$[141] have been reported and the spectra[141] of these complexes in acetonitrile have been given without interpretation. Unique high-spin dicobalt(II) complexes of the macrocyclic binucleating ligand **54** ($n = m = 3$) have been reported by Pilkington and Robson.[95] The square-pyramidal geometry of the cobalt(II) coordination spheres which was inferred from the spectral properties and magnetic behavior has been confirmed by an X-ray structure.[142]

A low-spin Co(II) complex ($\mu_B = 1.97$) of a sexadentate N_4S_2 macrocycle has been reported[90] but no assignments were given for the intense bands observed in the visible spectrum. The structures of the two known examples of cobalt(II) clathrochelates approximate trigonal prismatic co-ordination.[8,9,10,93,143,144] A detailed analysis[93] of the ligand field spectrum of the high-spin Co(II) complex of **53** is hampered by the presence of a Co(I) impurity exhibiting intense bands in the same spectral region.

4.2. Macrocyclic Cobalt(III) Complexes

Ligand field spectra provide a primary means for the characterization of cobalt(III) (diamagnetic, low-spin $3d^6$) complexes of macrocyclic ligands. Calculation of ligand field parameters for diacidotetramine cobalt(III) complexes usually follows the treatment of Wentworth and Piper[15] for analysing the splitting of the two transitions of octahedral cobalt(III) species ($^1T_{1g} \leftarrow {}^1A_{1g}$; $^1T_{2g} \leftarrow {}^1A_{1g}$) which occurs under the influence of a low-symmetry ligand field (Figure 3). For *cis*-folded macrocyclic tetramine complexes of Co(III)[3,7,22,145-148] (Table 8) the anticipated splitting for a C_{2v} donor atom array is not observed and the two bands are assigned on the basis of the octahedral model. The average ligand field strength (Dq^{av}) for the complex can be calculated from

$$Dq^{av} = (\nu_{1 T_{1g} \leftarrow {}^1A_{1g}} + C)/10$$

where C is given the value 3800 cm^{-1}. The ligand field strength of the macro-cyclic ligand can be estimated from Dq^{av} by the principle of averaged environment if Dq values for the nonmacrocyclic ligands are known.

The range of values for Dq^{av} for the two series *cis*-[Co([12–16]aneN$_4$)-CO$_3$]$^+$ and *cis*-[Co([12–14]aneN$_4$)Cl$_2$]$^+$ (Table 8) is very small, indicating only a slight, nonsystematic effect of ring size on the ligand field strength of folded macrocyclic ligands. The tendency for macrocyclic tetramine ligands to form folded complexes[7] is [12]aneN$_4$, *cis* only; [13]aneN$_4$ and [14]aneN$_4$,

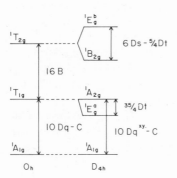

FIGURE 3. Ligand field diagram for a low-spin d^6 ion in O_h and D_{4h} symmetries.

TABLE 8. Absorption Spectra of Some Cobalt(III) Complexes with Macrocyclic Ligands in a *cis*-Folded Configuration

Complex	Ligand	$\lambda_{max}(\varepsilon)^a$	Dq^{av} (cm^{-1})	Ref.
[Co([12]aneN$_4$)CO$_3$]ClO$_4$	**1**	530(280); 368(210)^b	2267	22
[Co([13]aneN$_4$)CO$_3$]ClO$_4$	**3**	501(178); 356(133)^b	2376	7
[Co([14]aneN$_4$)CO$_3$]ClO$_4$	**6**	520(154); 365(140)^b	2303	7
[Co([15]aneN$_4$)CO$_3$]ClO$_4$	**12**	528(138); 366(182)^b	2274	7
[Co([16]aneN$_4$)CO$_3$]ClO$_4$	**13**	540(146); 375(197)^b	2232	7
[Co(*trans*-Me$_6$[14]aneN$_4$)CO$_3$]$^{+\ c,d}$	**10**	549^e; 386^e	2160	3
[Co(Me$_6$[14]4,11-dieneN$_4$)CO$_3$]$^{+\ c,d}$	**17**	515^e; 373^e	2280	3
[Co([12]aneN$_4$)Cl$_2$]Cl	**1**	560(185); 390(165)^f	2166^g	22
[Co([13]aneN$_4$)Cl$_2$]ClO$_4$	**3**	540(124)^{c,h}	2232	7
[Co([14]aneN$_4$)Cl$_2$]Cl	**6**	558(105)^{b,h}	2172^g	145
[Co(Me$_6$[14]4,11-dieneN$_4$)en]Cl(ClO$_4$)$_2$	**17**	495(133)^{h,i}	2400^g	146
[Co(Me$_6$[14]4,11-dieneN$_4$)acac](ClO$_4$)$_2$	**17**	522(181)^{h,j}	2296^g	146
[Co(5,12-Me$_2$-7,14-Ph$_2$-[14]4,11-dieneN$_4$)en](ClO$_4$)$_3$		478(166)^{h,i}	2472^g	146
[Co(5,12-Me$_2$-7,14-Ph$_2$-[14]4,11-dieneN$_4$)acac](ClO$_4$)$_2$		512(238)^{h,j}	2333^g	146
[Co(*ms*-Me$_2$pyo[14]eneN$_4$)Cl$_2$]	**27**	549(160); 418(191)^k	2200^g	147
[Co(*ms*-Me$_2$pyo[14]eneN$_4$)Br$_2$]	**27**	578(144); 452(641)^k	2110^g	147
[Co(*ms*-Me$_2$pyo[14]eneN$_4$)CO$_3$]	**27**	518(292); 370(339)^b	2310^g	147
[Co([14]aneS$_4$)Cl$_2$]BF$_4$	**46**	533(654); 420^{e,l}	2256^g	148
[Co([14]aneS$_4$)Br$_2$]BF$_4$	**46**	550(640)^{h,l}	2198^g	148
[Co([14]aneS$_4$)(NO$_2$)$_2$]BF$_4$	**46**	470(814)^{h,l}	2508^g	148

^{a} In nm. ^{b} H$_2$O solution. ^{c} Methanol solution. ^{d} Anion not specified.
^{e} ε not reported. ^{f} 30% HCl. ^{g} Calculated from reported data.
^{h} Higher-energy band is masked. ^{i} In 10:1 DMSO: 20% aqueous HClO$_4$.
^{j} Acetonitrile solution. ^{k} DMSO solution. ^{l} Nitromethane solution.

both *cis* and *trans*; [15]aneN$_4$ and [16]aneN$_4$, *trans* only unless forced to adopt *cis* configuration by the presence of a bidentate ligand such as carbonate. Unsaturated 14-membered N$_4$ macrocycles with up to two double bonds, e.g., Me$_6$[14]4,11-dieneN$_4$[3,146] **17**, and Me$_2$pyo[14]eneN$_4$,[147] **27**, are also observed to form *cis* complexes. A series of cobalt(III) complexes with folded configuration have been reported for [14]aneS$_4$,[148] **46**, with Dq^{av} values comparable to the corresponding complexes of [14]aneN$_4$.

Trans macrocyclic N$_4$ complexes of cobalt(III)[3,7,43,54,131,146,147,149–151] (Table 9) exhibit a pronounced splitting of the $^1T_{1g} \leftarrow {}^1A_{1g}$ band into two components $^1E_g{}^a \leftarrow {}^1A_{1g}$ and $^1A_{2g} \leftarrow {}^1A_{1g}$, and a complete ligand field analysis may be based on a D_{4h} model.[15] For complexes where $Dq^{xy} > Dq^z$ the transition $^1E_g{}^a \leftarrow {}^1A_{1g}$ lies at lower energy. If only this transition is observed (for example, when the higher-energy bands are obscured by ligand or charge transfer transitions), then Dq^{xy} for the macrocyclic ligand can be estimated from the relationship

$$Dq^{xy} = \tfrac{1}{5}(\nu_1{}_{E_g{}^a \leftarrow {}^1A_{1g}} + C) - Dq^z$$

TABLE 9. Ligand Field Spectral Parameters for Representative *Trans* Complexes of Cobalt(III) with N₄ Macrocyclic Ligands

Compound	Ligand	$\nu_{max}(\epsilon)^a$			$Dq^{z\,b}$	Dt^b	$Dq^{xy\,b}$	Ref.
		$^1E_g{}^a \leftarrow {}^1A_{1g}$	$^1A_{2g} \leftarrow {}^1A_{1g}$	$^1B_{2g} + {}^1E_g{}^b \leftarrow {}^1A_{1g}$				
[Co([13]aneN₄)Cl₂]ClO₄	3	600(33)	422(144)^c	Masked	1345	803	2750	7
[Co([14]aneN₄)Cl₂]Cl·H₂O	6	617(35)	458(24)	387(56)^c	1437	642	2562	7, 149
[Co(2,3-Me₂[14]aneN₄)Cl₂]ClO₄	8	625(112)	455(112)	382(185)^c	1460^d	595	2500	150
[Co(2,3-Me₂[14]aneN₄)Br₂]ClO₄	8	667(41)	Masked	374(2320)^c	1277^d	688	2480	150
[Co(ms-trans-Me₆[14]aneN₄)Cl₂]ClO₄	10	637	488	412^{c,e}	1460^d	560	2440	3, 151
[Co(rac-trans-Me₆[14]aneN₄)Cl₂]ClO₄	10	645	483	412^{c,e}	1460^d	536	2400	3, 151
[Co(ms-5,12-Me₂[14]4,11-dieneN₄)Cl₂]ClO₄	—	616(47)	435(49)	320sh^f	1460^d	622^g	2550^g	43
[Co(rac-5,12-Me₂[14]4,11-dieneN₄)Cl₂]ClO₄	—	616(45)	435(58)	320sh^f	1460^d	622^g	2550^g	43
[Co(Me₂[14]1,3-dieneN₄)Cl₂]ClO₄	19	610(89)	Masked	397(672)^c	1460^d	640	2580	150
[Co(Me₂[14]1,3-dieneN₄)Br₂]ClO₄	19	635(99)	Masked	382(2240)^c	1277^d	775	2630	150
[Co(ms-Me₆[14]4,11-dieneN₄)Cl₂]ClO₄	17	629^{c,e,h}	Masked	Masked	1460^d	582^g	2480^h	3
[Co(rac-Me₆[14]4,11-dieneN₄)Cl₂]ClO₄	17	629	448	341^{c,e}	1460^d	582^g	2480^h	3
		637(39)^c	Masked	Masked	1460^d	560	2440^g	54

Complex								Ref.
[Co(*rac*-7,14-Me$_2$-5,7,12,14-Et$_4$[14]4,11-dieneN$_4$(Cl$_2$)]ClO$_4$	—	637(38)^c	Masked	Masked	1460^a	560^g	2440^g	54
[Co(5,12-Me$_2$-7,14-Ph$_2$[14]4,11-dieneN$_4$)Cl$_2$]ClO$_4$	—	620(64)^f	Masked	Masked	1460^a	611^g	2530^g	146
[Co(Me$_6$[14]1,4,8,11-tetraeneN$_4$)Cl$_2$]ClO$_4$	62	599(52)^f	Masked	Masked	1460^a	674	2640	131
[Co(Me$_6$[14]1,4,8,11-tetraeneN$_4$)Br$_2$]ClO$_4$	62	633(87)^f	Masked	Masked	1277^a	792	2663	131
[Co(Me$_4$[14]1,3,8,10-tetraeneN$_4$Cl$_2$]PF$_6$	20	575(46)	429(144)	397(1100)^i	1460^a	754	2780	150
[Co(Me$_4$[14]1,3,8,10-tetraeneN$_4$)Br$_2$]PF$_6$	20	592(80)	Masked	392(2680)^i	1277^a	903	2860	150
[Co([15]aneN$_4$)Cl$_2$]NO$_3$ (Isomer I)^j	12	655(40)	520(69)	422(106)^c	1511	452	2303	7
[Co([15]aneN$_4$)Cl$_2$]Cl (Isomer II)^j	12	640(27)	490(38)	415(62)^c	1465	546	2421	7
[Co([16]aneN$_4$)Cl$_2$]ClO$_4$ (Isomer I)^j	13	679(57)	535(34)	430(99)^c	1457	452	2249	7
[Co([16]aneN$_4$)Cl$_2$]ClO$_4$ (Isomer II)^j	13	662(41)	510(35)	419(76)^c	1441	514	2341	7
[Co(*ms*-Me$_2$pyo[14]eneN$_4$)Cl$_2$]ClO$_4$ (α-Isomer)^j	27	617(44)	472(181)	386(148)^c	1500	570	2500^k	147
[Co(*ms*-Me$_2$pyo[14]eneN$_4$)Cl$_2$]ClO$_4$·H$_2$O (β-Isomer)^j	27	610(41)	465(166)	392(148)^c	1540	550	2500^k	147

^{a} In nm. ^{b} In cm^{-1}. ^{c} Methanol solution. ^{d} Assumed value. ^{e} ε values not reported. ^{f} Acetonitrile solution.
^{g} Calculated from spectral data given in this reference. ^{h} Differing values given in reference 147 are in error.[3] ^{i} Acetone solution.
^{j} Configurational isomer about secondary amine donors. ^{k} Average value for a series of complexes.

where a value of Dq^z is taken from other studies, and $C = 3800 \text{ cm}^{-1}$. This method has been used most extensively to determine the values of Dq^{xy} given in Table 9. The parameter Dt is then evaluated from the expression

$$Dt = (4/7)(Dq^{xy} - Dq^z)$$

When both low-symmetry components derived from the first octahedral band are observed a complete analysis[7] includes an independent evaluation of Dq^{xy}, Dt, and Dq^z according to

$$Dq^{xy} = (1/10)(\nu_{1_{A_{2g}\leftarrow {}^1A_{1g}}} + C)$$
$$Dt = (-4/35)(\nu_{1_{E_g{}^a\leftarrow {}^1A_{1g}}} - \nu_{1_{A_{2g}\leftarrow {}^1A_{1g}}})$$
$$Dq^z = Dq^{xy} - (7/4)Dt$$

Extensive correlations of Dq^{xy} have been made with macrocyclic ring size, number of substituents, and degree of unsaturation. For the complexes $[Co([13–16]aneN_4)Cl_2]^+$, Dq^{xy} decreases in the order $[13]aneN_4 \gg [14]aneN_4 > [15]aneN_4 > [16]aneN_4$.[7] The high value of $[13]aneN_4$ is attributed to a constrictive effect as indicated by strain energy calculations.[7] Conversely, the 15- and 16-membered macrocycles show a dilative effect with an opening too large to adopt normal Co–N bond lengths without ring strain.[7]

The presence in the macrocyclic ligand of two isolated double bonds (i.e., the $[14]4,11$-dieneN$_4$ ligands) produces no significant increase in Dq^{xy} and the introduction of a single di-imine unit (Me$_2[14]1,3$-dieneN$_4$) yields only a slight increase in Dq^{xy} as measured for the cobalt(III) complexes (Table 9). Four isolated azomethine linkages (Me$_6[14]1,4,8,11$-tetraeneN$_4$) or two di-imine units (Me$_4[14]1,3,8,10$-tetraeneN$_4$) produce significant increases in the macrocyclic ligand field strength (Table 9). A slight decrease in Dq^{xy} is noted for increasing methyl substitution in the following series: $[14]aneN_4$ (2562 cm^{-1}); $2,3$-Me$_2[14]aneN_4$ (2500 cm^{-1}); ms-$trans$-Me$_6[14]aneN_4$ (2440 cm^{-1}); and rac-$trans$-Me$_6[14]aneN_4$ (2400 cm^{-1}). For the ligand $[14]$-aneS$_4$,[148] Dq^{xy} has been evaluated as 2420 cm^{-1} indicating that the thiaether donors can exert an in-plane ligand field strength comparable to those of nitrogen donor ligands. Other tetradentate macrocyclic ligands for which Dq^{xy} has been estimated for cobalt(III) complexes are: TAAB, 31, (2563 cm^{-1})[152] and Me$_2$pyo$[14]$trieneN$_4$, 25, (2820 cm^{-1}).[153] A value of $Dq^{av} = 2520 \text{ cm}^{-1}$ has been calculated for the pentadentate ligand, pyeneN$_5$.[88]

For certain Co(III) complexes with highly unsaturated macrocyclic ligands no ligand field analyses are possible owing to intense intraligand or charge transfer transitions. A series of cobalt(III) complexes, $[Co(L)(py)_2]^+$, where L represents the unsaturated, dinegative ligands 34, 36, and 37, and $[Co(L')(py)_2]^{2+}$, where L' represents the uninegative ligand 38, have been reported by Holm and co-workers.[64] The spectra of these complexes in the visible and ultraviolet region consist of intense bands which are attributed to the highly delocalized ligand chromophore. Cobalt(III) complexes of 38 are

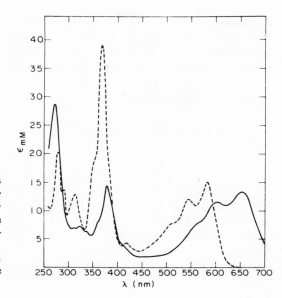

FIGURE 4. Electronic spectra of [Co(Ph$_2$[15]hexaenatoN$_4$)(CN)$_2$] in 9:1 vol./vol. DMSO-H$_2$O (—); dicyanocobalamin in 9:1 vol./vol. DMSO-H$_2$O containing 0.01 M KCN (---). Reprinted from reference 64 with permission. Copyright by The American Chemical Society.

especially noteworthy since they represent close analogs of the Co(III)–corrin moiety in cobalamins and related B$_{12}$ derivatives. Figure 4 illustrates the close agreement between the spectra of [Co(Ph$_2$[15]hexaenatoN$_4$)(CN)$_2$] and dicyanocobalamin. Related cobalt(III) complexes of dibenzocorrin analogs[72,154] have also been prepared but no details of their spectral properties have been reported.

A few examples of cobalt(III) complexes with macrocyclic ligands containing other than four donor atoms have been reported. The complex [Co(TRI)$_2$]$^{3+}$, where TRI represents the trimeric condensate of *o*-aminobenzaldehyde, has been synthesized as both the *meso*- and *dl*-isomers.[155]

60

Resolution of the racemic forms allowed circular dichroism studies[155,156] of this unique optically active complex. The absolute configuration of one of the enantiomers was established by X-ray diffraction studies.[156] The first reported clathrochelate was the Co(III) complex **60**.[10,144] No ligand field spectral analysis was possible since an intense band in the near ultraviolet region obscured the $d \rightarrow d$ transitions.

4.3. Cobalt(I) Macrocyclic Complexes

The spectra of chemically or electrochemically produced cobalt(I) derivatives of $Me_6[14]4,11$-dieneN$_4$, **17**, and $trans$-$Me_6[14]$aneN$_4$, **10**, have been reported.[141] The intense band at 679 nm ($\varepsilon = 15,950$) for [Co(Me$_6$[14]-4,11-dieneN$_4$)]$^+$ is assigned as a Co $\rightarrow$ ligand charge transfer transition. In the tetramine complex, lacking the azomethine unsaturation, the Co $\rightarrow$ L charge transfer transition occurs at 393 nm. A similar spectrum for [Co(Me$_6$[14]4,11-dieneN$_4$)]$^+$ was obtained in a pulse radiolysis study.[157]

5. Iron Complexes

5.1. Low-Spin (S = 0) Iron(II) Macrocycles

Six-coordinate iron(II) macrocyclic complexes are most commonly low spin.[158–164] These complexes exhibit magnetic moments in the range 0.5–0.7 B.M. attributed to temperature-independent paramagnetism (TIP). The ligand field spectra of six-coordinate spin-paired d^6 Fe(II) complexes can be interpreted with the crystal field diagram originally derived for complexes of Co(III) (Figure 3). Very often the unsaturated ligands which are usually required to produce low-spin Fe(II) also produce low-energy charge transfer transitions which obscure the $d \rightarrow d$ bands. Certain macrocyclic ligands have the advantage of yielding the low-spin configuration with clear observation of the ligand field transitions. Even here, however, sufficient spectral details to allow calculation of Dq^{xy} are available for only a few compounds.[158–161]

The reported spectra for selected low-spin, six-coordinate Fe(II) complexes[158–161] are given in Table 10. For those complexes described as tetragonal, the two lowest-energy bands are assigned as transitions to the split $^1T_{1g}$ state, i.e., $^1E_g{}^a \leftarrow {}^1A_{1g}$ and $^1A_{2g} \leftarrow {}^1A_{1g}$. The ordering of these transitions depends on the relative values of Dq^{xy} and Dq^z. For $Dq^{xy} > Dq^z$, $^1E_g{}^a$ lies below $^1A_{2g}$ and for $Dq^{xy} < Dq^z$ (as in the case of CN$^-$ derivatives), $^1A_{2g}$ is lowest. Those complexes where Dq^{xy} and Dq^z are close in value are described as pseudo-octahedral with the lowest-energy transition given as $^1T_{1g} \leftarrow {}^1A_{1g}$. When not obscured by ligand transitions, the $^1T_{2g} \leftarrow {}^1A_{1g}$ transition is observed at higher energy and is not split significantly in the presence of a tetragonal field.

TABLE 10. Representative Ligand Field Spectra for Tetragonal, Low-Spin Iron(II) Macrocycles

Compound	Ligand	$^1E_g^a \leftarrow {}^1A_{1g}$	$^1A_{2g} \leftarrow {}^1A_{1g}$	$^1B_{2g} + {}^1E_g^b \leftarrow {}^1A_{1g}$	$Dq^{z\,b}$	$Dq^{xy\,b}$	Ref.
[Fe([13]aneN₄)(CN)₂]	3	434(78)	532(18)	358(71)^c	3057	2208	158
[Fe([14]aneN₄)(CN)₂]	6	474(49)	589(5.1)	367(110)^c	2847	2029	158
[Fe([15]aneN₄)(CN)₂]	12	511(27)	661(6.2)	389(64)^c	2732	1842	158
[Fe(*ms-trans*-Me₆[14]aneN₄)(CH₃CN)₂](BF₄)₂	10	599(44)^d		376(47)^e	—	~2100	159
[Fe(Me₆[14]4,11-dieneN₄)(CH₃CN)₂](ClO₄)₂	17	510(85)^{d,e}		Masked		~2000	160
[Fe(Me₂[14]1,4,8,11-tetraeneN₄)(CH₃CN)₂](PF₆)₂		730(46)	469(105)^e	Masked	939	2460	161
[Fe(Me₂[14]1,4,8,11-tetraeneN₄)(SCN)₂]		630(235)	495(sh)^f	Masked	1490	2350	161
[Fe(Me₂[14]1,4,8,11-tetraeneN₄(py)₂](PF₆)₂)		805(580)	500(493)^g	Masked	814	2330	161

a In nm.   *b* In cm⁻¹.   *c* Methanol solution.   *d* Low-energy band is not split. Dq^{xy} estimated.   *e* Acetonitrile solution.   *f* Nitromethane solution.   *g* THF solution.

For the complexes given in Table 10 sufficient spectral details were observed to allow a calculation (or estimation) of Dq^{xy}. The *trans*-dicyano derivatives of [13–15]aneN$_4$[158] exhibit a decrease in Dq^{xy} with increasing ring size. This trend is attributed to a constrictive effect of the smaller macrocycles.[158] The value of Dq^z for CN$^-$ also varies as a function of the macrocyclic ring size. Dq^{xy} for the Me$_2$[14]1,4,8,11-tetraeneN$_4$ ligand[161] is significantly higher than the values for the saturated ligands, as is commonly observed. The small variations in the value of Dq^{xy} for the tetraene ligand with various axial groups are a good indication of the reliability of the calculation. The complexes [Fe(*ms-trans*-Me$_6$[14]aneN$_4$)(CH$_3$CN)$_2$]$^{2+}$ [159] and [Fe(Me$_6$[14]4,11-dieneN$_4$)(CH$_3$CN)$_2$]$^{2+}$ [160] (Table 10) do not exhibit a splitting of the lowest-energy octahedral band and the values of Dq^{xy} are estimated. The spectrum of [Fe(Me$_6$[14]aneN$_4$)(CH$_3$CN)$_2$]$^{2+}$ has been reported elsewhere without interpretation.[162]

Low-spin tetragonal iron(II) complexes with the ligands Me$_6$[14]1,4,11-trieneN$_4$, **61**, Me$_6$[14]1,4,8,11-tetraeneN$_4$, **62**, and Me$_6$[14]1,3,7,11-tetraeneN$_4$, **63**, are produced by oxidative dehydrogenation of [Fe(Me$_6$[14]4,11-dieneN$_4$)-

61   62   63

64   65

(CH$_3$CN)$_2$]$^{2+}$ (ligand **17**) and [Fe(Me$_6$[14]4,14-dieneN$_4$)(CH$_3$CN$_2$)]$^{2+}$ (ligand **18**).[163] The $d \to d$ bands for [Fe(Me$_6$[14]1,4,11-trieneN$_4$)(CH$_3$CN)$_2$]$^{2+}$ and [Fe(Me$_6$[1,4,8,11-tetraeneN$_4$)(CH$_3$CN)]$^{2+}$ occur at energies similar to that for the parent diene complex, but the more highly unsaturated complexes exhibit an enhanced charge transfer band. An intense low-energy absorption at 560 nm (3560) in the complex Fe(Me$_6$[14]1,3,7,11-tetraeneN$_4$)(CH$_3$CN)$_2$]$^{2+}$

is characteristic of the charge transfer band of iron(II) complexes of α-diimines.[163]

Oxidative dehydrogenation of [Fe(*ms-trans*-Me$_6$[14]aneN$_4$)(CH$_3$CN)$_2$]$^{2+}$ (ligand **10**) leads to iron(II) complexes of the unsaturated ligands Me$_6$[14]1,3,8-trieneN$_4$, **64**, and Me$_6$[14]1,3,8,10-tetraeneN$_4$, **65**.[164] The series of low-spin complexes, [Fe(L)A$_2$]Y$_2$ and [Fe(L)X$_2$], L = **64** or **65**, isolated for these ligands all exhibit an intense absorption band in the visible region which is assigned a charge transfer ($t_{2g} \rightarrow \pi^*$) within the iron(II)-diimine chromophore. The energy of the charge transfer band for [Fe(Me$_6$[14]1,3,8,10-tetraeneN$_4$)]$^{2+}$ varies as follows for different axial ligands: NO$_2$$^-$ (15.0 kK), CN$^-$ (15.6 kK), CH$_3$CN (17.3 kK), CH$_3$NC (19.5 kK), and HNC (20.0 kK). Increased π-bonding capability of the axial ligands lowers the energy of the t_{2g} level and increases the energy of the $t_{2g} \rightarrow \pi^*$ transition.[164] A related series of six-coordinate low-spin iron(II) complexes with Me$_4$[14]1,3,8,10-tetraeneN$_4$, **20**, including a unique carbon monoxide adduct, [Fe(L)(CH$_3$CN)(CO)](PF$_6$)$_2$, have been prepared but electronic spectral data were not reported.[165] The spectra of a series of low-spin iron(II) complexes with a macrocyclic ligand derived from 2,6-diacetylpyridine and hydrazine are interpreted as arising from an extensively delocalized system rather than isolated α-diimine chromophores.[166] The spectrum of low-spin [Fe(*ms*-Me$_2$pyo[14]eneN$_4$)(NCS)$_2$] has been reported without interpretation.[167]

5.2. High-Spin (S = 2) Iron(II) Macrocycles

Numerous examples of both five- and six-coordinate high-spin iron(II) complexes with macrocyclic ligands have been reported.[158–160,163,167] Spectral features for complexes with these two geometries (Tables 11a and 11b) are similar, with two weak bands in the near-infrared region assigned as quintet–quintet transitions to the components of the split ^5E_g (octahedral) band (Figure 5).[17] The splitting of the $^5T_{2g}$ level in a low symmetry field is assumed to be small (~ 500 cm^{-1}) and $^5B_{2g}$ is assumed to be the ground state based on results for Fe(py)$_4$X$_2$ complexes,[168,169] where X$^-$ = Cl$^-$, Br$^-$, or I$^-$. The energy of the $^5B_{1g} \leftarrow ^5B_{2g}(^5B_1 \leftarrow ^5B_2)$ transition can be taken as $10Dq^{xy}$, whereas the $^5A_{1g} \leftarrow ^5B_{2g}(^5A_1 \leftarrow ^5B_2)$ transition reflects the ligand field strength of the axial groups. For the macrocyclic complexes, the observed splitting of the two components derived from ^5E_g varies from ~ 2000 to ~ 9000 cm^{-1}. The upper ranges of these splittings are the largest reported for tetragonal iron(II) complexes.

The value of Dq^{xy} for a given ligand varies somewhat as a function of the axial donors. Dq^{xy} for [16]aneN$_4$ varies from 1234 cm^{-1} for two axial acetonitrile molecules to 1108 cm^{-1} for two axial thiocyanate ions but is always weaker than [15]aneN$_4$ [1406 cm^{-1} (CH$_3$CN); 1259 cm^{-1} (NCS$^-$)]. The higher Dq^{xy} value for smaller ring size is similar to the trend observed

TABLE 11a. Representative Ligand Field Spectra for Five-Coordinate (C_{4v}), High-Spin Iron(II) Macrocycles

Compound	Ligand	$\nu_{max}(\varepsilon)^a$ $^5A_1 \leftarrow {}^5E/{}^5B_2$	$^5B_1 \leftarrow {}^5E/{}^5B_2$	Dq^{xy} ^{b}	μ_B	Ref.
[Fe(Me$_6$[14]4,11-dieneN$_4$)Cl]ClO$_4$	**17**	2128(7)	820(5)^c	1220	5.01	160
[Fe(Me$_6$[14]4,11-dieneN$_4$)Br]ClO$_4$	**17**	~2000(~5)	820(4.9)^c	1220	5.11	160
[Fe(Me$_6$[14]4,11-dieneN$_4$)I]ClO$_4$	**17**	~2200(~4)	820(4)^c	1220	5.15	163
[Fe(Me$_6$[14]1,4,8,11-tetraeneN$_4$)Cl]ClO$_4$	**62**	2150(3)	860(3)^c	1162	5.00	160
[Fe(Me$_6$[14]1,4,8,11-tetraeneN$_4$)Br]ClO$_4$	**62**	2100(3)	858(3)^c	1166	5.14	163
[Fe(Me$_6$[14]1,4,8,11-tetraeneN$_4$)I]ClO$_4$	**62**	~2000	858(5)^c	1165	5.10	163
[Fe(*ms*-Me$_2$pyo[14]eneN$_4$)Cl]PF$_6$	**27**	~2040(~3)	926(4)^c	1080	4.95	167
[Fe(*ms*-Me$_2$pyo[14]eneN$_4$)Br]PF$_6$	**27**	~2200(~4)	909(5)^c	1100	5.11	167
[Fe(*ms*-Me$_2$pyo[14]eneN$_4$)I]PF$_6$	**27**	~2380(~5)	877(5)^c	1140	5.20	167
[Fe(*ms*-Me$_2$pyo[14]eneN$_4$)OAc]PF$_6$	**27**	—	862(10)^d	1160	5.11	167

^{a} In nm. ^{b} In cm^{-1}. Evaluated from position of $B_1{}^5 \leftarrow {}^5E/{}^5B_2(C_{4v})$ band.
^{c} Nitromethane solution. ^{d} Methanol solution.

TABLE 11b. Representative Ligand Field Spectra for Six-Coordinate (D_{4h}), High-Spin Iron(II) Macrocycles

Compound	Ligand	$\nu_{max}(\varepsilon)^a$ $^5A_{1g} \leftarrow {}^5E_g/{}^5B_{2g}$	$^5B_{1g} \leftarrow {}^5E_g{}^5B_{2g}$	Dq^{xy} ^{b}	μ_B	Ref.
trans-[Fe([15]aneN$_4$)(NCS)$_2$]	**12**	960(6.1)	794(5.2)^d	1259	5.32	15
trans-[Fe([15]aneN$_4$)(CH$_3$CN)$_2$](PF$_6$)$_2$	**12**	1040(3.6)	711(3.9)^e	1406	5.52	15
trans-[Fe([16]aneN$_4$)(NCS)$_2$]	**13**	1100(1.3)	903(4.6)^d	1108	5.55	15
trans-[Fe([16]aneN$_4$)(CH$_3$CN)$_2$](PF$_6$)$_2$	**13**	1080(3.1)	810(4.4)^e	1234	5.52	15
trans-[Fe(*ms*-Me$_6$[14]aneN$_4$)Cl$_2$]	**10**	1820(~1)	714(1.9)^d	1400	5.58	15
trans-[Fe(*ms*-Me$_6$[14]aneN$_4$)Br$_2$]	**10**	~2000	709(4.8)^d	1410	5.47	1
trans-[Fe(*ms*-Me$_6$[14]aneN$_4$)I$_2$]	**10**	—	704(7)^d	1420	5.37	1
trans-[Fe(*ms*-Me$_6$[14]aneN$_4$)(OAc)$_2$]	**10**	935(1)	714(1.9)^d	1400	5.70	1
trans-[Fe(*ms*-Me$_2$pyo[14]eneN$_4$)(N$_3$)$_2$]	**27**	—	855(55)^c	1170	5.50	1

^{a} In nm. ^{b} In cm^{-1}. Evaluated from position of $^5B_{1g} \leftarrow {}^5E_g/{}^5B_{2g}(D_{4h})$ band.
^{c} Nitromethane solution. ^{d} CHCl$_3$ solution. ^{e} Acetonitrile solution.

for the low-spin, six-coordinate iron(II) complexes. The saturated ligand *ms*-Me$_6$[14]aneN$_4$, **10**, exhibits larger values for Dq^{xy} in the series of six-coordinate species than the unsaturated ligands Me$_6$[14]4,11-dieneN$_4$, **17**, and Me$_6$[14]1,4,8,11-tetraeneN$_4$, **62**, in the five-coordinate complexes. Since increasing unsaturation normally increases Dq^{xy}, this reversed trend is

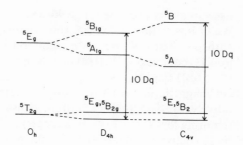

FIGURE 5. Ligand field diagram for high-spin d^6 ion in O_h, D_{4h}, and C_{4v} symmetries.

probably due to the iron atom being out of the plane of the macrocycle in the five-coordinate complexes.

5.3. Intermediate Spin (S = 1) Iron(II) Macrocycles

Two macrocyclic complexes, [Fe(Ph$_2$[14]tetraenatoN$_4$)], ligand **34**, and [Fe(Ph$_2$[16]tetraenatoN$_4$)], ligand **37**, are among the few well-characterized examples of four-coordinate, intermediate-spin iron(II) complexes.[170] Ambient temperature magnetic susceptibilities for the two complexes are 3.26 and 4.54 B.M., respectively. Both compounds follow Curie–Weiss behavior over the range 17–295 K, which confirms the intermediate-spin ($S = 1$) ground state and rules out a spin equilibrium between $S = 0$ and $S = 2$ states. Other synthetic macrocycles apparently exhibit the $S = 1$ ground state with [Fe(Me$_4$[14]tetraenatoN$_4$)][66] (ligand **66**) and [Fe(Me$_4$-

66   **67**

bzo$_2$[14]hexaenatoN$_4$)][171] (ligand **67**) having room temperature moments in the range 3.6–4.0 B.M. This intermediate-spin configuration has been well established for Fe(II) phthalocyanine [$\mu_B = 3.89$ (23°C)][172,173] and is likely for Fe(II) tetraphenylporphyrin [$\mu_B = 4.4$–5.0 (25°C)].[174–177] The intermediate-spin ($S = 1$) ground state may be a common occurrence in planar [Fe^{II}N$_4$] complexes. The electronic spectral bands for [Fe(Me$_4$bzo$_2$[14]-hexaenatoN$_4$] have been reported[171] without interpretation.

Intermediate-spin ground states have also been observed for [Fe([15]-aneN$_4$)(NO$_2$)](PF$_6$), [$\mu_B = 3.36$ (25°C)],[158] which is proposed to be six-coordinate with bidentate coordination of nitrite. Magnetic susceptibility measurements over the range 96–352 K may be approximated by the Curie–Weiss law with $\theta = 39.88°$, thus ruling out a spin-state equilibrium. Two authentic cases of spin-state equilibria have been reported for macrocyclic iron(II) complexes, [Fe([14]dieneN$_4$)phen](ClO$_4$)$_2$[160] [$\mu_B = 3.7$ (25°C)], and [Fe(Me$_6$[14]aneN$_4$)(NCS)$_2$][159] [$\mu_B = 1.5$ (25°C)]. The temperature-dependent magnetic moments for both of these complexes are interpreted quantitatively to yield the equilibrium constant for the conversion $^1A_{1g} \rightleftharpoons {}^5T_{2g}$ (ignoring low-symmetry components) as a function of temperature and thereby the activation parameters for the process.

5.4. Low-Spin (S = ½) Iron(III) Macrocycles

Examples of macrocyclic complexes of iron(III) exhibiting all three possible ground states ($S = \frac{1}{2}, \frac{3}{2}, \frac{5}{2}$) have been reported. In one unique series of five-coordinate iron(III) macrocycles containing a coordinated thiolate anion examples of all three spin states were found to occur.[170] For low-spin iron(III) macrocycles in both five- and six-coordinate geometries ligand field analyses of the $d \rightarrow d$ spectra have not been carried out and these complexes have been characterized principally by magnetic susceptibilities, EPR, and Mössbauer measurements. EPR and μ_B data for some low-spin iron(III) complexes with Ph$_2$[14–16]tetraenatoN$_4$ ligands (**34, 36, 37**) are shown in Table 12. Further examples of low-spin iron(III) complexes and the observed values of μ_B at ambient temperature are [Fe(*ms-trans*-Me$_6$[14]aneN$_4$)-(CH$_3$CN)$_2$](BF$_4$)$_3$ (2.19),[162] [Fe(Me$_6$[14]4,11-dieneN$_4$)Cl$_2$]ClO$_4$ (2.30),[160] Fe(Me$_6$[14]4,11-dieneN$_4$)(CH$_3$CN)$_2$](ClO$_4$)$_3$, (2.19),[160] [Fe(Me$_6$[14]4,11-dieneN$_4$)(NCS)$_2$]BPh$_4$ (2.14),[160] [Fe(Me$_6$[14]1,3,8-trieneN$_4$)Cl$_2$]PF$_6$ (2.13),[164] [Fe(Me$_6$[14]1,3,8,10-tetraeneN$_4$)Cl$_2$]PF$_6$ (2.06),[164] and [Fe(Me$_6$-[14]1,4,8,11-tetraeneN$_4$)Cl$_2$]ClO$_4$ (2.29).[163]

TABLE 12. EPR Data for Some Low-Spin, Five- and Six-Coordinate Iron(III) Macrocycles[a]

Compound	Ligand	μ_B [b]	g values[c]		
[Fe(Ph$_2$[14]tetraenatoN$_4$)(SPh)]	**34**	1.95	2.11	2.04	2.01[d]
[Fe(Ph$_2$[14]tetraenatoN$_4$)(SPh)(py)]	**34**	1.82	2.13	2.03	2.00[d]
[Fe(Ph$_2$[14]tetraenatoN$_4$)(py)$_2$]$^+$	**34**	—	2.08	2.02	2.00[e]
[Fe(Ph$_2$[15]tetraenatoN$_4$)(SPh)(py)]	**36**	1.87	2.19, 2.16	2.06	1.99[d]
[Fe(Ph$_2$[15]tetraenatoN$_4$)(py)$_2$]$^+$	**36**	—	2.09	2.05	1.99[e]
[Fe(Ph$_2$[16]tetraenatoN$_4$)(SPh)(py)]	**37**	—	2.41	2.10	1.96[d]
[Fe(Ph$_2$[16]tetraenatoN$_4$)(SCH$_2$Ph)(py)]	**37**	—	2.32	2.10	1.97[d]

[a] Reference 170.　　[b] At 295 K.　　[c] At ~95 K.　　[d] Toluene glass.
[e] 1/1(vol./vol.) DMF–CH$_2$Cl$_2$ glass.

5.5. High-Spin (S = $\frac{5}{2}$) and Intermediate-Spin (S = $\frac{3}{2}$) Iron(III) Macrocycles

A few examples of five- and six-coordinate high-spin iron(III) complexes of macrocyclic ligands have been reported.[159,167,170,171] These complexes invariably contain halide ions or similar weak axial donors. Owing to their spin-forbidden nature, the $d \rightarrow d$ transitions in high-spin iron(III) macrocyclic complexes have not been observed. Six-coordinate high-spin iron(III) complexes of ms-Me$_2$pyo[14]eneN$_4$, **27**,[167] with axial chloride or bromide ions and the complex Fe[($meso$-Me$_6$[14]aneN$_4$)(Cl)$_2$]ClO$_4$[159] have been reported. The five-coordinate complex [FeLCl], L = Me$_4$bzo$_2$[14]hexaenato-N$_4$, **67**, which results from the interaction of the highly reactive four-coordinate iron(II) species [FeL] ($S = 1$) with CHCl$_3$, is observed to be high spin.[171] The three five-coordinate derivatives, [Fe(Ph$_2$[16]tetraenatoN$_4$)Br], [Fe(Ph$_2$-[16]tetraenatoN$_4$)RCOO], and [Fe(Ph$_2$[16]tetraenatoN$_4$)(PhCH$_2$S)] are high spin, in contrast to corresponding derivatives of the [14]- and [15]-ring analogs, which are intermediate or low spin.[170] This represents a further example of a larger macrocyclic ring providing a weaker ligand field. The complexes [Fe(Ph$_2$[14]tetraenatoN$_4$)X] and [Fe(Ph$_2$[15]tetraenatoN$_4$)X], where X = halide, RCO$_2^-$, are presently the only definite examples of iron(III) complexes having the intermediate spin state $S = \frac{3}{2}$.[170] The complexes exhibit values of μ_B = 3.8–4.1 and accurately follow the Curie–Weiss law from 17 to 295 K.[170]

5.6. Other Iron-Containing Macrocycles

Oxo-bridged iron(III) dimers [N$_4$Fe^{III}-O-Fe^{III}N$_4$]$^{n+}$ have been obtained for several of the tetradentate N$_4$ macrocyclic complexes described above.[160,170,171] These μ-oxo species yield values for μ_B in the range 1.8–2.1 and exhibit a characteristic infrared stretching frequency for Fe-O-Fe in the range 850–890 cm^{-1}. The reactive iron(I) species, [Fe(Me$_6$[14]1,3,8,10-tetraeneN$_4$)]$^+$, ligand **65**, and the corresponding hydride, [HFe(Me$_6$[14]1,3,8,10-tetraeneN$_4$)(CH$_3$CN)], and alkyl derivatives, [RFe(Me$_6$[14]1,3,8,10-tetraeneN$_4$)], R = CH$_3$ or C$_6$H$_5$, have been prepared and characterized by magnetic measurements and EPR.[178] These low-spin $3d^7$ complexes gave values for μ_B = 2.1–2.3 and displayed rhombic EPR spectra.[178]

A unique series of high-spin iron(III) complexes, [Fe(N$_5$)X$_2$]Y, has been obtained with the planar pentadentate ligands **68** and **69**.[11,179,180] The pentagonal bipyramidal geometry deduced from physical properties was confirmed by X-ray crystallography.[11,181] Low-spin iron(II) complexes have been reported for an N$_4$S$_2$ and two N$_4$O$_2$ sexadentate macrocyclic ligands,[90] an N$_2$S$_3$ pentadentate macrocycle,[182] and for several clathrochelate ligands.[8,9,93,183]

6. Manganese Complexes

6.1. Macrocyclic Complexes of Manganese(II)

Manganese(II) complexes have been reported with two tetradentate macrocyclic ligands, ms-trans-Me$_6$[14]aneN$_4$,[184] **10**, and Me$_4$Bzo$_2$[14]hexa-enatoN$_4$,[185] **67**, and several pentadentate macrocyclic ligands, **68**,[186] and

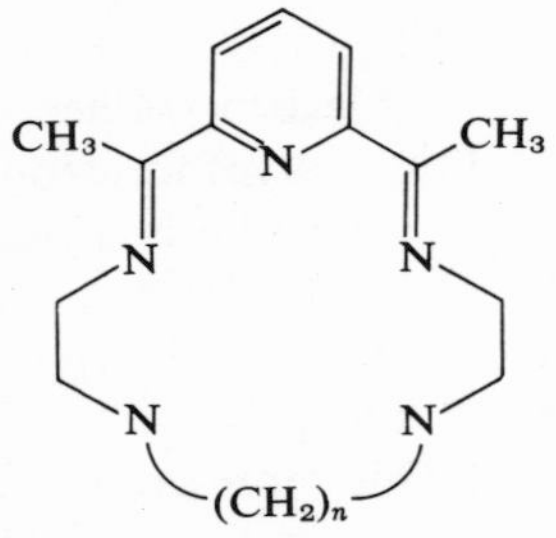

68 $n = 2$
69 $n = 3$

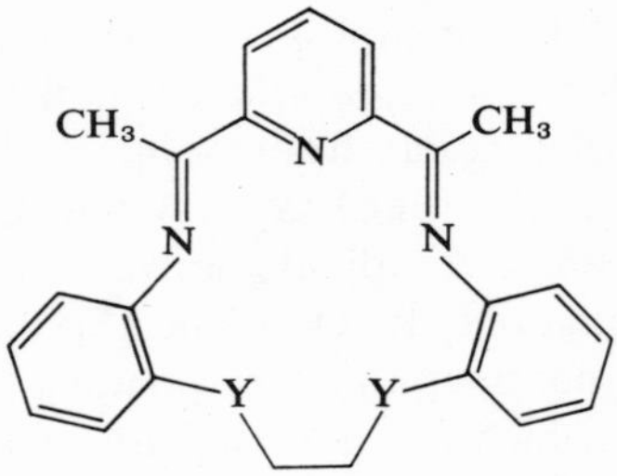

70 Y = NH
71 Y = O
72 Y = S

70–72.[187] All of these complexes are found to have a high-spin ($S = \frac{5}{2}$) configuration with $\mu_B = 5.7$–6.0. The spin-forbidden electronic transitions of Mn(II) are expected to be extremely weak in intensity and are not observed. The complexes of the dianionic ligand **67** are obtained as bis-triethylamine or bis-tri(n-propyl)amine adducts which exhibit intense ($\varepsilon \sim 10^3$–10^4) bands in the visible region of the spectrum.[185] These bands are attributed to transitions within the highly delocalized ligand framework since comparable bands are shown by the corresponding Zn(II) derivatives. EPR has provided minimal characterization of the Mn(II) macrocycles. The two complexes of **67** yield only a broad resonance at $g \sim 2.0$ for solid state samples at room temperature.[185] Measurements in toluene were not reproducible, indicating decomposition of the samples in solution.

Complexes of ms-trans-Me$_6$[14]aneN$_4$ were isolated as oxygen-sensitive CF$_3$SO$_3^-$ or CH$_3$COO$^-$ salts and were studied by EPR.[184] The major component of the room-temperature solid state EPR spectra of these complexes was a strong absorption at $g = 4.00$ (CF$_3$SO$_3^-$ derivative) or $g = 4.77$ (CH$_3$COO$^-$ derivative) together with four or five weaker bands at higher field strengths. A semiquantitative analysis[188,189] of these spectra indicates that the Mn(II) ion is experiencing only a small tetragonal distortion ($D = 0.1$–0.2 cm^{-1}).

The Mn(II) complex of **68** was isolated as an anhydrous chloride salt, as well as the corresponding mono- and hexahydrates.[186] All three forms of the complex were assumed to have a pentagonal bipyramidal geometry, which seems reasonable in light of the crystal structure reported for the Mn(II)

complex of **70** as a perchlorate salt.[187] The EPR spectra of the three compounds were studied in detail.[190] The hexahydrate gave a typical five-line spectrum but showed an unusual angular dependence, which indicated significant mixing of certain $|M\rangle$ levels. The monohydrate and the anhydrous complex are both fully high spin ($S = \frac{5}{2}$) from magnetic studies, but exhibit unusual one-line EPR spectra with $g \sim 2.0$. This result is attributed to a greater ligand field strength in the latter two forms of the complex.

6.2. Macrocyclic Complexes of Manganese(III)

A series of Mn(III) complexes with the stoichiometry [MnLX$_2$]Y, where X = Cl$^-$, Br$^-$, or NCS$^-$ and Y = X, PF$_6{}^-$, or BF$_4{}^-$, have been reported for *trans*-Me$_6$[14]aneN$_4$ in both the *meso* form[184] and one enantiomer of the racemic form.[27] All of the complexes are high spin ($S = 2$) and exhibit magnetic moments in the range 4.89–5.01 B.M. The ligand field spectra of the complexes are consistent with a tetragonally distorted d^4 system,[191] where for $Dq^{xy} > Dq^z$, the expected order of the transitions is $^5A_{1g} \leftarrow {}^5B_{1g}$, $^5B_{2g} \leftarrow {}^5B_{1g}$, and $^5E_g \leftarrow {}^5B_{1g}$. Only the lowest-energy transition at ~ 870 nm is resolved in the absorption spectra; the remaining bands appear as shoulders on an intense transition centered at ~ 286 nm. In the circular dichroism spectrum[27] of the optically active derivatives these shoulders are resolved as definite bands, and, from the position of the $^5B_{2g} \leftarrow {}^5B_{1g}$ transition, Dq^{xy} is reported to be 2300 cm^{-1}. This value is considerably higher than Dq^{xy} reported for Schiff-base complexes of Mn(III) (~ 1600 cm^{-1})[192] but compares favorably with values of Dq^{xy} for N$_4$ macrocyclic ligands about cobalt-(III) (cf. Table 9).

A series of complexes[185] MnLX, where L = **67** and X = Cl$^-$, Br$^-$, NCS$^-$, or N$_3{}^-$, have been reported to be high spin except for the azide derivative, where a slightly reduced value of μ_B (4.78 B.M.) suggests antiferromagnetic coupling between Mn(III) ions. The spectra of these complexes exhibit intense bands ($\varepsilon = 10^3$–10^4) in the region 660–330 nm, which are attributed to ligand transitions.

References

1. D. H. Busch, *Helv. Chim. Acta*, Fasciculus extraordinarius Alfred Werner, 174 (1967).
2. D. H. Busch, *Rec. Chem. Progr.* **25**, 107 (1964).
3. N. F. Curtis, *Coord. Chem. Rev.* **3**, 3 (1968).
4. D. H. Busch, K. Farmery, V. Goedken, V. Katovic, A. C. Melnyk, C. R. Sperati, and N. Tokel, *Adv. Chem. Ser.* **100**, 44 (1971).
5. L. Y. Martin, L. J. DeHayes, L. J. Zompa, and D. H. Busch, *J. Am. Chem. Soc.* **96**, 4046 (1974).

6. L. Y. Martin, C. R. Sperati, and D. H. Busch, *J. Am. Chem. Soc.* **99**, 2968 (1977).
7. Y. Hung, L. Y. Martin, S. C. Jackels, A. M. Tait, and D. H. Busch, *J. Am. Chem. Soc.* **99**, 4029 (1977).
8. J. E. Parks, B. E. Wagner, and R. H. Holm, *J. Am. Chem. Soc.* **92**, 3500 (1970).
9. E. Larsen, G. N. LaMar, B. E. Wagner, J. E. Parks, and R. H. Holm, *Inorg. Chem.* **11**, 2652 (1972).
10. D. R. Boston and N. J. Rose, *J. Am. Chem. Soc.* **90**, 6859 (1968).
11. M. G. B. Drew, A. H. bin Othman, P. D. A. McIlroy, and S. M. Nelson, *J. Chem. Soc. Dalton Trans.* **1975**, 2507 (1975).
12. F. P. Hinz and D. W. Margerum, *J. Am. Chem. Soc.* **96**, 4993 (1974).
13. F. P. Hinz and D. W. Margerum, *Inorg. Chem.* **13**, 2941 (1974).
14. E. R. Dockal, L. L. Diaddario, M. D. Glick, and D. B. Rorabacher, *J. Am. Chem. Soc.* **99**, 4530 (1977).
15. R. A. D. Wentworth and T. S. Piper, *Inorg. Chem.* **4**, 709 (1965).
16. D. A. Rowley and R. S. Drago, *Inorg. Chem.* **6**, 1092 (1967).
17. A. B. P. Lever, *Coord. Chem. Rev.* **3**, 119 (1968).
18. A. B. P. Lever, *Inorganic Electronic Spectroscopy*, Elsevier, Amsterdam (1968).
19. G. Maki, *J. Chem. Phys.* **28**, 651 (1958).
20. R. Smierciak, J. Passariello, and E. L. Blinn, *Inorg. Chem.* **16**, 2646 (1977).
21. G. A. Kalligeros and E. L. Blinn, *Inorg. Chem.* **11**, 1145 (1972).
22. J. P. Collman and P. W. Schneider, *Inorg. Chem.* **5**, 1380 (1966).
23. M. S. Holtman and S. C. Cummings, *Inorg. Chem.* **15**, 660 (1976).
24. B. Bosnich, M. L. Tobe, and G. A. Webb, *Inorg. Chem.* **4**, 1109 (1965).
25. N. F. Curtis, *J. Chem. Soc.* **1964**, 2644 (1964).
26. D. C. Olson and J. Vasilevskis, *Inorg. Chem.* **8**, 1611 (1969).
27. P. S. Bryan and J. C. Dabrowiak, *Inorg. Chem.* **14**, 299 (1975).
28. R. Buxtorf and T. A. Kaden, *Helv. Chim. Acta* **57**, 1035 (1974).
29. F. Wagner and E. K. Barefield, *Inorg. Chem.* **15**, 408 (1976).
30. E. K. Barefield and F. Wagner, *Inorg. Chem.* **12**, 2435 (1973).
31. E. M. Arnett, *Acc. Chem. Res.* **6**, 404 (1973).
32. F. Wagner, M. T. Mocella, M. J. D'Aniello, Jr., A. H-J. Wang, and E. K. Barefield, *J. Am. Chem. Soc.* **96**, 2625 (1974).
33. R. Buxtorf, W. Steinmann, and T. A. Kaden, *Chimia* **28**, 15 (1974).
34. M. Ciampolini and N. Nardi, *Inorg. Chem.* **5**, 41 (1966).
35. M. J. D'Aniello, Jr., M. T. Mocella, F. Wagner, E. K. Barefield, and I. C. Paul, *J. Am. Chem. Soc.* **97**, 192 (1975).
36. F. V. Lovecchio, E. S. Gore, and D. H. Busch, *J. Am. Chem. Soc.* **96**, 3109 (1974).
37. N. F. Curtis, *J. Chem. Soc.* **1965**, 924 (1965).
38. N. F. Curtis, *J. Chem. Soc.* (*A*) **1968**, 1584 (1968).
39. N. F. Curtis and D. A. House, *J. Chem. Soc.* (*A*) **1967**, 537 (1967).
40. W. H. Elfring, Jr. and N. J. Rose, *Inorg. Chem.* **14**, 2759 (1975).
41. S. C. Cummings and R. E. Sievers, *Inorg. Chem.* **9**, 1131 (1970).
42. N. F. Curtis and T. N. Milestone, *Aust. J. Chem.* **28**, 275 (1975).
43. R. W. Hay and G. A. Lawrence, *J. Chem. Soc. Dalton Trans.* **1975**, 1466 (1975).
44. J. G. Martin and S. C. Cummings, *Inorg. Chem.* **12**, 1477 (1973).
45. D. F. Cook, N. F. Curtis, and R. W. Hay, *J. Chem. Soc. Dalton Trans.* **1973**, 1160 (1973).
46. E. K. Barefield, F. V. Lovecchio, N. E. Tokel, E. Ochiai, and D. H. Busch, *Inorg. Chem.* **11**, 283 (1972).
47. E. Ochiai and D. H. Busch, *Inorg. Chem.* **8**, 1798 (1969).
48. J. F. Myers and N. J. Rose, *Inorg. Chem.* **12**, 1238 (1973).

49. D. A. House and N. F. Curtis, *J. Am. Chem. Soc.* **86**, 223 (1964).
50. J. D. Curry, Ph.D. Thesis, Ohio State University, 1964.
51. J. L. Karn and D. H. Busch, *Nature* **211**, 160 (1966).
52. L. G. Warner, N. J. Rose, and D. H. Busch, *J. Am. Chem. Soc.* **90**, 6938 (1968).
53. E. K. Barefield and D. H. Busch, *Inorg. Chem.* **10**, 108 (1971).
54. N. F. Curtis, *Aust. J. Chem.* **27**, 71 (1974).
55. J. L. Karn and D. H. Busch, *Inorg. Chem.* **8**, 1149 (1969).
56. G. A. Melson and D. H. Busch, *Proc. Chem. Soc.* **1963**, 223 (1963).
57. G. A. Melson and D. H. Busch, *J. Am. Chem. Soc.* **86**, 4834 (1964).
58. G. A. Melson and D. H. Busch, *J. Am. Chem. Soc.* **86**, 4830 (1964).
59. V. Katović, L. T. Taylor, F. L. Urbach, W. H. White, and D. H. Busch, *Inorg. Chem.* **11**, 479 (1972).
60. L. T. Taylor, F. L. Urbach, and D. H. Busch, *J. Am. Chem. Soc.* **91**, 1072 (1969).
61. V. Katović, L. T. Taylor, and D. H. Busch, *Inorg. Chem.* **10**, 458 (1971).
62. J. L. Karn, Ph.D. Thesis, Ohio State University, 1966.
63. S. C. Tang, S. Koch, G. N. Weinstein, R. W. Lane, and R. H. Holm, *Inorg. Chem.* **12**, 2589 (1973).
64. S. C. Tang and R. H. Holm, *J. Am. Chem. Soc.* **97**, 3359 (1975).
65. T. J. Truex and R. H. Holm, *J. Am. Chem. Soc.* **93**, 285 (1971).
66. T. J. Truex and R. H. Holm, *J. Am. Chem. Soc.* **94**, 4529 (1972).
67. E.-G. Jäger, *Z. Anorg. Allg. Chem.* **364**, 177 (1969).
68. E.-G. Jäger, *Z. Chem.* **8**, 30 (1968).
69. E. Jäger, *Z. Chem.* **8**, 392 (1968).
70. M. Green and P. A. Tasker, *Inorg. Chim. Acta* **5**, 65 (1971).
71. W. R. Pangratz, Ph.D. Thesis, Case Western Reserve University, 1974.
72. D. St. C. Black and A. J. Hartshorn, *J. Chem. Soc.* (*D*) **1972**, 706 (1972).
73. D. St. C. Black and P. W. Kortt, *Aust. J. Chem.* **25**, 281 (1972).
74. C. M. Kerwin and G. A. Melson, *Inorg. Chem.* **12**, 2410 (1973).
75. G. R. Brubaker and D. H. Busch, *Inorg. Chem.* **5**, 2114 (1966).
76. F. L. Urbach and D. H. Busch, *Inorg. Chem.* **12**, 408 (1973).
77. M. Ciampolini, *Inorg. Chem.* **5**, 35 (1966).
78. W. Rosen and D. H. Busch, *J. Am. Chem. Soc.* **91**, 4694 (1969).
79. G. F. Smith and D. W. Margerum, *J. Chem. Soc. Chem. Commun.* **1975**, 807 (1975).
80. J. Riker-Nappier and D. W. Meek, *J. Chem. Soc. Chem. Commun.* **1974**, 442 (1975).
81. L. G. Armstrong and L. F. Lindoy, *Inorg. Chem.* **14**, 1322 (1975).
82. R. W. Kluiber and G. Sasso, *Inorg. Chem. Acta* **4**, 226 (1970).
83. D. L. Johnston and W. DeW. Horrocks, Jr., *Inorg. Chem.* **10**, 687 (1971).
84. R. A. Lalancette, D. J. Macchia, and W. F. Furey, *Inorg. Chem.* **15**, 548 (1976).
85. G. A. Melson and D. H. Busch, *J. Am. Chem. Soc.* **87**, 1706 (1965).
86. L. T. Taylor, S. C. Vergez, and D. H. Busch, *J. Am. Chem. Soc.* **88**, 3170 (1966).
87. L. T. Taylor and D. H. Busch, *J. Am. Chem. Soc.* **89**, 5372 (1967).
88. M. C. Rakowski, M. Rycheck, and D. H. Busch, *Inorg. Chem.* **14**, 1194 (1975).
89. R. Louis, B. Metz, and R. Weiss, *Acta Crystallogr. Sec. B* **30**, 774 (1974).
90. E. B. Fleischer and P. A. Tasker, *J. Am. Chem. Soc.* **92**, 7072 (1970).
91. D. St. C. Black and I. A. McLean, *Chem. Commun.* **1968**, 1004 (1968).
92. L. F. Lindoy and D. H. Busch, *J. Am. Chem. Soc.* **91**, 4690 (1969).
93. J. E. Parks, B. E. Wagner, and R. H. Holm, *Inorg. Chem.* **10**, 2472 (1971).
94. G. A. Melson and D. H. Busch, *J. Am. Chem. Soc.* **86**, 4830 (1964).
95. N. H. Pilkington and R. Robson, *Aust. J. Chem.* **23**, 2225 (1970).
96. H. Okawa and S. Kida, *Bull. Chem. Soc. Japan* **45**, 1759 (1972).
97. H. Okawa, T. Tokii, Y. Muto, and S. Kida, *Bull. Chem. Soc. Japan* **46**, 2464 (1973).

98. N. F. Curtis and D. F. Cook, *Chem. Commun.* **1967**, 962 (1967).
99. E. S. Gore and D. H. Busch, *Inorg. Chem.* **12**, 1 (1973).
100. A. H. Maki, N. Edelstein, A. Davison, and R. H. Holm, *J. Am. Chem. Soc.* **86**, 4580 (1964).
101. M. Millar and R. H. Holm, *J. Am. Chem. Soc.* **97**, 6052 (1975).
102. E. K. Barefield and M. T. Mocella, *J. Am. Chem. Soc.* **97**, 4238 (1975).
103. J. G. Martin, R. M. C. Wei, and S. C. Cummings, *Inorg. Chem.* **11**, 475 (1972).
104. J. M. Palmer, E. Papaconstantinou, and J. F. Endicott, *Inorg. Chem.* **8**, 1516 (1969).
105. Y. M. Curtis and N. F. Curtis, *Aust. J. Chem.* **19**, 609 (1966).
106. M. M. Blight and N. F. Curtis, *J. Chem. Soc.* **1962**, 3016.
107. N. F. Curtis, *J. Chem. Soc. Dalton Trans.* **1973**, 863 (1973).
108. D. E. Fenton and S. E. Gayda, *J. Chem. Soc. Chem. Commun.* **1974**, 960 (1974).
109. H. Okawa and S. Kida, *Inorg. Nucl. Chem. Lett.* **7**, 751 (1971).
110. T. E. Jones, D. B. Rorabacher, and L. A. Ochrymowycz, *J. Am. Chem. Soc.* **97**, 7485 (1975).
111. R. Malkin and B. G. Malmström, *Adv. Enzymol.* **33**, 177 (1970).
112. R. Malkin, in: *Inorganic Biochemistry* (G. L. Eichhorn, ed.), p. 689, Elsevier Publishers, New York (1973).
113. E. I. Solomon, J. W. Hare, and H. B. Gray, *Proc. Natl. Acad. Sci. USA* **73**, 1389 (1976).
114. M. D. Glick, D. P. Gavel, L. L. Diaddario, and D. B. Rorabacher, *Inorg. Chem.* **15**, 1190 (1976).
115. P. H. Davis, L. K. White, and R. L. Belford, *Inorg. Chem.* **14**, 1753 (1975).
116. E. Billig, R. Williams, I. Bernal, J. H. Waters, and H. B. Gray, *Inorg. Chem.* **3**, 663 (1964).
117. M. J. Weeks and J. P. Fackler, *Inorg. Chem.* **7**, 2548 (1968).
118. R. W. Stotz and R. C. Stoufer, *J. Chem. Soc. D* **1970**, 1682 (1970).
119. H. Okawa, M. Honda, and S. Kida, *Chem. Lett.* **1972**, 1027 (1972).
120. D. C. Olson and J. Vasilevskis, *Inorg. Chem.* **10**, 463 (1971).
121. A. M. Tait, M. Z. Hoffman, and E. Hayon, *Inorg. Chem.* **15**, 934 (1976).
122. N. E. Tokel, V. Katović, K. Farmery, L. B. Anderson, and D. H. Busch, *J. Am. Chem. Soc.* **92**, 400 (1970).
123. K. M. Long and D. H. Busch, *Inorg. Chem.* **9**, 505 (1970).
124. M. J. Norgett, J. H. M. Thornley, and L. M. Venanzi, *J. Chem. Soc. (A)* **1967**, 540 (1967).
125. G. Dyer and D. W. Meek, *J. Am. Chem. Soc.* **89**, 3983 (1967).
126. M. J. Norgett, J. H. M. Thornley, and L. M. Venanzi, *Coord. Chem. Rev.* **2**, 99 (1967).
127. B. N. Figgis and R. S. Nyholm, *J. Chem. Soc.* **1959**, 338 (1959).
128. J. L. Love and H. K. J. Powell, *Inorg. Nucl. Chem. Lett.* **3**, 113 (1967).
129. S. C. Cummings, G. A. Melson, and D. H. Busch, *Inorg. Nucl. Chem. Lett.* **1**, 69 (1965).
130. P. R. Ireland and W. T. Robinson, *J. Chem. Soc. (A)* **1970**, 663 (1970).
131. A. M. Tait and D. H. Busch, *Inorg. Chem.* **15**, 197 (1976).
132. K. G. Caulton, *Inorg. Chem.* **7**, 392 (1968).
133. S. Cockle, H. A. O. Hill, S. Ridsdale, and R. J. P. Williams, *J. Chem. Soc. Dalton Trans.* **1972**, 297 (1972).
134. G. N. Schrauzer and L.-P. Lee, *J. Am. Chem. Soc.* **90**, 6541 (1968).
135. F. L. Urbach, R. D. Bereman, J. A. Topich, M. Hariharan, and B. J. Kalbacher, *J. Am. Chem. Soc.* **96**, 5063 (1974).
136. M. A. Hitchman, *Inorg. Chem.* **16**, 1985 (1977).

137. J. A. Topich, Ph.D. Thesis, Case Western Reserve University, 1974.
138. B. R. McGarvey, *Can. J. Chem.* **53**, 2498 (1975).
139. V. Malatesta and B. R. McGarvey, *Can. J. Chem.* **53**, 3791 (1975).
140. N. Sadasivan and John F. Endicott, *J. Am. Chem. Soc.* **88**, 5468 (1966).
141. J. Vasilevskis and D. C. Olson, *Inorg. Chem.* **10**, 1228 (1971).
142. B. F. Hoskins and G. A. Williams, *Aust. J. Chem.* **28**, 2607 (1975).
143. D. R. Boston and N. J. Rose, *Abstracts, 157th National Meeting of the American Chemical Society, Minneapolis, Minnesota, April 1969*, No. INOR 096.
144. G. A. Zakrzewski, C. A. Ghilardi, and E. C. Lingafelter, *J. Am. Chem. Soc.* **93**, 4411 (1971).
145. C. K. Poon and M. L. Tobe, *J. Chem. Soc. A* **1968**, 1549 (1968).
146. N. F. Curtis, *J. Chem. Soc. Dalton Trans.* **1973**, 1212 (1973).
147. E. Ochiai and D. H. Busch, *Inorg. Chem.* **8**, 1474 (1969).
148. K. Travis and D. H. Busch, *Inorg. Chem.* **13**, 2591 (1974).
149. B. Bosnich, C. K. Poon, and M. L. Tobe, *Inorg. Chem.* **4**, 1102 (1965).
150. S. C. Jackels, K. Farmery, E. K. Barefield, N. J. Rose, and D. H. Busch, *Inorg. Chem.* **11**, 2893 (1972).
151. N. Sadasivan, J. A. Kernohan, and J. F. Endicott, *Inorg. Chem.* **6**, 770 (1967).
152. S. C. Cummings and D. H. Busch, *Inorg. Chem.* **10**, 1220 (1971).
153. K. M. Long and D. H. Busch, *J. Coord. Chem.* **4**, 113 (1974).
154. D. St. C. Black and A. J. Hartshorn, *Tetrahedron Lett.* **1974**, 2157 (1974).
155. S. C. Cummings and D. H. Busch, *J. Am. Chem. Soc.* **92**, 1924 (1970).
156. R. M. Wing and R. Eiss, *J. Am. Chem. Soc.* **92**, 1929 (1970).
157. A. M. Tait, M. Z. Hoffman, and E. Hayon, *J. Am. Chem. Soc.* **98**, 86 (1976).
158. D. D. Watkins, Jr., D. P. Riley, J. A. Stone, and D. H. Busch, *Inorg. Chem.* **15**, 387 (1976).
159. J. C. Dabrowiak, P. H. Merrell, and D. H. Busch, *Inorg. Chem.* **11**, 1979 (1972).
160. V. L. Goedken, P. H. Merrell, and D. H. Busch, *J. Am. Chem. Soc.* **94**, 3397 (1972).
161. D. P. Riley, J. A. Stone, and D. H. Busch, *J. Am. Chem. Soc.* **98**, 1752 (1976).
162. D. C. Olson and J. Vasilevskis, *Inorg. Chem.* **11**, 980 (1972).
163. V. L. Goedken and D. H. Busch, *J. Am. Chem. Soc.* **94**, 7355 (1972).
164. J. C. Dabrowiak and D. H. Busch, *Inorg. Chem.* **14**, 1881 (1975).
165. D. A. Baldwin, R. M. Pfeiffer, D. W. Reichgott, and N. J. Rose, *J. Am. Chem. Soc.* **95**, 5152 (1973).
166. V. L. Goedken, Y.-A. Park, S.-M. Peng, and J. M. Norris, *J. Am. Chem. Soc.* **96**, 7693 (1974).
167. D. P. Riley, P. H. Merrell, J. A. Stone, and D. H. Busch, *Inorg. Chem.* **14**, 490 (1975).
168. D. M. L. Goodgame, M. Goodgame, M. A. Hitchman, and M. J. Weeks, *Inorg. Chem.* **5**, 635 (1966).
169. P. B. Merrithew, P. G. Rasmussen, and D. H. Vincent, *Inorg. Chem.* **10**, 1401 (1971).
170. S. Koch, R. H. Holm, and R. B. Frankel, *J. Am. Chem. Soc.* **97**, 6714 (1975).
171. V. L. Goedken and Y.-A. Park, *J. Chem. Soc. Chem. Commun.* **1975**, 214 (1975).
172. C. G. Barraclough, R. L. Martin, S. Mitra, and R. C. Sherwood, *J. Chem. Phys.* **53**, 1643 (1970).
173. B. W. Dale, *Mol. Phys.* **28**, 503 (1974).
174. J. P. Collman and C. A. Reed, *J. Am. Chem. Soc.* **95**, 2048 (1973).
175. H. Kobayashi and Y. Yanagawa, *Bull. Chem. Soc. Jpn.* **45**, 450 (1972).
176. S. M. Husain and J. G. Jones, *Inorg. Nucl. Chem. Lett.* **10**, 105 (1974).
177. J. P. Collman, J. L. Hoard, N. Kim, G. Lang, and C. A. Reed, *J. Am. Chem. Soc.* **97**, 2676 (1975).

178. M. C. Rakowski and D. H. Busch, *J. Am. Chem. Soc.* **97**, 2570 (1975).

179. S. M. Nelson and D. H. Busch, *Inorg. Chem.* **8**, 1859 (1969).

180. S. M. Nelson, P. Bryan, and D. H. Busch, *Chem. Commun.* **1966**, 641 (1966).

181. E. Fleischer and S. Hawkinson, *J. Am. Chem. Soc.* **89**, 720 (1967).

182. D. St. C. Black and I. A. McLean, *Inorg. Nucl. Chem. Lett.* **6**, 675 (1970).

183. S. C. Jackels and N. J. Rose, *Inorg. Chem.* **12**, 1232 (1973).

184. P. S. Bryan and J. C. Dabrowiak, *Inorg. Chem.* **14**, 296 (1975).

185. D. R. Neves and J. C. Dabrowiak, *Inorg. Chem.* **15**, 129 (1976).

186. M. D. Alexander, A. Van Heuvelen, and H. G. Hamilton, Jr., *Inorg. Nucl. Chem. Lett.* **6**, 445 (1970).

187. N. W. Alcock, D. C. Liles, M. McPartlin, and P. A. Tasker, *J. Chem. Soc. Chem. Commun.* **1974**, 727 (1974).

188. R. D. Dowsing and J. F. Gibson, *J. Chem. Phys.* **50**, 294 (1969).

189. R. D. Dowsing, J. F. Gibson, M. Goodgame, and P. J. Hayward, *J. Chem. Soc. A* **1969**, 187 (1969).

190. A. Van Heuvelen, M. D. Lundeen, H. G. Hamilton, Jr., and M. D. Alexander, *J. Chem. Phys.* **50**, 489 (1969).

191. T. S. Davis, J. P. Fackler, and M. J. Weeks, *Inorg. Chem.* **7**, 1994 (1968).

192. L. J. Boucher and D. R. Herrington, *Inorg. Chem.* **13**, 1105 (1974).

6

Chemical Reactivity in Constrained Systems

John F. Endicott and Bill Durham

1. Introduction

The reactions of complexes containing macrocyclic ligands are at least as varied as those of simpler transition-metal complexes. The binding of a metal to a macrocyclic ligand does introduce constraints on the kind of reaction that can occur. The steric constraints imposed by such binding can affect the electronic structure of the metal, the number and lability of binding sites, the affinity of the metal for other ligands, etc. In many instances the ligand plays a very active role in the reactions observed. Thus the macrocyclic ligand not only modulates the metal-ion reactivity by means of steric constraints, but the ligand may also provide complementary sites for reaction.

In order to keep this chapter tractable and to minimize duplication of material appearing elsewhere in this book, we have concerned ourselves mostly with reaction of complexes with tetraaza (N_4) macrocyclic ligands. Since these complexes are in many ways intermediate between the classical Werner-type coordination complexes and metallo enzymes, we have occasionally drawn material from these areas for contrast and comparison.

2. Predominantly Metal-Centered Reactions

Any reaction classification scheme is necessarily arbitrary. We have attempted to organize the available material into classical reaction categories and to distinguish, as much as is reasonable, reactions that result in changes largely localized on the metal from reactions that modify the ligand or both

John F. Endicott and Bill Durham • Department of Chemistry, Wayne State University, Detroit, Michigan 48202.

the ligand and metal. For various reasons, photochemical reactions are considered in an altogether separate section. The primary focus of this section is on simple "classical" reactions which directly involve the metal center. We have adopted this organizational scheme as a matter of convenience. We have not treated the categories as totally exclusive.

Equatorial coordination of a metal by a cyclic tetradentate ligand can result in species with properties strikingly different from those of coordination complexes with related monodentate or linear polydentate ligands. For example, such coordination of a metal to a cyclic ligand tends to make the equatorial positions relatively inert to substitution, and any reaction that involves changes of metal–ligand internuclear distances tends to be constrained to occur only at the axial coordination sites. Since the metal-equatorial-ligand bond distances and angles are determined largely by the ligand stereochemistry, the relative energies of electronic states can be altered and "tuned" by varying the "ligand field" (e.g., hole size or nature of the donor atom) through a homologous series of macrocyclic ligands. Reactions may be induced by preparing ligands which act as "templates" for certain reactants or which hold potentially reactive groups in promixity to a reactive coordination site. The possibility of "tuning" the reactivity of a metal by variations in the nature and stereochemistry of the donor atoms of a relatively "innocent" macrocyclic ligand is one of the most dominant themes of the work reviewed in this chapter.

2.1. Coordinative Lability

Mechanistic studies of the replacement of coordinated ligands by solvent species have long been of interest to inorganic chemists, and major aspects of this subject have frequently been reviewed.[1-14] Poon[6] has reviewed the literature through 1971 pertinent to substitution reactions of complexes containing macrocyclic ligands. Research activity in this area has continued to center on solvolysis reactions, especially acid and base hydrolyses.

2.1.1. Axial Substitution in trans-Co(N₄)XY Complexes

The overwhelming majority of studies of this class of reactions to date have been of cobalt(III) complexes. For most macrocyclic ligands axial substitution proceeds without an accompanying stereochemical change as in Eq. (1):

$$(1)$$

One may use the kinetic data obtained from reactions of this type to examine the *trans*-directing influence of ligands Y, to examine the labilizing influence of the equatorial (N_4) ligands, to examine the effects of the electronic structure, and to provide contrasts with those reactions in which there is stereochemical change. This is the simplest kind of reaction common to all six-coordinate complexes. Even among complexes of a single metal, e.g., cobalt(III), the lifetimes for these reactions can span 10–12 orders of magnitude, dependent on (N_4) and Y. The mechanistic problems of axial lability in these systems are common to substrates as diverse as the classical Werner complexes and the substituted cobalamins. With synthetic macrocyclic ligands one should in principle be able to investigate the effects of successive modifications of the equatorial ligand system, to isolate the ligand features that most affect lability, and to systematically span the range of interactions and reactivities between complexes with ammine and corrin or porphyrin ligand systems.

2.1.1.1. Acid Hydrolyses of Cobalt(III) Complexes. Poon and co-workers[6,7,15–27] have been systematically investigating the effects of equatorial (N_4) ligand substitution and axial (Y) ligand *trans* influence on the lability of acido ligands X. These investigators have found linear free energy correlations of the Hammett type[22,25] and inferred that the labilizing influences of equatorial (N_4) and axial (Y) ligands are additive. These and other studies have led to the identification of two distinct labilizing influences of the equatorial (N_4) ligands[6,16–31]: (i) a steric effect due to crowding of axial ligand positions by substituents of the equatorial ligand, and (ii) an electronic effect varying with the hybridization of the ligating nitrogen atoms. For example, hydrolysis rates increase in the order Co([14]aneN$_4$)XY < Co(Me$_2$[14]1,3-dieneN$_4$)XY < Co(Me$_4$[14]tetraeneN$_4$)-XY (X = Cl or Br and for a particular Y = Cl, Br, N$_3$, or NCS), with the differences being mostly attributable to the enthalpy of activation.[16,18] Poon and co-workers attribute the electronic labilizing influence of the equatorial ligands to an increase in polarizability or "softness" of the central metal with increasing unsaturation of the equatorial ligand.[6,16–27] For most Co(N$_4$)XY complexes, the hydrolysis rate is faster for X = Br than for X = Cl, and the ratio of rate constants, k_{Br}/k_{Cl}, decreases systematically with increases in unsaturation of the equatorial ligand. This last reactivity difference is largely a $\Delta S^{\ddagger}$ controlled effect for the aquation of chloro and bromo complexes (see Table 1) and may arise from a solvation interaction.

As is fairly common for reactions in solution,[8] there is a rough correlation of $\Delta H^{\ddagger}$ and $\Delta S^{\ddagger}$ for the hydrolysis reactions of type (1). Unfortunately, the interpretation of correlations of activation parameters is complicated by the strong correlation of errors in $\Delta H^{\ddagger}$ and $\Delta S^{\ddagger}$.[32] Somewhat more reliable is a correlation of $\Delta G^{\ddagger}$ with $\Delta H^{\ddagger}$.[33] For simple

TABLE 1. Axial Lability of Cobalt(III) Complexes Containing the Macrocyclic Ligands $trans$-$Co^{III}LXY + H_2O \xrightarrow{k_1} trans$-$Co^{III}L(Y)OH_2 + X$

L	Y	X	$\Delta G^{\ddagger\,a}$ kJ M^{-1}	$\Delta H^{\ddagger}$ kJ M^{-1}	$\Delta S^{\ddagger}$ kJ M^{-1} deg^{-1}	Ref.
[14]aneN₄	Cl⁻	Cl⁻	107	103	−13	6
	OH⁻	Cl⁻	85	79	−21	6
	NO₂⁻	Cl⁻	98	86	−38	19
	CN⁻	Cl⁻	109	98	−38	19
	NH₃	Cl⁻	111	101	−35	20
	NCS⁻	Cl⁻	123	144	+71	16
	NO₂⁻	Br⁻	91	87	−13	17
	N₃⁻	Cl⁻	104	99	−17	21
	Br⁻	Br⁻	100			cited in 6
	NCS⁻	Br⁻	118	149	+105	16
	N₃⁻	Br⁻	98	99	+4.2	18
	Br⁻	Cl⁻	106	94	−39	7
	Cl⁻	Br⁻	101	90	−39	7
ms-(5,12)-Me₆[14]aneN₄	Cl⁻	Cl⁻	91 (93)	115	+71	25
	NCS⁻	Cl⁻	108	115	+21	19
	CN⁻	Cl⁻	93	111	+59	19
	N₃⁻	Cl⁻	86	81	−17	19
	NO₂⁻	Cl⁻	81	93	+40	19
	NCS⁻	Br⁻	103	126	+79	19
	N₃⁻	Br⁻	80			19
	Cl⁻	Br⁻	84	98	+49	7
	Br⁻	Br⁻	81			25
	Br⁻	Cl⁻	89	80	−29	7
Me₆[14]4,11-dieneN₄	Cl⁻	Cl⁻	82			25
	Br⁻	Br⁻	81			25
	NCS⁻	Cl⁻	106	111	+17	20
	NCS⁻	Br⁻	105	126	+75	20
	CN⁻	Cl⁻	88	82	−21	7
	CN⁻	Br⁻	84	78	−21	7
	N₃⁻	Cl⁻	83	81	−4	20
	N₃⁻	Br⁻	80	89	+23	20
	NO₂⁻	Cl⁻	93			27
Me₂[14]4,11-dieneN₄	Cl⁻	Cl⁻	87			27
	NO₂⁻	Cl⁻	93			27
Me₄[14]tetraeneN₄	Cl⁻	Cl⁻	83.6, 84.3	83.3	−24	7, 26
	Br⁻	Br⁻	79			26
	NCS⁻	Cl⁻	106	120	+46	16
	NCS⁻	Br⁻	106	118	+40	16
	N₃⁻	Cl⁻	85	87	+6	18
	N₃⁻	Br⁻	84	86	+7	18
	NO₂⁻	Cl⁻	94	91	−9	21
	NO₂⁻	Br⁻	93	74	−63	21

TABLE 1 (continued)

L	Y	X	$\Delta G^{\ddagger a}$ kJ M^{-1}	$\Delta H^{\ddagger}$ kJ M^{-1}	$\Delta S^{\ddagger}$ kJ M^{-1} deg^{-1}	Ref.
Me$_2$[14]1,3-dieneN$_4$	Cl$^-$	Cl$^-$	94.9, 93.7	111	+59	26, 7
	Br$^-$	Br$^-$	85	118	+113	26, 7
	NCS$^-$	Cl$^-$	109	122	+42	16
	NCS$^-$	Br$^-$	109	136	88	16
	N$_3^-$	Cl$^-$	89	108	+62	18
	N$_3^-$	Br$^-$	88	91	+9	18
	Cl$^-$	Br$^-$	94	98	+14	7
	Br$^-$	Cl$^-$	92	86	−22	7
	NO$_2^-$	Cl$^-$	92	80	−42	21
	NO$_2^-$	Br$^-$	90	81	−29	21
13]aneN$_4$	Cl$^-$	Cl$^-$	94	123	53	31
[15]aneN$_4$ (isomer I)	Cl$^-$	Cl$^-$	91	94	31	31
[15]aneN$_4$ (isomer II)	Cl$^-$	Cl$^-$	85	76	−61	31
[16]aneN$_4$ (isomer I)	Cl$^-$	Cl$^-$	71			31

^{a} At 25°C: $\Delta G^{\ddagger} = RT (29.46 - nk)$, $RT = 2.49$ kJ mol^{-1}.

macrocyclic complexes a plot of $\Delta G^{\ddagger}$ vs. $\Delta H^{\ddagger}$ (Figure 1) indicates the following:

(i) There is a very rough correlation of the form $\Delta G^{\ddagger} = a + b\Delta H^{\ddagger}$, where b varies from approximately 1 to 0, depending on the equatorial (N$_4$) ligand.

(ii) Points for some combinations of axial ligands [especially Co(N$_4$)-(NCS)Cl^{2+} and Co(N$_4$)(NCS)Br^{2+}] deviate very significantly from such a correlation.

(iii) There is a considerable scatter of the data points with regard to any correlation.

Similar plots of available data for the *bis*-ethylenediamine and tetraammine analogs are more scattered (possibly because the data are from a larger number of laboratories), but do not indicate any unusual aspect of the activation parameters for systems undergoing stereochemical change in the course of solvolysis.[30] The preponderance of evidence indicates that acid hydrolyses of cobalt(III) complexes are largely dissociative, and a strong correlation with *net* stereochemical change would not be expected (for a contrasting point of view see reference 34).

Hung and Busch[31] have recently reported that hydrolyses of *trans*-Co([*m*]aneN$_4$)Cl$_2^+$ (*m* = 13, 15, 16) complexes are much more rapid than the hydrolysis rate of *trans*-Co([14]aneN$_4$)Cl$_2^+$. Through this whole series of cyclic tetraamine ligands it is observed that there are parallel variations in

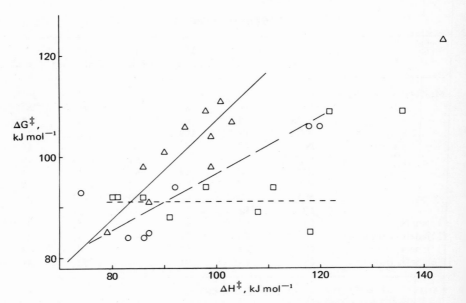

FIGURE 1. Correlation of activation parameters for acid hydrolyses of *trans*-Co(N₄)-XY⁺ complexes. Approximate correlation lines [excluding Co(N₄)(NCS)X⁺ data]: solid line for N₄ = [14]aneN₄ (△); horizontal dashed line for N₄ = Me₂[14]1,3-dieneN₄ (□); sloped dashed line for Me₄[14]-tetraeneN₄ (○).

$\Delta G^{\ddagger}$ and $\Delta H^{\ddagger}$, but that there is no stereochemical change accompanying substitution. It is postulated that the dominant feature affecting these variations in reactivity is release of strain energy in the nearly dissociative transition state. Calculated strain energies of the equatorial ligands were used to support these arguments.

The linear free-energy correlations found by Poon and co-workers have been rather limited in their range. Over an extended range of data $\Delta G^{\ddagger}$-[Co^{III}(N₄)XY] does appear to be approximately correlated with {$\Delta G^{\ddagger}$-[Co^{III}([14]aneN₄)XY] $-\Delta G^{\ddagger}$(Co([14]aneN₄)Cl₂⁺} (Figure 2); however, the correlation is not linear over the full range of data and there are several large deviations. While the scatter of points is appreciable, the correlation of hydrolysis rates does seem to support the contention of Poon and co-workers that labilizing influences are approximately additive.

The hydrolyses of Co^{III}([14]aneN₄)XY are relatively more sensitive to the leaving group (X) for the fastest reactions (smallest $\Delta G^{\ddagger}$) than is found for other Co^{III}(N₄)XY complexes. This is manifested in Figure 2 by the apparent curvature for small $\Delta G^{\ddagger}$. This and the observation that $\Delta G^{\ddagger}$ is generally larger for Co^{III}([14]aneN₄)XCl than for other Co^{III}(N₄)XCl complexes suggests that $\Delta G^{\ddagger}$ may have a minimum value of less than 80 kJ mol^{-1} for this

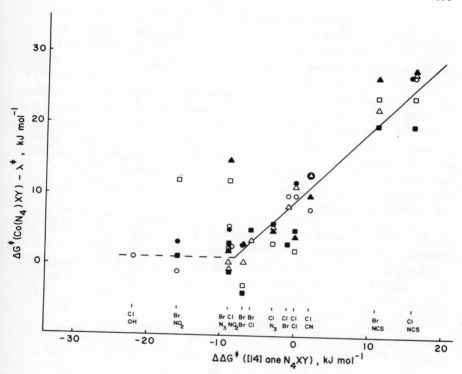

Key to Data Points

N_4	$\lambda^{\ddagger}$ (kJ mol^{-1})	Symbol	Reference
$Me_5[14]4,11$-dieneN$_4$	78	▲	Table 1
$Me_4[14]$tetraeneN$_4$	82	□	Table 1
$Me_2[14]1,3$-dieneN$_4$	89	■	Table 1
$Me_6[14]$aneN$_4$	81.4	△	Table 1
$(en)_2$	88	○	2, 4, 11
$(NH_3)_4$	77	●	2, 4, 11

FIGURE 2. Free-energy correlation for acid hydrolyses of $Co^{III}(N_4)XY$ complexes and $Co^{III}([14]aneN_4)XY$ complexes. $\Delta G^{\ddagger}$ from $k = (kT/h) \exp(-\Delta G^{\ddagger}/RT)$; $\Delta\Delta G^{\ddagger}([14]aneN_4$-$XY) = \Delta G^{\ddagger}(Co^{III}([14]aneN_4)XY) - \Delta G^{\ddagger}(Co^{III}([14]aneN_4)Cl_2)$; $\lambda^{\ddagger}$ is a "scale factor" adjusted to give a reasonable fit.

class of complexes. If $\Delta G^{\ddagger} \sim \Delta G_{coul} + \Delta G_{LFSE} + \Delta G_v + \Delta G_c$ (where ΔG_{coul} is the electrostatic free energy change associated with the separation of Cl$^-$ from $Co^{III}(N_4)Y$, ΔG_{LFSE} is change in ligand field stabilization energy associated with formation of a five-coordinate intermediate, ΔG_v is the free energy change associated with the increased volume of $Co^{III}(N_4)Y + Cl^-$, and ΔG_c is the covalent component of the Co–Cl bond), then using literature values of ligand field[35] and other parameters[36,37] and a change of 80 pm in the Co–Cl distance coordinated and free Cl^-, $\Delta G_{min}^{\ddagger} \sim \Delta G_{coul} + \Delta G_{LFSE} + \Delta G_v \sim (30 + 18 + 6)\ kJ\ mol^{-1} \sim 54\ kJ\ mol^{-1}$. A value of $\Delta G_c \sim 35\ kJ\ mol^{-1}$ [38] would give a reasonable intermediate value for $\Delta G^{\ddagger}$. Relaxation of strain in the equatorial ligand on formation of $Co^{III}(N_4)Y$ would be expected to contribute to ΔG_v ("relaxation" of strain corresponds to an increase in volume), while both the equatorial ligand and Y could modify ΔG_c and ΔG_{LFSE}. While much of the information about hydrolysis reactions is consistent with a linear superposition of energy terms, the approach to the limiting minimum value of $\Delta G^{\ddagger}$ would introduce nonlinearities in linear free energy correlations.

Recently German and Dogonadze have applied a general theory of reaction rates to solvolysis reactions of coordination complexes.[39,40] In this approach the free energy of activation for a dissociative reaction may be represented as a function[40] of the solvent reorganizational free energy ΔG_s, the reorganizational free energy associated with the departing ligand, ΔG_r^x, and the free energy change associated with the solvolysis reaction, ΔG^0 (see Figure 3):

$$\Delta G^{\ddagger} = (\Delta G_s + \Delta G_r^x + \Delta G^0)^2 / 4\Delta G_s \qquad (2)$$

Equation (2) leads to a linear free energy expression of the Hammett type for values of ΔG_s sufficiently large that $|\Delta G_s| > |\Delta G_r^x + \Delta G^0|$. More specifically, in this limit, for any particular equatorial ligand (N_4) we have

$$\Delta\Delta G^{\ddagger}(N_4XY) = \frac{1}{4}\Delta\Delta G_s(N_4XY) + \frac{\Delta\Delta G_r^x(N_4XY) + \Delta\Delta G^0(N_4XY)}{2} \qquad (3)$$

where $\Delta\Delta G_i(N_4XY) = \Delta G_i[Co(N_4)XY] - \Delta G_i[Co(N_4)Cl_2]$ for any free-energy quantity. Since ΔG_r^x should be substantially independent of (N_4) and Y, the linear free energy correlation

$$\Delta\Delta G^{\ddagger}(N_4XY) \simeq \tfrac{1}{4}[\Delta\Delta G_s(N_4XY) - \Delta\Delta G_s(N_4'XY]$$
$$+ \tfrac{1}{2}[\Delta\Delta G^0(N_4XY) - \Delta\Delta G^0(N_4'XY)] + \Delta\Delta G^{\ddagger}(N_4'XY) \qquad (4)$$

is obtained for hydrolysis reactions of families of complexes with different equatorial ligands (N_4) and (N_4'). If a strict superposition principle were applicable so that $\Delta G_i = \delta G_i + \delta G_i(N_4) + \delta G_i(X) + \delta G_i(Y)$, then the first two terms of (4) should be identically zero. This is not true, which suggests small residual differences in the solvation interaction terms.

While (2) provides a convenient context for the discussion of hydrolysis reactions, its applicability has not been established. As Swaddle has noted,[8] the assumption of parabolic potential surfaces which is involved in the

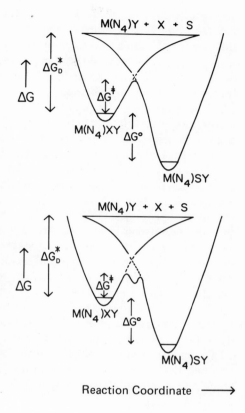

FIGURE 3. Qualitative representations of free-energy changes in solvolysis reactions showing the effect of formation of a metastable intermediate. For a purely dissociative reaction $\Delta G^{\ddagger}$ would approach the difference in the free-energy content of $\{M(N_4)Y + X + S\}$ and the ground state of the reactant system ($\Delta G_D{}^*$), where $M(N_4)Y$ has the internal internuclear coordinates of this moiety in $M(N_4)XY$. For a dissociative interchange mechanism in $\Delta G^{\ddagger} < \Delta G_D{}^*$, and for an associative mechanism $\Delta G^{\ddagger} \ll \Delta G_D{}^*$. Relaxation of strain in $M(N_4)Y$ would tend to decrease $\Delta G^{\ddagger}$, in the limit to form a metastable intermediate.

derivation of (2) cannot be strictly correct in a process that involves bond breaking [Figure 3 (top)]. Furthermore, formation of a metastable five-coordinate intermediate [Figure 3 (bottom)] would necessarily result in some modifications of (2). Numerical estimates above do indicate that ΔG_s is probably not large compared to $\Delta G_r{}^x + \Delta G^0$. The German and Dogonadze approach does emphasize that the solvent environment may play a profound role in affecting reactivity patterns, a possibility that is often neglected in mechanistic rationalizations, but seems required by the approximate correlations in Figure 1. The approximate dependence on the equatorial ligand of the correlation between $\Delta G^{\ddagger}$ and $\Delta H^{\ddagger}$ suggests that the equatorial ligand plays a significant role in determining the solvent reorganizational energy. Certainly one would expect ΔH_s to be large compared to ΔG_s. Since many of the mechanistic arguments about the relative rates of solvolytic reactions are primarily enthalpic (e.g., arguments about bond strength, ligand field stabilization energies, etc.), the significance of these contributions will remain ambiguous until the contributions of the solvent are better understood. In any case, attempts to "tune" solvolytic reactivity by varying substituents on the equatorial ligand, and to some extent the *trans* ligand, are currently at a highly empirical level and the effects observed could in principle have more to do with relaxation of strain in the dissociative transition state than with the transmission of electronic effects.

The relaxation of equatorial ligand strain is very likely an important factor in the solvolytic reactions of macrocyclic complexes. However, this relaxation process must have two principal components: (a) relief of internal strain in the isolated molecule as the bond to the leaving group is lengthened, and (b) work against the solvent as the relaxing ligand sweeps out a relatively large volume. Neither factor is easy to quantitatively evaluate, but the strain energy calculations of Busch and co-workers provide a basis for examining the internal processes. Only qualitative points can be made about the involvement of the solvent. First of all it should be noted that involvement of solvent in the formation of the transition state necessarily limits the usefulness of distinctions between associative and dissociative solvolytic reactions. Secondly there could be some notable differences between the solvolytic reactions that are stereoretentive and those that are not since relaxation of the equatorial ligand in the direction of the leaving group works against the entering ligand in stereoretentive reactions but assists the solvolysis process if there are accompanying *trans* ↔ *cis* isomerizations. In such highly substituted macrocyclic ligands as $Me_6[14]aneN_4$, the *cis* configurations tend to be much more strained than the *trans* configurations, and the relaxation of strain in the transition state could give the reactions an increased associative component (see Section 2.1.2).

2.1.1.2. Base Hydrolyses of Cobalt(III) Complexes. All known cobalt(III) ammine, and primary or secondary amine complexes, exhibit base-dependent hydrolysis pathways.[2,4–8,11,12,37] The observed hydrolysis rate constant for such systems may be represented $k_{obs} = k_{H_2O} + k_{OH}[OH^-]$. For *trans*-$Co(N_4)XY$ complexes the ratio k_{OH}/k_{H_2O} varies between $10^5\ M^{-1}$ and $10^{15}\ M^{-1}$, depending on (N_4), X, and Y. Effects of such magnitude are observed only in cobalt complexes that contain N–H protons, and the base hydrolysis effect has been attributed to formation of a very labile intermediate cobalt–amido complex,

$$Co^{III}(N_4)XY + OH^- \underset{k_{-5}}{\overset{k_5}{\rightleftharpoons}} Co^{III}(N_4-H^+)XY + H_2O \qquad (5)$$

$$Co^{III}(N_4-H^+)XY \xrightarrow{k_6} Co^{III}(N_4-H^+)Y + X \qquad (6)$$

$$Co^{III}(N_4-H^+)Y + H_2O \xrightarrow{k_7} Co^{III}(N_4)(OH^-)Y \qquad (7)$$

where $k_7 \gg k_6$. A stationary-state treatment of Eqs. (5)–(7) gives $k_{OH} = k_5 k_6/(k_{-5} + k_6)$. Values of k_{OH} span a range of nearly 10^6 for complexes with macrocyclic ligands. Most systems lie at the lower end of this range and their values of k_{OH} vary with X and Y; this is generally interpreted as corresponding to the condition $k_{-5} \gg k_6$. When very large values of k_{OH} $[(10^5–10^8)\ M^{-1}\ s^{-1}]$ are observed the hydrolysis rate constants appear to be less sensitive to changes in X and Y, and general base catalysis of the reactions is observed. Thus the reactions with very large rate constants are believed to approach the conditions for a limiting S_N1 mechanism, with $k_6 > k_{-5}$. Most macrocyclic amine complexes appear to approach this limit.

Similar base hydrolyses can dominate the substitutional behavior of cobalt(III) complexes containing β-diimine ligands. Thus $Co(Me_2[14]1,11\text{-}$ diene)XY^+ complexes are unusually labile, with the rates of hydrolysis being inversely proportional to acid concentration even for $[H^+] > 0.1\ M$.[43] The relatively acidic methylene group of the β-diimine moiety has a $pK_a \sim 6$ in these complexes (to be compared to $pK_a \sim 12$ for a coordinated amine), and some resonance forms of the $(Me_2[14]1,11\text{-}dienato(1-)N_4)$ ligand place the charge on the coordinated nitrogen atoms. Hence there is a formal similarity to the base hydrolyses of amine complexes discussed above. Furthermore, one may argue that the $Co(bzo_2[14]hexaenato(2-)N_4)Cl$ complex characterized by Goedken and co-workers[44] models the "transition state" in base hydrolysis reactions.

Many complexes with equatorial ligands that contain extensive unsaturation undergo degradative reactions in basic solution, with the possible involvement of intermediates containing radicals in the equatorial ligand (see Section 3 below).

2.1.2. Substitution in cis-Co(N₄)XY Complexes

There have been very few systematic studies of hydrolyses of *cis*-$Co^{III}(N_4)XY$ complexes. Such studies would seem essential in order to separate specific *trans*-labilization effects from more general effects, such as electronic, steric, solvation, etc.

Hung and Busch[31] have examined the stereoretentive solvolyses of *cis*-$Co([m]aneN_4)Cl_2^+$ complexes ($m = 12$ and 13) and compared their work with earlier studies on *cis*-$Co([14]aneN_4)Cl_2^6$ and some related systems. The *cis* complexes hydrolyze more rapidly than the *trans* analogs, but do not exhibit as large a range of reactivity. These systems exhibit a rather large range of values of $\Delta H^{\ddagger}$ without much variation in $\Delta G^{\ddagger}$.

There have been a handful of studies of hydrolysis reactions of complexes containing macrocyclic ligands in a folded "*cis*" configuration. Of the systems investigated to date, a reasonably large, stable family of *cis*-cobalt(III) complexes has been reported only with the [14]aneN₄ ligand. The *cis* configuration is "forced" on other macrocyclic ligands by bidentate ligands such as 1,2-diaminoethane, oxalate, carbonate, or nitrate.[45–47]

Endicott and co-workers[48,49] have investigated the hydrolytic reactions of $Co(N_4)CO_3^+$ complexes for N_4: $Me_6[14]4,11\text{-dieneN}_4$, $rac\text{-}(5,12)\text{-}Me_6[14]\text{-aneN}_4$, and $ms\text{-}(5,14)\text{-}Me_6[14]\text{aneN}_4$. In basic solutions the carbonate is displaced stepwise, forming the $trans\text{-}Co(N_4)(OH)_2^+$ product from an intermediate which is probably $trans\text{-}Co(N_4)(OH)CO_3$.[42]

The $Co(Me_6[14]4,11\text{-dieneN}_4)CO_3^+$ complex is relatively reactive in acidic solution. The hydrolysis proceeds smoothly to $trans\text{-}Co(Me_6[14]4,11\text{-dieneN}_4)(OH_2)_2^{3+}$. So far, attempts to trap cis intermediates have failed in this system.[48,49]

Cobalt(III) complexes of the $rac\text{-}(5,12)\text{-}Me_6[14]\text{aneN}_4$ and $ms\text{-}(5,14)\text{-}Me_6[14]\text{aneN}_4$ ligands are remarkable for their high affinity for CO_3^{2-}. Acid hydrolyses are exceptionally slow, complex, and reversible under most conditions. The situation is complicated by the large number of conformational isomers of these complexes and conformational isomerization reactions under hydrolytic conditions.[50,51] In fact, the $trans\text{-}Co(N_4)(OH_2)_2^{3+}$ complexes are converted to $Co(N_4)CO_3^+$ species even in acidic solution.[49] Carbonate hydrolyses in concentrated $HClO_4$ have permitted the isolation and characterization of $cis\text{-}Co(N_4)(OH_2)_2^{3+}$ complexes.[49,50] These cis-diaquo complexes are metastable in acidic aqueous solution, and an equilibrium mixture of conformational isomers is achieved slowly.[51] The stability of the $Co(N_4)CO_3^+$ relative to $cis\text{-}Co^{III}(N_4)XY$ complexes has been attributed to steric effects that arise largely from crowding of the coordination site by methyl groups of the N_4 ligand.[48,49]

Dasgupta[52] has reported that the acid hydrolysis of $Co([14]\text{aneN}_4)CO_3^+$ proceeds only slightly faster than acid hydrolysis of $Co(Me_6[14]4,11\text{-dieneN}_4)CO_3^+$. It was proposed that the slowness resulted from a combination of steric crowding and the electronic effect proposed by Francis and Jordan.[53] The hydrolysis product in perchloric acid solutions was found to be $cis\text{-}Co([14]\text{aneN}_4)(OH_2)_2^{3+}$.

Very complex anation reactions have been found of the $cis\text{-}Co(N_4)(OH_2)_2^{3+}$ complexes with $N_4 = rac\text{-}(5,12)\text{-}Me_6[14]\text{aneN}_4$.[54] Preliminary studies had suggested a nucleophilic pathway in the formation of $trans\text{-}Co(N_4)Cl_2^+$,[48] but the results of a thorough study of these systems are more consistent with very rapid rapid initial formation of $cis\text{-}Co(N_4)ClOH_2^{2+}$, the generation of a metastable $cis\text{-}Co(N_4)Cl_2^+$ complex, and the formation of $trans\text{-}Co(N_4)Cl_2^+$ from a five-coordinate intermediate, possibly in a scavenging reaction.[54] The thorough kinetic study of these systems is complicated by the difficulty of maintaining the reaction solutions isomerically pure.

Most aquohydroxy and dihydroxy complexes of cobalt(III) react rapidly with CO_2. Dasgupta and Harris[55] have thoroughly investigated the reaction of $cis\text{-}Co([14]\text{aneN}_4)(OH_2)^+$ and $cis\text{-}Co([14]\text{aneN}_4)(OH)OH_2^{2+}$ with CO_2. The reaction has been found to proceed stepwise with a relatively slow ring-closing step.

2.1.3. Axial Substitution Coupled with Electrophilic Attack on Ligand Groups

Poon and co-workers[27] have very recently discovered an extraordinary reaction in which the second step in the acid hydrolysis of a *trans*-$Co(N_4)Cl_2^+$ complex is accompanied by the displacement of a proton from an *N*-methyl group and formation of a cobalt–alkyl product.[56] These authors suggest that the initial, normal halide hydrolysis step

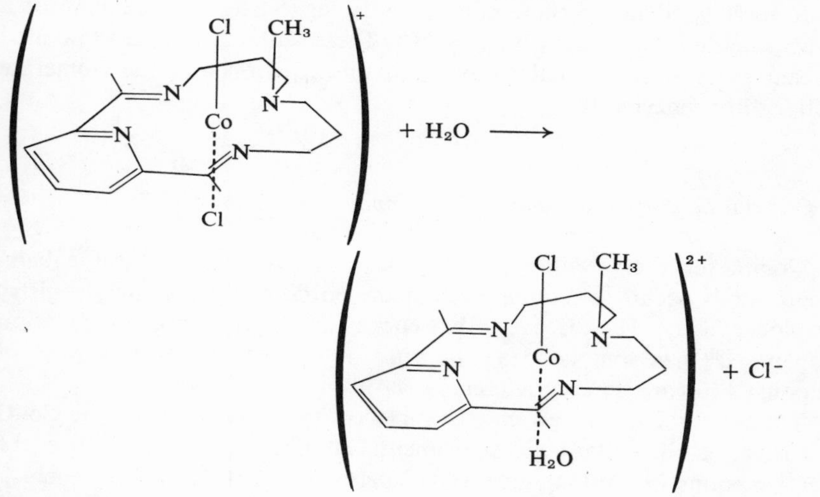

$$+ H_2O \longrightarrow$$

$$+ Cl^- \qquad (8)$$

1

involves the halide remote from the *N*-methyl group and that the alkylation is associated with substitution at the position adjacent to the *N*-methyl group:

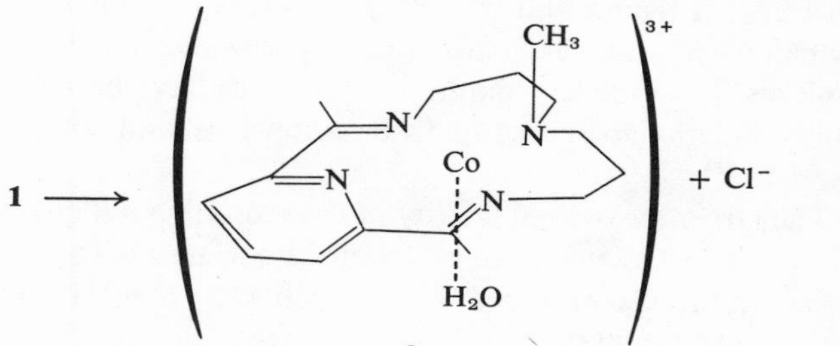

$$\mathbf{1} \longrightarrow$$

$$+ Cl^- \qquad (9)$$

2

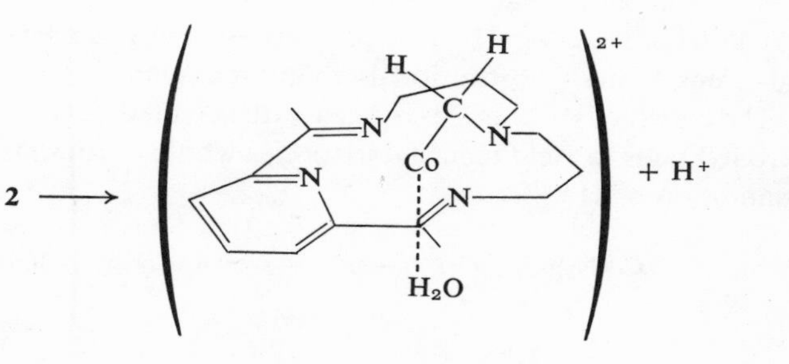

$$\mathbf{2} \longrightarrow$$

$$+ H^+ \qquad (10)$$

Steps indicated by (9) and (10) may or may not be concerted; however, electrophilic displacement of H^+ by a five-coordinate cobalt(III) intermediate seems an attractive mechanistic possibility.

Poon and co-workers[27] have also found a stepwise formation of $Co(2,6,11\text{-}Me_3pyo[14]trieneN_4)(NCS)Cl^+$ and $Co(2,6,11\text{-}Me_3pyo[14]triene\text{-}N_4)(NCS)_2{}^+$ when the hydrolysis of $Co(2,6,11\text{-}Me_3pyo[14]trieneN_4)Cl_2{}^+$ is carried out in excess thiocyanate. Failure to observe the formation of the cobalt–alkyl bond under these conditions is suggestive of scavenging of a five-coordinate intermediate(2) by NCS^-. These workers have also pointed out that there is some analogy between this reaction and the isomerase reactions of coenzyme B_{12}.

2.1.4. Axial Lability in Organo–Cobalt Complexes

Complexes that contain macrocyclic equatorial ligands and cobalt–carbon bonds occupy a unique place at the borders of inorganic chemistry and biochemistry. The discovery that coenzyme B_{12} is such a coordination complex[57,58] has stimulated a vast amount of research on this class of compounds (for reviews see references 58–70).

While both axial groups may be replaced, the ligand *trans* to the Co–C bond is very labile under most experimental conditions.[71] For example, the NCS^- anations of $Co([14]aneN_4)(OH_2)CH_3^{2+}$ and $Co(Me_4[14]tetraeneN_4)\text{-}(OH_2)CH_3^{2+}$ are similar in magnitude, implying a $Co\text{-}OH_2$ bond-breaking rate $k_w \sim 10^2 \ sec^{-1}$ [72] (assuming a dissociative mechanism, $k_{obs} \sim K_{os}k_w$ with $K_{os} \sim 1 \ M^{-1}$, where K_{os} is the equilibrium constant for outer-sphere association of ligand and complex).[3,10] This value is intermediate between the analogous water exchange rates of low-spin cobalt(III) and cobalt(II) complexes. Qualitatively similar observations have been made for the alkyl-aquobis(dimethylglyoximato)cobalt complexes and analogous B_{12} derivatives.[71,73–75]

The larger *trans*-axial lability of the $Co(N_4)(X)(alkyl)$ complexes than of their cobalt(III) analogs can be readily associated with a longer Co–X bond length in the alkyl complexes. Thus the $Co\text{-}OH_2$ bond is approximately 20 pm longer in $Co(N_4)(OH_2)CH_3^{2+}$ complexes than in $Co(N_4)(OH_2)_2^{3+}$ complexes,[72] but more than 14 pm shorter in the alkyl complexes than in the $Co(N_4)(OH_2)_2^{2+}$ analogs.[76] The qualitative correlation between bond length, bond strength, and substitutional lability is evident.

The cobalt–alkyl complexes react with solvated electrons e_{aq}^- at diffusion-controlled rates to yield metastable species whose decomposition involves the consumption of H^+ [77]:

$$Co(N_4)(OH_2)CH_3^{2+} + e_{aq}^- \longrightarrow Co(N_4)(OH_2)CH_3^+ \tag{11}$$

For $Co(Me_4[14]tetraeneN_4)(OH_2)CH_3^{2+}$ there is an initial increase in conductivity whose magnitude is consistent with formation of e_{aq}^- and H^+ in the radiolytic pulse, with e_{aq}^- being scavenged by $Co(N_4)(OH_2)CH_3^{2+}$. The subsequent conductivity change implies stoichiometric consumption of H^+ with a pH-independent decay time, $\tau_{15} \simeq 4.5 \times 10^{-4}$ sec, as in (12).

$$Co(N_4)(OH_2)CH_3^+ + H^+ \longrightarrow Co(N_4)(OH_2)_2^{2+} + CH_4 \qquad (12)$$

When the radiolysis is performed in D_2O (containing ~ 1 M t-butanol) the product gases contain a large amount of CH_3D, qualitatively consistent with (12).[78]

Except in such reduced systems, the cobalt–methyl bond seems to be thermally stable. For example, the $Co(Me_6[14]4,11$-dieneN$_4)(OH_2)CH_3^{2+}$ complexes can be reversibly dehydrated by concentrated H_2SO_4 and concentrated $HClO_4$, but the $Co-CH_3$ bond remains intact.[79] However, the alkyl ligands are readily displaced from organo–cobalt complexes by a variety of electrophiles.[60,62,69,80–85] Most of the quantitative studies have used cobalt complexes with oximate ligands. The rates of these reactions seem faster when the equatorial ligand is an oximate rather than a neutral macrocycle. For example, Hg^{2+} is methylated 10^5 times faster by $CH_3Co(DMG)_2$-OH_2 than by $CH_3Co(Me_4[14]tetraeneN_4)OH_2^{2+}$. Variations in the equatorial ligands are reported to result in $\sim 10^8$-fold variations in the rates of methylation of mercury.[86] This reactivity difference,[86] and differences in rates of *trans*-axial substitution,[71] have been attributed to increases in the amount of electron density, donated to the metal by the equatorial ligand. Pertinent to the same point, $CH_3Co(DMG)_2OH_2$ can be oxidized at a less positive potential than $CH_3Co(Me_4[14]tetraeneN_4)OH_2^{2+}$.[87]

2.1.5. Axial Substitution in $Co^{II}(N_4)$ Complexes

Addition of an electron to $Co^{III}(N_4)X_2$ complexes results in tetragonally distorted, low-spin cobalt(II) complex when N_4 is a 14-membered macrocyclic ligand.[76] The cobalt(II) complexes are conveniently described in terms of a single electron occupying an axial d_{z^2} orbital of the cobalt(III) complex, with a concomitant lengthening of the $Co^{II}-X$ bonds. Differences as large as 50 pm have been observed between the $Co^{II}-X$ and $Co^{III}-X$ bond lengths.[76] The axial distortion can be accompanied by an increase of axial lability by a factor of more than 10^{12} (Table 2).[76,88,89] Although the second hydrolysis constant k_{II} is very difficult to determine for most $Co^{III}(N_4)X_2$ complexes it is clear that both $Co^{II}-X$ bonds are greatly labilized with respect to their cobalt(III) counterparts. Similar tetragonal distortions and axial labilizations occur when $Co(CN)_6^{3-}$ is reduced and are associated with the low-spin (doublet) cobalt(II) electronic configuration.[76]

TABLE 2. Contrasting Axial Lability of Low-Spin Cobalt(III) and Cobalt(II) Complexes[a]

N$_4$	X	Co^{III}(N$_4$)X$_2$		Ref.	Co^{II}(N$_4$)X$_2$		Ref.
		k_I (sec^{-1})	k_{II} (sec^{-1})		k_I' (sec^{-1})	k_{II}' (sec^{-1})	
Me$_6$[14]4,11-dieneN$_4$	NH$_3$	3×10^{-6} ^{b}		29	$>10^6$	5×10^4	76
	CN	$<10^{-6}$ ^{d}			$>10^6$	1.0×10^4	76
Me$_4$[14]tetraeneN$_4$	NH$_3$	8×10^{-7} ^{b}		29	$>3 \times 10^5$	$>3 \times 10^5$	76
	CN	$<10^{-6}$ ^{d}			$>10^6$	1.0×10^3	76

^{a} Rate constants for stepwise replacement of X by H$_2$O in acidic aqueous solution at 25°C, except as indicated.
^{b} 68.5°C.
^{c} Not reported; very small rate constant.
^{d} Less than 10% decomposition in 24 hr.

2.1.6. Substitution at Other Metal Centers

Substitution at metals other than cobalt in complexes containing macrocyclic ligands have rarely been studied. The d^9-copper(II) complexes and the low-spin (singlet) d^8-nickel(II) complexes show very little evidence of axial interactions.[48,50,89,90] The Cu(Me$_2$pyo[14]trieneN$_4$)$^{2+}$ complex has been shown to exchange axially coordinated ligands with free solvent at a rate too fast to resolve by NMR relaxation techniques ($k > 5 \times 10^4$ sec^{-1}).[89]

The iron(II) and (III) complexes with 14-membered quadridentate macrocyclic ligands tend to be low spin (singlet and doublet ground states, respectively). Structural data on Fe(Me$_4$[14]tetraeneN$_4$)(NCCH$_3$)$_2^{2+}$ [91] and Fe(Me$_4$[14]tetraeneN$_4$)(OH$_8$)$_2^{3+}$ [92] indicate similar metal–ligand bond lengths The iron(III) complexes seem to equilibrate with the medium on times of the order of a minute or less,[93] but no systematic studies have been reported and the systems are complicated by very rapid redox reactions.

2.2. Oxidation–Reduction Reactions of Simple Stoichiometry

A rather bewildering range of chemical reactions is conveniently classified as "oxidation–reduction" reactions. The simplest of these are reactions in which a single electron is transferred from one metal center to another. Such reactions have been the subject of intensive study and there are many useful reviews (e.g., references 94–102). Complexes with quadridentate macrocyclic ligands have been useful in "stabilizing" unusual oxidation states of various metals and in systematic studies of the factors determining the rates of electron transfer reactions. The present section reviews both the thermodynamic and mechanistic studies.

2.2.1. Electrochemical Studies

Electrochemical studies of macrocyclic complexes have developed in parallel with synthetic studies, each type of study reinforcing the other.

Interest in electrochemical studies has usually been directed toward (1) generation and identification of complexes with metals in unusual oxidation states, (2) investigations of systems with electron density delocalized onto the ligands, and (3) determinations of the stabilities of various complexes as reflected in their redox potentials.

In studies of metals in unusual oxidation states it is critically important to establish the nature of the product formed in the electrolysis, and, if the measurements are to be useful, the reversibility of the process. In Table 3 only the number of electrons transferred have been indicated with no mention of whether it is to ligand- or metal-centered orbitals. In the majority of cases the first oxidation wave and to a lesser extent the first reduction wave correspond to M(III) and M(I), respectively[103]; however, there are exceptions.[104] Lovecchio et al.[103] have studied a very large number of nickel complexes and have used ESR to show that Ni(III) is the first oxidation product. Naturally, protic solvents will greatly affect the electrochemistry, most especially ligand-centered redox reactions that involve hydrogen, e.g., the reduction of imine bonds. An especially good example of such effects is the Ni(Me_2pyo[14]trieneN$_4$)$^{2+}$ [105] system, which was studied in both methanol and acetonitrile. The stepwise reduction in methanol of the series of related nickel complexes with various degrees of saturation strongly suggests that the imine functions are reduced, resulting eventually in the same saturated Ni(I) product. The electrochemistry in acetonitrile is very different and much simpler. The EPR spectrum of the first reduction product of Ni(Me_2pyo[14]-trieneN$_4$)$^{2+}$ in acetonitrile indicates that the electron density resides mainly on the ligand.[103]

The effect of structural parameters on redox potentials has been extensively investigated and some general trends are becoming apparent. The large amount of work by Busch and co-workers[103,105-107] is very informative, especially since it contains enough well-characterized electrochemical reactions to make meaningful generalizations. Using Ni(ms-(4,12)-Me$_6$[14]-aneN$_4$)$^{2+}$ as a reference point, the effects of various structural factors on the II $\rightarrow$ III potential appear to be separable and additive. The following are estimates of the contributions of various structural factors on the Ni(II)(ms-(5,12)-Me$_6$[14]aneN$_4$)$^{2+}$ II $\rightarrow$ III potential ($+0.87$ V): (1) [14] $\rightarrow$ [15]-membered ring $\Delta E_{1/2} = +225$ mV; (2) 14 $\rightarrow$ 16-membered rings, $+375$ mV; (3) two axial methyl groups in six-membered rings, $+183$ mV; (4) delocalized charge, -430 mV; (5) isolated imine, $+43$ mV; (6) α-diimine $+170$ mV. The estimates are based on the $E_{1/2}$'s of 24 complexes with a correlation coefficient between observed and calculated potentials of 0.984.

While the data are less extensive, very similar substituent effects seem to occur in the iron(II)–iron(III) couples as in the nickel(II)–nickel(III) couples.[106,119] One notable difference is a relatively greater sensitivity of the iron(II)–iron(III) couples to an α-diimine moiety ($\Delta E_{1/2} \sim 300$ mV). In contrast the half-wave potentials of cobalt(II)–cobalt(III) couples seem relatively

TABLE 3. Half-Wave Potentials for Reduction and Oxidation of Macrocyclic Complexes

Complex	$E_{1/2}$ (total number of electrons transferred)[a]						Solvent[b]	Electrode	Reference electrode	Ref.
	−2	−1	+1	+2	+3	+4				
Ni(ms-Me$_2$pyo[14]eneN$_4$)$^{2+}$		+0.89[c,e]	−1.53r				CH$_3$CN	sPt[e]	Ag/Ag$^+$[f]	105
Ni[Me$_2$pyo[14]dieneN$_4$)$^{2+}$		+0.93r	−1.25r	−1.88r			CH$_3$CN	sPt	Ag/Ag$^+$	105
Ni(Me$_2$pyo[14]trieneN$_4$)$^{2+}$		+1.03r	−0.96r	−1.55r			CH$_3$CN	sPt	Ag/Ag$^+$	105
Ni(Me$_2$pyo[14]tetraeneN$_4$)$^{2+}$		+1.05r	−0.84r	−1.45r			CH$_3$CN	sPt	Ag/Ag$^+$	105
Ni(ms-Me$_2$pyo[14]eneN$_4$)$^{2+}$			−1.15r				CH$_3$OH	dme[g]	sce[h]	105
Ni(Me$_2$pyo[14]dieneN$_4$)$^{2+}$			−0.65r				CH$_3$OH	dme	sce	105
Ni(Me$_2$pyo[14]trieneN$_4$)$^{2+}$			−0.52r	−0.88	−1.16	(−0.92, −1.17)	CH$_3$OH	dme	sce	105
Ni(Me$_2$pyo[14]tetraeneN$_4$)$^{2+}$					−0.92i	−1.09i	CH$_3$OH	dme	sce	105
Ni(Me$_6$[14]4,11-dieneN$_4$)$^{2+}$		(+1.03, +1.76)	(−1.22, −1.21)				CH$_3$CN	sPt	sce	108
Ni(Me$_6$[14]4,14-dieneN$_4$)$^{2+}$		(+1.12, +1.73)	−1.25r				CH$_3$CN	sPt	sce	108
Ni(Me$_8$[14]4,11-dieneN$_4$)$^{2+}$		+1.92	(−1.19, −1.16)				CH$_3$CN	sPt	sce	108
Ni(Me$_2$[13]4,7-dieneN$_4$)$^{2+}$		+0.69[j](A)[k]	(−2.26, −2.15)				DMF	sPt	sce	108
Ni(Me$_2$[14]4,7-dieneN$_4$)$^{2+}$		+0.61i(A)	−2.00t				DMF	dme	sce	108
Ni(ms-(4,12)-Me$_6$[14]aneN$_4$)$^{2+}$		(+1.21, +1.24)	(−1.24, −1.23)				CH$_3$CN	sPt	sce	108
Ni(rac-(5,12)-Me$_6$[14]aneN$_4$)$^{2+}$		(+1.28, +1.80)	(−1.25, −1.24)				CH$_3$CN	sPt	sce	108
Ni(ms-(5,12)-Me$_8$[14]aneN$_4$)$^{2+}$		(+1.36, +1.42)	(−1.21, −1.19)				CH$_3$CN	sPt	sce	108
Ni(rac-(5,14)-Me$_6$[14]aneN$_4$)$^{2+}$		+0.876r	−1.584r				CH$_3$CN	sPt	Ag/Ag$^+$	111
Ni(Me$_6$[14]4,11-dieneN$_4$)$^{2+}$		+0.856r	−1.519r				CH$_3$CN	sPt	Ag/Ag$^+$	111
Ni(Me$_6$[14]4,14-dieneN$_4$)$^{2+}$		(+0.79, +1.63)p[l]	−1.558r				CH$_3$CN	sPt	Ag/Ag$^+$	111
Ni([13]aneN$_4$)$^{2+}$		(+0.86, +1.55)p	−1.584r				CH$_3$CN[d]	sPt	Ag/Ag$^+$	112
Ni([14]aneN$_4$)$^{2+}$		(+0.7 → +0.9)ip	−1.70r				CH$_3$CN	sPt	Ag/Ag$^+$	112
Ni(Me$_2$[14]aneN$_4$)$^{2+}$		+0.67r	−1.73r				CH$_3$CN	sPt	Ag/Ag$^+$	112
Ni(Me$_4$[14]aneN$_4$)$^{2+}$		+0.68r	−1.66r				CH$_3$CN	sPt	Ag/Ag$^+$	112
Ni(ms-(4,12)-Me$_6$[14]aneN$_4$)$^{2+}$		+0.71r	−1.57r				CH$_3$CN	sPt	Ag/Ag$^+$	112
Ni(Me$_6$[14]4,9,11-tetraeneN$_4$)$^{2+}$		+0.82r	−1.57r	−2.0i(C)[m]			CH$_3$CN	sPt	Ag/Ag$^+$	112
Ni([15]aneN$_4$)$^{2+}$		+0.98r	−1.35r				CH$_3$CN	sPt	Ag/Ag$^+$	112
Ni(Me$_6$[16]aneN$_4$)$^{2+}$		+1.05r	−1.5i(C)				CH$_3$CN	sPt	Ag/Ag$^+$	112
Ni(Me$_6$[16]4,12-dieneN$_4$)$^{2+}$		+0.90r	−1.40r				CH$_3$CN	sPt	Ag/Ag$^+$	112
Ni(Me$_6$[16]1,4,12-trieneN$_4$)$^{2+}$		+1.3i(A)	−1.37r				CH$_3$CN	sPt	Ag/Ag$^+$	112
Ni(Me$_2$[14]1,3-dieneN$_4$)$^{2+}$		+1.3r	−1.30r				CH$_3$CN	sPt	Ag/Ag$^+$	112
Ni(Me$_6$[14]1,3,7,11-tetraeneN$_4$)$^{2+}$		+1.30	−1.16r	−1.62r			CH$_3$CN	sPt	Ag/Ag$^+$	112
Ni(Me$_4$[14]1,3,9,10-tetraeneN$_4$)$^{2+}$		+1.05r	−0.76r	−1.15r			CH$_3$CN	sPt	Ag/Ag$^+$	112
Ni(Me$_2$[14]4,7-dieneN$_4$)$^{2+}$		+1.00r	−0.82r				CH$_3$CN	sPt	Ag/Ag$^+$	112
Ni(Me$_2$[13]4,7-dienato(1−)N$_4$)$^{+}$		+0.72r	−2.30r				CH$_3$CN	sPt	Ag/Ag$^+$	112
Ni(Me$_2$[14]4,7-dienato(1−)N$_4$)$^{+}$		+0.27i(A)	−2.34r				CH$_3$CN	sPt	Ag/Ag$^+$	112
Ni(bzo$_4$[16]octaeneN$_4$)$^{2+}$		+0.23i(A)	−0.64nr(C)	−0.93nr(C)	−2.4i(C)		CH$_3$CN	rPt	Ag/Ag$^+$	109
(Bzo$_4$[16]octaeneN$_4$)$^{2+}$		+1.24n[n](A)	−0.43i(C)	−0.62r		−1.57i(C)	MeOH	dme	sce	109

Compound / X						Solvent	Electrode	Ref. elec.	Ref.
Ni²⁺...α,1,8]tetraeneN₄)²⁺			...(C)	−0.97nr(C)	−1.54i(C)	MeOH^a	dme	sce	109
Ni(Me₄[14]hexaeneN₄)²⁺			+0.11nr(C)	−0.70r		MeOH	dme	sce	109
Ni(Me₄[14]2.7,9,14-tetraeneN₄)²⁺			+0.11nr(C)			CH₃CN	rPt	sce	104
Ni^II(Me₆(X)₂[15]4,7,12,15-tetraenato(2−)N₄)						DMF^a	rPt	sce	104
X = —CH₂CH₂αC₅H₅N	+0.15i	−0.44r	−2.91nr			DMF	sPt	Ag/Ag⁺	112
—CH₂—CH₂—C(=O)—CH₃	+0.20i	−0.40nr	−2.87nr			DMF	sPt	Ag/Ag⁺	112
—CH₂—CH₂—C(=O)—C₂H₅	+0.29i	−0.39nr	−2.92nr			DMF	sPt	Ag/Ag⁺	112
—C(=O)—CH₃ , H	+0.39i	−0.36i	−2.83nr			DMF	sPt	Ag/Ag⁺	112
—C(=O)—CH₃	+0.51i	−0.14i	−2.64nr			DMF	sPt	Ag/Ag⁺	112
—C(=O)—NHαC₁₀H₇	+0.55i	+0.03r	−2.59nr			DMF	sPt	Ag/Ag⁺	112
—C(=O)—C₆H₄-p-CH₃	+0.63i	+0.19r	−2.44nr			DMF	sPt	Ag/Ag⁺	112
X = —C(=O)—C₆H₅	+0.64i	+0.21r	−2.42nr			DMF^a	sPt	Ag/Ag⁺	112
—C(=O)—(CH₂)₂—C(=O)—OC₂₇H₄₅	+0.66i	+0.22r	−2.50nr			DMF	sPt	Ag/Ag⁺	112
—C(=O)—(CH₂)... CH₃	+0.64i	+0.22nr	−2.52nr			DMF	sPt	Ag/Ag⁺	112
—NNC₆H₄-p-NO₂	+0.51	+0.305i	−1.53nr	−2.22i		DMF	sPt	Ag/Ag⁺	112

continued overleaf

Note: Footnotes to Table are on page 414.

TABLE 3 (continued)

Complex	−2	−1	+1	+2	+3	+4	Solvent[b]	Electrode	Reference electrode	Ref.
(structure) O=C–C6H4-p-NO2, NO2	+0.79i	+0.31nr	−1.47nr	−2.15nr			DMF	sPt	Ag/Ag+	112
Ni^II(Me2(CH3CO)2[14]tetraenato(2−)N4)	+1.04i		−2.00nr	−2.61			DMF	sP	Ag/Ag+	112
Ni^II(Me4(CH3CO)2[14]tetraenato(2−)N4)	+0.97i(A)	+0.42r					CH3CN[d]	sPt	Ag/Ag+	103
Ni^II(Me4(CH3CO)2[15]tetraenato(2−)N4)	+0.98i(A)	+0.25r					CH3CN	sPt	Ag/Ag+	103
Ni^II(Me6(CH3CO)2[15]tetraenato(2−)N4)	+0.92i(A)	+.26r, +0.27r					CH3CN	sPt	Ag/Ag+	103
Ni([16]aneN4)2+	+0.96i(A)	+0.28r, +0.90r					CH3CN[w]	sPt	Ag/Ag+	103
trans-Co[(13]aneN4)Cl2+			−0.66r				CH3CN	rPt	Ag/Ag+	107
trans-Co[(14]aneN4)Cl2+			−0.69r				CH3CN	rPt	Ag/Ag+	107
trans-Co[(15]aneN4)Cl2+ (1)o			−0.38r				CH3CN	rPt	Ag/Ag+	107
trans-Co[(15]aneN4)Cl2+ (2)p			−0.47nr				CH3CN	rPt	Ag/Ag+	107
trans-Co[(16]aneN4)Cl2+ (1)			−0.15i				CH3CN[w]	rPt	Ag/Ag+	107
(2)			−0.11i				CH3CN	rPt	Ag/Ag+	107
cis-Co[[12]aneN4)Cl2+			−0.49i				CH3CN	rPt	Ag/Ag+	107
cis-Co[(13]aneN4)Cl2+			−0.68i				CH3CN	rPt	Ag/Ag+	107
cis-Co[(15]aneN4)Cl2+			−0.54i				CH3CN	rPt	Ag/Ag+	107
Co^II(Me6[14]4,11-dieneN4)		+0.126r	−1.76(d)a, −1.658)p				CH3CN	dme	Ag/Ag+	113
Co^II(Me6[14]4,11-dieneN4)		(+0.010, +0.240)	(−1.693, −1.658)p	−2.7(d)	−2.52(d)		CH3CN	sPt	Ag/Ag+	113
Co^III(ms-5,12)-Me6[14]aneN4)S2			+0.078r	−2.03r			CH3CN	dme	Ag/Ag+	113
Co^III(ms-5,12)-Me6[14]aneN4)S2			(+0.021, +0.42)p	−2.037r			CH3CN	sPt	Ag/Ag+	113
Co^II(bzo4[16]octaeneN4)(ClO4)2		+0.51	−0.85r	−1.22	−1.85	−1.32	MeOH[r]	rPt		109
Co^II(bzo4[16]octaeneN4)(ClO4)2			−0.54	−0.89	−1.15		CH3CN	dme		109
Co(ms-5,12)-Me6[14]aneN4)Cl3			−0.14r	−1.94			CH3CN	sPt	sce	108
Co(ms-5,12)-Me6[14]aneN4)Br3			+0.06r				CH3CN	sPt	sce	108
Co(ms-5,12)-Me6[14]aneN4)(CN)2+			−1.00r				DMF	sPt	sce	108
Co(rac-5,12)-Me6[14]aneN4)Cl2+			−0.17r				CH3CN	sPt	sce	108
Co^III(Me6[14]4,11-dieneN4)Cl2+ X = Br−			+0.17r	−1.45r			CH3CN	sPt	sce	108
Cl−			(−0.20, −0.17)	(−1.55, −1.47)			CH3CN	sPt	sce	108
N3−			(−0.45, −0.42)	(−1.53, −1.50)			CH3CN	sPt	sce	108
H2O			(+0.32, −0.23)				CH3CN	sPt	sce	108

Header spanning columns −2 through +4: $E_{1/2}$ (total number of electrons transferred)[a]

Complex	E_1	E_2	E_3	E_4	E_5	Solvent	W.E.	Ref. E.	Ref.
X = NCS⁻		(+0.44, +0.59)	(−1.46, −1.45)			CH_3CN	sPt	sce	108
NO₂⁻		−0.48i(C)	(−1.66, −1.64)			CH_3CN	sPt	sce	108
NO₂⁻		(−0.40, −0.38)				CH_3CN	hde[s]	sce	108
CN⁻		−0.10r	(−1.78, −1.68)			CH_3CN	sPt	sce	108
NH₃		−0.91r	(−1.45, −1.43)			CH_3CN	sPt	sce	108
$Co^{II}(Me_6[14]4,11\text{-diene}N_4)(ClO_4)_2$		(+0.40, +0.72)	(−1.39, −1.36)			CH_3CN	sPt	sce	108
$Co^{II}(ms\text{-}(5,12)\text{-}Me_6[14]aneN_4)(ClO_4)_2$		(+0.37, +0.51)	(−1.76, −1.63)			CH_3CN	sPt	sce	108
$Co^{II}(Me_6[14]4,11\text{-diene}N_4)(ClO_4)_2$		(+0.53, +0.75)	−1.36r			CH_3CN	sPt	sce	108
$Cu^{II}(Me_6[14]4,11\text{-diene}N_4)^{2+}$		−0.93r	(−1.4 to −1.5)			CH_3CN	dme	Ag/Ag^+(0.1)	114
$Cu(Me_6[14]4,11\text{-diene}N_4)^{2+}$			−1.018nr(C)			CH_3CN	dme	Ag/Ag^+(0.1)	114
$Cu(ms\text{-}(5,12)\text{-}Me_6[14]aneN_4)^{2+}$		(+1.33, +1.180)	(−1.023, −0.773)	−2.0		CH_3CN	sPt	Ag/Ag^+(0.1)	114
$Cu(ms\text{-}(5,12)\text{-}Me_6[14]aneN_4)^{2+}$		(+1.119, +1.197)	(−1.161, −1.0)p	−2.20p		CH_3CN	sPt	Ag/Ag^+(0.1)	114
$Cu(Me_6[14]4,11\text{-diene}N_4)^{3+}$	1.43i(C)		−0.57i(C)			H_2O[t]	dme	sce	90
$Cu(ms\text{-}(5,12)\text{-}Me_6[14]aneN_4)^{3+}$			−0.62i(C)			H_2O	dme	sce	90
$\alpha\text{-}Cu^{II}(Me_6[14]4,11\text{-diene}N_4)^{2+}$		(+1.52, +1.60)	(−0.73, −0.53)			CH_3CN	sPt	sce	108
$\beta\text{-}Cu^{II}(Me_6[14]4,11\text{-diene}N_4)^{2+}$		(+1.52, +1.64)	(−0.70, −0.55)			CH_3CN	sPt	sce	108
$Cu(ms\text{-}(5,12)\text{-}Me_6[14]aneN_4)^{3+}$		(+1.54, +1.69)	−0.82			CH_3CN	sPt	sce	108
$Cu(Bzo_4[16]octaeneN_4)^{2+}$		+1.23	−0.23nr(C)	−1.49nr(C)		CH_3CN	rPt	Ag/Ag^+	109
$Cu(Bzo_4[16]octaeneN_4)^{2+}$			+0.06			CH_3CN	dme	sce	109
$Cu(Bzo_4(CH_3O)_2[16]hexaenato(2-)N_4)$			−0.21	−1.00	−1.00i	MeOH	dme	sce	109
$Cu^{II}(Me_4[14]hexaeneN_4)$			−0.04r	0.50nr(C)	−1.52i	CH_3CN[d]	rPt	sce	104
$Cu^{II}(Me_4[14]2,7,9,14\text{-tetraene}N_4)$			+0.17nr(C)			DMF[d]	rPt	sce	104
$Cu([14]aneS_4)^{2+}$			+0.689r			H_2O[t] / CH_3OH[b]	sPt	sce	115
$Mn^{III}(Bzo_2[14]hexaenato(2-)N_4)X$	+1.07					CH_3CN	sPt	sce	117
X = Cl⁻		+1.28	+0.562r	−0.681nr					
Br⁻		+1.02	+0.578r	−0.665nr					
NCS⁻		+1.20	+0.635nr	−0.658nr					
N₃⁻		(+0.15, +0.16)	+0.67i	+0.39i	+0.315i				
$Fe(Me_6[14]4,11\text{-diene}N_4)^{2+}$		0.18r		−1.86i(C)	−0.720nr	CH_3CN	sPt	sce	108
$Fe(ms\text{-}(5,12)\text{-}Me_6[14]aneN_4)^{2+}$		0.38r				CH_3CN	sPt	sce	108
$Fe(ms\text{-}(5,12)\text{-}[14]aneN_4)^{2+}$		0.27r				CH_3CN	sPt	Ag/Ag^+	118
$Fe(Me_6[14]aneN_4)^{2+}$		0.44r				CH_3CN	sPt	Ag/Ag^+	118
$Fe(Me_6[14]4,11\text{-diene}N_4)^{2+}$		0.51r				CH_3CN	sPt	Ag/Ag^+	118
$Fe^{II}(Me_6[14]1,4,8,11\text{-tetraene}N_4)^{2+}$		0.59r				CH_3CN	sPt	Ag/Ag^+	118
$Fe^{II}(Me_6[14]1,4,11\text{-triene}N_4)^{2+}$		0.76r				CH_3CN	sPt	Ag/Ag^+	118
$Fe^{II}(Me_6[14]1,3,8\text{-triene}N_4)^{2+}$		0.72r				CH_3CN	sPt	Ag/Ag^+	118
$Fe^{II}(Me_6[14]1,3,5,11\text{-tetraene}N_4)$		0.89r	−0.80r	−1.41r	−1.83r	CH_3CN	sPt	Ag/Ag^+	118
$Fe^{II}(Me_6[14]1,3,7,10\text{-tetraene}N_4)$		0.89r	−0.80r	−1.41r	−1.83r	CH_3CN	sPt	Ag/Ag^+	118
$Fe^{II}(Me_6[14]1,3,8,10\text{-tetraene}N_4)$		0.34r	−2.25i(C)			CH_3CN	sPt	Ag/Ag^+	118
$Fe^{II}(ms\text{-}(5,12)Me_6[14]aneN_4)$	+0.548					CH_3CN	sPt	Ag/Ag^+	119
$[FeMe_6[14]1,3,8,10\text{-tetraene}N_4)C_6H_5]$	−0.16r	−0.80r				CH_3CN[d]	sPt	Ag/Ag^+	120
$[FeMe_6[14]1,3,8,10\text{-tetraene}N_4)H]$	−0.01nr(A)	−0.73nr(A)				CH_3CN[d]	sPt	Ag/Ag^+	120

continued overleaf

TABLE 3 (continued)

Complex	$E_{1/2}$ (total number of electrons transferred)[a]						Solvent[b]	Electrode	Reference electrode	Ref.
	−2	−1	+1	+2	+3	+4				
Ag^{II}(ms-(5,12)-Me_6[14]janeN_4)		+0.617nr(A)		−0.283i(C)			CH_3CN	rPt	Ag/Ag$^+$	121
Ag^{II}([14]janeN_4)		+0.71nr(A)					CH_3CN	rPt	Ag/Ag$^+$	121
Ag^{III}([14]janeN_4)					−0.27i(C)		CH_3CN	rPt	Ag/Ag$^+$	121
Ag^{II}(1,4,8,11-Me_4[14]janeN_4)			+0.68nr(A)				CH_3CN	rPt	Ag/Ag$^+$	12
Ag^{II}(Me_6[14]4,11-dieneN_4)		+0.96nr(1)					CH_3CN	rPt	Ag/Ag$^+$	121
Zn^{II}(Me_4[14]2,7,9,14-tetraeneN_4)		+0.63nr(A)	+0.23i	−0.37(C)			CH_3CN^d	sPt	sce	104
Zn^{II}(Me_6[14]4,11-dieneN_4)				−1.32i(C)			H_2O^f	dme	sce	108
Hg([14]janeN_4)$_2^{2+}$		+1.6i					$CH_3CN^{4,u,z}$	sPt	sce	112

a The potentials in parentheses are the cathodic and anodic waves, respectively, for couples that are not reversible and for which information was available.
b Supporting electrolyte tetraethylammonium perchlorate except as indicated; 25°C.
c Reversible or very nearly as reported by investigator; usually of the order of $E_{3/4} - E_{1/4} = 56 \pm 4$.
d Tetra-n-butylammonium perchlorate electrolyte.
e Stationary platinum electrode.
f $Ag^+ = (0.1\ M)$ in CH_3CN
g Dropping mercury electrode.
h Saturated calomel (aqueous electrode).
i Number of electrons transferred either more than indicated or unknown.
j Irreversible.
k Anodic wave (cathodic wave not reported).
l Peak potentials.
m Cathodic wave (anodic wave not reported).
n Nearly reversible; 50 mV ≤ $|E_{3/4} - E_{1/4}|$ ≤ 80 mV or as reported by investigator.
o Isomer I.
p Isomer II.
q Distorted wave.
r KNO_3 electrolyte.
s Hanging drop mercury electrode.
t pH = 6; citric acid.
u KCl electrolyte.
v 80% CH_3OH, 20% H_2O.
w Tetra-n-butylammonium tetrafluoroborate electrolyte.
x −40°C.

insensitive to the degree of ligand unsaturation, but they respond to variations in axial ligation[108] and strain in the macrocyclic ligand.[107–109] Cobalt(II)–cobalt(I) couples have been found more responsive to equatorial ligand unsaturation than are their isoelectronic nickel(II)–nickel(III) analogs.[106]

Change of the ligating atoms can very greatly alter the potentials of metallo complexes. A most striking example is the greater than one volt anodic shift in the half-wave potential of the copper(II)–copper(I) couple when one goes from the tetraaza $Cu(ms-(5,12)-Me_6[14]aneN_4)^{2+}$ complex to the tetrathioether complex $Cu([14]aneS_4)^{2+}$.[115] The presence of donor sulfur atoms in the coordination sphere and coordination stereochemistry[116] both are significant factors in determining the copper(II)–copper(I) redox potentials. The sensitivity of the copper(II)–copper(I) couple to the presence of ligating sulfur has implications in the chemistry of blue copper proteins.

There are several ligand systems [e.g., $M(bzo_4[16]octaeneN_4)$,[109] $M(Me_4[14]hexaeneN_4)$,[104,110] $M(Me_4[14]tetraenatoN_4)$,[106] $M(bzo_2[14]-hexaenato(2-)N_4)$,[117] and many metalloporphyrinato complexes] for which oxidation or reduction of the unsaturated equatorial ligand is commonly observed. For example, $Ni^{II}(Me_4[14]hexaeneN_4)$ shows two one-electron oxidations. These have been shown by EPR to be consistent with the following series:

$$[Ni^{II}L^{2-}(16\pi)]^0 \longrightarrow [Ni^{II}L\cdot^-(15\pi)]^+ \longrightarrow [NiL^0(14\pi)]^{2+}$$

The relative differences between the two $E_{1/2}$'s is nearly constant for Ni(II), Cu(II), and Pd(II), which is consistent with ligand-centered reactions. The ligand $(Bzo_4[16]octaeneN_4)$, when coordinated to metals, tends to be reduced in two successive one-electron steps to the dianionic ligand $(Bzo_4[16]-septaenato(2-)N_4)$. Both cases just discussed have been described as porphyrin analogs, the end products of the electrochemistry containing highly delocalized π-electron density.

There have been several determinations of $Co^{III}(N_4)X_2/Co^{II}(N_4)X_2$ couples under relatively reversible conditions of conventional potentiometry.[108,123,124] Data are now available for several $trans$-$Co(N_4)(OH_2)_2^{3+,2+}$ couples (Table 4).

2.2.2. Electron Transfer Reactions

It is convenient to define two limiting classes of simple bimolecular electron transfer reactions[94–99,101,125]: (a) "outer-sphere" reactions in which the electron is transferred in a collision complex of the oxidant and reductant, but in which there is only very weak coupling of the reactant molecules in this collision complex, and (b) "inner-sphere" reactions in which the oxidant and reductant share a ligand in the collision complex. These limiting reaction pathways are related to one another as indicated in Scheme 1:

$$A^{III}X_6 + B^{II}L_6 \xrightleftharpoons{K_o} (A^{III}X_6, B^{II}L_6) \xrightleftharpoons[k_{-e}]{k_e} (A^{II}X_6, B^{III}L_6) \xrightleftharpoons[K'_{-o-}]{K_o'} A^{II}X_6 + B^{III}L_6 \qquad (13)$$

$$\qquad\qquad\qquad\qquad\quad \text{OSPC} \qquad\qquad\qquad\qquad \text{OSSC}$$

$$k_1 \big\Vert k_{-1}$$

$$(X_5A^{III} - X - B^{II}L_5) + L \qquad\qquad\qquad (14)$$

$$\text{ISPC}$$

$$k_e' \big\Vert k'_{-e}$$

$$A^{II}X_5S + B^{III}L_5X \xrightleftharpoons[k_{-d}]{S, k_d} (X_5A^{II} - X - B^{III}L_5) \qquad\qquad (15)$$

$$\text{ISSC}$$

(or $A^{II}X_6 + B^{III}L_5S$)

Scheme 1

Here Eq. (13) illustrates the outer-sphere pathway and (14) and (15) illustrate the inner-sphere pathway. In either pathway the first step must involve bringing the reactants together to form an outer-sphere (or ion-pair) complex; this species can be considered the precursor complex for the outer-sphere pathway (OSPC). Transfer of an electron in the OSPC results in formation of a similar ion pair with the charge redistributed: the outer-sphere successor complex (OSSC). If one of the reactant partners is sufficiently labile then a coordinated ligand may be displaced so that some ligand is shared by the two metals in the "inner-sphere precursor complex" (ISPC). For systems in which the electron transfer step is rate determining the observed rate constant

TABLE 4. Formal Potentials for $Co(N_4)(OH_2)_2^{3+,2+}$ Couples ($25°C$; $0.1\ M$ ionic strength, $NaClO_4/HClO_4$)

N_4	E^f (volts) vs. NHE[a]
$Me_4[14]$tetraeneN_4	0.564
$Me_6[14]4,11$-dieneN_4	0.564
$Me_2[14]1,11$-dieneN_4-13-one	0.600
Me_2pyo$[14]$trieneN_4	0.567
$[14]$aneN_4	0.45 ± 0.02[b]

[a] Based on reference 167 and M. Weaver, private communication.
[b] Based on quasireversible cyclic voltammetry in $1\ M\ HClO_4$ using a gold electrode (B. Durham, private communication).

for the outer-sphere pathway is given by $k_{obs} = K_o k_e$, while for the inner-sphere pathway

$$k'_{obs} = \frac{1}{1 + k_{-1}[L]/k_e'} \, k_1 K_o$$

(assuming the OSPC to be in equilibrium with the reactants and a stationary state in the ISPC). The outer-sphere pathway will be observed if $k_e > K_1 k_e'$ ($K_1 = k_1/k_{-1}[L]$) and if $k_e > k_1$.

The outer-sphere limit is conceptually the simplest and it is this limit that has attracted the greatest theoretical interest. The most frequently used theoretical formulation is that due to Marcus.[94-99] This approach assumes that the OSPC in which the metal centers have identical coordination environments (i.e., identical bond lengths and solvation) has a unit probability for electron transfer, while the contributions of OSPC's with other environments may be neglected. Taking $k_e = \nu \exp(-\Delta G^{\ddagger}/RT)$, where ν is the intrinsic frequency of forming products from reactants in the transition state, the free energy of activation may be represented as a linear superposition of the work terms involved in generating the OSPC with the critical environment from the equilibrium species; i.e., $\Delta G^{\ddagger} = \Delta G_r + \Delta G_o + \Delta G_i$, where ΔG_r is the Coulombic work involved in bringing the reactants together, ΔG_o is the work involved in repolarization of the solvent (e.g., from a dipolar ground-state OSPC to a transition state OSPC solvated as if there were no dipole moment), and ΔG_i is the free-energy change associated with bond compression and expansion necessary to make the metal–ligand bond lengths the same at both metal centers. With the further assumption that the interaction between donor and acceptor orbitals is weak, Marcus obtained the free energy correlation

$$\Delta G^{\ddagger}_{ab} = \frac{\lambda_{ab}}{4} + \frac{\Delta G^0_{ab}}{2} + \frac{(\Delta G^0_{ab})^2}{4\lambda_{ab}} \tag{16}$$

In this expression the subscripts denote reactants a and b, ΔG^0_{ab} is the standard free-energy change for the reaction conditions, and λ_{ab} is a reorganizational parameter with inner-sphere $[(\lambda_{ab})_i = 4(\Delta G_{ab})_i]$ and outer-sphere $[(\lambda_{ab})_o = 4(\Delta G_{ab})_o]$ components; λ_{ab} is generally taken as the average of the reorganizational parameters for the self-exchange electron transfer reactions of the oxidized and reduced forms of a and b, respectively:

$$\lambda_{ab} = (\tfrac{1}{2}\lambda_{aa} + \lambda_{bb}) \tag{17}$$

Equation (16) is applicable to a very large variety of reactions,[97,126,127] not just to outer-sphere electron transfer reactions, and it has been derived by using a variety of approaches and assumptions.[97,100,126] Whether λ_{ab} is a simple average of λ_{aa} and λ_{bb} has also been the subject of some concern.[97,99,126]

Applications of (16) have most often involved relative rate comparisons using the equation

$$k_{ab} = (k_{aa}k_{bb}K_{ab}f_{ab})^{1,2} \tag{18}$$

where the k_{ij} are rate constants for cross reactions $(i \neq j)$ or self-exchange reactions $(i = j)$, K_{ab} is the equilibrium constant for the reaction, and

$$\log f_{ab} = \frac{(\log K_{ab})^2}{4 \log (k_{aa}k_{bb}/Z^2)}$$

Equations (16) and (18) have been found of varying applicability to a large number of reactions for which $-\Delta G_{ab}^0 \leqslant \lambda_{ab}$. There have been relatively few systematic investigations of the Marcus model. There have been some experimental[128,129] and theoretical[130-133] indications that Eqs. (16) and (18) are only useful when $|\Delta G_{ab}^0| \leqslant \lambda_{ab}$. Even in this "normal" region there has frequently been some discrepancy between the expected and observed values of λ_{ab}.[134-143] These discrepancies are of some interest and may be an indication that the classical transition state model employed by Marcus is not exactly appropriate, even in the "normal" region.

Pretty much the same model has been used by Hush to describe inner-sphere reactions in which the metal centers are weakly coupled. This approach has been reasonably successful in describing intervalence transitions in binuclear complexes containing metals in different oxidation states[144-146] and the related electron transfer reactions.[147-150] The mixed-valence binuclear complexes may conveniently be regarded as models of the weak interaction limit of inner-sphere electron transfer reactions.

The role of the bridging ligand in the more classical inner-sphere reactions in which the reactant metals are coupled by means of some ligand such as a halide has been more elusive. Various models and theories have been proposed that range from molecular orbital arguments involving "direct," "double,"[151] and "superexchange"[152] interactions to the suggestion that the bridging ligand may effect perturbations of reactant and product potential energy surfaces which allow the otherwise "nonadiabatic" electron transfer processes to become "adiabatic."[153] Haim has recently published a coherent and useful review of experimental and theoretical investigations of the role of the bridging ligand.[101]

Obviously only a small fraction of the literature on electron transfer reactions has involved complexes with macrocyclic ligands. However, owing to their relatively slow rates of solvolysis, macrocyclic ligands do permit a systematic variation of the theoretically important parameters (e.g., ΔG_{ab}^0 and/or metal–ligand bond lengths), and many of the electron transfer studies using macrocyclic complexes have been directed toward the investigation of various models discussed above.

2.2.2.1. Outer-Sphere Reactions. Cobaltammine complexes have long occupied a unique position in the study of electron transfer reactions. Among the reasons for this are the following: (1) A large variety of substitution inert cobalt(III) complexes may be prepared so that one can systematically investigate the effect of variations of reactivity with alterations of the ligand; (2) cobalt(II) complexes equilibrate rapidly with the solvent so that back reactions are unimportant in acidic solutions; and (3) many of the reactions are slow enough to be followed using conventional spectrophotometry. These last two features pose some problems with regard to the investigation of theoretical models such as those mentioned above. Thus the lability of cobalt(II) complexes makes it difficult to obtain the crucial self-exchange rates and thermodynamic information about many of the cobalt(III)–cobalt(II) couples of interest. Furthermore, some of the rate constants for outer-sphere electron transfer reactions involving cobalt complexes have been somewhat mystifying; e.g., the very small ($< 10^{-9} \ M^{-1} \sec^{-1}$ at 65°C) self-exchange rate constant for the $Co(NH_3)_6^{3+,2+}$ couple[154] has not been readily explained by the Marcus model[97,100,138] and may indicate a strongly non-adiabatic pathway,[131] although the change in spin multiplicity has frequently been cited.[100,138,154–156] Very recently, Sargeson and co-workers[157] have found a relatively large self-exchange rate ($k_{exch} \sim 5 \ M^{-1} \sec^{-1}$) between the low-spin cobalt(III) and high-spin cobalt(II) complexes of the encapsulating ligand (S)-(1,3,6,8,10,13,16,19-octaazabicyclo[6.6.6]eicosane (abbreviated: sepulchrate). This is about 10^5 times faster than the self-exchange rate for the $Co(en)_3$ parent complexes and demonstrates that the spin constraint cannot be the major rate-determining feature in the cobalt system.

Tetraazamacrocyclic complexes of cobalt provide very useful comparisons to the ammine complexes since the $Co^{II}(N_4)$ species are reasonably stable in acidic aqueous solution, and since most of the cobalt(II) complexes studied to-date have low-spin (doublet) ground states,[76] so the $Co^{III}(N_4)$–$Co^{II}(N_4)$ couples do not involve a spin change. The rates of electron transfer reactions are again relatively slow.[123,124,139–142]

Low-spin cobalt(II) complexes are generally tetragonally distorted, five- or six-coordinate species with the unpaired electron in an orbital which is predominantly d_{z^2} in character.[76] Thus the low-spin tetraazacobalt(II) complexes are best formulated as $Co^{II}(N_4)(OH_2)_2$ complexes in aqueous solutions, with very weak bonds to the axial ligands (see Section 2.1.5). The axial Co^{II}–OH_2 bond lengths are from 30 to 50 pm longer in these complexes than are the corresponding Co^{III}–OH_2 bond lengths; in contrast, the Co^{II}– and Co^{III}–equatorial-nitrogen bond lengths are all about 190–196 pm, with cobalt–imine bonds being somewhat shorter than cobalt–amine bonds and with very similar Co^{II}–N and Co^{III}–N bond lengths. The cobalt(III)–cobalt(II) couples involving the smallest distortions tend to have the largest rate constants for electron transfer.[123,124,139-142,158-160] The Co(sepulchrate)$^{3+,2+}$ system of Sargeson and co-workers[157] is a much more constrained system in which the environments of the cobalt centers are more nearly similar (the bond length difference, [Co^{III}–N] − [Co^{II}–N], is 18 pm[161]; also compare the related encapsulated systems, references 162 and 163), and this system exhibits a relatively large self-exchange rate.

In order to circumvent some of the problems characteristic of the measurement of self-exchange rates using various tracer techniques, Durham and Endicott[123] have investigated the self-exchange rates of four well-characterized Co(N$_4$)(OH$_2$)$_3^{3+,2+}$ couples by determining the cross reactions between pairs of these closely related cobalt complexes. These reactions are conveniently followed spectrophotometrically since the complexes differ in color. Furthermore, since the standard reduction potentials (E^0 vs. NHE) of the Co(N$_4$)(OH$_2$)$_3^{2+,2+}$ couples are all about 0.6 V, the cross reaction rate constants are reasonably rigorously regarded as geometric means of the corresponding self-exchange rate constants (i.e., $\Delta G_{ab}^{\ddagger} \simeq (\Delta G_{aa}^{\ddagger} + \Delta G_{bb}^{\ddagger})/2$); Eq. (18) provides a further rationalization of this approach and the basis for some minor refinement of the data. The results (Table 5) indicate a very strong correlation between k_{exch} and the difference in axial bond lengths of the cobalt(III) and cobalt(II) complexes. More quantitatively, since $\Delta G_{exch}^{\ddagger} = \Delta G_r + \Delta G_o + \Delta G_i$, and since these very large complexes are similar in size, ΔG_o and ΔG_r cannot vary much ($\sim$ 19 and 4.4 kJ mol^{-1}, respectively), and the variations in $\Delta G_{exch}^{\ddagger}$ can only be ascribable to ΔG_i. Using a simple harmonic model, $\Delta G_i = \sum_j k_j (\Delta X_j)^2 = k_{III}(r^{\ddagger} - r_{III})^2 + k_{II}(r_{II} - r^{\ddagger})^2$, where $r^{\ddagger}$ is the critical transition state Co–OH$_2$ bond distance (the same at both centers), and r_{II} and k_{II}, and r_{III} and k_{III} are the Co–OH$_2$ bond lengths and force constants of the cobalt(II) and cobalt(III) complexes, respectively (note that the only bond-length changes are to axial ligands). Using reasonable estimates of force

TABLE 5. Constants for Self-Exchange Electron Transfer Rate and Related Structural and Thermodynamic Parameters for Co(N$_4$)(OH$_2$)$_3^{3+}$/Co(N$_4$)(OH$_2$)$_2^{2+}$ Couples at 25°C, Ionic Strength = 1.0

N$_4$	E^0 (volts)^a	Co^{III}–OH$_2$ (pm)^b	Co^{II}–OH$_2$ (pm)^b	$k_{exch}{}^a$ (M^{-1} sec^{-1})
Me$_6$[14]4,11-dieneN$_4$	0.56	196	248	3×10^{-5}
Me$_2$[14]1,11-dieneN$_4$-13-one	0.60	196	228, 239	4×10^{-3}
Me$_4$[14]tetraeneN$_4$	0.56	197	229	6×10^{-2}
Me$_2$pyo[14]trieneN$_4$	0.57	—	—	0.12

^{a} See Table 4.
^{b} From references 76, 158–160.

constants Durham[123] has found ΔG_i to vary systematically with r_{II} from 32.6 to 50 kJ mol^{-1} for the first three systems in Table 5, thus accounting reasonably quantitatively for the variations in $\Delta G^{\ddagger}_{exch}$. Using data for several other outer-sphere reactions involving these $Co(N_4)(OH_2)_2^{3+,2+}$ couples, Durham has also shown that $\lambda_{ab} \simeq (0.4 \pm 0.1)(\lambda_{aa} + \lambda_{bb})$.

A further test of Eq. (16) may be provided using data for various outer-sphere reactions of the $Co(Me_6[14]diene(N_4)(OH_2)_2^{3+,2+}$ and $Co(Me_4[14]$-tetraeneN$_4)(OH_2)_2^{3+,2+}$ couples. Equation (16) may be rewritten $4\Delta G^{\ddagger}_{ab}/\lambda_{ab} = (1 + \Delta G^0_{ab}/\lambda_{ab})^2$, so that $2(\Delta G^{\ddagger}_{ab}/\lambda_{ab})^{1/2}$ should be proportional to $\Delta G^0_{ab}/\lambda_{ab}$ with an intercept of 1.0 and a slope of 1.0 as indicated by the solid line in Figure 4. Given the considerable uncertainties in much of the literature data for self-exchange reactions, hence in values of $\lambda_{ab} = \frac{1}{2}(\lambda_{aa} + \lambda_{bb})$, the agreement seems generally satisfactory. However, there are some problems: (1) The Co^{3+} oxidations of $Co(N_4)(OH_2)_2^{2+}$ complexes are much slower than expected, apparently owing to a failure in the relation $\lambda_{ab} = \frac{1}{2}(\lambda_{aa} + \lambda_{bb})$, and (2) the oxidations of $Co(N_4)(OH_2)_2^{2+}$ substrates by polypyridyl complexes of iron(III) and ruthenium(III) are all somewhat slower than expected.

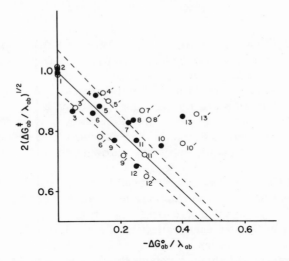

FIGURE 4. Marcus-type, reduced free-energy correlation for several outer-sphere reactions involving $Co(N_4)(OH_2)_2^{3+,2+}$ couples. Solid points, oxidations of $Co(Me_6[14]4,11$-dieneN$_4)(OH_2)_2^{2+}$ or reductions of $Co(Me_6[14]4,11$-dieneN$_4)(OH_2)_2^{3+}$; open circles (primed numbers), oxidation of $Co(Me_4[14]tetraeneN_4)(OH_2)_2^{2+}$ or reductions of $Co(Me_4[14]tetraeneN_4)(OH_2)_2^{3+}$. Counterreagents: 1 and 1′, $Co(Me_4[14]tetraeneN_4)$-$(OH_2)_2^{2+}$ with $Co(Me_6[14]4,11$-dieneN$_4)(OH_2)_2^{2+}$; 2, Co(edta)$^-$; 3 Fe^{3+}; 4, Fe(Me$_4$-phen)$_3^{3+}$; 5, Fe(Me$_2$phen)$_3^{3+}$; 6, Ru(NH$_3$)$_5$py^{2+}; 7, Fe(phen)$_3^{3+}$; 8, Fe(bipy)$_3^{3+}$; 9, Ru(NH$_3$)$_6^{2+}$; 10, Ru(bipy)$_3^{3+}$; 11, U^{2+}; 12, Mn^{3+}; 13, Co^{3+}. Data corrected for 1 M ionic strength, +6 charge type. Data taken from reference 123 and references cited therein.

These studies with macrocyclic complexes clearly demonstrate the effect on electron transfer rates of large changes in bond length across the reaction coordinate. Furthermore, it appears that the free energy correlation (16) holds reasonably well for a very wide range of electron transfer reactions, as do the somewhat more primitive assumptions that $\lambda_{ab} = \frac{1}{2}(\lambda_{aa} + \lambda_{bb})$ and $\Delta G^{\ddagger} = \Delta G_r + \Delta G_i + \Delta G_o$.

The often observed failure of (16) to correctly predict the intercepts of free energy correlations warrants some more careful consideration. The contribution to ΔG_i and ΔG_o are largely enthalpic. The failure of (16) to predict intercepts has led some workers to define "effective" λ_{aa} values to be used in the appropriate cross reactions. This approach seems to imply that reorganizational energies (or nonadiabatic contributions) are different in cross- and self-exchange electron transfer reactions. Such an inference can be partially examined using data for cross reactions involving $Co(N_4)$-$(OH_2)_2^{3+,2+}$ and other couples.

Equation (19) may readily be obtained from (16)[164]:

$$\Delta H_{ab}^{\ddagger} = \frac{\Delta H_{aa}^{\ddagger} + \Delta H_{bb}^{\ddagger}}{2}(1 - 4\alpha^2) + \frac{\Delta H_{ab}^0}{2}(1 + 2\alpha) \qquad (19)$$

where $\alpha = (\Delta G_{ab}^0)/4(\Delta G_{aa}^{\ddagger} + \Delta G_{bb}^{\ddagger})$. Since we expect $\Delta G_{ii}^{\ddagger} \simeq \Delta H_{ii}^{\ddagger} +$ (work terms, $\Delta G_{ab}^{\ddagger} \simeq$ (constant) $+ (\Delta H_{aa}^{\ddagger} + \Delta H_{bb}^{\ddagger})/2 + \Delta G_{ab}^0/2 + (\Delta G_{ab}^0)^2/4\lambda_{ab}$. Thus a plot of $\{\Delta G_{ab}^{\ddagger} - \Delta G_{ab}^0/2 - (\Delta G_{ab}^0)^2/4\lambda_{ab}\}(1 - 4\alpha^2)$ vs. $\Delta H_{ab}^{\ddagger} - (\Delta H_{ab}^0/2)(1 + 2\alpha)$ should be linear with unit slope and an intercept dependent largely on Coulombic work terms. Figure 5 is such a plot. Much of the scatter in Figure 5 probably arises from differences in ionic strength, ionic composition, work terms. Certainly the predominant contribution to variations in $\Delta G^{\ddagger}$ are enthalpic, once the contributions of ΔG^0 and ΔH^0 are properly taken into account (see Table 6). The deviations of reactions involving the Co^{3+}/Co^{2+} couple are huge. Figure 5 indicates that the large barriers to electron transfer in reactions involving the Co^{3+}/Co^{2+} couple are predominantly enthalpic. There does appear to be some compensating entropy effect in the free energy function. Figure 5 also demonstrates that reactions of macrocyclic and polypyridyl cobalt complexes do not have unusually large nonadiabatic components. The unusual behavior of the hexaaquo ions has been frequently discussed (e.g., see reference 142). Our analysis indicates that the kinetic peculiarities of these systems are largely attributable to enthalpic components, with relatively small entropic components attributable, perhaps, to nonadiabatic effects (small transmission coefficients).

Actually there is reason to expect some compensation between entropic and enthalpic components of $\Delta G^{\ddagger}$ in electron transfer reactions. Thus it seems very likely that electrons can be "thermally" activated to transfer between

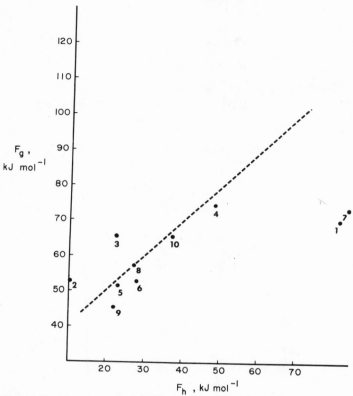

FIGURE 5. Interrelationship of activation parameters for outer-sphere electron transfer reactions: $F_g = \{\Delta G_{ab}^{\ddagger} - (\Delta G_{ab}^0/2 - [(\Delta G_{ab}^0)^2/4\lambda_{ab}])\}(1 - 4\alpha^2)$ vs. $F_h = \Delta H_{ab}^{\ddagger} - (\Delta H_{ab}^0/2)(1 + 2\alpha)$; $\alpha = \Delta G_{ab}^0/4(\Delta G_{aa}^{\ddagger} + \Delta G_{bb}^{\ddagger})$. Values of $\Delta G^{\ddagger}$, ΔG^0 at 25°C; ΔH^0 estimated from data in Table 6 and references 142, 156, 165–167. The solid line is drawn with the theoretical unit slope. 1, $Co^{3+}/Co(Me_4[14]\text{-tetraeneN}_4)(OH_2)_2^{2+}$; 2, $Co(Me_4[14]\text{-tetraeneN}_4)(OH_2)_2^{3+}/Ru(NH_3)_5py^{2+}$; 3, $Co(Me_6[14]dieneN_4)(OH_2)_2^{3+}/Ru(NH_3)_5py^{2+}$; 4, $Co(Me_2[14]diene\text{-}oneN_4)(OH_2)_2^{3+}/Co(Me_4[14]tetraeneN_4)(OH_2)_2^{2+}$; 5, $Fe^{3+}/Ru(NH_3)_6^{3+}$; 6, $Co(phen)_3^{3+}/Ru(NH_3)_6^{2+}$; 7, Co^{3+}/Fe^{2+}; 8, $Ru(NH_3)_6^{3+}/V^{2+}$; 9, $Ru(NH_3)_5py^{3+}/V^{2+}$; 10, $Co(phen)_3^{3+}/V^{2+}$.

reactant molecules with a large number of different nuclear configurations. If the ith allowed configuration of nuclear coordinates (including solvent) has a free-energy content g_j, and if p_j is the probability of electron transfer for this configuration, then the second-order rate constant for electron transfer may be represented as

$$k = \sum_j Z_j p_j \exp(-\Delta G_j/RT) \qquad (20)$$

TABLE 6. Activation Parameters and Related Quantities for Reactions Involving $Co(N_4)(OH_2)_2^{3+,2+}$ Couples[a]

Reductant/oxidant	$\Delta S^0_{red}/\Delta S^0_{ox}$ [b]	$-\Delta H^0_{ab}$ [c]	$-\Delta G^0_{ab}$ [c]	$\Delta G^{\ddagger}_{ab}$	$\Delta H^{\ddagger}_{gb}$ [c]	Ionic strength
$Co(Me_4[14]tetraeneN_4)(OH_2)_2^{2+}/Co^{3+}$	96/200	100	134	48 ± 1	50 ± 4	0.1
$Ru(NH_3)_5py^{2+}/Co(Me_4[14]tetraeneN_4)(OH_2)_2^{3+}$	62[d]/96	14	24	41 ± 1	4 ± 4	0.1
$Ru(NH_3)_5py^{2+}/Co(Me_6[14](4,11\text{-}dieneN_4)(OH_2)_2^{3+}$	62[d]/188	−13	24	54 ± 1	29 ± 2	0.1
$Co(Me_2[14](1,11)dieneN_4\text{-}13one)(OH_2)_2^{2+}/Co(Me_4[14]tetraeneN_4)(OH_2)_2^{3+}$	138/96	12	−3	72 ± 2	40 ± 4[e]	1.0

a From references 123 except as noted.
b kJ mol⁻¹ K⁻¹; red = reductant, ox = oxidant; values from reference 167.
c kJ mol⁻¹.
d Estimated value.
e R. Sriram and J. F. Endicott, unpublished work.

where $\Delta G_j = g_j - g_0$, g_0 is the free energy content of the "ground state," and Z_j is the frequency of collision at the particular internuclear separation and with the particular orientation. In this formulation the $\Delta G_j = G_{r,j} + G_{o,j} + G_{1,j}$ may be generated as before. One would anticipate that the maximum $p_j = p_m$ would occur for the classical limit described by Marcus, and that the predominant terms contributing to (20) would be of comparable magnitude, i.e., $Z_k p_k \exp(-\Delta G_k/RT) \sim Z_m p_m \exp(-\Delta G_m RT)$. Such a situation would preserve most of the features of (16) while allowing for peculiarities in values of the intercepts and in $\Delta H^{\ddagger}$ (p_j would be expected to act like a tunneling probability and to be only very weakly dependent on temperature). Such a formulation has qualitative features of many of the more recent theoretical approaches to the theory of electron transfer processes (e.g., references 130–133).

There have been almost no systematic studies of electron transfer reactions of macrocyclic complexes containing metals other than cobalt. Studies involving $Fe^{II}(N_4)X_2/Fe^{III}(N_4)X_2$ couples should be nicely complementary to those described above since the metal–ligand distances are nearly identical for iron(II) and iron(III), at least for $N_4 = Me_4[14]$tetraeneN_4.[91,168] Conversely, the copper(II)–copper(I)–cyclic-polythioether complexes should be very interesting owing to the very large changes in metal–ligand geometry which appear to occur.[169,170]

2.2.2.2. Inner-Sphere Reactions. For many combinations of reactants the inner-sphere pathway has a much lower activation barrier than the corresponding outer-sphere pathway. This appears to be especially the case when the electron is exchanged between antibonding, $*\sigma_m$, metal-centered orbitals. Table 7 compares the oxidations of $Co(N_4)(OH_2)_2^{2+}$ complexes by $M^{III}L_6$ and $M^{III}L_5Cl$ complexes. The differences in driving force for these pairs of reactants would be expected to account for no more than about a ten-fold variation in rate constant. [In the $Co(edta)^-/Co(hedta)Cl^-$ comparison the chloro complex could be even a poorer oxidant than $Co(edta)^-$ since the $Co(edta)^{2-}$ and $Co(hedta)H_2O^{0,-}$ couples have $E^0 \sim 0.6$[171] and 0.38 V[172], respectively.] Thus for these systems there is a 20 kJ mol^{-1}, or more, ΔG^0 independent advantage of the pathway involved in reactions of the chloro complexes. This kinetic advantage of the inner-sphere pathway is characteristic of $*\sigma_{M-donor}/*\sigma_{M-acceptor}$ systems.

The magnitude of this kinetic advantage is unequivocally demonstrated in the self-exchange reactions

$$Co(N_4)(OH_2)_2^{2+} + *Co(N_4)(OH_2)_2^{3+} \rightleftharpoons Co(N_4)(OH_2)_2^{3+} + *Co(N_4)(OH_2)_2^{2+} \quad (16)$$

$$Co(N_4)(OH_2)_2^{2+} + *Co(N_4)(OH_2)Cl^{2+} \rightleftharpoons Co(N_4)(OH_2)Cl^{2+} + *Co(N_4)(OH_2)_2^{2+} \quad (17)$$

For example, for $(N_4) = (Me_4[14]$tetraene$N_4)$ the rate constant for the outer-sphere pathway, (21), is 0.07 M^{-1} sec^{-1} (25°C, $\mu = 1.0$),[123] while the rate constant for the inner-sphere pathway, (22), is (1.0×10^5) M^{-1} sec^{-1} (25°C,

TABLE 7. Reactivity of $Co(N_4)(OH_2)_2^{2+}$ Toward Inner-Sphere (k_i) and Outer-Sphere (k_o) Oxidants[a]

Oxidants	$Co(Me_6[14]4,11-dieneN_4)(OH_2)_2^{2+}$	$Co(Me_4[14]tetraeneN_4)(OH_2)^{2+}$
Co^{3+}	66 ± 2 (3.0)	337 ± 23 (3.0)
$CoCl^{2+}$	$(1.6 \pm 0.2) \times 10^6$ (1.0)	$(1.5 \pm 0.1) \times 10^7$ (1.0)
Ratio[b]	$> 2 \times 10^4$	$> 4 \times 10^4$
Mn^{3+}	$(1.7 \pm 0.1) \times 10^4$ (3.0)	$(1.7 \pm 0.2) \times 10^5$ (3.0)
$MnCl^{2+}$	$(5.3 \pm 0.2) \times 10^6$ (1.0)	$(2.4 \pm 0.6) \times 10^7$ (1.0)
Ratio[b]	$> 5 \times 10^2$	$> 1 \times 10^2$
$Co(edta)^-$	$\sim 3 \times 10^{-2}$ (1.0)	$\sim 3 \times 10^{-2}$ (1.0)
$Co(edtaH)Cl^-$	$(5.5 \pm 0.5) \times 10^2$	$(9 \pm 1) \times 10^3$ (1.0)
Ratio[b]	$\sim 2 \times 10^4$	$\sim 3 \times 10^5$
$Co(N_4)(OH_2)_2^{3+}$	3×10^{-5} (1.0)	6×10^{-2} (1.0)
$Co(N_4)Cl_2^+$ [c]	99 (1.0)	8×10^5 (0.1)
Ratio[b]	$\sim 5 \times 10^4$	$\sim 2 \times 10^5$

[a] Second-order rate constants, $M^{-1} sec^{-1}$; 25°C; ionic strength in parentheses taken from reference 123.
[b] k_i/k_o corrected for ionic strength and charge type.
[c] D. P. Rillema and J. F. Endicott, unpublished work.

$\mu = 0.1$).[173] At the same ionic strengths the inner-sphere pathway would have nearly a 39 kJ mol^{-1} advantage in activation free energy over the outer-sphere pathway. However, Durham *et al.*[123,173] have demonstrated that the halide-bridged, inner-sphere reactions have rate patterns consistent with (16) and very likely determined by "Franck–Condon" reorganizational factors (e.g., see Figure 6). By comparisons of the variations of reactivity in self-exchange-like cross reactions, reactions in which the reduced and oxidized complexes (Co^{II} and Co^{III}) in (21) and (22) have different macrocyclic (N_4) ligands, it has been shown that most of the intrinsic (free-energy-independent) kinetic advantage of the inner-sphere pathway arises from a much smaller contribution of the inner-sphere reorganizational term $(\lambda_{ab}[I.S.])_i$; for Cl^--bridged reactions, as in (22), it has been shown that $(\lambda_{ab}[I.S.])_i \simeq 0.5(\lambda_{ab}[O.S.])_i$.[173] In addition there is a reduction of the outer-sphere (or solvent) reorganizational barrier, with $(\lambda_{ab}[I.S.])_o$ approaching zero for the chloride-bridged pathway, and there may be a very small difference in adiabaticity for the inner-sphere and outer-sphere reactions.[173]

Some variations in rate may be achieved by varying the bridging ligand in (22); thus free-energy-independent reaction rates tend to increase as the bridging ligand is changed from Cl^- to N_3^- to Br^-. These rate variations have been examined in some detail, both in cross reactions and self-exchange reactions, and appear to be a consequence of variations in $(\lambda[I.S.])_i$.[174]

Additional features of the $Co(N_4)(OH_2)_2^{2+}$ reductions of complexes with potential bridging ligands are illustrated in Figure 6. The reaction rates

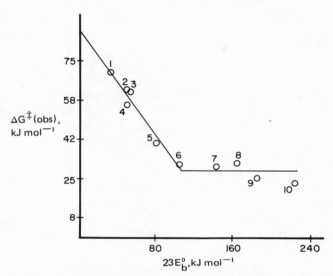

FIGURE 6. Plot of $\Delta G^{\ddagger}$ (obs) vs. contribution ($23E^0$) of oxidant to ΔG^0 for apparently inner-sphere oxidation of $Co(Me_6[14]4,11\text{-}dieneN_4)(OH_2)_2^{2+}$. Oxidants: 1, $Co(NH_3)_5\text{-}Cl^{2+}$; 4, $Co(Me_6[14]4,11\text{-}dieneN_4)Co(edtaH)Cl^-$; 4, $Co(Me_6[14]4,11\text{-}dieneN_4)Cl_2^+$; 5, $Co(Me_6[14]4,11\text{-}dieneN_4)Br_2^+$; 6, I_2^-; 7, $MnCl^{2+}$; 8, $CoCl^{2+}$; 9, Br_2^-; 10, Cl_2^-; 25°C; corrected to 0.1 M ionic strength, +4 charge type. From reference 123.

increase with ΔG^0, until a limiting rate is achieved for relatively exergonic reactions. The free energy dependence cannot be made really quantitative in these cases owing to the lack of appropriate self-exchange parameters. However, the slope in Figure 6 is consistent with (16). Similar oxidations of $Co(Me_4[14]tetraeneN_4)(OH_2)_2^{2+}$ also appear to be free energy dependent and are about 100 times faster than the corresponding reactions in Figure 6.

The limiting rates observed for highly exergonic oxidations of $Co(N_4)\text{-}(OH_2)_2^{2+}$ reactions are one to two orders of magnitude less than the expected diffusion limit. Consequently, it is believed that these rates are limited by the rate of substitution for water coordinated to cobalt(II) by $XM^{III}L_5$, i.e., for sufficiently large ΔG^0, $k_{obs} = K_o k_w$, where k_w is the water exchange rate on $Co(N_4)(OH_2)_2^{2+}$. From these observations k_w is estimated to be approximately 10^9 sec^{-1}. A similar value may be inferred from the rate ($k_{obs} \sim 10^8 \ M^{-1}\text{sec}^{-1}$) of $Co(N_4)(OH_2)_2^{2+}$ scavenging for methyl radicals.[79,175]

The smaller values of $(\lambda_{ab}[I.S.])_i$ than $(\lambda_{ab}[O.S.])_i$ are likely to be the result of several factors. First the Co–Cl force constants are smaller than Co–OH$_2$ force constants.[176] Secondly, the very weak Co^{II}–$(Cl\text{-}M^{III}L_5)$ bond would be predicted[76] to result in a decrease of the *trans*-Co^{II}–OH$_2$ bond length and a corresponding diminution of the reorganizational barrier. Thirdly, coordination of Co^{II} to Cl should weaken the M^{III}–Cl bond. Finally, the stretching of Co^{III}–Cl bond correlates to the motion compressing the Co^{II}–Cl bond.[98,177]

Espenson and co-workers[85,178,179] have reported on an interesting class of inner-sphere reactions in which the "bridging" group is an al^{l}yl ligand. These reactions are remarkable in that there is no electron pair on the alkyl ligand that may be donated to the second metal, and consequently one would anticipate that "precursor complexes" in this case should not be well defined nor distinguishable from the transition states for electron transfer. These reactions are relatively slow and preliminary observations indicate that "methyl-bridged" reactions analogous to (22) seem to have $(\lambda_{ab}[I.S.])_i \simeq (\lambda_{ab}[O.S.])_i$.[174] This seems somewhat strange since the Co–OH_2 bond lengths change much less for the alkyl than for the diaquo complexes. However, the data currently available are not adequate to assess the reorganizational barrier due to inversion of the methyl group being transferred or of the contributions of nonadiabatic and steric factors.

Finally, we wish to emphasize that the estimation of the free energy change, ΔG^0_{et}, for the electron transfer step of inner-sphere reactions is not in any way a simple matter (see Scheme 2).

The first two terms in Scheme 2 will be similar for inner-sphere and outer-sphere reactions. Thus $\Delta G^0_{et}(I.S.) \simeq \Delta G^0_{et}(O.S.)$ only if the entropies associated with ligand loss are similar for the oxidant and reductant species. An interesting feature of Scheme 2 is that it emphasizes that a major component of the driving force for inner-sphere electron (or group) transfer

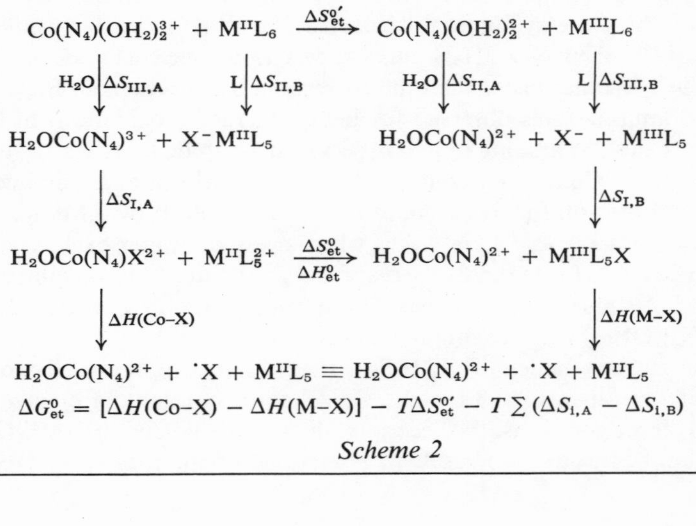

$$Co(N_4)(OH_2)_2^{3+} + M^{II}L_6 \xrightarrow{\Delta S^{0\prime}_{et}} Co(N_4)(OH_2)_2^{2+} + M^{III}L_6$$

$$\downarrow^{H_2O}{\Delta S_{III,A}} \qquad \downarrow^{L}{\Delta S_{II,B}} \qquad\qquad \downarrow^{H_2O}{\Delta S_{II,A}} \qquad \downarrow^{L}{\Delta S_{III,B}}$$

$$H_2OCo(N_4)^{3+} + X^-M^{II}L_5 \qquad\qquad H_2OCo(N_4)^{2+} + X^- + M^{III}L_5$$

$$\downarrow{\Delta S_{I,A}} \qquad\qquad\qquad\qquad \downarrow{\Delta S_{I,B}}$$

$$H_2OCo(N_4)X^{2+} + M^{II}L_5^{2+} \xrightarrow[\Delta H^0_{et}]{\Delta S^0_{et}} H_2OCo(N_4)^{2+} + M^{III}L_5X$$

$$\downarrow{\Delta H(Co-X)} \qquad\qquad\qquad\qquad \downarrow{\Delta H(M-X)}$$

$$H_2OCo(N_4)^{2+} + {}^{\cdot}X + M^{II}L_5 \equiv H_2OCo(N_4)^{2+} + {}^{\cdot}X + M^{II}L_5$$

$$\Delta G^0_{et} = [\Delta H(Co-X) - \Delta H(M-X)] - T\Delta S^{0\prime}_{et} - T\sum(\Delta S_{I,A} - \Delta S_{I,B})$$

Scheme 2

reactions is the difference in metal–ligand homolytic bond energies. The experimental determination of these quantities is in progress.[38,167,180]

2.2.3. Reactions with Dioxygen

During the past ten years the reactions of dioxygen with transition-metal complexes have been extensively studied, and there have been several excellent reviews of various aspects of these reactions.[181–187] Much of the available information may be discussed in terms of the reactions

$$ML_4XY + O_2 \underset{k_{-s}}{\overset{k_s}{\rightleftharpoons}} ML_4XO_2 + Y \tag{23}$$

$$ML_4XO_2 + ML_4XY \underset{k_{-d}}{\overset{k_d}{\rightleftharpoons}} (XL_4M)_2O_2 + Y \tag{24}$$

Many biologically important, reversible oxygen carriers (e.g., hemoglobin) are constrained so that (24) is not significant; however, the chemistry of many of the synthetic inorganic oxygen carriers is "complicated" by formation of μ-peroxo species, (24). A very large number of cobalt(II) complexes function as reversible oxygen carriers [relatively little contribution of (24)] at low temperatures, in relatively nonpolar solvent media.[183,184,188–190] The magnitude of K_s $(= k_s/k_{-s})$ has been found to be determined by the thermodynamics of the $Co^{III,II}$ couple for several systems.[188,189] Coordination apparently activates O_2 to facilitate oxidation of organic substrates in oxygenases and oxidases.

Despite the interest and activity there have been few systematic studies of reaction of O_2 with synthetic complexes containing macrocyclic ligands. It turns out that only a few $Co^{II}(N_4)$ complexes are very reactive toward O_2 in acidic aqueous (perchlorate) solutions. The irreversible oxidations of cobalt(II) substrates are markedly faster in halide solutions. There has been one report of a reasonably systematic study of such effects using a cobalt–oxime substrate.[191]

It has been known for some time that $Co([14]aneN_4)(OH_2)_2^{2+}$ reacts rapidly with O_2 in acidic aqueous solution to form the μ-peroxo complex $[Co([14]aneN_4)OH_2]_2O_2^{4+}$.[192] Current studies[174] indicate that $K_s k_d \sim 10^{10} M^{-1} sec^{-1}$ for this system. Solutions of this μ-peroxo complex have both oxidizing and reducing properties; for example, oxidizing toward Fe^{2+} and reducing toward Fe^{3+}. Both oxidizing and reducing reactions approach a first-order limit, corresponding to $k_{-d} \simeq 0.7 sec^{-1}$ ($\mu = 0.1$, perchlorate media). It is postulated that the reducing agents intercept very small stationary-state concentrations of a very reactive monomeric superoxo complex.[174] The much greater apparent reactivity of $Co([14]aneN_4)(OH_2)_2^{2+}$ than $Co(Me_4[14]tetraeneN_4)(OH_2)_2^{2+}$ toward dioxygen could be related to the

0.15-V difference in the potentials of the $Co^{III,II}$ couples (Table 4); however, quantitative kinetic comparisons are not yet possible.

It has been found that cobalt(II) complexes coordinated to β-diimine moieties mediate the oxygenation of the central carbon of the imine system [160,193,194] (see also Section 3 below). The reaction of $Co(Me_2[14]1,11\text{-}dieneN_4)(OH_2)_2^{2+}$ with O_2 is very rapid and apparently proceeds through several stages. [160,194] Similar reactions occur much more slowly with copper(II), nickel(II), and cobalt(III) β-diimine complexes. In the case of cobalt(III) complexes, the rates of oxygenation increase as the $Co^{III,II}$ couple becomes more oxidizing. For all these systems, oxygenation of the central carbon atom of the ligand appears to occur in competition with dimerization of the macrocyclic complexes. Both reactions are believed to proceed through radical intermediates, with radical–radical coupling occurring in competition with dioxygen scavenging for the radical intermediate. [194]

Related reactions of macrocyclic complexes are discussed in Sections 3 and 4 below.

3. Reactions of the Macrocyclic Ligands

In the course of the development of synthetic routes to macrocyclic compounds a variety of interesting and useful reactions of the ligands have been discovered. The remarkable resistance to total decomposition of many macrocyclic complexes has made this area of study particularly fruitful. Very powerful reagents are often necessary to carry out some of these reactions, and more fragile metal complexes would generally be destroyed. One important feature of macrocyclic ligands is their ability to stabilize unusual oxidation states such as Ni(III) or Cu(III). This property appears to play an important role in mediating reactions of the ligands. In the subsections which follow, more general reactions are discussed. Most of the data presently available is synthetic in nature, but it does provide clues as to the mechanisms of these reactions.

3.1. Oxidative Dehydrogenations

Oxidative dehydrogenations have been very useful in the synthesis of new complexes. The reactions considered here are oxidations of a coordinated ligand that result in the formation of a double bond in the ligand, often a $C{=}N$ bond. This class of reactions is not unique to macrocyclic ligands and has been shown to occur in some metal ethylenediamine complexes. [195–198]

Two of the better-characterized systems serve to illustrate typical reaction pathways[199]:

$$(25)$$

$$(26)$$

Both of these reactions may be carried out in steps, and the nickel(III), monoene, diene, or triene intermediates[200,201] may be isolated. The reactions of iron and nickel systems seem to be reasonably general, and several new macrocycles have been generated using such approaches.[118,199–202] The Fe(II) systems have the additional advantage that in the favorable cases the ligand may be removed from the metal. This enables one to prepare a variety of metallo complexes.[203] The different products obtained in Eqs. (25) and (26) should be noted. Busch and co-workers have shown[118,202,204] that unsaturation tends to occur in the five-membered rings in Fe(II) complexes. The propensity of iron complexes to form α-diimine linkages has been associated with the relative stability of the iron–(α-diimine) moiety and attributed to the partial aromatic nature of the five-membered rings.[202] Steric factors, such as the ring size, the number of double bonds, and metal center size may also be important.

A stepwise reaction scheme is useful in discussing the kind of ligand reactions such as (26):

$$(27)$$

$$\text{M}^{\text{III}}(\text{structure}) + \text{B} \;\rightleftharpoons\; \text{M}^{\text{III}}(\text{structure}) + \text{HB}^+ \tag{28}$$

$$\text{M}^{\text{III}}(\text{structure}) \;\rightleftharpoons\; \text{M}^{\text{II}}(\text{structure}) \tag{29}$$

$$\text{M}^{\text{II}}(\cdot\,N\text{-Me}_6[14]\text{aneN}_4)$$

$$\text{M}^{\text{II}}(\cdot\,N\text{-Me}_6[14]\text{aneN}_4) + \text{Ox} \;\rightleftharpoons\; \text{M}^{\text{III}}(\cdot\,N\text{-Me}_6[14]\text{aneN}_4) + \text{Red} \tag{30}$$

$$\text{M}^{\text{III}}(\cdot\,N\text{-Me}_6[14]\text{aneN}_4) + \text{B} \;\rightleftharpoons\; \text{M}^{\text{II}}(\text{structure}) + \text{HB}^+ \tag{31}$$

In Eqs. (27)–(31) we have used the $\text{Me}_6[14]\text{aneN}_4$ ligand for convenience, since it is the most studied example, to illustrate many important features of this reaction class. Barefield and Mocella[205] have reported evidence for a nickel(II) complex with a radical ligand that can be generated from the reaction of $\text{Ni}^{\text{III}}(\text{Me}_6[14]\text{aneN}_4)$ with water or other basic solvents. Whether such radical–ligand complexes could ever become the predominant solution species is subject to question (see Section 4.1 below). Nevertheless they do seem to be very plausible intermediates in this class of reactions. The dehydrogenation reaction proceeds in nitric acid media.[200,204,206] However, there is evidence for the reversibility of the acid–base reactions,[205] somewhat complicated by evidence that the position of net equilibrium of (28) and (29) depends on the nature of the acid.[207] It seems likely that the observation of oxidative dehydrogenation reactions depends both on favorable overall thermodynamics (i.e., $K_{31}K_{32}K_{33}K_{34}K_{35} \gg 1$) and the rate of attainment of equilibrium in each required step. Thus the apparent stability of the Fe–α-diimine moiety may be manifested in the net reaction efficiency, while kinetic features may account for the more common observation of this reaction for M=Fe than for M=Co in view of the very large reorganizational barriers expected in the oxidations of cobalt(II) complexes (see Section 2.2).

It should also be noted that the oxidant in (30) could either be an "external" species (O_2, HNO_3), $M^{III}(Me_6[14]aneN_4)$, or $M^{II}(\cdot N\text{-}Me_6[14]aneN_4)$. For example, Welsh and Henry[207] have found $Ni(\cdot N\text{-}Me_6[14]aneN_4)$ to be an efficient oxidant for cyclohexanone with the initial step likely to be hydrogen abstraction, thus a radical disproportionation could provide an alternate pathway[205] to the final product without consuming additional base [as in Eq. (31)].

In a closely related reaction, the oxidation of $Ru(bipy)_2en^{2+}$ has been shown to require four electrons to produce the $Ru(bipy)_2(diimine)$ product.[198]

Electrochemical oxidation[208] in acetonitrile, or Ce^{IV} oxidation[209] in aqueous sulfuric acid, produces $Cu^{III}(Me_6[14]4,11\text{-}dieneN_4)$ from the copper(II) complex. This copper(III) species decomposes spontaneously to produce, at least in part, a copper(II) product with increased unsaturation.[208] In aqueous sulfuric acid this decomposition is complete within a few seconds.[209]

Tang and Holm[210] have oxidatively dehydrogenated the copper(II), nickel(II), and cobalt(II) complexes of $Ph_2[15]tetraenato(2-)N_4$ and the copper(II) and nickel(II) complexes of $Ph_2[16]tetraenato(2-)N_4$ to the corresponding $Ph_2[15 \text{ or } 16]hexaenato(2-)N_4$ complexes by using 2,3-dichloro 5,6-dicyano-*o*-benzoquinone. This dehydrogenating agent has been successfully used with many organic compounds.[211] The reaction produces both C–C unsaturation and imines. Iron(III) chloride with excess pyridine has also been used successfully with cobalt(II), nickel(II), and copper(II) complexes of dibenzocorromin.[212] The only available mechanistic data is for organic compounds[213-215] and these seem to indicate abstraction of a hydride ion or hydrogen atom transfer.[211] No indications of oxidations of the metal center have been noted except in the cobalt(II) cases where the products are isolated as cobalt(III) complexes. This is an expected result considering that excess oxidant is always present with either pyridine or CN^- to stabilize Co(III) complexes. Busch and co-workers[199] have also been successful at introducing C–C unsaturation using Br_2 as the oxidant with $Ni(Me_2[13]1,10\text{-}dienato(-)N_4)^+$ and $Ni(Me_2[14]1,11\text{-}dienato(-)N_4)^+$.

The ligands of complexes containing β-diimine moieties are relatively easily oxidized.[194] Thus $M(Me_2[14]1,11\text{-}dienato(-)N_4)^{n+}$ complexes exhibit a wave ~ 0.4–0.8 V vs. NHE, more or less independent of the metal and attributable to one-electron oxidation of the ligands:

$$\tag{32}$$

The radical species may be intercepted by scavengers such as O_2 (see above) or may dimerize:

$$D_a \qquad\qquad (3$$

The oxidant in (32) may be either external (O_2, $FeCl_3$, etc.) or internal [Co^{III}, Ni^{III}, etc.; see also Eq. (29)]. In fact, the cobalt(II) dimer is formed very rapidly when $Co(Me_2[14]1,11\text{-dieneN}_4)Cl_2{}^+$ is mixed with base.[194] It further appears that the dimeric β-diimine complexes D_a may be dehydrogenated under very mild conditions ($FeCl_3$, O_2, etc.) to form a second dimer[194,216]:

$$D_a \xrightarrow{\text{[OX]}} \qquad\qquad D_b \qquad\qquad (34)$$

The hydrogenation of D_b and the dehydrogenation of D_a both occur under mild conditions in aqueous solution so that these complexes have the potential of functioning as two to four equivalent redox reagents. Electrochemical oxidations to generate C–C single-bonded dimers of related β-diimine complexes have also been reported.[217]

A number of other, possibly related, reactions have been observed which merit comment. These reactions show that base-promoted reactions that involve reduction of the metal are not unique to macrocyclic complexes nor do such reactions always involve deprotonation of relatively saturated ligands. For example, cyanocobalamin is reduced by heating in 0.4% NaOH. The reaction is complex but it suggests that dehydrogenation occurs on the B ring of the corrin.[218] Similarly, $Co^{III}(Me_4[14]tetraeneN_4)$ and $Co^{III}\text{-}$($Me_2pyo[14]trieneN_4$) are rapidly reduced in the presence of a variety of bases (OH^-, py, diethylamine, diphenylamine) and in a variety of solvents (H_2O, CH_3CN).[123] The reaction rate is dependent on base strength. The products have not been clearly characterized but are very similar to the cobalt(II) complexes with the original ligands, and there is some evidence for the formation of additional unsaturation in the ligands (NMR spectra show evidence of vinyl protons and new IR bands were observed at 1650 cm^{-1}).[123]

In a further example, (tetraphenylporphyrinato)iron(III) is reduced by CN^- or piperidine[219] with no apparent change in the porphyrin ring,[220] at least in that part of the iron(II) product that was examined. There is appreciable evidence that ·CN radicals and piperidine radicals are among the respective oxidation products.[219]

The analogous base-promoted reductions of the polypyridyl complexes, $M(bipy)_3^{3+}$ and $M(phen)_3^{3+}$ (M=Ru, Fe, Os), have been more carefully investigated.[221,222] The base-promoted reductions of the ruthenium(III) complexes are remarkable in that they are chemiluminescent with light energy equivalent to more than 2 eV released in the formation of the ruthenium(II) products.[223] This is reminiscent of the energy required to produce radicals such as ·CN in the iron porphyrin system, and demonstrates that these reactions must involve some highly exoergic step. Reactions of the iron and ruthenium *tris*-(bipyridyl) complexes have been shown to involve the buildup and decay of the concentration of an intermediate, perhaps free radical species.[221,222] Mechanisms proposed for these reactions are analogous to (28)–(31) but involve addition of the base to an α-carbon and a resulting formation of an amide–M^{III} species. Disproportionation reactions could in principle form stable molecular products [e.g., $Ru(bipy)_3^{2+}$ and H_2O_2] from the ligand-radical–M^{II} intermediates without necessitating prohibitively endoergic steps involving direct formation of ·OH, ·CN, etc. In this regard, Gillard[224] has postulated that metalloimine complexes are much more reactive than is commonly supposed, and that the polypyridyl complexes of highly oxidized metals may be in equilibrium with complexes in which the imine moieties are partly hydrolyzed.[225]

3.2. Hydrogenation

A large number of macrocyclic complexes containing unsaturated ligands have been successfully hydrogenated. The most commonly used reagent has been $NaBH_4$,[108,200,225,226] but H_2/Pt[227–229] and Al–Ni[230] alloy in NaOH solutions have also been used extensively. There are also some reports on electrochemical reductions.[41,45]

Curtis[41,200,225,226] has reported on the reductions of $Ni(Me_6[14]4,11$-dieneN$_4)^{2+}$. The ultimate product was the saturated complex $Ni(Me_6[14]ane$-N$_4)^{2+}$. Two isomers were formed in these reductions, rac-(5,12)-$Me_6[14]aneN_4$ and ms-(5,12)-$Me_6[14]aneN_4$, with the percent yield of each isomer depending on the method of reduction. Curtis[41,45] reports 22% of the rac-(5,12) isomer when reduction is carried out using Ni–Al alloy in basic solution, 33% using H_2/Pt, and 40% using $NaBH_4$.

Similar reductions were attempted with $Cu(Me_6[14]4,11$-dieneN$_4)^{2+}$, but decomposition was the predominant reaction.[41,45,227] Attempts to reduce $Co(Me_6[14]$-4,11-dieneN$_4)X_2^+$ generally result in reduction of the metal center but no reduction of the unsaturated ligand.[41,45] A notable exception is $Co(Me_6[14]4,11\text{-dieneN}_4)(CN)_2{}^+$, which is readily reduced to $Co(Me_6[14]\text{-aneN}_4)(CN)_2{}^+$ with $NaBH_4$.[47]

Some reductions have also been carried out using hypophosphorous acid. Tait and Busch[231] have shown that hypophosphorous acid reductions are specific for conjugated diimines. In all the cases thus far investigated only one of the imines was reduced in each diimine. Isolated imines are not attacked. For example, the complex $Co(Me_4[14]\text{tetraeneN}_4)Br_2{}^+$ was reduced to $Co(Me_4[14]1,8\text{-dieneN}_4)Br^{2+}$ after workup with HBr. Excess hypophosphorous acid was required, which may be a result of the reduction of the metal center that appears to occur during the reaction. Reactions with $Co(Me_6[14]4,11\text{-dieneN}_4)Br_2{}^+$ resulted in some decomposition and no ligand reduction. Hypophosphorous acid reductions were also employed by Curtis[45] with the complex $Ni(Me_6[14]1,3,7,11\text{-tetraeneN}_4)^{2+}$ to yield $Ni(Me_6[14]1,7,11\text{-trieneN}_4)^{2+}$.

3.3. Substitutions into the Macrocyclic Ligand

The γ position of six-membered, diiminato chelate rings has been found to be a very reactive, nucleophilic center. This reactivity has been employed to introduce a large number of substituents into macrocyclic ligands. Among reagents used successfully in these reactions are dioxygen,[160,193] nitriles, acetylenes, HNO_3,[232] benzoyl chloride, and methyl vinyl chloride.[112] When these reactions are performed on iron and cobalt complexes there is evidence that the metal center participates in the reaction. At least some of the oxygenation reactions involve ligand radicals (see Section 2.2.3).[39]

A large number of derivatives of the Ni complex in Eq. (1) have been produced by what have been proposed[233] to be electrophilic attacks on the γ carbon in a manner similar to substitution in ordinary aromatic rings:

$$\tag{35}$$

The carbonyl function is not necessary in all cases,[103] but nitrations were carried out as shown. In fact these rings are so highly activated that they will add to species like methyl vinyl ketone, ethylacrylate, and 2-vinylpyridine in a Michael addition.

The carbonyl side chain of the first complex in Eq. (35) reacts with methylfluorosulfonate in methylene chloride to form **3-I**. The reaction is reversed as H_2O is added in acetonitrile.[234] Complex **3-I** can be further

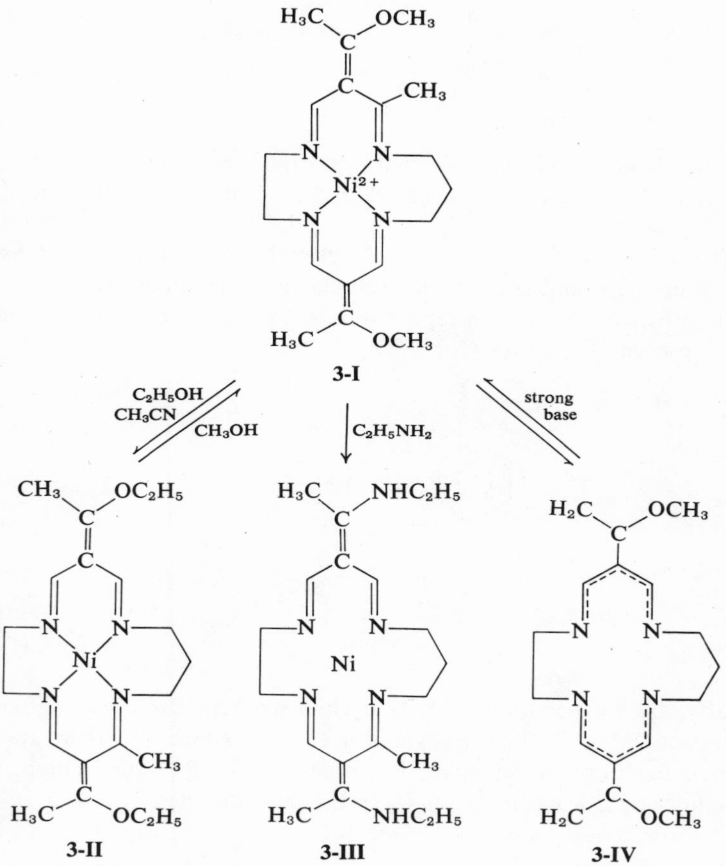

H_3C OCH_3
C
C CH_3
N   N
Ni^{2+}
N   N
H_3C OCH_3

3-I

C_2H_5OH
CH_3CN CH_3OH $C_2H_5NH_2$ strong base

CH_3 OC_2H_5
C
C
N   N
Ni
N   N
CH_3
C
H_3C OC_2H_5

3-II

H_3C NHC_2H_5
C
C
N   N
Ni
N   N
CH_3
C
H_3C NHC_2H_5

3-III

H_2C OCH_3
C
N   N
N   N
C
H_2C OCH_3

3-IV

modified by reactions characteristic of carboxylic acids as shown. Reaction of **3-I** with sodium ethoxide gave the final product.

Recently a series of reactions have been discovered which suggest that the metal center may be important in some of the electrophilic attacks at the γ carbon. For example, in two cases nitriles have reacted with the originally tetradentate ligand to form the sexa- or pentadentate ligands **4**[233] and **5**.[233] Analogous reactions have been observed with alkyl acetylenes.[193] It is impossible to tell at which time in the reaction the nitrogen coordinates to the

4   **5**

metal, but it is entirely feasible that coordination prior to reaction aids in positioning the reagents for the attack and it could stabilize the intermediate and end product.

The reactions of β-diimine complexes with dioxygen have been discussed in Sections 2.2.3 and 3.1 above. Related metal-mediated oxygenations of organic substrates have been reported; e.g., reaction (36) proceeds smoothly in the presence of Cu^{2+} or Co^{2+} [235]:

$$\tag{36}$$

Similarly, the oxygenation of *bis*-2-pyridyl methane requires only catalytic amounts of Cu^{2+}.[236] It has not yet been established whether there are radical mediated reactions, as postulated for macrocyclic β-diimine complexes, or simple nucleophilic additions of O_2 to the deprotonated ligand.

3.4. N-Alkylations

Barefield and co-workers[237,238] have successfully alkylated $Ni^{II}([14]aneN_4)$, $Ni^{II}(Me_6[14]aneN_4)$, and $Ni(Me_2[14]1,11\text{-dienato}(-)N_4)^+$ using KOH or $CH_3S(O)CH_2Na$ in DMSO followed by treatment with the appropriate alkyl halide. The stepwise formation of the singly and the doubly deprotonated metal complexes is clearly indicated by the changes in the spectra when stoichiometric amounts of base are added. The doubly deprotonated form of $Ni^{II}([14]aneN_4)$ is isolatable and stable in the absence of moisture. The deprotonations would indicate that a two-step procedure is necessary for total alkylation, but this is not the case with $Ni(Me_6[14]aneN_4)$. A variety of different alkyl halides may be used, including dihalides, those such as α,α'-dibromo-*o*-xylene which alkylate nitrogens of a single chelate ring. Similar reactions of $BrCH_2CO_2C_2H_5$ give hexacoordinate ligands. Partial and mixed alkylations are possible. It should be noted that the stereochemistry of these products is usually different from that produced by alkylation of the isolated ligand. The permethylation procedure just described for $Ni([14]aneN_4)$ results in the configuration referred to as the *trans*-III form, which is an overall chair configuration with two adjacent methyls up and two methyls down. The product with all methyls on one side may be produced by coordinating the already tetramethylated ligand.[238]

3.5. Additions

Some early work with the Cu(II) and Ni(II) complexes of the cyclo-tetrameric Schiff base of *o*-aminobenzaldehyde ($bzo_4[16]octaeneN_4$, TAAB) shows another possible site for nucleophilic attack. Both amines such as ammonia, dimethylamine, and 1,3-diaminopropane, and alkoxides will add to the carbons in the imine groups,[239,240] giving an α-amino ether with alkoxides. Generally, two molecules of nucleophile will add per macrocycle. Nucleophiles with bridging capabilities gave the complex 7, with X = S or NCH_3 and M = Ni or Cu.

7

4. Reactions Involving Free Radicals, Unusual Oxidation States, and Excited States

Studies of reactions involving metastable species of exceptional reactivity are particularly fascinating, and some such studies have involved complexes with macrocyclic ligands. Most of these studies have taken advantage of the constraints imposed on macrocyclic complexes to limit the reaction modes available to very reactive intermediates. The resulting reactivity patterns often provide useful new insights into a larger area of complex chemistry. A total contextual analysis of such studies is far beyond the scope of this chapter. The following paragraphs are offered as limited samplings of the possibilities in several fascinating areas of chemistry.

4.1. Free Radical Reactions

Some of the above discussion has been concerned with reactions which are inferred to be mediated by very small concentrations of free radical intermediates. In this section we discuss investigations that generate and directly study the reactions of free radical species.

Aqueous free radicals span a wide range of reactivities, both of oxidizing and reducing capacities.[241,242] Of the reducing radicals, e_{aq}^- reacts at diffusion-limited rates with most complexes to produce reduced species (e.g., see Table 8). The radicals $CO_2{}^-$, $(CH_3)_2\dot{C}OH$ [or $(CH_3)_2\dot{C}O^-$], and $\cdot CH_2O^-$ are more discriminating, exhibiting a considerable range of reactivity depending on the metal and the ligand; note that the basic forms of these radicals are much more powerful reducing agents than the acidic forms (Table 8 and references 175, 241, 244). These reactions are particularly interesting as a means of generating $M^I(N_4)$ species.

The dihalogen radical anions $X_2{}^-$ ($X = Cl$, Br, or I) are strongly oxidizing species that could in principle be used to generate relatively uncommon high-oxidation-state species such as $Ni^{III}(N_4)$ or $Cu^{III}(N_4)$ (e.g., see Section 2.2 and references 241, 242, 246–249). There have been relatively few studies of such reactions. Reactions of $X_2{}^-$ radicals with $Co^{II}(N_4)$ complexes have been investigated by flash photolysis techniques[246] and appear to occur at rates limited by the rate of axial $Co-OH_2$ bond breaking (see Section 2.2.2) to produce the corresponding $Co^{III}(N_4)(OH_2)X$ complexes. Very recently, Ferraudi and collaborators[249] and Meyerstein *et al.*[250] have investigated several $X_2{}^-/Ni(N_4)^{2+}$ reactions ($X = Cl$, Br, NCS; $N_4 = Me_6[14]aneN_4$, $Me_6[14]4,11\text{-}dieneN_4$, and $Me_6[14]1,4,8,11\text{-}tetraeneN_4$). Products of these reactions have been identified as metastable $Ni^{III}(N_4)X$ complexes; these primary products are reported to decay at high pH, through $Ni^{II}(N_4\text{-}radical)$ species, to dehydrogenated macrocyclic (N_4)nickel(II) products.[249] These authors provide evidence that the long-lived, reactive intermediates observed at high pH in $Ni^{III}(N_4)$ systems (e.g., see the discussion in Section 3.1 above and reference 205) are more likely to be distorted nickel(III) complexes than nickel(II) radical complexes as proposed by Barefield and Mocella.[205]

Many radicals are not conveniently categorized as "oxidizing" or "reducing." Notable among these are $\cdot H$ and $\cdot CH_3$. The reactions of H atoms with macrocyclic copper(II)[243] and cobalt(II)[244] complexes have been observed. These reactions are regarded as resulting in metallohydride

TABLE 8. Reactions of Some Aqueous Free Radicals with $M(N_4)X_2$ Complexes

Complex/radicals	$10^{-9} \times k\ (M^{-1}\ sec^{-1})$ [a]							
	$e^-(aq)$	CO_2^-	$(CH_3)_2\dot{C}O^-$	$\cdot CH_2O^-$	$\cdot CH_3$	I_2^-	Br_2^-	Cl_2^-
$Cu(Me_6[14]4,11\text{-dieneN}_4)^{2+}$	$50(<10^{-3})^b$	$2.3(0.1)^b$	$0.9(<10^{-2})^b$	$0.9(<10^{-2})^b$				
$Ni(Me_6[14]4,11\text{-dieneN}_4)^{2+}$	$78(<10^{-3})$	$6.7(0.1)^b$						
$Ni(Me_6[14]aneN_4)^{2+}$	$56(<10^{-3})^b$	$5.7(0.1)^b$						
$Co(Me_6[14]4,11\text{-dieneN}_4)(OH_2)_2^{2+}$	$44(<10^{-3})^b$				$0.5^f, 0.7^g$	$7.2(0.3)^e$	$9.8(0.01)^c$	$9.6(0.01)^c$
$Co(Me_6[14]4,14\text{-dieneN}_4)(OH_2)_2^{2+}$	$34\ (<10^{-3})^d$						$3.4(0.01)^c$	$2.0(0.01)^c$
$Co(Me_4[14]tetraeneN_4)(OH_2)_2^{2+}$	$49(<10^{-3})^d$	$4.7(0.1)^d$	$(5.5)^{d,h}$				$1.4(0.2)^e$	$1.0(1.0)^e$
$Co(Me_6[14]4,11\text{-dieneN}_4)(OH_2)_2^{3+}$		$0.8^{f,j}$	$(0.2)^{f,h}$		0.1^g	$\sim 1^i$	$\sim 1^i$	$\sim 1^i$
$Co(Me_4[14]tetraeneN_4)(OH_2)_2^{3+}$		$6.4^{f,j}$	$(1.9)^{f,h}$					
B_{12r}		4.7^k			2^k			

[a] 25°C; ionic strength in parenthesis.   [b] Reference 243.   [c] Reference 249.   [d] Reference 175.   [e] Reference 246.   [f] Reference 244.
[g] Reference 79.   [h] Reaction of $(CH_3)_2\dot{C}O^-$ in $HClO_4$. Ionic strength not specified.   [i] Second-order decays similar to $Co(Me_6[14]4,11\text{-dieneN}_4)(OH_2)_2^{2+} + X_2^-$ when these species were generated by flash photolysis of $Co(N_4)X_2^+$. Reactivity of $Co(Me_4[14]tetraeneN_4)(OH_2)_2^{2+}$ precluded an accurate determination of rate constants. (S. D. Malone and J. F. Endicott, unpublished work.)   [j] Ionic strength not specified; apparently less than 10^{-2}.   [k] Reference 245.

products; however, products have not been isolated or further characterized. It is possible that some of this chemistry is complicated by hydrogen atom abstraction from the organic ligand and that at least some of the product species observed in pulse radiolysis experiments contain the metal coordinated to a radical ligand rather than a metallohydride complex. In this regard it is interesting that the $\cdot$H atom and $\cdot$OH radical reactions with $Co(Me_6[14]4,11\text{-}dieneN_4)(OH_2)_2^{2+}$ result in products with many similar absorption features.

Methyl radical reactions have also been observed with many $Co^{II}(N_4)$ substrates to give $trans\text{-}Co(N_4)(OH_2)CH_3^{2+}$. Products of these reactions have been isolated and characterized for $N_4 = [14]aneN_4$, $Me_6[14]4,11\text{-}dieneN_4$, and $Me_4[14]tetraeneN_4$.[79] A minor ($\sim 5\text{-}10\%$) side reaction of $\cdot CH_3$ with $Co(Me_4[14]tetraeneN_4)(OH_2)_2^{2+}$ has been found to involve H-atom abstraction from an α-methylene carbon.[251] Several of these reactions[79,251] and the $\cdot CH_3$ reactions with B_{12r}[252] and $bis\text{-}(dimethylglyoximate)cobalt(II)$[252] have been investigated by means of flash photolysis and found to occur at rates smaller than the expected diffusion limit. These reactions are believed to be limited in rate by the rate of $Co\text{-}OH_2$ bond breaking in the respective $Co^{II}(N_4)(OH_2)_2$ complexes.

Elroi and Meyerstein[253] have recently reported on the reactions of $\cdot CH_2OH$, $CH_3\dot{C}HOH$, $HO\dot{C}HCH_2OH$, and $\cdot CH_2CHO$ and related species with $Co(Me_6[14]4,11\text{-}dieneN_4)(OH_2)_2^{2+}$. The rates of formation of organocobalt complexes were all found to be somewhat smaller than values reported for the reaction of this cobalt(II) complex with $\cdot CH_3$,[79,244] ranging from less than $10^7 \ M^{-1} \sec^{-1}$ for $\cdot CH_2C(CH_3)_2OH$ to $7 \times 10^7 \ M^{-1} \sec^{-1}$ for $\cdot CH_2OH$. The organocobalt adducts were found to be unstable with lifetimes of no more than a few seconds. For the adducts containing CH_2OH, CH_2CHO, and $CH(OH)CH_2OH$, the ultimate decomposition products contained approximately stoichiometric amounts of aldehydes. While different mechanisms are proposed for decomposition of several of the organocobalt intermediates, it is proposed that the $Co(Me_6[14]4,11\text{-}diene\text{-}N_4)(OH_2)CH(OH)CH_2OH^{2+}$ complex dehydrates and rearranges to form $Co(Me_6[14]4,11\text{-}dieneN_4)(OH_2)CH_2CHO^{2+}$. This is suggestive of the cobalamin-dependent chemistry of diol-dehydrase[62-70] and the proposed rearrangement of the coordinated organocobalt diol intermediate.[64,67,68] The observation of addition to rather than reduction of the cobalt center by these strongly reducing $R\dot{C}OH$ radicals may reflect the relatively negative potential of the $Co^{II,I}(Me_6[14]4,11\text{-}dieneN_4)$ couple.

4.2. Complexes Containing Metals in Unusual Oxidation States

The thermodynamics of generation of macrocyclic complexes with unusual oxidation states is discussed in Section 2.2.1. There have been a few studies of the reactions of these species. This is clearly a complex and important area worthy of much more, careful attention.

$M^I(N_4)$ Systems

Most of the tetraazametallo(I) complexes tend to be powerful and labile reducing agents. A point of some interest in the study of these systems is to design a system that makes the M^I state accessible at reasonable potentials, but for which the reorganizational barriers for redox reactions are small. One would expect relatively small reorganizational barriers for low-spin d^7–d^8, d^8–d^9, and d^9–d^{10} systems, provided the N_4 equatorial ligand system maintains a nearly planar microsymmetry through the series.

For the systems studied to date the $Ni^{II,I}$ couples have tended to be relatively reducing, thus accessible as transient species in reactions of powerful reducing radicals such as e_{aq}^- [244] but not easily attainable in an aqueous chemical system. The nickel(I) species do react with acids, but more slowly than do the analogous cobalt(I) complexes.[243] The reactions of $Ni^I(Me_6[14]$-4,11-dieneN$_4)^{2+}$ and $Ni^I(Me_6[14]aneN_4)$ with several metallo complexes [Co(bipy)$_3^{3+}$, Cr(bipy)$_3^{3+}$, Fe(bipy)$_3^{3+}$, Ru(NH$_3$)$_6^{3+}$, and Co(en)$_3^{3+}$][243] are more rapid than one might anticipate using (16), possibly owing to relatively small contributions of the quadratic term and too small a reorganizational term (i.e., λ_{ab} appears to be less than $\lambda_{bb}/2$). Unfortunately, not enough reactions have been examined with these substrates to permit definitive conclusions.

The tetraazacopper(I) complexes are potentially of more general chemical interest since they may be generated under relatively mild conditions (see Section 2.2.1) and their reactions with acids are not extremely rapid [e.g., the rate constant for $H^+ + Cu(Me_6[14]4,11$-dieneN$_4)^+$ is $5 \times 10^6\ M^{-1}$ sec^{-1}].[243] The electron transfer reactions of $Cu(Me_6[14]4,11$-dieneN$_4)^+$ are between two and four orders of magnitude slower than corresponding reactions of the nickel(II) analog, more or less as one would expect from (16) and the differences in $Ni^{II,I}$ and $Cu^{II,I}$ potentials.

The chemistry of tetraazacobalt(I) species has been extensively investigated in connection with cobalamin and related model compounds. The cobalt(I) oxidation state may be generated at reasonable potentials when the equatorial ligand contains extensive unsaturation (see Section 2.2.1). Hoffman and co-workers[175,244] used strongly reducing radicals to generate cobalt(I)–$Me_6[14]4,11$-dieneN$_4$, –$Me_6[14]aneN_4$, –$Me_6[14]4,14$-dieneN$_4$, and –$Me_4[14]$-tetraeneN$_4$ complexes. These cobalt(I) complexes all react rapidly with acids, presumably forming hydrido complexes; the rate constants for reactions with H^+ vary from $1.6 \times 10^5\ M^{-1}$ sec^{-1} ($Me_4[14]$tetraeneN$_4$) to $3.1 \times 10^9\ M^{-1}$ sec^{-1} ($Me_6[14]4,11$-dieneN$_4$). In fact, the relatively saturated $Co^I(N_4)$ complexes are very powerful and labile reducing agents. Very rapid rates of reaction are observed even with very weak oxidants; e.g., even the reaction with CH_3I may proceed by electron transfer.[79,175] In contrast, $Co^I(Me_4[14]$-tetraeneN$_4$) is a reactive but discriminating reductant, with many of its reactions being "too slow" to follow using the pulse radiolysis technique. Unfortunately this reagent is a bit hard to handle and there is little systematic information available about its reaction patterns. In common with the Ni^I and Cu^I complexes, it appears that the intrinsic component of the activation barrier to electron transfer reactions is unexpectedly small. As a consequence the $(Me_4[14]tetraeneN_4)Co^{II,I}$ couple is capable of playing a catalytic role in electron transfer reactions.[254]

A rather different approach to the generation of species with unusual oxidation states has been taken by Allred and co-workers[122,255] and by Barefield.[121] These investigators have found that cyclic tetramine ligands induce the disproportionation of Ag^+ in water to form $Ag^0 + Ag(N_4)^{2+}$. This species may be further oxidized to $Ag(N_4)^{3+}$, and the silver(III) complexes are not stable in water.[79,121] On the basis of the trends in gaseous ionization potentials, Allred *et al.*[122] predicted that it should be possible to prepare $Hg^{III}(N_4)$ and $Cd^{III}(N_4)$ complexes. Cyclic voltammetry indicated the formation of Hg^{III} species. This species proved too unstable to isolate, and EPR flow experiments in propionitrile at $-78°C$ indicated a 5-sec half-life for the decay of $Hg^{III}([14]aneN_4)$.

Macrocyclic complexes containing metals in rare oxidation states, either highly reduced or oxidized, could in principle have some potential as catalytic reagents or as selective redox agents. Thus Welsh and Henry[207] have investigated the use of the $Ni^{III}(N_4)$ complexes and the metastable $Ni^{II}(\cdot N_4)$ radical species (see Section 3) as oxidants for organic substrates such as cyclohexanol. As noted in Section 3, reactions of these oxidizing species with water seem to lead to ligand degradation so they are not likely to be useful in cycles involving water decomposition (e.g., see references 221 and 256).

The $M^I(N_4)$ complexes have received even less attention as potentially useful chemical reagents. Their reactivity toward H^+ is quite intriguing and their use as hydrogenating agents deserves some attention. Sutin and Creutz[221] and Balzani and co-workers[256] have pointed out that reducing species such as hydrido complexes could have a very important role in the photochemical degradation of water. Utilization of $M^I(N_4)$ complexes in this capacity has not received much attention, but such an application in homogeneous solutions is bound to be complicated by the relatively large absorptivity of these species and their $M^{II}(N_4)$ parents in the visible spectral region. Attempts to use the highly reducing excited states of $Ru(bipy)_3^{2+}$ and $Ru(bipy)_2(CN)_2$ to generate $M^I(N_4)$ species from $Cu(Me_6[14]4,11-dieneN_4)^{2+}$ and $Co(Me_4[14]$-tetraeneN$_4)(OH_2)_2^{2+}$ failed of their purpose owing to the interfering absorbencies of the $M^{II}(N_4)$ acceptor species.[257] However, the more powerful reducing agent $Ru(bipy)_3^+$ has been shown to reduce $Co(Me_6[14]4,11-dieneN_4)(OH_2)_2^{2+}$ in water, followed by evolution of hydrogen, as described by

$$Ru(bipy)_3^{2+} + h\nu \longrightarrow Ru(bipy)_3^{2+} \tag{37}$$

$$*Ru(bipy)_3^{2+} + Red \longrightarrow Ru(bipy)_3^{+} + Ox \tag{38}$$

$$Ru(bipy)_3^{+} + Co(N_4)(OH_2)_2^{2+} \longrightarrow Ru(bipy)_3^{2+} + Co^{(I)}(N_4) \tag{39}$$

$$Co^{(I)}(N_4) + H^+ \longrightarrow [Co(N_4)H] \xrightarrow{\ H^+\ } Co(N_4)(OH_2)_2^{3+} + H_2 \tag{40}$$

$$Co(N_4)(OH_2)_2^{3+} + Red \longrightarrow Co(N_4)(OH_2)_2^{2+} + Ox \tag{41}$$

where "Red" = Eu^{2+} or ascorbate.[258] This work illustrates potential of the photochemical generation of very reactive intermediates. By manipulating the $Co^{(II),(I)}$ potential with substituents on the macrocyclic (N_4) ligand it should be possible to accomplish direct $*Ru(bipy)_3^{2+}$ reduction of $Co^{(II)}(N_4)$, storing appreciable amounts of the excitation energy in H_2 and $Ru(bipy)_3^{3+}$ or a species oxidized by $Ru(bipy)_3^{3+}$ (e.g., ascorbate to dehydroascorbate, or water to O_2). Other macrocyclic complexes could be designed to mediate light-driven hydrogenations, perhaps by making the formation of $Co(N_4)$-$(OH_2)_2^{3+}$ thermodynamically difficult. Studies of such systems are very important areas for further research.

4.3. Photochemical Reactions

It has been conventional to classify photochemical studies of coordination complexes by the nature of the photoactive states generated following the absorption of radiation. Thus it is the practice to distinguish ligand field photochemistry (largely metal-centered states) from charge transfer photochemistry (states involving transfer of electron density between metal and ligand orbitals), and each of these categories tends to be distinguished from ligand-centered photochemistry (states involving predominantly ligand orbitals).[259,260] The high density of states and configurational mixing characteristic of transition-metal complexes greatly reduces the exclusivity of these categories, but they do provide a convenient device for organizing information about the gross photochemical features. Studies involving macrocyclic complexes are very fragmentary.

4.3.1. Ligand Field Photochemistry

Macrocyclic ligands, usually [14]aneN$_4$, have been used to introduce steric constraints on possible excited state distortions in mechanistic studies of ligand field photochemistry. Adamson and co-workers[261-264] have made a number of careful and interesting comparisons of the photochemistry of *trans*-M(en)$_2$Cl$_2^+$ and *trans*-M([14]aneN$_4$)Cl$_2^+$ complexes for M = Cr, Co, and Rh. Acidoamine complexes of chromium(III) are notoriously photosensitive. Quantum yields for aquation are commonly 0.2–0.4, and the photosolvolysis is usually accompanied by a *trans* → *cis* isomerization.[265-267] In marked contrast, ligand field irradiation of *trans*-Cr([14]aneN$_4$)Cl$_2$$^+$ results in very small ($\sim 3 \times 10^{-4}$) yields of stereoretentive solvolysis.[261,263] The behavior of linear and macrocyclic amine complexes of cobalt(III) and rhodium(III) does not present such a dramatic contrast. These contrasts in photochemical behavior imply that the distortions of the photoactive ligand field quartet states (configuration: $\pi_M{}^2\sigma_M{}^*$) in chromium(III) are of different symmetry types than the distortions characteristic of the photoactive states in the d^6 complexes (probably excited state configuration: $\pi_M{}^5\sigma_M{}^*$).

Selan and Rumfeld[268,269] have made some careful and critical comparisons of the ligand field photochemistry of *cis*- and *trans*-Rh([14]aneN$_4$)X$_2$$^+$ (X = Cl, Br, I) and of *cis*-Rh(en)$_2$Cl$_2$$^+$ in aqueous solutions. Consistent with predictions of a molecular orbital(MO)-based photochemical model,[270-273] irradiations of *cis*-Rh(en)$_2$Cl$_2$$^+$ result predominantly in Rh(en)(enH$^+$)-(OH$_2$)Cl$_2^{2+}$, a pathway that is repressed in the [14]aneN$_4$ complexes. For the *trans*-Rh([14]aneN$_4$)X$_2$$^+$ complexes the aquation yield increases in the order ϕ(Cl) < ϕ(Br) < ϕ(I), while the opposite order is observed for the *cis* complexes. Many features of the photochemistry of these systems seem consistent with the MO models, but some anomalies of the *cis*-Rh([14]aneN$_4$)X$_2$$^+$ complexes have been attributed to nontetragonal distortions in these species.

Photosubstitutional behavior is also observed for visible irradiations of Fe$^{(II)}$(Me$_4$[14]tetraeneN$_4$)XY complexes (X = NCS$^-$, CO, acetonitrile, imidazole).[274,275]

The excited states populated in these complexes may include some states with charge transfer to ligand character as well as ligand field states. As yet, a careful wavelength dependence of the product yields has not been reported. Molecular orbital models have been extended to account for the final product distributions in terms of the relative populations of metal-centered d orbitals.[274] Net product yields tend to be smaller for photolyses of Fe$^{(II)}$-(Me$_4$pyo$_2$[14]hexaeneN$_4$)XY than for Fe$^{(II)}$(Me$_4$[14]tetraeneN$_4$)XY, and this has been attributed to lower energy metal-to-ligand charge transfer than d–d states in the former, an orbital ordering that is thought to be reversed in the Me$_4$[14]tetraeneN$_4$ complexes of iron(II).[275] However, conventional and laser flash photolysis studies have shown that these systems have a rich transient chemistry and that the final product distributions depend on the reactions of metastable intermediate species.[276] A very reactive intermediate species with a lifetime of $\sim 10^{-7}$ sec in acetonitrile solutions has been tentatively identified as a five-coordinate Fe$^{(II)}$(Me$_4$[14]tetraeneN$_4$)X species; this species reacts with CO with a specific rate constant of $\sim 10^{10}$ M^{-1} sec^{-1} in acetonitrile solutions, in competition with reaction with the solvent.[276]

Many specific mechanistic inferences drawn from studies such as the above must be regarded as quite tentative since these interpretations tend to assume that the only alterations in excited-state behavior that occur when one changes ligands are due only to changing parameters in the reaction channels leading to products. In fact, excited states decay by means of competitive reactive (leading to photoproducts) and nonreactive (leading to unchanged substrate) channels. Consideration of a single decay channel, to the exclusion of others, is bound to lead to erroneous conclusions (see also references 277–279). Certainly the intervention and reactivity of metastable reaction intermediates would make the overall observed product distributions and yields rather weak evidence for such theories.

4.3.2. Charge Transfer Photochemistry

Malone and Endicott[246,280] have studied the photoredox chemistry of trans-$Co(Me_6[14]4,11\text{-dieneN}_4)X_2^+$ and trans-$Co(Me_4[14]\text{tetraeneN}_4)X_2^+$ (X = Cl, Br) complexes. The flash photolysis technique has been used to demonstrate the formation of X_2^- radicals and very rapid recombination reactions (see Section 2.2.2.2 and 4.1). Owing to the efficient recombination reactions it is difficult to determine meaningful quantum yields for photoredox processes in these systems. One rather surprising observation was that flash photolytic irradiation of trans-$Co(Me_4[14]\text{tetraeneN}_4)I_2^+$ did not lead to detectable radicals.[280] Not too surprisingly, there was evidence that this complex was reduced by I_2^- generated from flash photolysis of the $\{Co(NH_3)_6^{3+}, I^-\}$ ion pair.[246,280]

Reichgott and Rose[281] have reported on the $Fe^{(III)}(Me_4[14]\text{tetraeneN}_4)$ mediated photooxidation of methanol. The iron(III) species believed to be present in methanol solutions is $Fe(Me_4[14]\text{tetraeneN}_4)(OCH_3)(CH_3OH)^{2+}$. This species slowly (in a few days) oxidizes methanol in the dark to formaldehyde and other products. When exposed to sunlight the degassed system produces 0.47 mole CH_2O per mole of iron present. When O_2 is present in solution the photoassisted reaction proceeds in a cyclic manner until all the oxygen is consumed:

$$2Fe(Me_4[14]\text{tetraeneN}_4)(CH_3OH)_2^{2+} + \tfrac{1}{2}O_2 \xrightarrow{CH_3OH}$$

$$2Fe(Me_4[14]\text{tetraeneN}_4)(OCH_3)CH_3OH^{2+} + H_2O \quad (42)$$

$$2Fe(Me_4[14]\text{tetraeneN}_4)(OCH_3)CH_3OH^{2+} + CH_3OH \xrightarrow{\text{sunlight}}_{CH_3OH}$$

$$2Fe(Me_4[14]\text{tetraeneN}_4)(CH_3OH)_2^{2+} + CH_2O \quad (43)$$

The process, which very likely involves the $\cdot CH_2OH$ radical (with a possible $CH_3O\cdot$ precursor) intermediate, can be recycled many times. There is eventual degradation of the metal complex, and prolonged exposure leads to formation of an unidentified yellow product.

Ferraudi and Srisankar[282] have reported a unique photoredox reaction of dimeric cobalt(II)–sulfophthalocyanine complexes: Flash photolyses indicate a photoinduced disproportionation into cobalt(III) and cobalt(I)-sulfophthalocyanines. In the absence of scavengers the cobalt(I) and cobalt(III) species recombine to form the original cobalt(II) substrate. This behavior stands in contrast to photolyses of cobalt(III), copper(II), and iron(II) sulfophthalocyanines which result in formation of radical-ligand intermediates.[282,283]

4.3.3. Photostimulated Reactions of the Macrocyclic Ligand

The absorption spectra of many macrocyclic complexes exhibit features that are assigned as metal-to-ligand charge transfer (CTTL) or internal ligand transitions. There have been few investigations of such systems.

One example of the kind of behavior that might be common in such systems is the observation that $Cu(Me_6[14]4,11\text{-dieneN}_4)^{2+}$ decomposes in some solvent media when irradiated in its near ultraviolet CTTL absorption bands.[284] Two pathways are observed depending on the solvent. In water the orange substrate is photodecomposed to a blue product, which has been isolated and characterized as the copper(II) complex of 14-amino-4,4,9,11,11-pentamethyl-5,8,12-triazatetradeca-8,en-2-one, formed by adding a water molecule to an imine function of the substrate:

$$\text{8} + h\nu \longrightarrow \text{*CT} \qquad (44)$$

$$\text{*CT} \longrightarrow \text{8 or 9} \qquad (45)$$

$$\text{9} \xrightarrow{\text{H}^+} \text{8 or 10} \qquad (46)$$

$$10 + H_2O \longrightarrow \quad \left[\text{Cu}^{II} \text{ macrocyclic complex} \right] + H^+ \tag{47}$$

$$\longrightarrow \quad \left[\text{Cu}^{II} \text{ macrocyclic complex with } NH_2 \text{ and } C{=}O \right] \tag{48}$$

In deaerated methanolic solutions the observed reaction products are $Cu(Me_6[14]4,11\text{-dieneN}_4)^+$ and CH_2O. Flash photolysis experiments have shown that the copper(I) complex is generated over a period of about 10^{-4} sec, apparently in a secondary radical reaction. In view of the CTTL nature of the photoactive absorption region, and the discussion in Section 3.1, the postulated intermediate (or CTTL excited state), **9** is conveniently visualized as a copper(III) species with excess electron density on an adjacent nitrogen, i.e.,

$$\text{Cu}^{II}{-}\text{N}{=}\text{C}{<} + h\nu \longrightarrow \left[\text{Cu}^{III}{-}\text{N}{-}\text{C}{<} \right]$$

which may either be scavenged by methanol or H^+ (in water) to form the precursors to the observed products.

Ultraviolet irradiations also result in decomposition of $Cu(Me_2[13]\text{-}10,12\text{-dienato}(1-)N_4)^+$ in alcoholic media.[285] One of the predominant products has been identified as $Cu(Me_2[13]1\text{-monoeneN}_4)^{2+}$. A mechanism similar to that described above for photolyses of $Cu(Me_6[14]4,11\text{-dieneN}_4)^{2+}$ has been proposed.[285] While it does seem likely that the optical transitions of this complex include some metal-to-ligand charge transfer transitions, we have noted above the relative ease of oxidation of β-diimine moieties and would as a consequence also expect some low-energy charge transfer to metal chemistry to occur in these systems. Since not all the photoproducts could be characterized,[285] nor transient intermediates detected, the mechanism proposed for the photodecomposition of $Cu(Me_2[13]10,12\text{-dienato}(1-)N_4)^+$ is at best an interesting working hypothesis.

4.4. Photochemistry of Cobalt–Alkyl Complexes

There have been a large number of studies of the products resulting from light absorption by organocobalamin and organocobaloxime complexes.[59–63,69,70,251,252,285–300] There have been relatively few careful determinations of quantum yields.[251,289,292,294,298] The principal photochemical process in these systems is homolysis of the cobalt–alkyl bond. It appears that this process occurs efficiently, at least in the methyl–cobalt complexes, even for very low excitation energies.[251,252,297,298] In relatively nonpolar media there is evidence that some new process occurs in the alkyl cobaloximes —a process not involving cobalt–alkyl homolysis.[293–295]

One of the most interesting synthetic organocobalt complexes is $Co([14]aneN_4)(OH_2)CH_3^{2+}$. This complex has a remarkably "simple" absorption spectrum with weak absorption bands at 478 nm ($\varepsilon = 82\ M^{-1}\ cm^{-1}$) and 370 nm ($\varepsilon = 103\ M^{-1}\ cm^{-1}$), a weak ($\varepsilon \sim 10^2\ M^{-1}\ cm^{-1}$) shoulder at 280 nm, and an intense band ($\varepsilon = 2.4 \times 10^4\ M^{-1}\ cm^{-1}$) at 205 nm.[79] This contrasts to cobalt–alkyl complexes with extensively unsaturated equatorial ligands for which one observes transitions with large absorptivities in the visible region, and is very similar to the absorption spectrum of $Co(NH_3)_6^{3+}$. However, homolyses in $Co(NH_3)_6^{3+}$ are deep ultraviolet processes, while the threshold for $CoCH_3$ homolysis in $Co([14]aneN_4)(OH_2)CH_3^{2+}$ corresponds to $\lambda > 540$ nm.[251,298]

The low-energy threshold required for $Co–CH_3$ homolysis implies a relatively weak bond: ~ 200 kJ mol^{-1} for $Co–CH_3$ compared to ~ 273 kJ mole^{-1} for $Co–Cl$ in $Co(NH_3)_5Cl^{2+}$.[251,298,301] The bond energy and the spectroscopy suggest the correlation diagram in Figure 7 and the assignment of the lowest energy transition as excitation of an electron from the $Co–CH_3$ bonding orbital (Ψ_B) to the "$d_{x^2-y^2}$" orbital.[251] It has been suggested that the (Ψ_B–Ψ_{AB}) triplet state may lie at relatively low energies and mediate the photohomolysis.[219]

Cobalt(N_4)–alkyl bond energies do vary somewhat with the nature of the equatorial ligand. Thus the largest homolysis energies have been found for $Co(Me_4[14]tetraeneN_4)(OH_2)CH_3^{2+}$ and $Co(DMG)_2(OH_2)CH_3$ (about 208 kJ mole^{-1}), and the smallest to date for $Co(Me_6[14]4,11\text{-dieneN}_4)(OH_2)$-$CH_3^{2+}$ (less than 180 kJ mole^{-1}).[302] It is interesting to note that the last of these complexes is the cationic complex most easily dehydrated.[79] This is consistent with a 3-center–4-electron model proposed for the axial bonding interaction in cobalt–alkyl systems.[251]

The dependence of photohomolysis quantum yields on excited-state processes on the one hand, and solvent cage recombination processes on the other, have been the focus of considerable discussions and controversy (e.g., see references 241, 259, 267, 303, 304). Flash phololysis studies of alkyl cobalamins using 530-nm excitation pulses of a few picoseconds duration have demonstrated that both excited state and solvent cage recombination processes contribute to the net quantum yields of homolysis products.[305] This study has found the primary quantum yield (at $\leqslant 10$ psec) of radicals to be $\phi_0 \simeq 0.3$ for both methyl and adenosyl cobalamin, whereas the net yields are ~ 0.2 and ~ 0.1 in continuous photolysis studies,[297] or after about 1 nsec.[305] The difference in net yields arises because the methyl radical

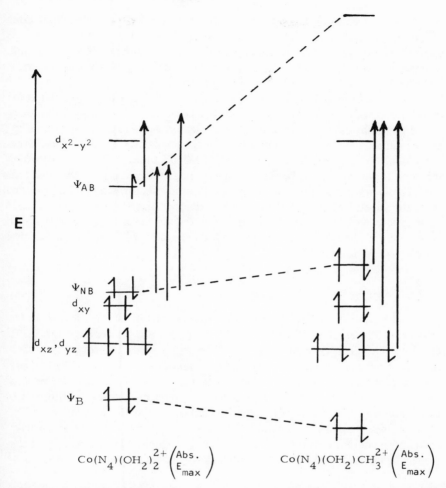

FIGURE 7. Correlation of orbital energies of cobalt(II) and methyl–cobalt[14]aneN$_4$ complex. Vertical arrows indicate energy of the absorption maxima for observed electronic transitions. Orbital energies have been selected for maximum consistency of observed transition energies and allowing 0.3–0.5 μm^{-1} for Franck–Condon contributions. A three-center–4-electron bonding model has been used to describe the axial interactions ($\Psi_{AB}, \Psi_{NB}, \Psi_B$). From reference 251.

diffuses out of the solvent cage more rapidly than the adenosyl radical ($k_d \simeq 3 \times 10^9$ and 0.7×10^9 sec^{-1}, respectively); cage recombination seems to be similar in both cases ($k_r \simeq 1.4 \times 10^9$ sec^{-1}).[305] Since the net quantum yields are nearly independent of excitation energy for these complexes, the picosecond studies imply that there is considerable thermalization of the excess excitation energy during the very short ($\leqslant 5$ psec) excited-state lifetime, and about 70% of the excited species relax directly to the ground state. These studies demonstrate clearly the strong tendency toward recombination of cobalt(II) with radical species, and they suggest that the radical, cobalt(II) pairs generated by photolysis are thermalized.

Since the biological functioning of coenzyme B_{12} has been postulated to involve thermal homolysis of the cobalt–carbon bond[64–67,79,70,306] studies of the transient chemistry in $Co^{(II)}(N_4)$–radical systems could have a bearing on the details of molecular functioning of the coenzyme. In most complexes the Co–C bond energies seem to run 150–210 kJ mol^{-1}[251,302]; however, thermal homolysis might be feasible if bonding is weakened in the position *trans* to the organo group. It does appear that radicals formed from molecules that are substrates (or similar to substrates) of the B_{12} enzyme undergo appropriate molecular transformations in the presence of cobalt(II) complexes.[253,300,305] It is to be hoped that further study will resolve some of the problems and ambiguities.

ACKNOWLEDGMENTS

The authors wish to thank Professor D. H. Busch, Professor J. H. Espenson, Professor M. Weaver, Dr. C. K. Poon, Dr. R. W. Hay, Dr. B. T. Golding, Dr. G. Navon, and Dr. G. J. Ferraudi, who generously provided them with very useful manuscripts well in advance of publication. The authors are grateful to Dr. C. K. Poon for a critical reading of the manuscript and for many helpful comments; Professor R. G. Wilkins and Dr. N. Sutin also made some helpful comments. We also acknowledge the assistance of Dr. C. L. Wong, Dr. J. M. Ciskowski, Dr. R. Sriram, and Dr. J. A. Switzer with many of the details. Support by the U. S. Public Health Service of some of the research efforts described is gratefully acknowledged.

References

1. H. Taube, *Chem. Rev.* **50**, 69 (1952).
2. F. Basolo and R. G. Pearson, *Mechanisms of Inorganic Reactions*, 2nd ed., Wiley, New York (1967).
3. R. G. Wilkins, *Acc. Chem. Res.* **3**, 408 (1970).
4. C. H. Langford and V. S. Sastri, *M. T. P. Int. Rev. Sci. Inorg. Chem. Ser.* [*1*] **9**, 203 (1972).
5. T. P. Dasgupta, *M. T. P. Int. Rev. Sci. Inorg. Chem. Ser.* [*2*] **9**, 63 (1974).
6. C.-K. Poon, *Coord. Chem. Rev.* **10**, 1 (1973).
7. C.-L. Wong, Ph.D. dissertation, University of Hong Kong, 1976.
8. T. W. Swaddle, *Coord. Chem. Rev.* **14**, 217 (1974).
9. A. V. Ablov and N. M. Samus, *Coord. Chem. Rev.* **17**, 253 (1975).

10. M. Eigen and R. G. Wilkins, *Adv. Chem. Ser.* **49**, 555 (1965).
11. J. Burgess, senior reporter, Inorganic Reaction Mechanisms, in: Vol. 1, 1970; Vol. 2, 1972; Vol. 3, 1975, Specialist Periodical Reports, The Chemical Society, Burlington House, London.
12. C. H. Langford and H. B. Gray, *Ligand Substitution Processes*, Benjamin, New York (1966).
13. J. O. Edwards, F. Monacelli, and G. Ontaggi, *Inorg. Chim. Acta* **11**, 47 (1974).
14. M. L. Tobe, *Acc. Chem. Res.* **3**, 377 (1970).
15. B. Bosnich, C.-K. Poon, and M. L. Tobe, *Inorg. Chem.* **5**, 1514 (1966).
16. C.-K. Poon and C.-L. Wong, *Inorg. Chem.* **15**, 1573 (1976).
17. C. K. Lui and C.-K. Poon, *J. Chem. Soc. Dalton Trans.* **1972**, 216 (1972).
18. C.-K. Poon and C.-L. Wong, *J. Chem. Soc. Dalton Trans.* **1976**, 966 (1976).
19. K. S. Mok and C.-K. Poon, *Inorg. Chem.* **10**, 225 (1971).
20. W.-K. Lee and C.-K. Poon, *Inorg. Chem.* **12**, 2016 (1973).
21. C.-K. Poon and C. L. Wong, *J. Chem. Soc. Dalton Trans.* **1977**, 523 (1977).
22. W.-K. Chau, W.-K. Lee, and C.-K. Poon, *J. Chem. Soc. Dalton Trans.* **1974**, 2419 (1974).
23. W.-K. Lee and C.-K. Poon, *J. Chem. Soc. Dalton Trans.* **1974**, 2423 (1974).
24. C.-K. Poon and H.-W. Tong, *J. Chem. Soc. Dalton Trans.* **1974**, 1 (1974).
25. W.-K. Lee and C.-K. Poon, *Inorg. Chem.* **13**, 2526 (1974).
26. P. W. Mak and C.-K. Poon, *Inorg. Chem.* **15**, 1949 (1976).
27. C.-K. Poon, W. K. Wan, and S. S. T. Liao, *J. Chem. Soc. Dalton Trans.* **1977**, 1247 (1977).
28. J. A. Kernohan and J. F. Endicott, *Inorg. Chem.* **9**, 1504 (1970).
29. D. P. Rillema, J. F. Endicott, and J. R. Barber, *J. Am. Chem. Soc.* **95**, 6987 (1973).
30. R. W. Hay and G. A. Lawrence, *J. Chem. Soc. Dalton Trans.* **1975**, 1556 (1975), **1976** 1086 (1976).
31. Y. Hung and D. H. Busch, *J. Am. Chem. Soc.* **99**, 4977 (1977).
32. R. R. Krug, W. G. Hunter, and R. A. Grieger, *J. Phys. Chem.* **80**, 2335 (1976).
33. R. R. Krug, W. G. Hunter, and R. A. Grieger, *J. Phys. Chem.* **80**, 2341 (1976).
34. M. L. Tobe, *Inorg. Chem.* **7**, 1260 (1968).
35. R. A. D. Wentworth and T. S. Piper, *Inorg. Chem.* **4**, 709 (1965).
36. W. Latimer, *Oxidation Potential*, Prentice-Hall, Englewood Cliffs, N.J. (1952).
37. E. A. Moelwyn-Hughes, *Physical Chemistry*, pp. 910–911, Pergamon, New York (1953).
38. J. F. Endicott, *Inorg. Chem.* **16**, 494 (1977).
39. R. R. Dogonadze, in: *Molecular Processes at Electrodes* (N. S. Hush, ed.), Chapter 3, p. 135, Wiley-Interscience, New York (1971).
40. E. D. German and R. R. Dogonadze, *J. Inorg. Nucl. Chem.* **34**, 3196 (1972).
41. E. Ahmed, M. L. Tucker, and M. L. Tobe, *Inorg. Chem.* **14**, 1 (1975).
42. P. L. Kendall and G. A. Lawrance, *Aust. J. Chem.* **30**, 1841 (1977).
43. J. A. Switzer, Ph. D. dissertation, Wayne State University, 1979.
44. M. C. Weiss, B. Bursten, S.-M. Peng, and V. I. Goedken, *J. Am. Chem. Soc.* **98**, 8021 (1976).
45. N. F. Curtis, *Coord. Chem. Rev.* **3**, 3 (1968).
46. P. O. Whimp and N. F. Curtis, *J. Chem. Soc. A* **1966**, 1827 (1966).
47. N. Sadasivan, J. A. Kernohan, and J. F. Endicott, *Inorg. Chem.* **6**, 770 (1967).
48. J. A. Kernohan and J. F. Endicott, *J. Am. Chem. Soc.* **91**, 6977 (1969).
49. J. F. Endicott, N. A. P. Kane-Maguire, D. P. Rillema, and T. S. Roche, *Inorg. Chem.* **12**, 1818 (1973).
50. P. O. Whimp, M. F. Bailey, and N. F. Curtis, *J. Chem. Soc. A* **1971**, 3397 (1971).

51. N. A. P. Kane-Maguire, J. F. Endicott, and D. P. Rillema, *Inorg. Chim. Acta* **6**, 443 (1972).
52. T. P. Dasgupta, *Inorg. Chim. Acta* **20**, 33 (1976).
53. D. J. Francis and R. B. Jordan, *Inorg. Chem.* **11**, 461 (1972).
54. N. A. P. Kane-Maguire and J. F. Endicott, unpublished work.
55. T. P. Dasgupta and G. M. Harris, *J. Am. Chem. Soc.* **99**, 2490 (1977).
56. X-ray structural work by R. Bau, cited in reference 27.
57. P. G. Lenhart and D. C. Hodgkin, *Nature (London)* **192**, 937 (1961).
58. D. C. Hodgkin, *Science* **150**, 979 (1965).
59. R. Bonnett, *Chem. Rev.* **63**, 573 (1963).
60. J. M. Pratt, *The Inorganic Chemistry of Vitamin B_{12}*, Academic Press, New York (1972).
61. J. M. Pratt and D. J. Craig, *Adv. Organomet. Chem.* **11**, 332 (1973).
62. D. G. Brown, *Prog. Inorg. Chem.* **18**, 177 (1973).
63. H. A. O. Hill, in: *Inorganic Biochemistry* (G. Eichhorn, ed.), Chapter 30, p. 1067, Elsevier, New York (1973).
64. R. H. Ables and D. Dolphin, *Acc. Chem. Res.* **9**, 114 (1976).
65. B. M. Babior (Ed.), *Cobalamin*, Wiley-Interscience, New York (1975).
66. R. H. Ables, in: *Biological Aspects of Inorganic Chemistry* (A. W. Addison, W. R. Cullen, D. Dolphin, and B. R. James, eds.), Wiley-Interscience, New York (1977).
67. B. M. Babior, *Acc. Chem. Res.* **8**, 376 (1975).
68. J. S. Krouwer and B. M. Babior, *Mol. Cell. Biochem.* **15**, 89 (1977).
69. G. N. Schrauzer, *Angew. Chem. Int. Ed. Engl.* **15**, 417 (1976); *Angew. Chem. Int. Ed. Engl.* **16**, 233 (1977).
70. B. T. Golding, in: *Comprehensive Organic Chemistry* (E. Haslam, ed.), Vol. 5, Pergamon Press, New York (1978).
71. R. J. Guschl and T. L. Brown, *Inorg. Chem.* **13**, 959 (1974).
72. J. F. Endicott, D. J. Halko, W. M. Butler, and M. D. Glick, manuscript in preparation; see also Proceedings XVII International Conference on Coordination Chemistry, Dublin, Ireland, August 1974, paper 3.43.
73. J. H. Espenson and R. Russell, *Inorg. Chem.* **13**, 7 (1974).
74. D. Thusius, *J. Am. Chem. Soc.* **93**, 2629 (1971).
75. G. C. Hayward, H. A. O. Hill, J. M. Pratt, and R. J. P. Williams, *J. Chem. Soc. A* **1971**, 196 (1971).
76. J. F. Endicott, J. Lilie, J. M. Kuszaj, B. S. Ramaswamy, W. G. Schmonsees, M. G. Simic, M. D. Glick, and D. P. Rillema, *J. Am. Chem. Soc.* **99**, 429 (1977).
77. J. F. Endicott and J. Lilie, unpublished observations.
78. J. F. Endicott and G. J. Ferraudi, unpublished observations.
79. T. M. Roche and J. F. Endicott, *Inorg. Chem.* **13**, 1575 (1974).
80. J. M. Wood, F. S. Kennedy, and C. G. Rosen, *Nature (London)* **220**, 173 (1968).
81. J. M. Wood and D. G. Brown, *Struct. Bonding (Berlin)* **11**, 47 (1971).
82. W. P. Ridley, L. J. Dizikes, and J. M. Wood, *Science* **197**, 329 (1977).
83. L. J. Dizikes, W. P. Ridley, and J. M. Wood, *J. Am. Chem. Soc.* **100**, 1010 (1978).
84. J. Halpern and J. P. Maher, *J. Am. Chem. Soc.* **86**, 2311 (1964).
85. Y.-T. Fanchiang, W. P. Ridley, and J. M. Wood, *J. Am. Chem. Soc.* **101**, 1442 (1979).
86. J. H. Espenson, W. R. Bushey, and M. E. Chmielewski, *Inorg. Chem.* **14**, 1302 (1975).
87. J. A. Switzer and J. F. Endicott, unpublished work.
88. S. F. Lincoln and R. J. West, *Aust. J. Chem.* **27**, 97 (1974).
89. L. L. Rusnak and R. B. Jordan, *Inorg. Chem.* **15**, 709 (1976).

90. J. M. Palmer, E. Papaconstantinou, and J. F. Endicott, *Inorg. Chem.* **8**, 1516 (1969).
91. H. W. Smith, G. W. Svetich, and E. L. Lingafelter, American Crystallographic Association Abstracts, 18, Winter, 1973.
92. T. W. Anderson, M. D. Glick, and J. F. Endicott, unpublished observations.
93. S. C. Pyke and J. F. Endicott, unpublished observations.
94. N. Sutin, in: *Inorganic Biochemistry* (G. L. Eichhorn, ed.), Chapter 19, p. 611, Elsevier, New York (1973).
95. R. G. Linck, *M. T. P. Int. Rev. Sci. Ser. I* **9**, 303 (1972).
96. R. G. Linck, *M. T. P. Int. Rev. Sci. Ser. 2* **9**, 173 (1974).
97. R. G. Linck, *Surv. Progr. Chem.* **7**, 89 (1976).
98. H. Taube, *Electron Transfer of Complex Ions in Solution*, Academic Press, New York (1970).
99. W. L. Reynolds and R. Lumry, *Mechanisms of Electron Transfer*, Ronald Press, New York (1966).
100. R. A. Marcus, *Ann. Rev. Phys. Chem.* **15**, 155 (1963).
101. A. Haim, *Acc. Chem. Res.* **8**, 264 (1975).
102. L. E. Bennett, *Progr. Inorg. Chem.* **18**, 1 (1972).
103. F. V. Lovecchio, E. S. Gore, and D. H. Busch, *J. Am. Chem. Soc.* **96**, 3110 (1974).
104. T. J. Truex and R. H. Holm, *J. Am. Chem. Soc.* **94**, 4529 (1972).
105. E. Kent Barefield, F. V. Lovecchio, N. E. Tokel, E. Ochiai, and D. H. Busch, *Inorg. Chem.* **11**, 283 (1972).
106. D. H. Busch, D. G. Pillsbury, F. V. Lovecchio, A. M. Tait, Y. Hung, S. Jackels, M. D. Rakowski, W. P. Schammel, and L. Y. Martin, in: *Electrochemical Studies of Biological Systems* (D. T. Sawyer, ed.), p. 32, A. C. S. Symposium Series, No. 38, American Chemical Society, Washington, D.C. (1977).
107. Y. Hung, L. Y. Martin, S. S. Jackels, A. M. Tait, and D. H. Busch, *J. Am. Chem. Soc.* **99**, 4029 (1977).
108. D. P. Rillema, J. F. Endicott, and E. Papaconstantinou, *Inorg. Chem.* **10**, 1739 (1971).
109. N. Takvoryan, K. Farmery, V. Katovic, F. V. Lovecchio, E. S. Gore, L. B. Anderson, and D. H. Busch, *J. Am. Chem. Soc.* **96**, 731 (1977).
110. M. Millar and R. H. Holm, *J. Am. Chem. Soc.* **97**, 6052 (1975).
111. D. C. Olson and J. Vasilevskis, *Inorg. Chem.* **8**, 1611 (1969).
112. D. G. Pillsbury and D. H. Busch, *J. Am. Chem. Soc.* **98**, 7836 (1976).
113. J. Vasilevskis and D. C. Olson, *Inorg. Chem.* **10**, 1228 (1971).
114. D. C. Olson and J. Vasilevskis, *Inorg. Chem.* **10**, 463 (1971).
115. E. R. Dockal, T. E. Jones, W. F. Sokol, R. J. Engerer, D. B. Rorabacher, and L. A. Ochrymowycz, *J. Am. Chem. Soc.* **98**, 4322 (1976).
116. H. Yokoi and A. W. Addison, *Inorg. Chem.* **16**, 1341 (1977).
117. V. F. C. McElroy, J. C. Dabrowiak, and D. J. Macero, *Inorg. Chem.* **16**, 947 (1977).
118. J. C. Dabrowiak, F. V. Lovecchio, V. L. Goedken, and D. H. Busch, *J. Am. Chem. Soc.* **94**, 5502 (1972).
119. D. C. Olson and J. Vasilevskis, *Inorg. Chem.* **11**, 980 (1972).
120. M. C. Rakowski and D. H. Busch, *J. Am. Chem. Soc.* **97**, 2570 (1975).
121. E. K. Barefield and M. T. Mocella, *Inorg. Chem.* **12**, 2829 (1973).
122. R. L. Deming, A. L. Allred, A. R. Dahl, A. W. Herlinger, and M. O. Kestner, *J. Am. Chem. Soc.* **98**, 4132 (1976).
123. B. Durham, Ph.D. dissertation, Wayne State University, 1977.
124. M. P. Liteplo and J. F. Endicott, *Inorg. Chem.* **10**, 1420 (1971).
125. H. Taube, *Can. J. Chem.* **37**, 129 (1959).
126. J. Murdoch, *J. Am. Chem. Soc.* **94**, 4410 (1972).

127. R. A. Marcus, *J. Phys. Chem.* **72**, 891 (1968).
128. B. S. Brunschwig and N. Sutin, *J. Am. Chem. Soc.* **100**, 7568 (1978).
129. C. Creutz and N. Sutin, *J. Am. Chem. Soc.* **99**, 291 (1977).
130. S. Efrima and M. Bixon, *J. Chem. Phys.* **64**, 3639 (1976).
131. E. Buhks, M. Bixon, J. Jortner, and G. Navon, *Inorg. Chem.*, submitted for publication.
132. R. P. VanDuyne and S. F. Fischer, *Chem. Phys.* **5**, 183 (1974).
133. G. J. Hoytink, in: *Chemiluminescence and Bioluminescence*, M. J. Cormier, D. M. Hercules, and J. Lee (Eds.), pp. 147–168, Plenum Press, New York (1973).
134. M. H. Ford-Smith and N. Sutin, *J. Am. Chem. Soc.* **83**, 1830 (1961).
135. G. Dulz and N. Sutin, *Inorg. Chem.* **2**, 917 (1963).
136. R. J. Campion, N. Purdie, and N. Sutin, *Inorg. Chem.* **3**, 1091 (1964).
137. R. J. Campion, C. F. Deek, P. King, and A. L. Wahl, *Inorg. Chem.* **6**, 672 (1967).
138. H. C. Stynes and J. A. Ibers, *Inorg. Chem.* **10**, 2304 (1971).
139. D. P. Rillema, J. F. Endicott, and R. C. Patel, *J. Am. Chem. Soc.* **94**, 394 (1972).
140. D. P. Rillema and J. F. Endicott, *Inorg. Chem.* **11**, 2361 (1972).
141. D. P. Rillema and J. F. Endicott, *J. Am. Chem. Soc.* **94**, 8711 (1972).
142. M. Chou, C. Creutz, and N. Sutin, *J. Am. Chem. Soc.* **99**, 5615 (1977).
143. A. Ekstrom, A. B. McLaren, and E. Smythe, *Inorg. Chem.* **14**, 2899 (1975).
144. C. Creutz and H. Taube, *J. Am. Chem. Soc.* **91**, 3988 (1969).
145. R. W. Collahan, T. J. Meyer, and G. M. Brown, *J. Am. Chem. Soc.* **96**, 7829 (1976).
146. R. C. Young, F. R. Keene, and T. J. Meyer, *J. Am. Chem. Soc.* **98**, 286 (1976).
147. S. Isied and H. Taube, *J. Am. Chem. Soc.* **95**, 8198 (1973).
148. D. Gaswick and A. Haim, *J. Am. Chem. Soc.* **96**, 7845 (1974).
149. H. Fischer, G. M. Tom, and H. Taube, *J. Am. Chem. Soc.* **98**, 5512 (1976).
150. K. Rieder and H. Taube, *J. Am. Chem. Soc.* **99**, 7891 (1977).
151. J. Halpern and L. E. Orgel, *Discuss. Faraday Soc.* **29**, 32 (1960).
152. P. George and J. S. Griffith, *Enzymes* **1**, 347 (1959).
153. R. A. Marcus, *Discuss. Faraday Soc.* **29**, 32 (1960).
154. N. S. Biradar, D. R. Stranks, and M. S. Vaidya, *Trans. Faraday Soc.* **58**, 2421 (1962).
155. J. F. Endicott and H. Taube, *J. Am. Chem. Soc.* **86**, 1686 (1964).
156. T. J. Meyer and H. Taube, *Inorg. Chem.* **7**, 2369 (1968).
157. I. I. Creaser, J. Mac B. Harrowfield, A. J. Herlt, A. M. Sargeson, J. Springborg, R. J. Geue, and M. R. Snow, *J. Am. Chem. Soc.* **99**, 3181 (1977).
158. M. D. Glick, J. M. Kuszaj, and J. F. Endicott, *J. Am. Chem. Soc.* **95**, 5097 (1973).
159. M. D. Glick, W. G. Schmonsees, and J. F. Endicott, *J. Am. Chem. Soc.* **96**, 5661 (1974).
160. B. Durham, T. J. Anderson, J. A. Switzer, J. F. Endicott, and M. D. Glick, *Inorg. Chem.* **16**, 271 (1977).
161. A. M. Sargeson, private communication.
162. D. R. Boston and N. J. Rose, *J. Am. Chem. Soc.* **95**, 4163 (1973).
163. G. A. Zabrzewski, C. Ghilardi, and E. C. Lingafelter, *J. Am. Chem. Soc.* **93**, 4411 (1971).
164. R. A. Marcus and N. Sutin, *Inorg. Chem.* **14**, 213 (1975).
165. L. E. Bennett and J. C. Sheppard, *J. Phys. Chem.* **66**, 1275 (1962).
166. T. J. Conocchioli, G. H. Nancollas, and N. Sutin, *J. Am. Chem. Soc.* **86**, 1453 (1964).
167. E. L. Yee, R. J. Cave, K. L. Guyer, P. D. Tyma, and M. J. Weaver, *J. Am. Chem. Soc.* **101**, 1131 (1979).

168. T. J. Anderson, M. D. Glick, S. C. Pyke, and J. F. Endicott, unpublished observations.
169. M. D. Glick, D. P. Gavel, L. L. Diaddario, and D. B. Rorabacher, *Inorg. Chem.* **15**, 1190 (1976).
170. E. R. Dockel, L. L. Diaddario, M. D. Glick, and D. B. Rorabacher, *J. Am. Chem. Soc.* **99**, 4530 (1977).
171. D. P. Fay and N. Sutin, *Inorg. Chem.* **9**, 1291 (1970).
172. R. G. Wilkins and R. Yelin, *J. Am. Chem. Soc.* **89**, 5496 (1967).
173. B. Durham, J. F. Endicott, C. L. Wong, and D. P. Rillema, *J. Am. Chem. Soc.* **101**, 847 (1979).
174. J. F. Endicott and C. L. Wong, work in progress.
175. A. M. Tait, M. Z. Hoffman, and E. Hayon, *J. Am. Chem. Soc.* **98**, 86 (1976).
176. K. Nakamoto, *Infrared Spectra of Inorganic and Coordination Compounds*, 2nd ed., Wiley-Interscience, New York (1970).
177. J. A. Stritar and H. Taube, *Inorg. Chem.* **8**, 2281 (1969).
178. J. H. Espenson and J. S. Shveima, *J. Am. Chem. Soc.* **95**, 4468 (1973).
179. J. H. Espenson and T. D. Sellers, *J. Am. Chem. Soc.* **96**, 94 (1974).
180. J. M. Ciskowski, C. L. Wong, and J. F. Endicott, work in propress.
181. J. Wilshire and D. T. Sawyer, *Acc. Chem. Res.* **12**, 105 (1979).
182. R. G. Wilkins, *Adv. Chem. Ser.* **100**, 111 (1971).
183. F. Basolo, B. M. Hoffman, and J. A. Ibers, *Acc. Chem. Res.* **8**, 384 (1975).
184. G. McLendon and A. E. Martell, *Coord. Chem. Rev.* **19**, 1 (1976).
185. J. Valentine, *Chem. Rev.* **73**, 235 (1973).
186. J. P. Collman, *Acc. Chem. Res.* **10**, 265 (1977).
187. L. Vaska, *Acc. Chem. Res.* **9**, 175 (1976).
188. M. J. Carter, L. M. Englehardt, D. P. Rillema, and F. Basolo, *Chem. Soc. Chem. Commun.* **1973**, 810 (1973).
189. M. J. Carter, D. P. Rillema, and F. Basolo, *J. Am. Chem. Soc.* **96**, 392 (1974).
190. G. McClendon and M. Mason, *Inorg. Chem.* **17**, 362 (1978).
191. R. Dreos, G. Tauzhor, G. Costa, and M. Green, *J. Chem. Soc.* **1975**, 2329 (1975).
192. B. Bosnich, C. K. Poon, and M. L. Tobe, *Inorg. Chem.* **5**, 1514 (1966).
193. M. C. Weiss and V. L. Goedken, *J. Am. Chem. Soc.* **98**, 3389 (1976).
194. J. A. Switzer and J. F. Endicott, work in progress.
195. V. L. Goedken, *J. Chem. Soc. Chem. Commun.* **1972**, 207 (1972).
196. B. C. Lane, J. E. Lester, and F. Basolo, *J. Chem. Soc. Chem. Commun.* **1971**, 1618 (1971).
197. D. F. Mahoney and J. K. Beattie, *Inorg. Chem.* **12**, 2561 (1973).
198. G. M. Brown, T. R. Weaver, F. R. Keene, and T. J. Meyer, *Inorg. Chem.* **15**, 190 (1976).
199. C. J. Hipp, L. F. Lindoy, and D. H. Busch, *Inorg. Chem.* **11**, 1988 (1972).
200. N. F. Curtis, *Chem. Commun.* **1966**, 882 (1966).
201. E. K. Barefield and D. H. Busch, *Inorg. Chem.* **10**, 108 (1970).
202. J. C. Dabrowiak and D. H. Busch, *Inorg. Chem.* **14**, 1881 (1975).
203. A. M. Tait and D. H. Busch, *Inorg. Chem.* **15**, 197 (1976).
204. V. L. Goedken and D. H. Busch, *J. Am. Chem. Soc.* **94**, 7355 (1972).
205. E. K. Barefield and M. T. Mocella, *J. Am. Chem. Soc.* **97**, 4238 (1975).
206. N. F. Curtis, *Coord. Chem. Rev.* **3**, 1 (1968).
207. ACS, Chicago, Ill., August 1975, INOR 149; private communication from P. M. Henry.

208. D. C. Olson and J. Vasilevskis, *Inorg. Chem.* **10**, 463 (1971).
209. J. M. Palmer and J. F. Endicott, unpublished observations.
210. S. C. Tang and R. H. Holm, *J. Am. Chem. Soc.* **97**, 3359 (1975).
211. H. O. House, in: *Modern Synthetic Reactions*, 2nd ed. (W. A. Benjamin, ed.), pp. 37–44, New York (1971).
212. D. Black and A. J. Hartshorn, *Tetrahedron Lett.* **25**, 2157 (1974).
213. E. A. Braude, L. M. Jackman, R. P. Linstead, and J. S. Shannon, *J. Chem. Soc.* **1960**, 3116 (1960).
214. E. A. Brande, L. M. Jackman, R. P. Linstead, and G. Lowe, *J. Chem. Soc.* **1960**, 3123, 3133 (1960).
215. R. Mechoulom, B. Yaginitinsky, and Y. Gaoni, *J. Am. Chem. Soc.* **90**, 2418 (1968).
216. J. A. Cunningham and R. E. Sievers, *J. Am. Chem. Soc.* **95**, 7183 (1973).
217. F. Z. McElroy and J. C. Dabrowiak, *J. Am. Chem. Soc.* **98**, 7112 (1976).
218. R. Yamada, T. Kato, S. Shimizu, and S. Fukai, *Biochim. Biophys. Acta* **117**, 13. (1966).
219. J. DelGaudio and G. N. Lamar, *J. Am. Chem. Soc.* **98**, 3014 (1976).
220. L. J. Radonovich, A. Bloom, and J. L. Hoard, *J. Am. Chem. Soc.* **94**, 2073 (1972).
221. C. Creutz and N. Sutin, *Proc. Natl. Acad. Sci.* **72**, 2858 (1975).
222. G. Nord and O. Wernberg, *J. Chem. Soc. Dalton Trans.* **1975**, 845 (1975).
223. D. M. Hercules, *Acc. Chem. Res.* **2**, 301 (1969).
224. R. D. Gillard, *Coord. Chem. Rev.* **16**, 67 (1975).
225. N.F. Curtis, *J. Chem. Soc. A* **1971**, 2834 (1971).
226. N. F. Curtis, *J. Chem. Soc. C* **1967**, 1979 (1967).
227. N. F. Curtis, *J. Chem. Soc.* **1964**, 2644 (1964).
228. J. L. Karn and D. H. Busch, *Inorg. Chem.* **8**, 1149 (1969).
229. V. Katovic, L. T. Taylor, F. L. Urbach, W. H. White, and D. H. Busch, *Inorg. Chem.* **11**, 479 (1972).
230. L. G. Warner and D. H. Busch, *J. Am. Chem. Soc.* **91**, 4092 (1969).
231. A. M. Tait and D. H. Busch, *Inorg. Chem.* **16**, 966 (1977).
232. C. J. Hipp and D. H. Busch, *Inorg. Chem.* **12**, 894 (1973).
233. K. Bowman, D. P. Riley, D. H. Busch, and P. W. R. Corfield, *J. Am. Chem. Soc.* **97**, 5036 (1975).
234. P. W. R. Corfield, J. D. Mokren, C. J. Hipp, and D. H. Busch, *J. Am. Chem. Soc.* **95**, 4466 (1973).
235. C. A. Sprecher and A. D. Zuberbuhler, *Angew. Chem.* **89**, 185 (1977).
236. F. L. Urbach, U. Knopp, and A. D. Zuberbuhler, *Helv. Chim. Acta* **61**, 1097 (1978).
237. F. Wagner and D. K. Barefield, *Inorg. Chem.* **15**, 408 (1976).
238. E. K. Barefield and F. Wagner, *Inorg. Chem.* **12**, 2435 (1973).
239. L. T. Taylor, F. L. Urbach, and D. H. Busch, *J. Am. Chem. Soc.* **91**, 1972 (1969).
240. V. Katovic, L. T. Taylor, and D. H. Busch, *J. Am. Chem. Soc.* **91**, 2122 (1969).
241. J. F. Endicott, in: *Concepts in Inorganic Photochemistry* (A. W. Adamson and P. D. Fleischauer, eds.), Chapter 3, p. 81, Wiley-Interscience, New York (1978).
242. V. M. Berdnokov and N. M. Bazhim, *Russ. J. Phys. Chem.* **44**, 395 (1970).
243. A. M. Tait, M. Z. Hoffman, and E. Hayon, *Inorg. Chem.* **15**, 934 (1976).
244. A. M. Tait, Z. M. Hoffman, and E. Hayon, *Int. J. Radiat. Phys. Chem.* **8**, 691 (1976).
245. R. Blackburn, A. Y. Erkol, G. O. Phillips, and A. J. Swallow, *J. Chem. Soc. Faraday Trans. I* **70**, 1693 (1974).
246. S. D. Malone and J. F. Endicott, *J. Phys. Chem.* **76**, 2223 (1972).
247. A. T. Thornton and G. S. Laurence, *J. Chem. Soc. Dalton Trans.* **1973**, 1632 (1973).

248. A. T. Thornton and G. S. Laurence, *J. Chem. Soc. Dalton Trans.* **1973**, 804 (1973).
249. P. Maruthamuthu, L. K. Patterson, and G. Ferraudi, *Inorg. Chem.* **17**, 3157 (1978).
250. M. Jaacobi, D. Meyerstein, and J. Lilie, *Inorg. Chem.* **18**, 429 (1979).
251. C. Y. Mok and J. F. Endicott, *J. Am. Chem. Soc.* **100**, 123 (1978).
252. J. F. Endicott and G. J. Ferraudi, *J. Am. Chem. Soc.* **99**, 243 (1977).
253. H. Elroi and D. Meyerstein, *J. Am. Chem. Soc.* **100**, 5540 (1978).
254. D. P. Rillema and J. F. Endicott, *Inorg. Chem.* **15**, 1459 (1976).
255. M. O. Kestner and A. L. Allred, *J. Am. Chem. Soc.* **94**, 7189 (1972).
256. V. Balzani, L. Moggi, M. F. Manfrin, F. Bolletta, and M. G. Gleria, *Science* **189**, 852 (1975).
257. G. J. Ferraudi and J. F. Endicott, unpublished work.
258. G. M. Brown, C. Creutz, B. Brunschwig, J. F. Endicott, and N. Sutin, *J. Am. Chem. Soc.* **101**, 1298 (1979).
259. V. Balzani and V. Carassiti, *Photochemistry of Coordination Compounds*, Academic Press, New York (1970).
260. A. W. Adamson and P. D. Fleischauer (eds.), *Concepts in Inorganic Photochemistry*, Wiley-Interscience, New York (1975).
261. C. Kutal and A. W. Adamson, *J. Am. Chem. Soc.* **93**, 5581 (1971).
262. C. Kutal and A. W. Adamson, *Inorg. Chem.* **12**, 1454 (1973).
263. C. Kutal and A. W. Adamson, *Inorg. Chem.* **12**, 1990 (1973).
264. R. A. Pribush, C.-K. Poon, C. M. Bruce, and A. W. Adamson, *J. Am. Chem. Soc.* **96**, 3027 (1974).
265. A. D. Kirk, *J. Am. Chem. Soc.* **93**, 283 (1971).
266. E. Zinato, Chapter 4 in reference 260.
267. P. D. Fleischauer, A. W. Adamson, and G. Sartori, *Progr. Inorg. Chem.* **17**, 1 (1972).
268. J. Sellan and R. Rumfeldt, *Can. J. Chem.* **54**, 519 (1976).
269. J. Sellan and R. Rumfeldt, *Can. J. Chem.* **54**, 1061 (1976).
270. J. I. Zink, *J. Am. Chem. Soc.* **94**, 8039 (1972).
271. J. I. Zink, *Inorg. Chem.* **12**, 1018 (1973).
272. J. I. Zink, *Inorg. Chem.* **12**, 1957 (1973).
273. L. G. Vanquickenborne and A. Ceulemans, *J. Am. Chem. Soc.* **100**, 475 (1978).
274. M. J. Incorvia and J. I. Zink, *Inorg. Chem.* **16**, 3161 (1977).
275. M. J. Incorvia and J. I. Zink, *Inorg. Chem.* **17**, 2250 (1978).
276. G. J. Ferraudi, *Inorg. Chem.*, submitted for publication.
277. P. C. Ford, *Inorg. Chem.* **14**, 440 (1975).
278. J. F. Endicott and G. J. Ferraudi, *J. Phys. Chem.* **80**, 949 (1976).
279. M. A. Bergkamp, R. J. Watts, P. C. Ford, J. Brannon, and D. Magde, *Chem. Phys. Lett.* **59**, 125 (1978).
280. S. D. Malone and J. F. Endicott, unpublished work.
281. D. W. Reichgott and N. J. Rose, *J. Am. Chem. Soc.* **99**, 1813 (1977).
282. G. Ferraudi and E. V. Srisankar, *Inorg. Chem.* **17**, 3164 (1978).
283. G. Ferraudi, *Inorg. Chem.*, submitted for publication.
284. G. J. Ferraudi and J. F. Endicott, *Inorg. Chem.* **16**, 2762 (1977).
285. G. Ferraudi, *Inorg. Chem.* **17**, 1741 (1978).
286. H. P. C. Hogenkamp, *Biochemistry* **5**, 417 (1966).
287. H. P. C. Hogenkamp, D. J. Vergamini, and N. A. Matwiyoff, *J. Chem. Soc. Dalton Trans.* **1975**, 2628 (1975).
288. J. M. Pratt, *J. Chem. Soc.* **1964**, 5154 (1964).
289. J. M. Pratt and B. R. D. Whitear, *J. Chem. Soc.* **1971**, 252 (1971).

290. G. N. Schrauzer, *Acc. Chem. Res.* **1**, 97 (1968).
291. G. N. Schrauzer, J. W. Sibert, and R. J. Windgassen, *J. Am. Chem. Soc.* **90**, 6681 (1968).
292. G. N. Schrauzer, L. P. Lee, and J. W. Sibert, *J. Am. Chem. Soc.* **92**, 2997 (1970).
293. C. Giannotti and J. R. Bolton, *J. Organomet. Chem.* **80**, 379 (1974).
294. C. Giannotti, G. Merle, C. Fontaine, and J. R. Bolton, *J. Organomet. Chem.* **91**, 357 (1975).
295. C. Giannotti, C. Fontaine, and B. Septe, *J. Organomet. Chem.* **71**, 107 (1974).
296. C. Giannotti and J. R. Bolton, *J. Organomet. Chem.* **110**, 383 (1976).
297. R. T. Taylor, L. Smucker, M. L. Hanna, and J. Gill, *Arch. Biochim. Biophys.* **156**, 521 (1973).
298. C. Y. Mok and J. F. Endicott, *J. Am. Chem. Soc.* **99**, 1276 (1977).
299. B. T. Golding, T. J. Kemp, P. J. Sellers, and E. Nocchi, *J. Chem. Soc. Dalton Trans.* **1977**, 1266 (1977).
300. H. Flohr, W. Pannhorst, and J. Retey, *Angew. Chem. Int. End.* **15**, 561 (1976).
301. J. F. Endicott, *Inorg. Chem.* **16**, 494 (1977).
302. J. M. Ciskowski and J. F. Endicott, work in progress.
303. A. W. Adamson, *Discuss. Faraday Soc.* **29**, 163 (1960).
304. J. F. Endicott, G. J. Ferraudi, and J. R. Barber, *J. Phys. Chem.* **79**, 630 (1975).
305. T. Netzel and J. F. Endicott, *J. Am. Chem. Soc.*, in press.
306. T. Toraya, K. Ushio, S. Fubui, and H. P. C. Hogenkamp, *J. Biol. Chem.* **252**, 963 (1977).

Metal Complexes of Phthalocyanines

Lawrence J. Boucher

1. Introduction

The special nature of the macrocycle phthalocyanine and its metal complexes has been known for over 40 years.[1-3] Since then the unique physical and chemical properties of this class of coordination compounds have been exploited from both the practical as well as the theoretical point of view. Thus, metallophthalocyanines have important uses as commercial dyes, optical and electrical materials, and catalysts. However, this review will focus primarily on the metallophthalocyanines as coordination compounds and will discuss their physical and chemical properties. Since the literature prior to 1965 is more than adequately reviewed by a monograph,[1] a brief review,[2] and a more extensive exemplary review,[3] this paper will stress work done in the past ten years or so. Unfortunately, aside from a detailed discussion of the catalytic activity of the metallophthalocyanines, no practical uses of the materials will be discussed. This, of course, is not due to perceived unimportance or to the "applied" nature of this very active phase of phthalocyanine research. It is rather a lack of space and, more importantly, a lack of the necessary expertise that leads to this exclusion.

In recent times a compelling reason to study the detailed coordination chemistry of metallophthalocyanines has arisen. The great activity in bioinorganic chemistry has led to a veritable avalanche of work on metalloporphyrins and other naturally occurring macrocyclic coordination compounds. Thus the metallophthalocyanines are being examined in earnest and their properties compared to those of the porphyrin complexes in hopes of arriving at an understanding of how the naturally occurring macrocyclic

Lawrence J. Boucher · Department of Chemistry, Western Kentucky University, Bowling Green, Kentucky 42101.

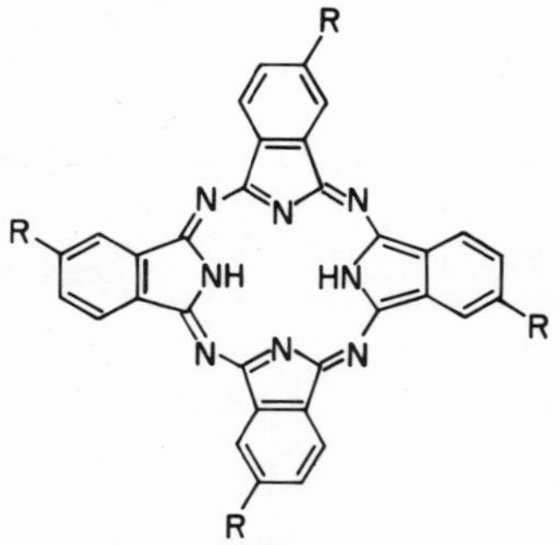

R	ABBRV.
H	PC
SO_3^-	$(PCTS)^{4-}$
$SO_2NHC_{18}H_{37}$	PCTO

FIGURE 1. Structural representation for phthalocyanine and its 4,4′,4″,4‴-tetrasubstituted derivatives.

porphyrin ligand affects the properties of the metal. Nevertheless, comparisons will be made to porphyrin complexes in this review only where it is particularly relevant to understanding the metallophthalocyanines.

The very nature of a critical review of any expanding area of chemistry requires hard and quite arbitrary decisions to be made about what topics are to be discussed and to what depth they are to be covered. This review has not attempted to be encyclopedic in its coverage but instead has quite naturally tried to be somewhat discriminating. The topics covered and the papers cited represent the more important and fruitful current areas of phthalocyanine research, in the reviewer's judgment.

A word about nomenclature is in order at this point. The metal-free phthalocyanine, H_2PC, can be deprotonated to yield the dianion PC,[4] which then may bind with a metal ion to form a complex, e.g., divalent metals form complexes of the type [M(PC)]. In this review PC will be used as a shorthand notation for the phthalocyanine ligand system as well as for its dianion. The most common derivative of the basic macrocycle is the 4,4′,4″,4‴-tetrasulfonate, abbreviated $(PCTS)^{4-}$; again note that the charge refers to the four sulfonate groups appended to the benz portion of the macrocycle. The two negative charges on the PC portion are understood. Figure 1 shows a representation of the most common phthalocyanine derivatives. Another point of terminology used here is the reference to the nitrogen atoms linking the pyrrole rings as aza nitrogens, while the others are called inner nitrogens. Other abbreviations used are listed at the end of the review.

2. Molecular Structure

Because of their rich coordination chemistry, i.e., the existence of complexes of a large number of metals and metalloids in a variety of oxidation states and coordination numbers, crystal and molecular structure determina-

TABLE 1. Selected Structural Data from X-Ray Single Crystal Structure on Metallophthalocyanines

Coordination sphere	Distance (pm)			Reference
	M–N	M–X	Δ	
[Ni(PC)]	183			6
β-[Cu(PC)]	194			7
α-[Pt(PC)]	198			8
γ-[Pt(PC)]	198			8
[Mg(PC)(H₂O)]	204	202(O)^a	50^b	11
[Zn(PC)(NH₂C₆H₁₁)]	206	218(N)	48	12
[Fe(PC)(γ-pic)₂]	192	200(N)	0	14
[Co(PC)(γ-pic)₂]	191	230(N)	0	15
[Mn₂(PC)₂(py)₂O]	196	171(O)	0	24
		215(N)		
[Si(PC)₂(OSi(CH₃)₃)₂]	192	168(O)	0	17
[Sn(PC)]	225		111	21
[Sn(PC)₂]	235		135	22
[Sn(PC)Cl₂]	205	245(Cl)	0	16
[U(PC)₂]	243		140	23

^{a} Bond length to axial ligand. Letters in parentheses show which atom the metal is bonded to.
^{b} Distance metal is out of best plane of the PC N donors (see Figure 2).

tions of PC complexes have been reported in the last ten years with increasing frequency. Selected structural data from this work are gathered in Table 1. Here we will only be concerned with the metal coordination sphere geometry and bond lengths. In general, structural parameters for the PC ligand and intermolecular packing in the solid will not be discussed. For this purpose the reader is referred to the references cited and to a recent discussion.[5] Typical coordination geometries are shown in Figure 2.

The first crystal structure of a PC complex was done by Robertson with

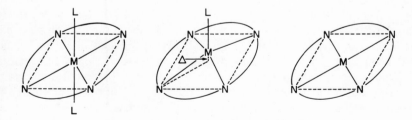

FIGURE 2. Idealized coordination spheres for metallophthalocyanines: 6-coordinate tetragonal (left), 5-coordinate square pyramidal (center), 4-coordinate square planar (right). The ring with four N donors represents the phthalocyanine ring.

[Ni(PC)].[6] The nominal coordination number of the square-planar Ni(II) as well as the Cu(II) derivatives is four. Nonetheless it has been proposed that there is a significant, albeit weak, intermolecular interaction between the metal in one molecule and an aza nitrogen in the molecules above and below it at a distance of 228 and 328 pm, respectively. The molecules themselves are planar to within 3 pm. Thus, the complexes of Ni(II) and Cu(II) as well as other first-row transition divalent derivatives [M(PC)] may be viewed as tetragonal six-coordinate structures in the solid. On the other hand, the polymorphs of the Pt(II) complex show axial Pt–N distances of 357 and 370 pm in the α and γ forms,[8] respectively. Thus these solids contain true square-planar complexes, as expected from the coordination properties of Pt(II). It is interesting to note that the Cu–N and Ni–N bond lengths are 13 and 4 pm shorter than the respective bond lengths in a representative [Cu(TPP)][9] and Ni(II) mesoporphyrin IX dimethylester.[10]

Typical five-coordinate structures have been found for the metal coordination sphere in [Mg(PC)H$_2$O]·2py*[11] and [Zn(PC)(NH$_2$C$_6$H$_{11}$)]-$\frac{1}{2}$NH$_2$C$_6$H$_{11}$.[12] The coordination sphere can best be described as square pyramidal with the metal atom above the best plane of the four PC nitrogens, by a distance Δ, toward the fifth axial ligand. The Mg–N and Mg–O bond lengths are 3 and 8 pm, respectively, shorter than for the analogous bonds in [Mg(TPP)H$_2$O].[13] On the other hand, Δ is larger for the PC complex by 23 pm. The Zn–N(PC) bond lengths are only 1 pm shorter than the [Zn(TPPy)Py] while Δ is significantly larger, by 15 pm, for the PC complex. Like the Mg(II) complex, the Zn(II) complex shows a ruffled nonplanar PC ring system with pairs of square-pyramidal molecules stacked with their basal planes parallel to each other at a distance of 324 pm and 326 pm, respectively.

The six-coordinate structures for [Fe(PC)(γ-pic)$_2$]·2(γ-pic),[14] [Co(PC)(γ-pic)$_2$]·2(γ-pic),[15] [Sn(PC)Cl$_2$],[16] and [Si(PC)(OSi(CH$_3$)$_3$)$_2$][17] show a tetragonal coordination geometry with the metal in the plane of the four PC nitrogen donors and with two equivalent axial bonds at a different distance from the in-plane bonds. The Co(II) and Fe(II) structures show shorter M–N(PC) bonds by 8 pm than for the analogous porphyrin complexes [Fe(TPP)(pip)$_2$][18] and [Co(TPP)(γ-pic)$_2$].[19] In the case of the Co(II) complex the axial Co–N bond length is 9 pm shorter in the PC complex than the porphyrin derivative. The PC ligand is planar for the Co(II), Fe(II), and Si(IV) complexes, while it is substantially crumpled in a stepped deformation (nonplanar) for [Sn(PC)Cl$_2$]. This effect results from the accommodation of the large Sn(IV) atom in the plane of the four PC nitrogen donors. This causes an expansion and distortion of the PC ligand. In fact, the Sn–N bond is the longest in-plane metal–N(PC) bond known. Nonetheless this bond is

* See list of abbreviations at the end of this chapter.

still 4 pm shorter than for the corresponding porphyrin complex [Sn(OEP)Cl$_2$].[20] On the other hand the Sn–Cl axial bond length is the same for the two complexes.

Several unusual structural types are noted for the PC derivatives that are yet to be found for the porphyrins. The molecular structure of the Sn(II) complex, [Sn(PC)], can be described as Ψ-square pyramidal,[21] i.e., the four N(PC) donor nitrogens occupy the basal plane while the metal atom is out of this plane by 111 pm and the fifth axial position is occupied by a stereochemically active lone pair of electrons. The PC ligand is distorted from planarity and is saucer shaped, being domed toward the metal. The Sn–N bond length is quite long. In fact, the overlarge size of the Sn(II) ion may account for the rather unusual out-of-plane "four-coordinate" structure. The coordination geometry about the metal in the Sn(IV) complex [Sn(PC)$_2$][22] can be described as showing a roughly square antiprismatic arrangement of the eight nitrogen donor atoms about the metal. The Sn–N bond length is fairly long in the eight-coordinate structure especially in comparison to that in [Sn(PC)Cl$_2$]. The metal atom is substantially out of the plane and the PC ligands adopt a domed configuration (toward the metal) in [Sn(PC)$_2$]. A similar eight-coordinate structure is noted for the U(IV) complex [U(PC)$_2$].[23]

The molecular structure of the dimer [Mn(PC)py]$_2$O·2py[24] is noteworthy because of its rather short Mn–O bond, which is comparable to that of the strong Si–O bond. This may be indicative of a strong Mn–O–Mn interaction in the Mn(III) complex. Another interesting feature of the structure is the fact that the Mn(III) atom is in the plane of the macrocycle, unlike the common out-of-plane structures for the porphyrin derivatives.[25] The Mn(III) structure is the only crystal structure done on a trivalent metal–PC complex. It would be interesting to see whether the five-coordinate square-pyramidal structure that is so common for porphyrins holds for the PC complexes like [M(PC)X].

A recent report on the crystal structure of the supposed [UO$_2$(PC)] complex shows that this material is actually a macrocyclic derivative of UO$_2^{2+}$ which contains five iminoisoindoline rings![5] Finally, a neutron diffraction study of metal-free phthalocyanine, H$_2$PC, has been reported.[26] The H$_2$PC molecules are planar and there is a half hydrogen atom noted on each inner nitrogen.

A comparison of the selected structural parameters of the PC and analogous porphyrin metal complexes points up several important differences between the two classes of macrocyclic compounds. In the four-coordinate square-planar and six-coordinate tetragonal structures, the in-plane M–N bonds are shorter for the PC compounds by 7 ± 3 pm. However, this difference is smaller, 2 ± 1 pm in the less stereochemically demanding five-coordinate square-pyramidal structure. On the other hand, the out-of-plane displacement is significantly larger, 15–23 pm greater for this type of PC complex when compared to the porphyrin analogs. Finally, the axial bonds in the PC complexes are generally equal to or somewhat shorter in the PC than in the porphyrin complexes. Although there can be substantial non-planar distortions when the metal coordination properties require it, in general the PC macrocycle shows less distortion from planarity than the porphyrins. Thus the phthalocyanine macrocycle is a rather rigid macrocycle with a smaller hole size than the porphyrins. These properties apparently lead to both enhanced metal–PC σ and π bonding in the PC complexes (see below). Another important consequence is the dramatic response in bond length changes to d-orbital population. For example, adding an electron to the vacant strongly σ antibonding $d_{x^2-y^2}$ orbital in going from Ni(II) to Cu(II) increases the in-plane M–N(PC) bond lengths by 11 pm. Similarly, adding an electron to the vacant and strongly axial σ antibonding orbital d_{z^2} in going from Fe(II) to Co(II) increases the axial bond length M–N by 30 pm!

It must, however, also be recognized that the coordination properties of the two macrocycles, PC and porphyrins, differ in degree and not in kind and that the coordination characteristics of the individual metal are an equally important determining structural factor for both macrocycles. Thus, there are many parallels in structural characteristics between porphyrins and phthalocyanines. For example five-coordinate structures are preferred for Mg(II) and Zn(II), pseudo-four-coordinate structures are preferred for Cu(II) and Ni(II), while six-coordinate structures are common (but not exclusive for the porphyrins) for Co(II) and Fe(II).

3. Electronic Structure

Theoretical molecular orbital studies have been carried out for both [M(PC)] [27–29] and metal-free H_2PC. [30,31] Extended Hückel calculations on [M(PC)], M = Mg(II), Zn(II), Ni(II), Fe(II), and Mn(II) are particularly noteworthy. [27] The results not only are able to reproduce the PC ligand $\pi \rightarrow \pi^*$ electronic spectra, but also give rational assignments for extra absorption bands as charge transfer transitions. Of equal importance, the calculations give the ground-state d-orbital configuration of the transition-metal ions as well as the electronic distribution in the molecule. The latter data can, of course, be interpreted in terms of metal–PC σ and π bonding. The qualitative positions of the PC and metal electronic energy levels are shown in Figure 3 for a typical case Ni(PC).

The M(PC) complexes are intensely colored and show a number of strong characteristic absorptions in the visible and ultraviolet region. The electronic spectra arise primarily from $\pi \rightarrow \pi^*$ transitions within the delocalized PC ring system. The frequencies of these transitions are generally independent of

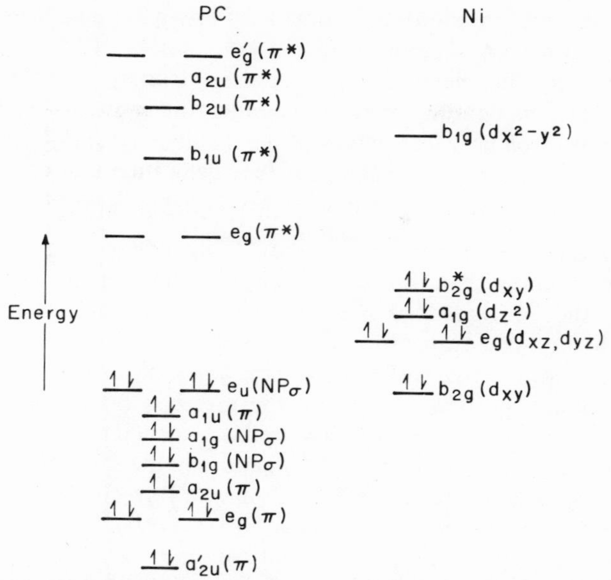

FIGURE 3. Qualitative molecular orbital energy levels of [Ni(PC)] (includes only highest filled and lowest empty bonding and antibonding MOs). The PC ligand orbitals are on the left while the Ni(II) d orbitals are on the right.

the metal ion. Table 2 gives the approximate energies of the absorptions along with the molecular orbital (MO) designation for the transition.[32] Further, the $Np_\sigma \rightarrow e_g$ (π^*) PC transition is predicted to occur near the B band and thus is responsible for the broadening of the absorption maxima in this region. Additional absorptions that are seen in the spectra of the [M(PC)] complexes have been assigned to metal to ligand charge transfer transitions. There are several reasons why M(PC) complexes in general have richer electronic spectra than for porphyrins: greater delocalization of metal orbitals in PC complexes which would enhance the intensity of charge transfer transitions involving $d\pi$

TABLE 2. Typical [M(PC)] Ligand $\pi \rightarrow \pi^*$ Transitions

Label	Transition	Maximum (nm)	Relative intensity
Q	$a_{1u} \rightarrow e_g$	658	1
B	$a_{2u} \rightarrow e_g$	320	4
N	$a'_{2u} \rightarrow e_g$	275	1
L	$a_{1u} \rightarrow e_g'$	245	0.1

metal orbitals, and an additional number of low-lying empty π^* orbitals to give a greater number of allowed transitions.

As expected, the metal $d_{x^2-y^2}$ orbital is strongly σ antibonding with respect to the sigma bonding pairs on the inner donor nitrogens. There is also significant out-of-plane π interaction of the metal $d\pi$ (d_{xz}, d_{yz}) and PC e_g (π^*) orbitals. Thus the strong metal–ligand covalent interactions are synergic, PC $\overset{\pi}{\underset{\sigma}{\rightleftharpoons}}$ M, in nature. In general the calculations also confirm the long-held notion that the smaller hole size in the PC ligands gives rise to a greater covalent interaction than for the porphyrins. The enhanced ligand field strength of the PC ligands is predicted to give unusual intermediate spin states for the first-row transition metals, as shown in Table 3.

The calculations give electronic charge populations which show that the magnitude of the positive charge on the metal is in the order Mg(II) > Zn(II) > Co(II) > Cu(II) > Ni(II) > Fe(II) > Mn(II). Of course, the magnitude of the negative charge on the PC ligand is in the same order. The charge distribution reflects the extent of PC σ donation [low for Mg(II) and Zn(II)] and π back donation [higher for Co(II), Ni(II), Fe(II), and Mn(II)].

The predicted ground-state d-orbital configuration and spin state for a number of divalent metals are given in Table 3. Calculations indicate that in addition to the σ and π interactions of the b_{1g}, a_{1g}, e_g metal orbitals with the appropriate PC orbitals, there is substantial σ interaction of the usually nonbonding b_{2g} metal orbital with the b_{2g} aza nitrogen lone-pair orbital for all the metals but Mn(II) in Table 3. Thus the metal configuration given in the table includes the b_{2g} and b_{2g}^* orbitals that arise from this interaction. Although the intermediate or low-spin states that are predicted by the theory are observed, the predicted orbital configurations do not agree with those obtained experimentally. The problem here seems to lie in the overestimation of the b_{2g} metal–aza nitrogen interaction.

TABLE 3. Ground-State d-Orbital Configuration and Spin State from Molecular Orbital Calculations for [M(PC)]

M(II)	Configuration	S
Mn	$b_{2g}^2 e_g{}^3 a_{1g}^1 (b_{2g}^*)^1$	3/2
Fe	$b_{2g}^2 e_g{}^4 a_{1g}^1 (b_{2g}^*)^1$	1
Co	$b_{2g}^2 e_g{}^4 a_{1g}^2 (b_{2g}^*)^1$	1/2
Ni	$b_{2g}^2 e_g{}^4 a_{1g}^2 (b_{2g}^*)^2$	0
Cu	$b_{2g}^2 e_g{}^4 a_{1g}^2 b_{1g}^1$	1/2

4. Spectral Properties

A large variety of physical techniques have been used to gather information about the molecular and electronic structure of metallophthalocyanines. Vapor phase spectra in the 200–800 nm region have been measured for [M(PC)], where M = Mg(II), Ti(IV)O, V(IV)O, Cr(II), Fe(II), Co(II), Ni(II), Cu(II), Zn(II), Sn(IV)Cl$_2$, and Pb(II).[32] While the spectra are usually blue shifted from DMSO solution spectra they are not greatly dependent on the metal. Unfortunately only [VO(PC)], [Co(PC)], [Ni(PC)], [Cu(PC)], and [Zn(PC)] are completely stable in the gas phase at elevated temperatures. In general these spectra are accurately reproduced by the molecular orbital calculations as described above. MCD spectra of DMSO and chlorobenzene solutions of [Co(PC)], [Ni(PC)], [Zn(PC)], and [Fe(PC)] confirm the assignment of the major absorptions to $\pi \rightarrow \pi^*$ ligand transitions.[33]

The near-infrared absorption spectra of [M(PC)] M = Cu(II), Ni(PC), Zn(II), Fe(II), Mn(II), and Cr(II) have been measured in the vapor phase and solution.[34] The spectra are generally rich with weak to medium intensity bands whose intensity and frequency are metal dependent. The band near ~1000 nm in most of the spectra is associated with decomposition of the macrocycle (irradiation damage) while a band near 1660 nm arises from a C–H stretching overtone of the PC ligand. Other bands in the region 900–2000 nm are strongly metal dependent intramolecular transitions for only the first transition-metal series. While the bands do not appear to be *d–d* metal transitions they may be charge transfer or more likely $n(\text{aza}) \rightarrow \pi^*$ PC transitions. The fact that these bands are metal dependent may be taken as evidence that the metal ions are interacting with the aza nitrogen, as suggested by the theoretical calculations. The electronic emission spectral properties of a number of [M(PC)] complexes has been surveyed at low temperature.[35-37] While [M(PC)], M = Mg(II), Zn(II), Cd(II), Pd(II) show the expected mirror image fluorescence, phosphorescence spectra (from the molecular triplet) are obtained for M = Pt(II), Pd(II), Cu(II), Zn(II), Cd(II), Mg(II), and V(IV)O. The Cu(II) complex does not show fluorescence.

X-ray photoelectron spectroscopy gives information about the charge distribution in [M(PC)] complexes. Thus, the PC N_{1s} binding energies have been determined for a series of complexes where M = Ti(IV)O, V(IV)O, Mn(II), Fe(II), Co(II), Ni(II), Cu(II), and Zn(II).[38] The energy is not markedly dependent on the metal and the two types of nitrogen atoms in the PC, inner and aza, are not resolved. The metal $2P_{3/2}$ binding energies have also been determined and the values are found to be higher for the PC ligands than for the corresponding porphyrins.[38,39] This is due to the higher positive charge on the metal, which may arise from an enhanced M $\rightarrow$ PC π charge donation. The measured heats of sublimation,[40] which are in the range −69 to −49 kcal mol^{-1}, are highest for Fe(II) and Co(II). The values are thought to reflect the contribution of the metal to the σ, π system of the PC ligand, i.e., intermolecular interactions with neighboring molecules in the solid are governed by the residual positive charge on the metal and negative charge in the PC ligand particularly that which is localized on the aza nitrogen.

Electron spin resonance measurements shed light on the electronic structure for the paramagnetic [M(PC)] complexes. For example these measurements for [Cu(PC)] show that the unpaired electron is in a b_{1g} metal orbital and that there is significant σ and π bonding of the metal b_{1g} and e_g orbitals with the PC sigma and $e_g(\pi^*)$ orbitals while the b_{2g} metal orbital is nonbonding.[41] In fact detailed analysis of the ESR spectra, as well as measurements of the paramagnetic anisotropy of the solid,[42] gives the d-orbital energy order $b_{1g} \gg a_{1g} > e_g > b_{2g}$ and corresponding configuration $b_{2g}^2 e_g^4 a_{1g}^2 b_{1g}^1$. For [Co(PC)] the ESR and magnetic susceptibility measurements indicate that the odd electron is in an a_{1g} orbital.[43,44] The marked solvent dependence of the ESR spectra is in accord with this result.[45] While the d-orbital configuration as obtained from experiment agrees exactly with that of MO theory for Cu(II), there is a sharp variance in the case of Co(II), i.e., the experiment shows the odd electron to be in an a_{1g}-type orbital while MO theory predicts the half-filled orbital to be of a b_{2g} type. On the other hand there is agreement between the ESR experiment and theory with respect to [VO(PC)] complex, i.e., the unpaired electron is in a b_{2g} metal d orbital.[46,47] The extent of the in-plane interaction of the metal orbital with the PC orbitals, however, has been estimated to be substantially higher from experiment than from theory. Nonetheless both agree that the orbital is not nonbonding as previously supposed.[48] Even though there are detailed disagreements between theory and experiment for these [M(PC)] complexes, the ESR spectra do indicate, like the MO theory, that there is a greater covalent interaction in PC complexes than in the porphyrin complexes. Finally, ESR spectra of $Na_4[VO(PCTS)]$ and $Na_4[Cu(PCTS)]$ in water-DMF at 77 K allow the characterization of a dimeric form of these complexes. Thus the spectra, which arise from a triplet state formation via the interaction of two $S = \frac{1}{2}$ paramagnetic ions, can be analyzed and spin Hamiltonian derived.[49]

Infrared spectra of [M(PC)] complexes have been measured in the 4000–400 cm^{-1} region and tentative assignments made.[50,51] As expected, the spectra are quite complex and show many metal-independent ligand absorptions. One attractive use of IR spectroscopy continues to be the identification of the various solid polymorphic solid modifications of [M(PC)].[52] Mass spectra of a number of [M(PC)] derivatives have been measured and the remarkable stability of the molecular ions that form noted.[53,54] In general, the spectra of [M(PC)] show mainly the molecular ions [M(PC)]$^+$ and [M(PC)]$^{2+}$. Metal dissociation and fragmentation of the PC ring are not major processes when M = Pt(II), Zn(II), Fe(II), Co(II), Cu(II), and Ni(II). For Mn(II), however, extensive fragmentation is noted, indicating an especial instability of $[Mn(PC)]^+$ and $[Mn(PC)]^{2+}$. The mass spectra of some trivalent metal complexes [M(PC)Cl], M = In(III), Al(III), Mn(III) mainly show the stable molecular ions $[M(PC)Cl]^+$, $[M(PC)Cl]^{2+}$ in addition to $[M(PC)]^+$, and $[M(PC)]^{2+}$. The spectrum of $[Nb(PC)Br_3]$ shows $[Nb(PC)Br_2]^+$ and $[Nb(PC)Br]^+$. On the other hand the spectra of [M(PC)Cl], M = Sc(III), Yb(III), and $[Ta(PC)Br_3]$ show extensive fragmentation and no parent molecular ion containing the metal. Thus the stability of the various higher-valent complexes seems to be quite dependent on the metal. The gas-phase ionization potentials of some divalent metal complexes have been measured by electron impact mass spectral measurements.[53] The values are generally in the range 7.22–7.46 eV. It is most likely that the electron being removed is from the $a_{1u}(\pi)$ or $e_u(NP_\sigma)$ orbitals of the PC ligand. The reputed high-temperature stability of [M(PC)] complexes is consistent with the ease of determining mass spectra and gas-phase electronic spectra. Nevertheless, most [M(PC)] complexes do show signs of thermal decomposition, albeit small, at above 300°C. Thus the narrow ESR signal at the free-electron g value ($g \approx 2.003$) that is found for [M(PC)] preparations even after sublimation has been ascribed to decomposition and the formation of defects in the solids.[55] Irradiation damage also increases the ESR signal as well as the formation of near-IR electronic absorptions. Oxygen does not seem to be a major factor in the origin of the decomposition.

5. Synthesis of New Derivatives

The standard methods of preparation as reviewed by Lever [3] have been applied to effect the synthesis of a PC derivative of almost every metal in the Periodic Chart. Nonetheless, new and improved synthesis of derivatives especially with metals in higher oxidation states are still being reported. For example a variation of the phthalonitrile condensation method has been used to prepare M(III) complexes of the type [M(PC)Cl] [54]:

$$(1)$$

M = Al(III), Sc(III), In(III), Yb(III)

A variation of the analogous condensation reaction with *o*-cyanobenzamide has also been used to prepare other M(III) derivatives [50]:

$$(2)$$

Using reaction (1), complexes of the type [M(PC)ClL] [where L = o-cyano-benzamide or phthalonitrile and M = Ru(III), Os(III), Ir(III)] were prepared as well as the compounds [Rh(PC)Cl] and [OsO₂(PC)].[50] Reaction (1) was also used to synthesize two trivalent rare earth complexes: [Yb(ClPC)Cl]·2H₂O and H[Gd(PC)₂].[56]

A cleaner method of preparing selected [M(PC)] derivatives involves metal exchange with a preformed PC complex [54] or acid–base reaction with the metal-free H₂PC [57]:

$$MBr_5 + [Li_2(PC)] \xrightarrow{\text{solvent}} [M(PC)Br_3] + 2LiBr, \quad M = Ta(V), Nb(V) \quad (3)$$

$$2CH_3HgN[Si(CH_3)_3]_2 + H_2PC \xrightarrow{\text{solvent}} [(CH_3Hg)_2 PC] + 2HN[Si(CH_3)_3]_2 \quad (4)$$

Lower-valent complexes may also be cleanly prepared by reduction of the higher-valent complex:

$$[M(PC)ClL] + XS\ aniline \xrightarrow{\Delta} [M(PC)(aniline)_2], \quad M = Ru(II), Ir(II) \quad (5)$$

The synthesis of the PC derivatives of Cr, Mn, Fe, Co, Si, Ge, and Sn are detailed in later reactions of this review.

Magnetic moments, electronic and ESR spectra were used to aid in fully characterizing these novel complexes. In general all of the above complexes that were formed via the self-condensation of the nitrile or benzamide using the metal as a template are formed in low yield. Further, purification procedures require repeated extraction with a variety of solvents and, if it is sufficiently soluble, recrystallization. The condensation reactions are obviously quite complex and involve a redox reaction. It is interesting to note that it has been shown that the formation of [Ni(PC)] proceeds via a stepwise condensation of 1,3-diaminoisoindoline ligands coordinated to Ni(II).[58]

6. Redox Reactions

The PC complexes of divalent metal ions can be chemically and electrochemically reduced. The reaction of dilithium benzophenone or sodium in THF with [M(PC)] can yield a series of anions of the type [M(PC)]$^{n-}$, $n = 1$ to 4 and where M = Mn(II), Fe(II), Co(II), Ni(II), Cu(II), Zn(II), Mg(II), and Al(III)Cl.[59–62,47] The reduction products of the [Mg(PC)], which are illustrative of most of the other reduction products, have been characterized by magnetic susceptibility (of the Li$^+$ salt),[59] electron spin resonance,[60] electronic spectra,[61] and MCD measurements.[62] The data indicate that the electrons added in forming the anions are occupying the lowest empty antibonding π orbital of the ligand. Thus the reduced species are Mg(II) complexes of: $n = 1$, PC$^-$, $e_g(\pi^*)^1$, $n = 2$, PC^{2-}, $e_g(\pi^*)^2$, $n = 3$, PC^{3-}, $e_g(\pi^*)^3$, $n = 4$, PC^{4-}, $e_g(\pi^*)^4$. The [Mg(PC)]$^-$ species is paramagnetic and shows an

ESR signal near that of the free electron, $g = 2.0027$. Major electronic absorptions are seen at 961, 637, 562, and 340 nm. The dianion is diamagnetic and shows characteristic absorptions at 520 and 336 nm. The species $[Mg(PC)]^{3-}$ is again paramagnetic, $g = 2.0030$ and has electronic absorptions at 1110, 588, and 339 nm. Finally, the tetraanion is diamagnetic and shows electronic absorption bands at 840, 621, and 305 nm. The electronic absorptions of the reduced species are ascribed to PC^{n-} $e_g(\pi^*)^n \rightarrow b_{1u}(\pi^*)$ and $b_{2u}(\pi^*)$ transition for the low-energy bands (> 625 nm) while the high-energy bands correspond to the expected $\pi \rightarrow \pi^*$ transitions.

The pattern of reduction of [Zn(PC)], [Al(PC)Cl], and [Ni(PC)] is quite analogous to that of [Mg(PC)], i.e., PC ligand reduction and no metal reduction. Conversely the reduction products of [Cu(PC)] do not conform to this general pattern.[63] Physical evidence indicates that the first reduction product $[Cu(PC)]^-$ is a Cu(II) complex of the reduced PC^- ligand. However, the next reduction yields a Cu(I), $3d^{10}$ complex of PC^-. Subsequent reduction seems to give rise to the loss of Cu. The reduction products of [Cr(PC)], [Mn(PC)], [Fe(PC)], and [Co(PC)] will be discussed in the respective sections devoted to each particular element.

Electrochemical measurements on DMF solutions of [Mg(PC)], [Zn(PC)], [Cd(PC)], and [Al(PC)Cl] give the reduction potentials for the first and second reduction.[64] The first reduction appears near -0.9 V (vs. SCE) while the second reduction is near -1.4 V for the [Mg(PC)]. The order for ease of reduction is Al(III)Cl > Zn(II) > Mg(II) > Cd(II). Surprisingly the Al(III) complex is significantly easier to reduce ($+0.28$ V) than the others. ESR and optical spectra show that the first reduction for all the complexes corresponds to the expected formation of PC^-.[65] It is interesting to note that the PC complex is easier to reduce, by $+0.4$ V, than the corresponding porphyrin complex. This, of course, reflects the increase in stability of the $e_g(\pi^*)$ orbital in the more delocalized PC macrocycle.

Polarographic reduction of the anions $[M(PCTS)]^{4-}$ M = Co(II), Ni(II), Cu(II) in DMSO, has been studied.[66] The first three reduction waves are noted for [Ni(PC)] at -0.17, -1.16, and -1.93 V (vs. SCE). The first wave is slightly more negative for [Cu(PC)] while the second and third waves are more positive. ESR and visible electronic spectra of the first and second reduction species for both [Ni(PC)] and [Cu(PC)] show that they are M(II) complexes of the reduced ligand, PC^- and PC^{2-}, respectively. This result for $[Cu(PCTS)]^{4-}$ is in contradistinction to that for [Cu(PC)], which indicates a Cu(II) $\rightarrow$ Cu(I) reduction. Thus the PC and $PCTS^{4-}$ ligands differ in their ability to stabilize the lower-valent state of Cu.

Electrochemical oxidation of [M(PC)] M = Cd(II), Mg(II), Zn(II), Al(III)Cl in DMF has also been studied.[64] The oxidation potentials for the first and second oxidations of [Mg(PC)] are $+0.61$ and 1.26 V (vs. SCE). The order for ease of oxidation is Cd(II) > Mg(II) > Zn(II) > Al(III), which is opposite to the reduction sequence. ESR and electronic spectra show[65] that the mono cations formed, $[M(PC)]^+$, are M(II) complexes of the oxidized PC ligand, PC^+ $(a_{1u})^1$. The chemical oxidation[67] of $[Zn(PC)]$ with $SOCl_2$ yields the solid complex of the PC^+ ligand $[Zn(PC)Cl]_2$, which shows strong electronic absorptions at 740, 550, 446, 379, and 323 nm. Although the spectra have not been analyzed, these bands are most likely due to the $\pi \rightarrow \pi^*$ transitions of the PC^+ macrocycle. The electrochemical oxidation of transition-metal $[M(PC)]$ complexes in chloronaphthalene has also been measured.[68] The order of ease of oxidation is Mn(II) > Fe(II) > Zn(II) > Co(II) > Cu(II) > Ni(II). In the case of Mn(II), Fe(II), and Co(II) the metal is being oxidized, M(II) → M(III), while for Zn(II), Cu(II), and Ni(II) the ligand PC is being oxidized. The ease of oxidation of PC ligands is significantly greater (~ 0.25) than for the corresponding porphyrin complexes. Thus the interval between the first oxidation and first reduction potentials is ~ 1.6 V for PC ligands while it is ~ 2.2 V for the porphyrins. The complexes of the PC ligand are both easier to oxidize and to reduce. This is presumably a consequence of the enhanced delocalization in PC macrocycle.

7. Aggregation of Complexes

The metal complexes of the tetrasulfonic acid derivative of PC, $PCTS^{4-}$, have been widely studied because the materials have appreciable solubility in aqueous solution, especially in the basic pH range, as well as in other solvents of high dielectric constant. Unfortunately, even the $[M(PCTS)]^{4-}$ complexes tend to form associated species at concentrations greater than 10^{-5} M. A driving force for self-association is thought to be the π–π interaction between the planar conjugated ring systems. If the concentration range examined is closely restricted, 10^{-5}–10^{-7} M, then the simple monomer–dimer equilibrium can be studied and thermodynamic as well as kinetic data for the dimerization can be determined. The approximate equilibrium constant for dimerization is in the range $K_D = 10^5$–10^7 M^{-1}, and is in the order Cu(II) > Fe(III) > V(IV)O $\geqslant$ Zn(II) > Co(II).[69]

More accurate thermodynamic information as well as kinetic data have been determined for the Co(II)[70] and V(IV)O[71] complexes. The relevant data is collected in Table 4. Kinetic measurements indicate that the dimerization reaction is apparently second order in both cases. Although the equilibrium constants for the two metal complex dimerizations are within an order of magnitude of each other, both the rates of the forward and the back reactions are 10^3–10^4 times faster for the V(IV)O complex. The small difference in ionic strength for the two systems, 0.006 vs. $\sim 10^{-4}$, cannot give rise to the large rate difference. Thus, the lability of the equilibrium must be a function of the metal ion. A simple-minded rationale of the rate data would take

TABLE 4. Thermodynamic and Kinetic Parameters
for the Equilibrium

$$2[M(PCTS)]^{4-} \underset{k_b}{\overset{k_f}{\rightleftharpoons}} \{[M(PCTS)]_2\}^{8-} \text{ at } 58°C$$

	Co(II)[70]	V(IV)O[71]
K_D	$2.05 \pm 0.05 \times 10^5\ M^{-1}$	$1.9 \times 10^6\ M^{-1}$
k_f	$5.36 \pm 0.15 \times 10^2\ M^{-1}\,\mathrm{s}^{-1}$	$8.4 \pm 0.8 \times 10^6\ M^{-1}\,\mathrm{s}^{-1}$
k_b	$2.8 \pm 0.10 \times 10^{-3}\ M^{-1}\,\mathrm{s}^{-1}$	$4.5 \pm 0.2\ \mathrm{s}^{-1}$
$\Delta H°$	$-14 \pm 0.9\ \mathrm{kcal\ mol}^{-1}$	
$\Delta G°$	$-8 \pm 0.02\ \mathrm{kcal\ mol}^{-1}$	
$\Delta S°$	$-18 \pm 3\ \mathrm{eu}$	
$\Delta H^{\ddagger}$	$-13.3 \pm 0.63\ \mathrm{kcal\ mol}^{-1}$	$4.9 \pm 1.5\ \mathrm{kcal\ mol}^{-1}$
$\Delta G_f^{\ddagger}$	$15.3 \pm 1.3\ \mathrm{kcal\ mol}^{-1}$	
$\Delta S_f^{\ddagger}$	$86.2 \pm 2\ \mathrm{eu}$	$-12\ \mathrm{eu}$
$\Delta H_f^{\ddagger}$	$-0.66 \pm 0.66\ \mathrm{kcal\ mol}^{-1}$	$13.6 \pm 1.5\ \mathrm{kcal\ mol}^{-1}$
$\Delta G_b^{\ddagger}$	$23.3 \pm 1.3\ \mathrm{kcal\ mol}^{-1}$	
$\Delta S_b^{\ddagger}$	$-72.4 \pm 2\ \mathrm{eu}$	$-14\ \mathrm{eu}$

into account the difference in the strength of the axial interaction for Co(II) and V(IV)O. In the former case the metal is in the plane of the PC ligand with two axial H_2O ligands which are relatively strongly bound. Conversely, the V(IV)O complex most likely has the metal out of the plane of the PC ligand toward the strongly bound axial O^{2-} ligand. Thus, the sixth axial ligand will not be strongly bound. ESR measurements on the vanadyl dimer in DMF-H_2O indicate that the two $S = \frac{1}{2}$ metal ions are coupled to form a triplet state.[49] By analysis of the triplet spectra and use of the magnetic dipole–dipole interaction model, the metal–metal separation is ~ 450 pm in the dimer. This relatively short distance requires that the sixth axial ligand must be lost in forming the dimer. In the case of the vanadyl complex, this process should be quite facile and most likely not rate determining in the formation of the dimers. On the other hand the loss of axial ligands may not be facile, and this process may contribute to the overall complex rate of dimerization for Co(II). A similar argument can be put forth for the dedimerization reaction. It is relatively slow, especially for Co(II), because the rate-determining step may involve separating the two planar ligands, which are held together by both PC–PC (π) interactions and perhaps metal–aza nitrogen intermolecular interactions. The out-of-plane structure for the vanadyl dimer would preclude the latter interaction and thus the only significant interaction here would be the PC–PC interaction.

The influence of solvent and added solute on the nature of the species in aqueous solutions of $[Co(PCTS)]^{4-}$ has been monitored by electronic spectral measurements.[72] In general the Q band shifts to higher energy by 900 cm^{-1}

and broadens in forming the dimer. The equilibrium constant determined, $K_D = 2.7 \pm 0.3 \times 10^5\,M^{-1}$ at 20°C, $\mu = 5 \times 10^{-3}$, and 20% (vol./vol.) methanol–water agrees quite well with the value obtained from the kinetic study above. Evidence is also given that the formation of the dimer requires 12 molecules of water to stabilize it, presumably via hydrogen bonding to the sulfonate groups.

ESR spectra of DMF water solutions of $[Cu(PCTS)]^{4-}$ can be analyzed in the same way as for $[VO(PCTS)]^{4-}$ to yield a Cu–Cu separation in the dimer of 430 pm. Similarly the Co(II) dimer shows a distance of 450 ± 50 pm.[73] Mixed dimers of Co(II), Ni(II), Cu(II), Zn(II) have also been examined and the Cu–Co separation is 450 ± 50 pm. Further, the Cu–Ni dimer is the most stable of this type formed while the least stable is the Co–Zn dimer. In general, the solvent order for depolymerization at room temperature in mixed aqueous solution mixtures is found to be DMSO > DMF > alcohols > acetone $\geqslant$ acetonitrile. The solvent effect may arise by a removal of the specifically bound H_2O molecules (hydrogen bonded to $-SO_3^-$ groups) as suggested by earlier work, and replacement with good charge solvating molecules.

The equilibrium constant for dimerization of metal complexes of another soluble derivative of PC, PCTO, has also been measured spectrophotometrically for several metals in a range of nonaqueous solvents. Equilibrium constant data are collected in Table 5.[74,75] The metal dependence is striking and the magnitude of K_D is in the order $V(IV)O \geqslant Zn(II) > Cu(II)$. The metal order for K_D in nonaqueous solvents for $[M(PCTO)]$ is at variance with that of $[M(PCTS)]^{4-}$ in aqueous solution. Thus specific solvation, and hydrogen bonding effects, as postulated for Co(II) may stabilize the charged dimers (in addition to other interactions) in aqueous solutions. The order of decreasing $[Cu(PCTO)]$ dimerization in several solvents is CCl_4 > benzene > toluene > chloroform > dioxane > THF. The solvent dependence of dimer stability is quite large and reflects the dielectric constant and solvating power of the solvent, i.e., the higher the solvating power of the solvent, the higher the solubility of the $[M(PCTO)]$ and the lower stability of the dimer.

TABLE 5. Dimerization Constants for $[M(PCTO)]$ in Several Solvents at 25°C[74,75]

M	Solvent	$K_D\,(M^{-1})$
Cu(II)	CCl_4	$2.97 \pm 0.02 \times 10^6$
Cu(II)	C_6H_6	$1.58 \pm 0.09 \times 10^4$
Cu(II)	THF	$1.41 \pm 0.50 \times 10^2$
Zn(II)	C_6H_6	$1.09 \pm 0.23 \times 10^6$
V(IV)O	C_6H_6	$2.01 \pm 0.49 \times 10^6$
V(IV)O	THF	$1.07 \pm 0.01 \times 10^4$

In addition to the other derivatives, the PC complexes undoubtedly aggregate in solvents where they have appreciable solubility. Thus, by monitoring the visible spectra it is found that [Fe(PC)] forms dimers [76] in DMSO, in the concentration range 5×10^{-4} to 5×10^{-6}. The rate of dedimerization is found to be $k = 1 \times 10^{-3}$ s^{-1} at 28°C. This slow reaction is comparable to that of [Co(PCTS)]$^{4-}$ for presumably the same reasons. Dimer formation is so persistent that a dimeric form of [Cu(PC)] has been isolated in the solid state. [77]

8. Chromium Complexes

A convenient synthesis of β[Cr(PC)], in 40% yield, has been reported. [78] The reaction involves the condensation of phthalonitrile in the presence of Cr(CO)$_6$ in refluxing chloronaphthalene:

$$4 \; \text{(benzene ring with two CN groups)} + \text{Cr(CO)}_6 \xrightarrow{\Delta} \beta[\text{Cr(PC)}] + 6\text{CO} \qquad (6)$$

This preparation is especially attractive since a pure product precipitates out of the reaction mixture and no subsequent sublimation or extraction is necessary for purification. In fact, controlled sublimation of β[Cr(PC)] at 300–350°C gives the polymorph α[Cr(PC)]. [79] The β form of the [Cr(PC)] is quite reactive toward oxygen and exposure of it to air rapidly and quantitatively yields a green solid with the empirical formula [Cr(PC)OH]. This Cr(III) complex has been fully characterized but no detailed study of this oxidation (oxygenation) reaction has been carried out. [80] This complex has been used to synthesize a number of new Cr(III) derivatives [78]:

$$[\text{Cr(PC)OH}] + \text{HO-P(=O)-R}_2 \xrightarrow[\text{CH}_3\text{OH}]{\Delta} [\text{Cr(PC)(CH}_3\text{OH)}_2](\text{O}_2\text{PR}_2)$$
$$R = \text{CH}_3, \text{C}_6\text{H}_5 \qquad (7)$$

$$[\text{Cr(PC)(CH}_3\text{OH)}_2](\text{O}_2\text{PR}_2) \xrightarrow{\Delta} [\text{Cr(PC)(CH}_3\text{OH)O}_2\text{PR}_2] + \text{CH}_3\text{OH} \qquad (8)$$

$$[\text{Cr(PC)OH}] + \text{HO-P(=O)(CH}_3)(\text{C}_6\text{H}_5) \xrightarrow[\text{CHCl}_3]{\Delta} [\text{Cr(PC)(H}_2\text{O)O}_2\text{PCH}_3\text{C}_6\text{H}_5] \qquad (9)$$

The phosphinate derivatives are all six-coordinate Cr(III) complexes. Attempts to remove the Lewis-base donors by heating *in vacuo* to form either five-coordinate complexes or polymeric solids with bridging phosphinates leads only to decomposition.

The reaction of solid $\beta[Cr(PC)]$ with gaseous NO at ambient temperatures yields a 1:1 adduct[81]:

$$\beta[Cr(PC)] + NO \rightleftharpoons [Cr(PC)NO] \qquad (10)$$

Adduct formation, which is most facile for Cr(II), can be reversed by heating of the solid *in vacuo* at 280°C for several hours. The IR spectrum[82] of the solid shows a $\nu(NO)$ at 1690 cm^{-1}. This indicates that the complex can be formulated as a Cr(III) complex of NO$^-$. The solid readily adds pyridine:

$$[Cr(PC)NO] + py \longrightarrow [Cr(PC)(NO(py)]$$

$$[Cr(PC)] + py + NO \longrightarrow \qquad \qquad \uparrow \qquad (11)$$

The same complex, which shows a $\nu(NO)$ of 1680 cm^{-1}, can also be formed by bubbling NO through a pyridine solution of [Cr(PC)]. The solid adduct, [Cr(PC)NO], reacts with oxygen to give a material with a composition of [Cr(PC)(NO)OH] and $\nu(NO) = 1690$ cm^{-1}. No further characterization of this supposed Cr(IV) complex has been reported. Finally, the modification $\alpha[Cr(PC)]$ does not react either O$_2$ or NO in the solid state. This, of course, is in contrast to the reactivity of $\beta[Cr(PC)]$. This has been ascribed to the difference in solid state structure for the two forms. Although the interplanar spacing between PC planes remains 340 pm, the metal–metal distance increases from 340 to 480 pm in going from the α to the β form. Thus, the metal is less accessible to the small molecules, O$_2$ or NO, and no binding in the solid state occurs for the α form.

The reduction of [Cr(PC)] with Na in THF[47] yields an anion [Cr(PC)]$^-$, which shows an ESR spectrum consistent with low-spin Cr(I) with a d-orbital configuration of $e_g{}^4b_{2g}^1$. In this case the e_g level is depressed below the b_{2g} level by a back π bonding Cr(I) $d\pi$ to PCπ^* interaction. This is consistent with the π donor properties of low-valent Cr atoms. Reduction of [Cr(PC)] with Na in HMPA yields a different species than for the THF solution. ESR spectra are consistent with low-spin Cr(I) with d-orbital configuration $b_{2g}^2 e_g{}^3$. Here the axially bound, π donor, solvent molecule raises the energy of the metal e_g orbital. Further reduction of [Cr(PC)] or [Cr(PC)ClH$_2$O][83] with dilithium benzophenone in THF yields the green crystals, Li$_2$[Cr(PC)]·6THF. The magnetic moment of the solid suggests that there are two unpaired electrons in the material. This would be consistent with a complex of Cr(I) and PC$^-$: $e_g{}^4b_{2g}^1$, $e_g(\pi^*)^1$. Oxidation of [Cr(PC)][67] with SOCl$_2$ yields the tetragonal complex [Cr(PC)Cl$_2$]. The magnetic moment, electronic and IR spectra and oxidative titration support the formulation Cr(III) (PC$^+$). Thus the PC ligand is oxidized by the removal of an electron from the a_{1u} (π) orbital of the PC macrocycle. No evidence is obtained for the existence or formation of Cr(IV) or higher metal oxidation state.

9. Manganese Complexes

The magnetic properties of solid [Mn(PC)] have been studied as a function of temperature.[84] The susceptibility measurements can be interpreted in terms of an unusual intermediate spin state, $S = 3/2$, for the d^5 Mn(II) with an $^4A_{2g}$ ground state and a d-orbital configuration, $b_{2g}^2 e_g^2 a_{1g}^1$. Once again the experimental d-level ordering is at variance with the theoretical.[27] Magnetic resonance measurements on the [Mn(PC)] solid also give information about exchange interactions and long-range ordering in the solid.[85,86] [Mn(PC)] is square planar in the solid and it readily forms six-coordinate tetragonal complexes in solution and in the solid state[87]:

$$[Mn(PC)] + 2py \rightleftharpoons [Mn(PC)py_2] \qquad (12)$$

Heating this adduct at 100°C *in vacuo* reverses the process. The electronic spectrum of [Mn(PC)] is changed by the formation of the dipyridinate and the Mn(II) is converted to the low-spin, $S = \frac{1}{2}$ form with reversal of the d orbital to give the configuration $e_g^4 b_{2g}^1$. This may be attributed to axial σ donation of the pyridine, which destabilizes the metal a_{1g} orbital, and π donation of the Mn $d\pi$ to the π^* of the PC, which stabilizes the metal e_g orbital. The electronic spectrum of [Mn(PC)] in noncoordinating solvents or solid films is rather unusual and shows a number of additional prominent absorptions at 910 and 460 nm in addition to the Q band (720 nm) and B band (350 nm). Formation of the dipyridine complex enhances the intensity of all the bands and shifts the Q and B bands to 680 and 330 nm, respectively, and the additional bands to 900 and 465 nm. Although the additional bands have not been as yet assigned on the basis of theoretical or experimental work, it is likely that they arise from metal to ligand $e_g \rightarrow e_g(\pi^*)$ charge transfers. The prominence of the additional bands may then be related to an appreciable Mn $\rightarrow$ PC(π) interaction.

While [Mn(PC)] alone is not sensitive to oxygen, the presence of donor molecules like pyridine promotes irreversible oxidation of the Mn(II) to Mn(III) according to the overall scheme[87]

$$2[Mn(PC)py_2] + \tfrac{1}{2}O_2 + H_2O \longrightarrow 2[Mn(PC)(OH)py] \qquad (13)$$

$$2[Mn(PC)(OH)py] \longrightarrow [Mn_2(PC)_2(py)_2O] + H_2O \qquad (14)$$

The X-ray crystal structure of the product shows that the dimer has a linear py–Mn–O–Mn–py axial arrangement with the metal in the plane of the equatorial PC ligand.[24] The short Mn–O–Mn bond and strong metal–metal interaction results in a substantial antiferromagnetic coupling of the two d^4 metal atoms. Although the complex is weakly paramagnetic at room temperature, it has been suggested that the material is inherently diamagnetic with some paramagnetic impurity in the solid.[88] Heating of the dimer produces a solid lacking pyridine, $[Mn_2(PC)_2O]$. Conversely, dimers with pyridine being replaced by other Lewis bases such as N-methylimidazole, 3-picoline, and piperidine can readily be prepared by passing oxygen through a solution of $[Mn(PC)]$ in the neat amine. The physical properties of these new dimers are quite similar to those of the pyridine complex and it is reasonable to assume that they all have analogous structures. An interesting point brought out by this work is the inability to unambiguously assign any infrared absorption that arises from the ν(Mn–O–Mn) asymmetric stretch. The band must be quite weak since even ^{18}O substitution does not lead to the appearance of any new bands. On the other hand, the dimers do show strong visible absorptions in the region 633–617 nm, which are characteristic since the monomeric Mn(III) or Mn(II) complexes do not show strong absorptions in this region. Although the possibility of the formation of a peroxo intermediate has been eliminated, the detailed mechanism of oxygenation of $[Mn(PC)]$ in pyridine and subsequent formation of dimer has not been worked out.[88]

The reaction of $MnSO_4$ and sodium-4 sulfophthalic acid in air yields a manganese complex of $PCTS^{4-}$. Although first formulated as a Mn(II) complex,[89] later work supports its formulation as a Mn(III) species, $Na_4[Mn(PCTS)OHH_2O]$.[90] The low room-temperature magnetic moment of the solid and especially of the aqueous solution of the complex indicates that the monomeric hydroxide complex is partially dimerized. In fact the formation of a μ-oxo dimanganese(III) species occurs slowly at 25°C in 1:1 DMF, H_2O at pH 9–12.[90] This postulate is supported by the appearance of the characteristic dimer electronic absorption band at 625 nm and by the lowering of the solution magnetic moment to ~ 1.9 B.M. The mechanism of dimerization is suggested to be

$$\underbrace{[Mn(PCTS)OHH_2O]^{4-}}_{A} \rightleftharpoons \underbrace{[Mn(PCTS)(OH)_2]^{5-}}_{B} + H^+ \qquad (15)$$

$$A + B \underset{k_3}{\overset{k_1}{\rightleftharpoons}} \underbrace{[Mn_2(PCTS)_2OH(H_2O)]^{9-}}_{C} + H_2O \qquad (16)$$

$$2B \xrightarrow{k_2} \underbrace{[Mn_2(PCTS)_2(OH)_2O]^{10-}}_{[B_2]} + H_2O \qquad (17)$$

$$[B_2] + H^+ \longrightarrow C \qquad (18)$$

By analysis of the rapid spectral changes, from a Q band at 714 nm characteristic of A to one at 680 nm characteristic of B upon addition of base, the pK$_a$ of equilibrium (15) is found to be 10.3. Near pH = 12 and $\mu = 0.01$, $k_2 = 1.1 \times 10^2\ M^{-1}\,s^{-1}$ and $k_3 = 6.4 \times 10^{-4}\ M^{-1}\,s^{-1}$. At long times and higher pH, loss of the Mn(III) complex is noted. The intimate mechanism of the dimerization has not been worked out. Presumably the rate-determining step would be the loss of one axially bound solvent molecule. This would correspond to a dissociative mechanism (case II) as found for [Fe(PC)] (see below). The next step would be a rapid formation of the dimer.

A number of other Mn(III) complexes have been prepared by the oxidation of [Mn(PC)]. Exposure of solid [Mn(PC)] to anhydrous formic acid slowly gives [Mn(PC)O$_2$CH(HCOOH)].[91] If the six-coordinate complex is heated at 110°C, the five-coordinate [Mn(PC)O$_2$CH] species forms. Similarly, oxidation of [Mn(PC)] with SOCl$_2$ or SOBr$_2$ yields [Mn(PC)Cl] or [Mn(PC)Br].[67] All the Mn(III) complexes show high-spin d^4 magnetic moments, $\mu = 4.9$ B.M. Unlike the general observation for all other PC complexes, the ground-state d-orbital configuration in the spin-free state is $b_{2g}^1 e_g^2 a_{1g}^1$. In spite of this the electronic spectra of [Mn(PC)X] are typical of PC complexes with no extra bands and a Q band at 704 nm and B band at 360 nm in pyridine. The reaction of solid [Mn(PC)] with gaseous NO slowly yields the 1:1 adduct [Mn(PC)NO].[81] Addition of pyridine gives [Mn(PC)NO(Py)], which can also be formed by the passage of NO through a pyridine solution of [Mn(PC)]. The infrared spectra of the solids [Mn(PC)NO] and [Mn(PC)NOPy] show ν(NO) at 1760 cm^{-1} and 1737 cm^{-1}, respectively. These values indicate that the materials can be formulated as Mn(III)-NO$^-$ complexes.[82] Reaction of [Mn(PC)NO] with oxygen slowly yields a material with the composition [Mn(PC)NO(OH)]. Aside from the observation that the IR spectra shows a ν(NO) at the same frequency as the starting material this interesting solid has not been fully characterized. The apparent oxidation state, from the merger evidence presented, is Mn(IV). However, it is interesting to note that attempts to further oxidize [Mn(PC)Cl] with SOCl$_2$ did not yield a Mn(IV) complex.[67]

The reduction of [Mn(PC)] in THF with dilithium benzophenone yields a series of solids with the composition Li$_n$[Mn(PC)]mTHF.[92] Similar anions [Mn(PC)]$^{n-}$, $n = 1$–5, can be prepared by reduction with sodium in THF. From electronic spectral data and magnetic moments, it is clear that the first reduction product [Mn(PC)]$^-$ is a Mn(I) complex[61] with the d-orbital configuration $b_{2g}^2 e_g^3 a_{1g}^1$. On the other hand, the second reduction leads to a Mn(II) complex of the reduced ligand, PC^{2-}, i.e., $b_{2g}^2 e_g^2 a_{1g}^1$, $e_g(\pi^*)^2$. Similar reasoning gives the configuration of [Mn(PC)]$^{3-}$ as $b_{3g}^2 e_g^2 a_{1g}^1 (e_g^*)^3$ or Mn(II)PC^{3-}. Experimentally, no ESR spectrum is seen for the reduced [Mn(PC)]$^{n-}$ species.[60] From magnetic moments the formation of the species [Mn(PC)]$^{4-}$ and [Mn(PC)]$^{5-}$ seems to be that of unusual high-spin Mn(II) or Mn(I) complexes of PC^{4-} or high-spin Mn(I) complexes of PC^{3-} or PC^{4-}, respectively.

10. Iron Complexes

Several useful preparative techniques involving iron PC derivatives have been worked out. For example, a new rapid synthesis of [Fe(PC)] has been reported which utilizes the condensation of phthalonitrile in the presence of $Fe(CO)_5$ in refluxing chlorophthalene.[78] The yield of pure product, which precipitates out of the reaction mixture in 30 min, is 50%. Using the same procedure with tetrafluorophthalonitrile under N_2, a 75% yield of perfluoro-[Fe(PC)] is formed.[93] Like [Fe(PC)] the perfluoro derivative shows an intermediate spin, $S = 1$ at 33°C in acetone, while it is diamagnetic in donor solvents like DMSO and pyridine.

Detailed physical studies on [Fe(PC)] have been carried out. Low-temperature magnetic moments show that in the solid, the Fe(II) is in an intermediate spin state, $S = 1$. The data are consistent with the ground state $^3B_{2g}$ and the d-orbital configuration is $e_g^4 b_{2g}^1 a_{1g}^1$.[94] The zero-field splitting parameter is $D = 64\ cm^{-1}$ and $g_\parallel = g_\perp = 2.74$. Magneto circular dichroism[33] spectra of [Fe(PC)] in dichlorobenzene solution are interpreted to support the ground state $^3A_{2g}$ ($S = 1$) and the configuration $a_{1g}^2 b_{2g}^2 e_g^2$. Extra bands in the near-IR spectra at 800 and 714 nm are assigned to metal–ligand charge transfer $e_g \to b_{1u}, b_{2u}(\pi^*)$ transitions. Mössbauer spectra of a powdered sample of [Fe(PC)] in a magnetic field have been measured in the temperature range 4.2–100 K.[95] Interpretation of the data gives the ground state, $^3E_g: b_{2g}^2 e_g^3 a_{1g}^1$ with g (in all three directions) > 2.0. Thus, while all the measurements agree on the $S = 1$ intermediate spin state they differ on the ground state d-orbital configuration. The magnetic susceptibility and Mössbauer measurements are on the solid at low temperature, but the MCD spectra were taken on solutions albeit in a solvent of low coordinating ability. Thus it may be possible that the d-orbital configuration is different in the solid and solution. An interesting study that may be relevant here is the report of the Mössbauer spectra of the α and β forms of [Fe(PC)].[96] The solid state structures, with different intermolecular interactions, gives rise to observable differences in the Mössbauer parameters. These can be interpreted in terms of an increase in the intermolecular Fe–aza N interaction in going from the α to the β form.

The formation of six-coordinate complexes of Fe(II) readily occurs upon addition of the suitable donor ligand to solutions or slurries of [Fe(PC)]. Complexes of the type $[Fe(PC)L_2]$, where L is a nitrogen donor ligand have been prepared and characterized. Mixed ligand complexes [Fe(PC)LL′] with a variety of ligands are also known and gathered in Table 6. The stereochemistry of this class of tetragonal complexes is typified by the structure of $[Fe(PC)(\gamma\text{-pic})_2]$ (see Table 1). By application of an induced shift calculation with NMR proton resonances of the coordinated alkylamines of $[Fe(PC)(n\text{-}C_4H_9NH_2)_2]$ an Fe–N axial bond length of 194 pm is calculated for the complex in $CDCl_3$ solution.[97] Since this distance is comparable to that determined in the crystal structure of the analogous Fe(II) complex, it is reasonable to assume that the same tetragonal structure is maintained in the solid and in solution for $[Fe(PC)L_2]$.

TABLE 6. [Fe(PC)LL'] Complexes Isolated and Characterized

L	L'	Reference
$n\text{-BuNH}_2$	$n\text{-BuNH}_2$	97
py	py	98
im	im	98
pip	pip	98
$C_6H_5CH_2NC$	$C_6H_5CH_2NC$	99
$C_6H_5CH_2NC$	py	99
$C_6H_5CH_2NC$	pip	99
$C_6H_5CH_2NC$	CH_3im	99
C_6H_5NO	$n\text{-BuNH}_2$	100
$p\text{-}CH_3C_6H_4NO$	$n\text{-BuNH}_2$	100
$p\text{-}(CH_3)_2CHC_6H_4NO$	$n\text{-BuNH}_2$	100
$P(OC_6H_5)_3$	$n\text{-BuNH}_2$	100
$p\text{-}CH_3C_6H_4NO$	CH_3im	100
$C_6H_{11}NC$	CH_3im	100
CH_3im	CH_3im	100
$C_6H_5CH_2NH_2$	$C_6H_5CH_2NH_2$	100
$P(OC_2H_5)_3$	$P(OC_2H_5)_3$	100
$C_6H_{11}NC$	$C_6H_{11}NC$	100
$n\text{-}C_4H_9NC$	$n\text{-}C_4H_9NC$	100
$C_6H_{11}O_3P$	$C_6H_{11}O_3P$	100

The complexes are diamagnetic with the apparent d-orbital configuration $b_{2g}^2 e_g^4$. As expected, addition of axial ligands raises the energy of the a_{1g} orbital sufficiently to cause spin pairing in b_{2g} and e_g orbitals. The electronic spectra and MCD spectra show only the normal π–π^* ligand transitions,[33] i.e., no extra absorptions due to charge transfer transitions are noted. In order to probe the nature of axial interactions in the [Fe(PC)L$_2$] complexes, Mössbauer and MCD spectra have been studied. The Mössbauer parameters for L = py, im, BuNH$_2$, pip, and [Fe(PC)(CN)$_2$]$^-$ are interpreted in terms of a tetragonal structure with the in-plane ligand field greater than the axial field.[98] The axial interactions involve moderate σ bonds and weak π bonds (only with π acceptor ligands). The approximate σ donor order is CN$^-$ > pip > BuNH$_2$ > py and the π acceptor order is CN$^-$ > py > im > BuNH$_2$ ~ pip. Low-temperature Mössbauer measurements have been made for [Fe(PC)py$_2$].[101,102] The large quadrupole splitting in comparison to those of the isostructural porphyrin complexes has been ascribed to a greater in-plane π interaction with the PC ligand system. By analysis of the MCD spectra for [Fe(PC)(CN$_2$)]$^-$ and [Fe(PC)L$_2$], L = im, py, DMSO, orbital reduction factors can be calculated and related to the e_g(metal) $\rightarrow$ $e_g(\pi^*)$PC in-plane π interaction.[103] This interaction in the complexes is in the order

DMSO > py > im > CN⁻. The implication is that the axial ligand e_g(metal) → L(π^*) interaction is in the opposite order. This, of course, agrees with the expected π acceptor properties of the axial ligand.

The study of the thermodynamics and kinetics of axial ligand exchange for [Fe(PC)L₂] has proven to be quite fruitful in recent years. Equilibrium constants can be conveniently determined using visible spectrophotometric techniques. The data, log K_1 and log K_2, for various systems are tabulated in Table 7.

$$[Fe(PC)L_2] + L' \overset{K_1}{\rightleftharpoons} [Fe(PC)LL'] + L \tag{19}$$

$$[Fe(PC)LL'] + L' \overset{K_2}{\rightleftharpoons} [Fe(PC)L_2'] + L \tag{20}$$

For the substituted pyridines the order of the strength of binding of the ligand to [Fe(PC)] is X-py: 4NH₂ > 4CH₃ > H > 4CH₃CO > 4CN > 2-CH₃. This follows the pK_{BH^+} for the ligands.[104] Thus, within a restricted donor set, the strength of binding of the ligand follows the intrinsic Lewis basicity of the ligand (σ bonding). However, when ligands of the same donor atom but different binding characteristics are compared the order is im > pip > py > CH₃im, and then other effects have to be considered. The placement of imidazole ahead of piperidine must reflect a π acceptor interaction of the imidazole with the Fe(II). However, the binding of the poor sigma donor and good π acceptor, CO, is not favored even when in competition with only good sigma donors like piperidine. Thus, sigma bonding effects are a good deal more important than π-bonding effects with axial ligands. In fact CO binding to [Fe(PC)] is several orders of magnitude weaker than for an Fe(II) porphyrin complex where axial π bonding is considered important.[105] The relative unimportance of axial π bonding in [Fe(PC)] may reflect the signifi-

TABLE 7. Equilibrium Constant Data for Ligand, L',
Binding to [Fe(PC)L₂]

L	L'	log K_1	log K_2	Reference
DMSO[a]	im	5.0	3.8	104
DMSO[a]	4NH₂py	3.8	3.8	104
DMSO[a]	4CH₃py	2.3	1.4	104
DMSO[a]	py	2.0	0.4	104
DMSO[a]	4 acetpy	1.7	−0.4	104
DMSO[a]	4CNpy	1.0	0.6	104
DMSO[a]	2CH₃py	0.2	−0.2	104
im[b]	CO	−1.5		105
pip[b]	CO	−0.1		105
py[b]	CO	−0.4		105
2CH₃im[b]	CO	0.9		105

[a] 25°C, in DMSO.
[b] 23°C, in toluene.

cant in-plane π interaction of the metal with the PC ligand which tends to saturate the π-bonding capacity of the metal. Finally, the relatively weak binding of $2CH_3py$ and CH_3im to [Fe(PC)] indicates the importance of steric hindrance in dictating ligand binding for this system.

Kinetic data for ligand exchange reactions are collected in Tables 8 and 9.

$$[Fe(PC)LL'] + L'' \rightleftharpoons [Fe(PC)LL''] + L' \tag{21}$$

All the reactions seem to follow the common dissociative mechanism

$$[Fe(PC)LL'] \underset{k_{-1}}{\overset{k_{+1}}{\rightleftharpoons}} [Fe(PC)L] + L' \tag{22}$$

$$[Fe(PC)L] + L'' \underset{k_{-2}}{\overset{k_2}{\rightleftharpoons}} [Fe(PC)LL''] \tag{23}$$

The rate law, neglecting the k_{-2} term, is

$$\text{rate} = \frac{k_2 k_1 [Fe(PC)LL'][L'']}{k_{-1}[L'] + k_2[L'']}$$

The limiting cases are then

$$\text{I.} \quad k_2[L''] \gg k_{-1}[L'], \quad \text{rate} = k_1[Fe(PC)LL']$$

$$\text{II.} \quad k_{-1}[L'] \gg k_2[L''], \quad \text{rate} = \frac{k_1 k_2}{k_{-1}[L']} [Fe(PC)LL'][L'']$$

The kinetic measurements can be made at a variety of temperatures, using rapid mixing, stop flow, or variable-temperature NMR measurements.[100] The reactions of [Fe(PC)(DMSO)$_2$] follow case II kinetics, while case I applies for the tolylnitroso and isocyanide complexes. The full rate law is seen for the CO reactions, however.

TABLE 8. Kinetic Parameters for Ligand Exchange Reactions of [Fe(PC)LL'']

L	L'	L''	$k_1 (s^{-1})$	$k_{-2} (s^{-1})$	k_{-1}/k_2	Reference
im	im	CO^a	2.6×10^{-3}	2.0×10^{-2}	4.2	105
pip	pip	CO^a	5.0×10^{-1}	1.3×10^{-1}	3.3	105
py	py	CO^a	1.2×10^{-1}	9.0×10^{-2}	3.4	105
$2CH_3im$	$2CH_3im$	CO^a	3.3×10^1	7.0×10^{-1}	5.8	105
$BuNH_2$	tolNO	$tolNO^b$	1.3×10^1			100
py	CH_3CH_2NC	py^c	1.98×10^{-4}			99
pip	CH_3CH_2NC	pip^c	5.15×10^{-4}			99
CH_3im	CH_3CH_2NC	CH_3im^c	9.2×10^{-5}			99
ϕCH_2NC	ϕCH_2NC	py^d	0.1			99

^{a} 23°C, in toluene.
^{b} 25°C, in CDCl$_3$.
^{c} 30°C, in toluene.
^{d} 20°C, in toluene.

TABLE 9. Activation Parameters for Ligand Exchange Reactions of [Fe(PC)LL']

L	L'	L"	k_{obs} $(M^{-1}s^{-1})$	$\Delta H^{\ddagger}$ (kcal mol^{-1})	$\Delta S^{\ddagger}$ (eu)	Reference
DMSO	DMSO	im[a]	3.1	17.3 ± 0.5	1.8	106, 107
DMSO	DMSO	n-CH$_2$im[a]	6.9	18.2 ± 0.5	6.2	108
DMSO	DMSO	n-Buim[a]	7.9	18.3 ± 0.5	7.1	108
BuNH$_2$	tolNO	tolNO[b]	—	20.1 ± 0.2	14.2 ± 0.2	100
im	CO	im[c]	—	27	15	105

[a] 25°C, in DMSO.
[b] 25°C, in CDCl$_3$.
[c] 25°C, in toluene.

The kinetic data show that the activation parameters are independent of the entering group as expected for a dissociative reaction. Further, the rate constant for dissociation of CO, k_{-2}, for a series of complexes with a different *trans* ligand L, does not vary greatly, indicating a small nonleaving group or *trans* effect. However, the leaving group effect is substantial, with the relative order of reactivity CH$_3$im > tolNO > pip > py > CO > im > ϕCH$_2$NC spanning the range from 1 for the least reactive to 10^4 for the most reactive. The dissociative character of the reaction and common mechanism is also supported by the general parallel between the strength of binding of the ligand as measured by K_{eq} and the rate of ligand exchange. One of the most reactive [Fe(PC)] complexes found involves the dissociation of a nitroso ligand C$_7$H$_7$NO. NMR measurements give evidence for ligand binding to the Fe(II) via the nitrogen atom. Thus the enhanced reactivity here can be ascribed to a steric acceleration of ligand dissociation which is related to the bulky phenyl substituent. Similar steric arguments can be made for the rapid dissociation of CH$_3$im. Of course, the observation of steric acceleration adds support to the assignment of the ligand exchange reaction to a dissociative mechanism.

The nature of the five-coordinate intermediate in the dissociative mechanism is of some interest. The values for the rate constant ratio k_{-1}/k_2 indicate that the intermediate shows some discrimination between nucleophiles, i.e., preferring reaction with the better nucleophile amine than with the CO ligand. This may be taken as evidence for the formation of an intermediate of finite lifetime. The stereochemistry of the intermediate can be one of two possibilities; both are square pyramidal, one with the metal in the plane of the PC ligand and the other with the metal above the plane toward the lone axial ligand. Although there is precedent for the latter geometry in the Mg(II) and Zn(II) five-coordinate PC complexes, the former structure is favored for the intermediate. The reasoning here is based on the low- and intermediate-spin state of the in-plane six-coordinate [Fe(PC)L$_2$] and four-coordinate [Fe(PC)] species, respectively. An out-of-plane intermediate would imply a change of spin state to give a high-spin Fe(II) complex which would be kinetically labile and chemically unstable.

It is interesting to note that the rate of ligand dissociation is only about 1/40th as fast for [Fe(PC)] as for the corresponding low-spin Fe(II) porphyrin.[105] Although still rapid in comparison to the reactions of low-spin Fe(II) complexes, the PC derivative does react significantly slower than the porphyrin complexes. This may be due to the higher positive charge on the metal in the PC complexes when compared to the porphyrins, which leads to stronger axial bonding. Of course, the kinetic consequence of a stronger axial bond for a dissociative reaction are clear: the stronger the bond, the slower the reaction.

Another interesting kinetic result is the observation that the dissociation of benzylisocyanide (Table 9) is accelerated (10^3 times) by visible light. In fact, the equilibrium[99] shifts to the left in the light but shifts to the right in the dark.

$$[Fe(PC)L_2] + C_6H_5CH_2NC \rightleftharpoons [Fe(PC)L(C_6H_5CH_2NO)] + L \tag{24}$$
$$L = py, pip, CH_3im$$

The condensation reaction of sodium 4-sulfophthalic acid with ferrous sulfate in air yields a solid which was originally formulated as Na[Fe(PCTSH$_3$)]·3H$_2$O,[89] an Fe(II) complex, but now appears to be a mixture of the Fe(III) complexes: Na$_3$[Fe(PCTS)]·7H$_2$O and Na$_4$[Fe(PCTS)OH]·2H$_2$O.[69] The magnetic moment of the solid gives a value of $\mu = 3.03$ B.M. at room temperature while aqueous solution moments give a value of $\mu = 1.80$ B.M.[89] The latter moment would be consistent with a low-spin tetragonal Fe(III) complex with axial H$_2$O or OH$^-$ ligands. The higher moment in the solid may indicate an intermediate-spin state with the metal in lower coordination number. Of course, the low moments may also arise from partial formation of a μ-oxo-*bis*-Fe(III) dimer much like that of the well-known Mn(III) PC case. In fact a spectral study has shown that dimerization and further polymerization of the Fe(III) complex occurs in aqueous solution in the pH range 6–9. At higher pH decomposition slowly occurs. Unfortunately, no subsequent study of the dimerization has been carried out as yet.[69] In particular it has not been shown whether the latter dimerization is of the π–π type or of the hydrolysis (μ-oxo) type.

Heating of Na$_4$[Fe(PCTS)OH]·2H$_2$O at 100°C in N$_2$ slowly yields a solid which has a magnetic moment of $\mu = 4.80$ B.M., which is consistent with a high-spin Fe(II) complex with the configuration $b_{1g}^2 e_g^2 a_{1g}^1 b_{1g}^1$. This, of course, is an unusual configuration, being one of the few PC complexes that is a high-spin form with the strongly antibonding orbital singly occupied.

Exposure to air of the latter solid yields the original Fe(III) complex. This apparent reversible oxygenation has also been noted in a separate study. The formation of a green solution occurs when the metal-free ligand H_2PCTS^{4-} is added to an aqueous solution of Fe(II) at pH 6–7.[109] The visible spectrum is typical of a divalent [M(PC)] with a sharp Q band at 671 nm and a B band at 330 nm. Further, the complex shows a strong band at 425 nm, which is characteristic of a low-spin tetragonal Fe(II) complex[33] such as $[Fe(PCTS)(H_2O)_2]^{4-}$. The pK of the coordinated water is found to be ~8.0. Reversible oxygenation is noted for the aqueous solution, pH = 6.5 at room temperature with the stoichiometry

$$\underbrace{[Fe(PCTS)(H_2O)_2]^{4-}}_{A} + O_2 \xrightarrow{K_1} \underbrace{[Fe(PCTS)(H_2O)_2O_2]^{4-}}_{B} + H_2O \quad (25)$$

$$A + B \xrightarrow{K_2} \{[Fe(PCTS)(H_2O)]^{4-}\}_2O_2 + H_2O \quad (26)$$

Upon addition of oxygen the spectrum of the starting material quickly disappears and new bands at 637 and 344 nm that are characteristic of the oxygenation product grow in. Deoxygenation can only be accomplished slowly at ~70°C by passing nitrogen through the solution. Although the authors prefer to view the oxygenation product as an adduct of Fe(II), the visible spectrum of this material is similar to that of $[Fe(PCTS)OH]^{4-}$ formed from $FeCl_3$[109] and $H_2(PCTS)^{4-}$ or from the air oxidation during the condensation reaction of Fe(II) with sodium 4-sulfophthalic acid.[69] Thus the oxygenation product can best be formulated as $Fe(III)-O_2^{2-}-Fe(III)$, a peroxo-Fe(III) dimer. Unfortunately, no magnetic susceptibility measurements were made of these solutions nor were solid materials isolated and characterized. However, by a combination of spectrophotometric and manometric techniques, the equilibrium constants K_1 and K_2 have been estimated to be $2 \times 10^4 \ M^{-1}$ and $4 \times 10^7 \ M^{-1}$, respectively, at pH 6.5 and 20°C.

The apparent ease of oxidation and high stability of the oxygenation product for $[Fe(PCTS)(H_2O)_2]^{4-}$ is in contradiction to observations for [Fe(PC)]. Thus [Fe(PC)] in donor or nondonor solvents, as well as in the solid, does not react with oxygen at room temperature. However, the surface of the solid does weakly chemisorb oxygen at low temperatures with desorption occurring around $-115°C$. The activation energy for desorption is 9.8 kcal mol^{-1} and the kinetics first order.[110] The data here constitute evidence for lack of O–O bond rupture during the chemisorption process.

The existence of the Fe(III) oxidation state for PC-type ligand is somewhat controversial. After long periods of time, the solid β[Fe(PC)] reacts with gaseous NO to give [Fe(PC)NO] at room temperature.[81] The 1:1 adduct may be dissociated by heating at 250°C in a nitrogen stream. The infrared spectrum[82] of the solid adduct shows a strong ν(NO) at 1737 cm^{-1}, which is consistent with its formulation as an Fe(III)–NO$^-$ complex. The relative instability of the adduct is demonstrated by the observation that the NO is easily displaced by pyridine vapor. Surprisingly, however, [Fe(PC)NO] is stable in air and does not react with oxygen. The polymorph α[Fe(PC)] does not react with NO, presumably for the same structural reasons as detailed for α[Cr(PC)].[78] Unfortunately, no physical studies (magnetic, Mössbauer, or electronic absorption) have been carried out on [Fe(PC)NO].

Oxidation of [Fe(PC)] with $SOCl_2$ yields [Fe(PC)Cl_2].[67] Spectral studies are consistent with the formulation of the complex as an Fe(III) complex of the oxidized ligand (PC$^+$). It is interesting to note that the Fe(III) here is in a low-spin form, $S = \frac{1}{2}$, with the apparent configuration $b_{2g}^2(e_g)^3$. The reaction of aqueous HCl with [Fe(PC)] in air is purported to yield an Fe(III) complex, [Fe(PC)Cl]. However, recent work supports the notion that the major product of this reaction is in fact an HCl adduct of [Fe(PC)], i.e., [Fe(PCH)Cl].[111] In order to eliminate any side reactions which may occur with water, the interaction of solid [Fe(PC)] with gaseous HCl was followed by mass sorption measurements. At room temperature the 4:1 adduct is formed, [Fe(PC)]·4HCl. By heating at 120°C or by treating with wet N_2, this solid is converted to [Fe(PC)]·HCl. Magnetic moment, $\mu = 3.69$ B.M., and Mössbauer spectral parameters are consistent with the formulation of the adduct as Fe(II) complex of the azaprotonated ligand, HPC$^+$, with an axially bound Cl$^-$ ligand. The acid–base adduct can also be reversibly formed in DMSO or DMA solution by reaction of [Fe(PC)] with HCl gas.[112] Visible absorption spectral measurements show isobestic points and an equilibrium is demonstrated here by the observation that dilution of the adduct solution reverses the process. The adduct is diamagnetic in DMSO,[113] which is consistent with the presence of low-spin tetragonal Fe(II) in this solvent, e.g., [Fe(PCH)(DMSO)$_2$]$^+$·Cl. Adduct formation has also been demonstrated for the reaction of gaseous HBr and [Fe(PC)] in DMA and DMSO.[114] Infrared spectra of the solids [Fe(PCH)Br] and [Fe(PCH)Cl] show for IR bands at 310 cm^{-1} and 224 cm^{-1} for ν(M–X) of the axially bound ligand in the solid. The adducts undergo slow dissociation in DMSO and DMA with concomitant spectral and magnetic susceptibility changes. For the Cl$^-$ complex the solution remains diamagnetic while for the Br$^-$ derivative the adduct is paramagnetic. The following reaction scheme appears to be consistent with the data collected:

$$[Fe(PC)(DMSO)_2] + HX \underset{\text{fast}}{\overset{\text{fast}}{\rightleftharpoons}} [Fe(PCH)(DMSO)_2]^+ + X^- \qquad (27)$$

$$[Fe(PCH)(DMSO)_2]^+ + X^- \underset{\text{slow}}{\overset{\text{fast}}{\rightleftharpoons}} [Fe(PCH)(DMSO)X] + DMSO \qquad (28)$$

For Br$^-$ the mixed complex is intermediate spin while for the Cl$^-$ it is low spin. This difference in spin state may reflect a difference in molecular structure or may reflect the decrease in axial ligand field (with stabilization of the

a_{1g} orbital) in going from the Cl^- to the Br^- complex. The kinetic and structural behavior of the Fe(II) complexes of the protonated PC ligand follow quite reasonably from the prior discussion of [Fe(PC)] complexes.

The reduction of [Fe(PC)] in the THF with dilithium benzophenone yields solids with the general formula $Li_n[Fe(PC)] \cdot mTHF$,[59] $n = 1$–4. The first three anions can also be generated by reaction of [Fe(PC)] with sodium. Magnetic susceptibility[59] and ESR measurements[47] are consistent with the following formulations and electronic configurations:

$$[Fe(PC)]^- : \quad Fe(I), \; b_{2g}^2 e_g^4 a_{1g}^1$$

$$[Fe(PC)]^{2-} : \quad Fe(0), \; b_{2g}^2 e_g^4 a_{1g}^2$$

$$[Fe(PC)]^{3-} : \quad Fe(0), \; b_{2g} e_g^4 a_{1g}^2; \quad PC^-, \; e_g(\pi^*)^1$$

Unfortunately, electronic spectra[61] do not support the postulated reduction to Fe(0) but indicate that reduction of the PC ligand occurs. Thus there is compelling evidence only for the existence of one low-valent-state iron, Fe(I).

Several of the reduced species react with organic molecules to give interesting and novel organometallic derivatives[115]:

$$Li_2[Fe(PC)] \cdot 6THF + CH_3I \xrightarrow{\;THF\;} Li[Fe(PC)CH_3] \cdot 6THF \qquad (29)$$

The diamagnetic carbanion-Fe(II) derivative is formed in 75% yield. An analogous phenyl derivative can also be prepared. The alkyl derivative undergoes a series of characteristic reactions:

$$[Fe(PC)CH_3]$$

$$\phi_3CCl \uparrow$$

$$CH_3I + [Fe(PC)] + I^- \;\xleftarrow{\;I_2\;}\; [Fe(PC)CH_3]^- \xrightarrow{\;\Delta\;} \tfrac{1}{2}C_2H_6 + [Fe(PC)]^- \quad (30)$$

$$H^+ \downarrow 2py$$

$$[Fe(PC)(py)_2] + CH_4$$

The chemistry of the Fe(III) alkyl complex has not been explored. The Fe(II) alkyl derivative reacts with both homolytic (thermal) and heterolytic cleavage of Fe–C bond. It is interesting to note that the oxidizing agent is important in determining whether the Fe(III) alkyl compound forms. On the other hand, the oxidation of the phenyl derivative can be accomplished with I_2:

$$[Fe(PC)C_6H_5]^- + \tfrac{1}{2}I_2 \;\longrightarrow\; [Fe(PC)C_6H_5] + I^- \qquad (31)$$

Thus, the aryl anion is able to stabilize the higher oxidation state of the metal to a greater extent than the alkyl anion. Unfortunately, there is no detailed physical characterization of this interesting new class of organometallic compounds. Both the alkyl and aryl derivatives of Fe(II) and Fe(III) react with styrene in the presence of Pd(II) salts to yield $C_6H_5CH=CHR$.[116] The order of reactivity here is Fe(II) > Fe(III) and $CH_3^- > C_6H_5^-$.

11. Cobalt Complexes

The square-planar [Co(PC)] readily adds one or two donor molecules that bind in the axial coordination positions[117]:

$$[Co(PC)] + L \rightleftharpoons [Co(PC)L] \tag{32}$$

$$[Co(PC)] + 2L \rightleftharpoons [Co(PC)L_2] \tag{33}$$

The most common material that forms is a six-coordinate tetragonal complex, $[Co(PC)L_2]$. Structural data for a typical complex of this type are given in Table 1. It is evident that the axial bond is quite long and this axial binding is presumed not to be strong. In fact solid $[Co(PC)py_2]$ and $[Co(PC)(\gamma\text{-pic})_2]$ readily lose both axial ligands in the temperature ranges 80–120°C and 90–130°C, respectively. No evidence is seen for the stepwise loss of L with intermediate formation of the 1:1 adduct. Spectral data of these complexes in donor solvents are consistent with a tetragonal structure similar to that of the solid state. In fact, the g values determined from the ESR spectra of the paramagnetic $S = \frac{1}{2}$ system are sensitive to axial interaction since the single unpaired electron is in an a_{1g} orbital.[45,117] The extent of axial interaction is in the order expected from their donor strength to metal centers: quin > γ-pic > py > $2CH_3$quin > β-pic > isoquin. Similarly the order im > py > DMSO is found for $[Co(PCTS)]^{4-}$ in DMSO.[43] In nondonor solvents the complex $[Co(PC)L_2]$ dissociate after a few seconds and [Co(PC)] precipitates out of solution. Analysis of the ESR spectra of the α and β polymorphs of solid [Co(PC)] indicates that intermolecular Co(II)–N(aza) PC axial interactions are greater in the α form than in the β form as expected from their structural differences in the solid state.[118] Consideration of the MCD spectra of $[Co(PC)L_2]$, L = DMSO, py, and $[Co(PC)(CN)_2]^-$, in DMSO lead to the conclusion that the visible absorption bands are due to PC ligand $\pi \rightarrow \pi^*$ transitions and none can be assigned to charge transfer transitions.[33] Under special conditions the five-coordinate complex [Co(PC)L] can be formed from [Co(PC)], i.e., crystallization of the solid from hot neat solution of the base at 100°C. This is in the temperature region where the 2:1 adduct readily loses axial ligands.[117] Conversion of [Co(PC)L] to $[Co(PC)L_2]$ only occurs quite slowly (days) in a slurry of the solid in the neat donor at ambient temperature. The 1:1 adduct is more thermally stable than the 2:1 adduct since it takes a higher temperature to dissociate the axial ligand, e.g., [Co(PC)py] and [Co(PC)γ-pic] lose base in the temperature range 130–160°C and 150–180°C, respectively. It is interesting to note that [Zn(PC)L] loses pyridine and γ-picoline at 40–80°C higher temperature.[119] This again attests to the overall weakness of axial binding in the Co(II) PC complexes. ESR spectra of [Co(PC)L] formed initially are still consistent with a low-spin form with the odd electron in the a_{1g} orbital. Again the ESR parameters indicate that the axial interactions are in the order [Co(PC)L$_2$] > [Co(PC)L] > [Co(PC)]. Thus, the spectral properties of the 1:1 adduct are consistent with a five-coordinate, square-pyramidal structure with the donor molecule in the axial position. Whether the Co(II) is in the plane of the PC ligand or above it toward the base like other five-coordinate [M(PC)] structures is not known.

Recently, the synthesis by the same procedure as above of the complex [Co(PC)β-pic] has been reported.[120] This solid is a low-spin complex that loses the axial ligand in the 130°–180°C temperature range. If, however, the solid is slurried in neat base at ambient temperatures for 15 days the magnetic moment increases to $\mu = 4.89$ B.M. This moment indicates that the new material is a high-spin Co(II) form. Variable-temperature magnetic moments suggest that the ground state of the high-spin complex has a contribution from an orbital singlet. The ESR spectrum of the high-spin complex shows three g values, $g_1 = 5.55$, $g_2 = 2.78$, and $g_3 = 1.54$, which suggest the d-orbital configuration $b_{2g}^2 e_g^3 a_{1g}^1 b_{1g}^1$. The attainment of the high-spin form of Co(II) is highly unusual, especially for the PC derivative. It is well-known, for example, that square-planar [Ni(PC)] is diamagnetic and addition of donor molecules does not give rise to the formation of the paramagnetic high-spin *bis* adduct as is noted for the porphyrin complexes.[27] It is also interesting to note that the Co(II) porphyrins do not give high-spin forms in either the 1:1 or 2:1 adducts.[121] Thus some special interaction must occur in [Co(PC)β-pic] that increases the axial interaction (and/or decreases the in-plane interaction) to the point that the high-spin form is favored. It has been suggested that a slow dimerization occurs in the solid state with the two five-coordinate square-pyramidal molecules lining up with the two square bases parallel. In this configuration, if the Co(II) is in the plane of the PC, the metal from each molecule can, of course, interact with an aza nitrogen atom in the next molecule.

Although [Co(PC)] does not form an oxygen adduct even in donor solvents at room temperature, if the temperature is lowered sufficiently (liquid N$_2$) then reversible oxygenation is observed, presumably in the sequence[122]

$$[Co(PC)(HMP)_2] \rightleftharpoons [Co(PC)HMP] + HMP \qquad (34)$$

$$[Co(PC)(HMP)] + O_2 \rightleftharpoons [Co(PC)(HMP)O_2] \qquad (35)$$

Oxygenation of [Co(PC)γ-pic] in HMP also gives a 1:1 adduct:

$$[Co(PC)\gamma\text{-pic}] + O_2 \rightleftharpoons [Co(PC)(\gamma\text{-pic})O_2] \tag{36}$$

Upon oxygenation the ESR signal typical of Co(II) rapidly changes to that of the oxygen adduct. The spectrum of the latter species is consistent with the formulation Co(III)·O_2^-. The additional axial base seems to influence the stability since ESR intensities indicate that the adduct with γ-pic is more stable (formed to a greater extent) than the adduct with a *trans* HMP donor.

Oxygen binding is also noted with [Co(PC)L$_2$], L = py, γ-pic in CH$_2$Cl$_2$ at 115 K.[117] Dissociation of one donor must, of course, occur prior to oxygenation since addition of free base to the solution decreases the concentration of the oxygen adduct. Further, the complexes [Co(PC)L$_2$] are not sensitive to oxygen in the solid state even at low temperature.

The reversible formation of a 1:1 oxygen adduct[123] is also noted for methanol solutions of the Co(II) complex [Co(PCTS)(H$_2$O)$_2$]$^{4-}$ at $-84°$C:

$$[Co(PCTS)(H_2O)_2]^{4-} + O_2 \rightleftharpoons [Co(PCTS)(H_2O)O_2]^{4-} + H_2O \tag{37}$$

Again the ESR spectrum is consistent with the formulation Co(III)·O_2^-. At room temperature no oxygenation of [Co(PCTS)]$^{4-}$ occurs unless the pH of the aqueous solution is raised to greater than 12.[124] Prolonged exposure to high pH, however, destroys the complex. The adduct shows a Q band at 670 nm while the Co(II) species has a slightly less intense and broader Q band at 710 nm.

Monometric measurements show that the oxygen adduct that slowly forms at high pH at 25°C is a 2:1 complex[125]:

$$\underbrace{[Co(PCTS)(H_2O)OH]^{5-}}_{A} + O_2 \rightleftharpoons \underbrace{[Co(PCTS)(HO)O_2]^{5-}}_{B} + H_2O \tag{38}$$

$$A + B \rightleftharpoons \{[Co(PCTS)(HO)]^{5-}\}_2O_2 + H_2O \tag{39}$$

Analysis of the rate data supports the notion that the adduct formation occurs in two stages. The rate of oxygenation is first order in Co(II) and oxygen. Thus, the first step, equilibrium (38), is the rate-determining step with a second-order rate constant at pH 13 of 3.24 M^{-1} s^{-1}.

Presumably, the intimate mechanism of the reaction would be analogous to that of [Fe(PC)], i.e., a dissociative mechanism (case II) where the rate-determining step is the dissociation of an axial water ligand. Also, the oxygenation rate is dependent on the OH$^-$ concentration, i.e., the rate increases as the OH$^-$ concentration increases. This suggests that prior axial coordination of an OH$^-$ to the Co(II) is needed to activate the metal center. The adduct immediately releases oxygen when H$^+$ is added. No hydrogen peroxide is formed. The acid presumably rapidly neutralizes the axial OH$^-$ and this subsequently destabilizes the oxygen adduct.

The visible absorption spectrum of the 2:1 adduct shows a Q-band maximum at 671 nm, which is the frequency of the absorption for the 1:1 adduct. Since ESR measurements indicate that the 1:1 adduct is a Co(III) complex, a similar conclusion for the 2:1 adduct can be made based on the visible spectral data. Thus the 2:1 adduct can be formulated as a Co(III)O_2^{2-} Co(III) (μ-peroxo) dimer. Unfortunately, no magnetic measurements have been made on the dimer to confirm that they are diamagnetic.

The Co(II) complex, Na$_4$[Co(PCTS)(H$_2$O)$_2$] was prepared by the standard condensation reaction of sodium 4-sulfophthalic acid utilizing CoSO$_4$ as the template.[89] An aqueous solution, at neutral pH, of the complex can only be slowly air-oxidized in the presence of strong field ligands like CN$^-$ or imidazole. Thus the spin-paired six-coordinate Co(III) complexes [Co(PCTS)(im)$_2$]$^{3-}$ and [Co(PCTS)H$_2$O(X)]$^{4-}$, X = OH$^-$, (CN$^-$) were prepared and characterized.[126]

In addition to reacting with oxygen, solid [Co(PC)] slowly reacts with gaseous NO to form an air-stable 1:1 complex [Co(PC)NO].[81] The infrared spectrum of the solid shows a ν(NO) at 1705 cm^{-1}, which is consistent with the formulation Co(III) NO$^-$.[82] The NO can be displaced from the complex by addition of pyridine vapor or by heating the solid at 200°C in a nitrogen stream. A remarkable stability of [Co(PC)NO] is exemplified by the fact that the complex can be recovered unchanged from concentrated sulfuric by precipitation with water! Oxidation of [Co(PC)][67] with SOCl$_2$ or SOBr$_2$ yields [Co(PC)X$_2$] X = Cl$^-$, Br$^-$. Spectral and chemical properties indicate that the materials are Co(III) complexes of the oxidized, PC$^+$, ligand. Previous work on the air oxidation of [Co(PC)] in concentrated HCl or HBr may have also yielded the same species. It is interesting to note that the oxidation potential for [Co(PCTS)]$^{4-}$ in DMSO is approximately +0.46 V (vs. SCE) while the corresponding ligand oxidation is +1.09 V.[66] Therefore, it should be possible with suitable conditions to generate only the Co(III) complexes.

The chemical reduction of [Co(PC)] in THF with dilithium benzophenone gives the series of solids, Li$_n$[Co(PC)]·m THF, where n = 1 to 5.[59] Magnetic moments, ESR spectra and electronic absorption spectra support the formulations for the anions[47,61,62]:

$$[\text{Co(PC)}]^- : \quad \text{Co(I)}b_{2g}^2 e_g^4 a_{1g}^2 : \text{PC}$$

$$[\text{Co(PC)}]^{2-} : \quad \text{Co(I)}b_{2g}^2 e_g^4 a_{1g}^2 : \text{PC}^- e_g(\pi^*)^1$$

$$[\text{Co(PC)}]^{3-} : \quad \text{Co(II)}b_{2g}^2 e_g^4 a_{1g}^1 : \text{PC}^{3-} e_g(\pi^*)^3$$

$$[\text{Co(PC)}]^{4-} : \quad \text{Co(I)}b_{2g}^2 e_g^4 a_{1g}^2 : \text{PC}^{3-} e_g(\pi^*)^3$$

$$[\text{Co(PC)}]^{5-} : \quad \text{Co(I)}b_{2g}^2 e_g^4 a_{2g}^2 : \text{PC}^{4-} e_g(\pi^*)^4$$

Electrochemical and spectral measurements on DMSO solutions of [Co(PCTS)]$^{4-}$ yield similar results for the first two anions.[66] The polarographic half-wave potential for the Co(II) → Co(I) reduction is −0.547 V (vs. SCE) while the next reduction, to form PC$^-$, gives a much more negative potential, -1.35 V. This latter potential is about 0.2 V more negative than the corresponding reduction potential for the isoelectronic $[Ni(PC)]^{4-}$ complex. The difficulty of reduction of the PC ligand in the Co(I) complex reflects the high negative charge in the macrocycle. This charge buildup could occur from extensive charge donation from the low-valent metal to PC ligand via $e_g \rightarrow e_g(\pi^*)$ bonding.

The relative stability of the low-valent Co(I) oxidation state is quite unique among [M(PC)] complexes. It can also be generated from $[Co(PC)]^{(127)}$ in DMA by reduction with BH_4^- and from $[Co(PCTS)]^{4-}$ in neutral water with hydrazine or sodium sulfide.[128] The electronic spectrum of the d^8, diamagnetic complex shows a characteristic asymmetric absorption band at 467 nm in addition to the normal PC, $\pi \rightarrow \pi^*$ bands at 705 and 310 nm. On the basis of MCD measurements this band has been assigned to the overlap of the two metal to ligand charge transfer transitions $e_g \rightarrow b_{1u}, b_{2u} (\pi^*)$.[33]

The Co(I) complex is quite reactive and is a powerful nucleophile. Typical reactions result in oxidative addition to give Co(III) complexes:

$$[Co(PC)]^- + CH_3I \xrightarrow{\text{DMA}} [Co(PC)CH_3] + I^- \qquad (40)$$

A similar reaction occurs for $[Co(PCTS)]^{4-}$ in aqueous solution.[127]

Solid n-propyl and phenyl derivative can also be prepared by the reaction of n-C_3H_7I or C_6H_5I with $Li[Co(PC)]$ in THF.[115] All the alkyl derivatives are diamagnetic and generally their NMR spectra show proton resonance for the axially bound alkyl groups. The electronic spectrum no longer shows the characteristic Co(I) band and gives absorptions different from those of [Co(PC)]. The complexes are most likely six coordinate in donor solvents and addition of CN^- to $[Co(PC)CH_3]$ in DMA gives a spectral change which can be interpreted as due to the formation of $[Co(PC)CH_3CN]^-$. The alkyl derivatives are light sensitive and are photolyzed to give [Co(PC)]'and alkane. The complex $[Co(PC)CH_3] \cdot 2THF$ also turns out to be a good alkylating agent in methanol and THF. The reaction with styrene to yield $C_6H_5CH = CHCH_3$ is promoted by Pd(II) salts.[116] However, the Co(II) complex $Li[Co(PC)(C_6H_5)_2] \cdot 8THF$ is more reactive in this reaction than the Co(III) derivatives. The Co(I) complex also can be irreversibly oxidized by reaction with oxygen in DMF[129]:

$$[Co(PCTS)(DMF)_2]^{5-} + O_2 \longrightarrow [Co(PCTS)(DMF)O_2]^{5-} + DMF \qquad (41)$$

Visible and ESR spectra are consistent with the formulation of this complex as a $Co(II) \cdot O_2^-$ material. Interestingly enough the $Co(I)$–PC^--type dianion reacts with oxygen in two stages:

$$[Co(PCTS)(DMF)_2]^{6-} + O_2 \longrightarrow [Co(PCTS)(DMF)O_2]^{6-} + DMF \qquad (42)$$

$$[Co(PCTS)(DMF)O_2]^{6-} + O_2 \rightleftharpoons [Co(PCTS)(O_2)_2]^{6-} + DMF \qquad (43)$$

ESR and electronic spectral evidence indicates that the first irreversible oxidation yields a $Co(I) \cdot O_2^-$ complex while the second reversible oxygenation gives a $Co(II)(O_2^-)_2$ complex.

12. Group IV Metal Complexes

Synthetic and structural work on the group IV metalloid (Si, Ge, and Sn) complexes of phthalocyanine have been particularly active in the past 10 years. The complexes contain the metal in the M(II) or M(IV) oxidation state. The higher oxidation state materials are most common and have been well characterized. They are six-coordinate tetragonal complexes with structures analogous to those in Table 1. In addition to their intrinsic chemical value, a number of interesting applications of these complexes have also been noted.

A new synthesis, in good yield, of $[Si(PC)Cl_2]$ has been reported, which involves the condensation of *o*-cyanobenzamide or 1,3-diiminoisoindoline with $SiCl_4$ under anhydrous conditions.[130] This complex can be used as the starting material for the synthesis of a host of other derivatives in high yield. For example, the synthesis of pseudohalogen[131] and oxy derivatives[132,133] can be easily accomplished:

$$[Si(PC)Cl_2] + 2AgX \xrightarrow{\Delta} [Si(PC)X_2] + 2AgCl \tag{44}$$
$$X = NCO^-, NCS^-, NCSe^-$$

$$[Si(PC)Cl_2] + HOR \longrightarrow [Si(PC)(OR)_2] + 2HCl \tag{45}$$
$$R = CH_3, H, Si(C_3H_7)_3$$

IR spectra of the pseudohalogen derivatives indicate that the axial anions are N-bonded.

By using a procedure similar to that used for the dichloride complex derivatives of the type $[Si(PC)RCl]$, where $R = CH_3, C_3H_7, C_6H_5$, etc., can be prepared from $RSiCl_3$.[132,133] The silicon–carbon bonds in the alkyl and aryl derivatives are quite stable to hydrolysis since treatment of the methyl derivative with concentrated sulfuric acid or hydrofluoric acid yields only $[Si(PC)CH_3X]$, $X = OH^-, F^-$, respectively. The materials are, however, light sensitive and photolytically decomposed by visible light. Reactions of the hydroxo derivative yield a number of new complexes, e.g.,

$$[Si(PC)(CH_3)OH] + HOR \longrightarrow [Si(PC)(CH_3)OR] \tag{46}$$
$$R = C_6H_5, C_6H_4Cl, Si(C_3H_7)_3, SiCH_3[OSiCH_3(OCH_3)_2]_2, Si(CH_3)_2[OSiCH_3(OCH_3)_2]$$

Bis-siloxane derivatives have also been synthesized via the useful synthetic procedure

$$[Si(PC)(OH)_2] + HY \longrightarrow [Si(PC)(OY)_2] + H_2 \tag{47}$$
$$Y = Si(CH_3)_2OSi(CH_3)_3, Si(CH_3)[OSi(CH_3)_3]_2, Si[OSi(CH_3)_3]_3$$

Mixed *bis*-siloxane derivatives can also be prepared. All of the Si(IV) complexes have been fully and carefully characterized via elemental analyses and electronic, IR, and NMR spectroscopy.

The X-ray crystal structure[17] of $\{Si(PC)[OSi(CH_3)_3]_2\}$ shows the Si–O–Si bond angle in the solid to be 157°. Induced shift calculations based on proton NMR data yield a solution bond angle of 162°.[17] From a structural point of view the mono- and *bis*-siloxane derivatives of Si(IV)PC are important since a methodology has been worked up, using their proton NMR data, by which solution conformations of the siloxane groups (and their alkyl substituents) can be inferred. This approach combined with the development of detailed synthetic schemes for making the complexes $[Si(PC)X_2]$, where X is an oligomer of a siloxane polymer, could lead to a rational construction of thermally stable silicon-containing polymers of the type $\{Si(PC)-O-Si(R)_2\}_n O$. Along this line, several interesting macromolecular products of this type have been recently prepared and partially characterized.[134] Initial work has involved the formation of model monomeric compounds:

$$[Si(PC)(OH)_2] + C_6H_5XH \longrightarrow [Si(PC)(C_6H_5X)_2] + 2H_2O$$
$$C_6H_5X = C_6H_5CO_2{}^-, C_6H_5O^-, C_6H_5S^- \tag{48}$$

When difunctional reagents of the same type are used, a series of low molecular weight, thermally stable polymers are formed:

$$\{Si(PC)-O-R\}_n O, \qquad R = -CH_2CH_2, \ -\langle O \rangle-, \ \overset{O}{\underset{\|}{C}}(CH_2)_4\overset{O}{\underset{\|}{C}}, \ \text{etc.}$$

Unlike silicon, a divalent Ge derivative has been prepared via the $BH_4{}^-$ reduction of $[Ge(PC)Cl_2]$ in 77% yield.[135] While $[Ge(PC)]$ is light sensitive it is surprisingly not oxidized by Br_2, I_2, or H_2O_2 and does not react with oxygen. The synthesis of a number of derivatives of Ge(IV) in high yield proceeds in a fashion similar to that of silicon[136]:

$$[Ge(PC)Cl_2] + 2H_2O \longrightarrow [Ge(PC)(OH)_2] + 2HCl \tag{49}$$

$$[Ge(PC)(OH)_2] + HX \longrightarrow [Ge(PC)X_2] + 2H_2O$$
$$X = F^-, Br^-, OD^-, OSi(C_2H_5)[OSiCH_3OSi(CH_3)_3]_2 \tag{50}$$

The synthesis of the pseudohalogen derivatives follows the same procedure as the silicon derivatives.[131] The complexes $[Ge(PC)X_2]$ have been fully characterized both chemically and spectrally.

The synthesis of low-molecular-weight polymers of the type $\{Ge(PC-R-O\}_n$ has been carried out in a way analogous to those of silicon.[134] The formation of an intractable polymeric solid presumably of high molecular weight can be done by heating $[Ge(PC)(OH)_2]$ to 350°C:

$$n\text{-}[Ge(PC)(OH)_2] \xrightarrow{\Delta} \{Ge(PC)-O\}_n + nH_2O \tag{51}$$

It is interesting that an analogous sequence occurs to form a Ti(IV) polymer[137]:

$$[Ti(PC)Cl_2] + 2H_2O \xrightarrow{-2HCl} [Ti(PC)(OH)_2] \xrightarrow[-H_2O]{\Delta} \{Ti(PC)-O\}_n \quad (52)$$

The synthesis of the starting material is accomplished in almost quantitative yield from $TiCl_3$ and phthalonitrile[137]:

$$2TiCl_3 + 4C_6H_4(CN)_2 \xrightarrow{\Delta} [Ti(PC)Cl_2] + TiCl_4 \quad (53)$$

In this exemplary reaction the metal is a reducing agent while the nitrile is the oxidizing agent. The reaction is forced to the right by distilling the volatile $TiCl_4$ off. The pure product is collected from the cooled chloronaphthalene solvent. Low-molecular-weight polymers of Ti(IV) may also be made by reactions with phosphites:

$$[Ti(PC)Cl_2] + C_6C_5PO(OH)_2 \longrightarrow \{Ti(PC)OPO(C_6H_5)-O\}_n \quad (54)$$

Both the divalent and tetravalent states of [Sn(PC)] complexes are known. In fact two forms of Sn(IV) have been characterized and crystal structures determined: the common six-coordinate structure for [Sn(PC)Cl₂], and the unusual eight-coordinate square-antiprismatic *bis*-(PC) structure for [Sn(PC)₂]. The divalent complex [Sn(PC)] is rather unique since it shows a domed structure (see Table 1 and following discussion). The ^{119}Sn Mössbauer spectrum of this material is consistent with the postulate that a stereochemically active lone-electron pair is localized in an axial position.[21]

The synthesis of a number of six-coordinate Sn(IV) complexes has been worked out, e.g., [Sn(PC)X₂], X = OH^-, F^-, Cl^-, Br^-, I^-, etc.[138] The reaction of the dihydroxide with a silanol is a model for a general reaction to yield polymers and other interesting materials[139]:

$$[Sn(PC)(OH)_2] + 2HO-Si(C_2H_5)_3 \xrightarrow{-2H_2O} \{Sn(PC)[OSi(C_2H_5)_3]_2\} \quad (55)$$

Spectral characterization of the tetragonal Sn(IV) derivatives indicates they are quite similar in structure and bonding to the corresponding Si(IV) and Ge(IV) derivatives. Mössbauer spectroscopy of a series of complexes [Sn(PC)X₂][140] reveals a linear relationship between the isomer shift and the electronegativity of the axial anion. This can be interpreted by assuming that the charge density on the Sn(IV) is chiefly determined by σ bonds of the axial anion. It is also postulated that π donation from PC to Sn is appreciable and that charge donation from the axial anions retards this interaction. The PC → M π bonding should be expected for high oxidation (IV) state complexes. X-ray photoelectron spectra of the Cl^- and F^- complex support this charge distribution argument.[39]

13. Catalytic Activity

The [M(PC)] complexes have been extensively used as both homogeneous and heterogeneous catalysts in a wide range of chemical reactions.[141-144] Although not exclusively so, the [M(PC)] function as redox catalysts, i.e., the complex changes oxidation state during the catalytic cycle. Further, it is assumed that at least one of the reactants interacts with the catalyst via axial coordination to metal or via peripheral bonding to the PC ligand. In many cases the catalytic activity of the [M(PC)] materials is quite dependent on the particular metal and the total oxidation state of the complex. To a smaller degree, substitution on the PC backbone also modulates the reactivity of the catalyst. Additional ligands (Lewis bases) also can enhance or retard the catalysts' activity. Finally, the physical state of the catalyst, as a solid or in homogeneous solution, can be of some importance. Thus, many variables may be manipulated in order to achieve maximum activity in a desired reaction. An additional feature of the [M(PC)] complex catalysts is their superior thermal stability, which allows high-temperature gas phase heterogeneous reactions to be carried out where both the reactants and products are volatile.

An important class of reactions that is catalyzed by [M(PC)] is the autoxidation of various organic molecules. The role of the catalyst in these reactions is generally thought to involve activation of molecular oxygen. An example of this type of reaction is the autoxidation of 2,6-di-*t*-butyl phenol [see formula (56)].[145] In the reaction a suspension of the catalyst in a suitable solvent and the phenol are subjected to moderate oxygen pressure, 3-4 atm, and temperature, 20-65°C. The order of metal activity is Co(II) $\geqslant$ Mn(II) $\geqslant$ Fe(II) $\geqslant$ Cu(II), while the DMF suspensions are more active than those in

$$\text{(56)}$$

DMSO or CH_3OH. The selectivity of the reaction also depends on the metal, solvent, and temperature. Thus while [Mn(PC)] give exclusively II, [Co(PC)] gives a I/II ratio of three, while for [Fe(PC)] it is 1/3. The extent of product I formation decreases in the solvent order DMF > CH_3OH > DMSO while it also decreases as the temperature is increased. As expected from the magnitude of their oxidation potentials, the oxidation of disubstituted phenols is more facile than monosubstituted phenols, which is in turn more facile than for phenol. The latter two classes of substrates as well as less sterically hindered 1,6-dialkylphenols tend to yield polymeric polyphenylene oxide type products. The [Fe(PC)] mediated stoichiometric autoxidation of 2,4-di-*t*-butyl phenol in methanol at atmospheric pressure yields *o*-dihydroxy coupled products, whereas 2,4,6-tri-*t*-butyl phenol gives mainly the 4-peroxo coupled product.[146] Under the same conditions, the 2,6-di-*t*-butyl phenol gives exclusively product I. While the reaction with [Fe(PC)] is slow, the Mn(II), Co(II), Ni(II), and Cu(II) complexes are nearly inactive. In addition, the reduction products, $[Co(PCTS)]^{n-}$, $n = 5, 6$, are also active in the autoxidation of polyhydric phenols such as catechol, hydroquinone, and pyrogallol to yield polymeric materials.[129] Conversely the oxidation of phenol and methylphenol does not occur. Apparently, electrochemically reduced [Ni(PCTS)] provides a similar catalytic system.

The [M(PC)] catalyzed autoxidation of the di-*t*-butyl phenols gives products similar to those of other one-electron oxidizing agents like Ag_2O or $[Fe(CN)_6]^{3-}$. In view of this and the known (or presumed) oxygen binding properties of [M(PC)] especially for Fe(II), Co(II), and Mn(II), it is reasonable to propose that the following general mechanism holds for the phenol autoxidations:

$$S + [M(PC)] + O_2 \rightleftharpoons [M(PC)(S)O_2] \tag{57}$$

$$[M(PC)(S)O_2] + RH \longrightarrow [M(PC)(S)O_2H] + R\cdot \tag{58}$$

$$[M(PC)(S)O_2H] \longrightarrow [M(PC)S] + HO_2\cdot \tag{59}$$

$$R' + HO_2\cdot \longrightarrow I + H_2O$$
$$2R' + \tfrac{1}{2}O_2 \longrightarrow II + H_2O \tag{60}$$

In reaction (57) the importance of axial solvent coordination in the activation of oxygen is supported by the fact that the reaction rate and selectivity are solvent sensitive. The metal-bound oxygen has been activated to such an extent that it is an efficient one-electron oxidizing agent, provided the substrate has a low-enough oxidation potential. Reaction (58) forms a phenoxy radical which has two principal resonance structures. In subsequent steps the phenoxy radical undergoes a series of reactions to yield a variety of products, which depends on the substituent pattern in the substrate. In the case of the sterically hindered 1,6 derivative, the possibilities are somewhat more restricted. The phenoxy radical can couple with the peroxy radical and eventually yield product I. In addition the phenoxy radical can couple with itself to form a dihydro derivative which is then oxidized (dehydrogenated) by molecular oxygen to yield product II.

The [M(PC)] is involved in at least three steps in the catalytic cycle. In reaction (57) the oxygen molecule is bound and activated, presumably by electron transfer to it from the metal. The active species is then the oxidized metal complex of the superoxide anion $M(III) \cdot O_2^-$. Of course, the extent (magnitude of equilibrium constant) to which this occurs is dependent on the particular metal. Further, in reaction (58), the reactivity (oxidizing power) of the bound superoxide anion is also dependent on the metal. Finally, the efficiency of the breakdown of the M(III) peroxy anion complex via a Haber–Weiss mechanism, in reaction (59), is metal dependent.[143] Although the catalaselike activity of [M(PC)] complexes has not been studied thoroughly, the $[Fe(PCTS)(H_2O)_2]^{4-}$ complex does decompose hydrogen peroxide quite readily. The exact mechanism is rather complicated but it does appear that a redox reaction of the metal is involved.[147] The low activity of the Cu(II) derivative may be due to the poor oxygen binding of [Cu(PC)]. In fact there is no evidence for oxygen binding of [Cu(PC)] complexes at ambient temperatures nor for the existence of high oxidation states like Cu(III), at least in the PC system.

An unsettled point that still stands relates to the physical state of the [M(PC)] catalyst in the phenol oxidations. The reaction mixture is described as being a slurry of solid catalyst in DMSO, etc. Undoubtedly some of the catalyst is in homogeneous solution. The important question then as to whether the solid or solution catalyst are both active and to what extent and with what selectivity is thus not answered. It is possible then that the variation in activity and selectivity may reflect [M(PC)] solubility and/or solid state effects.

Hydrocarbon autoxidation at ambient pressure to yield hydroperoxide is also promoted by solid [M(PC)] at 80–110°C. For example, cumene oxidation[148,149] activity is in the order Co(II) > Fe(II) > Mn(II) > Ni(II) > Zn(II) > Cu(II) > Mg(II). A reasonable mechanism would involve oxygen activation, hydrogen abstraction, and peroxide decomposition via steps (51)–(53). Subsequent steps in the chain reaction could be

$$R\cdot + O_2 \rightleftharpoons RO_2\cdot \qquad (61)$$

$$RO_2\cdot + RH \rightleftharpoons RO_2H + R\cdot \qquad (62)$$

$$R\cdot + \cdot O_2H \longrightarrow RO_2H \qquad (63)$$

The Mn(II), Fe(II), and Co(II) complexes are also good catalysts for the breakdown of the alkyl peroxide product giving the final product alcohol.

The mechanism here would presumably involve a Haber–Weiss-type process which features a M(II) → M(III) → M(II) change in metal oxidation state cycle. The [M(PC)] complexes with M = Cu(II), Mg(II), Zn(II) do not catalyze the decomposition of the peroxide.

When electron-donating substituents are present on the periphery of the PC ring in [Cu(PC)], then the rate of cumene autoxidation is enhanced.[149] For example, when R = OCH_3 or NO_2 (see Figure 1 for structure), the rate is some ten times faster than when R = Cl or Br. This observation supports a mechanism that involves a formal oxidation of the [Cu(PC)] since increased electron density in the macrocycle should enhance the ease of oxidation of the catalyst. It is unlikely that the metal center in the catalyst strongly binds and activates oxygen as it does for [Fe(PC)]. Thus, oxygen sorption by solid [Cu(PC)] is quite weak with desorption taking place at $-140°C$ with an activation energy of 7.5 kcal mol^{-1}. The corresponding values for [Fe(PC)] are $-115°C$ and 9.8 kcal mol^{-1}.[110] An alternate mechanism for oxygen activation may hold for [Cu(PC)] as well as for other metals like Mg(II), Zn(II), and possibly Ni(II) where metal center oxidation is not likely. It has been proposed that in these cases oxygen activation is accomplished by partial electron transfer from the PC ligand π orbital (see below) to the oxygen. Even though this pathway seems to be a less efficient way for oxygen activation, solid [Cu(PC)] catalyzes a variety of atmospheric pressure autoxidations. Thus the autoxidation of diphenylethane yields a peroxide, which subsequently decomposes to acetophenone and phenol.[150] Supported [Cu(PC)] on K_2CO_3, MgO, LiO_2, SiO_2, Al_2O_3 is active for the autoxidation of propylene to propyleneoxide.[151] On the other hand, α methylstyrene yields a peroxo derivative.[152] Conversely, the autoxidation of cyclohexene to cyclohexeneoxide is catalyzed by solid [Fe(PC)] but not by [Cu(PC)]. The opposite trend is noted for the tetraimido derivative

$$R = -\overset{\overset{\textstyle O}{\|}}{C}-NH-\overset{\overset{\textstyle O}{\|}}{C}-$$

i.e., Cu(II) is more active than Fe(II). Finally, both [Fe(PC)] and its tetraimido derivative catalyze the autoxidation of ethyleneacetal of acetaldehyde to form ethyleneglycol acetate.[153]

The autoxidation of sulfur-containing organic substrates has also been reported. For example, the oxidation (1 atm) of thiols yields disulfides in 2 M aqueous sodium hydroxide at 30°C:

$$CH_3CH_2SH + \tfrac{1}{2}O_2 \xrightarrow{\text{M(PC)}} (CH_3CH_2S)_2 + H_2O \qquad (64)$$

Here the order is Co(II) > Fe(II) > Cu(II).[154] Since it has been shown that at room temperature under basic conditions $[M(PCTS)]^{4-}$, M = Co(II), and Fe(II) bind oxygen in a 2-1 complex, it is reasonable to assume that an oxygen activation mechanism operates here as well. The order of catalytic activity for the metal is also consistent with this suggestion. Addition of the ligand CN^- to the Co(II) complex poisons the catalytic activity. This again suggests that reactant binding to the metal is required for catalytic activity. An attractive mechanism here is

$$RSH + OH^- \rightleftharpoons RS^- + H_2O \tag{65}$$

$$2[M(PC)OHH_2O]^- + O_2 \rightleftharpoons \{[M(PC)OH]_2^{2-} \cdot O_2\} \tag{66}$$

$$\{[M(PC)OH]_2^{2-} \cdot O_2\} + 2RS^- \rightleftharpoons \{[M(PC)SR]_2^{2-} \cdot O_2\} \tag{67}$$

$$2H_2O + \{[M(PC)SR]_2^{2-} O_2\} \longrightarrow [M(PC)] + H_2O_2 + 2RS\cdot + 2OH^- \tag{68}$$

$$2RS\cdot \longrightarrow (RS)_2 \tag{69}$$

The H_2O_2 could then be decomposed via a Haber–Weiss mechanism. Thiol autoxidation activity is also noted with $[Co(PCTS)]^{4-}$ in concentrated base.[155] When the formation of the 2:1 $Co\cdot O_2$ complex is prevented by isolating $[Co(PC)]$ on a polymer matrix, the catalytic activity increases fivefold. This may arise because the activity of $[Co(PC)(O_2)SR]^-$ is inherently greater than the 2:1 complex.

In the oxygen activation mechanism $[M(PC)]$ acts as an electron donor to oxygen, e.g.,

$$[M(PC)] + O_2 \rightleftharpoons [M(PC)]^+ \cdot O_2^- \tag{70}$$

Another general mechanism appears to occur in a number of other autoxidations, i.e., the catalyst acts as an electron acceptor. The selective gas-phase autoxidation of 2-propanol over solid $[M(PC)]$ at 100–400°C, atmospheric pressure, yields acetone[156]:

$$CH_3CHOHCH_3 + \tfrac{1}{2}O_2 \xrightarrow{[M(PC)]} CH_3COCH_3 + H_2O \tag{71}$$

The order of catalytic activity here is Cu(II) > Fe(II) > Ni(II) > Cr(II) > Mn(II) > Co(II). The observed order is roughly opposite to that for the phenol autoxidation. With this in mind it is reasonable to speculate that the autoxidation of 2-propanol follows a different pathway from that involving the activation of oxygen outlined previously. An attractive mechanism would involve a proton transfer from the alcohol to the aza nitrogen of the PC to form $[Cu(PCH)OCH(CH_3)_2]$. This could be followed by a hydride transfer from the axially bound alkoxy anion to another aza nitrogen to produce acetone and $[Cu(PCH_2)]$. The latter species would be the protonated form of the doubly reduced anion $[Cu(PC)]^{2-}$. $[Cu(PCH_2)]$ then could be oxidized by air to yield the catalyst and water. This final step may involve axial coordination of the oxygen to the metal prior to hydrogen transfer from the PCH_2.

The gase phase autoxidation of 2-propanethiol at 350°C and atmospheric pressure is catalyzed by solid $[Cu(PC)]$.[157,158] It is assumed that this reaction proceeds in a similar way to that of 2-propanol. Thus the initial reaction would yield $[Cu(PCH)SCH(CH_3)_2]$, which then breaks down with homolytic cleavage of the Cu–S bond to give $[Cu(PCH)] + \frac{1}{2}[(CH_3)_2CHS]_2$. The complex $[Cu(PCH)]$, a protonated form of the reduced species $[Cu(PC)]^-$, is then oxidized by oxygen to regenerate catalyst. During the course of this reaction (and in the absence of O_2), the 2-propanethiol also loses H_2S to form propylene. The liberated H_2S binds to $[Cu(PC)]$ to block further catalytic activity. However, when oxygen is admitted the H_2S is oxidized to SO_2 and H_2O. This interesting result shows that $[Cu(PC)]$ can activate a small molecule like H_2S.

The autoxidation at ambient temperature and pressure of hydrazine in aqueous base, pH = 13, to form nitrogen is strongly catalyzed by $[Co(PCTS)]^{4-}$ but little or no activity is noted with Fe(III), Mn(III), Cu(II), and for all the $[M(PC)]$ at pH < 12[159]:

$$N_2H_4 + O_2 \longrightarrow N_2 + 2H_2O \qquad (72)$$

In the case of the Mn(III) and Fe(III) complexes, hydrazine reduced the metal to at least the M(II) level. Subsequent reaction with oxygen leads to decolorization and presumably destruction of the ligand. Similarly, hydrazine reacts with the Co(II) complex to apparently form a Co(I) species since the visible spectrum shows an absorption characteristic of that oxidation state. Subsequent bubbling of oxygen through the solution gives a spectrum typical of the Co(II) adduct of molecular oxygen. The investigators prefer, however, to interpret the spectral changes as indicating that a ternary complex, $[Co(PCTS)(N_2H_4)O_2]^{4-}$, forms from the three reactants. The ternary complex breaks down to yield the products. The kinetics of the overall process can be described by a Michaelis–Menten rate law with $k = 140$ and $K = 160$ for 1 M aqueous NaOH at 25°C. The approximate activation energy is 8 kcal mol^{-1}. The intimate mechanism of the reaction must be complicated since it involves a series of four-proton and four-electron transfers to give the observed products. Hydroxylamine reacts with oxygen in the presence of $[Co(PCTS)]^{4-}$ to yield nitrogen, dinitrogen oxide, and nitrite ion only at pH > 10. The oxygen is reduced to water and hydrogen peroxide.[160] Again the catalytic role of the metal is thought to be to bring the reactants into the coordination sphere of the metal and then to transfer electrons from the reductant to the oxidant. Analysis of the Michaelis–Menten kinetics indicates the autoxidation of hydroxylamine is about 70 times faster than the autoxidation of hydrazine. Finally, the hydrogen peroxide that is formed initially reacts with hydroxylamine to yield nitrogen and dinitrogen oxide. The hydrogen peroxide reaction rate is about 1–2% of the oxygen reaction rate.

The electrocatalytic reaction of oxygen on graphite or carbon black electrodes[161,162] impregnated or coated with $[M(PC)]$ or with suspensions of the catalysts surrounding platinum electrodes[163] has been extensively studied. The order of catalytic activity in sulfuric acid medium is Fe(II) >

Co(II) > Ni(II) ⩾ Cu(II). The catalytic order and other evidence suggests that the electrocatalytic reaction proceeds via formation of the 1:1 oxygen-[M(PC)] adduct on the electrode surface.[164] The activation of the oxygen lowers the potential necessary to reduce it to hydrogen peroxide:

$$[M(PC)O_2] \xrightarrow[2H^+]{2e^-} [M(PC)] + H_2O_2 \qquad (73)$$

The reduction of the hydrogen peroxide is subsequently strongly catalyzed by [Fe(PC)]. Addition of ligands like CN^- poisons the catalyst presumably by displacing the bound oxygen from the metal. Further, electron-withdrawing groups on the periphery of the PC ligand depress the catalytic activity.[165] This is consistent with the first step in the electrocatalytic cycle being an oxidation of the [M(PC)].

An interesting gas phase oxidation–reduction reaction which is catalyzed by solid [M(PC)] at 250°C is[68]

The metal order for catalytic activity is Ru(II) > Mn(II) > Fe(II) > Co(II) > Pd(II) > Zn(II) > Cu(II) > Ni(II). This roughly corresponds to that for the ease of electrochemical oxidation of the [M(PC)] in chloronaphthalene. Thus, the initial, rate-determining step in the catalytic cycle is thought to be the oxidation of [M(PC)] by nitrobenzene. A similar reactivity pattern is observed when oxygen is used as the oxidant. The oxidized catalyst abstracts hydrogen atoms from the substrate dien to give benzene. It is interesting to note that those metals that give metal oxidation M(II) → M(III), i.e., Mn(II), Ru(II), Fe(II), Co(II) are more active than those that give PC ligand oxidation, e.g., Cu(II). Furthermore, the metal complexes of the tetrafluoro PC ligand (R = F) show decreased catalytic activity. The limited number of electron-withdrawing F substituents make catalyst oxidation more difficult when it is acting as an electron donor. Conversely, with perfluoro PC complexes, the great electron withdrawal of the substituents changes the nature of the catalytic action here. Thus, the catalyst acts as an electron acceptor and abstracts hydrogen atoms from the dien yielding [M(PCH₂)] and benzene even in the absence of oxygen. Subsequent oxidation of the reduced catalyst completes the cycle.

At this point it is of interest to compare the catalytic activity in oxygen activation reactions of PC complexes with those of the porphyrins. The electrochemical reduction of oxygen is facilitated by porphyrin complexes more than for the PC complexes for both Cu(II) and Co(II) from pH 0 to 14.[165] The catalytic activity of metalloporphyrins in reaction (74) is also higher than for PC complexes for Co(II), Ni(II), Cu(II), and Zn(II).[68] Both of these results correlate well with the ease of oxidation of the chelates. Thus it is easier to oxidize the porphyrin complexes than for PC complexes. This is especially true for the case where the metal is being oxidized, M(II) → M(III).

The intermediacy of reduced [M(PC)] derivatives has been proposed, and a growing number of useful catalytic systems developed for a range of reactions. For example, the hydrogenation of acetylene is catalyzed by alkaline solutions (pH 9–10) of $[Co(PCTS)]^{4-}$ at ambient temperature and pressure. The catalytic cycle is thought to be[166]

$$[Co(PCTS)(H_2O)_2]^{4-} \xrightarrow[\text{H}^+]{2e^-} [Co(PCTSH)(H_2O)_2]^{5-} \tag{75}$$

$$HC{\equiv}CH \Big\downarrow -H_2O$$

$$C_2H_4 + OH^- \xleftarrow{\ 2H_2O\ } [Co(PCTS)CH{=}CH_2(H_2O)]^{5-}$$

The apparent activation energy for the reduction is 8 kcal mol^{-1} and CO and O_2 inhibit the reaction completely. The first step in the process is the reduction of the Co(II) complex to a Co(I) complex of the protonated singly reduced form of PCTS^{4-}. The site of the ligand protonation is on an aza nitrogen since the IR spectrum shows a $\sim\nu$(N–H) at ~ 3000 cm^{-1}. The next step is the axial binding of the acetylene (presumably via a π complex) to the Co(I). Subsequently there is a hydride transfer from the macrocycle to the substrate. This results in the formation of an ethylene anion complex of Co(II). Hydrolysis of this species yields the product and regenerates the catalyst. Visible absorption spectral changes during the reaction are consistent with the oxidation state changes shown in the catalytic cycle. The reduction can also be carried out at a platinum electrode above -1.0 V (vs. SCE). The activity of several metals falls in the order Co(II) > Fe(II) $\sim$ Ni(II) > Cu(II). In this example, the catalyst is clearly acting as an electron acceptor.

Atmospheric pressure hydrogenation of allyl alcohol is facilitated over solid [Ni(PC)] at 100°C to form propanol. At 150–300°C, the product is predominantly propane.[167] Propionaldehyde forms propanol and then propane at 100–250°C while acrolein reacts quite rapidly at 100°C to form propanol, propionaldehyde, and propane. Presumably the hydrogenation reaction occurs via an acceptor mechanism with the formation of the reduced [M(PC)] complexes, i.e., dissociative absorption of hydrogen by [Ni(PC)], axial binding of the substrate to the nickel in [Ni(PCH$_2$)], then hydride–proton (or hydrogen atom) transfer and finally product dissociation. A variation of this mechanism most likely obtained for the [M(PC)] mediated HCO$_2$H decomposition.[168] The formic acid dissociatively adsorbs on the solid [M(PC)] to yield the complex [M(PCH)HCO$_2$]. Here the formate anion is axially bound to the metal while the proton is bound to an aza nitrogen of the

PC ligand. At elevated temperature, this complex undergoes a hydride transfer from the HCO_2^- to another aza nitrogen in the PC with the liberation of CO_2. The $[M(PCH_2)]$ then loses H_2 to regenerate the catalyst. The catalytic activity of $[M(PC)]$ is in the order $Cu(II) > Ni(II) > Co(II) \geqslant Fe(II) > Zn(II) \gg Mn(II)$. The actual product distribution which is a mixture of CO, H_2O, CO_2, and H_2, results from the water–gas-shift reaction. The inactivity of the latter two materials is linked to the fact that the $[Zn(PC)] \cdot HCOOH$ adduct is quite stable and can be isolated and that a stable Mn(III) salt, $[Mn(PC)HCO_2]$, forms from the oxidation of $[Mn(PC)]$.

The electroreduction of CO_2 in aqueous solution to form $C_2O_4^{2-}$ is catalyzed by $[M(PC)]$ impregnated on a graphite electrode.[169] The active catalyst is thought to be $[M(PC)]^{2-}$ and the order of activity is $Co(II) \sim Ni(II) \gg Cu(II) \sim Fe(II) > Mn(II) \sim Pd(II)$. It is suggested that the active complexes contain electrons in both the metal d_{z^2} and ligand $e_g(\pi^*)$ level. The catalytic reaction is thus seen to involve an axial coordination of CO_2 with subsequent electron transfer from the reduced $[M(PC)]$ to the substrate. The $\cdot CO_2^-$ then dimerizes to form the product.

The breadth and importance of acceptor catalysis can be judged by the number of reports concerning the catalytic activity of films of reduced $[M(PC)]$ catalysts,[170] $[M(PC)]^{n-}$. For example, solid films of $Na_2[M(PC)]$ at 170–240°C have been shown to catalyze the gas-phase reduction of CO with H_2 to yield methane and higher hydrocarbons.[171] The order of catalytic activity for $[M(PC)]^{2-}$ is $Fe(II) > Co(II) \geqslant Pt(II)$. Carbon monoxide is sorbed by the film in the temperature range 25–200°C but is only desorbed at higher temperature. The hydrogenation of propylene is also catalyzed by film of $Li_n[M(PC)]$. While $[M(PC)]$ is not active the anions like $[Fe(PC)]^{n-}$, $n = 3, 4$; $[Ni(PC)]^{n-}$, $n = 1, 2, 3, 4$; $[Co(PC)]^{n-}$, $n = 2, 3, 4, 5$; $[Zn(PC)]^{n-}$, $n = 1–4$ show good activity with the higher n values being the more active. In general, the complexes that are most active possess d^8 metal configuration and with the reduced PC ligand with one or more electrons in the $e_g(\pi^*)$ level.[172] Reaction of a film of $[Fe(PC)]^{4-}$ with H_2 gives a species with visible absorption peaks typical of $[Fe(PC)]^{3-}$ and $[Fe(PC)]^{2-}$. Evacuation of this film at 40°C reverses the process. The film, $Na_2[Fe(PC)]$, also catalyzes the reduction of nitrogen to yield ammonia.[173] Lower activity was noted with Mo(II), Ti(II), Co(II) while a negligible amount of NH_3 was formed with Cu(II), Pt(II), Ni(II), and Zn(II).

Related, but simpler, reactions involving hydrogen exchange are also mediated by films of reduced $[M(PC)]$ complexes.[174] The rapid H_2–D_2 exchange reaction is markedly dependent on the metal derivative and on the level of reduction of the film.[175–176] That is, the activity is generally greatest for $[M(PC)]^{n-}$ when $n = 4$ and it decreased as n decreases. Thus, the population of the $e_g(\pi^*)$ level of the PC is important in dictating reactivity here. The activity is dependent on the metal being in the order $Ni(II) > Co(II) >$

Fe(II) > Mn(II) ~ Mg(II), Cu(II), Zn(II). The exchange reaction is thought to occur via both the π-conjugated systems of the ligand and the metal site. Consistent with this, the reactivity of the latter site is poisoned by ligands like CO.

All the reactions of hydrogen mediated by the films of reduced [M(PC)] most likely proceed via a common mechanism. The first step may be a dissociative adsorption of hydrogen by $[M(PC)]^{n-}$ with the hydrogen associated with aza nitrogens. Higher electron density in the PC $e_g(\pi^*)$ orbital apparently facilitates this process. The substrate, acetylene, CO, N_2, then binds to the metal in the axial coordination position. Binding of the soft ligands is facilitated by the fact fact the metal is most likely in a low-valent state. The reaction proceeds via a series of hydride–proton or hydrogen atom transfers to the substrate. Readsorption of molecular hydrogen can occur and then the reaction proceeds until the final product is formed and dissociates from the metal coordination site.

It is obvious that the variety of catalytic activity noted for [M(PC)] derivatives is linked to the fact that the catalyst can act as an electron acceptor or donor and that reactant molecules can interact with at least two sites on the catalyst, i.e., the metal via the axial coordination site and with PC ligand via the aza nitrogen site. The large number of oxidation states readily available to the catalyst is also a major factor in the catalytic chemistry of [M(PC)].

The catalytic activity of a series of so-called sheetlike polymeric polyphthalocyanine metal complexes has been surveyed and in many cases found to be superior to that of the solid monomeric [M(PC)] complexes. The most obvious advantage of using the polymeric materials is the increase in surface area since they are in general microcrystalline. The synthesis of the materials is accomplished by carrying out the metal-mediated condensation reactions with tetrafunctional benzene derivatives like 1,2,4,5-tetracyanobenzene or pyromellitic dianhydride. In general, the products of the condensation reaction are a mixture which is difficult to work up. Polymeric materials are still quite impure even after extensive extraction with a variety of solvents. Finally, it is difficult to reproducibly prepare the polymeric solids from one laboratory to the next. In fact the only pure materials that can be isolated are a monomeric form of [M(PC)] with imide or carboxylic acid substituents on the peripheral sites of the ligand.[177] The lack of reproducibility in catalyst preparation and the likelihood that the polymeric solid has a number of active sites of both a metal and nonmetal-containing type makes it difficult to assign the genesis of catalytic activity in these materials. Nonetheless a number of studies have been reported. Thus mixed Cu(II), Fe(II) poly-PC complexes are good catalysts for the autoxidation of neat acetaldehyde ethyleneacetal at 30°C and 1 atm pressure.[178] The autoxidation of cumene is catalyzed by mixed and mono poly[M(PC)] in the order Cu(II)–Fe(II) > Cu(II)–Co(II) > Cu(II)–Mn(II) > Cu(II) > Co(II) > Cr(II) > Fe(II)–Co(II) > Mn(II) > Fe(II).[179,180] Surprisingly, neglecting the mixed metal complexes, the order is opposite to that of the monomeric [M(PC)] solids. This may arise from a change of mechanism in going from the monomeric [M(PC)] where oxygen activation by metal coordination is most prominent to one in the polymeric solids where ligand oxidation becomes more important. In addition, the site of oxygen adsorption may not be a metal ion but some defect in the organic sheet structure. This, however, is not consistent with the observation that the addition of large amounts of pyridine deactivates the catalyst presumably by occupying all the metal coordination sites to form poly[M(PC)(py)$_2$]. On the other hand, small amounts of pyridine activate the catalyst, perhaps by making oxygen binding more likely by prior coordination of one py to the metal site. It is interesting to note that the Cu(II), Fe(II), and mixed polymeric solid absorbs oxygen at room temperature and the oxygen can only be desorbed at 100–120°C.[179] The mixed complex also shows a second population of oxygen molecules that are desorbed at 300°C! This result is in contradiction to the results for the monomeric [M(PC)], which desorb near −120°C.[110] Since the latter work has been interpreted from the point of view of a nondissociative adsorption of oxygen, the results with the polymeric materials suggest a dissociative adsorption of oxygen. Whether the oxygen atoms on the surface of the latter catalyst are localized on metal or organic ligand sites or both is not known. The mixed Cu(II)–Fe(II) poly[M(PC)] is also found to be active in the autoxidation of acrolein.[181] Again, oxidation is enhanced by small amounts of pyridine. The chain reaction here is initiated by an aldehyde hydrogen abstraction by the oxygenated poly[M(PC)]. The general conclusion to be drawn from the autoxidation work on cumene and acrolein is that the polymeric catalysts appear to be fundamentally different from the monomeric catalysts. The origin of the difference is at this time uncertain. The oxidative dehydrogenation of alcohols over metal polyphthalocyanines at 60–150°C yields an aldehyde or ketone.[182] The caltaytic order for 2-propanol is Cu(II) > Fe(II) > Fe(II), Cu(II) > Fe(II), Pd(II) ≥ Fe(II), Mn(II). The dehydrogenation of ethanol and methanol is less facile. All the poly[M(PC)] are active while monomeric [Fe(PC)] and metal-free polyPC are not active. The proposed mechanism of the catalytic reaction involves oxygen binding to the metal site, activation of the oxygen, and then hydrogen abstraction from the substrate. The difference in reactivity of the monomeric and polymeric [M(PC)] in this reaction is ascribed to an unspecified "polymer" effect. Needless to say, there are serious reservations, based on catalyst purity, etc., that render mechanistic arguments quite speculative.

14. Comparison of Chemistry of Chromium, Manganese, Iron, and Cobalt Complexes

The coordination chemistry of the first-row transition metals, Cr, Mn, Fe, Co, generally has many common features, and trends in structural type and reactivity can be linked to the d-orbital population. Within the restrictions imposed by this ligand system, it is expected that the metallophthalocyanines will fit the same pattern. It is therefore of interest to compare here the properties of the Cr, Mn, Fe, and Co phthalocyanine derivatives.

All of the divalent metal complexes, [M(PC)], are square planar in the solid state and they show an intermediate spin state, i.e., the d electrons are distributed among the four lowest orbitals, b_{2g}, e_g, and a_{1g} so as to give the maximum free spin. The high-energy, sigma antibonding, b_{1g} level is not occupied. When the materials form the common tetragonal complex [M(PC)L$_2$] by reaction with Lewis bases, they adopt a low-spin state. In this case the a_{1g} orbital is not populated because it is destabilized by sigma inter-action with the axial ligands. The only one of the divalent metals to form a five-coordinate complex, [M(PC)L], is Co(II). This can be related to the fact that this metal has an electron in the antibonding a_{1g} orbital even in the low-spin form.

The ease of reaction of [M(PC)] toward oxygen varies strongly with the metal and is in the order Cr(II) > Mn(II) > Fe(II) > Co(II). Thus, Cr(II) is rapidly and irreversibly oxidized by air at ambient temperature to form a monomeric Cr(III) complex, [Cr(PC)OH]. The presence of a Lewis base, L, is required before [Mn(PC)] is irreversibly air oxidized to yield a dimeric Mn(III) complex, [Mn$_2$PC$_2$L$_2$O]. On the other hand, both [Fe(PC)] and [Co(PC)L] reversibly bind oxygen at low temperature in the form, $M(III) \cdot O_2^-$. At room temperature and strongly basic aqueous solution only [Co(PCTS)]$^{4-}$ and [Fe(PCTS)]$^{4-}$ are reversibly oxidized to form dimers of the form, $M(III) \cdot O_2^{2-} \cdot M(III)$. It thus appears that here the formation of the M(III) oxidation state is dependent on the charge on the PC ligand. In fact the Co(II)–Co(III) oxidation potential is 0.3 V less favorable for [Co(PC)] than for [Co(PCTS)]$^{4-}$ [161] In a like manner, [Co(PCH$_4$)]$^{4+}$ in concentrated sulfuric acid is still more difficult to oxidize. Finally, the order of reactivity to oxygen of [M(PC)] parallels the ease of electrochemical oxidation.[143]

The reactivity of [M(PC)] towards NO, to give M(III)NO$^-$, also follows the order Cr(II) > Mn(II) > Fe(II) > Co(II).[80] The reaction can be reversed by heating the solids in a nitrogen stream. The stability of [M(PC)NO] to removal of NO by this method or by ligand displacement with pyridine is the opposite of the oxidation order. The solids [VO(PC)], [Cu(PC)], and [Ni(PC)] do not react with NO.[80] This is surprising since [VO(PCTS)]$^{4-}$ is slowly oxidized by O$_2$ in aqueous base.[183] Here, of course, the charged ligand may be an important factor.

In general, the stability of the trivalent oxidation state in the PC com-plexes is greatest for Cr(III) and least for Co(III). For example, a large num-ber of six-coordinate Cr(III) complexes like [Cr(PC)XL] or [Cr(PC)L$_2$]X are known and five-coordinate structures are rare. Although Mn(III) is also stable and its complexes numerous, the majority of them are five coordinate.

The formation of a five-coordinate structure may be linked to the fact that the Mn(III) complexes are high spin and have one of the unpaired electrons occupying the a_{1g} orbital. The only example of a low-spin form, the dimer, is six coordinate.

For Fe and Co there are only a few substantiated reports of M(III) complexes and these are mainly for the PCTS^{4-} derivatives. All of these complexes appear to be six coordinate and low spin. No authenticated complexes of Cr, Mn, Fe, or Co in the high oxidation state M(IV) have been reported.

Reduction of all of the complexes, [M(PC)], yields materials with the metal in the low oxidation state, M(I). The stability of this oxidation state appears to be greatest for Co(I). In this case the reduction Co(II) $\rightarrow$ Co(I) is also dependent on the PC derivative and is in the order[161] $[Co(PCH_4)]^{4+} >$ $[Co(PC)] > [Co(PCTS)]^{4-}$.

The catalytic activity of [M(PC)] also varies with the metal and the order depends on how the complex is functioning in the catalytic cycle. When the complex is acting as an electron donor, e.g., in the autoxidation of organic substrates or electrochemical reduction of oxygen, the order of catalytic activity is Mn(II) > Fe(II) > Co(II). Unfortunately, little work has been done on Cr(II) catalysis because they are so easily oxidized and thus difficult to work with. When the catalyst is functioning as an electron donor as in the reduction (hydrogenation) of substrates then the order of activity is Co(II) > Fe(II) > Mn(II) > Cr(II). Both of the catalytic activity orders agree with the ease of oxidation and presumably reduction of the [M(PC)] complex, respectively. Thus it is reasonable to conclude that the lower oxidation state, M(I), becomes more stable going from Cr to Co while the higher oxidation state, M(III), is more stable in going in the opposite direction, i.e., Co to Cr. This difference in oxidation state stability may reflect the change in d-orbital population and the change in relative position of the metal and PC, σ, and π orbitals. Further theoretical and experimental consideration of these observations seem warranted.

Abbreviations

TPP	Dianion of $\alpha,\beta,\gamma,\delta,$ tetraphenylporphine	im	Imidazole
		CH$_3$im	2-Methylimidazole
TTpy	Dianion of $\alpha,\beta,\gamma,\delta,$ tetrapyridylporphine	quin	Quinoline
		isoquin	Isoquinoline
OEP	Dianion of octaethylporphyrin	CH$_3$quin	2-Methylquinoline
ClPC	Monochlorophthalocyanine	DMSO	Dimethylsulfoxide
py	Pyridine	DMF	Dimethylformamide
γ-pic	γ-Picoline, 4-methyl pyridine	DMA	Dimethylacetamide
β-pic	β-Picoline, 3-methyl pyridine	HMP	Hexamethylphosphoramide
pip	Piperidine	tolNO	p-Nitrosotoluene
BuNH$_2$	n-Butylamine		

References

1. F. H. Moser and A. L. Thomas, *Phthalocyanine Compounds*, Reinhold Publishing Corp., New York (1963).
2. F. H. Moser and A. L. Thomas, *J. Chem. Ed.* **41**, 245 (1964).
3. A. B. P. Lever, *Adv. Inorg. Radiochem.* **7**, 28 (1965).
4. D. D. Ledson and M. V. Twigg, *Inorg. Chim. Acta* **13**, 43 (1975).
5. V. W. Day, T. J. Marks, and W. A. Wachter, *J. Am. Chem. Soc.* **97**, 4519 (1975).
6. J. M. Robertson, *J. Chem. Soc.* **1936**, 1195 (1936).
7. C. J. Brown, *J. Chem. Soc.* (*A*) **1968**, 2488 (1968).
8. C. J. Brown, *J. Chem. Soc.* (*A*) **1968**, 2494 (1968).
9. E. B. Fleischer, C. K. Miller, and L. E. Webb, *J. Am. Chem. Soc.* **86**, 2342 (1964).
10. T. A. Hamor, W. S. Caughey, and J. L. Hoard, *J. Am. Chem. Soc.* **87**, 2305 (1965).
11. M. S. Fischer, D. H. Templeton, A. Zalkin, and M. Calvin, *J. Am. Chem. Soc.* **93**, 2622 (1971).
12. T. Kobayashi, T. Ashida, N. Uyeda, E. Suito, and M. Kakudo, *Bull. Chem. Soc. Japan* **44**, 2095 (1971).
13. R. Timkovich and A. Tulinsky, *J. Am. Chem. Soc.* **91**, 4430 (1969).
14. T. Kobayashi, F. Kurokawa, T. Ushida, W. A. U. Yeda, and E. Suito, *Chem. Commun.*, 1631 (1971).
15. F. Cariati, F. Morazzoni, and M. Zocchi, *Inorg. Chim. Acta* **14**, L31 (1975).
16. D. Rogers and R. S. Osborn, *Chem. Commun.*, 840 (1971).
17. J. R. Mooney, C. K. Choy, K. Knox, and M. Kenney, *J. Am. Chem. Soc.* **97**, 3033 (1975).
18. L. J. Radonavich, A. Bloom, and J. L. Hoard, *J. Am. Chem. Soc.* **94**, 2073 (1975).
19. J. A. Ibers, J. W. Lauher, and R. G. Little, *Acta Crystallogr.* **B30**, 268 (1974).
20. D. L. Cullen and E. F. Meyer, *Chem. Commun.*, 616 (1971).
21. M. K. Friedel, B. F. Hoskins, R. L. Martin, and S. A. Mason, *Chem. Commun.*, 400 (1970).
22. W. E. Bennett, D. E. Broberg, and N. C. Baenziger, *Inorg. Chem.* **12**, 930 (1973).
23. A. Gieren and W. Hoppe, *Chem. Commun.*, 413 (1971).
24. L. H. Vogt, A. Zalkin, and D. H. Templeton, *Inorg. Chem.* **6**, 1725 (1967).
25. V. W. Day, B. R. Stultz, E. L. Tasset, R. S. Marianelli, and L. J. Boucher, *Inorg. Nucl. Chem. Lett.* **11**, 505 (1975).
26. B. F. Hoskins, S. A. Mason, and J. C. B. White, *Chem. Commun.*, 554 (1969).
27. A. M. Schaffer, M. Gouterman, and E. R. Davidson, *Theoret. Chim. Acta* **30**, 9 (1973).
28. S. C. Mathur and J. Singh, *Int. J. Quan. Chem.* **6**, 57 (1972); **6**, 747 (1972); **8**, 79 (1974).
29. I. Chen, M. Abkowitz, and J. H. Sharp, *J. Chem. Phys.* **50**, 2237 (1969).
30. A. Henriksson and M. Sundbom, *Theoret. Chim. Acta* **27**, 213 (1972).
31. I. Chen, *J. Mol. Spectrosc.* **23**, 131 (1967).
32. L. Edwards and M. Gouterman, *J. Mol. Spectrosc.* **33**, 292 (1970).
33. M. J. Stillman and A. J. Thomson, *Trans. Faraday Soc. II* **70**, 790 (1974).
34. P. E. Fielding and A. G. McKay, *Aust. J. Chem.* **28**, 1445 (1975).
35. D. Eastwood, L. Edwards, M. Gouterman, and J. Steinfeld, *J. Mol. Spectrosc.* **20**, 381 (1966).
36. L. Bajema, M. Gouterman, and B. Meyer, *J. Mol. Spectrosc.* **27**, 225 (1968).
37. P. S. Vincent, E. M. Voight, and K. E. Rieckoff, *J. Chem. Phys.* **55**, 4131 (1971).
38. Y. Niwa, H. Kobayashi, and T. Tsuchiya, *Inorg. Chem.* **13**, 2891 (1974).
39. M. V. Zeller and R. G. Hayes, *J. Am. Chem. Soc.* **95**, 3854 (1973).

40. A. G. McKay, *Aust. J. Chem.* **26**, 2425 (1973).
41. C. M. Guzy, J. B. Raynor, and M. C. R. Symons, *J. Chem. Soc.* (*A*) **1969**, 2299 (1969).
42. R. L. Martin and S. Mitra, *Inorg. Chem.* **9**, 182 (1970).
43. L. D. Rollman and S. I. Chan, *Inorg. Chem.* **10**, 1978 (1971).
44. R. L. Martin and S. Mitra, *Chem. Phys. Lett.* **3**, 183 (1969).
45. J. M. Assour, *J. Am. Chem. Soc.* **87**, 4701 (1965).
46. M. Sato and T. Kwan, *J. Chem. Phys.* **50**, 558 (1969).
47. C. M. Guzy, J. B. Raynor, L. P. Stodulski, and M. C. R. Symons, *J. Chem. Soc.* (*A*) **1969**, 997 (1969).
48. J. M. Assour, J. Goldmacher, and S. E. Harrison, *J. Chem. Phys.* **43**, 159 (1965).
49. P. D. W. Boyd and T. D. Smith, *J. Chem. Soc. Dalton Trans.* **1972**, 839.
50. I. M. Keen and B. W. Malerbi, *J. Inorg. Nucl. Chem.* **27**, 1311 (1965).
51. H. F. Shurvell and L. Penzuti, *Can. J. Chem.* **44**, 125 (1966).
52. B. I. Knudsen, *Acta Chem. Scand.* **20**, 1344 (1966).
53. D. D. Eley, D. J. Hazeldine, and T. F. Palmer, *Trans. Faraday Soc. II* **70**, 1808 (1974).
54. K. Varmuza, G. Maresch, and A. Meller, *Mh. Chem.* **105**, 327 (1974).
55. J. F. Boas, P. E. Fielding, and A. G. McKay, *Aust. J. Chem.* **27**, 7 (1974).
56. A. G. McKay, J. F. Boas, and G. J. Troup, *Aust. J. Chem.* **27**, 955 (1974).
57. P. Clare and F. Glocking, *Inorg. Chim. Acta* **14**, L12 (1975).
58. T. H. Hurley, M. A. Robinson, and S. I. Trotz, *Inorg. Chem.* **6**, 389 (1967).
59. R. Taube, *Z. Chem.* **6**, 8 (1966).
60. D. W. Clack, N. S. Hush, and J. R. Yandle, *Chem. Phys. Lett.* **1**, 157 (1967).
61. D. W. Clack and Y. R. Yandle, *Inorg. Chem.* **11**, 1738 (1972).
62. R. E. Linder, J. R. Rowlands, and N. S. Hush, *Mol. Phys.* **21**, 417 (1971).
63. R. Taube and H. Arfert, *Z. Naturforsch.* **22b**, 219 (1967).
64. D. Lexa and M. Reix, *J. Chim. Phys.* **71**, 511 (1974).
65. D. Lexa and M. Reix, *J. Chim. Phys.* **71**, 517 (1974).
66. L. D. Rollman and R. T. Iwamato, *J. Am. Chem. Soc.* **90**, 1455 (1968).
67. J. F. Meyers, G. W. Rayner-Canham, and A. B. P. Lever, *Inorg. Chem.* **14**, 461 (1975).
68. J. Manassen and A. Bar-Ilan, *J. Catal.* **17**, 86 (1970).
69. H. Sigel, P. Waldmeier, and B. Prijs, *Inorg. Nucl. Chem. Lett.* **7**, 161 (1971).
70. Z. A. Schelly, R. D. Farina, and E. M. Eyring, *J. Phys. Chem.* **74**, 617 (1970); Z. A. Schelly, D. J. Harward, P. Hemmes, and E. M. Eyring, *J. Phys. Chem.* **74**, 3040 (1970).
71. R. D. Farina, D. J. Halko, and J. H. Swinehart, *J. Phys. Chem.* **76**, 2343 (1972).
72. E. W. Abel, J. M. Pratt, and R. Whelan, *J. Chem. Soc. Dalton Trans.* **1976**, 509 (1976).
73. J. A. DeBolfo, T. D. Smith, J. F. Boas, and J. R. Pilbrow, *Trans. Faraday Soc. II* **72**, 481 (1976).
74. A. R. Monahan, J. A. Brado, and A. F. DeLuca, *J. Phys. Chem.* **76**, 446, 1994 (1972).
75. M. Abkowitz and A. R. Monahan, *J. Chem. Phys.* **58**, 2281 (1973).
76. J. G. Jones and M. V. Twigg, *Inorg. Nucl. Chem. Lett.* **8**, 305 (1972).
77. J. H. Sharp and M. Abkowitz, *J. Phys. Chem.* **77**, 477 (1973).
78. E. G. Melone, L. R. O'Cone, and B. P. Block, *Inorg. Chem.* **6**, 424 (1967).
79. C. Ercolani, C. Neri, and P. Porta, *Inorg. Chim. Acta* **1**, 415 (1967).
80. A. B. P. Lever, *J. Chem. Soc.* **1968**, 1821 (1968).
81. C. Ercolani and C. Neri, *J. Chem. Soc.* **1967**, 1715.

82. C. Ercolani, C. Neri, and G. Sartori, *J. Chem. Soc. (A)* **1968**, 2133 (1968).
83. R. Taube and K. L. Lunkenheimer, *Z. Naturforsch.* **19b**, 653 (1964).
84. C. G. Barraclough, R. L. Martin, S. Mitra, and R. C. Sherwood, *J. Chem. Phys.* **53**, 1638 (1970).
85. H. Miyoshi, H. Ohya-Nishiguchi, and Y. Deguchi, *Bull. Chem. Soc. Japan* **46**, 2724 (1973).
86. H. Miyoshi, *Bull. Chem. Soc. Japan* **47**, 561 (1974).
87. A. Yamamoto, L. K. Phillips, and M. Calvin, *Inorg. Chem.* **7**, 847 (1968).
88. G. W. Rayner-Canham and A. B. P. Lever, *Inorg. Nucl. Chem. Lett.* **9**, 513 (1973).
89. J. H. Weber and D. H. Busch, *Inorg. Chem.* **4**, 469 (1965).
90. K. Fenkart and C. H. Brubaker, Jr., *Inorg. Nucl. Chem. Lett.* **4**, 335 (1968); *J. Inorg. Nucl. Chem.* **30**, 3245 (1968).
91. H. Hanke, *Z. Anorg. Allg. Chem.* **355**, 160 (1967).
92. R. Taube, H. Munke, and J. Petersen, *Z. Anorg. Allg. Chem.* **390**, 257 (1972).
93. J. G. Jones and M. V. Twigg, *Inorg. Chem.* **8**, 2018 (1969).
94. C. G. Barraclough, R. L. Martin, S. Metra, and R. C. Sherwood, *J. Chem. Phys.* **5**, 1643 (1970).
95. B. W. Dale, *Mol. Phys.* **28**, 503 (1974).
96. T. S. Srivastava, J. L. Przylinski, and A. Nath, *Inorg. Chem.* **13**, 1562 (1974).
97. J. E. Maskasky, J. R. Mooney, and M. E. Kenney, *J. Am. Chem. Soc.* **94**, 2132 (1972).
98. B. W. Dale, R. J. P. Williams, P. R. Edwards, and C. E. Johnson, *Trans. Faraday Soc.* **64**, 620 (1968).
99. D. V. Stynes, *J. Am. Chem. Soc.* **96**, 5943 (1974).
100. J. J. Watkins and A. L. Balch, *Inorg. Chem.* **14**, 2720 (1975).
101. A. Hudson and H. J. Witfield, *Inorg. Chem.* **6**, 1120 (1967).
102. T. H. Moss and A. B. Robinson, *Inorg. Chem.* **7**, 1692 (1968).
103. M. J. Stillman and A. J. Thomson, *Trans. Faraday Soc. II* **70**, 805 (1974).
104. J. G. Jones and M. V. Twigg, *Inorg. Chim. Acta* **10**, 103 (1974).
105. D. V. Stynes and B. R. James, *J. Am. Chem. Soc.* **96**, 2733 (1974).
106. J. G. Jones and M. V. Twigg, *Inorg. Nucl. Chem. Lett.* **5**, 333 (1969).
107. H. P. Bennetto, J. G. Jones, and M. V. Twigg, *Inorg. Chim. Acta* **4**, 180 (1970).
108. J. G. Jones and M. V. Twigg, *Inorg. Chem. Acta* **12**, L15 (1975); *Inorg. Chem.* **8**, 2121 (1969).
109. D. Vonderschmitt, K. Bernauer, and S. Fallab, *Helv. Chim. Acta* **48**, 951 (1965).
110. J.-P. Contour, P. Lenfant, and A. K. Vigh, *J. Catal.* **9**, 8 (1973).
111. L. L. Dickens and J. C. Fanning, *Inorg. Nucl. Chem. Lett.* **12**, 1 (1976).
112. J. G. Jones and M. V. Twigg, *Inorg. Nucl. Chem. Lett.* **6**, 245 (1970).
113. J. G. Jones and M. V. Twigg, *J. Chem. Soc. (A)* **1970**, 1546 (1970).
114. J. G. Jones and M. V. Twigg, *Inorg. Chim. Acta* **4**, 602 (1970).
115. R. Taube, H. Drevs, and T. Duc Hiep, *Z. Chem.* **9**, 115 (1969).
116. M. E. Volpin, R. Taube, H. Drevs, L. G. Volkova, I. Y. A. Levitin, and T. M. Ushakova, *J. Organomet. Chem.* **39**, C79 (1972).
117. F. Cariati, D. Galizzioli, F. Morazzoni, and C. Busetto, *J. Chem. Soc. Dalton Trans.* **1975**, 556 (1975).
118. J. N. Assour and W. K. Kahn, *J. Am. Chem. Soc.* **87**, 207 (1965).
119. T. Kobayashi, N. Uyeda, and E. Suito, *J. Phys. Chem.* **72**, 2446 (1968).
120. F. Cariati, F. Morazzoni, and C. Busetto, *J. Chem. Soc. Dalton Trans.* **1976**, 496 (1976).
121. F. A. Walker, *J. Am. Chem. Soc.* **92**, 4235 (1970).

122. C. Busetto, F. Cariati, D. Galizzioli, and F. Morazzoni, *Gaz. Chim. Ital.* **104**, 161 (1974).
123. E. W. Abel, J. M. Pratt, and R. Whelan, *Chem. Commun.*, 449 (1971).
124. L. C. Gruen and R. J. Blagrove, *Aust. J. Chem.* **26**, 319 (1973).
125. D. M. Wagnerová, E. Schwertnerova, and J. Veprek-Siska, *Coll. Czech. Chem. Commun.* **39**, 1980 (1974).
126. J. H. Weber and D. H. Busch, *Inorg. Chem.* **4**, 473 (1965).
127. P. Day, H. A. O. Hill, and M. G. Price, *J. Chem. Soc.* (*A*) **1968**, 90 (1968).
128. D. H. Busch, J. H. Weber, D. H. Williams, and N. J. Rose, *J. Am. Chem. Soc.* **86** 5161 (1964).
129. S. Mechitsuka, M. Ichikawa, and K. Tamaru, *Chem. Commun.* 360 (1975).
130. M. K. Lowey, A. J. Starshak, J. N. Esposito, P. C. Krueger, and M. E. Kenney, *Inorg. Chem.* **4**, 128 (1965).
131. A. J. Starshak, R. D. Joyner, and M. E. Kenney, *Inorg. Chem.* **5**, 331 (1966).
132. J. N. Esposito, J. E. Floyd, and M. E. Kenney, *Inorg. Chem.* **5**, 1979 (1966).
133. A. R. Kane, J. F. Sullivan, D. H. Kenny, and M. E. Kenney, *Inorg. Chem.* **9**, 1445 (1970).
134. G. Meyer and D. Wohrle, *Die Macromol. Chem.* **175**, 714 (1974).
135. R. L. Stover, C. L. Thrall, and R. D. Joyner, *Inorg. Chem.* **10**, 2335 (1971).
136. J. N. Esposito, L. E. Sutton, and M. E. Kenney, *Inorg. Chem.* **6**, 1116 (1967); A. R. Kane, R. G. Yalman, and M. E. Kenney, *Inorg. Chem.* **7**, 2588 (1968).
137. B. P. Block and E. G. Melone, *Inorg. Chem.* **4**, 111 (1965).
138. W. J. Kroenke and M. E. Kenney, *Inorg. Chem.* **3**, 251, 696 (1964).
139. L. E. Sutton and M. E. Kenney, *Inorg. Chem.* **6**, 1869 (1967).
140. M. O'Rourke and C. Curran, *J. Am. Chem. Soc.* **92**, 1501 (1970).
141. H. W. Krause, *Fortschr. Chem. Forsch.* **6**, 327 (1966).
142. W. Hanke, *Z. Chem.* **9**, 1 (1969).
143. J. Manassen, *Catal. Rev.* **9**, 223 (1974).
144. J. Manassen, *Fortschr. Chem. Forsch.* **25**, 1 (1972).
145. V. M. Kothari and J. J. Tazuma, *J. Catal.* **41**, 180 (1976).
146. M. Tada and T. Katsu, *Bull. Chem. Soc. Japan* **45**, 2558 (1972).
147. P. Waldmeier and H. Sigel, *Inorg. Chim. Acta* **5**, 659 (1971).
148. H. Kropf, *Angew. Chem. Int. Ed.* **11**, 239 (1972).
149. H. Kropf and Hd. Hoffman, *Tetrahedron Lett.*, 659 (1967).
150. H. Kropf, W. Gebert, and K. Franke, *Tetrahedron Lett.*, 5527 (1968).
151. V. Ragaini and R. Saravalle, *Rect. Kinet. Catal. Lett.*, **1**, 271 (1974).
152. Y. Kamiya, *Tetrahedron Lett.* 4965 (1968).
153. D. J. Baker, D. R. Boston, and J. C. Bailar, Jr., *J. Inorg. Nucl. Chem.* **35**, 153 (1973).
154. C. F. Cullis and D. L. Trimm, *Discuss. Faraday Soc.* **28**, 144 (1968).
155. T. A. M. M. Maas, M. Kuijer, and J. Zwart, *Chem. Commun.* 86 (1976).
156. F. Steinbach and K. Hiltner, *Z. Phys. Chem. N. F.* **83**, 126 (1973).
157. F. Steinbach and H. Schmidt, *J. Catal.* **29**, 515 (1973).
158. F. Steinbach and H. H. Schmidt, *J. Catal.* **39**, 190 (1975).
159. D. M. Wagnerová, E. Schwertnerova, and J. Veprek-Siska, *Coll. Czech. Chem. Commun.* **38**, 756 (1973).
160. D. M. Wagnerová, E. Schwertnerova, and J. Veprek-Siska, *Coll. Czech. Chem. Commun.* **39**, 3036 (1974).
161. F. Beck, *Ber. Bunsenges. Phys. Chem.* **77**, 353 (1973).
162. M. Savy, P. Andro, C. Bernard, and G. Magner, *Electrochim. Acta* **18**, 191 (1973).
163. H. Alt, H. Binder, and G. Sandstede, *J. Catal.* **28**, 8 (1973).
164. W. Beyer and F. von Sturm, *Angew. Chem. Int. Ed.* **11**, 140 (1972).

165. J. Manassen, *J. Catal.* **33**, 133 (1974).
166. M. Ichikawa, R. Sonoda, and S. Meshitsuka, *Chem. Lett.* 709 (1973).
167. H. Kropf and D. J. Witt, *Z. Phys. Chem. N. F.* **76**, 331 (1971).
168. W. Hanke and D. Gutschick, *Z. Anorg. Allg. Chem.* **366**, 201 (1969).
169. S. Meshitsuka, M. Ichikawa, and K. Tamaru, *Chem. Comm.* 158 (1974).
170. K. Tamaru, *Advan. Catal.* **20**, 327 (1969); *Catal. Rev.* **4**, 161 (1971).
171. M. Ichikawa, M. Sudo, R. Soma, T. Onishi, and K. Tamaru, *J. Am. Chem. Soc.* **91**, 1538 (1969).
172. M. Ichikawa, M. Soma, T. Onishi, and K. Tamaru, *Bull. J. Chem. Soc. Jpn.* **41**, 1739 (1968).
173. M. Sudo, M. Ichikawa, M. Soma, T. Onishi, and K. Tamaru, *J. Phys. Chem.* **72**, 1174 (1968).
174. M. Ichikawa, M. Soma, T. Onishi, and K. Tamaru, *J. Phys. Chem.* **70**, 2069 (1966).
175. M. Ichikawa, M. Soma, T. Onishi, and K. Tamaru, *Trans. Faraday Soc.* **64**, 1215 (1968).
176. S. Naito, M. Ichikawa, and K. Tamaru, *Trans. Faraday Soc.* **68**, 1451 (1972).
177. D. R. Boston and J. C. Bailar, Jr., *Inorg. Chem.* **11**, 1578 (1972).
178. H. Inoue, Y. Koda, and E. Imoto, *Bull. J. Chem. Soc. Jpn.* **40**, 184 (1967).
179. T. Hara, Y. Ohkatsu, and T. Osa, *Bull. Chem. Soc. Jpn.* **48**, 85 (1975).
180. T. Hara, Y. Ohkatsu, and T. Osa, *Chem. Lett.* **1973**, 953 (1973).
181. T. Hara, Y. Ohkatsu, and T. Osa, *Chem. Lett.* **1973**, 103 (1973).
182. S. Naito and K. Tamaru, *Z. Phys. Chem. N. F.* **94**, 150 (1975).
183. E. Schwertnerova, D. M. Wagnerová, and J. Veprek-Siska, *Z. Chem.* **14**, 311 (1974).

8

Coordination Chemistry of Porphyrins

Lawrence J. Boucher

1. Introduction

The current interest in the coordination chemistry of synthetic macrocyclic ligands stems directly from the fact that among the most important biological compounds that contain metal ions are those with naturally occurring macrocyclic ligands, e.g., the porphyrins, the chlorophylls, and the corrins. These ligands are generally bound to iron, magnesium, and cobalt, respectively, in the biological systems. Work on the biological compounds has been extensive and many reviews and monographs give the recent results.[1-6] Model compound studies of both the synthetic and naturally occurring macrocyclic ligand complexes have, for the most part, been useful in arriving at an enhanced understanding of the function of the metal in the biological system.[7-10] The recent literature on metalloporphyrins is vast and there are many excellent reviews on various aspects of the subject.[11-13] This review will only summarize the most important parts of the various topics and will reference pertinent reviews or exemplary articles. Details of experiment and interpretation, although of great interest, will generally not be given here.

2. Synthesis

The porphyrin ligands that are used to form metal complexes can be divided into two general classes. The first class, the porphins (or alternately called porphines), do not have substituents on the pyrrole rings but generally have substituents on the methine (bridge) carbon of the macrocycle. The

Lawrence J. Boucher · Department of Chemistry, Western Kentucky University, Bowling Green, Kentucky 42101.

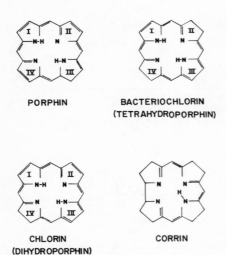

PORPHIN   BACTERIOCHLORIN
           (TETRAHYDROPORPHIN)

CHLORIN   CORRIN   *FIGURE 1.* Structural representation of
(DIHYDROPORPHIN)   tetrapyrrole macrocyclic ligands.

parent macrocyclc porphin is shown in Figure 1, while the popular *meso*-substituted tetraphenylporphins TPP are shown in Figure 2 with the nomenclature commonly used. The second class, the porphyrins, have substituents on the pyrrole rings but generally do not have any on the methine carbons. Structures and nomenclature for these are given in Figure 3. The porphins are produced synthetically,[14] while the porphyrins are generally isolated from natural sources directly[15] or are derived from the naturally occurring

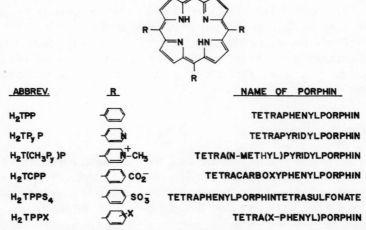

ABBREV.	R	NAME OF PORPHIN
H_2TPP	⟨phenyl⟩	TETRAPHENYLPORPHIN
H_2TP$_y$P	⟨pyridyl⟩	TETRAPYRIDYLPORPHIN
H_2T(CH$_3$P$_y$)P	⟨N$^+$-CH$_3$⟩	TETRA(N-METHYL)PYRIDYLPORPHIN
H_2TCPP	⟨phenyl⟩CO$_2^-$	TETRACARBOXYPHENYLPORPHIN
H_2TPPS$_4$	⟨phenyl⟩SO$_3^-$	TETRAPHENYLPORPHINTETRASULFONATE
H_2TPPX	⟨phenyl⟩X	TETRA(X-PHENYL)PORPHIN

FIGURE 2. Structural representation and nomenclature for α-, β-, γ-, δ-*meso*-tetraphenyl-porphin derivatives. X = *p*-Cl, *p*-CH$_3$, *m*-CH$_3$, *o*-CH$_3$, *p*-CN, *p*-NO$_2$, *p*-OH, *p*-OCH$_3$, *p*-OC$_2$H$_5$, *p*-*n*C$_3$H$_7$.

macrocycle by simple transformation of one or two of the ring substituents.[16] An exception here is the much used octaethylporphyrin, which is synthesized in the laboratory.[17]

Strictly speaking the other porphyrins may be totally synthesized as well although the yields are low and the procedures tedious and time consuming.[18] The general term used here for both classes of materials will simply be porphyrins or Por. Of course the various porphyrins differ from each other in both chemical and physical properties. From the point of view of their utility as ligands, relevant properties are their solubility (and of their complexes as well) and donor strength. The TPP derivatives without charged groups are generally soluble only in nonaqueous solvents. On the other hand, those with charged groups or pyridine ring have some solubility in basic

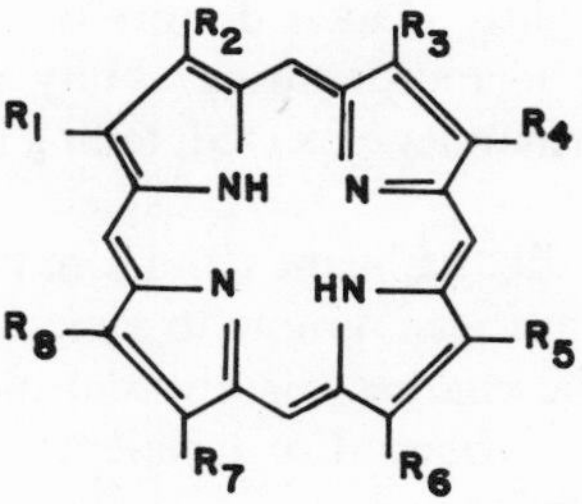

ABBREV.	R_1	R_2	R_3	R_4	R_5	R_6	R_7	R_8
H_2POR	H	H	H	H	H	H	H	H
H_2OEP	C_2H_5	C_2H_5	C_2H_5	C_2H_5	C_2H_5	C_2H_5	C_2H_5	C_2H_5
H_2ETIO	CH_3	C_2H_5	CH_3	C_2H_5	CH_3	C_2H_5	CH_3	C_2H_5
H_2MESO	CH_3	C_2H_5	CH_3	C_2H_5	CH_3	P	P	CH_3
H_2DEUT	CH_3	H	CH_3	H	CH_3	P	P	CH_3
H_2PROT	CH_3	V	CH_3	V	CH_3	P	P	CH_3
H_2PEN	CH_3	E	CH_3	E	CH_3	N	N	CH_3
H_2HAEM	CH_3	B	CH_3	B	CH_3	P	P	CH_3

$P = CH_2CH_2CO_2CH_3$, $V = CH = CH_2$, $E = CH_2CH_2NHCH_2CH_2NH_2$

$N = CH_2CH_2CONHCH_2CH_2NH_2$, $B = CHOHCH_3$

ABBREV.	R_1	R_2	R_3	R_4	R_5	R_6	R_7	R_8
H_2COP	CH_3	C	CH_3	C	CH_3	C	CH_3	C
H_2URO	A	C	A	C	A	C	A	C

$C = CH_2CH_2CO_2H$, $A = CH_2CO_2H$

FIGURE 3. Structural representation and nomenclature for porphins.

or acid aqueous solution. Similarly, the porphyrins shown in Figure 3 have appreciable solubility in nonaqueous solvents only when the methyl esters are present. Conversely, solubility in neutral or basic solution is noted for those materials with carboxylic substituents and in neutral acidic solution for those with amine substituents. The donor strength of the porphyrins is related to the presence of electron-attracting or -withdrawing substituents on the macrocycle, i.e., the greater the electron-donating power of the substituents the better the donating ability of the porphyrin. The order of donor strength of the porphyrins estimated from pK_3 measurements is[19]

OEP > Etio > *Meso* > HAEM > TPP > Deut > Prot

The position of PEN is expected to be about the same as *Meso*, while that of URO and COP should be a little lower than *Meso*. The carboxylic acid derivatives should be slightly weaker donors than the methyl esters. Within the TPP derivatives the ligand-donating ability varies with the substituent on the phenyl ring in the manner expected, taking into account both inductive and conjugative effects.[20]

The porphyrins are dibasic ligands and lose two protons upon coordination of the four pyrrole nitrogen donors to a metal. The dinegative charge on the ligand will satisfy the charges for divalent metals, and thus inner complexes, [M(Por)], can be formed. For trivalent and tetravalent metals additional anionic ligands are required to form complexes like [M(Por)X] and [M(Por)X$_2$]. The ease and method of formation of metal complexes of porphyrins depend on the particular metal. Several general procedures have been developed for synthesis of metalloporphyrins.[21] One approach is typified by the reaction of a divalent metal salt in refluxing dimethyl formamide or glacial acetic acid:

$$M^{2+} + H_2Por \longrightarrow [M(Por)] + 2H^+ \tag{1}$$

Using basic anions to reduce the H^+ concentration can enhance the reaction. Subsequent exposure to oxygen may also give the trivalent oxidation state of the metal:

$$2[M(Por)] + 2H^+ + \tfrac{1}{2}O_2 \longrightarrow 2[M(Por)]^+ + H_2O \tag{2}$$

Subsequent addition of a coordinating anion X^- will give [M(Por)X], etc. The first-row transition elements Cr, Fe, Mn, Co, Ni, Cu give the M(II) complex as in reaction (1). In the presence of oxygen the higher-valent complexes can be formed for Cr(III), Fe(III), Mn(III), Co(III), Sc(III), Ti(IV)O, and V(IV)O. The second- and third-row early transition elements give complexes with the metal in higher oxidation states, e.g., Zr(IV), Hf(IV), Mo(IV), Mo(V), Re(V). Conversely, the latter transition elements give divalent and trivalent metal complexes, e.g., Ru(II), Ru(III), Os(II), Rh(II), Rh(III), Ir(III),

Pd(II), Pt(II), Ag(II), Ag(III), Au(III). The Group IVB porphyrin complexes of Si(IV), Ge(IV), Sn(II), Sn(IV), Pb(II), Pb(IV) are generally prepared by the reaction of the metalloid halide MX_4 with the porphyrin in a nonprotic solvent. This latter method may also be used with the higher-valent metal halides of the second and third transition series.[22]

A recent method of metalloporphyrin synthesis involves a ligand exchange reaction with the formation of a volatile product like acetylacetone which drives the reaction to completion[23]:

$$ML_2 + H_2Por \longrightarrow [M(Por)] + 2HL \tag{3}$$

The novel use of organometallic reagents has also allowed the clean formation of a number of metalloporphyrins[24]:

$$M(CO)_n + H_2Por \longrightarrow [M(Por)] + nCO + H_2 \tag{4}$$

Many unusual complexes where the porphyrin acts as a bidentate or tridentate ligand can be prepared in this way.

The kinetics and mechanism of metalloporphyrin formation continues to be the subject of wide interest.[25] It is evident that the rates of formation of metalloporphyrins vary widely. Unfortunately, while rate laws have been determined, unambiguous detailed mechanistic information is still lacking. It is likely that the rate of metal insertion into the porphyrin macrocycle is governed to a high degree by the dissociation of the ligands and/or solvent molecules from the first coordination sphere of the metal. Other contributions to the rate include the proton abstraction from the porphyrin and the out-of-plane deformation of the porphyrin.

3. Structure

The electronic structure of metalloporphyrins has been the subject of an extensive research effort both from the theoretical and experimental point of view. The methods of molecular orbital calculations have been applied at various levels of approximation and the results used to rationalize the known spectral properties of metalloporphyrins.[26] However, the predictive power of the calculations has been low. The molecular orbital studies do substantiate the qualitative suppositions about bonding in metalloporphyrins. For example, the porphyrin dianion is a good sigma donor with the pyrrole nitrogen atoms interacting strongly with the metal atoms' $d_{x^2-y^2}$ and d_{z^2} orbitals. In addition to this an out-of-plane Π interaction of the porphyrin highest occupied Π MO or lowest unoccupied Π^* MO with the metal $d_{xz} d_{yz}$ is envisioned. This type of interaction varies considerably with the metal, while in-plane Π bonding with the metal d_{xy} orbital is not important.

A qualitative molecular orbital energy level diagram for the relevant

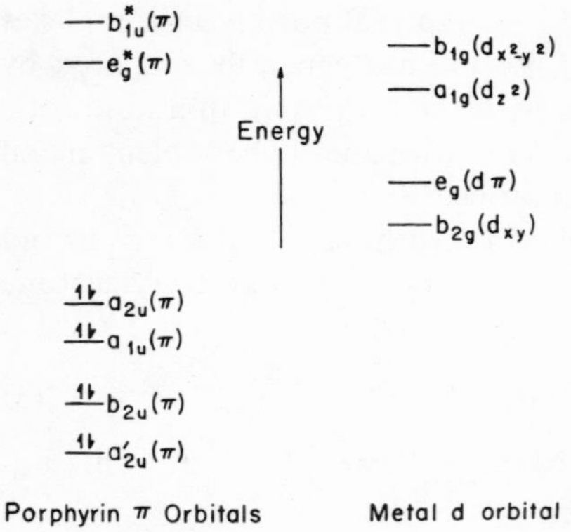

$$—b^*_{1u}(\pi)$$
$$—e^*_g(\pi)$$

$$—b_{1g}(d_{x^2_-y^2})$$
$$—a_{1g}(d_{z^2})$$

Energy

$$—e_g(d\,\pi)$$
$$—b_{2g}(d_{xy})$$

$$\underline{\text{⇈}}\,a_{2u}(\pi)$$
$$\underline{\text{⇈}}\,a_{1u}(\pi)$$

$$\underline{\text{⇈}}\,b_{2u}(\pi)$$
$$\underline{\text{⇈}}\,a'_{2u}(\pi)$$

Porphyrin π Orbitals   Metal d orbital

FIGURE 4. Qualitative molecular orbital energy level scheme for metalloporphyrins.

porphyrin and metal d orbitals is given in Figure 4. For a typical transition metal in a square-planar complex, the highest-energy orbital is the strongly sigma antibonding $d_{x^2-y^2}$, while the more weakly sigma antibonding d_{z^2} falls substantially below this in energy. Conversely, the calculations show that in a six-coordinate structure the energy of the d_{z^2} orbital is increased and the gap between it and the $d_{x^2-y^2}$ orbital is reduced. For a five-coordinate complex the situation is intermediate, with both the $d_{x^2-y^2}$ and d_{z^2} orbitals being somewhat destabilized. The Π interactions are generally less in the five-coordinate (out-of-plane) structure than in the four- or six-coordinate complexes.

The d-orbital distribution and spin states for three common molecular geometries of metalloporphyrins of a number of first-row transition elements can be determined with a simple qualitative ligand field model.[27] For example, the four-coordinate square-planar complex of Ni(II), [Ni(Por)], is diamagnetic, while the six-coordinate, tetragonal complex, [Ni(Por)L_2], is paramagnetic $(S = 1)$.[19] The respective d-orbital distributions here are $(d_{xy})^2(d_{xz}d_{yz})^4(d_{z^2})^2$ and $(d_{xy})^2(d_{xz}d_{yz})^4(d_{z^1})^1(d_{x^2-y^2})^1$. Thus the increase in the ligand field along the z axis by the addition of two Lewis-base ligands destabilizes the metal d_{z^2} orbital and substantially reduces the gap between the d_{z^2} and $d_{x^2-y^2}$ orbital. The case with Fe is quite interesting and biologically relevant. The observations are schematically shown in Figure 5. The five-coordinate Fe(II) and Fe(III) complexes are high spin with d-orbital occupations: $(d_{xy})^2(d_{xz}, d_{yz})^2(d_{z^2})^1(d_{x^2-y^2})^1$ and $(d_{xy})^1(d_{xz}, d_{yz})^2(d_{z^2})^1(d_{x^2-y^2})^1$. In this case the metal is out of the plane of the porphyrin donors, with the result that the $d_{x^2-y^2}$ orbital is somewhat stabilized and partially occupied. Addition of a sixth ligand along the z axis *trans* to the fifth ligand increases the ligand field along this direction, and as a result the d_{z^2} orbital is destabilized. In addition the metal moves further into the plane of the porphyrin, with the result that the $d_{x^2-y^2}$ orbital is substantially destabilized. The result is that both complexes are low spin with the d-orbital distributions:

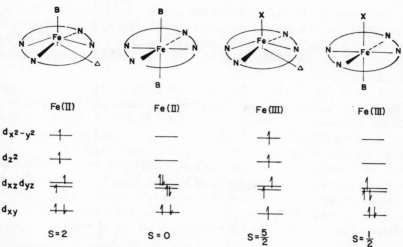

	Fe(II)	Fe(II)	Fe(III)	Fe(III)
$d_{x^2-y^2}$	⊣	—	⊣	—
d_{z^2}	⊣	—	⊣	—
$d_{xz}d_{yz}$	⇅ ⇅	⇅ ⇅	⊣ ⊣	⇅ ⊣
d_{xy}	⇅	⇅	⊣	⇅
	S = 2	S = 0	$S = \frac{5}{2}$	$S = \frac{1}{2}$

FIGURE 5. Schematic representation for the change in spin state of iron porphyrins with coordination number.

$(d_{xy})^2(d_{xz}, d_{yz})^4$ and $(d_{xy})^2(d_{xz}, d_{yz})^3$. The Fe porphyrins represent an intermediate behavior between the early transition elements and the latter elements. Thus the Cr(II), Cr(III), Mn(II), Mn(III) are high spin, while Co(II) and Co(III) are low spin. As expected the second and third transition series metalloporphyrins are all low spin.

In general the electron spin resonance spectra of the paramagnetic metalloporphyrins are in agreement with the results of calculations, at least with respect to the placement of unpaired electrons in d orbitals.[28] Unfortunately, agreement of orbital coefficients (degree of covalency) is not good.[29] Other spectroscopic techniques have been applied to filling out the details of metalloporphyrin electronic structure, e.g., NMR,[30] Mössbauer,[31] resonance Raman,[32] and electronic absorption spectra.[33] The interpretation of the experimental results agree with, for the most part, the description of the metalloporphyrins as being tetragonal complexes with strong in-plane sigma bonding and weaker axial bonding. Small but important Π-bonding effects of the metal to the porphyrin and to the axial ligands are superimposed upon the sigma bonding.

The characteristic, intense color of metalloporphyrins is due to the presence of several allowed Π → Π* ligand transitions, $a_{1u} \to e_g^*$ and $a_{2u} \to e_g^*$, in the visible region. The pattern, energy, and intensity of the electronic transition has been adequately explained by molecular orbital calculations notably by M. Gouterman and his group.[33] Initially, the four-orbital model focused on the occupied Π molecular orbitals a_{1u}, a_{2u} and on the lowest empty Π^* molecular orbital e_g^*, but the latter work has been extended to take into account the sigma orbitals as well. A typical electronic spectrum of a metalloporphyrin is shown in Figure 6. The intense band near 400 nm is the so-called Soret band, while two weaker bands near 590 nm are the α and β bands.

In general the visible spectrum is not sensitive to the metal coordinated to the porphyrin. This has been interpreted as evidence against a significant Π interaction between the metal and porphyrin Π orbitals. However, several complexes, e.g., Cr(III), Mn(III), and Fe(III), show medium to low intensity charge transfer absorptions that are notably metal dependent.[27] In these cases, the spectra and molecular orbital studies have been interpreted in terms of a significant metalloporphyrin Π interaction. The details of bonding in these complexes, however, continues to be subject to some discussion even though a number of spectroscopic techniques have been applied to the problem. The situation with the electronic structure of Mn(III) porphyrins

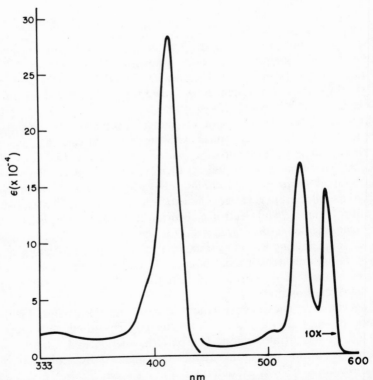

FIGURE 6. Electronic absorption spectrum of zinc(II)etioporphyrin in pyridine.

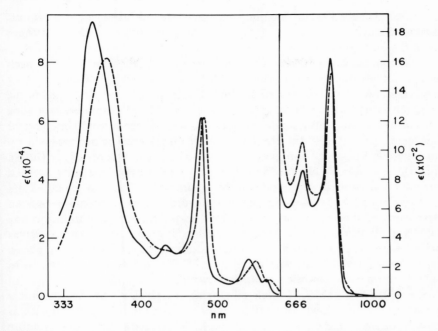

FIGURE 7. Electronic absorption spectrum of chloromanganese(III)etioporphyrin in chloroform.

illustrates this point. The unusual electronic spectrum of the [Mn(Por)X] complexes, which is shown in Figure 7, was interpreted and tentative assignments of the observed transition made to ligand $\Pi \rightarrow \Pi^*$ transitions and ligand-to-metal $\Pi \rightarrow d$ charge transfers.[34] Qualitative bonding schemes were proposed that included strong porphyrin-to-metal sigma donation and a back Π donation from the $d\Pi$ metal orbitals to the ligand $e_g{}^*(\Pi)$ orbital. This is shown in Figure 8. The resonance Raman spectra of a number of [Mn(Por)X] complexes have been measured and assignments made.[35] Excitation within the different absorption bands gives different resonance Raman patterns and intensities. The anomalous absorption at 455 nm can be assigned to a charge transfer transition mixed with a ligand $\Pi \rightarrow \Pi^*$ transition. On the other

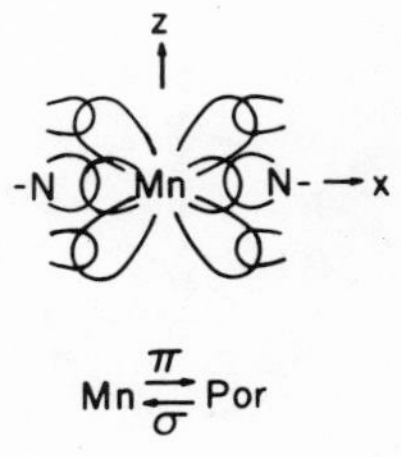

FIGURE 8. Schematic representation of the proposed manganese(III)porphyrin bonding. σ: $\mathrm{Mn}(d_{x^2-y^2})$–Por(Nσ). π: $\mathrm{Mn}(d_{xz}, d_{yz})$–Por($\pi^*$).

hand, bands near 590 and 560 nm are assigned to $\Pi \rightarrow \Pi^*$ ligand transitions. Linear dichroism spectra of manganese(III) porphyrins are also consistent with these assignments.[36]

The molecular structure of metalloporphyrins in the solid state has been extensively studied by the method of X-ray diffraction.[37,38] The pioneering and careful work of J. L. Hoard on single-crystal X-ray crystallography of the metalloporphyrins has been particularly noteworthy. While there is a wide variation in detail, only a small number of structural types seem to be common for metalloporphyrins. Representation of the three stereochemistries is shown in Figure 9. The four-coordinate square-planar structure is found for Cu(II), Ni(II), and presumably for Ag(II), Pd(II), and Pt(II). The only unusual structural feature of these complexes is the somewhat shortened metal–nitrogen bond. The five-coordinate structure shows a square-pyramidal arrangement of ligand atoms with the four porphyrin nitrogen atoms forming the basal plane and the fifth ligand occupying an axial position perpendicular to the base. The metal atom is positioned above the plane in the direction toward the axial ligand. The magnitude of Δ generally varies from 10 to 70 pm and depends on the metal, porphyrin, and axial ligand. The divalent metals Co, Zn, Mg, Mn, Fe (high-spin form), and Cr show the five-coordinate structure [M(Por)L], with the fifth ligand L being a Lewis-base donor such as water, alkylamine, heterocyclic amine, etc. The square-pyramidal structure is also common for complexes of trivalent metals [M(Por)X], where M is Mn Fe, Cr, and Co, and X is an anionic ligand. The most common six-coordinate structure, [M(Por)L₂], can best be described as tetragonal with the metal atom in the plane of the four porphyrin nitrogen donor atoms and the axial bonds generally being longer than the in-plane bonds. Divalent metal such as Fe in the low-spin form and Co can show this stereochemistry. Another variation of the six-coordinate structure, [M(Por)XL], is seen for trivalent

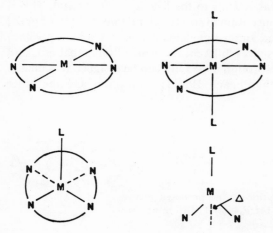

FIGURE 9. Common idealized stereochemistries for metallo-porphyrins.

metals such as Co, Fe (low-spin form), and Mn. For these structures, the metal is still out of the plane of the porphyrin donors toward the axial anion ligand. Although the porphyrin ligand is highly aromatic the macrocycle can be ruffled. The nonplanar conformation seen can have individual carbon and nitrogen atoms out of the best plane of the porphyrin by as much as 40 pm. The pyrrole rings, however, show local planarity. The nonplanar structures are thought to arise from crystal-packing forces, with the unstrained planar form being the most stable form in free solution. The bond angles and bond lengths in the macrocycle are generally in agreement with the delocalized, unsaturated nature of the ring system.

By way of example, it is instructive to consider the details of recent crystallographic work on manganese porphyrins. The results of seven X-ray crystallographic studies exemplify the structural changes noted for metallo-porphyrins in the various common stereochemistries and metal oxidation states. The relevant structural parameters for the various structures are sum-marized in Table 1. The five-coordinate square-pyramidal structure is seen for [Mn(TPP)Cl] and [Mn(TPP)N$_3$], while the six-coordinate structure is seen for [Mn(TPP)N$_3$(CH$_3$OH)], [Mn(TPP)Cl(Py)], and [Mn(TPP)NO(4CH$_3$-Py)]. For the Mn(III) complexes the change from five to six coordination does not appreciably affect the in-plane metal-porphyrin–nitrogen bond length, 202 $\pm$ 10 pm. On the other hand, the axial bond length to the anion increases from the five- to six-coordinate structure, while the out-of-plane displacement Δ decreases by 15 pm for the same change. Thus it appears that as a sixth, axial ligand is added the metal moves away from the other axial ligand toward the plane of the porphyrin ring system. With the formation of a sixth bond the fifth bond weakens, while the in-plane interaction is not significantly affected. There is therefore a delicate balance between the five- and six-coordinate structure with both stereochemistries possessing an appreciable stability. It is important to note that the bond to the anion is still the stronger of the axial bonds. This supports the notion that the axial bonds are quite ionic in nature.

TABLE 1. Bond Lengths and Out-of-Plane Displacement Δ (in pm) for Manganese Porphyrins [Mn(TPP)XL]

X	L	Mn–X	Mn–N	Mn–L	Δ	Reference
Cl	—	237	202	—	26	39
N$_3$	—	204	201	—	23	40
Cl	Py	247	201	244	12	41
N$_3$	CH$_3$OH	218	203	233	8	40
NO	4CH$_3$-Py	164	203	221	10	42
—	CH$_3$-Imid	—	209	213	56	43

The reduction of [Mn(Por)X] gives a five-coordinate square-pyramidal Mn(II) complex, [Mn(Por)L], where L is a Lewis-base donor. The major change in going from Mn(III) to Mn(II) is the dramatic lengthening of the metal-porphyrin–nitrogen bond length by 11 pm. This results in an even larger increase, 30 pm, in the out-of-plane displacement. Since both the five- and six-coordinate Mn(III) complexes and the Mn(II) complex are high spin, the structural change cannot be ascribed to a change in spin state. The apparent large increase in size of the manganese atom can, however, be related to the addition of one electron to the strongly antibonding $d_{x^2-y^2}$ orbital in going from Mn(III) to Mn(II). Thus the high-spin Mn(II) porphyrin complex can only form if the structure adopted has the metal appreciably out of the porphyrin plane. This structure tends to reduce the energy of the $d_{x^2-y^2}$ orbital. Finally, where multiple oxidation states are possible for the metal, bonding to the higher oxidation state is generally stronger than to the low oxidation state.

4. Reactions

Metalloporphyrins undergo a number of characteristic reactions that are common to other classes of coordination compounds.[44] The metal-centered reactions include: change in coordination number, structure, and spin state. The first type of reaction involves addition of ligands to the metal in its vacant axial coordination positions. The change in coordination number from four to five and five to six is illustrated in the following equilibria:

$$[M(Por)] + L \rightleftharpoons [M(Por)L] \tag{5}$$

$$[M(Por)L] + L \rightleftharpoons [M(Por)L_2] \tag{6}$$

In the case of square-planar complexes, like those of Ni(II), the *bis* complex forms most readily.[45] For five-coordinate complexes, like those of Mg(II), only the equilibrium for the addition of the sixth ligand is noted.[46] In general the equilibrium constants and free-energy changes for reactions (5) and (6) are dependent on the particular porphyrin, added ligand, and metal. Thus as the sigma donor strength of the ligand increases the equilibrium constant increases. There is also some evidence that the Π-acceptor property of the ligand is important here. Further, as the basicity of the porphyrin increases the ability to add axial ligands decreases, i.e., the increase in the donation of electron density to the metal atom from the porphyrin reduces its ability to accept electron density from the axial ligands.[45,46] By their very nature, ligation reactions involve a change in structure, i.e., from square planar to tetragonal and square pyramidal to tetragonal. In the latter case the metal atom may or may not move into the plane of the porphyrin upon addition of the sixth ligand. For the less electronegative metals like Zn(II) and Mg(II) it appears as if the metal is still out of the plane of the porphyrin nearer to one of the axial ligands. In addition to the change in molecular structure upon ligation there can be a change in electronic structure. For example, in the nickel complex the metal changes spin state on going from the four- to the six-coordinate complex.

The oxidation–reduction chemistry of the first-row transition metalloporphyrins is particularly rich.[22] Reaction of the metal center can give a number of metal oxidation states: M(I), M(II), M(III), and M(IV). The lowvalent state M(I), which can be formed both electrochemically and chemically from the M(II) complex, has been characterized for Fe(I)[47] and Co(I).[48] The higher-valent state M(IV), which can be generated by the oxidation of the M(III) complex both chemically and electrochemically, has been only partially characterized for Mn(IV)[49] and Fe(IV).[50] In both cases M(I) and M(IV) the structures of the unusual oxidation states appear to be similar to those of their isoelectronic neighbors in the periodic table. The more common oxidation states M(II) and M(III) can be interconverted quite easily, again both chemically and electrochemically, and have been fully characterized. The stability of the M(III) state is in the order Cr > Co > Mn > Fe. The stability of the higher oxidation state is dependent on both the porphyrin and axial ligand (anion). The better the donor strength of the porphyrin the more stable the higher oxidation state. Hard electronegative axial anions also tend to stabilize the higher oxidation state.[22,34]

In addition to the oxidation–reduction of the metal center in metalloporphyrins, the ligand itself is subject to oxidation and reduction.[22] In this type of reaction the electrons are added to and removed from the e_g^* and the a_{1u}, a_{2u} Π orbitals of the porphyrin. In general the porphyrin ligand undergoes two one-electron oxidations. For example, for the divalent metal complex [M(Por)] the species $[M(Por)]^+$, $[M(Por)]^{2+}$, $[M(Por)]^-$, and $[M(Por)]^{2-}$ are observed. These species have characteristic absorption spectra as well as electron spin resonance and magnetic properties.

Another important reaction that involves both oxidation and ligation is the reversible binding of oxygen to divalent Cr, Mn, Fe, and Co complexes at low temperature[51]:

$$[M(Por)L] + O_2 \rightleftharpoons [M(Por)LO_2] \tag{7}$$

The tendency to oxygen binding is in the order Cr > Mn > Fe > Co.

The oxygen adduct can be described as $M(III)O_2^-$ for Co, Cr, and Fe, while Mn appears to be best described as $Mn(IV)O_2^{2-}$. Reversible oxygen binding is found to be dependent on the fifth axial ligand and the porphyrin. Increasing electron donation to the metal leads to a greater extent of oxygen binding.

In addition to the thermally mediated reactions, metalloporphyrins are subject to photochemically induced ligation and oxidation–reduction reactions of the metal and ligands.[52] The various reactions mentioned above, ligation and oxidation–reduction, are generally quite rapid and a number of kinetics studies have been carried out.[53]

The nonbiological catalytic activity of metalloporphyrins has been explored in only a few reactions. In general the metalloporphyrin functions as a redox catalyst with either the metal center or porphyrin, or both, undergoing a change in oxidation state during the catalytic cycle. For example, the electrolytic dehydrogenation of cyclohexene and cyclohexadiene to form benzene is catalyzed by [Co(TPP)].[54] Under the electrolysis conditions of the experiment the catalyst is in the form of the radical anion of the Co(I) complex, $[Co(TPP)]^{2-}$. The catalytic cycle proposed involves an initial binding of the olefin to the soft Co(I) center via axial coordination. In subsequent steps hydrogen atoms are transferred to the porphyrin ligand to give a chlorin. Finally, electrolysis yields hydrogen and the regenerated catalyst. The catalytic activity of the complex is a function of solvent, being negligible in good coordinating solvents like pyridine. Apparently an easily displaced axial ligand is needed for activity. Also, it seems that a low oxidation state like Co(I) is needed since complexes of Ni(II), Pd(II), Pt(II), Cu(II), Mn(III)Cl, Fe(III)Cl, and V(IV)O are not active catalysts.

In several other reactions the metalloporphyrin also functions as an oxidation catalyst. In one case molecular oxygen is first activated by the divalent metalloporphyrin.[55] In a subsequent step the solid adduct $[M(TPP-OCH_3)O_2]$ abstracts hydrogen atoms from the substrate cyclohexadiene gas to eventually give H_2O and benzene. The order of catalytic activity, which is Co > Fe > Cu, parallels the oxygen-binding ability of the porphyrin. The electrocatalytic reduction of oxygen also appears to involve the prior formation of a metalloporphyrin adduct $[M(TPP)O_2]$.[56] Electrochemical reduction of this intermediate and proton addition gives the product H_2O_2. The order of catalytic activity for the acid-suspended solids is Co(II) > Fe(III) > Ni(II) $\approx$ Cu(II) $\approx$ O. A similar order is found in neutral or basic solution.[57] The gas-phase catalytic dehydrogenation of cyclohexadiene by nitrobenzene at 300°C has also been reported. The solid Co(TPP) is oxidized by nitrobenzene. The oxidized catalyst then abstracts hydrogen from the substrate and eventually passes it onto the nitrobenzene radical anion.[58]

Catalytic decomposition of H_2O_2 by Co(III) porphyrins has been extensively studied. In this case axial coordination of the peroxide anion and peroxide molecule occurs first. The next step is a complicated electron transfer via the cobalt from one axial ligand to the other.[59] Whether or not there is a transient formation of higher or lower oxidation states of the cobalt porphyrin is not known.

5. Chlorins and Corrins

The macrocyclic tetrapyrrolic class of compounds called chlorins are dihydro or reduced porphyrins. The tetrahydroporphyrin is called bacteriochlorin. The structural representations of these macrocyclics are shown in Figure 1. Again there are two general classes of chlorins. The first are those derived by the reduction of the synthetic porphyrins,[60] like H_2TPP and H_2OEP, or a corresponding metal complex.[61] The second class of chlorins contains those macrocycles that are derived from the naturally occurring chlorophyll molecule.[62] Simple transformation and degradations can give a number of chlorophyll-like ligands. The removal of Mg(II) from chlorophyll *a* gives metal-free pheophytin *a*, while the additional *trans* esterification with methanol gives methylpheophorbide, e.g., the phytol residue is replaced by a methyl group. The structural representation of a chlorophyll derivative is given in Figure 10. Pheophytin *b* and methylpheophorbide *b* are analogous to the "*a*" derivatives except that the ring II methyl group is replaced by a formyl group. Rupture of the exocyclic ring V gives a series of materials of which chlorin e_6 is perhaps best known (see Figure 11). Small amounts of chlorin derivatives are also found in the preparation of synthetic porphyrins.[63] The chlorins are substantially weaker donors than the porphyrins, and in general the chlorophyll derivatives are weaker donors than the synthetic chlorins with β substituents but stronger donors than those with methine substituents.[64,65] It appears that the reduced pyrrole ring acts as an electron-withdrawing center in the macrocycle. In comparison to the large number of metalloporphyrins known only a few metallochlorins have been characterized. For the methyl pheophorbide, pheophytin, and chlorin e_6 derivatives, only the first-row transition metal complexes have been prepared, e.g., Mg(II), Zn(II), Cu(II), Ni(II), Fe(II), Co(II), Hg(II), Cd(II), Fe(III), Mn(III), Pd(II).[64–69] In the case of the synthetic chlorins the complexes of Mg(II), Zn(II), Cd(II), Cu(II), Ag(II), Sn(II), Co(II), Pd(II), Mn(III) have been prepared.[64,70,71] The chlorin complexes of the divalent metals can be prepared in an analogous manner to that of the porphyrins. However, in the

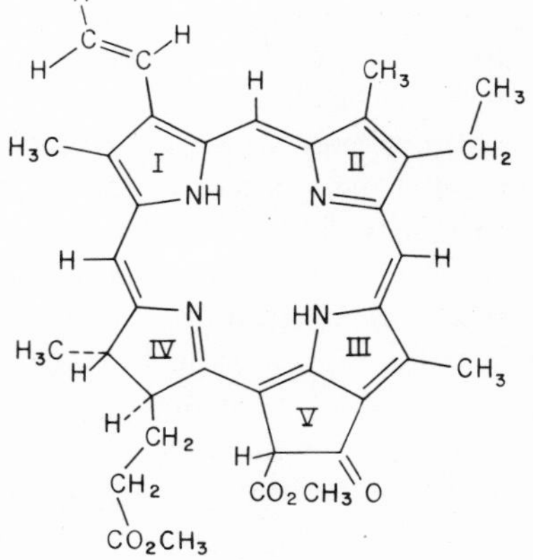

FIGURE 10. Structural representation of methyl pheophorbide *a*.

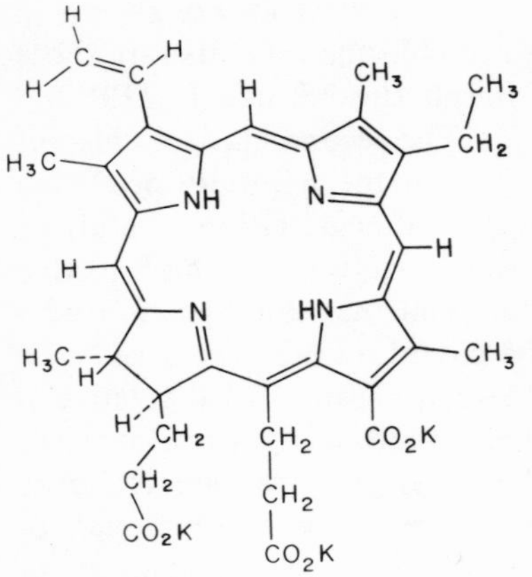

FIGURE 11. Structural representation of chlorin e_6.

preparation of the trivalent metal complexes some oxidation of the chlorin ligand occurs in air.[64]

The electronic spectrum and structure of the metallochlorins has been investigated by the method of molecular orbital calculations.[72] The theoretical work predicts the porphyrin a_{1u} and a_{2u} orbitals are raised in energy, while the empty e_g are lowered in energy and their degeneracy removed in the chlorins. This agrees with the observation that it is easier to oxidize and reduce the chlorin ligand Π system than that of the porphyrin.[65,73] The chlorin ligand appears to be a weaker ligand than the porphin. This is exemplified in many of its properties. For example, the ability of the metal center to accept additional ligands is enhanced over that of the porphyrins.[74] Secondly, the chlorin ligand stabilizes the M(III) oxidation state to a lower degree than the porphyrin.[64] Only a few X-ray structures of metallochlorins have been determined and these have been only for the Mg(II) complex of ethyl pheophorbide *a* and ethyl pheophorbide *b*.[75] The reduced donor strength of the chlorin is not reflected in the bond length here since the average Mg–N bond length is the same in the porphyrins and in the chlorin. However, the Mg–N bond to the reduced ring is substantially increased by 10 pm in the chlorin. Also the chlorin macrocycle is somewhat more distorted from a planar structure. Recent X-ray photoelectron[76] and Mössbauer[77] spectroscopic studies of metal complexes of macrocycles are also consistent with the notion that chlorins are weaker donors than the porphyrins.

The corrin system is found in nature as a Co(III) complex.[78] The structural representation of the vitamin B_{12} coenzyme is depicted in Figure 12. Vitamin B_{12}, which contains a cyano group instead of the adenine nucleoside, is an artifact produced during the isolation of the biologically active form. The corrin ring structure, stripped of substituents, is shown in Figure 1. Although a tetrapyrrolic macrocycle, the corrin ligand system

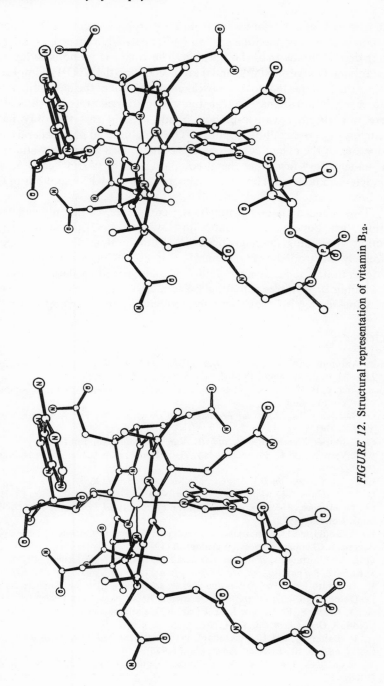

FIGURE 12. Structural representation of vitamin B_{12}.

differs in several important ways from the porphin system. Most importantly the corrin lacks one methine carbon and is partially reduced, i.e., it has five fewer double bonds than the porphin. The corrin is monobasic ligand, and two anionic charges (axial ligands) are needed for the Co(III) complex. Thus the corrin is expected to be a weaker donor than the porphin. Also the macrocycle is not planar even in the unconstrained state. In the coenzyme there is a direct cobalt-to-carbon bond, making this the only naturally occurring organometallic compound. The structure and physical and chemical properties of the cobalt corrin complexes have been extensively studied. Unfortunately, very few other metal complexes of the corrin ligand have been characterized in detail. However, much work on model compounds of Co(III) has been reported.[79-82]

Two reduced states of vitamin B_{12} complexes, commonly called cobalamines, are known: one, containing low-spin Co(II), vitamin B_{12r}, and another containing diamagnetic Co(I), vitamin B_{12s}. The latter species is extremely electrophilic and a number of alkyl Co(III) complexes can be made from it. The ready accessibility of a number of oxidation states allows the vitamin B_{12} complexes to function as catalysts in several enzyme reactions. These include methyl group transfer, reductions, and rearrangements.[83]

References

1. E. Antonini and M. Brunori, *Hemoglobin and Myoglobin in Their Reactions with Ligands*, North-Holland Publishing Co., Amsterdam (1971).
2. B. Chance, R. W. Estabrook, and T. Yonetani (eds.), *Hemoproteins*, Academic Press, New York (1966).
3. O. Hayaishi (ed.), *Molecular Mechanism of Oxygen Activation*, Academic Press, New York (1974).
4. G. S. Marks, *Heme and Chlorophylls*, Van Nostrand, London (1969).
5. L. P. Vernon and G. R. Seely (eds.), *The Chlorophylls*, Academic Press, New York (1966).
6. E. L. Smith, *Vitamin B_{12}*, 3rd ed., Methuen, London (1965).
7. D. R. Williams, *The Metals of Life*, Van Nostrand-Reinhold, London (1971).
8. M. N. Hughes, *The Inorganic Chemistry of Biological Processes*, John Wiley and Sons, London (1972).
9. R. F. Gould (ed.), *Bioinorganic Chemistry*, Advances in Chemistry Series No. 100, American Chemical Society, Washington (1971).
10. G. L. Eichhorn (ed.), *Inorganic Biochemistry*, Elsevier, Amsterdam (1973).
11. J. E. Falk, *Porphyrins and Metalloporphyrins*, Elsevier, Amsterdam (1964).
12. K. M. Smith (ed.), *Porphyrins and Metalloporphyrins*, Elsevier, Amsterdam (1975).
13. D. Dolphin (ed.), *The Porphyrins*, Academic Press, New York (1978).
14. A. D. Adler, F. R. Longo, J. D. Finarelli, J. Goldmacher, J. Assour, and L. Korsakoff, *J. Org. Chem.* **32**, 476 (1967).
15. A. R. Battersby and E. McDonald, in: *Porphyrins and Metalloporphyrins* (K. M. Smith, ed.), p. 61, Elsevier, Amsterdam (1975).
16. W. S. Caughey, J. O. Alben, W. Y. Fujimoto, and J. L. York, *J. Org. Chem.* **31**, 2631 (1966).

17. H. W. Whitlock, and R. Hanauer, *J. Org. Chem.* **33**, 2161 (1968).
18. G. W. Kenner and K. M. Smith, *Ann. N. Y. Acad. Sci.* **206**, 138 (1973).
19. W. S. Caughey, W. J. Fujimoto, and B. P. Johnson, *Biochemistry* **5**, 3830 (1966).
20. F. A. Walker, D. Beroiz, and K. M. Kadish, *J. Am. Chem. Soc.* **98**, 3484 (1976).
21. K. M. Smith, in: *Porphyrins and Metalloporphyrins* (K. M. Smith, ed.), p. 61, Elsevier, Amsterdam (1975).
22. J. H. Fuhrhop, *Struct. Bonding (Berlin)* **18**, 1 (1974).
23. J. W. Buchler, L. Puppe, K. Rohback, and H. H. Scheehage, *Ann. N. Y. Acad. Sci.* **206**, 116 (1973).
24. D. Ostfeld and M. Tsutsui, *Acc. Chem. Res.* **7**, 52 (1974).
25. W. Schneider, *Struct. Bonding (Berlin)* **23**, 123 (1975).
26. S. J. Chantrell, C. A. McAuliffe, R. W. Munn, and A. C. Pratl, *Coord. Chem. Rev.* **16**, 259 (1975).
27. M. Gouterman, L. K. Hanson, G. E. Khalil, W. R. Leemstra, and J. W. Buchler, *J. Chem. Phys.* **62**, 2343 (1975).
28. J. Subramanian, in: *Porphyrins and Metalloporphyrins* (K. M. Smith, ed.), p. 555, Elsevier, Amsterdam (1975).
29. P. W. Lau and W. C. Lin, *J. Inorg. Nucl. Chem.* **37**, 2389 (1975).
30. H. Scheer and J. J. Katz, in: *Porphyrins and Metalloporphyrins* (K. M. Smith, ed.), p. 399, Elsevier, Amsterdam (1975).
31. P. Hambright and A. J. Bearden, in: *Porphyrins and Metalloporphyrins* (K. M. Smith, ed.), p. 539, Elsevier, Amsterdam (1975).
32. T. G. Spiro, *Biochem. Biophys. Acta* **416**, 169 (1975).
33. A. J. McHugh, M. Gouterman, and C. Weiss, *Theor. Chim. Acta (Berlin)* **24**, 346 (1972).
34. L. J. Boucher, *Coord. Chem. Rev.* **7**, 289 (1972).
35. R. R. Gaughan, D. F. Schriver, and L. J. Boucher, *Proc. Natl. Acad. Sci.* **72**, 433 (1975); S. A. Asher and K. Sauer, *J. Chem. Phys.* **64**, 4115 (1976); J. A. Shelnutt, D. C. O'Shea, W-T. Yu, L. D. Cheun, and R. H. Felton, *J. Chem. Phys.* **64**, 1156 (1976).
36. R. Gale, R. D. Peacock, and B. Samori, *Chem. Phys. Lett.* **37**, 430 (1976).
37. J. L. Hoard, *Science* **174**, 1295 (1971).
38. E. B. Fleischer, *Acc. Chem. Res.* **3**, 105 (1970).
39. L. J. Radonovich and J. L. Hoard, private communication.
40. V. W. Day, B. R. Stults, E. L. Tasset, R. S. Marianelli, and L. J. Boucher, *Inorg. Nucl. Chem. Lett.* **11**, 505 (1975).
41. J. F. Kirner and W. R. Scheidt, *Inorg. Chem.* **14**, 2081 (1975).
42. P. L. Piculo, G. Rupprect, and W. R. Scheidt, *J. Am. Chem. Soc.* **96**, 5293 (1974).
43. B. Gonzalez, J. Kouba, S. Yee, C. A. Reed, J. F. Kirner, and W. R. Scheidt, *J. Am. Chem. Soc.* **97**, 3247 (1975).
44. P. Hambright, *Coord. Chem. Rev.* **6**, 247 (1971).
45. B. D. McLees and W. S. Caughey, *Biochemistry* **7**, 642 (1968).
46. C. B. Storm, A. H. Corwin, R. R. Arellano, M. Martz, and R. Weintraub, *J. Am. Chem. Soc.* **88**, 2525 (1966).
47. I. A. Cohen, D. Ostfeld, and B. Lichtenstein, *J. Am. Chem. Soc.* **94**, 4522 (1972).
48. H. Kobayashi, T. Hara, and Y. Kuizu, *Bull. Chem. Soc. Japan.* **45**, 2148 (1972).
49. M. Calvin and P. A. Loach, in: *Theory and Structure of Complex Compounds* (B. Jezska-Trzebiatouska, ed.), p. 13, Macmillan, New York (1964).
50. R. H. Felton, D. Dolphin, D. C. Borg, and J. Fajer, *J. Am. Chem. Soc.* **93**, 6332 (1971).
51. F. Basolo, B. M. Hoffman, and J. A. Ibers, *Acc. Chem. Res.* **8**, 384 (1975); B. M. Hoffman, C. J. Weschler, and F. Basolo, *J. Am. Chem. Soc.* **98**, 5493 (1976).

52. F. R. Hopf and D. G. Whitten, in: *Porphyrins and Metalloporphyrins* (K. M. Smith, ed.), p. 667, Elsevier, Amsterdam (1975).
53. P. Hambright, in: *Porphyrins and Metalloporphyrins* (K. M. Smith, ed.), Chap. 6, Elsevier, Amsterdam (1975).
54. H. Kageyama, M. Kidai, and Y. Uchida, *Bull. Chem. Soc. Japan* **45**, 2898 (1972); **46**, 2901 (1973).
55. J. Manassen, *Cat. Rev. Sci. Eng.* **9**, 223 (1974).
56. H. Alt, H. Binder, and G. Sandstede, *J. Catal.* **28**, 8 (1973).
57. J. Manassen, *J. Catal.* **33**, 133 (1974).
58. A. Bar Ilan and J. Manassen, *J. Catal.* **33**, 137 (1974).
59. H. Sigel and R. F. Pasternak, *J. Inorg. Nucl. Chem.* **37**, 1093 (1975).
60. H. W. Whitlock, Jr., R. Hanauer, M. Y. Oester, and B. K. Bower, *J. Am. Chem. Soc.* **91**, 7485 (1969).
61. H. H. Inhoffen, J. W. Buchler, and R. Thomas, *Tetrahedron Lett.*, 1141 (1969).
62. H. H. Svec, in: *The Chlorophylls* (L. P. Vernon and G. R. Seely, eds.), p. 22, Academic Press, New York (1966).
63. U. Eisner, A. Lichtarowrez, and R. P. Linstead, *J. Chem. Soc.*, 733 (1957).
64. L. J. Boucher and J. W. Klinehamer, *Bioinorg. Chem.* **2**, 231 (1973).
65. F. J. Ryan, R. A. Bambara, and P. A. Loach, *Bioorg. Chem.* **2**, 44 (1972).
66. L. J. Boucher and J. J. Katz, *J. Am. Chem. Soc.* **89**, 4703 (1967).
67. H. H. Inhoffen, G. Klotmann, and G. Jeckel, *Liebigs Ann.* **695**, 112 (1966).
68. L. J. Boucher, *J. Am. Chem. Soc.* **92**, 2725 (1970).
69. B. D. Berezin and A. N. Drobysheva, *Russ. J. Inorg. Chem.* **13**, 1401 (1968).
70. B. D. Berezin and N. I. Sosnikava, *Russ. J. Phys. Chem.* **39**, 719 (1965).
71. G. D. Dorough and F. M. Hueenkens, *J. Am. Chem. Soc.* **74**, 3974 (1952).
72. C. Weiss, *J. Mol. Spectrosc.* **44**, 37 (1972).
73. J. H. Fuhrhop, *Z. Naturforsch.* **25b**, 255 (1970).
74. J. R. Miller and G. D. Dorough, *J. Am. Chem. Soc.* **74**, 3977 (1952).
75. H. C. Chow, R. Serlin, and C. E. Strouse, *J. Am. Chem. Soc.* **97**, 7230 (1975); R. Serlin, H-C. Chow, and C. E. Strouse, *J. Am. Chem. Soc.* **97**, 7237 (1975).
76. D. H. Karweik and N. Winograd, *Inorg. Chem.* **15**, 2336 (1976).
77. J. C. Fanning, J. J. Jen, and A. J. Mouchet, *J. Inorg. Nucl. Chem.* **38**, 645 (1976).
78. R. H. Abeles and D. Dolphin, *Acc. Chem. Res.* **9**, 114 (1976).
79. G. N. Schrauzer, *Acc. Chem. Res.* **1**, 97 (1968); *Forschs. Chem. Org. Natust.* **31**, 583 (1974); *Adv. Chem. Ser.* **100**, 1 (1971).
80. A. Bigatto, G. Costa, G. Mestroni, G. Pellizer, A. Puxeddu, E. Reisenefer, L. Stefani, and G. Taurzer, *Coord. Chem. Rev.* **4**, 41 (1970).
81. D. G. Brown, *Prog. Inorg. Chem.* **18**, 187 (1973).
82. H. A. O. Hill, J. M. Pratt, and R. J. P. Williams, *Chem. Brit.*, 156 (1969).
83. T. C. Stadtman, *Science* **171**, 859 (1971).

9

Physicochemical Studies
of Crown and Cryptate Complexes*

Alexander I. Popov and Jean-Marie Lehn

1. Introduction

For many years coordination chemistry of alkali metal ions was completely ignored by chemists. Indeed, there was strong doubt that such coordination chemistry could even exist since it was universally accepted that the alkali ions in solutions were quite inert, i.e., they were highly resistant to solvolytic, redox, or complexation reactions. In fact, many studies of the more glamorous transition-metal complexes in solutions were often carried out in the presence of large concentrations of alkali salts so as to maintain a "constant ionic strength." In most cases no thought was given to the possibility that a competitive formation of an alkali ion complex may also take place.

It is only very recently that these Cinderellas of the Periodic Table were transformed into serious rivals of the transition-metal family. This transformation process was extremely rapid and in one decade the study of alkali complexes changed from a chemical oddity to a coherent discipline. In addition to a purely chemical interest in a new and unexplored domain of alkali ion coordination, this field of research receives a strong impetus from the biological role of Li^+, Na^+, and K^+ cations, for example, in an effective treatment of psychiatric disorders with lithium salts, the Na^+-K^+ "pump," etc.

---

* Dedicated to Professor J. D. Roberts on the occasion of his 60th birthday, 1978.

Alexander I. Popov · Department of Chemistry, Michigan State University, East Lansing, Michigan. *Jean-Marie Lehn* · Department of Chemistry, Université Louis Pasteur de Strasbourg, Strasbourg, France.

Only a little more than a decade ago it was shown that some "traditional" ligands, such as ethylenediaminetetraacetic acid, do form complexes with alkali ions.[1] In particular, Carr and Swartzfager have shown that in aqueous solutions the formation constants of potassium, sodium, and lithium ions with 1,2-diaminocyclohexane-N,N,N',N'-tetraacetic acid are 68, 4.6×10^4, and 1.3×10^6, respectively[2]:

$$\begin{array}{c}
HOOC-H_2C \diagdown \qquad\qquad\qquad CH_2-COOH \\
\qquad\qquad\qquad N-C-C-N \\
HOOC-H_2C \diagup \; H_2C \diagup \!\! {}^{H\; H} \!\! \diagdown CH_2 \!\! {}^{CH_2-COOH} \\
\qquad\qquad\qquad C-C \\
\qquad\qquad\qquad H_2 \; H_2
\end{array}$$

A new era in the coordination chemistry of alkali elements was inaugurated by the discovery of new types of ligands, macrocyclicpolyethers ("crowns") by Pedersen[3,4] in 1967. Pedersen reported syntheses of over 50 cyclic polyethers in which the size of the macrocyclic rings, the number of ether oxygens, and the number and type of substituent groups on the ring varied widely. For example, the polyether rings contained from 9 to 60 atoms, of which 3 to 20 were oxygen atoms. Structures of some typical crown ethers, their trivial names, and code designation are given in Figure 1.

Pedersen suggested useful and descriptive trivial names for these fascinating compounds. He proposed that the cyclic polyethers be called "crowns" and be identified by (a) the number and kind of substituent groups on the ring, (b) the total number of atoms in the ring, (c) the class name—crowns, and (d) the number of oxygens in the ring. Thus the trivial name of compound 5, of Figure 1, is dibenzo-18-crown-6, while according to the IUPAC nomenclature it is called 2,3,11,12-dibenzo-1,4,7,10,13,16-hexaoxa-cyclooctadeca-2,11-diene.

It was immediately recognized that these compounds have an unusual ability to form relatively strong complexes with alkali and alkaline earth ions. In fact, in the presence of some of these ligands alkali salts can be solubilized in nonpolar organic solvents. Perhaps the most spectacular effect was the dissolution of potassium permanganate in benzene by an 18-member poly-ether ring DC18C6.[4]

The discovery of the crown ethers was soon followed by synthesis of macrobicyclic polyethers containing three polyether strands joined by two bridgehead nitrogens.[5,6] These compounds have three-dimensional cavities which can accommodate a metal ion of suitable size and form an *inclusion* complex. Once again the IUPAC names for these compounds are quite cumbersome. Since it has been shown that complexation by a bicyclic ligand involves envelopment of the metal ion, these ligands are called [2]-*cryptands* ([2] indicating a bicyclic ligand), and the corresponding complexes, *cryptates*. Since bicyclic ligands may have a variable number of ether oxygens in each

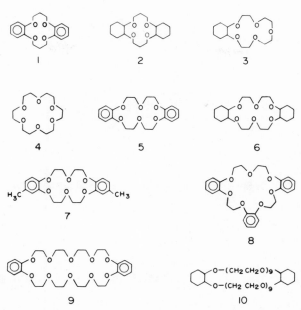

FIGURE 1. Structural formulas of some typical polyether crowns. The trivial names and codes are listed below.

1.	Dibenzo-14-crown-4	**DB14C4**
2.	Dicyclohexyl-14-crown-4	**DC14C4**
3.	Cyclohexyl-15-crown-5	**C15C5**
4.	18-crown-6	**18C6**
5.	Dibenzo-18-crown-6	**DB18C6**
6.	Dicyclohexyl-18-crown-6	**DC18C6**
7.	4,4′-dimethyldibenzo-18-crown-6	**4,4′DiMeDB18C6**
8.	Tribenzo-21-crown-7	**TB21C7**
9.	Dibenzo-30-crown-10	**DB30C10**
10.	Dicyclohexyl-60-crown-20	**DC60C20**

strand it is convenient to use the number and the distribution of oxygen atoms for coding the cryptands. As shown in Figure 2, a compound with two oxygen atoms on each hydrocarbon strand is cryptand 222, hereafter abbreviated C222. The IUPAC name for this compound is 4,7,13,16,21,24-hexaoxa-1,10-diazabicyclic(8,5,8)hexacosane. Trivial names and code designations of [2]-cryptands are shown in Figure 2.

Similar ligands containing three and four macrocycles ([3]-cryptands and [4]-cryptands) have been synthesized [7,38] as well. Typical structures are given in Figure 3.

Crown and cryptand ligands have been prepared in which some, or all, of the ether oxygens are replaced by nitrogen or sulfur atoms. As will be seen

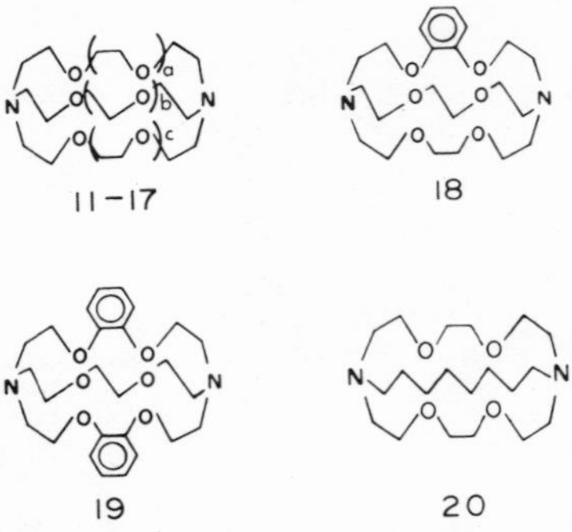

11–17   18

19   20

FIGURE 2. Structural formulas of some typical [2]-cryptands.

11. $a = b = c = 0$	Cryptand 111	C111
12. $a = b = 0, c = 1$	Cryptand 211	C211
13. $a = 0, b = c = 1$	Cryptand 221	C221
14. $a = b = c = 1$	Cryptand 222	C222
15. $a = b = 1, c = 2$	Cryptand 322	C322
16. $a = 1, b = c = 2$	Cryptand 332	C332
17. $a = b = c = 2$	Cryptand 333	C333
18. —	Monobenzo cryptand 222	C222B
19. —	Dibenzo cryptand 222	C222Bz
20. —	—	C22C$_8$

later, such substitutions can drastically modify the complexing abilities of the ligands.

Macrobicyclic compounds with three hydrocarbon chains attached to two nitrogen bridgeheads were synthesized by Simmons and Park.[8] The length of the chains varied from 6 to 10 carbon atoms.

Finally, certain naturally occurring antibiotic ionophores such as valinomycin, monensin, nigericin, etc. have been shown to complex alkali metal ions (for example, see references 9 and 10), but this is an extensive subject in its own right and will not be discussed in this chapter, which is focused primarily on the physicochemical studies of crown and cryptate complexes with metal ions.

Different topologies of the crown and cryptand ligands are shown in Figure 4, where Z represent bridgehead moieties. Crown ethers, schematically represented by A, have two-dimensional cavities that can accommodate an ion with spherical symmetry of proper dimension. The [2]-cryptand, B, as well as [3]-cryptands, C (cylindrical) and D (spherical), and the [4]-cryptand, E, all have three-dimensional cavities that are capable of forming an inclusion complex with a cation (in case D, also with an anion, see p. 557) or cations of proper size.

The task of presenting an adequate or even semiadequate description of the current state of this field of research is made very difficult by the rapid growth in the popularity of macrocyclic ligands. Numerous reviews have been

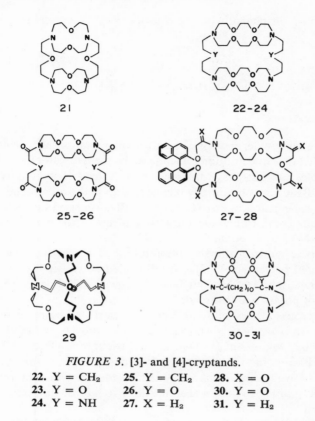

FIGURE 3. [3]- and [4]-cryptands.

22. Y = CH₂	25. Y = CH₂	28. X = O
23. Y = O	26. Y = O	30. Y = O
24. Y = NH	27. X = H₂	31. Y = H₂

published within the last 5 years.[11-15] However, advances in this field are so prolific and rapid that the reviews are outdated by the time of publication. This chapter is not an exception.

2. Synthetic Methods

2.1. Crown Polyethers

Aromatic crown polyethers were prepared by Pedersen by condensation

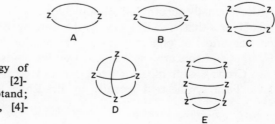

FIGURE 4. Schematic topology of macrocycles. A, crown; B, [2]-cryptand; C, cylindrical [3]-cryptand; D, spherical [3]-cryptand; E, [4]-cryptand.

reactions of a *bis*-phenol with a dichloride.[3,4,16] A typical reaction is given below:

$$2 \text{ (catechol)} + 2\text{Cl-(CH}_2)_2\text{-O-(CH}_2)_2\text{- Cl} \xrightarrow{\text{M}^+\text{OH}^-} \text{(dibenzo crown ether)}$$

The yield strongly depends on the nature of the cation M^+, which seems to indicate that the condensation is facilitated by the template effect of M^+. Thus, for example, if potassium hydroxide is used instead of sodium hydroxide the yield of DB18C6 is substantially increased.[17] On the other hand, a base with a large organic cation, n-Bu$_4$NOH, drastically decreased the yield. Potassium ion fits quite comfortably into the DB18C6 ring, while Na$^+$ ion is too small and n-Bu$_4$N$^+$ ion is too large.

Catalytic hydrogenation of aromatic hydrocarbons yields the corresponding saturated polyethers.[16] The unsubstituted polyether 18C6 was prepared by the condensation of hexaethylene glycol with 1,2-dimethoxyethane in the presence of potassium *t*-butoxide.[4]

Numerous authors followed Pedersen's work in developing new synthetic paths for crown ethers. For example, an improved synthesis of 18C6 is suggested by Gokel *et al.*,[18] who reacted triethylene glycol with triethylene glycol dichloride in 10% aqueous tetrahydrofuran. It is interesting to note that this crown easily forms a crystalline adduct with acetonitrile. Under gentle heat and high vacuum acetonitrile is volatilized and pure crown 18C6 is obtained.

Synthetic methods for the preparation of 12C4 and 15C5 have been developed by several workers.[19–21] For example, base condensation of ethylene glycol with 1,8-dichloro-3,6-dioxaoctane in dimethyl sulfoxide in the presence of lithium perchlorate gives a 13% yield of 12C4.[21]

A comprehensive review of crown syntheses has been published by Gokel and Durst.[15]

Numerous crown compounds have been prepared in which some of the ether oxygens have been replaced by sulfur atoms Pedersen prepared sulfur analogs of DB15C5, B15C5, B18C6, and DB18C6 containing two, and in the latter case, four sulfur atoms by condensing mercaptophenols or dithiols with terminally substituted ether dichlorides. In more recent years numerous mixed oxa–thia crowns were synthesized by Bradshaw and co-workers by condensation of oligoethylene glycol dichloride with a dimercaptan.[23–25] Synthesis of thia crowns was reviewed by Bradshaw and Hui in 1974.[26]

Crowns with mixed ether–ester and ether–ester–amide functions also have been synthesized recently by Bradshaw and co-workers.[27–29] Some typical examples are given below. The compounds were obtained by reacting dioxodioic acid chlorides with various oligoethylene glycols or with ethylene diamine:

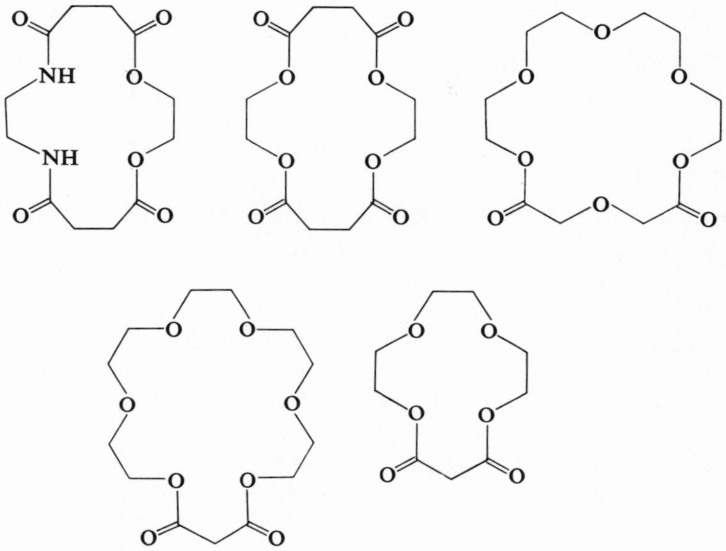

Synthesis of 4'-vinyl derivatives of B15C5, B18C6, and DB18C6 was first described by Kopolow *et al.*[30] A simple version of the synthesis was also published.[31] The two first monomers can be polymerized to give linear polymers:

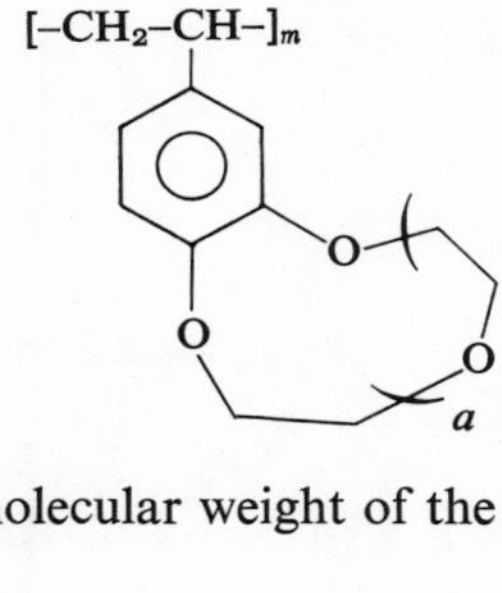

$a = 3$ for B15C5
$a = 4$ for B16C6

The average molecular weight of the polymers was found to be ∼40,000.

2.2. [2]-Cryptands

2.2.1. Diazamacrobicycles

Diazabicycloalkanes were prepared by Simmons and Park[8] by reacting a diazacycloalkane with a diacid chloride:

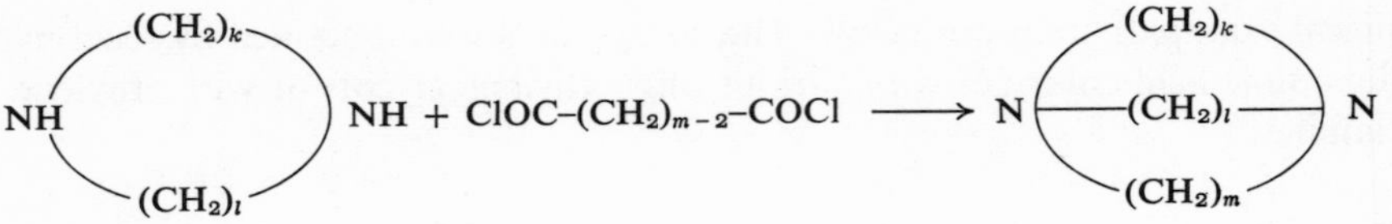

Bicyclic compounds were prepared with chains of equal length, $k = l = m = 7$–10, as well as with chains of unequal length, [6,6,4], [6,8,10], and [8,8,10].

2.2.2. Diazapolyoxamacrobicycles

The synthesis of diazapolyoxamacrobicyclic polyethers is more involved and tedious than the synthesis of crown ethers (see the reaction scheme[5] given in Figure 5). In the case of cryptand C222, a monocyclic diazapolyether is first formed by a high-dilution reaction between 1,8-diamino-3,6-dioxa-octane and triglycolyl dichloride in benzene solution. The dilactam is reduced with lithium aluminum hydride in tetrahydrofuran solution. The second cyclization is carried out with another molecule of triglycolyl dichloride; the resulting dilactam is reduced with diborane to a *bis*-(borane-amine) adduct of C222, which is treated by 6 *M* HCl to yield the dihydrochloride derivative. The product is dissolved in water and passed through an anion exchange

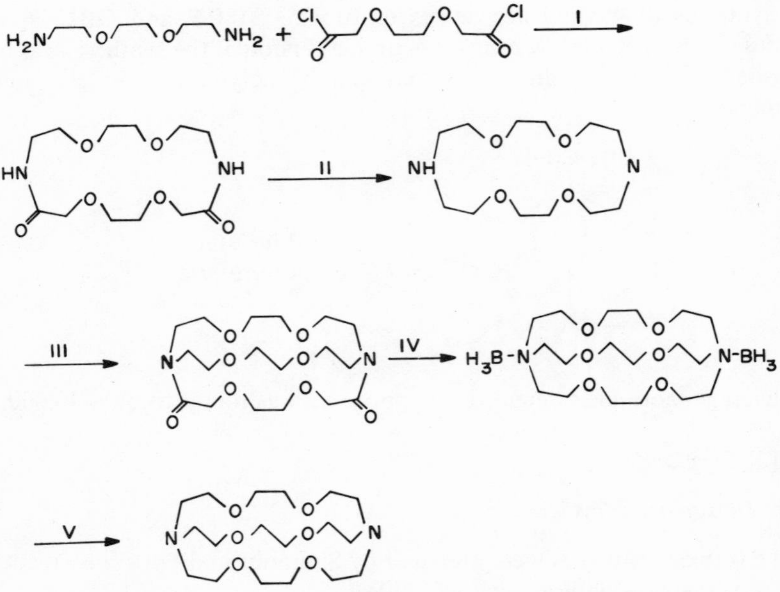

FIGURE 5. Reaction scheme for the synthesis of cryptand C222.

column in the hydroxide form to yield the free base C222. By using this technique and varying the number of ether oxygens, compounds have been prepared from C111 to C333.

The high-dilution technique requires very gradual (~ 8 h) mixing of reagents in order to avoid polymerization reactions. It has been shown by Dye *et al.* that the reaction can be carried out in less than one minute if the reagents are efficiently mixed in a suitable flow cell.[32]

2.2.3. Polythia and Polyoxa [2]-Cryptands

By substituting a dithia-diamine or a dithia dicarboxylic acid dichloride for the corresponding oxygen species, diazaoxathia macrobicyclic cryptands were prepared[33]:

$$y\text{---}\diagdown\diagup_{S}\diagdown\diagup_{S}\diagup\text{---}y \qquad y = COCl \text{ or } CH_2NH_2$$

In particular, analogs of C222 were synthesized that contained two sulfur atoms in one, two, or all three strands.

Similarly, by replacing oxygen atoms by N–R (R = H, CH_3, $COOCH_3$, or Ts) groups, cryptands containing nitrogen atoms in the hydrocarbon strands were produced.[34,35]

2.3. [3]- and [4]-Cryptands

Syntheses of cylindrical and spherical [3]-cryptands (compounds **21–31**, Figure 3) have been described in several publications.[7,36–39] The general scheme for the stepwise synthesis of cylindrical [3]- and [4]-cryptands is given in Figure 6. The reaction sequence for the synthesis of compound **25** is shown in Figure 7.

3. Metal–Cation Complexes: Preparation and Structure

3.1. Monocyclic Ligands (Crowns)

In his classical paper on the synthesis and properties of cyclic polyethers, Pedersen also described the preparation of over 50 crystalline 1:1 complexes of various crowns, mostly with alkali or alkaline earth salts.[4] In a subsequent paper, synthetic methods for the crown complexes are developed in more detail[40] and it was found that in cases where the size of the ring was smaller than the size of the complexed ion, crystalline complexes were often obtained which contained two molecules of ligand per metal ion. Pedersen postulated that in such cases "sandwich" complexes are formed in which the two ligand rings are held together by a small spherical ion. Complexes with

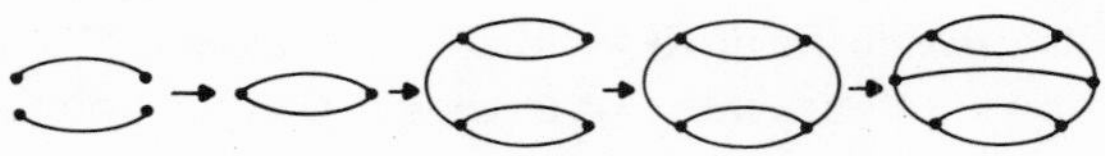

FIGURE 6. General scheme for the syntheses of [3]- and [4]-cryptands.

inverse "sandwich" arrangement M$^+$–Crown–M$^+$ were not observed in this study; however, their existence was demonstrated in later investigations.

Cesium complexes with B18C6, DB18C6, and DC18C6 were also obtained with the (Crown)$_3$(CsX)$_2$ stoichiometry. Extending the previous hypothesis, Pedersen thought it possible that these complexes have a "club sandwich" structure: Crown–Cs$^+$–Crown–Cs$^+$–Crown.

In principle, the preparation of solid crown complexes with alkali salts is quite simple. The desired metal salt and the ligand are dissolved in an appro-

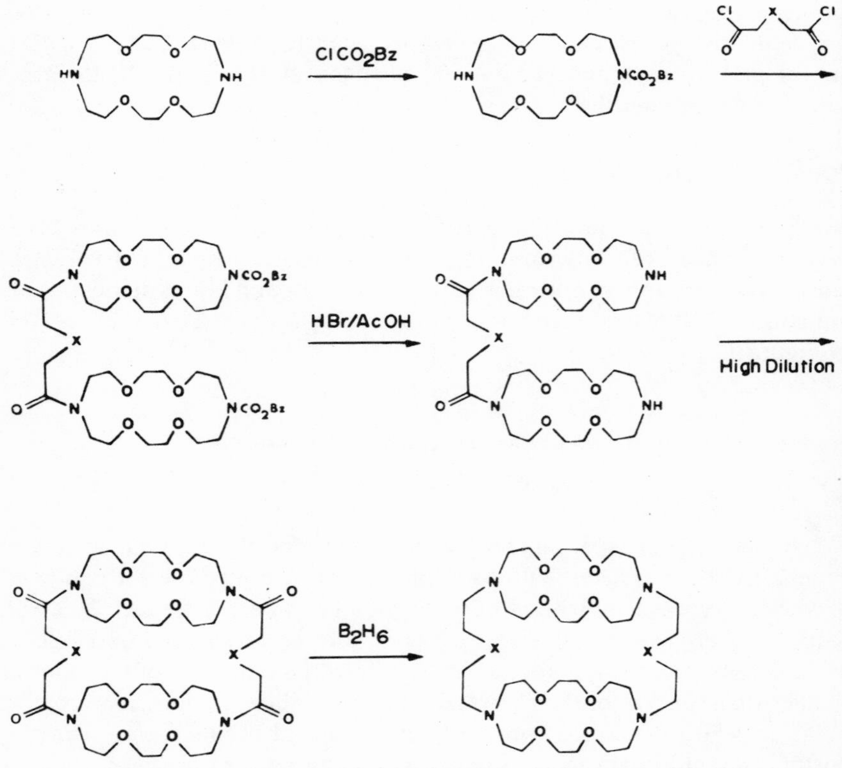

FIGURE 7. Synthesis of a cylindrical [3]-cryptand.

priate warm nonaqueous solvent, and on cooling, the complex precipitates out. In general, ethanol seems to be the solvent of choice although many others have been used. Obviously the formation of the crystalline complex is determined by relative solubilities of the reactants and the products in the given solvent. Equimolar mixtures of KSCN, Ca(SCN)$_2$, and B15C5 in ethylacetate give the calcium complex, while the same mixture in methanol yields the potassium complex.[41]

In a number of cases, although reactions were carried out in "anhydrous" solvents, the complexes obtained contained one or more molecules of water of crystallization. Poonia and Truter[42] pointed out that rigorous exclusion of water is necessary in order to obtain anhydrous complexes. Occasionally, but not frequently, a solvate other than the hydrate is formed. Parsons and Wingfield obtained calcium thiocyanate·B15C5 with methanol or pyridine of solvation when the reactions were carried out in these respective solvents.[41]

Parsons *et al.*[43] prepared a series of alkali tetraphenylborate complexes of B15C5, 18C6, DB18C6, DB24C8, and DB30C10. In most cases 1:1 complexes were obtained. However, with B18C6, sodium, potassium, and cesium complexes were prepared with L$_2$MBPh$_4$, stoichiometry (L = ligand). It is interesting to note that B15C5 gave 1:1 complexes only with NaBPh$_4$, and these complexes invariably contained one molecule of solvent (water or an alcohol) in the crystal lattice.

On the other hand, with a large ligand, DB30C10, both L·NaX and L(NaX)$_2$ crystalline complexes were obtained. In the latter case the crystals also contained one or two molecules of water although they were prepared in presumably anhydrous solvents, dichloromethane, methanol, and tetrahydrofuran. Alkali complexes of cryptands also often contain some water of crystallization.

Crystal structures of a number of crown complexes have been determined by many investigators. Pioneering work in this field was done by M. R. Truter and co-workers.[44] In general, the studies indicate that for the 1:1 complexes the cations are located at or near the center of the oxygen atom ring and all of the oxygen atoms participate in the coordination. The metal ion is also coordinated to water of hydration and/or to the anion in the direction perpendicular to the plane of the ring.

If the size of the cation is somewhat larger than the ring, the former can be situated above the plane of the oxygen atoms. Thus in the benzo-15-crown-5–sodium-iodide complex, B15C5·NaI·H$_2$O, the cation is 75 pm* out

---

* The unit picometer has been used in this chapter for the sake of consistency with the rest of this book, but the use of Angstrom units for internuclear distances would have been preferable according to the last conference of the International Union of Crystallographers in Warsaw, Poland, summer of 1978.

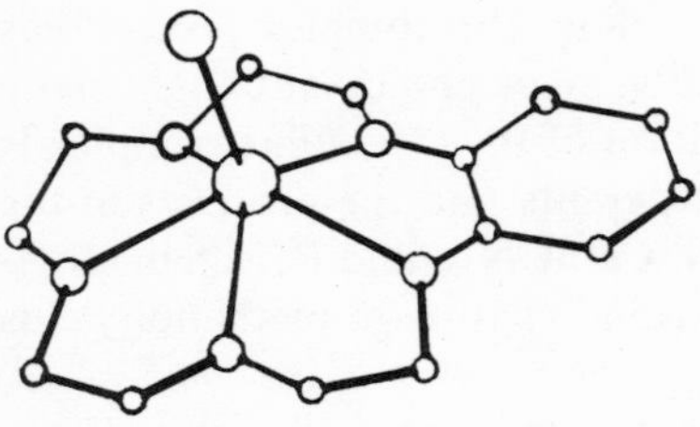

FIGURE 8. The structure of the benzo-15-crown-5 sodium iodide complex.[44] (Reproduced with permission.)

of the plane of the five oxygen atoms (Figure 8). It is also coordinated to the water molecule, which is hydrogen bonded to the anion.[45] The sodium-oxygen distances of 235–243 pm are very nearly equal to the sum of the sodium and oxygen crystallographic radii.

The structure of dibenzo-18-crown-6 (DB18C6) complex with rubidium thiocyanate was determined from an analysis of a mixed-cation complex, $DB18C6 \cdot Rb_{0.55}Na_{0.45}SCN$, and it is illustrated in Figure 9.[46,47] The Rb^+ ion is almost coplanar with the six oxygen atoms and is coordinated to the thiocyanate ion through the nitrogen.

A detailed study of 18C6 and its complexes has been carried out by Dobler, Dunitz, and co-workers.[48–53] In the free ligand the three OCH_2CH_2O subunits are clearly different. The mean carbon–carbon and oxygen–oxygen distances are both short compared to gaseous 1,4-dioxane.[48] In the sodium complex, $18C6 \cdot NaSCN \cdot H_2O$, the sodium ion is coordinated to six oxygen atoms of the ring, and to a water molecule. One oxygen atom of the ligand is out of the plane of the other five. On the other hand, the potassium thiocyanate complex $18C6 \cdot KSCN$ was obtained in anhydrous form; the K^+ ion

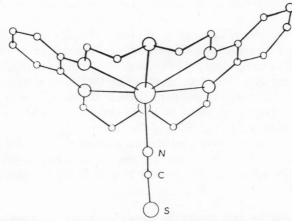

FIGURE 9. The structure of dibenzo-18-crown-6 complex with RbSCN.[46,47] (Reproduced with permission.)

occupies a crystallographic center of symmetry and is also coordinated to six oxygen atoms of the ligand.[50]

In the case of larger alkali cations, rubidium and cesium, the cation is displaced from the mean plane of the polyether ring, by ~ 120 pm for Rb^+ [51] and ~ 144 pm for Cs^+.[52] Finally, for the calcium thiocyanate complex $18C6 \cdot Ca(SCN)_2$, each cation is surrounded by a nearly planar hexagon of oxygen atoms.[53]

A crystalline complex of magnesium(II) chloride with a small crown, 12C4, with the composition $MgCl_2 \cdot 12C4 \cdot 6H_2O$, has an unusual structure. The metal ion is not found in, or near, the macrocyclic ring. Instead the ligand is hydrogen bonded to the water molecules coordinated to the Mg^{2+} ion.[54] The results indicate that there is little chance for the existence of a magnesium(II)–12C4 complex in aqueous solutions. In fact Mg^{2+} is known to have very high affinity for water and it does not readily undergo complexation reactions in aqueous solutions.

In contrast to magnesium, sodium chloride forms a sandwich complex with 12C4 with the stoichiometry $NaCl \cdot (12C4)_2 \cdot 5H_2O$.[55] The cation is insulated from the anion and the water molecules. A similar structure has been found for the sodium hydroxide complex $NaOH \cdot (12C4)_2 \cdot 8H_2O$.[56]

A structure intermediate between the Na^+ and Mg^{2+} cases is that of the calcium chloride complex $CaCl_2 \cdot 12C4 \cdot 8H_2O$, where the cation is octa coordinated to the four oxygen atoms of the crown on one side and four water molecules on the other.[57]

A copper(II) chloride complex of 12C4 was obtained in anhydrous form.[58] In this case the metal ion is hexa coordinated in the form of a distorted octahedron with the two chloride ions in *cis* position.

Another case of a crystalline crown complex in which the cation is well outside the ligand ring is that of the uranyl complex of 18C6. The compound was first prepared by Costes *et al.*[59,60] On the basis of infrared data on the solid adduct $UO_2(NO_3)_2 \cdot 18C6 \cdot 4H_2O$, the authors assumed that this was a *bona fide* complex with the cation in the center of the ring. However, a recent crystal structure determination of the dihydrate $UO_2(NO_3)_2 \cdot 18C6 \cdot 2H_2O$ by Bombieri *et al.*[61] indicates that the uranyl ion is coordinated to the two nitrate ions and to two oxygen atoms of the water molecules. There is no direct ligand–cation bonding, but the former is weakly hydrogen-bonded to the water molecules, which thus form a link between it and the cation.

In the above complexes and in similar 1:1 crown–alkali or crown–alkaline-earth complexes, the ligand molecules remain quite inflexible as long as the size of the ring and that of the ion are commensurate. On the other hand, if the size of the ring is considerably *larger* than the cation, the ligand may deform so as to bring all of its donor atoms into a bonding contact with the cation. An X-ray study of dibenzo-30-crown-10 (DB30C10) complex with

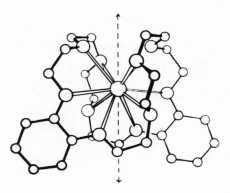

FIGURE 10. Three-dimensional crown complex dibenzo-30-crown-10·KI.[62] (Reproduced with permission.)

potassium iodide shows that this ligand is twisted around the potassium ion ("as seams of a tennis ball") so as to bring all ten oxygen atoms into coordinating positions (Figure 10).[62] The potassium–oxygen distances are in the range 285.0–293.1 pm, which again is close to the sum of the ionic radius of K^+ and the van der Waals radius of oxygen. There is no interaction between potassium ion and the iodide ion.

In the case of $(Crown)_2MX$ complexes, the X-ray investigations confirmed Pedersen's assumption of a "sandwich structure." For example, in the $(B15C5)_2KI$ complex the K^+ ion is sandwiched between two centrosymmetrically related crown molecules[63]; the five oxygen atoms of each ligand coordinated to the cation are approximately coplanar. The iodide ions occupy the spaces between the complexed cations at distances which preclude any cation–anion interaction.

It appears that structures of $3:2$ complexes such as $(DB18C6)_3(CsSCN)_2$ and $(DC18C6)_3(CsI_3)_2$ have not been determined as yet and the question of the "club sandwich" structure appears to still be unresolved.

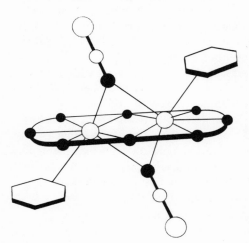

FIGURE 11. Dibenzo-24-crown-8·2KSCN complex.[64,65] (Reproduced with permission.)

TABLE 1. Reported Crystalline Structures
of Crown Complexes

Complex	Reference
18C6	48
DC18C6 "A"	81
DC18C6 "B"	81
12C4·MgCl$_2$·6H$_2$O	54
(12C4)$_2$NaCl·5H$_2$O	55
(12C4)$_2$NaOH·8H$_2$O	56
12C4·CoCl$_2$·8H$_2$O	57
12C4·CuCl$_2$	58
B15C5·NaI·H$_2$O	45
(B15C5)$_2$KI	63
18C6·NaSCN·H$_2$O	49
18C6·KSCN	50
18C6·RbSCN	51
18C6·CsSCN	52
18C6·Ca(SCN)$_2$	53
18C6·K tosylate	67
18C6·K ethyl acetate enolate	68
18C6·UO$_2$(NO$_3$)$_2$·2H$_2$O	61
DB18C6·RbSCN	46, 47
DB18C6·NaBr·2H$_2$O	66
DC18C6·NaBr·2H$_2$O	69
DC18C6·Ba(SCN)$_2$	80
DB24C8·2KSCN	64, 65
DB30ClO·KI	62
Trithio 12C4	82
Dithio 15C5	82
Dithio 18C6	82
Dithia 18C6·PdCl$_2$	86

Of particular interest is the structure of the complexes of two metal salts with one crown ether. The crystal structure of *bis*-(potassium thiocyanate)-dibenzo-24-crown-8 complex is shown in Figure 11.[64,65] It is seen that the complex does not have an "inverse sandwich" structure. The two cations are accommodated within the nearly planar ring of the eight oxygen atoms. Each cation is bound to five oxygens with two oxygen atoms shared by both cations. The cations are also bridged by the nitrogen atoms of thiocyanate anion.

Table 1 lists the crown complexes whose crystal structures have been reported in the literature.

3.2. Macropolycyclic Ligands (Cryptands)

3.2.1. Macrobicyclic Complexes—[2]-Cryptates

Pioneering work on the crystal structure of numerous diazapolyoxa-macrobicyclic complexes ([2]-cryptates) has been performed by R. Weiss and exo–exo   exo–endo   endo–endo

FIGURE 12. Different conformations of [2]-cryptands.

co-workers. In several cases, as with crowns, crystalline complexes contain water of crystallization. Proton magnetic resonance of diazabicycloalkanes $N[(CH_2)_{k,l,m}]_3N$ in solutions by Simmons and Park[8] showed that the molecules can exist in three different configurations, the *out–out* (*o,o*), in which the lone pair of the nitrogen atoms are directed away from the central cavity, the *out–in* (*o,i*), and the *in–in* (*i,i*), where the lone pairs are inside the cavity. These configurations are shown schematically in Figure 12; they have also been designated as *exo–exo*, *exo–endo*, and *endo–endo*, respectively.

Structures of the uncomplexed ligand C222 and of its *bis*-borohydride

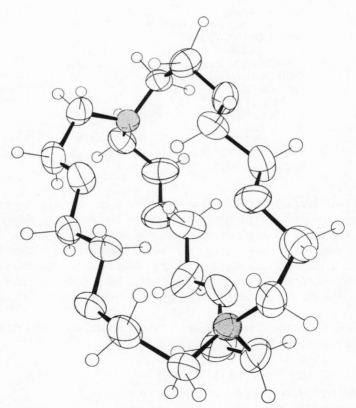

FIGURE 13. Crystal structure of cryptand C222.[70] (Reproduced with permission.)

derivative have been determined by Metz *et al.*[70] The free ligand has the *endo–endo* conformation (Figure 13) similar to that of the complexed ligand (see below). However, the conformation of the three bridges (e.g., around the central C–C bond) is different from that found in the complexed ligand. It should be remembered that the observed conformation in the crystalline state may be quite different from the one in solution.

The *bis*-(borane-amine) derivative (Figure 14), on the other hand, is *exo–exo* with the borohydride groups directed outside of the ligand cavity.

The structure of the smallest bicyclic cryptand C111 in the form of its mono(borane-amine), C111·BH_3, has been recently determined by Weiss and co-workers.[71] As expected, the ligand is in the *exo–endo* configuration with the BH_3 group being outside the cavity. The N–N distance is 494 pm as compared with 687 pm in uncomplexed C222.[70] The cavity of C111 is small and

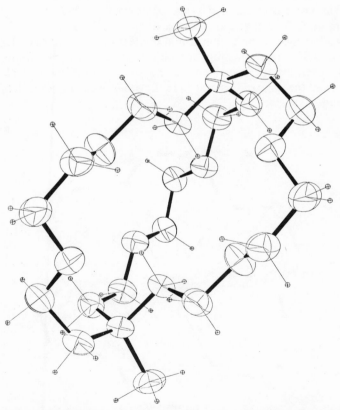

FIGURE 14. Crystal structure of *bis*-borohydride derivative of cryptand C222.[70] (Reproduced with permission.)

only Li$^+$ has been found to form a complex.[72] This ligand also forms efficiently shielded proton cryptates whose deprotonation kinetics are very slow.[72]

The metal cryptates in general have 1:1 stoichiometry and the ligand is in the *endo–endo* configuration. Alkali complexes of the cryptand C222 present a particularly interesting pattern in that the ligand shows considerable flexibility and can adapt its conformation to accommodate alkali ions from sodium through cesium.

The crystal structures of potassium, rubidium, and cesium complexes with cryptand-C222 were determined by Weiss *et al.*[73,74] Potassium iodide and rubidium and cesium thiocyanates were used as the alkali salts. In the two latter cases the crystals also contained one molecule of water of hydration.

The crystals belong to the monoclinic system and contain four molecules of the complex per unit cell. The cations are located in the center of the ligand cavity and are surrounded symmetrically by the eight heteroatoms of the ligand. The latter is in the *endo–endo* conformation. The cation–anion distances are quite long (695 pm in the case of the KI complex). It is evident, therefore, that there is no cation–anion contact in the solid complex.

In the case of the potassium complex (Figure 15) the cation–oxygen and cation–nitrogen distances are equal to the sum of the van der Waals radii of the atoms and the ionic radius of K$^+$. In the other two complexes the distances are somewhat shorter.

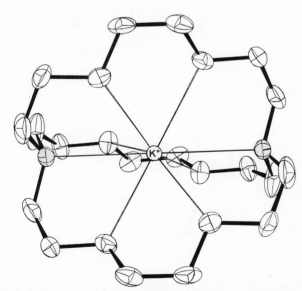

FIGURE 15. Potassium complex of cryptand C222.[74] (Reproduced with permission.)

The study shows that the cryptand cavity can adapt, within certain limits, to the ionic radius of the enclosed cation. The size of the cavity of C222 is close to that of the potassium and rubidium cations. Thus the formation of potassium and rubidium cryptates occurs without undue distortion of the ligand. On the other hand, introduction of cesium ion into the C222 cavity can only be accomplished by the adaptation of the ligand to a larger cation.[74] For example, for the K^+, Rb^+, and Cs^+ complexes we have the following variations in bond lengths and bond angles: N–N distances—575, 600, and 607 pm; O–O distances (for atoms on different chains)—426, 239, and 450 pm; the N–C–C–O dihedral angles are 54°, 57°, and 70°, respectively. These data indicate that the stabilities of the three complexes should be in the order $K^+ > Rb^+ > Cs^+$. This is indeed the order of the respective stability constants found in solutions. The heats of formation are in the order $Rb^+ \sim K^+ > Cs^+$. The crystal structure of the sodium-iodide–C222 complex, in general, is very similar to that of other alkali cryptates[75] except that the Na–N and Na–O bond distances of 275 and 257 pm are considerably larger than the sum of the Na^+ ionic radius and the van der Waals radii of N and O, respectively (252 and 242 pm). The ligand must undergo a considerable deformation in order to accommodate small Na^+ ions. As expected, the stability of the $C222 \cdot Na^+$ complex in solutions is lower than that of $C222 \cdot K^+$.

The crystal structure of the thallium(I) formate–C222 complex was found to be very similar to that of the corresponding potassium and rubidium complexes.[76] The Tl–O distance of 290.4 pm is very close to the sum of the ionic radius of Tl^+ and of the van der Waals radius for oxygen. On the other hand, the Tl–N distance is somewhat shorter than the respective sums of the two radii (294.5 vs. 302 pm), which could be due to a slightly covalent character of the Tl–N bond. It should be noted that these results are in accord with the superior stability of the thallium(I) cryptates.

In the case of the lead thiocyanate complex, the cation again occupies the center of the ligand cavity in the *endo–endo* form.[77] The cation is bonded to the ten heteroatoms *and* to the two anions, one through the nitrogen and the other through the sulfur atoms. The Pb–O and Pb–N (N of the macrocycle) bonds are longer than the sum of the ionic radius of Pb^{2+} and the van der Waals radii of the heteroatoms. The cation is bonded to O, N, and S through electrostatic interactions and, as in previous cases, the ligand adapts the dimensions of the cavity to the size of the cation. The distance between the two nitrogen atoms in the complex is 576 pm as compared to 687 pm in the free ligand.

The structure of the corresponding barium complex, $Ba(SCN)_2 \cdot C222 \cdot H_2O$, differs in that the cation interacts with the eight heteroatoms, with the water molecule, and with *one* of the anions (through N).[78,79] Since K^+ and Ba^{2+} ions have nearly the same ionic radii the structures of their C222 complexes are nearly the same. With a larger cryptand C322, a dihydrated complex $Ba(SCN)_2 \cdot C322 \cdot 2H_2O$ was obtained.[79] In this case the cation interacts with the nine heteroatoms of the ligand, with both water molecules, but not with the anions.

Mathieu and Weiss[83] determined the crystal structure of a cobalt(II) thiocyanate complex with C221, $[C221 \cdot Co]^{2+}[Co(SCN)_4]^{2-}$. The cobalt(II) ion is completely enclosed in the ligand cavity and it is coordinated to the seven heteroatoms of the C221. The C–C torsion angles vary from 36° to 52° indicating that the ligand is strained. It should be noted that the size of the Co^{2+} cation is much smaller than the size of the C221 cavity. The structure of the lithium-iodide–C211 complex is similar to that found for C222 alkali cryptates. The ligand conformation is again *endo–endo* and the cation fits snugly into the ligand cavity.

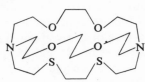

A very different structure was found for the palladium(II) chloride complex with a dithia analog of C222, hereafter designated C22S$_2$.[84] In contrast to the alkali complexes, the palladium ion is *outside* the cavity and is bound, in a square-planar configuration, to the two sulfur atoms and to the two chlorines. Nevertheless, the ligand is in the *endo–endo* configuration (N–N distance is 613.8 pm as in other complexes).

The strand with the two sulfur atoms is different from the two other strands in that the hydrogens attached to the carbon atoms *between* the two sulfurs are inside the cavity and form hydrogen bonds with oxygen and nitrogen atoms.

3.2.2. Macrotricyclic Complexes—[3]-Cryptates

The crystal structure of the silver nitrate complex with the [3]-cryptand, **21** (see Figure 3), was determined by Wiest and Weiss.[85] Slow evaporation of methanol–butanol solution of the ligand and of silver nitrate yielded crystals with stoichiometry **21** $\cdot 3AgNO_3$. Two of the silver atoms are located inside the molecular cavity of **21** with the Ag^+-Ag^+ distance of 388 pm. Each silver atom is bound to five heteroatoms of the ligand and to an oxygen atom of a nitrate ion. The third silver ion is located outside the ligand and is bound to all three NO_3^- anions, forming an $[Ag(NO_3)_3]^{2-}$ unit.

Mellinger *et al.*[87] determined the structure of a 2:1 complex of **22** $\cdot 2NaI$. The sodium ions are located in the two lateral cavities and each atom is bound to two nitrogen atoms and five oxygen atoms. The Na^+-Na^+ distance is 640 pm.

The spherically symmetrical [3]-cryptand, **29**, can form both cationic and

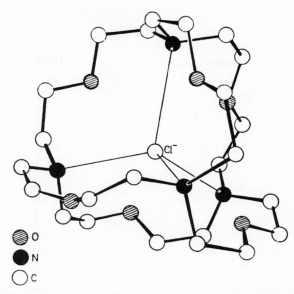

FIGURE 16. The X-ray structure of the complex cation $[ClLH_4]^{3+} 3Cl^- \cdot 7H_2O$.[88] (Reproduced with permission.)

anionic complexes. In the cationic complex, $29 \cdot NH_4I \cdot H_2O$, and the anionic complex, $29 \cdot 4HCl \cdot 7H_2O$, the ligand has all four nitrogens in the *endo* configuration. In both cases, the ligand encapsulates the species NH_4^+ *and* H_4Cl^{3+}.[88] The chloride ion of the second complex is in the center of the cavity and it is hydrogen bonded to the four protonated nitrogens of the ligand. Its structure is schematically shown in Figure 16.[88]

It is seen from the above brief discussion that one cannot speak of a "coordination number" of an alkali or an alkaline earth cation in a crown complex or in a cryptate. The number of metal–ligand bonds varies drastically with the size of the ring or cavity, the configuration of the ligand, the number and nature of the donor atoms, and the case of access of the anion and/or the solvent molecules "of solvation" to the complexed cation. Table 2 (p. 558) lists cryptates whose crystal structures have been reported in the literature.

4. Complexes in Solutions: Experimental Techniques

4.1. General Considerations

Detection of complexation reactions in solutions, the determination of the stability of the resulting complex or complexes, and the determination of

TABLE 2. Reported Crystalline Structures
of Cryptates

Complex	Reference
[C222]	70
[C222·2BH$_3$]	70
[C222·Rb$^+$]SCN$^-$·H$_2$O	73
[C222·Ba^{2+}]2SCN$^-$·H$_2$O	78, 180
[C322·Ba^{2+}]2SCN$^-$·H$_2$O	79, 180
[C211·Li$^+$]I$^-$	181
[C222·Na$^+$]I$^-$	75, 182
[C222·K$^+$]I$^-$	74, 182
[C222·Rb$^+$]SCN·H$_2$O	74
[C222·Cs$^+$]SCN$^-$·H$_2$O	74, 182
[C222·Ca^{2+}]2Br$^-$·3H$_2$O	183
[C222·Tl$^+$]HCOO$^-$·H$_2$O	76
[C221·Co^{2+}][Co(SCN)$_4^{2-}$]	83
[C222Pb^{2+}](SCN)$_2$	77
[C22S$_2$·Pd^{2+}]2Cl$^-$	84
[C222·Na$^+$]Na$^-$	184
[3]-cryptand **21**·3AgNO$_3$	85
[3]-cryptand **29**·NH$_4^+$I$^-$·H$_2$O	88
[3]-cryptand **29**·4HCl·7H$_2$O	88

enthalpy, entropy, or the kinetics of complex formation can be achieved by a variety of physicochemical measurements. Several texts describe these techniques in considerable detail.[89,90]

In most cases, experimental methods in solution chemistry have been developed for studies of aqueous solutions. Physicochemical (and especially electrochemical) techniques for studies of nonaqueous solutions have been developed to a much lesser extent. Techniques for precise determinations of formation constants greater than 10^5 in aprotic solvents are particularly lacking. Yet, due to the hydrophobic "skin" of macrocyclic complexes, they are usually much more soluble in nonaqueous solvents than in water. Consequently these complexes are particularly suited for the studies of solvent influence on complexation reactions.

In the studies of complexation constants of macrocyclic complexes it is a common (but not a universal) practice to ignore activity corrections since crowns and cryptands are neutral ligands and, therefore, the complexation reaction

$$M^n + L \rightleftharpoons ML^{n+}$$

does not involve separation of charges. The thermodynamic formation constant is

$$K_f = \frac{a_{ML^{n+}}}{a_{M^{n+}}a_L} = \frac{[ML^{n+}]}{[M^{n+}][L]}\frac{\gamma_{ML^{n+}}}{\gamma_{M^{n+}}\gamma_L} = K_c\frac{\gamma_{ML^{n+}}}{\gamma_{M^{n+}}\gamma_L}$$

where a_i's are activities of species i and γ_i's are the corresponding activity coefficients. In the calculations of formation constants it is usually assumed (often, tacitly) that the activity coefficient of the uncharged ligand, γ_L, is equal to unity *and* that $\gamma_{ML^{n+}} = \gamma_{M^{n+}}$. Thus the value of the concentration constant K_c should be very close to that of the thermodynamic constant. It should be remembered, however, that both of the above assumptions are *valid only in dilute solutions*. In fact, the ion-size parameter a in the Debye–Hückel equation is certain to have very different values for the free ion and the complex. In addition, at fairly high concentrations ($\sim 1\ M$) γ_L is no longer equal to unity.

Recent measurements of concentration equilibrium constants of crown complexes in methanol solutions as a function of ionic strength show that K_c values are quite constant up to an ionic strength of ~ 0.05 but begin to vary at higher ionic strengths.[91] It should be noted that activity effects are usually strongly magnified in nonaqueous solvents.

In order to determine the thermodynamics governing complexation reactions, it is desirable not only to determine the stability of a complex (i.e., ΔG^0), but also to divide it into the enthalpic and entropic contributions:

$$\Delta G^0 = \Delta H^0 - T\Delta S^0 \tag{1}$$

The most accurate method for the determination of the above thermodynamic parameters is to measure ΔH^0 calorimetrically and then to calculate the entropy of the complexation reaction from ΔH^0 and ΔG^0 obtained from precise measurement of $\log K_f$. Unfortunately such measurements are rare. By far the most common technique used for the determination of ΔH^0 and ΔS^0 is the measurement of $\log K_f$ as a function of temperature and obtaining the two above parameters from the slope and the intercept of a (presumably) linear van't Hoff plot. This method is straightforward but can be subject to serious errors, especially if the measurements are carried out in a narrow temperature range.

4.2. Electrochemical Techniques

4.2.1. Potentiometric Measurements

Being given a complexation reaction

$$M^{n+} + L \rightleftharpoons ML^{n+}$$

in any solvent, it is very simple to calculate the concentration formation constant of the complex if one can determine the concentration of the free metal ion (or of the ligand) in a solution containing known stoichiometric amounts of the ligand and the metal salt. Potentiometric determinations of the free metal ions by using ion-selective electrodes are the measurements of choice since, in principle, they are simple and can be carried out to very low concentrations of the free metal ion, thereby allowing the *determination of stability constants for very stable complexes*. It should be noted, however, that indicator electrodes used for such measurements are "cation selective" and not "cation specific." Consequently, small amounts of impurities may have serious effects if the studies are carried out in very dilute solutions.

An important limitation of potentiometric techniques is that they are rarely suited to nonaqueous solvents since the behavior of electrodes commonly becomes irreversible and erratic. Even in waterlike solvents such as the lower alcohols, selective ion electrodes must be carefully preconditioned before reliable results can be obtained.[92] The problem is strongly exacerbated in aprotic solvents where very few, if any, careful measurements with cation-selective electrodes (except for some measurements of "pH") are known.

A very careful and comprehensive potentiometric study of a large number of crown complexes with the alkali cations as well as with ammonium and silver ions is reported by Frensdorff.[92] Measurements were made in water and in methanol solutions. A silver–silver-chloride electrode was usually used as the reference except for the studies of the silver complexes, where it was replaced by the calomel electrode. In the case of methanol solutions, the indicator electrodes were pretreated by stepwise conditioning in aqueous methanol solutions with increasing amounts of methanol. Under these experimental conditions, the reproducibility of the EMF measurements was of the order of ± 1 mV and the uncertainty in the log K_f values was of the order of 0.04.

Cation-selective electrodes were also used extensively by Lehn and Sauvage for the determination of cryptate stability constants in water and in 95% methanol solutions.[93,94] Since [2]-cryptands are diprotic bases, their complexing abilities in aqueous solutions are strongly influenced by the pH of the medium. The following equilibria occur in aqueous solutions of cryptands and metal salts:

$$LH_2^{2+} \rightleftharpoons LH^+ + H^+ \tag{2}$$

$$LH^+ \rightleftharpoons L + H^+ \tag{3}$$

$$LM^{2+}2H^+ \rightleftharpoons LM^+H^+ + H^+ \tag{4}$$

$$LM^+H^+ \rightleftharpoons LM^+ + H^+ \tag{5}$$

$$L + M^+ \rightleftharpoons LM^+ \tag{6}$$

Titrations of the free ligands with an acid in the absence and in the presence of a metal ion were carried out and the data analyzed. Analysis of the titration curves in light of equilibria (2)–(6) yields the values of the complexation constants. It should be noted that equilibria (2), (4), and (5) involve separation of charges, and activity corrections may be important.

A similar technique was used by Anderegg[95] in the study of linear, monocyclic, and bicyclic diazapolyoxa complexes with several monovalent and divalent cations.

Recently, Gordon and Rock used the following cell without a liquid junction to determine the stability of the DC18C6 complex with Ba^{2+} ion[96]:

$$Hg-Pb|PbCO_{3(s)}, BaCO_{3(s)} |BaLCl_{2(aq)}, L_{(aq,sat)}|Hg_2Cl_{2(s)}-Hg_{(l)}-Hg_2Cl_{2(s)}$$

$$|BaCl_{2(aq)}|BaCO_{3(s)}, PbCO_{3(s)}|Pb-Hg \qquad (7)$$

where L is the ligand DC18C6.

The net cell reaction in the presence of excess solid ligand is given by

$$BaCL_{2(aq)} + L_{(s)} \rightleftharpoons BaLCl_{2(aq)} \qquad (8)$$

The potential of the cell is given by

$$E = E^0 - \frac{RI}{2F} \ln \frac{(BaL^{2+})(Cl^-)^2}{(Ba^{2+})(Cl^-)^2} \frac{\gamma_{BaL^{2+}}\gamma_{Cl^-}^2}{\gamma_{Ba^{2+}}\gamma_{Cl^-}^2} \qquad (9)$$

EMF measurements on the above cell combined with known solubilities of the ligands allow the determination of the BaL^{2+} formation constant.

The authors assume that the activity coefficients of Ba^{2+} and BaL^{2+} ions are essentially equal and that the complex is stable so that with excess of ligand nearly all of the Ba^{2+} ion is in the form of the complex; thus E becomes $\sim E^0$. The original paper should be consulted for details.

4.2.2. Electrical Conductance Measurements

In general, electrical conductance measurements are easier to carry out in nonaqueous (and especially aprotic) solvents than the potentiometric measurements since the vexing problem of obtaining a reversible electrode system in a given solvent is largely eliminated. The disadvantage of the conductance method is its sensitivity to the presence of even small amounts of conducting impurities.

The formation of a crown complex is an ion–molecule reaction, and therefore complexation results in a decrease in the mobility of the cation (due to increase in size) rather than in the formation or disappearance of charged species. Shchori *et al.*[97,98] used electrical conductance measurements to study the complexation of Na^+ by DB18C6 as well as its 4,4'-diamine and 4,4'-dinitro derivatives in dimethylformamide and dimethoxyethane solutions.

In a series of solutions containing a constant amount of a metal salt and varying concentrations of a crown, the concentration of *free* crown, (L), in any solution is given by the expression

$$(L)_{free} = (L)_{total} - (M^+)\frac{\Lambda_A - \Lambda}{\Lambda_A - \Lambda_B} \tag{10}$$

where Λ is the measured equivalent conductance, while Λ_A and Λ_B are the equivalent conductances of the solvated metal salt and of the complexed salt, respectively.

An iteration procedure is used to obtain the values of Λ_B and of K_f. In dimethoxyethane solutions, due to the relatively low dielectric constant of 7.2, the salt is largely ion paired, and in addition to the usual complexation constant K_f for the reaction

$$M^+ + L \; \overset{K_f}{\rightleftharpoons} \; ML^+$$

the following equilibria were considered:

$$M^+X^- + L \; \overset{K_{ip}}{\rightleftharpoons} \; ML^+X^-$$

$$M^+X^- \; \overset{K_d^{\,A}}{\rightleftharpoons} \; M^+ + X^-$$

$$ML^+X^- \; \overset{K_d^{\,B}}{\rightleftharpoons} \; ML^+ + X^+$$

and the ion-pair dissociation constants determined by the method of Fuoss and Accascina.[99] It is easily seen that $K_{ip}/K_f = K_d^A/K_d^B$. The values obtained are given in Table 3. It is interesting to note that since the values of K_f and K_{ip} in DME are very nearly the same, the free-energy change for the complexation reaction is essentially the same whether the reactant is the solvated sodium ion or the ion pair.

Matsuura et al.[100–101] also used electrical conductance measurements to study the formation constants of the DB18C6 with the alkali ions in dimethyl sulfoxide, dimethylformamide, and propylene carbonate solution. In this case the concentration of the ligand was kept constant and the concentration of the salt was changed. The authors devise the following equation for the equivalent conductance of an equimolar solution of an alkali salt MX and a crown L:

$$\Lambda_{obs} = \kappa/C = (1 - \alpha)\Lambda_{MX} + \alpha\Lambda_{MLX} \tag{11}$$

where κ is the observed specific conductance of the solution, C is the concentration of the salt and of the crown, and α is the fraction of the cation in the complexed form. Equation (11) is solved by iteration. First a value of α is estimated and the values of Λ_{MX} and Λ_{MLX} are calculated. The values of α are adjusted until the numerical value of the right-hand side of Eq. (11) equals κ_{obs}/C. Once the value of α is known, K_f can be easily calculated. It was pointed

TABLE 3. Thermodynamic Parameters for the Formation of Crown Complexes in DMF and DME Solutions[97,98]

L	MX	Solution	$\log K_f$	ΔH (kcal mol^{-1})	ΔS (cal mol^{-1} deg^{-1})	Reference
DB18C6	NaSCN	DMF	~2.8 (25°C)	−6.	−7.	97
4,4'-diaminoDB18C6	NaSCN	DMF	2.9 (20°C)	−6.	−7.	98
4,4'-dinitroDB18C6	NaSCN	DMF	1.99 (30°C)	—	—	98
DB18C6	NaBPh$_4$	DME	3.66 (30°C)	−3.9	+3.9	98
DB18C6	NaBPh$_4$	DME	$\log K_{\rm ip} = 3.62$ (30°C)	−3.4	+5.4	98
DB18C6	NaBPh$_4$	DME	$pK_d^{\,A} = 4.27$	−2.0	−26	98
DB18C6	NaBPh$_4$	DME	$pK_d^{\,B} = 4.22$	−2.5	−28	98

out by Jagur-Grodzinski[102] that since this technique monitored the change in the conductance of solutions with variable salt concentration but at constant crown concentration, the calculation of formation constant is based on small differences in the values of Λ_{MX} and Λ_{MLX} with salt concentration and to the difference in the degree of complexation. This technique makes the method very sensitive to small experimental errors. It should be noted that the results of Matsuura *et al.* show a curious lack of sensitivity of the complexation constant to the nature of the cation.

Precise conductance measurements were carried out by Evans *et al.*[103] in methanol and acetonitrile solutions of alkali salts with DB18C6 and DC18C6. The measurements were made on a series of solutions with constant salt concentration with varying concentration of the ligand. Again, the nonlinear least-squares iteration method was used to obtain the values of Λ_{MLX} and of K_f. It was assumed that at the salt concentrations used ($\leqslant 4 \times 10^{+3}\,M$), the formation of ion pairs could be neglected since the dielectric constants of the two solvents are quite high. In general, the addition of a crown lowers the equivalent conductance of a salt M^+X^- by about 10–15%. The results of these measurements are shown in Table 4. It is interesting to note that with DB18C6 the stabilities of the sodium and potassium complexes are reversed in going from methanol ($K_{f(K^+)} > K_{f(Na^+)}$) to acetonitrile ($K_{f(Na^+)} > K_{f(K^+)}$).

Electrical conductance measurements were also used by Boileau *et al.*[104a,b] to determine the ion-pair dissociation constants in the 20 to $-60°C$ temperature range of $MBPh_4$ salts ($M = Na^+$, K^+) in tetrahydrofuran solutions in the presence of cryptands C221 and C222. The conductance data were analyzed by the method of Fuoss[99] and of Justice *et al.*[105] The K_d values obtained by the two methods were not significantly different and were in the

TABLE 4. Formation Constants of Some Crown Complexes in Methanol and Acetonitrile Solutions at 25°C

Ligand	Cation	Solvent	$\log K_f$
DB18C6	Na^+	MeOH	4.16
DB18C6	K^+	MeOH	4.57
DB18C6	Na^+	MeCN	5.04
DB18C6	K^+	MeCN	4.83
DC18C6	Na^+	MeOH	4.05
DC18C6	K^+	MeOH	5.35
DC18C6	Cs^+	MeOH	3.85
DC18C6	Na^+	MeCN	5.20
DC18C6	K^+	MeCN	5.63
CC18C6	Cs^+	MeCN	4.26

10^{-5}–10^{-4} range. The van't Hoff plots for $K^+Ph_4B^-$, $[C221 \cdot Na^+]Ph_4B^-$, and $[C222 \cdot K^+]Ph_4B^-$ systems are shown in Figure 17a and they are obviously not linear. The variation in the ΔH^0 of the ion-pair dissociation with temperature is probably due to an equilibrium between contact and solvent-separated ion pairs. Similar plots for $[C222 \cdot Na^+]Ph_4B^-$ and $[C322 \cdot Cs^+]Ph_4B^-$ in tetrahydrofuran solutions, however, give linear van't Hoff plots (Figure 17b).[104b]

Ion-pair dissociation constants of DB18C6 complexes with sodium and potassium thiocyanates were determined in nitrobenzene and nitrobenzene–toluene mixture.[106] These data are given in Table 5. As expected, the ion-pair dissociation constant decreases with decreasing dielectric constant. It is interesting to compare these data with those for the conductance of sodium and potassium picrates in nitrobenzene. The ion-pair dissociation constants of these two salts are reported to be 2.8×10^{-5} and 6.86×10^{-4}, respectively, while the equivalent conductance values at infinite dilution are 32.30

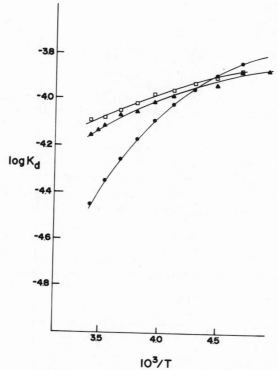

FIGURE 17a. Variation of $\log K_d$ with $1/T$ for KBPh$_4$ (●), KBPh$_4$ + C222 (□), and NaBPh$_4$ + C221 (▲), all in tetrahydrofuran solutions.[104a] (Reproduced with permission.)

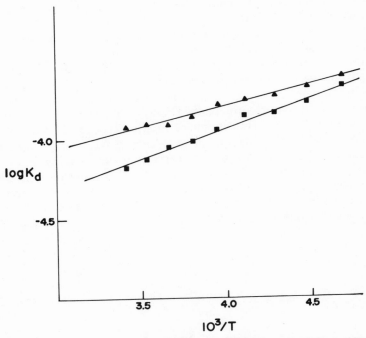

FIGURE 17b. Variation of $\log K_d$ with $1/T$ for CsBPh$_4$ + C222 (▲) and NaBPh$_4$ + C222 (■) in tetrahydrofuran solutions.[104b] (Reproduced with permission.)

and 33.81.[107] In comparing these data with those of Evans *et al.*[103] it is seen that the formation of a large crown cation increases the dissociation constant by more than one order of magnitude. It is surprising that the Λ_0 values seem to be unaffected by the formation of the crown complex, although it is difficult to estimate the influence of substituting the picrate ion for the thiocyanate.

TABLE 5. Ion-Pair Dissociation Constant of 18C6 Alkali Complexes in Nitrobenzene and Nitrobenzene–Toluene Mixtures[106]

Solvent	Dielectric Constant	$K\,(M^{-1})$	$\Lambda_0\,(\Omega^{-1}\,\mathrm{cm}^2\,\mathrm{mol}^{-1})$
PhNO$_2$	35	Na$^+$, 7×10^{-4}	35
		K$^+$, 1×10^{-2}	33
70% PhNO$_2$, 30% PhMe	23	Na$^+$, 6×10^{-5}	43
		K$^+$, 4×10^{-4}	45
50% PhNO$_2$, 50% PhMe	16	Na$^+$, 5×10^{-6}	63
		K$^+$, 1×10^{-4}	45

Conductometric titrations have been used by Pedersen and Frensdorff[11] to determine the stoichiometry of crown complexes. The results, illustrated in Figure 18, indicate two types of behavior. The addition of crown to a KCl solution in methanol, where the salt is nearly completely dissociated, will *decrease* the equivalent conductance due to the decreased mobility of the cation. With stable complexes there will be an abrupt break at the ligand-to-metal stoichiometric mole ratio since further addition of the ligand would not affect the equivalent conductance of the complexed salt.

On the other hand, since in nonpolar solvents alkali salts either can be solubilized as macrocyclic complexes, or, if in solution, exist predominantly as nonconducting ion pairs, the addition of a crown to a suspension of a metal salt in chloroform results in an *increase* in the conductance until the stoichiometric mole ratio is followed by a plateau.

Smid *et al.*[30,108] showed that the interactions of crown ether polymers with alkali salts in solution can also be followed conductometrically. However, the break at the stoichiometric mole ratio is not as sharp as with the monomer since there may be a problem of obtaining a fully cation-saturated polymer.

Electrical conductance measurements have also been used by Smid *et al.*[109] to determine the formation constants of 4'-substituted B15C5–sodium complexes and 4'-substituted B18C6 complexes in acetone solution. The treatment of data was similar to that used by Evans *et al.*[103]

Ion-pair dissociation constants of fluorenyl salts and their complexes were measured by Hogen-Esch and Smid as a function of temperature in tetrahydrofuran and tetrahydropyran solutions.[110]

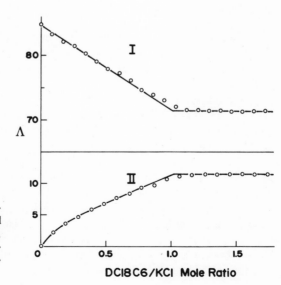

FIGURE 18. Equivalent conductance of the DC18C6–KCl system. I, Methanol; II, 90–10% chloroform–methanol mixture.[11] (Reproduced with permission.)

Conductometric measurements have been used recently[111] to follow interactions of a number of crown ethers with organic acids in 1,2-dichloroethane solutions. The formation constants of the proton complexes

$$H^+X^- + L \underset{}{\overset{K_f}{\rightleftharpoons}} LH^+X^-$$

and the ion-pair dissociation constant

$$LH^+X^- \underset{}{\overset{K_d}{\rightleftharpoons}} LH^+ + X^-$$

have been determined.

The protonation constants with picric and *p*-toluenesulfonic acids are of the order of 10^2–10^3. As expected for a 1,2-dichloroethane, the ion-pair dissociation constants are of the order of $\sim 10^{-5}$.

Association constants and electric dipole moments were obtained from *permittivity measurements* of dilute solutions of potassium *p*-toluenesulfonate (KTs), LiCl, DB18C6, cyclohexyl-15-crown-5, and of their mixtures in octanoic acid. The dipole moment of DB18C6–KTs is 8.27, and from the distance between the ionic electrical centers it is inferred that the K^+–Ts^- distance is stretched by 50 pm in the complex relative to that in the uncomplexed KTs ion pairs.[101]

4.2.3. Polarography, Cyclic Voltametry, and Other Electrochemical Techniques

Electroreduction of alkali metal and thallium(I) complexes of DC18C6 showed completely reversible electrode kinetics.[112] Electroreduction of cryptate complexes was first studied by Peter and Gross in propylene carbonate solutions.[113] Decomposition potentials of Na^+, K^+, and Ba^{2+} ions were shifted to negative potentials with the magnitude of the shift dependent on the stability of the complex. For example, the complexes with cryptand C222 are reduced at approximately the same potential as tetrabutylammonium bromide.

A polarographic study of the DB18C6·K^+ complex in the same solvent showed that the half-wave potential is at -1.970 V, i.e., about 130 mV more negative than the uncomplexed ion (vs. SCE), while for the cryptate C222·K^+, the $E_{1/2}$ value was -2.70 V.

The authors also studied C222·K^+ cryptate in propylene carbonate by coulometry, polarography at short drop time, pulse polarography, and cyclic voltametry.[114a] Coulometric studies indicate that the reduction of C222·K^+ cryptate requires 1.2 electrons. The reduction product is metallic potassium, which is not complexed by C222. The mechanism may involve dissociation of the complex in the electrical double layer and the reduction of the free K^+ ion, or the reduction takes place inside the cryptand cavity and the resulting neutral complex C222·K is extremely weak and rapidly dissociates. It was shown by Britz and Knittel[28] that in aqueous solutions the $C222 \cdot K^+$ cryptate is adsorbed at the dropping mercury electrode up to -1.70 V. The diffusion coefficient of the $C222 \cdot K^+$ is 6.9×10^{-6} cm^{-2} s^{-1}, which is about one-third the value for the free hydrated K^+ ion.

Electrochemical studies on C222 complexes with lithium, sodium, rubidium, and cesium ions were carried out in propylene carbonate solutions. The half-wave potentials for $C222 \cdot Na^+$ and $C222 \cdot Rb^+$ are -2.73 and -2.39 V vs. aqueous SCE. In all cases one-electron irreversible waves were obtained. It should be noted that the half-wave potentials for the sodium and potassium cryptates are the same despite the fact that the potassium complex should be the more stable one.

Louati and Gross studied electrochemical reduction of Mg^{2+} ion and of $C221 \cdot Mg^{2+}$ cryptate in propylene carbonate solutions.[114b] A polarographic study of uncomplexed Mg^{2+} ion gave half-wave potential of -1.62 ± 0.05 V (vs. SCE), which was independent of concentration.

A well-defined polarographic wave was obtained with the Mg^{2+} cryptate, but the half-wave potentials vary from ~ 2.15 to 2.38 V vs. SCE in the 4×10^{-4} to 1.025×10^{-3} M concentration range. Both polarographic and cyclic voltammetry studies indicate that the reduction of the solvated Mg^{2+} and of the $Mg^{2+} \cdot C221$ ions is polarographically irreversible and slow. Electron transfer is the rate-determining step. The reduction of the solvated Mg^{2+} ion is much slower than the reduction of the cryptate.

Agostiano *et al.*[115] polarographically determined formation constants of sodium, potassium, and cesium complexes with DC18C6 in lower alcohols. The values of K_f's were obtained in the usual manner from the shifts of the cationic half-wave potential upon the addition of the ligand.[88,89]

A polarographic study of thallium(I) in propylene carbonate gave the stability constant of the $C222 \cdot Tl^+$ complex as log $K_f = 9.0 \pm 0.3$.[116]

Recent cyclic voltammetry studies on the europium(II)–europium(III) couple in aqueous solutions in the presence of cryptand C221 indicate that in contrast to the aquated ions, the $[C221 \cdot Eu^{3+}]/[C221 \cdot Eu^{2+}]$ is an electro-chemically reversible redox couple at room temperature.[117] It is interesting to note that at the addition of the fluoride or the hydroxide ion, the cyclic voltammograms for the $C221 \cdot Eu^{3+}$ ion are shifted to more negative potentials. The data seem to indicate ion-pair association between those ions and the complexed cation. No evidence for such association was found for the europium(II) complex, presumably due to the lower charge on the cation.

4.3. Spectroscopic Techniques

4.3.1. Proton Magnetic Resonance (PMR)

Proton magnetic resonance is one of the most useful techniques for the studies of macrocyclic ligands and their complexes in solutions. Many of the unsubstituted crowns and cryptands have very simple PMR spectra which are sensitive to the conformation changes that the ligands undergo in complex formation. Examples of PMR studies are quite numerous and only typical examples will be cited here.

It has been shown that the formation of the sodium complex of DB24C8 in chloroform solutions results in a 0.1–0.15 ppm upfield shift of the CH_2–O protons, while the Ar–O–CH_2 and the aromatic protons are shifted downfield by 0.12–0.17 and 0.11–0.20 ppm, respectively.[118]

The proton chemical shifts of the ligands are strongly affected by the formation of contact ion pairs with aromatic anions in solvents of low polarity. An early paper of Wong et al.[119] shows that the PMR of 4,4'-dimethyl-DB18C6 is significantly affected by complexation. The spectrum of the free ligand contains CH_3 signal at 2.27 ppm, aromatic protons at 6.77 ppm, and polyether ring protons form a broad band centered at 4.05 ppm. The addition of an equimolar amount of sodium fluorenyl (Na^+Fl^-) to a THF solution of the ligand results in a strong upfield shift of the polyether ring protons—one set by 1.1 ppm, the other by 0.75 ppm. The aromatic protons of the crown were also shifted upfield by ~ 0.2 ppm, and there was very little effect on the methyl protons. Complexation with alkali ions should produce downfield shifts due to increased electronegativity of the ether oxygen; the upfield shift must result from the interaction of the ligand with the fluorenyl anion, i.e., to the contact ion-pair formation.

When 4,4'-dimethyl-DB18C6 is dissolved in pyridine the polyether protons show two bands centered at 4.15 and 3.95 ppm. Upon the addition of Na^+Fl^- the two peaks are shifted to 3.80 and 3.22 ppm, respectively. These data again indicate that an ion pair is formed between complexed cation and Fl^- ion.

The PMR signal of free C222 consists of a triplet due to N–CH_2 at 2.65 ppm, a triplet for N–CH_2–CH_2–O at 3.60 ppm, and a singlet for O–CH_2–

TABLE 6. Chemical Shifts of Proton Signals of
Free and Complexed C222 in Deutorochloroform
(in ppm)[121]

Compound	N–CH_2	N–CH_2–CH_2–O	O–CH_2–CH_2–O
C222	2.65	3.60	3.68
C222·Li^+	2.73	3.67	3.73
C222·Na^+	2.68	3.63	3.63
C222·K^+	2.57	3.57	3.62
C222·Rb^+	2.57	3.56	3.62
C222·Cs^+	2.55	3.53	3.62
C222·Ca^{2+}	2.93	—	3.89
C222·Ba^{2+}	2.80	3.87	3.92

C$\underline{H}_2$–O protons at 3.68 ppm.[120,121] Upon complexation, the latter proton resonances are not affected very much by the alkali metals but are shifted downfield by alkaline earths. The N–C$\underline{H}_2$ triplet moves upfield with increasing cationic radius. In general, the magnitude of the chemical shifts is 0.1–0.2 ppm. Some examples are given in Table 6. More complete data are given in reference 121.

Chemical shifts of the sulfur-substituted cryptand $C22S_2$ in deuterochloroform solutions are particularly sensitive to the complexation of transition-metal ions. The formation of the potassium cryptate affects

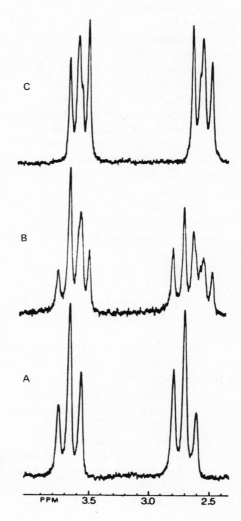

FIGURE 19. Proton magnetic resonance of the spherical [3]-cryptand **29** in CDCl$_3$ solutions. A, Free ligand; B, ligand + CsBr complex; C, **29**·CsBr complex.[38] (Reproduced with permission.)

primarily the N–C$\underline{H}_2$ resonance, while Ag$^+$ and Pb^{2+} cryptates show large shifts, particularly for the S–C$\underline{H}_2$–C$\underline{H}_2$–S singlet.[33]

The high symmetry of the spherical [3]-cryptand, **29**, results in a particularly simple PMR spectrum shown in Figure 19. It consists of two triplets ($J = 5.5$ Hz) for the C$\underline{H}_2$–N and C$\underline{H}_2$–O protons at 2.70 and 3.65 ppm, respectively.[38] The ligand must exist either in the tetra-*endo* (i_4), or tetra-*exo* (o_4) forms or as a mixture of all forms in rapid equilibrium. As seen in Figure 19, the addition of a metal ion results in a definite chemical shift of the proton resonance.

Proton NMR (as well as ^{13}C NMR) were used by Live and Chan[122] to elucidate the structures of several crown complexes in water, acetone, and chloroform solutions and in water–acetone mixtures. Detailed analysis of proton chemical shifts and of proton–proton vicinal coupling constants was made possible by using a high-field (52 kG) Fourier transform NMR spectrometer. Complexes of B18C6, DB18C6, and DB30C10 with Na$^+$, K$^+$, Cs$^+$, and Ba^{2+} were studied.

The addition of a potassium salt to DB18C6 in acetone resulted in two spectral changes; there is a change in coupling constants, indicating a small increase in OCH$_2$CH$_2$O dihedral angle and all proton resonances shift downfield. Very similar spectral changes were observed for potassium complexes of crown B18C6. Change in anion produced very little effect. The same general behavior was observed for the sodium complexes although smaller frequency shifts and more anion dependence was observed.

In the case of cesium complexes with DB18C6 a different behavior was observed. The addition of Cs$^+$ first produced an upfield shift until a minimum at a Cs$^+$/L mole ratio of ~ 0.5, and then a gradual downfield shift for the ether protons 1 and 2

$$AR–O–\overset{1}{C}H_2–\overset{2}{C}H_2–O$$

(Figure 20). The behavior clearly indicates the formation of L$_2$Cs and LCs$^+$ complexes. Similar results were obtained by other investigators using ^{133}Cs NMR (p. 585).

Not surprisingly, more anion dependence of chemical shifts was observed in deuterochloroform solution, where a considerably greater extent of ion pairing is expected.

For potassium and cesium complexes the behavior of a large crown DB30C10 in acetone and deuterochloroform solutions shows a great deal of similarity with that of DB18C6. In the case of the sodium ion the spectrum is quite different from that of the heavier alkali complexes. The results indicate that the structure of the Na$^+$ complex is different from the other two and that a conformational change of the ligand takes place upon complexation, since the ring of the DB30C10 crown ether is considerably larger than the diameter of the sodium ion.

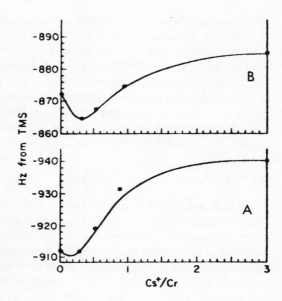

FIGURE 20. Proton chemical shifts on complexation of DB18C6 with Cs^+ in acetone solutions. A, H_1 proton; B, H_2 proton (see text).[122] (Reproduced with permission.)

Formation constants of several complexes were obtained from the analysis of the proton chemical shift dependence on the ligand/cation mole ratio. The results are given in Table 7. The method is analogous to the one described below.

PMR measurements at variable temperature were used to study diazabicyclo (k,l,m) alkanes.[8] With the dihydrochlorides there is a slow isomerization in solution from (o,o) to (i,i). The PMR spectra of the two isomers, observed with a 220 MHz spectrometer, are quite different. For diazabicyclo-

TABLE 7. Formation Constants of Crown Complexes Obtained from PMR Measurements[122]

Ligand	Cation	Solvent	$\log K_f$
DB18C6	K^+	Acetone	> 4.3
DB18C6	Na^+	Acetone	~ 4.6
B18C6	K^+	Acetone	> 4
B18C6	Na^+	Acetone	> 4
B18C6	K^+	D_2O	2
B18C6	Ba^{2+}	D_2O	3.70
DB30C10	Na^+	Acetone	2.54
DB30C10	K^+	Acetone	4.3
DB30C10	Cs^+	Acetone	4.23

(8.8.8) alkane the activation energy for the nitrogen inversion is 7.7 kcal mol^{-1} and the rate constant is 1.4×10^7 s^{-1}.

Equilibrium constants for $(o,o) \rightarrow (i,i)$ isomerization were obtained for alkanes with different lengths of the hydrocarbon chains. Measurements were made in a 50:50 mixture of trifluoroacetic acid and water. For $k = l = m = 8$, the equilibrium constant is greater than 100. It has a value of 24 for the (6.6.8) compound but is small for the rest. For example, in the case of (7.7.7) and (9.9.9) alkanes, it is <1. The equilibrium is established much more rapidly in MeOD than in D_2O or in D_2O–MeOD mixtures.[123]

4.3.2. Carbon-13 Magnetic Resonance

Carbon-13 NMR is a very useful adjunct to the PMR. With present day instrumentation and, particularly, with the use of Fourier transform spectroscopy, the task of obtaining spectra of unenriched samples is no longer a problem. The range of chemical shifts is much larger than for the proton, and the resonance frequencies are quite sensitive to small changes in the conformation and/or the chemical environment of the studied compound. Often ^{13}C measurements are made to confirm the results obtained with PMR. Thus the interpretation of Live and Chan of the PMR studies described above were supported by the ^{13}C measurements.[122]

Carbon-13 NMR has been used to study intramolecular cation exchange in a [3]-cryptand, **21**.[124] The free ligand shows the expected four ^{13}C resonances. On addition of alkali cations the four signals shift smoothly and level off at high M$^+$/**21** mole ratios. With alkali earth cations, however, the addition of a salt results in the appearance of a *new set* of four lines. The relative intensities of the two sets change as more salt is added, and the signal of the free ligand disappears when 1:1 mole ratio is reached. Further addition of the salt does not change the intensities of the new set of ^{13}C resonances which, obviously, must belong to the complex.

A variable-temperature study of the ^{13}C resonances of several metal complexes in D_2O showed that the **21**·M^{2+} complexes display an interconversion between two species in which the cation is located unsymmetrically in the ligand cavity. The coalescence temperature, exchange rates, and free energies of activation for this process were obtained and are shown in Table 8.

The protonation of cryptand **29** and the formation of anionic complexes can be followed very conveniently by ^{13}C NMR.[125] The formation of species **29**·2H$^+$ and **29**·4H$^+$·Cl$^-$ can be clearly identified.

Carbon-13 NMR was also used in the study of polynuclear complexes of the [3]-cryptand, **23**.[126] Figure 21 shows the spectra of the **23**·2Pb^{2+} and **23**·Pb^{2+}·Ag$^+$ complexes in the C–N region. It is seen that the resonances of the top and the bottom ring are different in the mixed complex and are

TABLE 8. Spectral Parameters, Coalescence Temperatures, and Free Energies of Activation for Intramolecular Cation Exchange in $M^{n+} \cdot 21$ Cryptates[124]

Compound	^{13}C signal[a]	$\Delta\nu$,[b] ± 2 Hz (temp. °C)	T_0 [c] +4°C	k_0 [d] (s^{-1})	$\Delta G_0^{\ddagger}$,[e] ± 0.3 (kcal mol^{-1})
$21 \cdot Ca^{2+} \cdot 2Cl^-$	C_R–N	48 (4)	40	107	15.4
	C_B–N	26 (4)	30	58	15.3
	C_R–N	21 (32)	~105	~ 47	~19.5
$21 \cdot Sr^{2+} \cdot 2Cl^-$	C_R–N	77 (5)	27	171	14.5
	C_B–N	26 (5)	15	58	14.5
$21 \cdot Ba^{2+} \cdot 2Cl^-$	C_R–N	70	< 3	>155	<13.3
$21 \cdot La^{3+} \cdot 3NO_3^-$	C_R–N	26 (32)	> 93	< 58	>18.6

[a] The R and B designate, respectively, carbon atoms in the 12-membered ring and in the two bridges.
[b] $\Delta\nu$—separation of the ^{13}C signals at temperatures indicated.
[c] T_0—coalescence temperature.
[d] k_0—exchange rate.
[e] $\Delta G_0^{\ddagger}$—free energy of activation at coalescence.

shifted with respect to the two homonuclear complexes. The spectrum of the mixed complexes also indicates that the *intra*molecular cation exchange is slow by the NMR time scale.

C-13 relaxation times were used to investigate the molecular motions in DB18C6, DC18C6, and in their K^+ complexes; the uncomplexed macrocycles display segmental mobility which is absent in the complexes; furthermore, the correlation times obtained indicate ion pairing between the anion and the complexed cation.[126b]

4.3.3. Nuclear Magnetic Resonance of Nuclei Other Than ^1H and ^{13}C

4.3.3.1. Introduction. The importance of the PMR and ^{13}C NMR in the elucidation of the solution behavior of the macrocyclic ligands and of their complexes has been amply illustrated by the examples discussed in Sections 4.3.1 and 4.3.2. In recent years, with the development of NMR instrumentation, NMR measurements on other nonmetallic and metallic nuclei possessing nuclear spins became possible, and their use for the studies of macrocyclic complexes is becoming increasingly popular.

Of course, in both ^1H and ^{13}C NMR measurements, the studied nuclei belong to the ligand (or, sometimes the solvent) and do not participate directly in the interaction with the metal ion. It is also possible to study the behavior of the ligands by looking at the resonances of the donor atoms themselves, such as oxygen (^{17}O) or nitrogen (^{14}N or ^{15}N).

The ^{17}O nucleus has a spin of $\frac{5}{2}$ and a natural abundance of $(3.7 \times 10^{-2})\%$. While ^{17}O NMR has been used in complexation studies, the low

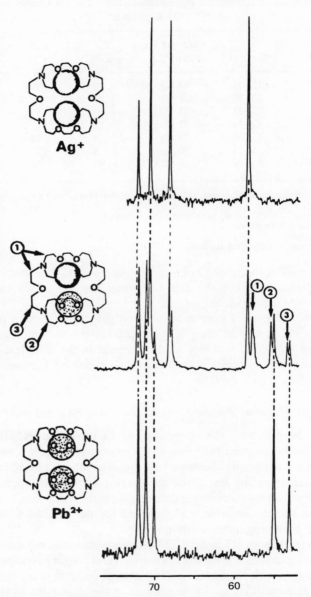

FIGURE 21. Carbon-13 NMR of the [3]-cryptand **23** complexes in CD_3OH-H_2O mixture (95–5%). Shifts are given in ppm with TMS as standard.[126] Unshaded circles on the left-hand side of the figure represent Ag^+; shaded circles represent Pb^{2+}. (Reproduced with permission.)

sensitivity as well as very low natural abundance seems to have deterred its use in the studies of macrocyclic complexes.

The nitrogen-14 nucleus, although present in nearly 100% natural abundance, has a spin $I = 1$, and a quadrupole moment $Q = 7.1 \times 10^{-2}$ barns. The ^{14}N resonances are usually very broad, and at this time it is not a convenient nucleus for the studies of polyaza macrocycles.

On the other hand, the nitrogen-15 nucleus has a spin $I = \frac{1}{2}$ and therefore a very narrow linewidth, but its natural abundance is only 0.1% and its sensitivity is only 0.001 of that of proton (at constant field). It is possible, however, to observe ^{15}N signals in natural abundance by using Fourier transform NMR and, preferably, a high-field magnet with a superconducting solenoid. The use of ^{15}N NMR in the studies of polyaza macrocyclics has been recently studied by Forster and Roberts.[127]

Before discussing the above studies, it should be noted that in this chapter we will designate paramagnetic (*downfield*) shifts as *negative* and diamagnetic (*upfield*) shifts as *positive*.

Foster and Roberts studied ^{15}N chemical shifts of cryptands C211, C221, and C222 with alkaline and alkaline earth ions as well as with Ag^+ and Tl^+. The direction and the magnitude of the chemical shifts (up to 10 ppm) varied with the ligand and the metal ion and, in general, depended on charge and ionic character of the metal ion as well as on the tightness of the fit of the ion in the cryptand cavity. These pioneering results appear to be quite promising, but additional work is needed before the utility of ^{15}N NMR for the studies of polyaza complexes is fully realized.

Since the macrocyclic polyethers and polyoxapolyaza ligands are particularly noted for their complexing abilities *vis à vis* the alkali cations, it is natural that the NMR of the alkali metal nuclei is being used quite intensively for the studies of alkali complexes.

The nuclear properties of the alkali elements are shown in Table 9. It is seen that all members of the alkali family have NMR-active isotopes, and that in all cases the nuclei have quadrupole moments.

Despite the quadrupole moment, the natural linewidths of ^{23}Na, ^{39}K, and especially ^{7}Li and ^{133}Cs are quite narrow (less than 1 Hz in the last two cases). The range of chemical shifts varies directly with the atomic number being on the order of ~ 10 ppm for ^{7}Li and several hundred ppm for ^{133}Cs.

Magnetic resonance studies of alkali salt solutions in water and in non-aqueous solvents have shown that this technique represents a very sensitive probe of the immediate chemical environment of the respective cations.[128,129]

The nucleus of a complexed ion usually resonates at a different frequency than the uncomplexed ion. Thus, if reaction conditions are adjusted so that both free and complexed metal ions are present in solution and if the exchange between the complexed and the free cation is slow on the NMR time scale, then two resonance lines will be observed corresponding to the

TABLE 9. Nuclear Properties of Alkali Metals

Isotope	NMR frequency in MHz at 14.2 kG	Natural abundance (%)	Spin	Sensitivity relative to ^1H at constant field
^{6}Li	8.896	7.63	1	0.0085
^{7}Li	23.497	92.57	3/2	0.294
^{23}Na	15.992	100.0	3/2	0.0927
^{39}K	2.821	93.08	3/2	0.00051
^{41}K	1.551	6.91	3/2	0.00008
^{85}Rb	5.838	72.8	5/2	0.010
^{87}Rb	19.783	27.2	3/2	0.177
^{133}Cs	7.931	100.0	7/2	0.047

two chemical environments. Since the integrated intensities of the signals are proportional to the concentrations of the two species, intensity measurements, in principle, can lead to the values of the stability constants. However, at the present state of NMR technology, the accuracy of intensity measurements in most cases is not sufficient for an accurate calculation of high formation constants. In addition, slow exchange is rather rare, in the great majority of cases the exchange between two sites is fast, and only one, population average, signal will be observed whose chemical shift is given by

$$\delta_{obs} = P_f \delta_f + P_c \delta_c \tag{12}$$

where P_f and P_c are the populations of the cation in the free and complexed state, while δ_f and δ_c are the corresponding chemical shifts.

If we assume that *only a 1:1 complex is formed*, we can derive the following equation for the observed chemical shift[130]:

$$\delta_{obs} = \{[KC_M{}^t - KC_L{}^t - 1] + [K^2(C_L{}^t)^2 + K^2(C_M{}^t)^2 - 2K^2C_M{}^tC_L{}^t$$
$$+ 2KC_L{}^t + 2KC_M{}^t + 1]^{1/2}\}\left\{\frac{\delta_f - \delta_c}{2KC_M{}^t}\right\} + \delta_c \tag{13}$$

where $C_M{}^t$ and $C_L{}^t$ are the total concentrations of the metal and of the ligand and K is the formation constant.

Usually δ_{obs} is measured as a function of ligand/metal ion mole ratio. Since δ_f is readily obtained from measurements on metal salt solutions without added ligand, Eq. (13) contains two unknown quantities K_f and δ_c. The experimental parameters δ_{obs}, δ_f, $C_M{}^t$, and $C_L{}^t$ are substituted into Eq. (13), and K and δ_c values are adjusted with a nonlinear least-squares program.

The above technique can be very useful for complexes with formation constants varying from less than unity to about 10^4–10^5. As with all spectroscopic techniques, NMR fails when it has to deal with very stable complexes.

4.3.3.2. Lithium-7 NMR. The effect of the addition of a cryptand to a lithium salt solution in a given solvent is very much dependent on the solvating ability of the medium. In the case of the lithium salts, the addition of the larger cryptands C222 and C221 produces no effect in a strongly solvating solvent such as dimethyl sulfoxide. In a nearly equally polar but much less solvating solvent—nitromethane—the addition of a large cryptand does result in a downfield shift, indicating complexation, but the exchange between

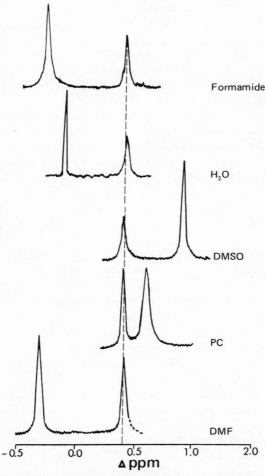

FIGURE 22. Lithium-7 NMR spectra of lithium-C211 cryptate in various solvents. Chemical shifts are measured versus aqueous LiClO$_4$ solution at infinite dilution.[129] (Reproduced with permission.)

the free and complexed lithium ion is fast on the NMR time scale, and only one population-average resonance signal is observed.[129] On the other hand, with the cryptand C211, whose cavity is close to the dimensions of the desolvated Li^+ ion, the addition of the ligand to a Li^+ salt produces not only a paramagnetic chemical shift but also, in the presence of an excess Li^+ ion, two NMR signals are observed corresponding to the resonances of the complexed and of the free ion. In this case the lithium ion is not readily released by the cryptand, and the exchange between the two cationic sites is slow.

It is interesting to note that the limiting chemical shift of the Li^+ ion in the C211 cavity is almost completely independent of the solvent (Figure 22). It is obvious, therefore, that the ion is effectively insulated from the medium by the ligand sheath.

Lithium-7 NMR has also been used for the determination of Li^+–C222 complex formation constants in water and in pyridine by previously described techniques. The values obtained were, respectively, $\log K_f^{H_2O} = 0.99 \pm 0.15$ and $\log K_f^{Py} = 2.94 \pm 0.10$.

The kinetics of the decomplexation reaction of $C211 \cdot Li^+$ complex have been studied in a number of solvents by temperature-dependent line-shape analysis.[131] The data are given in Table 10. The activation energy for the release of the Li^+ ion from the complex seems to be related to the donicity of the solvent as defined by Gutmann[132] rather than to the dielectric constant. The activation energy values vary from 14.1 kcal mol^{-1} in formamide to 21.3 kcal mol^{-1} in water. It is interesting to note that Shchori et al.[133] found that the activation energies for release of Na^+ from DB18C6 were independent of the solvent (see below).

4.3.3.3. Sodium-23 NMR. The sodium-23 resonance is considerably broader than that of lithium-7, and the quadrupolar interaction with the electric field gradient in the complexes may result in strong additional broadening. Thus, for example, the two-dimensional crown complexes result in a considerable electrical field gradient about the nucleus, and the addition of an equimolar amount of DC18C6 to Na^+ solution in nitromethane completely wiped out the ^{23}Na signal.[134]

The addition of diMeDB18C6 to a $NaBPh_4$ solution in tetrahydrofuran resulted in an upfield shift from 5.7 ppm (vs. 1.0 M aqueous NaCl) to 14.9 ppm. A very large increase was noted in the linewidth for the complexed Na^+ ion. The exchange was fast and only one ^{23}Na signal was observed.[135]

On the other hand, for the cryptate complexes, the line broadening is much less pronounced and the exchange is much slower. Ceraso and Dye[136] found that for the $C222 \cdot Na^+$ complex in ethylenediamine, two ^{23}Na signals can be observed at room temperature (Figure 23). Line-shape analysis gave the activation energy of 12.2 ± 1.1 kcal mol^{-1} for the decomplexation reaction. A more detailed study using Fourier transform ^{23}Na NMR gave

TABLE 10. Exchange Rates and Thermodynamic Parameters of Lithium Cryptate Exchange in Various Solvents[131]

Solvent	Donicity	Dielectric constant	E_a (kcal mol^{-1})	$k_b \times 10^3$ (s^{-1}, 298 K)	$\Delta H_0^\ddagger$ (kcal mol^{-1})	$\Delta S_0^\ddagger$ (cal K^{-1} mol^{-1})	$\Delta G_0^\ddagger$ (kcal mol^{-1}, 298 K)
				Cryptand 211			
Pyridine	33.1	12.3	19.6 (3.5)[a]	0.12 (0.24)	19.0 (3.4)	−12.5 (9.2)	22.7 (1.1)
Water	33.0	78.6	21.3 (1.2)	4.9 (2.0)	20.7 (1.1)	+ 0.4 (3.0)	$20.6 (0.2_5)$
Dimethyl sulfoxide	29.8	45.0	16.1 (0.6)	23.2 (5.4)	15.5 (0.6)	−13.8 (1.4)	$19.7 (0.1_4)$
Dimethyl formamide	26.6	36.1	16.0 (0.6)	13.0 (3.3)	15.4 (0.6)	−15.5 (1.4)	$20.0 (0.1_5)$
Formamide	24.0	111.0	14.1 (0.7)	7.4 (2.9)	13.5 (0.7)	−22.8 (1.8)	$20.8 (0.2_3)$
				Cryptand 221			
Pyridine	33.1	12.3	13.5 (0.4)	1230 (196)	12.9 (0.4)	−14.9 (0.9)	17.9 (0.1)

[a] Values in parentheses are standard deviations.

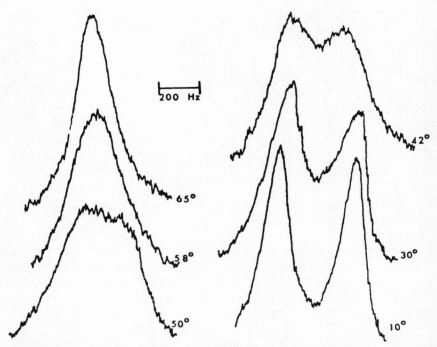

FIGURE 23. Temperature dependence of ^{23}Na nuclear magnetic resonance spectrum for a solution 0.4 *M* NaBr and 0.2 *M* C222 in ethylenediamine solution.[136] (Reproduced with permission.)

chemical shifts of free and cryptated sodium ions in several solvents (Table 11) as well as exchange rates and thermodynamic parameters of sodium cryptate (Table 12).[137]

The activation energies for the decomplexation reaction depend on the

TABLE 11. Chemical Shifts of ^{23}Na Resonance for Na$^+$ and Na$^+ \cdot$C222 Species in Different Solvents[137]

Salt	Solvent	δ_{Na^+} [a,b]	$\delta_{\text{Na}^+\text{C222}}$
NaI	H$_2$O	-0.1	8.6
NaBr	Ethylenediamine	-13.7	10.7
NaBPh$_4$	Tetrahydrofuran	7.5	12.0
NaBPh$_4$	Pyridine	-0.9	12.4

[a] Infinitely dilute aqueous Na$^+$ as reference.
[b] Positive values indicate upfield (diamagnetic) shifts.

TABLE 12. Exchange Rates and Thermodynamic Parameters of Sodium
Cryptate Exchange in Several Solvents[137]

Solvent	Donicity	k_{-1} (s^{-1}, 298 K)	$\Delta H_0^{\ddagger}$ (kcal mol^{-1})	$\Delta S_0^{\ddagger}$ (cal K^{-1} mol^{-1})	$\Delta G_0^{\ddagger}$ (kcal mol^{-1}, 298 K)
Pyridine	33.1	1.14 ± 0.09	13.6 ± 0.2	-12.6 ± 0.6	17.374 ± 0.004
Tetrahydrofuran	20.0	8.03 ± 0.27	13.8 ± 0.2	$- 8.1 \pm 0.6$	16.22 ± 0.02
Water	33.0	147.4 ± 2.6	16.1 ± 0.2	5.3 ± 0.8	14.49 ± 0.01
Ethylenediamine	55.0	165.0 ± 4.9	12.3 ± 0.2	$- 7.6 \pm 0.6$	14.44 ± 0.02

solvent used, although there is no close correlation with the donicity of the
solvent defined by Gutmann,[132] as was found in the case of the $Li^+ \cdot C211$
complexes.[131]

Shchori *et al.* used sodium-23 NMR for kinetic studies of Na^+ complexation by DB18C6, DiNO$_2$DB18C6, DiNH$_2$DB18C6, and DC18C6 in several
solvents. Considerable line broadening was observed upon complexation.
For example, in the Na^+–DB18C6 system in dimethylformamide solution the
resonance line of the complexed ion was about 25 times broader than the
resonance of the free ion.[97] Due to the broadness of the signal, it was
difficult to measure precisely the chemical shifts, but the data seem to
indicate that complexation by the crown does not result in a significant
chemical shift of the ^{23}Na resonance. Of course, the observation is valid only
for DMF solutions since it has been shown[134] that the ^{23}Na resonance
frequency is very much dependent on the solvent.

Line-shape analysis as a function of temperature gave a pseudo-first-
order rate constant for the decomplexation reaction

$$Na^+DB18C6 \rightleftharpoons Na^+ + DB18C6$$

of approximately 10^5 s^{-1} at zero ionic strength and 298 K. The activation
energy for the decomplexation step was found to be 12.6 ± 0.6 kcal mol^{-1}.

Subsequent studies on substituted DB18C6 and DC18C6 in methanol,
dimethylformamide, dimethoxyethane, and tetrahydrofuran solutions showed
that while the nature of the solvent strongly affects the stabilities of complexes,
there is a very small solvent effect on the activation energy for the decomplexation reaction.[133] It should be noted, however, that solvents used in the
above studies have approximately the same donicity.

Kintzinger and Lehn[138a] studied Na^+-cryptates with C222, C221,
C211, and C22S$_2$ in 95% methanol solutions by sodium-23 NMR. In all
cases when an excess of sodium ion was present, two signals were observed.
In the case of the $Na^+ \cdot C222$ complex, the free energy of activation for the

decomplexation reaction at 331 K is 15.4 kcal mol^{-1}, in agreement with the value obtained from PMR studies.[139] The ^{23}Na chemical shifts (vs. 0.25 M aqueous NaCl solution) varied drastically with the ligand. The values obtained were -11.15 ppm for Na$^+$C211, $+4.25$ ppm for Na$^+$C221, $+11.40$ ppm for Na$^+$C222, and $+6.20$ ppm for Na$^+$C22S$_2$. The calculated ^{23}Na nuclear quadrupole coupling constants χ obey a linear relationship as a function of the Na^{23} chemical shifts δ: $\chi = -0.05\delta + 1.65$ MHz. Thus, measuring Na^{23} shifts and relaxation times allows a detailed study of Na$^+$ solvation (field gradients, cation mobilities) in various solvents.[138b] The method should be extendable to other quadrupolar nuclei. It should be noted that the sign indication used in the above paper is opposite to the one in this chapter.

Both crowns and cryptands react with metal solutions in appropriate solvents to give the solvated alkali cation and an alkali anion CM$^+ \cdot$M$^-$. Sodium-23 NMR measurements on the Na$^+$C222$\cdot$Na$^-$ salt in methylamine, ethylamine, and tetrahydrofuran solutions show that for Na$^-$ the ^{23}Na resonance is shifted strongly upfield from the resonance of Na$^+$ ions (free or complexed) in solutions.[140,141] In the above three solvents the resonance of the Na$^-$ nucleus is at $\sim +63$ ppm from the reference (saturated aqueous NaCl solution) and it is approximately at the same frequency as the calculated value for the gaseous anion. The above results indicate that at least in the solvents studied, the anion is essentially nonsolvated. The exchange between the two sodium sites Na$^+$C222 and Na$^-$ is slow and both resonances can be observed.

The sodium anion salt was also obtained by complexing Na$^+$ with 18C6 in methylamine solutions. In this case the exchange rate between 18C6$\cdot$Na$^+$ and Na$^-$ is faster than in the case of cryptates and the resonances for both sodium species are much broader.

4.3.3.4. Potassium-39 NMR. The ^{39}K NMR measurements are difficult due to a low sensitivity of the nucleus and the low resonance frequency. It is only very recently that macrocyclic complexes of K$^+$ were investigated by potassium-39 NMR.

Preliminary studies of K$^+$-cryptand systems in nonaqueous solvents showed slow exchange between the free and the "cryptated" K$^+$ ion when C222 and C221 cryptands were used.[142]

4.3.3.5. Rubidium-87 NMR. Rubidium-87 resonance yields very broad lines since the nucleus has a very short relaxation time. The only ^{87}Rb NMR measurements involving crown or cryptate complexes are the studies of Dye *et al.*[141] on the C222$\cdot$Rb$^+ \cdot$Rb$^-$ salt in ethylamine and in tetrahydrofuran solutions. The Rb$^-$ anion resonates nearly 200 ppm upfield from the aqueous Rb$^+$ ion. The linewidth of the C222$\cdot$Rb$^+$ resonance was too broad to be observed.

4.3.3.6. Cesium-133 NMR. In contrast to ^{87}Rb, the linewidth of the ^{133}Cs resonance is very narrow (<1 Hz), while the sensitivity is relatively

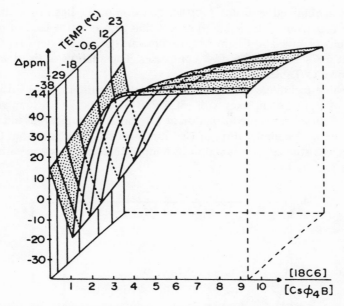

FIGURE 24. A three-dimensional plot of the ^{133}Cs chemical shift versus mole ratio and temperature (°C) for CsBPh$_4$ and 18C6 in pyridine solutions.[146] (Reproduced with permission.)

high and resonances of 0.001 M solutions can be observed at a field of 14.1 kG.[143,144]

Cesium-133 NMR studies of Cs$^+$–18C6 in pyridine solutions were carried out recently by Mei *et al.*[145,146] A plot of the ^{133}Cs chemical shift vs. the 18C6/Cs$^+$ mole ratio at different temperatures is shown in Figure 24. It is interesting to note that the resonance shifts linearly downfield until a 1:1

TABLE 13. Limiting Chemical Shift of the Cs$^+ \cdot 2(18C6)$ Complex and the Thermodynamic Parameters for the Reaction Cs$^+ \cdot 18C6 + 18C6 \rightleftharpoons Cs^+(18C6)_2$ in Pyridine^a [146]

Temp (K)	K_1 (min) (M^{-1})	K_2 (M^{-1})	δ_2 (ppm)
297	10^5	79 ± 2	47.8
285	10^6	121 ± 5	49.4
272	10^6	218 ± 14	49.9
255	10^6	432 ± 58	51.4
244	10^6	623 ± 35	51.9
235	5×10^6	1173 ± 160	51.2

^{a} $(\Delta G_2^0)_{298} = -2.58 \pm 0.02$ kcal mol^{-1}; $\Delta H_2^0 = -5.8 \pm 0.2$ kcal mol^{-1}; $\Delta S_2^0 = -10.7 \pm 0.6$ cal mol^{-1} deg^{-1}.

mole ratio is obtained, and upon further addition of the ligand produces an *upfield* shift. It seems reasonable to expect that just as in the case of proton NMR results of Live and Chan[122] this behavior indicates a stepwise formation of $18C6 \cdot Cs^+$ and $(18C6)_2Cs^+$ complexes. The analysis of the data gave results shown in Table 13.

The same technique was used for the determination of the formation constants of cesium complexes with 18C6, DB18C6, and DC18C6 in various solvents.[147] In all cases the formation of 1:1 and 2:1 complex was indicated.

The rate of decomplexation of Cs^+ ion from DC18C6 and from C222B complexes was studied as a function of temperature in propylene carbonate

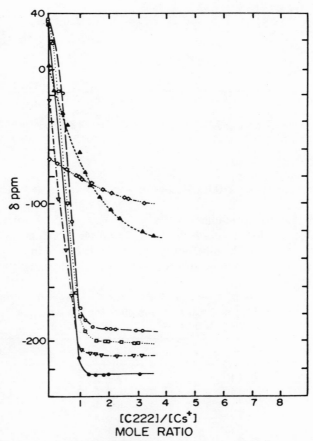

FIGURE 25. Cesium-133 chemical shifts as a function of $C222/Cs^+$ mole ratio in different solvents.[148] ●, pyridine; ▽, MeCN; □, acetone; ○, PC; ▲, DMF; ◇, DMSO. (Reproduced with permission.)

solutions. The activation energy for the decomplexation process was found to be 8.5 and 15 kcal mol^{-1}, respectively. The chemical shifts of Cs$^+$ ion and its 1:1 complex with 18C6 are strongly solvent dependent; however, the chemical shift of the 2:1 complex is independent of the solvent, indicating that in the sandwich complex, the Cs$^+$ ion is very effectively screened from the environment.

The addition of the cryptand C222 to a cesium solution in a nonaqueous solvent produced very strong paramagnetic shifts (Figure 25); in the case of pyridine, it goes to 250 ppm.[148] It is seen that in propylene carbonate, acetone, acetonitrile, and pyridine solutions the plots show a sharp break at 1:1 mole ratio, indicating the formation of a stable complex; on the other hand, in strongly solvating solvents, dimethyl formamide and dimethyl sulfoxide, only very weak complexes are formed.

It has been mentioned previously that cesium ion can be accommodated into the C222 cavity, albeit with some strain. A tight-fitting cation will result in a strong overlap of the ether oxygen electron pairs with the cation orbitals even though the individual electron-pair–cation interaction may be on the repulsive side of the binding curve; consequently the cationic chemical shifts will be large. On the other hand, in C222B·Cs$^+$ complex the paramagnetic shifts are of the order of 50–80 ppm. Since C222B has a smaller cavity than

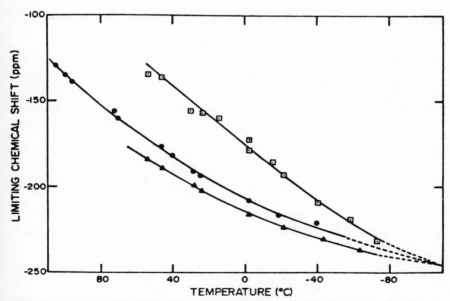

FIGURE 26. Plots of limiting ^{133}Cs chemical shifts for C222·Cs$^+$ complex in acetone, propylene carbonate, and dimethylformamide versus temperature.[149] ▲, acetone; ●, PC; ☐, DMF. (Reproduced with permission.)

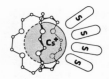

FIGURE 27. Schematic illustration of an "exclusive" complex.[148] (Reproduced with permission.)

C222, formation of exclusive complexes (see below) is a distinct possibility in which case the electron-pair–cation orbital overlap will be much smaller.

Variation of the ^{133}Cs chemical shift for the C222·Cs$^+$ complex as a function of temperature was studied in dimethylformamide, propylene carbonate, and acetone.[149] As seen in Figure 26, the chemical shifts converge at lower temperature, i.e., the shift is no longer solvent dependent. The results indicate the existence of two types of 1:1 complexes. At low temperatures we have predominantly an *inclusive* complex with the cation inside the ligand cavity and insulated from the solvent. At higher temperature, on the other hand, the cation is only partially accommodated inside the cavity (exclusive complex) (Figure 27) and is accessible to the solvent molecules.

Analysis of the data gave the enthalpy and the entropy for the complexation reactions

$$Cs^+ + C222 \rightleftharpoons C222 \cdot Cs^+_{exclusive}$$

$$C222 \cdot Cs^+_{exclusive} \rightleftharpoons C222 \cdot Cs^+_{inclusive}$$

The results, shown in Table 14, clearly show the expected strong influence of the solvent on the first reaction. The formation of an inclusive complex has little dependence on the solvent since the cation is already partially in the ligand cavity.

The spin-lattice relaxation time (T_1) of the ^{133}Cs was studied recently in methanol and dimethylformamide in the presence of 18C6.[150a] Plots of spin-lattice relaxation rate (T_1^{-1}) vs. ligand/cation mole ratio in dimethylformamide gave results very similar to those obtained for this system with ^{133}Cs chemical

TABLE 14. Thermodynamic Parameters (ΔH_1^0 and ΔS_1^0) for the Formation of the Exclusive Complex, Cs$^+$·C222, and for Its Conversion (ΔH_2^0 and ΔS_2^0) to the Inclusive Complex in Acetone, Propylene Carbonate, and N,N-Dimethylformamide[149]

Solvent	ΔH_1^0 (kcal mol^{-1})	ΔS_1^0 (cal mol^{-1} deg^{-1})	ΔH_2^0 (kcal mol^{-1})	ΔS_2^0 (cal mol^{-1} deg^{-1})
Acetone	$(-12.9)^a$	$(-26.8)^a$	-2.5	-5.6
Propylene carbonate^b	-8.6	-13.7	-2.9	-7.0
N,N-Dimethylformamide^b	-5.7	-11.2	-2.6	-7.6

^{a} The limited temperature range and small number of points obtained for acetone solutions provide only these crude estimates of ΔH_1^0 and ΔS_1^0.
^{b} Estimated uncertainties in the ΔH^0 and ΔS^0 values are ± 0.4 kcal mol^{-1} and ± 1.5 cal mol^{-1} deg^{-1}, respectively, for solutions in PC and DMF.

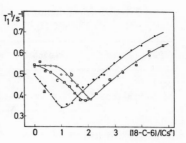

FIGURE 28. ^{133}Cs spin-lattice relaxation rate T_1^{-1} as a function of 18C6/Cs$^+$ mole ratio.$^{(150)}$ ●, 0.1 M CsI; ○, 0.1 M CsI + 0.1 M KI; □, 0.1 M RbI; solvent, DMF. (Reproduced with permission.)

shifts. T_1^{-1} decreased to mole ratio of 1:1 and then increased. In the presence of equimolar amounts of potassium or rubidium, the decrease was very gradual to a minimum value of 2:1 mole ratio (Figure 28). The behavior is consistent with a competitive formation of Rb$^+$ or K$^+$ complexes which are more stable than the cesium complex since the former two ions have a more favorable size.

4.3.3.7. Thallium-205 NMR. Thallium-205 NMR has been performed on the complexes of thallous ion Tl$^+$ with 18C6 and DB18C6.$^{(150b)}$ Large ^{205}Tl shifts occur on complexation. Competition experiments yield the sequence of stability constants of the alkali complexes with respect to Tl$^+$. Proton–thallium coupling constants have been observed in the proton spectrum of the C222·Tl$^+$ cryptate.$^{(139)}$

4.3.4. Electronic Spectroscopy

In general, ultraviolet-visible spectroscopy has somewhat limited usefulness for the studies of macrocyclic complexes. In his classical paper Pedersen reported changes in the ultraviolet spectrum of benzene-substituted crown ethers upon complexation.$^{(4)}$ For example, upon addition of potassium thiocyanate, DB18C6 in methanol shows a new absorption band about 6 nm to the larger wavelength side of the major peak. Similar results were obtained by Tusek *et al.*$^{(118)}$ Shchori *et al.*$^{(151)}$ determined the stability constants of several DB18C6 complexes in aqueous solutions by combining solubility data on the crown with spectrophotometric determination of the total complex concentration.

Chock$^{(152)}$ used a modification of the Benesi–Hildebrand method$^{(89,90)}$ to determine the formation constants of alkali complexes and of NH$_4^+$ and Tl$^+$ complexes in methanol solutions as a function of temperature and, consequently, obtained the enthalpy of complexation. The data are given in Table 15.

A recent paper reports the spectrophotometric determination of stepwise formation constants of C221, pyridine cryptands, and of some noncyclic polyethers.$^{(153)}$ It is claimed that the technique can be used for the determination of up to four successive formation constants. Spectrophotometry is,

TABLE 15. Formation Constant and Enthalpy of Complexation of
DB30C10 Complexes in Methanol Solutions[152]

Parameter	Na$^+$	K$^+$	Rb$^+$	Cs$^+$	NH$_4{}^+$	Tl$^+$
log K_f	2.11	4.57	4.64	4.23	2.43	4.51
$\Delta H_{25°}^0$, kcal mol^{-1}	-4	-11.5	-12.7	-11.2	-5.5	-11

however, an unreliable method for the determination of stepwise equilibrium constants.[89,90]

Reflectance spectra were used to characterize DB18C6 complexes of some potassium tetrahalo cobalt(II) salts.[154,155]

Hogen-Esch, Smid, and co-workers have shown in a series of elegant papers that the ultraviolet spectroscopy can be used very effectively in distinguishing between contact and solvent-separated ion pairs of alkali salts of carbanions.[156] Thus a sodium fluorenyl solution in tetrahydrofuran shows a narrow absorption band at 356 nm at room temperature which shifts to 373 nm band as the temperature is lowered. The two bands are assigned to a contact (tight) ion pair and a solvent-separated ion pair, respectively.

Spectrophotometric measurements show that the addition of a crown ether to a solution at room temperature converts the contact ion pair to a solvent-separated ion pair.[119] Similar results were obtained with alkali picrates in THF solutions[157] where the sodium, potassium, and cesium contact ion pairs have absorption maxima at 351, 357, and 362 nm, respectively. The addition of DiMeDB18C6 to a solution of sodium picrate shifts the maximum to 378 nm, indicating the formation of Na$^+$LPic$^-$ ion pair. On the other hand, the addition of 15C5 does not change the absorption spectrum. In this case it seems reasonable to assume that the tight ion pair persists and the structure of the complex is LNa$^+$Pic$^-$.

Addition of DiMe18C6 to difluorenylbarium in a 1:1 ratio in THF at 25°C produces an optical spectrum of equal fractions of separated and contact ion pairs; when, however, C222 is used as the complexing agent only the absorption peak of the separated ion pairs is observed.[158]

The perturbation of photoexcited states by cation complexation has been studied by the effect of alkali metal cations on the emission from crown ethers containing naphthalene units.[159]

4.3.5. Vibrational Spectroscopy

Initial studies on the infrared spectra of crown ethers and their complexes were carried out by Pedersen.[4]

Parsons *et al.*[42,43] measured the infrared spectra of numerous solid alkali and alkaline earth complexes with crown ethers in nujol mulls. They report that the infrared spectra of DB30C10 with sodium, potassium, and cesium tetraphenyl borate are very similar, indicating similar structure. These data seem to contradict the conclusions of Live and Chan[122] that the sodium–DB30C10 complex has a different structure from that of the heavier alkalies. It should be noted, however, that Live and Chan studied the complexes in solutions and there is no assurance that the solid complexes would have the same structure.

It was noted that the greatest change in the infrared spectrum of the crown ethers in going from the free ligand to the complex occurs in the 900–1200 cm^{-1} region.

Several complexes were obtained as mono- or polyhydrates. It was observed that for the monohydrates the ν_{OH} bands are sharp, indicating direct bonding of the water molecule to the alkali ion. On the other hand, in complexes with several water molecules, such as $18C6 \cdot 2NaBPh_4 \cdot 5H_2O$, the ν_{OH} band in the 3550–3400 cm^{-1} region is broad, indicating that some of the water molecules are located in the lattice.

In the case of the alkaline earth crown compounds the spectra of the perchlorate anion were informative. The infrared spectra of $Sr(ClO_4)_2 \cdot (B15C5)_2$ and $Ba(ClO_4)_2(B15C5)_2$ are identical and the asymmetric Cl–O stretching frequency is not perturbed, indicating that the anion is not ion paired to the cations, thus confirming the sandwich structure. On the other hand, the band is split in the $Ba(ClO_4)_2(B18C6)$ complex, indicating coordination of the anion to barium.

Conformational changes of macrobicyclic diazaalkanes and of C222 were studied by Lord and Siamwiza.[160] In the case of the cryptand the IR spectrum in deuterochloroform solution contains a shifted C–D stretching band due to C–D · · · N hydrogen bonding with the solvent; this indicates that the molecule is predominantly in the *exo–exo* conformation. On addition of Ba^{2+} this band disappears, indicating a change to the expected *endo–endo* form.

Izatt *et al.* prepared solid DC18C6 complexes with strong acids, $H_3O^+ClO_4 \cdot DC18C6$, $D_3O^+ClO_4^- \cdot DC18C6$, and $H_3O^+PF_6^- \cdot DC18C6$.[161] The vibrational spectra of the compounds in perfluorokerosene mull confirmed the presence of the hydronium ion.

It has been shown that solvated alkali cations undergo low-frequency vibrations in the solvent cages. The exact frequencies of such vibrations depend on the cation *and on the solvent*.[162] However, it has been shown by Tsatsas *et al.*[163] that the addition of DB18C6 to an alkali salt solution results in a vibrational band whose frequency depends on the metal ion but is *independent* of the solvent. Thus, for example, the vibrational frequencies of Na^+ ion in dimethyl sulfoxide and in pyridine cages are 202 and 181 cm^{-1}, respectively. Upon addition of DB18C6 these bands disappear and a new band appears at $214 \, \text{cm}^{-1}$. This band must represent the cation–crown vibration. In the case of K^+ ion, this vibration occurs at $168 \, \text{cm}^{-1}$.

Similar results were obtained by Cahen and Popov[164] with cryptands C211 and C222 (Figure 29). The Li^+-inside-C211 vibration is at $348 \, \text{cm}^{-1}$, while Na^+-inside-C222 vibration frequency is $234 \, \text{cm}^{-1}$. It is interesting to note that these two bands are Raman inactive. This fact confirms the strongly electrostatic nature of the cation–macrocycle bond.

Recently Gans *et al.*[165] studied Raman spectra of sodium and potassium cyanides in liquid ammonia in the presence of B15C5 and C222. The ν_{C-N} was observed between 2030 and $2080 \, \text{cm}^{-1}$, and the spectra are shown in Figure 30. Without the macrocyclic ligand, potassium cyanide solution shows two bands in the C–N stretching region corresponding to the free solvated anion and anion in the K^+CN^- ion pair at 2056 and $2054 \, \text{cm}^{-1}$, respectively. Since the molar intensities of the two bands can be assumed to be approximately

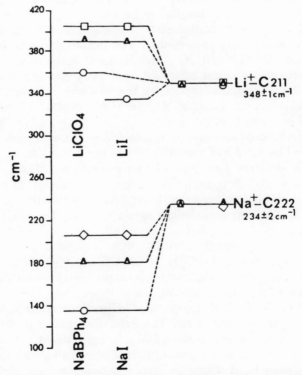

FIGURE 29. Comparison of ion-motion band frequencies for sodium and lithium salts and their 222 and 211 cryptates, respectively, in pyridine, acetonitrile, nitromethane, and dimethyl sulfoxide solutions.[164] △, pyridine; □, CH_3CH; ○, nitromethane; ◇, DMSO. (Reproduced with permission.)

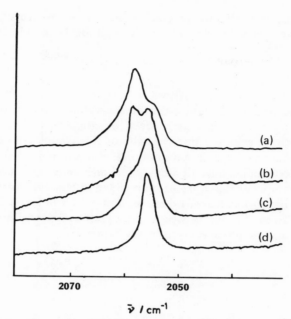

FIGURE 30. Raman spectra of KCN in liquid ammonia. (a) KCN, (b) KCN + one equivalent of B18C6, (c) KCN + two equivalences of B18C6, (d) KCN + one equivalent of C222.[165] (Reproduced with permission.)

equal, we see that KCN exists mostly in the form of contact ion pairs. The addition of one equivalent of B18C6 complexes some, but not all of the cation, and the concentrations of the two species are approximately equal (it should be remembered that liquid ammonia is a strongly solvating solvent). The concentration of ion pairs decreases further upon the addition of another equivalent of B18C6. Cryptand C222 reacts quantitatively with the potassium ion and the band for the ion pair is no longer visible.

4.3.6. Electron Spin Resonance–Nuclear Quadrupole Resonance

Radical anions of aromatic hydrocarbons (benzene, toluene, naphthalene, anthracene) have been prepared by dissolution of sodium or potassium metal using ligands like DCH18C6 or C222.[166–170] The addition of the ligand leads to more or less complete separation of the ion-paired species. With C222 and potassium in toluene the absence of proton hyperfine splittings indicates fast electron exchange between the toluene anion radical and toluene molecules.[167] On the other hand, hyperfine structure is still observed when DCH18C6 is used, indicating that the anions remain associated with the complexed cation to some extent.[167,168]

[35]Cl, [81]Br, and [127]I nuclear quadrupole resonance measurements have been performed on CdX_2 and HgX_2 (X = Cl, Br, I) complexes with 15C5 and 18C6.[171,172]

4.4. Extraction Studies

Since the formation of a crown or a cryptate complex results in the formation of a cation enclosed partially or completely in a hydrophobic sheath, it is to be expected that the salts of this complexed cation will be soluble in nonpolar organic solvents and also extractable from aqueous solutions into organic solvents immiscible with water. Frensdorff studied extractions of aqueous solutions of sodium and potassium picrates by crown solutions in dichloromethane, *n*-hexane, and 1,1,2-trifluoro-1,2,2-trichloroethane.[173] The equilibrium between aqueous salt and crown (indicated below by L) in the organic phase is given by the expression

$$M_{aq}^+ + Pic_{aq}^- + L_{or} \rightleftharpoons ML^+Pic_{or}^-$$

or

$$K_e = \frac{[ML^+Pic^-]_{or}}{[M^+]_{aq}[Pic^-]_{aq}[L]_{or}}$$

The ML^+Pic^- ion pair in the organic phase is partially dissociated:

$$ML^+Pic_{or}^- \rightleftharpoons ML_{or}^+ + Pic_{or}^-$$

and the dissociation constant is

$$K_d = \frac{[ML^+]_{or}[Pic^-]_{or}}{[ML^+Pic^-]_{or}} \tag{15}$$

The distribution of the free ligand is given by

$$P_e = \frac{[L]_{aq}}{[L]_{or}} \tag{16}$$

In all cases terms within brackets indicate activities.

Frensdorff used DC18C6 as a mixture of isomers and separated isomers A and B as well as DBz18C6 for the extraction studies. The results are given in Table 16.

It is seen that the extractibility of the cation varies with the stability of the complex. Thus with the same solvent, dichloromethane, the extraction equilibrium constant K_e is two orders of magnitude smaller for sodium than for potassium. The extracting ability of solvent varies in the order $CH_2Cl_2 > CF_2Cl–CFCl_2 > n$-$C_6H_{14}$, and it is seen that the ion-pair dissociation in the organic phase is so small that activity corrections can be neglected.

Somewhat similar results were obtained by Sadakane *et al.*,[174] who used

TABLE 16. Extraction Equilibria of Alkali Picrates with Crowns[173]

Crown	Cation	Solvent	P_e	K_e	K_d
DC18C6—mixed	K^+	$n\text{-}C_6H_{14}$	0.15	3.8×10^3	10^{-6}
DC18C6—mixed	K^+	$CF_2Cl\text{--}CFCl_2$	7×10^{-2}	1.9×10^4	10^{-6}
DC18C6-A	K^+	CH_2Cl_2	2.5×10^{-4}	4.0×10^6	2.5×10^{-5}
DC18C6-B	K^+	CH_2Cl_2	2.7×10^{-4}	1.1×10^6	2.4×10^{-5}
DC18C6—mixed	Na^+	CH_2Cl_2	2.6×10^{-4}	1.5×10^4	10^{-4}
DBz18C6	K^+	CH_2Cl_2	—	7.1×10^5	4.4×10^{-6}

DB18C6 as the ligand and benzene as the extracting solvent for alkali picrates. The extraction equilibrium constant [Eq. (14)] was determined in the 10–30°C temperature range and the enthalpy and entropy of extraction were calculated in the usual manner from the temperature dependence of the equilibrium constant. The results are given in Table 17.

Not unexpectedly, the most efficient extraction is found for K^+. It is seen that the extraction is controlled by the enthalpy but is adversely affected by the entropy change.

In the presence of a large excess of crown, evidence was obtained for the formation of the 2:1 complexes $(DB18C6)_2MPic$ for Rb^+ and Cs^+. The formation constant for the second step

$$(DB18C6)M^+Pic^- + DB18C6 \rightleftharpoons (DB18C6)_2M^+Pic^-$$

in the benzene phase was estimated to be 5×10^2 for Rb^+ and 3×10^2 for Cs^+. The authors conclude that countercurrent extraction of the alkali crown complexes may be a useful technique for the separation of alkali ions.

Selective transport of alkali metal cation through a liquid membrane by macrobicyclic ligands was studied by Kirch and Lehn.[175] Cryptands C222, C322, C333, and $C22C_8$ were tried. The alkali metal picrates used were transported with the aid of the above cryptands from an in to an out aqueous phase through a bulk liquid chloroform membrane. When the carrier is

TABLE 17. Thermodynamic Parameters for the Extraction of DB18C6 Alkali Complexes into Benzene[174]

Cation	K_e	ΔH^0 (kcal mol^{-1})	ΔS^0 (cal mol^{-1} deg^{-1})
Na^+	1.6×10^2	-8.9	-20
K^+	4.5×10^4	-16.3	-33.4
Rb^+	5.6×10^3	-16.4	-37.8
Cs^+	1.2×10^3	-16.0	-39.6

added to the membrane the alkali metal picrate from the in aqueous phase distributes into the chloroform and the cation and picrate concentration are found to increase in the out phase. It was found that carrier–cation pairs forming very stable cryptate complexes display efficient extractions of the salt into the organic phase. The rate of transport is of the same order of magnitude as observed for antibiotic-mediated transport across a bulk liquid membrane but, in contrast, the *relative* transport rates are not proportional to the complex stability and extraction efficiency.

4.5. Calorimetric Techniques

Calorimetric techniques are extremely useful for the determination of thermodynamic parameters of a complexation reaction. There are many examples in the literature, and some in this chapter, where the enthalpy and the entropy of a reaction are determined from the temperature dependence of the equilibrium constant. Good results necessitate many careful measurements in a decent temperature range.

The calorimetric method can also be used for the determination of equilibrium constants. If $\Delta H°$ for a reaction is known and q is the measured heat of the reaction for a given solution per kilogram of solvent, the concentration of the complex in molal units is $q/\Delta H°$ and, therefore,

$$K_{\text{conc}} = \frac{q/\Delta H°}{(C_{\text{M}}^{\text{tot}} - q/\Delta H°)(C_{\text{L}}^{\text{tot}} - q/\Delta H°)} \tag{17}$$

where $C_{\text{M}}^{\text{tot}}$ and $C_{\text{L}}^{\text{tot}}$ are the total concentrations of the metal ion and of the ligand, respectively.

Calorimetric techniques for the determination of complex formation constants are nearly ideal for nonaqueous solutions. Just as for spectroscopic techniques, the method becomes unreliable when the formation constants are larger than 10^4–10^5.

Calorimetric studies of numerous crown ether complexes were pioneered principally by Izatt and co-workers.[176-178] The results are given in Chapter 3. Enthalpies of complexations of DB18C6 complexes of alkali ions in dimethyl sulfoxide, acetone, tetrahydrofuran, and water were reported by Arnett and Moriarty.[179] Enthalpies of cryptate formations were studied by Anderegg[95] and by Kauffmann *et al.*[185]

4.6. Relaxation Techniques

The dynamics and kinetics of complex formation and dissociation have been studied by relaxation techniques (temperature jump, ultrasonic absorption)[152,186-190] and stopped-flow methods,[191,192] in addition to variable-temperature NMR measurements.[136,139] The rates vary greatly from one cation/ligand couple to another. For the crown complexes, they are several orders of magnitude slower than the rates of exchange of solvent molecules in the solvation shell. For the cryptates, the cation exchange rates are usually even slower than those of the crown complexes. Conformational changes of the ligand may accompany complexation.[152,187] The mechanism of formation of the $C211 \cdot Na^+$ and $C211 \cdot Ca^{2+}$ cryptates has been studied in detail by temperature-jump relaxation spectrometry.[187]

5. Conclusion

Physicochemical studies of crown and cryptate complexes have had a two-way impact. On the one hand, sophisticated physical methods have been of immense value in the detailed analysis of the features of this novel and fascinating class of compounds. Conversely, these complexes have provided a rich set of substrates for physicochemical investigations, presenting new physical properties and being amenable to study by a large variety of techniques.

Many new systems may be imagined and many will undoubtedly be synthesized, exposing novel aspects of molecular behavior to the scrutiny of a broad range of physicochemical means and ways. Such feedback should ensure high activity in this recent field of physical organic chemistry.

References

1. G. Anderegg, *Helv. Chim. Acta* **50**, 2333 (1967).
2. J. D. Carr and D. G. Swartzfager, *Anal. Chem.* **42**, 1238 (1970); *Anal. Chem.* **43**, 583, 1520 (1971).
3. C. J. Pedersen, *J. Am. Chem. Soc.* **89**, 2495 (1967).
4. C. J. Pedersen, *J. Am. Chem. Soc.* **89**, 7017 (1967).
5. B. Dietrich, J.-M. Lehn, and J.-P. Sauvage, *Tetrahedron Lett.*, 2885 (1969).
6. B. Dietrich, J.-M. Lehn, and J.-P. Sauvage, *Tetrahedron Lett.*, 2889 (1969).
7. J.-M. Lehn, J. Simon, and J. Wagner, *Nouv. J. Chim.* **1**, 77 (1977).
8. H. E. Simmons and C. H. Park, *J. Am. Chem. Soc.* **90**, 2428 (1968).
9. W. Simon, W. E. Morf, and P. Ch. Meier, *Struct. Bonding (Berlin)* **16**, 113 (1973).
10. P. G. Gertenbach and A. I. Popov, *J. Am. Chem. Soc.* **97**, 4738 (1975).
11. C. J. Pedersen and H. K. Frensdorff, *Angew. Chem. Int. Ed.* **11**, 16 (1972).
12. J.-M. Lehn, *Struct. Bonding (Berlin)* **16**, 1 (1973).
13. J. J. Christensen, D. J. Eatough, and R. M. Izatt, *Chem. Rev.* **74**, 351 (1974).
14. C. Kappenstein, *Bull. Soc. Chim. France*, 89 (1974).
15. G. W. Gokel and H. D. Durst, *Synthesis*, 168 (1976).
16. C. J. Pedersen, *J. Am. Chem. Soc.* **92**, 391 (1970).
17. R. N. Greene, *Tetrahedron Lett.*, 1793 (1972).

18. G. W. Gokel, D. J. Cram, C. L. Liotta, H. P. Harris, and F. L. Cook, *J. Org. Chem.* **39**, 2445 (1974).
19. F. A. L. Anet, J. Krane, J. Dale, K. Daasvatn, and P. O. Kristiansen, *Acta Chem. Scand.* **27**, 3395 (1973).
20. J. Dale and P. O. Kristiansen, *Acta Chem. Scand.* **26**, 1471 (1972).
21. F. L. Cook, T. C. Caruso, M. P. Byrne, C. W. Bowers, D. H. Speck, and C. L. Liotta, *Tetrahedron Lett.*, 4029 (1974).
22. C. J. Pedersen, *J. Org. Chem.* **36**, 254 (1971).
23. J. S. Bradshaw, J. Y. Hui, B. L. Haymore, J. J. Christensen, and R. M. Izatt, *J. Heterocycl. Chem.* **10**, 1 (1973).
24. J. S. Bradshaw, J. Y. Hui, Y. Chan, B. L. Haymore, R. M. Izatt, and J. J. Christensen, *J. Heterocycl. Chem.* **11**, 45 (1974).
25. J. S. Bradshaw, R. A. Reeder, M. D. Thompson, E. D. Flanders, R. L. Carruth, R. M. Izatt and J. J. Christensen, *J. Org. Chem.* **41**, 134 (1976).
26. J. S. Bradshaw and J. Y. K. Hui, *J. Heterocycl. Chem.* **11**, 649 (1974).
27. J. S. Bradshaw, L. D. Hansen, S. F. Nielsen, M. D. Thompson, R. A. Reeder, R. M. Izatt, and J. J. Christensen, *J. Chem. Soc. Chem. Commun.*, 874 (1975).
28. D. Britz and D. Knittel, *Electrochim. Acta* **20**, 891 (1975).
29. R. M. Izatt, J. D. Lamb, G. E. Maas, R. E. Asay, J. S. Bradshaw, and J. J. Christensen, *J. Am. Chem. Soc.* **99**, 2365 (1977).
30. S. Kopolow, T. E. Hogen-Esch, and J. Smid, *Macromolecules* **6**, 133 (1973).
31. J. Smid, B. El. Haj, T. Majewicz, A. Nonni, and R. Sinta, *Org. Prep. Proced. Int.* **8**, 193 (1976).
32. J. L. Dye, M. T. Lok, F. J. Tehan, J. M. Ceraso, and K. J. Voorhees, *J. Org. Chem.* **38**, 1773 (1973).
33. B. Dietrich, J.-M. Lehn and J.-P. Sauvage, *J. Chem. Soc. Chem. Commun.*, 1055 (1970).
34. J.-M. Lehn and F. Montavon, *Tetrahedron Lett.*, 4557 (1972).
35. J.-M. Lehn and F. Montavon, *Helv. Chim. Acta* **59**, 1566 (1976).
36. J.-M. Lehn, J. Simon, and J. Wagner, *Angew. Chem. Int. Ed.* **12**, 578 (1973).
37. J. Cheney, J.-M. Lehn, J.-P. Sauvage, and M. E. Stubbs, *J. Chem. Soc. Chem. Commun.*, 1100 (1972).
38. E. Graf and J.-M. Lehn, *J. Am. Chem. Soc.* **97**, 5022 (1975).
39. B. Dietrich, J.-M. Lehn, and J. Simon, *Angew. Chem. Int. Ed.* **13**, 406 (1974).
40. C. J. Pedersen, *J. Am. Chem. Soc.* **92**, 386 (1970).
41. D. G. Parsons and J. N. Wingfield, *Inorg. Chim. Acta* **18**, 263 (1976).
42. N. S. Poonia and M. R. Truter, *J. Chem. Soc. Dalton Trans.*, 2062 (1973).
43. D. G. Parsons, M. R. Truter, and J. N. Wingfield, *Inorg. Chim. Acta* **14**, 45 (1975).
44. M. R. Truter, *Struct. Bonding (Berlin)* **16**, 71 (1973).
45. M. A. Bush and M. R. Truter, *J. Chem. Soc. Perkin Trans. 2*, 341 (1971).
46. D. Bright and M. R. Truter, *Nature* **225**, 176 (1970).
47. D. Bright and M. R. Truter, *J. Chem. Soc.*, 1544 (1970).
48. J. D. Dunitz and P. Seiler, *Acta Crystallogr.* **B30**, 2739 (1974).
49. M. Dobler, J. D. Dunitz, and P. Seiler, *Acta Crystallogr.* **B30**, 2741 (1974).
50. P. Seiler, M. Dobler, and J. D. Dunitz, *Acta Crystallogr.* **B30**, 2744 (1974).
51. M. Dobler and R. P. Phizackerley, *Acta Crystallogr.* **B30**, 2746 (1974).
52. M. Dobler and R. P. Phizackerley, *Acta Crystallogr.* **B30**, 2748 (1974).
53. J. D. Dunitz and P. Seiler, *Acta Crystallogr.* **B30**, 2750 (1974).
54. M. A. Newman, E. C. Steiner, F. P. van Remoortere, and F. P. Boer, *Inorg. Chem.* **14**, 734 (1975).
55. F. P. van Remooretere and F. P. Boer, *Inorg. Chem.* **13**, 2071 (1974).

56. F. P. Boer, M. A. Neuman, F. P. van Remooretere, and E. C. Steiner, *Inorg. Chem.* **13**, 2826 (1974).
57. P. P. North, E. C. Steiner, F. P. van Remooretere, and F. P. Boer, *Acta Crystallogr.* **B32**, 370 (1976).
58. F. P. van Remooretere, F. P. Boer, and E. C. Steiner, *Acta Crystallogr.* **B31**, 1420 (1975).
59. R. M. Costes, G. Folcher, N. Keller, P. Plurien, and P. Rigny, *Inorg. Nucl. Chem. Lett.* **11**, 469 (1975).
60. R. M. Costes, G. Folcher, P. Plurien, and P. Rigny, *Inorg. Nucl. Chem. Lett.* **12**, 13 (1976).
61. G. Bombieri, G. DePaoli, A. Cassol, and A. Immirzi, *Inorg. Chim. Acta* **18**, L23 (1976).
62. M. A. Bush and M. R. Truter, *J. Chem. Soc. Perkin Trans.* 2, 345 (1972).
63. P. R. Mallinson and M. R. Truter, *J. Chem. Soc. Perkin Trans.* 2, 1818 (1972).
64. D. E. Fenton, M. Mercer, N. S. Poonia, and M. R. Truter, *J. Chem. Soc. Chem. Commun.*, 66 (1972).
65. M. Mercer and M. R. Truter, *J. Chem. Soc. Dalton Trans.*, 2469 (1973).
66. M. A. Bush and M. R. Truter, *J. Chem. Soc.*, 1440 (1971).
67. P. Groth, *Acta Chem. Scand.* **25**, 3189 (1971).
68. C. Riche, C. Pascard-Billy, C. Cambillau, and G. Bram, *J. Chem. Soc. Chem. Commun.*, 183 (1977).
69. M. Mercer and M. R. Truter, *J. Chem. Soc. Dalton Trans.*, 2215 (1973).
70. B. Metz, D. Moras, and R. Weiss, *J. Chem. Soc. Perkin Trans.* 2, 423 (1976).
71. R. Weiss, personal communication.
72. (a) J. Cheney and J.-M. Lehn, *J. Chem. Soc. Chem. Commun.*, 487 (1972).
    (b) J. Cheney, J. P. Kintzinger, and J.-M. Lehn, *Nouv. J. Chim.*, in press (1978).
73. B. Metz, D. Moras, and R. Weiss, *J. Chem. Soc. Chem. Commun.*, 217 (1970).
74. D. Moras, B. Metz, and R. Weiss, *Acta Crystallogr.* **B29**, 383, 388 (1973).
75. D. Moras and R. Weiss, *Acta Crystallogr.* **B29**, 396 (1973).
76. D. Moras and R. Weiss, *Acta Crystallogr.* **B29**, 1059 (1973).
77. B. Metz and R. Weiss, *Inorg. Chem.* **13**, 2094 (1974).
78. B. Metz, D. Moras, and R. Weiss, *Acta Crystallogr.* **B29**, 1382 (1973).
79. B. Metz, D. Moras, and R. Weiss, *Acta Crystallogr.* **B29**, 1388 (1973).
80. N. K. Dalley, D. E. Smith, R. M. Izatt, and J. J. Christensen, *J. Chem. Soc. Chem. Commun.*, 90 (1972).
81. N. K. Dalley, J. S. Smith, S. B. Larson, J. J. Christensen, and R. M. Izatt, *J. Chem. Soc. Chem. Commun.*, 43 (1975).
82. N. K. Dalley, J. S. Smith, S. B. Larson, K. L. Matheson, J. J. Christensen, and R. M. Izatt, *J. Chem. Soc. Chem. Commun.*, 84 (1975).
83. F. Mathieu and R. Weiss, *J. Chem. Soc. Chem. Commun.*, 816 (1973).
84. R. Louis, J. C. Thierry, and R. Weiss, *Acta Crystallogr.* **B30**, 753 (1974).
85. R. Wiest and R. Weiss, *J. Chem. Soc. Chem. Commun.*, 678 (1973).
86. B. Metz, D. Moras, and R. Weiss, *J. Inorg. Nucl. Chem.* **36**, 785 (1974).
87. M. Mellinçer, J. Fischer, and R. Weiss, *Angew. Chem.* **85**, 828 (1973).
88. B. Metz, J. M. Rosalky, and R. Weiss, *J. Chem. Soc. Chem. Commun.*, 533 (1976).
89. F. J. C. Rossotti and H. S. Rossotti, *The Determination of Stability Constants*, McGraw-Hill, New York (1961).
90. M. T. Beck, *Chemistry of Complex Equilibria*, Van Nostrand Reinhold Co., London (1970).
91. A. J. Smetana and A. I. Popov, *J. Chem. Therm.*, in press.

92. H. K. Frensdorff, *J. Am. Chem. Soc.* **93**, 600 (1971).
93. J.-M. Lehn and J.-P. Sauvage, *J. Chem. Soc. Chem. Comm.*, 440 (1971).
94. J.-M. Lehn and J.-P. Sauvage, *J. Am. Chem. Soc.* **97**, 6700 (1975).
95. G. Anderegg, *Helv. Chim. Acta* **58**, 1218 (1975).
96. A. Z. Gordon and P. A. Rock, *J. Electrochem. Soc.* **124**, 534 (1977).
97. E. Shchori, J. Jagur-Grodzinski, Z. Luz, and M. Shporer, *J. Am. Chem. Soc.* **93**, 7133 (1971).
98. E. Shchori and J. Jagur-Grodzinski, *Isr. J. Chem.* **11**, 243 (1973).
99. R. M. Fuoss and F. Accascina, *Electrolytic Conductance*, p. 229, Interscience, New York (1959).
100. (a) N. Matsuura, K. Umemoto, Y. Takeda, and A. Sasaki, *Bull. Chem. Soc. Japan* **49**, 1246 (1976). (b) N. Matsuura, K. Umemoto, Y. Takeda, and A. Sasaki, *Bull. Chem. Soc. Japan* **50**, 3078 (1977).
101. T. P. I and E. Grunwald, *J. Am. Chem. Soc.* **96**, 2879 (1974).
102. J. Jagur-Grodzinski, *Bull. Chem. Soc. Japan* **50**, 3077 (1977).
103. D. F. Evans, S. L. Wellington, J. A. Nadis, and E. L. Cussler, *J. Solution Chem.* **1**, 499 (1972).
104. (a) S. Boileau, P. Hemery, and J. C. Justice, *J. Solution Chem.* **4**, 873 (1975). (b) P. Hemery, S. Boileau, and P. Sigwalt, *J. Polym. Sci.* **52**, 189 (1975).
105. J. C. Justice, *Electrochim. Acta* **16**, 701 (1971).
106. P. R. Danesi, R. Chiarizia, C. Fabiani, and C. Domenichini, *J. Inorg. Nucl. Chem.* **38**, 1226 (1976).
107. C. R. Witschonke and C. A. Kraus, *J. Am. Chem. Soc.* **69**, 2472 (1967).
108. S. Kopolow, Z. Machacek, U. Takaki, and J. Smid, *J. Macromol. Sci. Chem.* **A7**, 1015 (1973).
109. R. Ungaro, B. El Haj, and J. Smid, *J. Am. Chem. Soc.* **98**, 5198 (1976).
110. T. E. Hogen-Esch and J. Smid, *J. Phys. Chem.* **79**, 233 (1975).
111. N. Nae and J. Jagur-Grodzinski, *J. Am. Chem. Soc.* **99**, 489 (1977); *J. Chem. Soc. Faraday Trans. 1*, 1951 (1977).
112. (a) J. Koryta and M. L. Mittal, *J. Electroanal. Chem.* **36**, app14 (1972). (b) L. Pospisil, M. L. Mittal, J. Kůta, and J. Koryta, *J. Electroanal. Chem.* **46**, 203 (1973).
113. F. Peter and M. Gross, *Compt. Rend. Acad. Sci. (Paris)* **277**, 907 (1973).
114. (a) F. Peter and M. Gross, *Electroanal. Chem.* **53**, 307 (1974). (b) A. Louati and M. Gross, *Electrochim. Acta* **21**, 7 (1976).
115. A. Agostiano, M. Caselli, and M. Della Monica, *J. Electroanal. Chem.* **74**, 95 (1976).
116. J. P. Gisselbrecht and M. Gross, *J. Electroanal. Chem.* **75**, 637 (1977).
117. O. A. Gansow, A. R. Kausar, K. M. Triplett, M. J. Weaver, and E. L. Yee, *J. Am. Chem. Soc.* **99**, 7087 (1977).
118. L. Tusek, H. Meider-Gorican, and P. R. Danesi, *Z. Naturforsch.* **31B**, 330 (1976).
119. K. H. Wong, G. Konizer, and J. Smid, *J. Am. Chem. Soc.* **92**, 666 (1970).
120. B. Dietrich, J.-M. Lehn, J.-P. Sauvage, and J. Blanzat, *Tetrahedron* **29**, 1629 (1973).
121. B. Dietrich, J.-M. Lehn, and J.-P. Sauvage, *Tetrahedron* **29**, 1647 (1973).
122. D. Live and S. I. Chan, *J. Amer. Chem. Soc.* **98**, 3769 (1976).
123. C. H. Park and H. E. Simmons, *J. Am. Chem. Soc.* **90**, 2429 (1968).
124. J.-M. Lehn and M. E. Stubbs, *J. Am. Chem. Soc.* **96**, 4011 (1974).
125. E. Graf and J.-M. Lehn, *J. Am. Chem. Soc.* **98**, 6403 (1976).
126. (a) J.-M. Lehn and J. Simon, *Helv. Chim. Acta* **60**, 141 (1977). (b) M.-C. Fedarko, *J. Magnetic Res.* **12**, 30 (1973).
127. H. G. Forster and J. D. Roberts, personal communication.

128. A. I. Popov, in: *Solute–Solvent Interactions* (J. F. Coetzee and C. D. Ritchie, eds.), Vol. 2, pp. 271–330, Marcel Dekker, Inc., New York (1976).
129. Y. M. Cahen, J. L. Dye, and A. I. Popov, *J. Phys. Chem.* **79**, 1289 (1975).
130. E. T. Roach, P. R. Handy, and A. I. Popov, *Inorg. Nucl. Chem. Lett.* **9**, 359 (1973).
131. Y. M. Cahen, J. L. Dye, and A. I. Popov, *J. Phys. Chem.* **79**, 1292 (1975).
132. V. Gutmann and E. Wychera, *Inorg. Nucl. Chem. Lett.* **2**, 257 (1966).
133. E. Shchori, J. Jagur-Grodzinski, and M. Shporer, *J. Am. Chem. Soc.* **95**, 3842 (1973).
134. R. H. Erlich, E. Roach, and A. I. Popov, *J. Am. Chem. Soc.* **92**, 4989 (1970).
135. A. M. Grotens, J. Smid, and E. deBoer, *J. Chem. Soc. Chem. Commun.*, 759 (1971).
136. J. M. Ceraso and J. L. Dye, *J. Am. Chem. Soc.* **95**, 4432 (1973).
137. J. M. Ceraso, P. B. Smith, J. S. Landers, and J. L. Dye, *J. Phys. Chem.* **81**, 760 (1977).
138. (a) J. P. Kintzinger and J.-M. Lehn, *J. Am. Chem. Soc.* **96**, 3313 (1974). (b) C. Detellier and P. Laszlo, *Bull. Soc. Chim. Belg.* **84**, 1087 (1975).
139. J.-M. Lehn, J.-P. Sauvage, and B. Dietrich, *J. Am. Chem. Soc.* **92**, 2916 (1970).
140. J. M. Ceraso and J. L. Dye, *J. Chem. Phys.* **61**, 1585 (1974).
141. J. L. Dye, C. W. Andrews, and J. M. Ceraso, *J. Phys. Chem.* **79**, 3076 (1975).
142. J. S. Shih and A. I. Popov, to be published.
143. W. J. DeWitte, R. C. Schoening, and A. I. Popov, *Inorg. Nucl. Chem. Lett.* **12**, 251 (1976).
144. W. J. DeWitte, L. Liu, E. Mei, J. L. Dye, and A. I. Popov, *J. Solution Chem.* **6**, 337 (1977).
145. E. Mei, J. L. Dye, and A. I. Popov, *J. Am. Chem. Soc.* **98**, 1619 (1976).
146. E. Mei, J. L. Dye, and A. I. Popov, *J. Am. Chem. Soc.* **99**, 5308 (1977).
147. E. Mei, A. I. Popov, and J. L. Dye, *J. Phys. Chem.* **81**, 1677 (1977).
148. E. Mei, L. Liu, J. L. Dye, and A. I. Popov, *J. Solution Chem.* **6**, 771 (1977).
149. E. Mei, A. I. Popov, and J. L. Dye, *J. Am. Chem. Soc.* **99**, 6532 (1977).
150. (a) F. W. Wehrli, *J. Mag. Reson.* **25**, 575 (1977). (b) C. Srivanavit, J. I. Zink, and J. J. Dechter, *J. Am. Chem. Soc.* **99**, 5876 (1977).
151. E. Shchori, N. Nae, and J. Jagur-Grodzinski, *J. Chem. Soc. Dalton Trans.* 2381 (1975).
152. P. B. Chock, *Proc. Natl. Acad. Sci. USA* **69**, 1939 (1972).
153. B. Tummler, G. Maass, E. Weber, W. Wehner, and F. Vogtle, *J. Am. Chem. Soc.* **99**, 4683 (1977).
154. P. C. L. Birkbeck, D. S. B. Grace, and T. M. Sheperd, *Inorg. Nucl. Chem. Lett.* **7**, 801 (1971).
155. A. C. L. Su and J. F. Weiher, *Inorg. Chem.* **7**, 176 (1968).
156. T. E. Hogen-Esch and J. Smid, *J. Am. Chem. Soc.* **87**, 669 (1965); *J. Am. Chem. Soc.* **88**, 307 (1966).
157. (a) K. H. Wong, M. Bourgoin, and J. Smid, *J. Chem. Soc. Chem. Commun.* 715. (b) M. Bourgoin, K. H. Wong, J. Y. Hui, and J. Smid, *J. Am. Chem. Soc.* **97**, 3462 (1975).
158. U. Takaki and J. Smid, *J. Am. Chem. Soc.* **96**, 2588 (1974).
159. L. R. Sousa and J. M. Larson, *J. Am. Chem. Soc.* **99**, 307 (1977).
160. R. C. Lord and M. N. Siamwiza, *Spectrochim. Acta* **31A**, 1381 (1975).
161. R. M. Izatt, B. L. Haymore, and J. J. Christensen, *J. Chem. Soc. Chem. Commun.* **1972**, 1308.
162. A. I. Popov, *Pure Appl. Chem.* **41**, 275 (1975).
163. A. T. Tsatsas, R. W. Stearns, and W. M. Risen, *J. Am. Chem. Soc.* **94**, 5247 (1972).

602 Alexander I. Popov and Jean-Marie Lehn

164. Y. M. Cahen and A. I. Popov, *J. Solution Chem.* **4**, 599 (1975).
165. P. Gans, J. B. Gill, and J. N. Towning, *J. Chem. Soc. Dalton Trans.*, 2202 (1977).
166. J. L. Dye, M. T. Lok, F. J. Tehan, R. B. Coolen, N. Papadakis, J. M. Ceraso, and M. G. DeBacker, *Ber. Bunsenges. Physik. Chem.* **75**, 659 (1971).
167. B. Kaempf, S. Raynal, A. Collet, F. Schué, S. Boileau, and J.-M. Lehn, *Angew. Chem. Int. Ed. Engl.* **13**, 611 (1974).
168. M. A. Komarynsky and S. I. Weissman, *J. Am. Chem. Soc.* **97**, 1589 (1975).
169. G. V. Nelson and A. von Zelewsky, *J. Am. Chem. Soc.* **97**, 6279 (1975).
170. G. F. Pedulli, A. Alberti, and M. Guerra, *J. Chem. Soc. Perkin Trans.* **2**, 1327 (1977).
171. G. Wulfsberg, *Inorg. Chem.* **15**, 1791 (1976).
172. G. Wulfsberg and A. Weiss, *J. Chem. Soc. Dalton Trans.*, 1640 (1977).
173. H. K. Frensdorff, *J. Am. Chem. Soc.* **93**, 4684 (1971).
174. A. Sadakane, T. Iwachido, and K. Toei, *Bull. Chem. Soc. Japan* **48**, 60 (1975).
175. M. Kirch and J.-M. Lehn, *Angew. Chem. Int. Ed.* **14**, 555 (1975).
176. R. M. Izatt, D. P. Nelson, J. H. Rytting, B. L. Haymore, and J. J. Christensen, *J. Am. Chem. Soc.* **93**, 1619 (1971).
177. R. M. Izatt, R. E. Terry, D. P. Nelson, Y. Chan, D. J. Eatough, J. S. Bradshaw, L. D. Hansen, and J. J. Christensen, *J. Am. Chem. Soc.* **98**, 7626 (1976).
178. R. M. Izatt, R. E. Terry, B. L. Haymore, L. D. Hansen, N. K. Dalley, A. G. Avondet, and J. J. Christensen, *J. Am. Chem. Soc.* **98**, 7620 (1976).
179. E. M. Arnett and T. C. Moriarty, *J. Am. Chem. Soc.* **93**, 4908 (1971).
180. B. Metz, D. Moras, and R. Weiss, *J. Am. Chem. Soc.* **93**, 1806 (1971).
181. D. Moras and R. Weiss, *Acta Crystallogr.* **B29**, 400 (1973).
182. B. Metz, D. Moras, and R. Weiss, *J. Chem. Soc. Chem. Commun.*, 444 (1971).
183. B. Metz, D. Moras, and R. Weiss, *Acta Crystallogr.* **B29**, 1377 (1973).
184. F. J. Tehan, B. L. Barnett, and J. L. Dye, *J. Am. Chem. Soc.* **96**, 7203 (1974).
185. E. Kauffmann, J.-M. Lehn, and J.-P. Sauvage, *Helv. Chim. Acta* **59**, 1099 (1976).
186. W. Burgermeister and R. Winkler-Oswatitsch, *Top. Curr. Chem.* **69**, 91 (1977).
187. E. Grell and I. Oberhäumer, in: *Molecular Biology, Biochemistry and Biophysics*, Vol. 24, *Chemical Relaxation in Molecular Biology*, p. 371, Springer, Heidelberg (1977).
188. K. Henco, B. Tümmler, and G. Maass, *Angew. Chem. Int. Ed. Engl.* **16**, 538 (1977).
189. G. W. Liesegang, M. M. Farrow, N. Purdie, and E. M. Eyring, *J. Am. Chem. Soc.* **98**, 6905 (1976).
190. L. J. Rodriguez, G. W. Liesegang, R. D. White, M. M. Farrow, N. Purdie, and E. M. Eyring, *J. Phys. Chem.* **81**, 2118 (1977).
191. B. G. Cox and H. Schneider, *J. Am. Chem. Soc.* **99**, 2809 (1977).
192. V. M. Loyola, R. Pizer, and R. G. Wilkins, *J. Am. Chem. Soc.* **99**, 7185 (1977).

10

Natural-Product Model Systems

Virgil L. Goedken

1. Introduction

What constitutes a model system? What is the importance of studies involving models? What is the proper approach in the use of model systems? These are some of the questions that will be addressed in this chapter before considering specific details of the use of macrocyclic ligands as models for certain naturally occurring metalloproteins and enzymes.

1.1. Model Systems—Criticisms, Objectives, and Definitions

The utility and nature of a model is related to the complexity of the system and the field of science under investigation. Some problems are so intractable that only a simplified approach is capable of yielding useful information that will assist in the understanding of the real system.

Those disciplines requiring a great number of mathematical manipulations, e.g., physics and quantum chemistry, have for many years resorted to methods of approximation that attempt to model situations not amenable to exact treatments because of their complexity. Over the past half-century, methods of approximation have become increasingly sophisticated such that in some instances it is possible to determine certain parameters more accurately by calculation, using approximate methods, than it is to determine them experimentally. The Bohr model of the atom yielded a crude description of atomic systems. Yet it led to increasingly sophisticated and useful models. Today, approximate numerical methods such as MINDO II[1] yield information about excited states and reaction intermediates of complicated molecules that cannot be determined experimentally even with today's advanced physical methods.

Virgil L. Goedken • Department of Chemistry, Florida State University, Tallahassee, Florida 32306.

The proper approach to the development and study of model systems introduces an age-old quandary, viz., that of the objective versus subjective philosophy. Purists argue vehemently in favor of the objective approach, of being concerned with facts without bias introduced by the prejudices of individual backgrounds. This has generally been regarded as fundamental to the scientific method. However, such arguments, when taken too far, frequently lead to logical absurdities. Even the study of native biological material when isolated from *in vivo* systems constitutes an alteration of the system. Indeed, the actual study of *in vivo* systems involves a certain amount of perturbation.

In actual practice, the approach of investigators to model systems in bioinorganic chemistry has been largely subjective, being conditioned by individual scientific backgrounds, individual interests, and personal idiosyncrasies. The approach to problem solving in this area has traditionally taken two paths, the biochemical approach and the inorganic approach. In recent years these paths appear to be merging. The biochemist's approach to the elucidation of the structure and function of metalloenzymes has frequently involved alterations of native material and investigation of the resultant changes—spectral, reactivity rates, etc. These alterations can take the form of modification of the ligand, substitution of one metal for another, or substituting one substrate for another until the active sites and/or its function can be identified and characterized. Substitution of cobalt(II) for zinc(II) in carboxypeptidase and examination of the magnetic and spectral features of the cobalt provided an indicator of the coordination environment of zinc(II) in the original system.[2]

From the viewpoint of synthetic inorganic chemists who tend to focus their attention on metal ions, the approach to developing model systems has been toward the construction of ligands and metal complexes which approximate, to some degree, the environment about certain metal ions of the active sites, and to explore the full range of complex reactivity in these systems. This generally takes the form of introducing the proper number of donor atoms, geometry, type of donor atom, steric requirements, etc. into the ligand. The complete investigation, however, involves the total characterization of the metal complexes and a comparison of their properties with the native systems.

It must be borne in mind by the critics that the flow of information of understanding cannot be one way, i.e., from an understanding of the model system and application to the natural system. Just as assuredly, an understanding of the functions of the natural system may indicate a role for the model system, or some derivative of it (or reveal some underlying principle) which may be applicable in a completely unrelated way to some other processes, reaction, property of industrial or theoretical interest. For example, studies of chlorophyll have revealed it to be a system of uncompromising complexity. Studies have led not only to a better understanding of chlorophyll, but to the nature of the excited state, and energy transfer processes in general.

What is the objective of the model-system approach? Relevant synthetic models, because of their simplicity, are sometimes useful in identifying critical chemical features among a large number of possible causes, in short-cutting years of experimental work, and in more easily reaching the core of the more complicated biological problems.[3] Unfortunately, the term "model system" has been so abused with respect to biological relevance in the past 15 years that it has become almost meaningless. It is therefore necessary for our purpose to examine and redefine the term. The definition of a model system depends, to a very real extent, on the degree of understanding of the structure and function of the system being modeled. Approximately 30% of all enzymes contain metal ions. The detailed structure of only a small fraction of these has been determined accurately enough to enable one to design a synthetic system which may be capable of functioning to some extent in a manner related to the natural system. Thus model systems frequently evolve in their degree of sophistication as the structure and function of the naturally occurring systems become better known.

The objectivist attempting to model metalloenzymes and metalloproteins would undoubtedly use an operational definition and attempt to emulate their function. Systems amenable to this approach are those that catalyze the conversion of hydrogen peroxide to oxygen and water, the oxidation of ethanol to acetaldehyde, and transport and storage of molecular oxygen. However, unless the mechanisms of these processes bear some similarities to those of natural systems, they should not be claimed to be model systems. To do so would add more confusion than understanding to the problem.

Realistically, one should expect the geometrical features of the model system to conform to some extent to those of the natural system. To develop a synthetic system that merely duplicates some physical property of the natural system, i.e., some electronic structural feature, or magnetic property such as a magnetic moment, in the absence of structural information is normally of limited merit. Although sometimes of heuristic value, such approaches normally aid very little our understanding of either the "model" or the "real system." This is especially true of features such as charge transfer absorptions that are insensitive to changes in coordination geometry of the metal, or the surrounding tertiary structure of a large complex system. Yet, in certain instances (see below) such studies are apparently of some value in discrediting existing theories and in formulating new ones.

For our purposes, we shall interpret as a valid model any system that modifies or isolates certain salient features pertinent to the real system. The examination of the full range of behavior inherent in certain metals and their environments within the constraints of coordination geometry and the type of ligand is viewed as being within the realm of a model-system investigation.

The constitution of a salient feature is also highly subjective, being frequently dependent on the field of the investigator. Features construed to be salient at early stages of investigation are frequently those that are unexpected, unique, or apparently diagnostic for a particular system. They are generally dependent upon the extent to which the structure and function of the system being modeled is understood and also upon the analytical methods available and/or appropriate to the study.

The truly impressive rates and high degree of specificity associated with most metalloenzymes has mistakenly led to the impression that undiscovered chemical reactions or features were present to permit this remarkable reactivity. Very detailed X-ray structural results during the past ten years, coupled with a variety of studies, many on model systems, have led to the conclusion that the impressive reactivity and specificity of enzymes are not the result of any particular feature, but rather are due to the accumulation of numerous favorable factors. These include proper tertiary protein structure yielding the correct size and shape of cavity for a specific substrate, for producing the optimum dielectric, the optimum charge distribution, hydrogen-bonding interactions, and in some instances providing interactions that lead to strain and activation of the substrates. The point to be gleaned from what might appear to be wandering discussion is that the most important role of models is the demonstration of the importance of individual contributing factors, rather than to duplicate all of the features of a given system.

Properties that appear novel or unusual when first encountered in biological systems frequently are later found in a large number of compounds. For example, the discovery of an unprecedented cobalt–carbon bond in vitamin B_{12}[4] resulted in immediate synthetic attempts to synthesize other cobalt(III)–carbon bond species. To date, over a thousand such complexes have been synthesized which, in actual fact, have aided very little our understanding of vitamin B_{12}. The mechanistic details of most of its functions remain as elusive as ever. Similarly, the discovery of five-coordinate deoxyhemoglobin provided the impetus for the synthesis and characterization of other five-coordinate iron(II) and iron(III) complexes of porphyrin and nonporphyrin ligands.[5–7] Cytochrome P-450 has as its most striking feature, the red shift of the Soret bands from the normal range down to 450 nm. The blue copper proteins possess a distinctive blue color of unusual intensity that is thought to be related to the type of donor atoms and coordination geometry about copper. Here the uniqueness criteria of the spectral features are of great importance. In the absence of other distinguishing properties, the danger that other complexes have similar properties but entirely different coordination geometry and ligands is ever present. Relying too heavily on one or two distinctive physical characteristics can lead to unwarranted and misleading conclusions.

1.2. Importance of X-Ray Structural Analyses

The importance of knowledge concerning the structural details of both naturally occurring metallosystems and of models is self-evident. Structure analyses not only reveal the coordination environment about the metal but also illuminate the role of the steric effects of the quaternary structure of metalloproteins and its involvement in specific substrate interactions. Much higher accuracy associated with models augments the cruder structures obtained from native systems.

These aspects are best exemplified with heme proteins. The gross details of the structure of hemoglobin and myoglobin have been revealed by single-crystal X-ray diffraction studies.[8] While these have been invaluable in elucidating the overall conformation of the protein side chains and placement of the heme with respect to the protein and to one another, X-ray studies of large biological molecules do not yield sufficient resolution to provide accurate details of bond distances and angles. While nonstructural parameters such as redox potentials provide some information upon which to initiate construction of model systems, more exact knowledge of metal–donor bond distances, placement of the metal with respect to the macrocyclic ligand plane, and the role of the quaternary structure and its interaction with axially located substrates are of considerable importance in designing useful model systems.

A large number of highly accurate crystal structure determinations of synthetic porphyrins, principally of $\alpha,\beta,\gamma,\delta$-tetraphenylporphyrin and its metal complexes, have yielded significant insight into the flexibility of the ligand (particularly the radius as defined by the nitrogen to center, N–Ct, distance), the metal–nitrogen distances, and the positioning of the metal with respect to the plane of the macrocycle.[9] Careful analysis of this structural data has provided considerable insight to better understand the oxygenation of hemoglobin, particularly with regard to the nature and origin of cooperative effects involved with binding of dioxygen.

Iron complexes that may be divided into four categories, iron(II) and iron(III) and the high-spin and low-spin electronic configurations of each, are realized in one or more of the naturally occurring hemes. Although structures of iron complexes representative of each of the four classes have not been determined, the details of the coordination geometry of all metal porphyrins follow well-defined structural principles that allow realistic extrapolation.

The rather striking differences in coordination number and bond parameters that are found in the structures correlate well with the oxidation state and spin state of the iron. Normally, that is in the absence of constraints, iron–nitrogen distances are approximately 20 pm longer in high-spin states

($\sim$198 pm), in agreement with electrons occupying the destabilizing $d_{x^2-y^2}$ and d_{z^2} orbitals in the high-spin configurations. The rigid constraints of macrocyclic ligands restrict the N–Ct distance such that displacement of the metal from the N_4 plane provides the best mechanism for achieving a minimum-energy configuration. The undistorted N–Ct radius for porphyrins is about 201 pm. While this radius is sufficient for unstrained centering of low-spin iron(II) and iron(III) in the plane, preferred high-spin iron(II) and iron(III) nitrogen distances are appreciably longer, approximately 220 pm. Consequently, high-spin iron(II) and iron(III) in iron porphyrins are displaced approximately 50 and 70 pm, respectively, from the N_4 plane. This displacement generally results in the loss of a sixth ligand and the formation of five-coordinate complexes.

These observations are critical to the understanding of oxygenation phenomena of myoglobin and particularly the cooperativity effects of dioxygen binding of hemoglobin. Thus Hoard et al.[9] pointed out that any interaction of high-spin five-coordinate iron(II) porphyrin with a sixth ligand to produce a spin change to low-spin iron(II) would simultaneously bring the iron into the N_4 coordination plane. The amplitude of the relative motion of the iron atom should exceed 50 pm and would be expected to be accompanied by motion of the surrounding protein framework. Perutz has proposed an overall mechanism of hemoglobin oxygenation that invokes the conformational changes attending oxygenation of a single heme as sufficient to initiate cooperative interactions in the hemoglobin molecules.[10] A number of more recent X-ray structural investigations of a five-coordinate imidazole iron(II) complex and the dioxygen complex confirm the earlier predictions of Hoard.[11,12]

1.3. Evolution of Models

The construction of a model system for a naturally occurring macrocyclic system proceeds through several levels of approximation. Considerable overlap exists in the processes outlined below; the separation is somewhat artificial and the division is for explanatory purposes only.

1. The first requirement is structural, i.e., the synthesis of a metal complex having the proper type of donor atoms fixed in the coordination sphere such that a reasonable approximation of the naturally occurring metal–ligand distances and inner coordination angles is reproduced. Thus to model chlorophyll, heme proteins, and corrinoid rings a tetraazamacrocyclic ligand provides the minimum requirements.

2. The next step is the development of synthetic methods to "coarse tune" the system by proper selection of donor–acceptor properties to achieve the proper electronic environment for the metal. This can be achieved by using a combination of techniques: (a) varying the size of the macrocyclic

ligand, i.e., by selecting a 13-, 14-, 15-, or 16-membered ligand, and (b) varying the types of unsaturation present in the chelate rings (α-diimines, isolated imines, and the number of charges and extent of its delocalization).

3. "Fine-tuning" of the oxidation–reduction potentials and other ground state as well as excited electronic features can be achieved by placing appropriate substituents on the periphery of the macrocyclic ring.

4. The final and most difficult task is that of providing the proper environment to enable it to function. The role of the surrounding tertiary structure varies from one metalloprotein or enzyme to another. The highly involved convolutions are largely determined by highly specific base-pairing hydrogen-bonding interactions. The importance of the environment is best illustrated with the heme proteins hemoglobin and myoglobin. Both reversibly coordinate dioxygen only in the presence of the globular protein, to which the heme functions are not chemically bound. In these systems, the protein serves three functions: (a) It solubilizes the normally water-insoluble hemes, (b) it places the heme in a hydrophobic cleft, and (c) it isolates the heme moieties from one another. This latter function is very important since it prevents the irreversible oxidation of iron(II) to iron(III) via a second-order reaction involving two iron centers.

2. Macrocyclic Complexes as Models

The past ten years has been a period of continued evolution of macrocyclic ligands having potential as vehicles for model studies. These studies have largely been devoted to three classes of synthetic macrocyclic ligands: (a) the synthetic porphyrins and their derivatives, (b) synthetic nonporphyrin tetraazamacrocyclic ligands, and (c) cyclic and polycyclic ethers that serve as models for naturally occurring discriminations of alkali metal and alkaline earth cations.

In a sense, it is premature to write this chapter at this time. The most complete investigations of macrocyclic systems have been more concerned with investigating the fundamental chemistry and the detailed physical characterization of the ground state, and to a minor extent, the excited state properties. Deliberate attempts to reproduce the critical aspects of a native system are relatively recent and reports are largely in the form of communications. Thus to a large extent, the usefulness of synthetic macrocyclic systems, especially the nonheme variety, remains untested.

2.1. Macrocyclic Ethers and Thiaethers in Model Systems

The discovery in 1964 that the antibiotic valinomycin[13] exhibited alkali specificity in rat liver mitochondria, and Pederson's observations[14] that macrocyclic polyethers form stable complexes with the alkali and alkaline earth cations was a starting point for a large number of studies in the area of

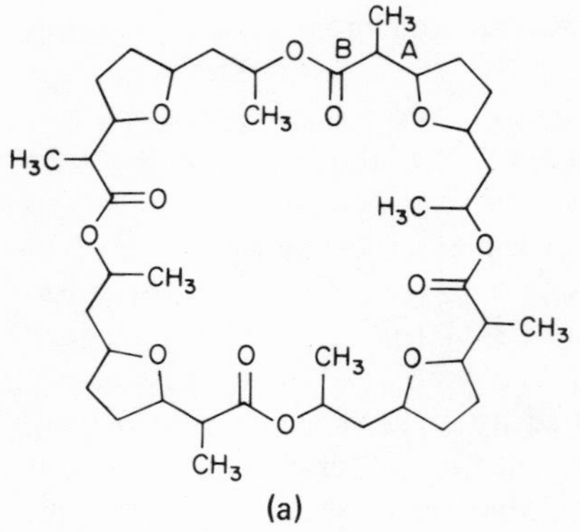

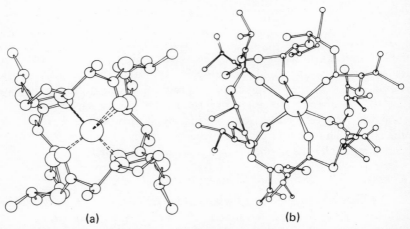

FIGURE 1. Macrocycles used in biophysical transport: (a) Valinomycin and (b) nonactin.

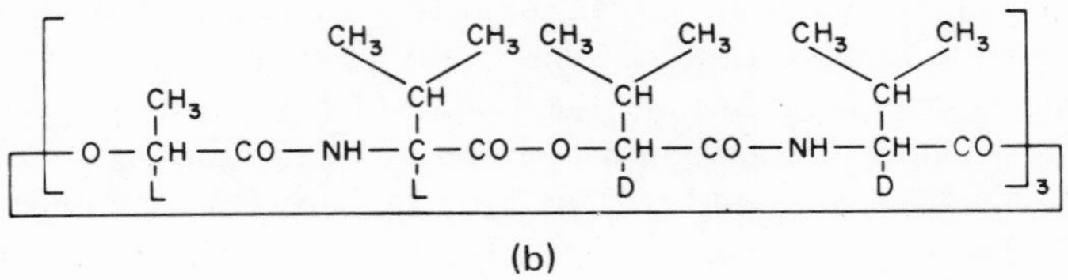

alkali and alkaline earth cation selectivity of biological and model systems. Valinomycin is a cyclododecadepsipeptide having a 36-atom ring with 12 carbonyl oxygen atoms (Figure 1a). Nonactin, a microtetrolide (Figure 1b), complexes alkali metal salts with stability constants in the order of Li < Na < Cs < Rb < K.[15] The crystal structure of the KNCS complex (Figure 2) consists of cubic coordination about the potassium ion.[16] It is

FIGURE 2. Crystal structures of the potassium complexes of (a) nonactin (reproduced from reference 15 with permission) and (b) valinomycin (reproduced from reference 16 with permission) illustrating the coordination environment in each case.

surrounded by four oxygen atoms from the tetrahydrofuran rings and four carbonyl oxygen atoms.

The naturally occurring antibiotics and the synthetic macrocyclic and macroheterobicyclic ligands have the following common features[17]: (1) One molecule can provide all the ligating atoms for the cation; (2) the cation is often anhydrous with the solvation sheath having been replaced by the oxygen atoms of the ligand; (3) the outside of the complex consists of aliphatic groups yielding lipophilic properties; (4) complex formation and dissociation are generally very rapid in solution (although dissociation rates for certain Sr^{2+} and Ba^{2+} complexes may be slow).[18]

The macrocyclic ethers and polythiaether ligands generally are not synthesizable by the metal template methods that have proven so useful for their nitrogen counterparts. Their synthesis generally involves brute force organic techniques, frequently necessitating high-dilution techniques, making them difficult to obtain. Yet because of their attractiveness as vehicles for the complexation of alkali and alkaline earth metals, and because they complement the nitrogen-occurring macrocycles, considerable effort and success has been devoted to them. The macrocyclic and polycyclic ether ligands have been found to duplicate some of the features of naturally occurring antibiotics, while copper complexes of the sulfur-containing macrocycles have been employed in studies designed to mimic some of the features of the blue copper proteins.

More than 60 neutral macrocyclic polyethers containing 4–20 oxygen atoms, in addition to nitrogen and sulfur atoms, have been synthesized (see Chapter 9). Many of these ligands form complexes in which the donor atoms of the polyether encircle the cation. The related macroheterobicyclic compounds of the type depicted in Figure 3 have remarkable complexing properties, as demonstrated by their stability constants (Table 1). The ligands may properly be construed as models for biological systems. Their amenability to discrimination among cations permits design of synthetic ligands for metal-ion specificity outside the range available from the naturally occurring ligands. Ligands of the macrobicyclic type (the cryptands or molecular sepulchers), because of their greater rigidity, display the best overall selectivity. They discriminate well against cations that are either smaller or larger than the preferred one. Some of the ligands already synthesized appear to have great

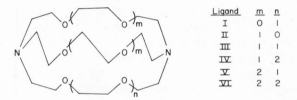

Ligand	m	n
I	0	1
II	1	0
III	1	1
IV	1	2
V	2	1
VI	2	2

FIGURE 3. General representation for various bicyclic cryptand ligands.

TABLE 1. Log K values for M$^+$—Macrobicyclic Ligand Complex
Formation in Aqueous Solution[a,b]

Ligand[c]	Cavity size (pm)	Li$^+$ (60)	Na$^+$ (95)	K$^+$ (133)	Rb$^+$ (148)	Cs$^+$ (169)	Ag$^+$ (126)
(I)	0.8	4.30	2.80	< 2	< 2	< 2	
(II)	1.15	2.50	5.40	3.95	2.55	< 2	10.6
(III)	1.4	< 2	3.90	5.40	4.35	< 2	9.60
(IV)	1.8	< 2	< 2	2.2	2.05	2.20	
(V)	2.1	< 2	< 2	< 2	≤ 0.7	< 2	
(VI)	2.4	< 2	< 2	< 2	≤ 0.5	< 2	

[a] Cation radii in picometers given in parentheses under each M+.
[b] Data from reference 19.
[c] Ligand structures are given in Figure 3.

potential for the selective removal of toxic metals, e.g., Hg^{2+} or Pb^{2+} or radio-active ^{85}Sr^{2+}, from biological systems, including humans. It has been demonstrated that the ligand whose Sr^{2+} complex has greater stability (log K = 13) than its Ca^{2+} complex (log K = 4.1) is moderately effective in removing ^{85}Sr^{2+} which had been injected into rats.[20]

The "blue" copper proteins have generated exceptional interest because of their characteristic intense absorption band in the 600-nm region,[21] a feature which is absent in low-molecular-weight copper(II) complexes. A number of recent investigations have identified the presence of one or more copper(II)–S bonds in these blue copper proteins. Rorabacher and co-workers have synthesized a large number of copper(II) complexes of sulfur macrocyclic ligands varying in size from [12]aneS$_3$ to the [21]aneS$_6$ ligand and examined their spectroscopic and redox properties.[22,23] A number of open-chain tetradentate ligands with donor atoms varying four nitrogen to four sulfur atoms were examined for comparison. In general, it was noted that both the macrocyclic and open-chain polythiaethers have an intense absorption band ($\varepsilon_{max} = 10^3$–10^4) in the 600-nm and near 400-nm regions. The single-crystal X-ray structure determination of [Cu([14]aneS$_4$)](ClO$_4$)$_2$ complex has revealed a "regular" structure with the copper centered in the plane of the four sulfur atoms (Cu–S bond lengths, 230 pm) with perchlorates occupying the axial positions (Cu–O distance, 265 pm).[24] Since the spectra of the [12]aneS$_4$ and [13]aneS$_4$ complexes exhibit virtually identical molar absorptivity values to the spectrum of the [14]aneS$_4$ complex, even though the former ligands are too small to permit planar coordination to copper(II), it appears that no specific coordination geometry about copper(II) is required for the generation of the characteristic absorption features. These observations discredit earlier hypotheses that the 600-nm band arose from a highly distorted ligand environment about the copper ion which was imposed by the protein superstructure.[25-29]

The markedly positive copper(II)–copper(I) formal electrode potentials ($+0.2$ to $+0.8$ V vs. SHE at pH 7) have also been recognized as a characteristic of the blue copper proteins, and are also a source of controversy.[21,30] Shifts of the copper(II)–copper(I) reduction potentials to more positive values could result from ligands that sterically or electronically destabilize tetragonal copper(II) and/or enhance the stabilization of copper(I). This led a number of workers to suggest that either tetrahedral or trigonal bipyramidal coordination was implied by the high redox potentials.[31-33] However, Rorabacher *et al.* in a systematic investigation of the redox properties of the series of polythiaether and polyaminothiaether complexes mentioned above found that the redox potentials were in the blue copper protein range and were independent of coordinative geometric constraints.[23] All the $E_{1/2}$ values for the polythiaether complexes S_3 to S_6 were in the range $+0.67$ to 0.90 V, and are among the highest values reported for the copper(II)–copper(I) couple. Since the potentials for the S_4 ligands of [13]aneS_4 and [14]aneS_4 are similar ($E_{1/2} = 0.674$ and 0.689 V, respectively), but the coordination geometries are grossly different (the 14-membered ligand has a regular tetragonal coordination, while that of the 13-membered ring is distorted), the large copper(II)–copper(I) potentials do not appear to be heavily dependent on distortions in the coordination geometry. Two open-chain tetrathiaether complexes examined for comparison exhibited the most positive potentials. However, reduction in the number of sulfur donor atoms to two or one was accompanied by a dramatic reduction (~ 0.5 V) in the potential. Elimination of the last sulfur upon going to an open-chain N_4 system (trien) results in a further decrease (~ 0.6 V) to negative potentials. It was also observed that the complex of 2,3,2-S_4, involving both thiaether and mercaptide sulfur donors, exhibited a potential similar to the corresponding open-chain tetrathiaether complexes and was very different from the complex of the corresponding diaminodithiaether analog, 2,3,3-NSSN. This was interpreted to imply that coordination of either mercaptide sulfur or thiaether sulfur may produce a similar influence on the copper(II)–copper(I) potential.

2.2. Synthetic Tetraazamacrocyclic Systems

The biologically occurring tetrapyrrole–metal complexes represent the ultimate in metal-ion control. This is accomplished through (1) intricate modifications of the basic tetrapyrrole framework, (2) peripheral substituents on the ligand, and (3) environmental effects. These influence metal–ligand geometry (bond distances and angles), electronic structure (ground and excited states), oxidation–reduction potentials, and access to coordination sites. Congruent with these observations, three components are necessary for most naturally occurring macrocyclic metal complexes to be biologically active: (1) the metal ion (frequently having more than one accessible oxidation state), (2) the macrocyclic ligand, and (3) the appropriate environment (convoluted protein, solvent, and axial ligands). To a large extent, the initial goal of modeling such systems has been the development of macrocyclic ligands which duplicate some essential features of naturally occurring tetrapyrrole ligands.

It is not possible to consider metal ions and ligands separately in most cases for a number of reasons. *In vivo* metal-ion–substrate interactions cannot occur in the absence of the coordinated and specially tuned tetrapyrrole ligands. In addition, the synthesis of many synthetic macrocyclic ligands is critically dependent upon the presence of a metal-ion template to direct the steric course of the reaction. Finally, some macrocyclic ligands are incapable of maintaining their identity in the absence of metal ions and either undergo decomposition or some type of internal rearrangement when the metal ion is removed.

The synthetic tetraaza macrocyclic ligands provide a natural framework from which to start construction of model systems. They contain certain features reminiscent and present in the heme, chlorophyll, and corrinoid ligands that are instantly recognizable, i.e., the presence of four donor nitrogen atoms more or less rigidly confined to a plane leaving two axial sites available for interaction with substrates. Furthermore, starting with a relatively few basic macrocyclic frameworks, it is possible to systematically alter the substituents and degree of saturation and conjugation within the limits of normal functional group reactivity contained therein. For many nonporphyrin macrocyclic ligands the presence of the coordinated metal greatly assists in these transformations. In addition, in recent years peripheral substituents have been placed on these ligands to modify the redox properties of the complexes, and in some cases to serve at least part of the function of the protein in the native structures. Thus, somewhere in the almost infinite variety of modifications that can be incorporated into synthetic tetraaza ligands, certain properties such as redox potential, proximity of low-lying states, ligand field strength, and coordination geometry may approach those found in natural environments.

The most important characteristics observed for most nonheme synthetic macrocyclic complexes was their kinetic stability toward ligand dissociation compared to their noncyclic counterparts.[34,36] A correlate of this behavior was their stability toward protonation in strongly acidic solutions and toward metal-ion precipitation under basic conditions, as well as their ability to stabilize unusual oxidation states.[36] These properties are also common to those of naturally occurring macrocyclic compounds. The observation that the same basic framework, i.e., tetrapyrrole derivatives are found repeatedly in compounds carrying out extremely diverse functions, e.g., light harvesting (chlorophyll), oxygen carrier (hemoglobin), electron transport (cytochromes), and modification of substrates (vitamin B_{12}) suggest the importance of the macrocyclic framework in devising models for these systems. The relative importance of the fused pyrrole rings to the overall stability and function of these systems also cannot be underestimated. The aromatic (26 π-electron) system of porphyrins leads to high thermodynamic stability, and the five-membered pyrrole rings produce a nearly flat (but with a small amount of ruffling) ligand that cannot be folded.[9]

In the absence of significant unsaturation, 16-membered rings have a N–Ct distance too large to yield highly stable first-row transition-metal complexes.[37] In the absence of fused five-membered rings, conjugated 16-membered rings tend to assume a marked saddle shape[38] that leads to reactivity on the macrocycle ligand because of activation from twisted double bonds.[39]

The kinetic and thermodynamic stability of the naturally occurring tetrapyrrole derivatives is extreme compared to other organic and biochemical substances. Porphyrins are frequently found in oil and oil shales as their nickel(II), or in some cases vanadium(II) complexes,[40] with the basic porphyrin framework remaining intact through millions of years. Since porphyrins of these metals are unknown in present biological systems and it is unlikely that they have existed in the past, metal exchange has probably taken place. Abelsonite, a newly discovered mineral, is a nickel(II) complex of an etioporphyrin derivative of chlorophyll.[41] Apparently, over eons of time, the magnesium and iron porphyrins have been replaced by metal ions which, although less abundant, yield more stable metalloporphyrins. The extraordinary stability of some synthetic nonheme macrocyclic ligand complexes has been covered in Chapter 3.

For the purposes of this chapter, we shall confine ourselves to tetraaza macrocyclic ligands and also include the synthetic porphyrins, phthalocyanines, and the *bis*-glyoximato ligand derivatives. The *bis*-glyoximato complexes, though lacking the high degree of stability toward dissociation and hydrolysis that is characteristic of other macrocyclic ligands, have a number of other features in common with them. The strength of the hydrogen bond is sufficient to force them to assume a square planar coordination geometry. With many metal ions, the strength of the α-diimine bond and the strength of the O$\cdots$H–O bond imparts high stability and enables studies to be conducted in mildly acidic media (pH 4–5). For example, many of the fundamental reactions of cobalt in the corrin ring system of vitamin B_{12} take place when cobalt is incorporated into *bis*-glyoximato complexes, tetradentate Schiff-base complexes, and macrocyclic ligands.[42] It is only logical to include those glyoximato complexes when their properties fall into line with other macrocyclic ligands.

Attention will be focused on 13-, 14-, 15-, and 16-membered ligands.

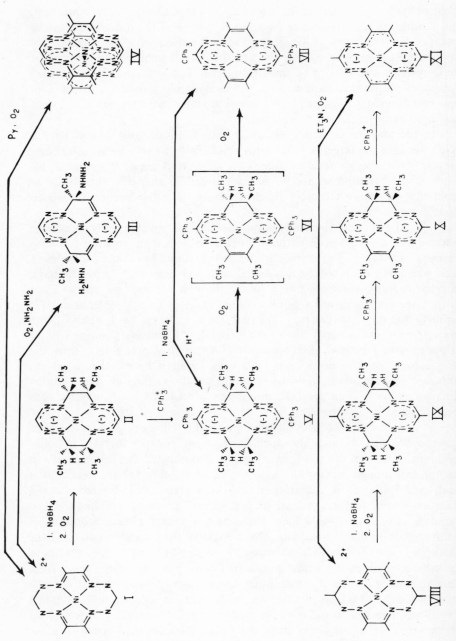

FIGURE 4. Reaction scheme illustrating the chemical transformations possible for the nickel(II) complex of the Me$_4$[14]tetraeneN$_4$(N$_4$)

Smaller ring sizes lead to folded conformations and obviate the need for inclusion in this chapter, whereas ring sizes larger than 16 exert a dilative effect on the metal–nitrogen interaction. For purposes of model studies, 13- to 16-membered macrocycles, in addition to their general robust behavior in acidic and basic media, have the general feature of four nitrogen atoms confined to an approximate square-planar configuration, thus leaving axial coordination sites available for interaction with other ligands and/or substrates. For each ring size, there exists a fully saturated member. Fourteen-membered ligands vary from fully saturated to fully conjugated dianionic systems.

2.3. Fundamental Studies of Synthetic Macrocyclic Ligand Complexes

An up-to-date, comprehensive review by Busch[43] has recently appeared. Therefore, only a few selected studies will be considered in this section.

2.3.1. Syntheses and Chemical Transformations: Examples

The systematic investigation of synthetic complexes containing ligand structures with different fused chelate ring size patterns (6-5-6-5, 6-5-6-6, 6-6-6-6) and degrees of π unsaturation provides a basis for the interpretation of certain stereochemical, electronic, and reactivity patterns of tetraaza-macrocyclic metal complexes. The octaazamacrocyclic ligand system *I* (Figure 4) is capable of undergoing a large number of facile chemical transformations that lead to different types of chelate rings and patterns of unsaturation.[44-46] The properties of the coordinated metal are highly dependent upon macrocyclic ring characteristics, i.e., α-diimine chelate rings versus diazonato chelate rings. Many of these complexes have been characterized by X-ray structural investigations. These 16 π-electron systems are formally antiaromatic, and total bond delocalization is not predicted. However, within the 6-5-6-5 chelate ring system, two limiting patterns of delocalization are represented by resonance isomers and are permissible (Figure 5). X-ray crystal structures have demonstrated that the particular resonance isomer is metal-ion dependent. The structure of an alkyl–cobalt(III)

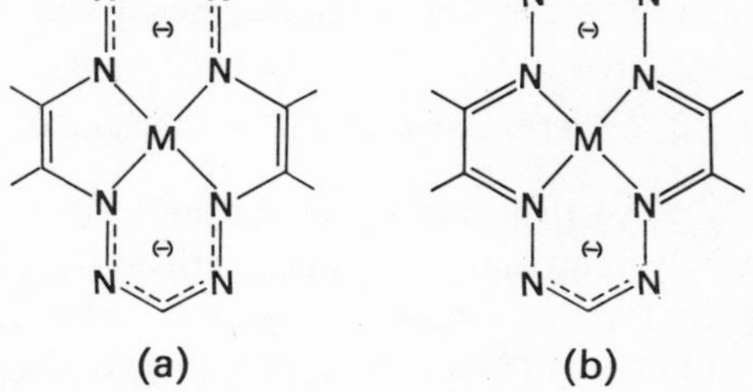

FIGURE 5. Two possible resonance isomers for the Me$_4$[14]-hexaeneatoN$_4$(N$_4$) ligand complexes.

(a)   (b)

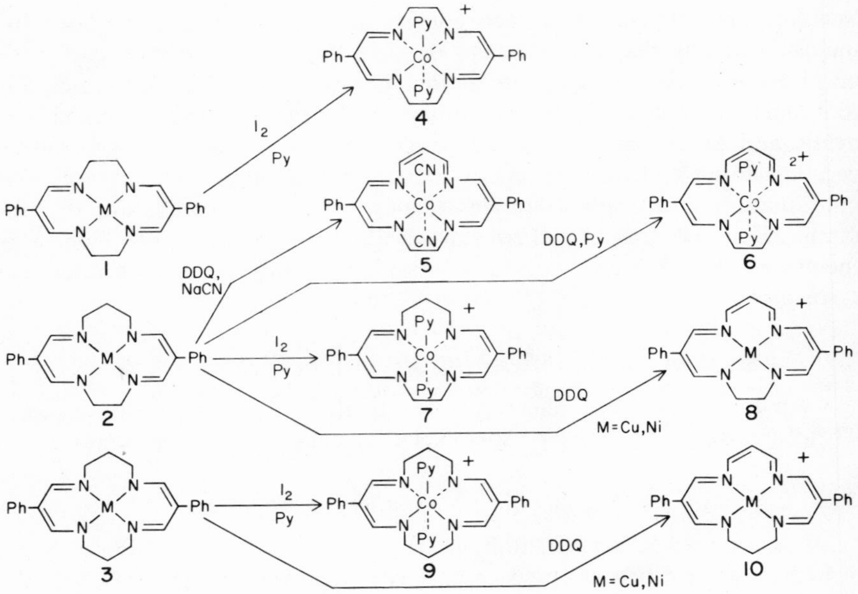

FIGURE 6. Scheme illustrating methods for increasing the conjugation of various macrocyclic rings (from reference 47).

complex has definitively demonstrated an α-diimine five-membered chelate ring and a three-atom delocalized allylic-type arrangement in the six-membered rings.[45] However, the nickel(II) structure contains the fully delocalized diazaonato rings of resonance form b.[46]

Tang and Holm, as another example, have demonstrated the feasibility of internal modifications of basic ring types.[47] Figure 6 illustrates the synthetic preparation of 12, 14, 15, and 16 π-electron systems. These are obtained from the 12-π precursor complexes via oxidative dehydrogenation. Structure **8** represents the synthesis of the 14-π corrin ring nucleus devoid of supporting pyrroline rings. The principal electronic differences between that species and vitamin B_{12}, its derivatives, and purely synthetic corrins occur in relative band intensities and low-energy shifts.[47] These complexes may be considered to contain the mildly perturbed corrin chromophore.

2.3.2. Spectroscopic and Electrochemical Studies

Investigations by Busch and co-workers over the past 15 years have amply demonstrated and quantified the effects of macrocyclic ring size, donor atom type, and substituents on the physical properties of the complexes.[48–52] Primary emphasis has been placed on determining ligand field strengths, spectroscopic properties, magnetic properties, redox properties, and compound stability. The effects of variation of macrocyclic ligand size on the in-plane ligand field (Dq^{xy}) have been demonstrated in a number of studies, but most clearly illustrated with an extensive series of cobalt(III) complexes of saturated macrocyclic ligands.[52] The order of increasing Dq^{xy} as a function of ring size is $13 > 14 > 15 > 16$ when the ligands occupy the planar coordination sites. For the *trans*-[Co(13–16)aneN$_4$]Cl$_2$ complexes Dq^{xy} varies by $\sim 20\%$, from 2750 to 2249 cm^{-1} for the 13- and 16- membered rings, respectively.

This study also illustrated the profound effects of incorporating various degrees of unsaturation on Dq^{xy}. The ligand field increases with the degree of unsaturation, e.g., Dq^{xy} varied from 1414 cm^{-1} for Me$_4$[14]aneN$_4$ to 1767 cm^{-1} for Me$_4$[14]1,3,8,10-tetraeneN$_4$. These changes are attributable to (1) the stronger interaction of imines with metals than secondary amines and (2) macrocyclic ring contraction accompanying the introduction of each double bond, leading to shorter metal–nitrogen distances and higher Dq values. Busch and co-workers have made similar observations with macrocyclic complexes of nickel(II).[51]

More impressive is the extent to which oxidation–reduction potentials and ligand field strengths can be controlled by varying the substituents on the periphery of the macrocyclic ring while holding the ring size and degree of unsaturation constant. Busch *et al.* have demonstrated that electron-withdrawing substituents (X and Y) are highly effective at producing more positive potentials[53] (see Figure 7). When $X = Y = NO_2$, $E_{1/2}$ for the Ni^{2+}/Ni^{3+} is 0.42 V (vs. Ag/AgNO$_3$) and decreases to -0.44 V for –CH$_2$CH$_2$–C$_5$H$_5$N, for a total range of 0.86 V. A total range of 1.80 V has been observed for the Ni^{2+}/Ni^{3+} couple, varying from 1.3 V (vs. Ag/AgNO$_3$) for saturated 16-membered rings to -0.5 V for the dianionic, 14-membered ligands.

In another ongoing study, the redox potentials of a large number of iron complexes having macrocycles of varying ring size and degrees of unsaturation were measured. A truly impressive range for the Fe(II)–Fe(III) couple

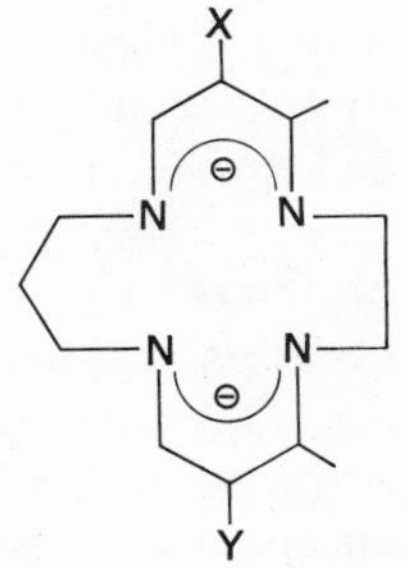

FIGURE 7. The dianionic ligand used by Busch *et al.* (reference 54) to demonstrate substituent effects on redox couples.

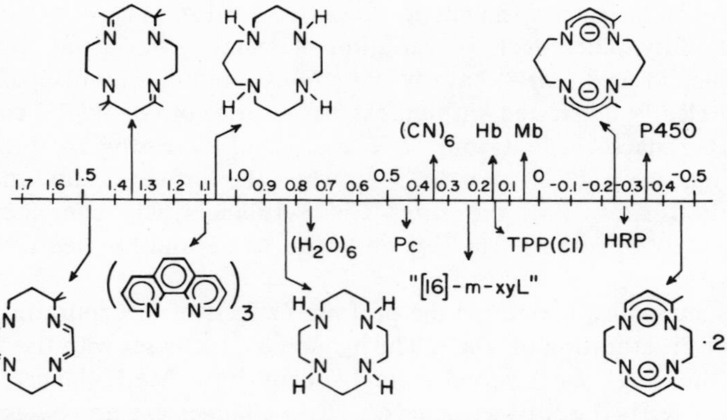

FIGURE 8. The range of redox potentials spanned by various macrocyclic complexes of iron (from reference 54).

was observed, from 1.48 V for the *bis*-α-diimine ligand systems to −0.46 V for the 14-membered dianionic ligand, "[16]-*m*-xyL". (See Figure 8 and Figure 13, Section 3.3.)[54] The synthesis of complexes having redox potentials close to natural heme systems is an important aspect in the development of nonheme models.

3. Modeling of Heme Proteins

Among the heme proteins, myoglobin and hemoglobin are the most obvious candidates for model-systems studies for a number of reasons. First, their function, that of transporting and storing dioxygen, is among the best understood of the metalloproteins. Secondly, their structures have been fully elucidated by single-crystal X-ray diffraction techniques. Third, a very wide variety of synthetic porphyrins and nonporphyrin macrocyclic ligands have been synthesized in recent years which are ideally suited for model studies.

Hemoglobin has a molecular weight of approximately 64,000 and has four subunits, each of which contains one heme group. Myoglobin contains only one heme group, and the structure of the surrounding protein is similar to that of the subunits of hemoglobin. The heme group in both hemoglobin and myoglobin is not covalent linked to the protein; rather it is lodged in a crevice created by the folds of the protein and linked through a coordinated histidine–nitrogen atom of the protein to the iron(II) of the heme. The deoxygenated forms of both hemoglobin and myoglobin are five coordinate.

The vast majority of metal complexes, including macrocyclic complexes and even isolated native heme, react with dioxygen irreversibly. From an analysis of the thermodynamic data obtained in aqueous solution, Ochiai has concluded that the μ-peroxo, $M-O_2-M$ type of complexes are generally favored over the superoxo type.[55] With the exception of naturally occurring heme proteins and a few specially designed synthetic ligands, iron(II) complexes are irreversibly oxidized to iron(III) species via the following scheme[56–58]:

$$Fe(II) + O_2 \; \xrightleftharpoons \; \begin{array}{c} Fe(II)-O_2 \\ or \\ Fe(III)-O_2{}^- \end{array} \; \xrightarrow{Fe(II)}$$

$$Fe(II)-O-O-Fe(II) \; \xrightarrow{\text{auto-oxidation}} \; Fe(III) \text{ species}$$

To circumvent this problem, investigators have employed metal ions other than iron(II) or have studied the coordination of small molecules such as carbon monoxide, which normally does not take part in redox reactions.

3.1. Studies Involving Metals Other than Iron

Simple replacement of iron(II) with cobalt(II) in many planar tetradentate ligands, both cyclic and noncyclic, leads to reversible oxygenation of the metal center in a large number of complexes.[59–61] Most of these have been shown to be diamagnetic peroxo-bridged species of the type $(B)(L)Co^{III}-(O_2)-Co^{III}(L)(B)$, where B stands for a nitrogenous base and L is the planar ligand. However, Calderazzo and co-workers made the important observation that Co(3-methoxysalen) in pyridine solution reacts with dioxygen to yield the monomeric Co(3-methoxysalen)(py)(O_2) complex.[60] Basolo and co-workers independently made a similar observation with the cobalt(II) complexes of N,N'-ethylenebis(acetylacetoniminato) ligand.[62,63] The significance of these observations obviously stemmed from the observation that oxyhemoglobin is a one:one dioxygen:iron complex and that no other dioxygen–metal complexes suitable as models had been found.

EPR studies by Hoffman and co-workers of the latter cobalt(II) complexes in both liquid and frozen solution clearly indicated monomeric complexes with only one interacting ^{59}Co nucleus.[64] Their analysis of the EPR spectra indicated approximately a 90% transfer of spin density from the cobalt(II) to dioxygen upon complex formation. This result, together with the observed O–O stretching frequency $\sim 1130 \text{ cm}^{-1}$, which is close to that of ionic $O_2{}^-$, strongly suggests that charge transfer occurs upon complex formation and that these monomeric dioxygen complexes are best described as formal $Co(III)-O_2{}^-$ species.[63,65] The acceptance of this formulation is not universal, however.[66–68] A number of single-crystal X-ray diffraction studies[69–72] have shown that the oxygen is end bonded with an angular Co–O–O bond, 117°–120°. The O–O distance, 127–130.2 pm, is also consistent with the O–O distance found in ionic superoxides, 128 pm.

Experiments with cobalt(II) complexes of protoporphyrin(IX) dimethyl ester showed that cooled toluene solutions of the porphyrin complex, in the presence of base, react with dioxygen according to the following equation[73-75]:

$$Co(porphyrin)(B) + O_2 \rightleftharpoons Co(porphyrin)(B)(O_2)$$

The EPR parameters of the resulting dioxygen complex were not qualitatively different from those of monomeric dioxygen complexes of Schiff-base cobalt(II) complexes.

This set the stage for experiments by Hoffman *et al.* with coboglobins [hemoglobin and myoglobin reconstituted with cobalt(II) rather than iron].[76] EPR studies showed that the cobalt atom in deoxycoboglobin is five coordinate as iron is in the natural proteins.[76] Exposure of the coboglobins to dioxygen yields the dioxygen coboglobins, and EPR studies again showed that the cobalt–dioxygen spectra parameters were not significantly different from those obtained from the earlier studies of the dioxygen complexes of protoporphyrin(IX) dimethyl ester.[77] Thermodynamic studies of dioxygen binding by coboglobin and comparisons with cobalt(II) protoporphyrin(IX) dimethyl ester reveal that the equilibrium constant for dioxygen binding by coboglobins is about 300 times greater than for protoporphyrin(IX).[77] This has been attributed to a hydrophobic pocket already organized to accommodate the strongly dipolar Co(III)–O_2 linkage. The difference in binding constants is largely entropic; in the cobalt(II) protoporphyrin complex dioxygen complexation from a neutral cobalt(II) porphyrin and a neutral oxygen molecule is probably accompanied by appreciable solvent reorganization and appears thermodynamically as a negative entropy contribution.[77,78]

Although the study of dioxygen complexation with metal ions not isoelectronic with iron(II) might appear to vitiate any comparison with dioxygen complexes of iron, recent results of dioxygen binding to chromium(II) porphyrins and iron(II) porphyrins (Section 3.3) indicate a common mode of metal–dioxygen interactions among a number of one:one dioxygen:metal complexes.

3.2. Iron(II) Carbon Monoxide Complexes

The difficulty of obtaining stable dioxygen complexes of synthetic iron(II) systems was undoubtedly responsible for most investigations of the carbon-monoxide-binding ability of iron(II) complexes of macrocyclic and other closely related ligands. These studies may be rationalized as appropriate for model studies through two arguments. First, many of the naturally occurring heme proteins as well as the isolated hemes, coordinate carbon monoxide very strongly with binding constants approximately 200 times greater than for dioxygen.[79] Equally important is the fact that carbon monoxide is a natural product of the catabolism of hemoglobin and myoglobin.[80] This endogenous source of carbon monoxide produces 10–50 ppm P_{CO} in mammalian tissue under equilibrium conditions,[81] leading to a normal "poisoning" of 4–5% of the hemoglobin.[82] Secondly, a casual examination of the reactivities of a number of macrocyclic complexes of iron(II) indicates that those which are very sensitive to oxidation by dioxygen also appear to coordinate carbon monoxide well, and vice versa.

The objectives in the studies of dioxygen versus carbon monoxide complexes have been somewhat different because of the relative ease of obtaining carbon monoxide as opposed to the difficulty of dioxygen complexes. The objectives with dioxygen complexes have been twofold: (1) the design of suitable ligands to stabilize the iron–dioxygen assemblage and (2) the interpretation of the physical parameters, principally infrared and X-ray data of the dioxygen complexes to ascertain the nature of the metal–dioxygen interactions. These have taken precedence over a detailed analysis of kinetic and thermodynamic parameters to establish the effects of various equatorial and axial ligands. Investigations in the latter areas will unquestionably be the subject of future research as more synthetic dioxygen complexes become available. The obtainment of iron(II)–carbon-monoxide complexes has been less troublesome, and considerable effort has been expended to evaluate the influence of various macrocyclic ligands and ligands *trans* to the carbon monoxide on the thermodynamic parameters of carbon monoxide complexation.

Although carbon monoxide is an exceedingly weak Lewis base, its strong π-acceptor properties make it a good ligand with low-valent transition metals. As mentioned, iron(II) in native hemoglobin and myoglobin has a very high affinity for one molecule of carbon monoxide, approximately 200 times higher than for dioxygen. This high affinity for carbon monoxide persists for isolated iron(II) hemes and for all iron(II) porphyrin complexes in general. The strength of these interactions has been viewed as quite remarkable considering that the myriad amine and Schiff-base complexes of iron(II) display an insignificant tendency to coordinate carbon monoxide at ambient temperatures and 1 atm of carbon monoxide pressure.

These observations raise a number of obvious questions. What features of the heme ligands are responsible for the very strong binding of carbon monoxide? What range of in-plane ligands will stabilize iron(II)–carbon-monoxide complexes? What are the values of the thermodynamic parameters associated with carbon monoxide binding in these complexes and how do they vary with the type of in-plane ligand?

A few iron(II)–carbon-monoxide complexes have been known for decades. These include $[Fe(CN)_5(CO)]^{2+}$ and $[Fe(phen)_2(CO)X]^{2+}$ (X = solvent or monodentate ligand). More recently, it has been found that iron(II) complexes of certain linear polydentate amines, for example,

1,5,8,12-tetraazadodecane, coordinate carbon monoxide quite strongly, as is evident from the low carbon monoxide stretching frequency, 1940 cm^{-1}, compared to that of free carbon monoxide, 2143 cm^{-1}.[83]

Like most other known CO complexes, the carbonyl derivatives of hemoglobin and myoglobin, as well as all other known iron(II) carbon monoxide complexes, are low spin. Although the M—C≡O angle is nominally linear in all synthetic transition-metal complexes, single-crystal diffraction studies indicate a nonlinear Fe—C≡O in three heme proteins investigated structurally.[84–86] This appears to be a consequence of steric hindrance imposed by the side chains of the amino acid residues of the surrounding protein.

The carbonyl-versus-deoxy-difference map of the monomeric insect hemoglobin, erythrocruorin, revealed that the carbon monoxide molecule is inclined to the heme plane with an Fe—C≡O angle of 145° ± 15°.[84] The crystalline three-dimensional structure of carbonyl myoglobin, determined by neutron diffraction studies, also had an Fe–CO angle of 135°.[86] It should be pointed out that the resolution of all these studies is insufficient to differentiate between a bent Fe—C≡O bond and a linear Fe—C≡O arrangement which is tilted with respect to the heme plane. Since the energy required to distort a bond angle is a function of the extent of deformation, it is reasonable to assume that the overall minimum-energy configuration contains both types of distortion.

Thus, in addition to the preceding questions posed for model systems, it may be asked how steric interactions effect the geometry and binding constants of coordinated carbon monoxide.

Many factors influence the binding of carbon monoxide (and other small molecules) to iron(II) complexes. Some of the more important factors, listed in order of decreasing importance, are: the nature of the in-plane ligand, the number and type of axial ligands, steric interactions, and the solvent medium. Ideally, these various factors may be evaluated to some extent by holding all variables except one constant during a series of measurements. However, experimental limitations such as solubility, reactivity of the iron complex with solvent, etc., severely limit the extent to which valid comparisons can be made. Thus it is difficult to make valid comparisons of thermodynamic parameters among the large number of variables (e.g., solvent, concentration of axial base, etc.) employed by different investigators on different systems.

What criteria are most useful for quantitatively assessing the "nature" of iron(II)–carbon-monoxide interactions? C≡O stretching frequency? Equilibrium constants? Rate of Fe—C≡O formation? Rate of Fe—C≡O dissociation? Enthalpy of formation? Fe—C≡O bond energy? Free energy of reaction? Each of the foregoing measurements provides only one indicator of either the strength or stability of Fe—C≡O bonds. No single measurement can adequately describe the total character of a metal–carbon-monoxide bond. Frequently, comparisons of a given property between two related compounds can be meaningless, as will be demonstrated shortly.

The carbonyl stretching frequency is one of the most easily measured parameters of carbon monoxide complexes. It provides one measure of the strength of the iron–carbon-monoxide interaction that can be used for comparison in very closely related compounds. Table 2 lists the carbonyl stretching frequencies for a wide variety of iron(II)–carbonyl complexes. For synthetic complexes having linear carbon monoxide coordination, it is observed that the CO stretching frequency decreases (indicating a stronger Fe–CO interaction) as the basicity of the ligand increases. These vary from about 2050 cm^{-1} for macrocycles having *bis-α*-diimine chelates down to 1950 cm^{-1} for ligands having pentanediiminato chelates. The CO stretching frequencies for porphyrins are observed about midway in these ranges; isolated hemes and porphyrins have CO stretching frequencies near 1970 cm^{-1} (with some

TABLE 2. Carbonyl Stretching Frequencies of Carbonyl Derivatives of a Model Compound (FeLBCO)

Planar ligand	Axial base	ν_{CO} (cm^{-1})	Condition	Reference
protoporphyrin IX	py	1970	KBr disk	87
hemoglobin		1951		88
TpivPP	*N*-MeIm	1969		89
Traylor's (Figure 9)		1963	in chloroform	90
		1950	in solid	
2,4-diethyl DP′	py	1973.0	in bromoform with 0.12 *M* py	90
DP′	py	1975.1	in bromoform with 0.12 *M* py	90
bis-DPGH	py	1996		91
bis-DPGH	*m*-chlorophyll			91
TIM	CH$_3$CN	2031		92
(3,2,3-tet)	CH$_3$CN	1975		93
[14]aneN$_4$	CH$_3$CN	1965		93
(Me$_6$[14]aneN$_4$)	CH$_3$CN	1978		93
(Me$_6$[14]4,11-dieneN$_4$)	CH$_3$CN	1974		93
(Me$_6$[14]1,4,8,11-tetraeneN$_4$)	CH$_3$CN	2010		93
(Me$_4$[14]1,3,8,10-tetraeneN$_4$)	CH$_3$CN	2048		93
(Me$_4$[14]decaeneN$_6$)	CH$_3$CN	2058		93
Me$_4$bzo$_2$[14]hexaeneatoN$_4$	CH$_3$CN	1930		93
[14]aneN$_4$]	CH$_3$CN	1930		93
(Me$_6$[14]4,11-dieneN$_4$)	Cl	1951		93
(Me$_6$[14]1,4,8,11-tetraeneN$_4$)	Cl	1983		93
(Me$_4$[14]1,3,8,10-tetraeneN$_4$)	py	2044		93
(tetren)	Cl	1940		93

dependence on axial base and solvent medium), while normal hemoglobin and myoglobin, with nonlinear carbonyl coordination, have lower frequencies, ~ 1950 cm^{-1}. Ligands containing α-diimine chelates compete for the d π-electron density of the metal and weaken the interaction between the metal and the carbonyl. Similar arguments can be made for ligands occupying positions *trans* to carbon monoxide. It is possible to envisage ligands *trans* to the carbon monoxide exhibiting either an antagonistic or a symbiotic *trans* effect. Strong σ donors might be expected to stabilize the iron(II)–carbon-monoxide bond, whereas π-acceptor ligands compete with carbon monoxide,[94] thus explaining the absence of *bis* carbon monoxide complexes of iron(II).

Two of the most common parameters given in the literature are the equilibrium constants and $P_{1/2}$, the pressure of carbon monoxide required to have 50% of the complex in the form of its carbon monoxide complex.

Differing coordination numbers of the precursor iron(II) complexes certainly invalidate comparison of most thermodynamic parameters of carbon monoxide complexation. Iron(II) phthalocyanine is four coordinate in non-donor solvents and deoxyhemoglobin and myoglobin are five coordinate. For these systems the formation of a carbon monoxide complex is a straight-forward association process. However, isolated hemes and most synthetic iron(II) macrocyclic complexes are six coordinate and dissociation of an axial ligand must occur prior to carbon monoxide coordination. Thus the form of the equilibrium constants differ as shown below:

$$K_6 = \frac{[FeLB(CO)][B]}{[FeLB_2][CO]}$$

$$K_5 = \frac{[FeLB(CO)]}{[FeLB][CO]}$$

For the six-coordinate precursor case, the equilibrium constant is strongly dependent upon the bond strength of the iron(II) axial base which dissociates, and is influenced by the nature of the interaction between the remaining base and iron(II), and to some extent, the effect of the base on the binding of carbon monoxide. Similarly, $P_{1/2}$ data may be used to obtain quantitative data and make comparisons of carbon monoxide affinities in a series of closely related systems under very similar experimental conditions. For six-coordinate precursor complexes, $P_{1/2}$ values are dependent upon the concentration of axial ligands present which tend to vary with the particular investigation.

Other important factors, not easily evaluated, are changes in spin state and accompanying structural changes (metal–ligand bond distances and angles) and their influence on thermodynamic parameters of complexation. For example, five-coordinate deoxyhemoglobin and myoglobin are high spin

($S = 2$), and iron(II) phthalocyanine has an intermediate ($S = 1$) spin state. Both become low spin ($S = 0$) and six coordinate upon coordination of carbon monoxide.

Caution must also be exercised in drawing conclusions from thermodynamic parameters when comparing the effects of various axial bases on the binding of carbon monoxide *trans* to it, especially when these arguments are based on the relative basicity and π-acceptor properties of these ligands. X-ray structural determinations of six-coordinate iron(II) porphyrins reveal that α hydrogens of ligands such as pyridine and piperidine have steric interactions with the porphyrin such that the iron(II)–nitrogen distances are longer than predicted.[95] Obviously, if steric interactions prevent normal approach of the base, arguments based on basicities alone can be misleading. For six-coordinate metal complexes of highly conjugated planar ligands, steric interactions with α hydrogens of piperidine or pyridine are likely to occur whenever the metal–nitrogen distance approaches 200 pm.

Tables 3 and 4 contain thermodynamic parameters of carbon monoxide complexation for porphyrin and nonporphyrin macrocycles, respectively. Several observations warrant special emphasis. The binding of carbon monoxide by synthetic five-coordinate porphyrins is much more effective than for native hemoglobin and myoglobin. In the native systems steric interactions lead to a tilting of the carbon monoxide and greatly reduce the ability of CO to coordinate. The binding in the picket fence porphyrins is so strong that reversibility cannot be observed. [All porphyrin and many nonporphyrin iron(II) carbonyl complexes are photosensitive to carbon monoxide ejection.] A significant observation has been made by Collman[100] regarding the profound influence of N-methylimidazole versus 2-methylimidazole on the affinity of iron(II): deuteroporphyrin complexes of the 2-methyl complex have only $\frac{1}{200}$ the affinity of those of the N-methyl base[101–103] because steric interactions in the former restrain the five-coordinate form.

Binding parameters of carbon monoxide to six-coordinate iron(II) complexes cannot be compared with those of five-coordinate complexes as previously mentioned. However, it can be seen that they span an enormous range. Complexes with α-diimine chelates have a low affinity for carbon monoxide. In general, these α-diimine complexes also have the lowest reactivity with dioxygen. Iron(II) complexes of dimethylglyoxime appear to be an exception; although they are extremely oxygen sensitive, they display little tendency to coordinate carbon monoxide at room temperature under 1 atm of carbon monoxide pressure. In general, the carbon monoxide binding constants increase with the expected basicity of the ligand. Of the nonporphyrin ligands, the dianionic ligand $Me_4bzo_2[14]hexaenatoN_4$ is the most basic. The binding for this system was so strong that reversibility could not be observed. For complexes without steric interactions, the carbon monoxide equilibrium constants parallel the Fe^{2+}/Fe^{3+} redox potentials.[93]

TABLE 3. Equilibrium Constants and $P_{1/2}$ Values of Carbon Monoxide Binding by Various Fe(II) Compounds[a]

Compound	Conditions	$(K_{CO})_6$ (Unitless)	$(K_{CO})_6$ (M/Torr)	$(K_{CO})_5$ M^{-1}	$P_{1/2}(CO)$ (Torr)	Reference
Hemoglobin					0.004(0.035)[b]	79
Fe(TpivPP)(N-MeIm)(CO)					Irreversible	89
Fe(II)PP(H$_2$O)$_2$	21.8–27.8°C in H$_2$O pH 9.6–10.5	1.87×10^7	23.2		2.4	96
Fe(II)PP(H$_2$O)$_2$ or Fe(II)PP(H$_2$O)(py)	[heme]$_{total}$ = 1.0×10^{-4} M, 21–24°C, ph 11, in H$_2$O					96
	[heme]$_{total}$ = 5×10^{-5} M					
	[py] = 5×10^{-5} M				0.8	96
	[py] = 1×10^{-4} M				0.4	96
	[py] = 5×10^{-4} M				0.29	
Fe(II)PP(pip)$_2$	At 23°C in piperidine	2.3×10^5	2.0		5.8	97
Fe(II)TPP(pip)$_2$	At 23°C in piperidine	1.5×10^4	0.13		75	97
Fe(II)MP'(py)$_2$	[heme]$_{total}$ = 3.5×10^{-5} M	10,300		[7,950]	12.5	98
	[py] = 1.3 M					
Fe(II)DP'(py)$_2$	At 20°C in CHCl$_3$(?)	4,600		[3,500]	28	98
Fe(II)PP'(py)$_2$	At 23°C in CHCl$_3$	3,800		[2,950]	33	98
T-2[c]				4.4×10^5	0.20	98

[heme]$_{total}$ = 5 × 10^{-5} M

Ligand				Ref
T-1^c		8.2 × 10^4	1.1	98
T-3^c		7.5 × 10^5	0.13	98

At 25°C in chlorobenzene

At 23°C

In toluene

Ligand				Ref
Fe(II)(DPGH)$_2$(py)$_2$	43			91
Fe(II)Pc(Im)$_2$	0.03			99
Fe(II)Pc(pip)$_2$	0.85			99
Fe(II)Pc(py)$_2$	0.37			99
Fe(II)Pc(2-MeIm)$_2$	7.8			99

Planar ligand (Fe(II) complexes)	Solvent (axial bases)			Ref
[14]aneN$_4$	CH$_3$CN	9.42 × 10^4	103	93
Me$_6$[14]aneN$_4$	CH$_3$CN	2.93 × 10^4	373	93
Me$_6$[14]4,11-dieneN$_4$	CH$_3$CN	3.68 × 10^5	30.0	93
Me$_6$[14]1,4,8,11-tetraeneN$_4$	CH$_3$CN	2.64 × 10^5	41.3	93
Me$_4$[14]1,3,8,10-tetraeneN$_8$	CH$_3$CN	2.27 × 10^4	481	93
Me$_4$[14]decaeneN$_6$	CH$_3$NO$_2$	8.66 × 10^2	12600	93
Me$_4$Bzo$_2$[14]hexaenatoN$_4$	CH$_3$CN	<2.40 × 10^9	<0.005	93

^{a} Abbreviations used: PP, protoporphyrin IX; PP', protoporphyrin IX dimethyl ester; TPP, α,β,γ,δ-tetraphenylporphyrin; DP', deuteroporphyrin dimethyl ester; MP', *meso*-porphyrin dimethyl ester; DPGH, diphenylglyoxime; Pc, phthalocyanine; py, pyridine; pip, piperidine; Im, imidazole; MeIm, methyl imidazole.

^{b} The first value is for the binding of the fourth carbon monoxide; the value in parentheses is for the overall equilibrium.

^{c} See Figure 7 for key to ligand structure.

TABLE 4. Thermodynamic Properties of O_2 Binding[a]

Substance	$P_{1/2}O_2$ (20°C)	ΔH[b]	ΔS[c]
Fe(TpivPP)(N-MeIm)	0.31	−65.27	−159
OxMb	0.55	−63	−155
Co(TpivPP)(N-MeIm)	100	−51.0	−160
	44	−54.8	−163
CoMb	54	−52.7	−155
Co(PP$_{IX}$DME)(N-MeIm)	1.4×10^4	−48.1	−188

[a] Data from reference 110b.
[b] $kJ \, mol^{-1}$.
[c] $J \, mol^{-1} \, deg^{-1}$.

Chang and Traylor have examined the effects of neighboring groups and proximal base on the binding of carbon monoxide to a number of *meso*-hemes with different appended bases.[98] Some of these *meso*-hemes are shown in Figure 9. The effect of having a pyridine covalently attached to the heme as in Figure 9 (structure **1**) is that it competes for carbon monoxide and leads to a dramatic reduction in the carbon-monoxide-binding equilibrium constant, from 4.4×10^5 to $8.2 \times 10^4 \, M^{-1}$. The complex denoted by Figure 9 (structure **3**) has an imidazole which occupies a position similar to its position in myoglobin, but cannot approach the iron itself. The approximate twofold

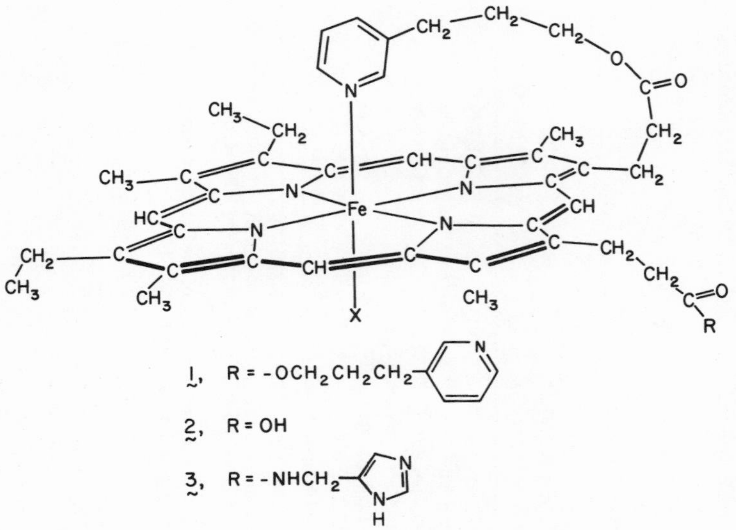

FIGURE 9. Meso-heme derivatives containing various appended bases. Reproduced from reference 98 and with permission.

increase in binding constant on going from 2 to 3 is consistent with the proposed hydrogen bonding between this imidazole and the carbon monoxide ligand.[98] They also demonstrated the difference between having a pyridine group or imidazole group occupying the fifth position on carbon monoxide and oxygen binding. Although the binding constants for carbon monoxide were similar for the two, the pyridine–*meso*-heme favored dioxygen binding by a factor of 3800.

The sensitivity of carbon monoxide binding to iron(II) on the nature of the macrocyclic ligand is perhaps best exemplified by the differing affinity of the phthalocyanine and porphyrin complexes by a factor of approximately 10^7 (Table 3). The porphyrin ring of hemes and phthalocyanines both may be described as containing 16-membered macrocyclic rings comprised of four pyrrole moieties, and similar carbon monoxide affinities might be expected. The enormous variation in the binding constants for carbon monoxide coordination has its origins in two structural differences of the two macrocycles. The pyrrole rings of phthalocyanines are fused together with nitrogen rather than carbon linkages and benzene rings are fused onto the pyrrole rings. Because C–N distances are approximately 10 pm shorter than C–C distances, the nitrogen-to-center distance for phthalocyanines is approximately 10 pm shorter than for porphyrins. Apparently, the phthalocyanine ligand itself nearly meets all the electronic requirements for iron(II). Four-coordinate iron(II) phthalocyanines are common, and interactions with axial bases tend to be relatively weak; they do not readily oxidize when exposed to dioxygen.

3.3. Dioxygen Complexes

The problem of preventing or inhibiting bimolecular interactions of iron centers involving coordinated dioxygen that lead to irreversible oxidation to iron(III) has been approached in a number of ways. One method is to immobilize the iron(II) centers via a substitutionally inert complex on a polymer matrix or other suitable solid support. Second, bulky substituents may be placed on the periphery of the ligand to block access to larger axial ligands but permit entry of small ligands such as dioxygen and to also prevent close approach of two iron centers. Closely related to this approach is the construction of a dome or arch over one side of the macrocyclic plane which can also produce a hydrophobic cavity and can serve to block access to larger axial ligands and prevents bimolecular interactions. Finally, lowering the temperature of solutions containing a variety of iron(II) porphyrin derivatives inhibits the rate of irreversible oxidation of the iron(II) sufficiently in some instances to enable characterization of the solution species and to perform kinetic and mechanistic studies of the dioxygenated species.

Wang,[104] in the 1950s, theorizing that an anhydrous space is necessary to prevent irreversible oxidation of iron(II) porphyrins, impregnated iron(II) porphyrins on polystyrene polymer. On the basis of spectroscopic evidence, he concluded that this system was capable of reversibly coordinating dioxygen. While this approach clearly illustrated the importance of keeping the iron centers separated, the method provided a crude approximation of native heme proteins that was not amenable to detailed characterization. Weber and Busch[105] demonstrated in 1965 that iron(II) phthalocyanines can add dioxygen reversibly in the solid state, which was also an indicator of the importance of keeping the iron centers separated. In an extension of these concepts, Chang and Traylor[106] synthesized simple heme imidazole compounds, having the same geometry as the heme imidazole complex of myoglobin, but with the imidazole covalently linked to one of the porphyrin pyrrole groups. This complex was found to reversibly bind oxygen in the solid state or when deposited on a polystyrene film. Their results were in agreement that factors such as the electronic nature of the base (imidazole), neighboring group effects of the basic group, and immobilization of the heme group are important in determining the reversibility of dioxygen binding. Interestingly, in a related study,[107] Collman and Reed observed that sufficient mobility of iron(II) porphyrins occurred in solvent-swollen, crosslinked polystyrene to permit irreversible oxidation of the iron.

The most important advances in our understanding of dioxygen complexes of iron(II) occurred in 1973 when two research groups independently synthesized iron(II) complexes capable of binding dioxygen reversibly. Baldwin and Huff[108] demonstrated the importance of providing steric interactions by synthesizing a nonporphyrin 14-membered ring macrocyclic ligand, an octaaza[14]annulene derivative. One of these utilized a dione derived from anthracene [Figure 10(a) and (b)], creating walls and a cavity 500 pm deep, to prevent close approach of the iron(II) centers. A considerable difference in the sensitivity of these iron(II) complexes to irreversible oxidation to iron(III) was noted when the nominally planar ligand utilizing cyclohexanone was substituted for the dione derived from anthracene. Whereas the first is oxidized immediately under all conditions, the latter anthracene derivative could be carried through a number of freeze–thaw cycles at $-78°C$ without appreciable iron oxidation taking place. Above $-50°C$, however, irreversible degradation of the oxygenated complex occurred. Although these results are impressive, analysis of the ligand structure reveals that it is an unlikely candidate for a good hemoglobin or myoglobin model. The reasons are summarized as follows. A fully unsaturated 14-membered ring of this type produces abnormally short iron–nitrogen distances, 183 pm, and probably favors a higher oxidation state which has shorter metal–nitrogen distances.[109] Secondly, the negative charge formally resides on a nitrogen atom coordinated to iron. Both aspects are expected to strongly stabilize the iron(III) state.

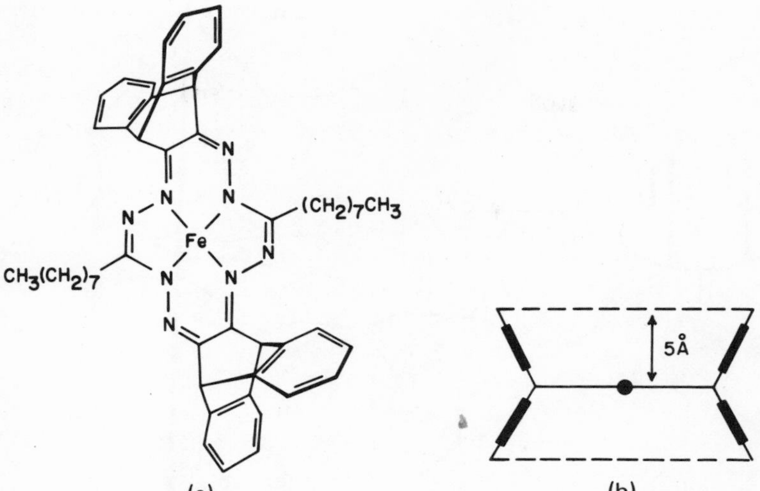

FIGURE 10. (a) The octaaza[14] annulene complex of iron designed to prevent close approach of iron centers. (b) Diagram illustrating well depth.

Nonetheless, this first observation of reversible oxygenation on a nonporphyrin iron complex in solution strongly suggested that other, better model systems could be developed.

Closely following this report, Collman and co-workers[110] described a modification of a tetraphenylporphyrin, using instead of benzaldehyde, *o*-nitrobenzaldehyde, and subsequently introduced pivalamide groups to produce the now famous "picket fence" porphyrin (Figure 11). The diamagnetic, six-coordinate complex $Fe[(TpivPP)(N-RIm)_2]$ is completely oxygenated in solution at 25°C under 1 atm of O_2, forming $Fe(TpivPP)$-$(N-RIm)(O_2)$, and found to be reversible. Total irreversible oxidation of the iron required longer than 12 hr at 25°C. A small excess of axial base was required, presumably to prevent coordination of dioxygen from the unencumbered side of the porphyrin. A single-crystal X-ray diffraction study of this complex, although of limited accuracy, has confirmed that dioxygen is bound to the iron in an angular fashion, Fe–O–O angle = 131°, O–O distance = 116(5) pm, as first proposed by Pauling[111] and later by Weiss.[112] Careful, detailed infrared studies of a number of these oxygen complexes with a variety of imidazole bases, together with comparisons with other well-characterized transition-metal complexes, have led to the conclusion that the bent dioxygen ligand is best viewed as bound superoxide, O_2^-.[89,110b] The O–O stretching frequencies of the picket fence porphyrins are in the range 1159–1163 cm^{-1}.[89] This is not far removed from the O–O stretching fre-

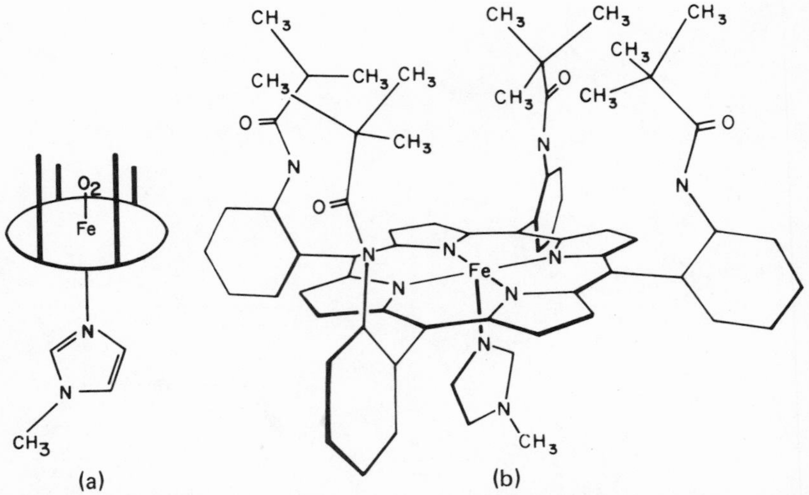

FIGURE 11. (a) Diagrammatic sketch of Collman's "picket fence" and (b) the pivalamide derivative of tetraphenyl porphyrin forming the "picket fence" porphyrin for reversible coordination of dioxygen (reproduced from reference 110 with permission).

quency of a variety of ionic superoxides that range from 1150 to 1100 cm^{-1}.[113-115] The O–O stretching frequencies of oxygenated hemoglobin and myoglobin are 1107 and 1103 cm^{-1}, respectively, virtually identical to that observed in coboglobin, 1106 cm^{-1}, which has cobalt(II) substituted for iron(II).

Solid-crystalline picket-fence oxygen complexes readily lose oxygen under vacuum, affording a high-spin deoxy derivative, a process that is reversed when the partial pressure of oxygen is increased.[110b] This remarkable solid–gas equilibrium enabled the determination of the enthalpy and entropy changes. The values of $P_{1/2}$, ΔH, and ΔS for the picket-fence porphyrin are very similar to those of myoglobins (Table 4). These data strongly support the utility of this synthetic complex for further model–system studies.

Reasoning along the same line, Baldwin and co-workers, in an elegant synthesis, prepared the capped iron(II) porphyrin depicted in Figure 12.[116,117] Exposure of this complex in pyridine to dioxygen at 25°C was accompanied by an immediate change in the spectrum, presumably a result of coordination of dioxygen. Deaeration by freeze–thawing restored the spectrum of the iron(II) complex, a process that could be repeated for many cycles without appreciable deterioration of the complex. The lifetime of the dioxygen adduct was about 20 hr. The stability of the dioxygen adduct was shown to be largely dependent on the nature and concentration of the coordinating base. Solutions of 5% 1-methylimidazole in benzene had approximately a 5-hr lifetime

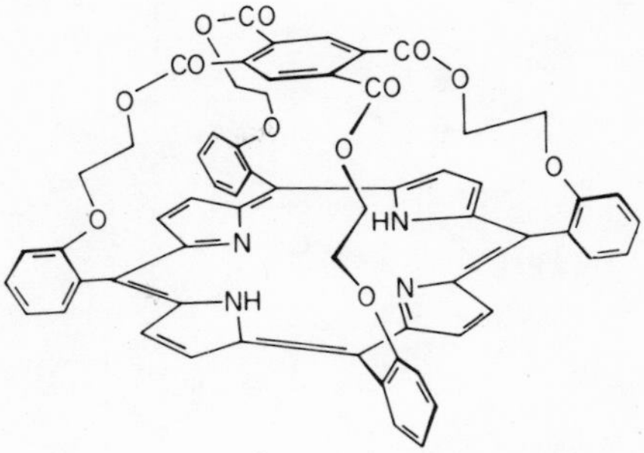

FIGURE 12. Domed porphyrin designed to form a projective cavity for reversible dioxygen coordination. Reproduced with permission from reference 116.

(25°C). In the absence of any axial base, the complex is instantaneously irreversibly oxidized by dioxygen, probably through oxygenation of the unprotected side.

Busch and co-workers have synthesized the first iron(II) complexes of neutral, nonporphyrin macrocyclic ligand which are capable of reversibly coordinating carbon monoxide.[54]* The framework selected for this ligand was one for which the redox potentials most closely matched those of hemoglobin. This 16-membered ligand (Figure 13) has access to one of the axial sites partially blocked by a *m*-xylyl group which is attached to the two six-membered chelate rings. The *N*-methylimidazole adduct has been characterized and preliminary evidence indicates that this species does interact with carbon monoxide and dioxygen under ambient conditions. Two other significant aspects to this complex are (1) the iron complex exists as a cation and thus has the potential for water solubility and (2) the Fe(II)/Fe(III) couple for this species has an $E_{1/2}$ of -0.36 V (vs. Ag/AgNO$_3$) in acetonitrile, and falls very close to that of hemoglobin (see Figure 9).

An imaginative approach to the design of linked hemes has been carried out by Traylor and co-workers.[118] In their model two hemes are covalently linked together with *meso*-1,2-di(3-pyridyl)ethylenediamine to give a molecule

* See note added in proof on page 649.

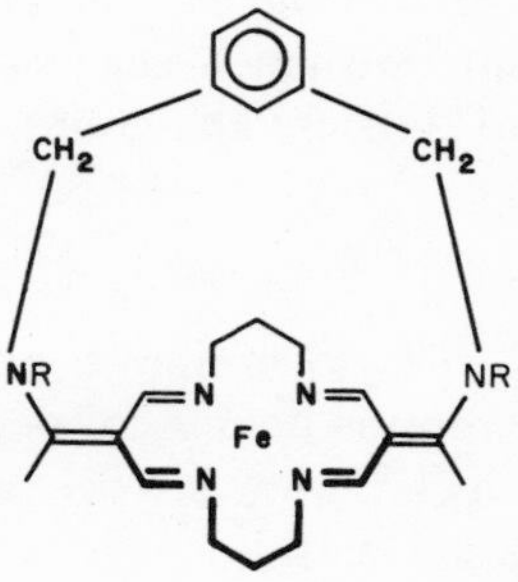

FIGURE 13. Neutral, nonporphyrin macrocycle with appended *m*-xylyl group to partially block access to one axial ligand site.

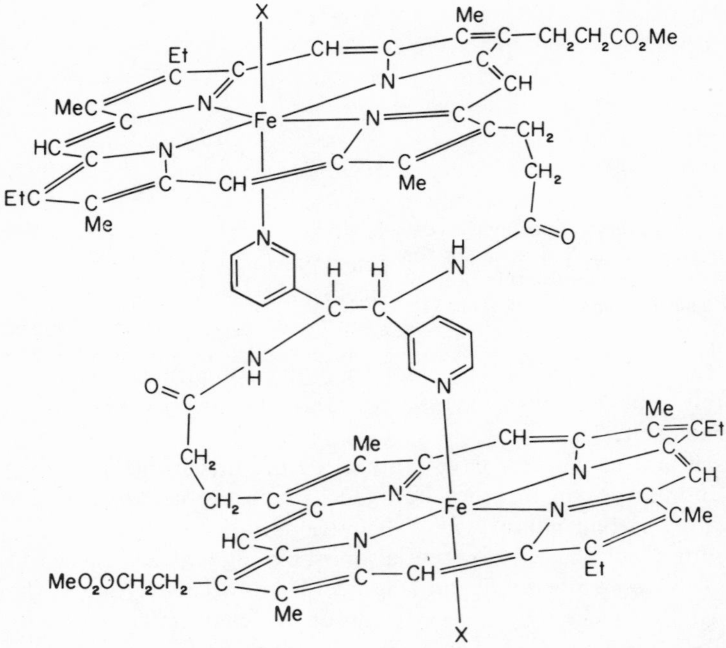

FIGURE 14. Covalently linked porphyrins designed for studying cooperativity effects. Reproduced with permission from reference 118.

having a center of symmetry as shown in Figure 14. For the five-coordinate forms of the heme rings, any change in the upper iron-containing ring conformation should affect the conformation of the lower ring. Although the five-coordinate species do reversibly bind oxygen in solution, no data are currently available for the rate constants for the two different iron atoms. However, it is important that carbon monoxide reacts with the complex with two rate constants. The first carbon monoxide reacts at a rate of 4×10^7 liter mol^{-1} s^{-1}, about ten times faster than the second carbonyl. Although the detailed mechanism has not been established, it is suggested that it proceeds through a combination of (a) base elimination to give a four-coordinate intermediate which reacts with carbon monoxide, followed by (b) reassociation of the base and coordination of the second carbonyl.

3.4. Cytochromes

Cytochromes c are defined as those cytochromes in which the heme group is bound covalently to the protein via the side chains of the porphyrin.[119] They are believed to play a role in electron transport processes. The prosthetic group and the protein are bound to one another not only via the thioether bridges, but also through coordination to the heme iron atom. The latter type of interaction is of major structural and functional importance for the molecule and has been a focal point of interest in cytochrome c research. X-ray diffraction studies of ferricytochrome c have firmly established that the coordination shell of iron is completed by the side chain of a histidine residue, and the second coordinated group is provided by the thioether of a methionine side chain.[120-122] Thus the mammalian-type cytochromes are mixed hemochromes containing sulfur and nitrogen bound to the axial heme positions. The heme group is imbedded in a crevicelike arrangement of the surrounding protein such that only an edge is exposed to the surroundings. The structure reveals that the hydrophobic regions meet the surface near the presumed binding sites for electron donor and acceptor molecules. A number of closely spaced aromatic groups lie approximately planar to one another; they may serve a function in the transfer of electrons to and from the heme groups. The heme groups of the c cytochromes profoundly influence the structure and properties of the protein component, and the latter greatly affects the behavior of the prosthetic group. The interplay between these two is not understood.

Cytochrome P-450 oxygenases catalyze the hydroxylation of C–H bonds in metabolism, hormone regulation, and drug detoxification.[123] These hemoproteins derive their name, in part, from the atypical Soret absorption which they exhibit near 450 nm when reduced and reacted with carbon monoxide. All other hemoproteins capable of interacting with CO in the reduced state have an absorption near 420 nm.

Four different states of iron in cytochrome P-450 have been identified in the catalytic cycle of cytochrome P-450 camphor hydroxylase: substrate-free low-spin $(S + \frac{1}{2})$ iron(III), substrate-bound high-spin $(S = \frac{5}{2})$ iron(III), deoxy high-spin $(S = 2)$ iron(II), and diamagnetic oxygenated iron(II). Thus considerable room exists for model studies of cytochrome P-450 both with respect to porphyrin and synthetic nonporphyrin macrocyclic ligands.[124]

No X-ray diffraction studies of cytochrome P-450 revealing the details of the coordination environment have been reported. However, the physical properties have been extensively studied, and certain features, particularly the unusual spectra of the iron(II) carbonyl complex, have aroused the interest of a number of coordination chemists. Model studies of cytochromes have focused on two aspects. One of these has centered around identification of the unusual axial ligands, which are essential to our understanding of their enzymic function. Second, a number of complexes containing the $Fe^{III}N_4SR$ unit have been synthesized to probe the properties of the donor atom arrangement and the ground state electronic features.

Some of the earliest clues to this identification came in 1969 and 1971 when it was shown that the EPR spectrum of low-spin cytochrome P-450 was similar to that obtained from the addition of mercaptoethanol or alkyl mercaptides to iron(III) hemoglobin or iron(III) myoglobin.[125] Other model studies with heminthiol complexes suggested that a necessary axial ligand for the iron in iron(III) cytochrome P-450 is a thiol, presumably from cystein.[125,127] Although the EPR spectra could be used to infer the nature of the axial ligand, this does not necessarily infer the nature of the axial ligand in the iron(II) carbonyl complex. Stern and Peisach in 1974[128] reported the successful preparation of heme carbon monoxide model compound having absorption maxima at 450 and 555 nm from hemin chloride, and demonstrated the dependence of these absorptions of iron(II) heme, sulfur as mercaptide, and carbon monoxide.

Collman et al_j, working with the synthetic porphyrin tetraphenyl porphyrin, have been successful in isolating a five-coordinate oxygen-sensitive iron(III) complex, $Fe(TPP)(SC_6H_5)$, which yielded an EPR spectrum similar to substrate-bound high-spin P-450.[129] This complex was rapidly reduced to the iron(II) complex $Fe(TPP)B_2$, and also formed the disulfide $(C_6H_5S)_2$. However, at low temperatures they demonstrated the existence of a metastable species, $Fe(TPP)(SC_6H_5)(B)$. This species was very unstable in solution at 25°C but could be dramatically stabilized by immobilization when the axial imidazole ligand was N-bonded to a cross-linked polystyrene. Similar observations have been made by Holm and co-workers.[130] These results could provide an explanation for the membrane stabilization of cytochrome P-450's. In related work they observed that their picket fence porphyrin $Fe(TpivPP)$ (Figure 11) reacts with sodium thiomethanolate and carbon monoxide to yield a Soret band at 449 nm.[131] The absorbance was quite sensitive to the polarity of the solvent and to the nature of the porphyrin. A significant observation in this work was the appearance of the carbon monoxide stretching frequency at 1945 cm^{-1}, lower than for any other known carbonyl complexes of iron porphyrins. The carbon monoxide stretching frequency for the native enzyme has not been reported.

Collman and co-workers also examined the magnetic circular dichroism of iron(II) carbonyl adducts of cytochromes P-450 and P-420.[132] Model systems derived from iron(II) porphyrins with sodium methylmercaptide and carbon monoxide in benzene had MCD and absorption spectra almost identical to those of cytochrome P-450, indicating that a mercaptide serves as the fifth ligand in the iron(II) carbonyl of P-450. The MCD spectra of iron(II) porphyrins and carbon monoxide with either propylmercaptan or N-methylimidazole as the remaining axial ligand were identical with that of P-420, making it impossible to make unambiguous assignments of the axial ligands in this case. Axial mercaptide bases are polarizable electron donors; when $trans$ to carbon monoxide, a good π acid, a net drift of electron density through the iron to the carbonyl is expected. This is confirmed by a shift of the carbon monoxide stretching frequency from 1964 to 1945 cm^{-1} when the axial base is changed from N-methylimidazole to methylmercaptide ion.

Conclusions similar to those of Collman *et al.* have been drawn by Chang and Dolphin in a related study.[133]

Holm and co-workers examined a series of [$Fe^{III}N_4SR$] coordination units, where N_4 represents the donor atoms of the 14-, 15-, and 16-membered ligands, Nos. 4, 7, and 9 in Figure 6.[134] Particularly significant is the fact that this series of complexes encompasses three spin states. From their studies the following conclusions were drawn: (1) Porphyrin complexes of $Fe^{III}N_4SR$ formulation are high spin and can only be isolated when more electronegative (aromatic) thiolate ligands occupy the axial position. (2) The magnetic properties of the [$Fe^{III}N_4R$] unit are dependent upon both the nature of R and the macrocyclic ligand. Increasing the size of the macrocyclic ring favors stabilization of high-spin states due to larger "core" size, i.e., ideal metal–nitrogen distances, and attendant decrease in ligand field strength.[135] (3) The uncommon quartet ground state is readily achieved in stable macro-cyclic complexes.

4. Binuclear Systems

Metal–metal interactions are critical to the function of a number of elaborately constructed biological systems. These include (a) hemerytherin, (b) superoxide dismutase, (c) chlorophyll dimers in photosynthetic units, (d) binuclear manganese complexes in the photosynthetic oxidation of water to give dioxygen, (e) cytochrome oxidase for the multielectron reduction of dioxygen, (f) monooxygenases for the activation of dioxygen by two-electron transfer, and (g) others, some yet to be discovered. The appreciation of the finesse with which these systems carry out their functions is yet another example of the stimulus provided by biological systems for the development of new ligand systems to test current theories and develop new ones. For binuclear systems, fundamental questions concern the nature of metal–metal interactions in general, and the nature of energy or electron transfer from one monomer unit to another in particular. Fundamental to understanding these phenomena are knowledge of the perturbation of the energy levels of one unit by the other, the rate of electron transfer, and the redox potential for electron transfer. These are largely determined by the nature of the macro-cyclic ligands from which the dimers are constructed, by the distance between ligands, effect of bridging ligands, influence of the spin state of the metal, etc.

The information derived from these investigations will undoubtedly extend beyond the original objectives. Studies involving related systems employing ruthenium–ruthenium interactions or molybdenum–molybdenum interactions could prove useful as models for the binding and reduction of dinitrogen. Dimer systems containing metals such as rhodium or iridium may have applications in organometallic chemistry.

Coupling of two macrocyclic ligands together serves two major purposes: (1) It provides a means of keeping two metal ions in the vicinity of one another and (2) it provides a rational starting point for controlling the coordination environment of metal ions, particularly those that are labile with all but macrocyclic ligands. The coupling of ligands can obviously be accomplished in many different ways. The simplest involves a single point of attachment between two ligands. Although the metal centers are held in close proximity, a considerable amount of undesired flexibility results with respect to metal–metal distances. More control over the metal–metal separation can be exercised with multiple attachments between two macrocyclic ligands which lead to more rigid frameworks. By varying the number and type of bridging atoms, metal–metal separations may be varied as desired. The two limiting types of metal–metal interactions (through-space interactions and through-bonding interactions, i.e., inductive effects) are dependent on the nature of the macrocyclic bridges, the presence or absence of ligands bridging the axial sites of the metals, and the particular metals involved.

4.1. Cofacial Diporphyrins

Most synthetic binuclear systems have employed a porphyrin nucleus upon which to construct another porphyrin molecule or set of nonporphyrin donor atoms. Attention with cofacial porphyrins has been twofold: (a) spectral properties of the complexes and their relationship to chlorophyll pair interactions and (b) metal–metal interactions.

Collman and co-workers have linked two tetraphenylporphyrin derivatives together to form "face-to-face" dimers [Figure 15(c)] and the more flexible dimers *exo–cis* DUBP and *endo–cis* DUPB [Figure 15(a),(b)].[136] Cobalt(II) and copper(II) have been successfully inserted into these ligands and the complexes characterized by EPR, visible spectra, and PMR. For both *endo–cis* DUBP and *trans*-DUPB complexes where the ligands have the face-to-face conformation, the EPR spectra of Cu(II) and Co(II) complexes show clear evidence for metal–metal interactions. The positions of the $\pi–\pi^*$ Soret peaks are shifted 15–17 nm to shorter wavelength, and have been interpreted in terms of porphyrin–porphyrin interactions. The internal pyrrole N–H's are shifted 1.4 ppm upfield for *endo–cis* DUBP and for *trans*-DUBP, compared to monomeric porphyrins, consistent with the ring current shifts expected in face-to-face porphyrins. The relatively recent nature of these investigations and their preliminary nature may not be of direct assistance in extensions to biological systems, but provide an important beginning to further extend and refine these investigations.

Chang, in some elegant syntheses, has prepared a number of cofacial diporphyrins that are bridged through two pyrrole groups, as illustrated in Figure 16.[137] The length of the chains bridging the two porphyrin units

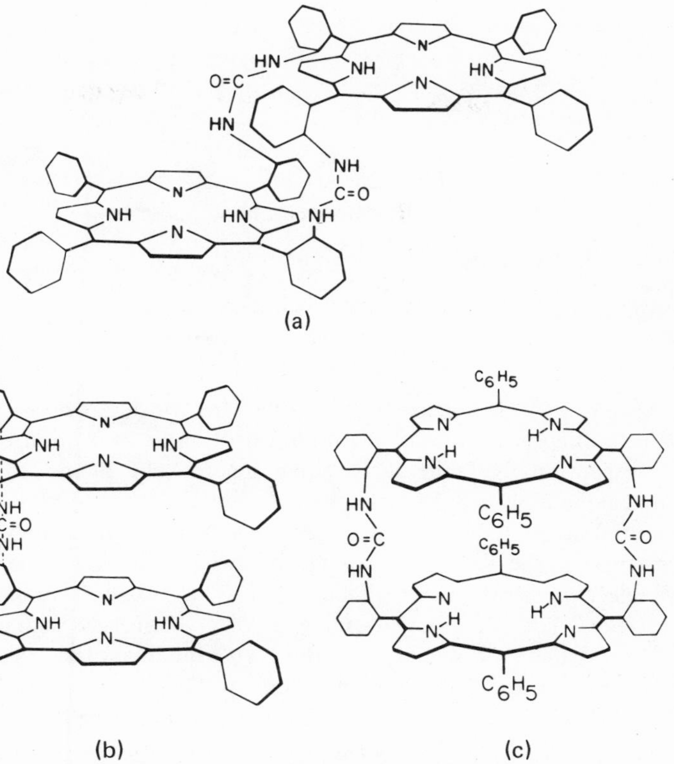

(a)

(b)   (c)

FIGURE 15. Dimeric porphyrins (a) exo–*cis* DUBP, (b) endo–*cis* DUPB, and (c) *trans*-DUPB face-to-face porphyrins. Reproduced with permission from reference 136.

was varied to achieve interplanar separations between 420 and 640 pm. The method of synthesis was such that only one or two different metals could be inserted. Diporphyrins have been prepared with $Cu-H_2$, Fe–Cu, $Mg-H_2$, Mg–Mg, and Fe–Mg occupying the core. The stability of most of the dimetal complexes appears to be similar to the monomers, with the exception of the Mg–Mg system for which one Mg is easily expelled to give the $Mg-H_2$ dimer. Excitation coupling between the two porphyrin rings produces a blue shift of the Soret bands, a red shift of the visible bands, and a red tail of the Soret band extending out to 500 nm. The magnitude of these shifts varied with interplanar separation and with the coordinated metals.

Partial electrolysis of the Mg–Mg diporphyrin yielded a solution thought to contain $Mg-Mg^+$ monocation radicals. EPR measurements on these solutions revealed a single line, $g = 2.003$, with an extremely narrow line-

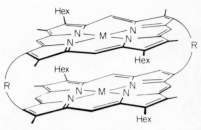

FIGURE 16. Cofacial porphyrins with variable interplanar separations. Reproduced with permission from reference 137.

$\underset{\sim}{1}$ R = $-CH_2CH_2CON(n-Bu)CH_2CH_2CH_2-$, d = 6.4 Å
$\underset{\sim}{2}$ R = $-CH_2CON(n-Bu)CH_2CH_2CH_2-$, d = 5.4 Å
$\underset{\sim}{3}$ R = $-CH_2CON(n-Bu)CH_2CH_2-$, d = 4.2 Å

width of only 1.05 G. The extremely narrow linewidth has been speculated to result from oligomerization of the monocation radicals in solution, a process that is exceptionally facile, with some macrocyclic ligands derived from *o*-phenylenediamine and 2,4-pentanedione.[138]

Multielectron reduction of dioxygen may be easily achieved in metal diporphyrins if the dioxygen can be sandwiched between the metals and receive electrons from them (Figure 17). The cobalt(II) complex of the diporphyrin **3** (Figure 16), 420-pm gap, reacts instantly with dioxygen to produce a diamagnetic species consistent with the formulation as a μ-peroxo [Co–O$_2$–Co] species. The EPR spectrum of the paramagnetic species obtained by oxidation with a trace of iodine consists of a well-defined isotropic spectrum comprised of 15 lines expected for a μ-superoxo dicobalt complex involving two equivalent ^{59}Co nuclei.

Addition of dioxygen to iron(II) porphyrins of the diporphyrin **3** resulted in instantaneous oxidation of the heme at $-45°C$. Kinetic measurements indicated that the rate of auto-oxidation of this *bis*-heme was at least 10^3 times faster than monomeric myoglobin. The rate-limiting step in monomeric iron(II) hemes is the formation of the μ-peroxo complex, but in **3** the favorable position of the two iron(II) atoms leads to a lowering of the barrier to dioxygen reduction and an increase in the oxidation rate.

Boxer and Closs[139] and Katz and co-workers[140] have synthesized dimeric chlorophyll molecules via area linkages that are held in a cofacial position by hydrogen-bonding interactions. Substantial spectral alterations

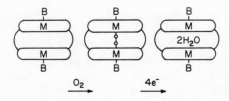

FIGURE 17. Schematic representation of dioxygen reduction with cofacial porphyrins. Reproduced with permission from reference 137.

FIGURE 18. Binuclear cryptand incorporating the tripod, nitrogen donor ligand, tren (from reference 146).

were observed. Ogoshi,[141] Kagen,[142] and others[143] have also synthesized cofacial porphyrins closely related to those of Chang and Collman. However, detailed studies of metal–metal interactions in these new ligand systems have not yet been reported.

Lehn and co-workers have synthesized a large number of elegant binuclear cryptates incorporating two macrocycles with cations coordinated to each macrocycle.[143–145] One of the most promising nitrogen systems for bioinorganic and catalysis developments incorporated two molecules of tren, $N(CH_2CH_2NH_2)_3$ (Figure 18).[146] Binuclear complexes of zinc(II), copper(II), and cobalt(II) have been isolated and characterized. The interaction distances are estimated to be approximately 450 pm. The magnetic moment of the copper complex, $\mu_{eff} = 2.0\,\mu_B$ per copper at 294 K and the EPR spectrum have been interpreted as indicating weak copper–copper interactions. Evidence has also been accumulated to indicate that substrates such as water, cyanide, and azide bridge the two metals as shown. Considerable potential exists in these systems for enhanced cooperativity effects between the two metals.

4.2. Unsymmetrical Binuclear Systems

Recent communications deal with two types of unsymmetrical binuclear complexes. Chang "crowned" a porphyrin nucleus with a crown ether (Figure 19) which is able to coordinate a transition metal and a Group 1A or Group 2A cation simultaneously.[147] The size of the cavity between the porphyrin and the crown ether was estimated to be about 600 pm from CPK models and probed by examining the properties of the iron(II) complex. In

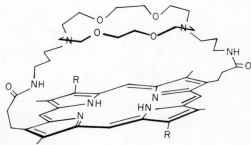

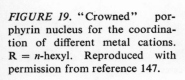

FIGURE 19. "Crowned" porphyrin nucleus for the coordination of different metal cations. R = *n*-hexyl. Reproduced with permission from reference 147.

the presence of imidazole the iron(II) complex binds dioxygen at room temperature with a half-life of 3 min, indicating that dioxygen as well as imidazole ligands are able to enter the cavity. In the presence of a larger axial base such as 1-triphenylmethylimidazole, the oxygenated species is stable at 25°C with a half-life greater than 1 hr and indicates that the N-base can coordinate only to the "free" side with the dioxygen in the cavity being protected from bimolecular oxidation by the crown ether. Preliminary NMR experiments indicated that Na and Cs cations are coordinated by the crown ether portion of the molecule.

The second type of unsymmetrical *bis*-metallosystem was reported almost simultaneously by Buckingham[148] and Elliott.[149] The basic ligand framework, illustrated in Figure 20, is derived from Collman's picket fence porphyrin by replacing the pivalyl groups with *meta*-substituted pyridyl functions. The pyridine groups, although free to rotate, can orient to form a near-planar array of donor atoms above the porphyrin plane. This arrangement allows the insertion of the same or different metal ions into the two different coordination sites. A number of mono- and *bis*-metal complexes have been characterized. The mono Cu(II), Ni(II), Fe(II) and Fe(III), and Cu(III) complexes have been isolated with the metal coordinated to the porphyrin. *Bis*-metal complexes that have been examined include Cu–Cu, Ni–Ni, Ni–Cu, Fe(III)–Cu, Fe(III)–Fe(III), and Cu(II)–Co(III), with the first and second metals given being coordinated to the porphyrin and the pyridine units, respectively. The metals coordinated to the pyridines are more labile and can generally be removed by treatment with dilute acid.

The visible spectra of the *bis*-metallosystems are dominated by typical monometalloporphyrin chromophore with little change observed when a

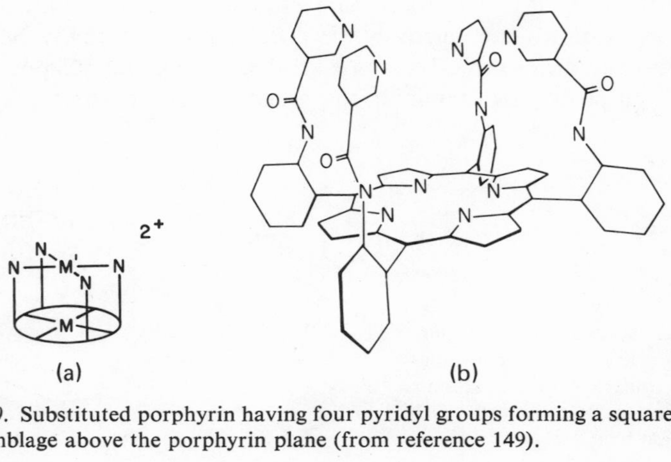

(a)   (b)

FIGURE 20. Substituted porphyrin having four pyridyl groups forming a square-planar donor assemblage above the porphyrin plane (from reference 149).

second metal is inserted into the cap. However, the EPR spectra of the *bis*-Cu system shows marked Cu⋯Cu interactions. The internuclear separation of 590 pm, estimated from the EPR parameters, was in good agreement with that estimated from Dreiding stereomodels. Addition of copper(II) to a solution of the mono-iron(III) complex results in the gradual disappearance of the iron signal with no new signals appearing due to copper(II) and demonstrates the strong coupling of the $S = \frac{5}{2}$ and $S = \frac{1}{2}$ spin systems in the $[Fe(III)–Cu(II)]^{2+}$ system.

The groups of Lehn[150] and Weiss[151] have synthesized binuclear

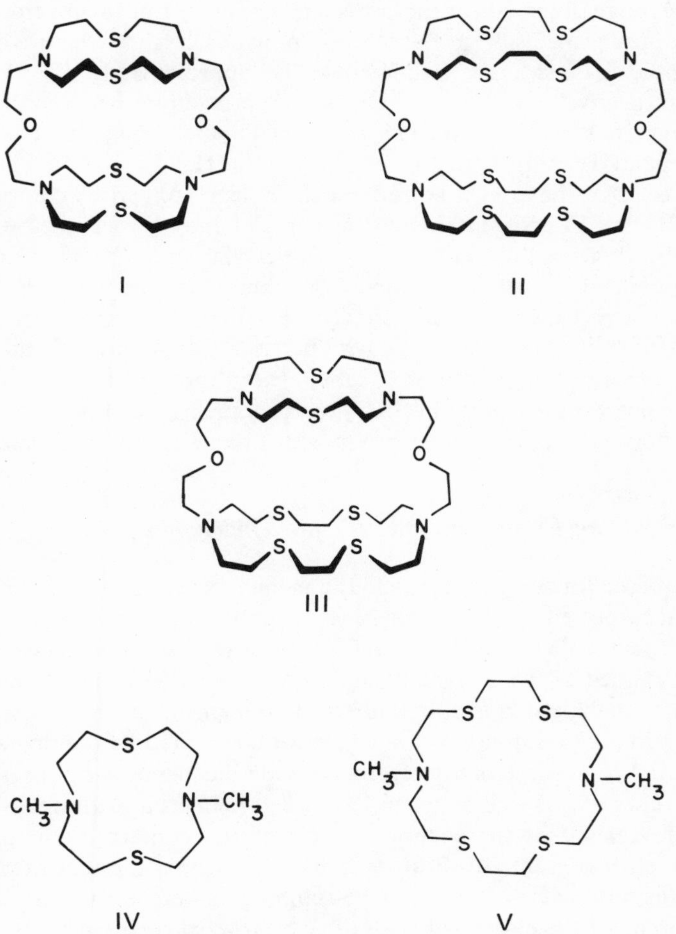

FIGURE 21. Heteromacrotricyclic ligands for forming symmetrical and unsymmetrical binuclear complexes.

copper(II) complexes of macrotricyclic ligands. Lehn *et al.* report a general synthetic procedure for the construction of macrotricycles which contain different macrocyclic subunits that have potential for stabilizing different cations or different oxidation states. The macrocycles, depicted in Figure 21, contain nitrogen- and sulfur-binding sites. The copper complexes of ligands I–III may be considered binuclear models of the copper proteins like the simple macrocyclic thiaether complexes IV and V are models for mononuclear copper systems. Electrochemical reduction of the copper(II) complex of I (Figure 21) indicates the reversible transfer of two electrons at + 445 mV vs. SHE electrode. The markedly positive potentials for this complex fall in the range observed for the copper(II) complexes of thiamacrocycles [22] and of copper proteins.[21,30] Based upon previous crystal structure data, the intercationic distances are estimated to be approximately 500, 700, and 500 pm, respectively, for I, II, and III. Binuclear reactions between the copper ions is suggested by the spectral shifts obtained upon adding KO_2 to solutions of the copper(II) complex of I or of dioxygen to II.

Weiss *et al.* have synthesized the binuclear copper(II) complex of I (Figure 21).[151] Violet solutions of this complex have a very intense absorption at 550 nm ($\varepsilon = 1200\ M^{-1}\ cm^{-1}$). The crystal structure of this complex reveals a copper–copper separation of 561 pm. This together with the EPR spectral data reveal little if any direct interaction between the two copper centers. However, there is a large cavity between the two copper cations that should permit insertion of diatomic substrates. Although there is no evidence for such a process the addition of KO_2 to the copper(II) complex or dioxygen to the copper(I) complex yields species with similar electronic spectra.

5. Comments on Vitamin B_{12} and Related Inorganic Systems

The search for the "antipernicious anaemia factor" started in 1926 when it was first suspected and lasted for 20 years, until it was isolated as crystalline vitamin B_{12} in 1948.[152,153] The availability of crystalline material, the fact that it contained a "heavy-metal" ion, and the number of atoms per asymmetric unit challenged crystallographers to attempt one of the first structural determinations of an important biochemical compound containing a transition metal. The upsurge in interest of not only biochemists, but organic and inorganic chemists as well, dates to 1961 when Lehnert and Hodgkin's X-ray analysis revealed that the coenzyme form of B_{12} contained a cobalt(III)–carbon bond[4] (Figure 22). Prior to 1961, it had been thought that higher-valent transition-metal σ-bonded carbon complexes were intrinsically unstable and apparently little effort had been directed toward their synthesis.

The challenge of the total synthesis of B_{12} attracted the best talent in the world (Eschenmoser in Switzerland, Cornforth in Britain, and Woodward in

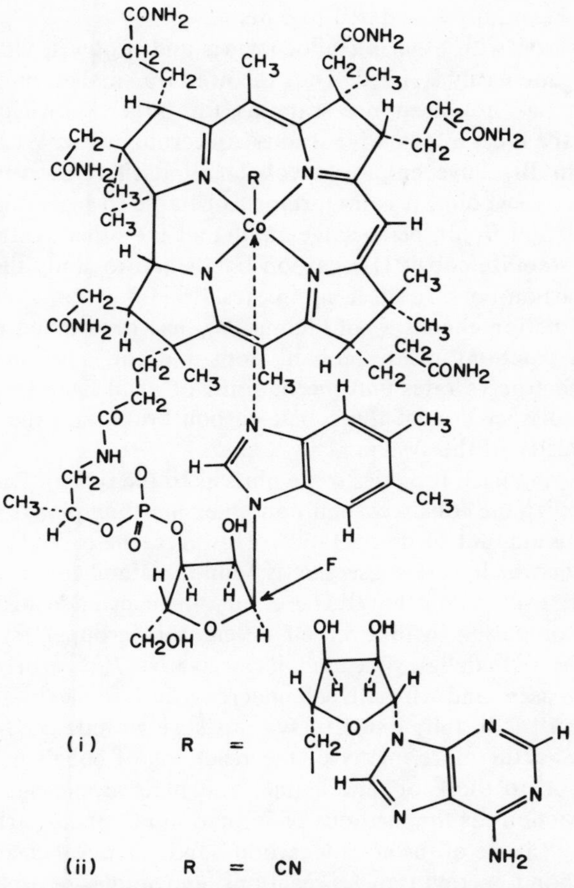

FIGURE 22. Structure of vitamin B_{12} coenzyme, 5′deoxyadenosylcobalamin.

the United States).[154,155] To the biochemist, the interest in vitamin B_{12} arose because of its ability to interact with a wide variety of substrates and to carry out some of the most remarkable chemical transformations known. The unexpected discovery of the cobalt(III)–carbon bond struck the inorganic and organometallic chemists as the most unusual aspect, and their initial efforts were largely relegated to exploring the range of systems which stabilized cobalt(III)–carbon bonds.

About a dozen enzymatic reactions are known that require corrinoids.[156] These have been classified into (1) reactions which involve the transfer of a methyl group and which do not require 5-deoxyadenosylcorrinoids and (2) reactions which appear to involve transfer and which require

5-deoxyadenosylcorrinoids. The mechanistic details for none of these enzymic reactions have been fully elucidated to date.

By comparison with other metalloenzymes and proteins, vitamin B_{12} is relatively small and easily isolated. Thus the need for smaller, more tractable model systems has not been necessary as for larger, unwieldy systems. Consequently, the most informative studies concerning the role and mechanisms of vitamin B_{12} have employed cobalamin itself. However, from an overall chemical viewpoint, it is important to be able to place the inorganic reactions of vitamin B_{12} in perspective among all the other synthetic ligand systems which stabilize cobalt(III)–carbon bonds and to study the reactivity of the cobalt carbon bond in these complexes.

The coordination chemistry of vitamin B_{12} has been taken to be those properties and reactions of the cobalt atom and the axial ligands. This includes such factors as rates and mechanisms of axial ligand substitution, formation and dissociation of the cobalt–carbon bond, and the oxidation–reduction chemistry of the system.

The corrin ring itself provides some hints as to the type of ligand system that might stabilize the cobalt–carbon bond, i.e., a strong planar ligand field with a moderate amount of unsaturation. However, the exceedingly diverse nature of ligands which were subsequently found to stabilize the cobalt(III)–carbon bond is truly astonishing. These vary from cyanide ligands, as in $Co(CN)_5R^{3-}$ complexes (with over 60 different R groups),[157] bidentate chelates such as dimethylgloxime and its derivatives,[158] nearly all tetradentate Schiff bases, and virtually all macrocyclic complexes—from fully saturated cyclam[159] to fully oxidized systems such as tetraaza[14]annulene ligands.[160,161] Furthermore, many of the reactions of cobalt in the corrin ring are common to those of tetradentate and macrocyclic ligands. These include such reactions as the methods of formation of cobalt–carbon bonds, photochemical cleavage of the cobalt–carbon bond, chemical cleavage of the cobalt–carbon bond, methyl transfer reactions, and studies of exchange rates of axial ligands. The dimethylglyoximato complexes of cobalt are probably the most extensively studied for the purposes of comparison with the cobalamin system. Despite the great disparity in the nature of the *bis*-dimethylglyoximato ligands compared to the corrin ring, a remarkable one-to-one correspondence of the inorganic chemistry of the two systems has been documented.[158] The inorganic chemistry of vitamin B_{12} has been comprehensively reviewed in a number of places,[156,162,163] as has the organometallic chemistry of cobalt(III) [157,164] Macrocyclic ligands have played a significant role in a number of these studies. However, there is no sharp line of demarcation separating fundamental studies of macrocyclic complexes of cobalt from those of the innumerable tetradentate Schiff-base and related ligands. Therefore, for a treatment of the reactivity of macrocyclic complexes of cobalt, the reader is referred to Chapter 6. For more detailed treatment, the reader is referred to the reviews cited above.

Note Added in Proof

Recent X-ray structural studies by Busch and co-workers have demonstrated the existence of two isomers of the ligand shown in Figure 13.[165] Alternate connections of the bridgehead nitrogen produces two isomers, a "lid-on" isomer for the ligand with R=H in which the *m*-xylyl cap projects directly over the metal and a "lid-off" isomer for the case of R=CH$_3$ in which the cap is displaced toward the rear. This isomerism creates cavities of different sizes and dimensions; the "lid-on" isomer is 7.5 Å high and 5.05 Å wide whereas the "lid-off" isomer is only 5.0 Å high but 7.34 Å wide. Small molecules such as CO and O$_2$ should be reasonably well accommodated by the former cavity and be more restricted in the latter. Preliminary equilibrium constants for carbon monoxide formation of the Fe(II) complexes of this ligand with CH$_3$CN as the axial base reveal that K for the "lid-on" isomer is three orders of magnitude greater than for the "lid-off" isomer.

In related studies, with macrocyclic ligands containing bridges, Busch and co-workers have investigated the oxygen-binding capabilities of a number of cationic cobalt(II) complexes of the macrocyclic ligands represented in Figure 23.[165] The bridging methylene chains function to limit access of bound O$_2$ molecules from solvent interactions and to prevent the formation of dimeric peroxo-bridged species. Extensive equilibrium data for the series of compounds in Figure 23a–e in acetonitrile solution have been obtained and demonstrate the profound effect of the bridging group on the oxygen-binding ability of the cobalt complexes. $K_{eq^2}^0$ for Figure 23e as an *N*-methylimidazole adduct in aqueous solution is 1.6 ± 0.1 torr^{-1} at 20°C, which may be compared with the $K_{eq^2}^0$ for cobalt myoglobin of 0.030 torr^{-1}. This 1:1 adduct for the Co(II) complex of Figure 23e is the most thermodynamically stable 1:1 adduct reported to-date. The pronounced effect of the bridging group is

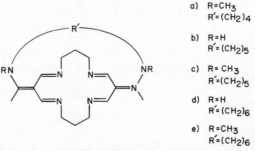

a) R=CH$_3$
   R'=(CH$_2$)$_4$

b) R=H
   R'=(CH$_2$)$_5$

c) R=CH$_3$
   R'=(CH$_2$)$_5$

d) R=H
   R'=(CH$_2$)$_6$

e) R=CH$_3$
   R'=(CH$_2$)$_6$

FIGURE 23. Neutral, nonporphyrin macrocycles with bridging —(CH$_2$)$_n$— groups that partially block access to one of the axial coordination sites.

demonstrated by the observation that the O_2-binding constant of Figure 23a is over five orders of magnitude smaller than that for Figure 23e.

References

1. M. J. S. Dewar, *Science* **187**, 1037 (1975); *Science* **190**, 591 (1975).
2. S. A. Latt, and B. L. Vallee, *Biochemistry* **10**, 4263 (1971), and references cited therein.
3. J. H. Wang, *Acc. Chem. Res.* **1**, 90 (1970).
4. P. G. Lenhert and D. C. Hodgkin, *Nature* **192**, 937 (1961).
5. P. H. Merrell, V. L. Goedken, D. H. Busch, and J. A. Stone, *J. Am. Chem. Soc.* **92**, 7590 (1970).
6. J. L. Hoard, G. H. Cohen, and M. D. Glick, *J. Am. Chem. Soc.* **89**, 1922 (1967).
7. V. L. Goedken, J. A. Molin-Case, and G. G. Christoph, *Inorg. Chem.* **12**, 2894 (1973).
8. M. F. Perutz, M. G. Rossman, A. F. Cullis, M. Muirhead, G. Will, and A. C. T. North, *Nature* **185**, 416 (1960); J. C. Kendrew, R. E. Dickerson, B. E. Strandberg, *et al.*, *Nature* **185**, 422 (1960).
9. J. L. Hoard, *Science* **174**, 1295 (1975); J. L. Hoard, in: *Porphyrins and Metalloporphyrins* (K. M. Smith, ed.), p. 317, Elsevier Scientific Publishing Company, New York (1975).
10. M. F. Perutz and L. F. TenEyck, *Cold Spring Harbor Symp. Quant. Biol.* **36**, 295 (1972).
11. J. L. Hoard and W. R. Scheidt, *Proc. Natl. Acad. Sci. USA* **70**, 3919 (1973); *Proc. Natl. Acad. Sci. USA* **70**, 1578 (1974).
12. J. P. Collman, R. R. Gagne, C. A. Reed, W. T. Robinson, and G. A. Rodley, *Proc. Natl. Acad. Sci. USA* **71**, 1326 (1974).
13. B. C. Pressman, *Fed. Proc. Fed. Am. Soc. Exp. Biol.* **27**, 1283 (1968), and references cited therein.
14. C. J. Pedersen, *J. Am. Chem. Soc.* **89**, 7017 (1967).
15. M. R. Truter, *Struct. Bonding (Berlin)*, **16**, 71 (1973).
16. B. T. Kilbourn, J. D. Dunitz, L. A. R. Pioda, and W. Simmon, *J. Mol. Biol.* **30**, 559 (1967); M. Dobler, J. D. Dunitz, and B. T. Kilbourn, *Helv. Chim. Acta* **52**, 2573 (1969).
17. J.-M. Lehn, *Struct. Bonding (Berlin)* **16**, 3 (1973).
18. J.-M. Lehn and J.-P. Sauvage, *J. Am. Chem. Soc.* **97**, 6700 (1975).
19. J.-M. Lehn, and J.-P. Sauvage, *Chem. Commun.*, 440 (1971).
20. W. H. Muller, *Naturwissenschaften* **57**, 248 (1970).
21. R. Malkin and B. G. Malmstrom, *Adv. Enzymol. Relat. Subj. Biochem.* **33**, 177 (1970); R. Malkin, in: *Inorganic Biochemistry* (G. L. Eichhorn, ed.), Vol. 2, p. 689, Elsevier, New York (1973).
22. T. E. Jones, B. D. Rorabacher, and L. A. Ochrymowycz, *J. Am. Chem. Soc.* **97**, 7485 (1975).
23. E. R. Dockal, T. E. Jones, W. F. Sokol, R. J. Engerer, D. B. Rorabacher, and L. A. Ochrymowycz, *J. Am. Chem. Soc.* **98**, 4322 (1976).
24. M. D. Glick, D. P. Gavel, L. L. Diaddario, and D. B. Rorabacher, *Inorg. Chem.* **15**, 1190 (1976).
25. R. Osterberg, *Coord. Chem. Rev.* **12**, 309 (1974).

26. B. L. Vallee and W. E. C. Wacker, *The Proteins*, 2nd ed., Vol. V, pp. 100–102 (H. Neurath, ed.), Academic Press, New York (1970).
27. H. B. Gray, *Adv. Chem. Ser.* **100**, 365 (1971).
28. O. Siiman, N. M. Young, and P. R. Carey, *J. Am. Chem. Soc.* **96**, 5583 (1974).
29. V. Miskowski, S. P. W. Tang, T. G. Spiro, E. Shapiro, and T. H. Moss, *Biochemistry* **14**, 1244 (1975).
30. A. Brill, R. B. Martin, and R. J. P. Williams, in: *Electronic Aspects of Biochemistry*, p. 519 (B. Pullman, ed.), Academic Press, New York (1964).
31. R. J. P. Williams, *Inorg. Chim. Acta Rev.* **5**, 137 (1971).
32. O. Siiman, N. M. Young, and P. R. Caray, *J. Am. Chem. Soc.* **98**, 744 (1976).
33. E. I. Solomon, P. J. Clendening, and H. B. Gray, *J. Am. Chem. Soc.* **97**, 3878 (1975).
34. N. F. Curtis, *Coord. Chem. Rev.* **1**, 3 (1968).
35. D. H. Busch, *Rec. Chem. Prog.* **25**, 107 (1964).
36. F. V. Lovecchio, E. S. Gore, and D. H. Busch, *J. Am. Chem. Soc.* **96**, 3109 (1974).
37. L. Y. Martin, L. J. DeHayes, L. J. Zompa, and D. H. Busch, *J. Am. Chem. Soc.* **96**, 4046 (1974).
38. S. W. Hawkinson and E. B. Fleischer, *Inorg. Chem.* **8**, 2402 (1969).
39. V. Katovic, L. T. Taylor, and D. H. Busch, *J. Am. Chem. Soc.* **27**, 2122 (1969).
40. J. V. Moore and H. N. Dunning, *Indus. Chem. Eng.* **47**, 1440 (1955); G. C. Hodgson and B. L. Baker, *Chem. Geol.* **2**, 197 (1967).
41. C. Milton, personal communication.
42. G. N. Schrauzer, *Acc. Chem. Res.* **1**, 97 (1968).
43. D. H. Busch, *Acc. Chem. Res.* **11**, 392 (1978).
44. G. Gordon, S-M. Peng, and V. L. Goedken, *Inorg. Chem.* **17**, 3578 (1978).
45. V. L. Goedken and S-M. Peng, *J. Chem. Soc. Chem. Commun.*, 258 (1975).
46. V. L. Goedken and G. Gordon, to be published.
47. S. C. Tang and R. H. Holm, *J. Am. Chem. Soc.* **97**, 3359 (1975).
48. D. H. Busch, *Helvetica Chimica Acta, Fasiculus extraordinarius, Alfred Werner Commemorative Volume*, p. 174 (1967).
49. D. H. Busch, *Adv. Chem. Ser.* **100**, 44 (1971).
50. A. M. Tait, F. V. Lovecchio, and D. H. Busch, *Inorg. Chem.* **16**, 2206 (1977).
51. L. Y. Martin, C. R. Sperati, and D. H. Busch, *J. Am. Chem. Soc.* **99**, 2968 (1977).
52. Y. Hung, L. Y. Martin, S. C. Jackels, A. M. Tait, and D. H. Busch, *J. Am. Chem. Soc.* **99**, 4029 (1977).
53. D. G. Pillsbury and D. H. Busch, *J. Am. Chem. Soc.* **98**, 7836 (1976).
54. D. H. Busch, D. G. Pillsbury, F. V. Lovecchio, A. M. Tait, Y. Hung, S. Jackels, M. C. Rakowski, W. P. Schammel, and Y. L. Martin, in: *Electrochemical Studies of Biological Systems* (D. T. Sawyer, ed.), ACS Symposium Series, Vol. 38, American Chemical Society, Washington, D.C. (1977); W. P. Schammel, Ph.D. Thesis, Ohio State University (1976).
55. E. I. Ochiai, *J. Inorg. Nucl. Chem.* **35**, 3375 (1973).
56. I. A. Cohen and W. S. Caughey, *Biochemistry* **7**, 636 (1968).
57. J. O. Alben, W. H. Fuchsman, C. A. Beaudreau, and W. S. Caughey, *Biochemistry* **7**, 624 (1968).
58. G. S. Hammond and C. H. S. Wu, *Adv. Chem. Ser.* **77**, 186 (1968).
59. A. E. Martell and M. Calvin, *Chemistry of the Metal Chelate Compounds*, pp. 336–357, Prentice-Hall, Englewood Cliffs, N.J. (1952).
60. C. Floriani and F. Calderazzo, *J. Chem. Soc. A* **1969**, 946 (1969).
61. G. Henrici-Olivé and S. Olivé, *Angew. Chem. Int. Ed. Engl.* **13**, 29 (1974).
62. A. L. Crumbliss and F. Basolo, *Science* **164**, 1168 (1969).
63. A. L. Crumbliss and F. Basolo, *J. Am. Chem. Soc.* **92**, 55 (1970).

64. B. M. Hoffman, D. L. Diemete, and F. Basolo, *J. Am. Chem. Soc.* **92**, 61 (1970).
65. B. M. Hoffman, T. Szymanski, and F. Basolo, *J. Am. Chem. Soc.* **97**, 673 (1975).
66. B. S. Tovrog, D. J. Kitko, and R. S. Drago, *J. Am. Chem. Soc.* **98**, 5144 (1976).
67. J. H. Burness, J. G. Dillard, and L. T. Taylor, *J. Am. Chem. Soc.* **97**, 6080 (1975).
68. W. A. Goddard III and B. D. Olafson, *Proc. Natl. Acad. Sci. USA* **72**, 2335 (1975).
69. G. A. Rodley and W. T. Robinson, *Nature* **235**, 438 (1972).
70. R. S. Gall, J. F. Rogers, W. P. Schaefer, and G. G. Christoph, *J. Am. Chem. Soc.* **98**, 5135 (1976).
71. A. Avdeef and W. P. Schaefer, *J. Am. Chem. Soc.* **98**, 5153 (1976).
72. R. S. Gall and W. P. Schaefer, *Inorg. Chem.* **15**, 2758 (1976).
73. F. Basolo, B. M. Hoffman, and J. A. Ibers, *Acc. Chem. Res.* **8**, 384 (1975).
74. H. C. Stynes and J. A. Ibers, *J. Am. Chem. Soc.* **94**, 1559 (1972).
75. F. A. Walker, *J. Am. Chem. Soc.* **92**, 4235 (1970); B. M. Hoffman and D. H. Petering, *Proc. Natl. Acad. Sci. USA* **67**, 637 (1970).
76. B. M. Hoffman, C. A. Spilburg, and D. H. Petering, *Cold Spring Harbor Symp. Quant. Biol.* **36**, 343 (1971).
77. C. A. Spilburg, B. M. Hoffman, and D. H. Petering, *J. Biol. Chem.* **247**, 4219 (1972).
78. R. Lumry and S. Rajender, *Biopolymers* **9**, 1125 (1970).
79. E. Antonini and M. Brunori, in: *Hemoglobin and Myoglobin in their Reactions with Ligands*, pp. 28 and 221–226, North-Holland Publishing Company, Amsterdam (1971).
80. T. Sjostrand, *Acta Physiol. Scand.* **26**, 328 (1952).
81. G. Metz and T. Sjostrand, *Acta Physiol. Scand.* **31**, 384 (1954).
82. P. O'Earra, in: *Porphyrins and Metalloporphyrins*, K. M. Smith (Ed.), pp. 123–153, Elsevier Scientific Publishing Co., New York.
83. L. R. Melby, *Inorg. Chem.* **9**, 2186 (1970).
84. R. Huber, O. Epp, and H. Formanek, *J. Mol. Biol.* **52**, 349 (1970).
85. E. A. Padlan and W. E. Love, *J. Biol. Chem.* **249**, 4067 (1975).
86. J. C. Norvell, A. C. Nunes, and B. P. Schoenborn, *Science* **190**, 568 (1975).
87. J. W. Wang, A. Nakahara, and E. B. Fleischer, *J. Am. Chem. Soc.* **80**, 1109 (1958).
88. W. S. Caughey, *Ann. N.Y. Acad. Sci.* **174**, 148 (1970).
89. J. P. Collman, J. I. Brauman, T. R. Halbert, and K. S. Suslick, *Proc. Natl. Acad. Sci. USA* **73**, 3333 (1976).
90. C. K. Chang and T. G. Traylor, *Proc. Natl. Acad. Sci. USA* **70**, 2647 (1973).
91. L. Vaska and T. Yamaji, *J. Am. Chem. Soc.* **93**, 6673 (1971).
92. D. A. Baldwin, R. M. Pfeiffer, P. W. Reichgott, and N. J. Rose, *J. Am. Chem. Soc.* **95**, 5152 (1973).
93. M. Suh, Ph.D. Thesis, The University of Chicago, 1976.
94. V. L. Goedken and S.-M. Peng, *J. Am. Chem. Soc.* **96**, 7826 (1974).
95. L. J. Radonovich, A. Bloom, and J. L. Hoard, *J. Am. Chem. Soc.* **94**, 2073 (1972).
96. A. Nakahara and J. H. Wang, *J. Am. Chem. Soc.* **80**, 6526 (1958).
97. D. V. Stynes, H. C. Stynes, and J. A. Ibers, *J. Am. Chem. Soc.* **95**, 8475 (1973).
98. C. K. Chang and T. G. Traylor, *J. Am. Chem. Soc.* **95**, 8475 (1973); C. K. Chang and T. G. Traylor, *J. Am. Chem. Soc.* **95**, 8477 (1973).
99. D. V. Stynes and B. R. James, *J. Am. Chem. Soc.* **96**, 2733 (1974).
100. J. P. Collman, *Acc. Chem. Res.* **10**, 265 (1977).
101. M. Rougee and D. Brault, *Biochem. Biophys. Res. Commun.* **55**, 1364 (1974).
102. M. Rougee and D. Brault, *Biochemistry* **14**, 4100 (1975).
103. F. Antonini and M. Brunori, *Hemoglobin and Myoglobin and Their Reactions with Ligands*, p. 23, North-Holland Publishing Company, Amsterdam (1971).

104. J. H. Wang, in: *The Oxygenases* (O. Hayaishi, ed.), Academic Press, New York (1962).
105. J. H. Weber and D. H. Busch, *Inorg. Chem.* **4**, 469 (1965).
106. C. K. Chang and T. G. Traylor, *Proc. Natl. Acad. Sci. USA* **70**, 2647 (1973).
107. J. P. Collman and C. A. Reed, *J. Am. Chem. Soc.* **95**, 2048 (1973).
108. J. E. Baldwin and J. Huff, *J. Am. Chem. Soc.* **95**, 5757 (1973).
109. R. G. Little, J. A. Ibers, and J. E. Baldwin, *J. Am. Chem. Soc.* **97**, 7049 (1975).
110. (a) J. P. Collman, R. R. Gagne, T. R. Halbert, J. C. Marchon, and C. A. Reed, *J. Am. Chem. Soc.* **95**, 7870 (1973). (b) For a complete review of this work, see J. P. Collman, *Acc. Chem. Res.* **10**, 265 (1977).
111. L. Pauling, *Stanford Med. Bull.* **6**, 215 (1948); L. Pauling, *Nature (London)* **203**, 182 (1964).
112. J. J. Weiss, *Nature (London)* **202**, 85 (1964); J. S. Griffith, *Proc. R. Soc. London Ser. A* **235**, 23 (1956).
113. G. Herzberg, *Molecular Spectra and Molecular Structure. I. Spectra of Diatomic Molecules*, p. 560, Van Nostrand Co., New York (1950).
114. R. Rolfe, W. Halzer, W. F. Murphy, and J. H. Berstein, *J. Chem. Phys.* **49**, 963 (1968).
115. J. C. Evans, *J. Chem. Soc. Chem. Commun.* **1969**, 682 (1969).
116. J. Almog, J. E. Baldwin, and J. Huff, *J. Am. Chem. Soc.* **97**, 227 (1975).
117. J. Almog, J. E. Baldwin, R. L. Dyer, and M. Peters, *J. Am. Chem. Soc.* **97**, 227 (1975).
118. T. G. Traylor, Y. Tatsuno, D. W. Powell, and J. B. Cannon, *J. Chem. Soc. Chem. Commun.* **1977**, 732 (1977).
119. M. Florkin and E. H. Stotz (eds.), *Comprehensive Biochemistry*, 2nd ed., p. 18. Vol. 13, Elsevier, Amsterdam, 1965.
120. E. Margoliash, W. M. Filch, and R. E. Dickerson, *Brookhaven Symp. Biol.* **21**, 259 (1968).
121. R. E. Dickerson, M. L. Kopka, J. E. Weinzierl, J. Varnum, D. Eisenberg, and E. Margoliash, *J. Biol. Chem.* **242**, 3015 (1967).
122. T. Takano, O. B. Kallai, R. Swanson, and R. E. Dickerson, *J. Biol. Chem.* **248**, 5234 (1973).
123. I. C. Gunsalus, J. R. Meeks, J. D. Lipscomb, P. Debrunner, and E. Munck, in: *Molecular Mechanisms of Oxygen Activation* (O. Hayaishi, ed.), p. 561, Academic Press, New York (1974).
124. V. Ullrich, *Angew. Chem. Int. Ed. Engl.* **11**, 701 (1972).
125. C. R. E. Jefcoate and J. L. Gaylor, *Biochemistry* **8**, 3464 (1969); W. E. Blumberg and J. Peisach, in: *Probes of Structure and Function of Macromolecules and Membranes* (B. Chance, T. Yonetani, and A. S. Mildvan, eds.), Vol. II, p. 215, Academic Press, New York (1971).
126. H. A. O. Hill, A. Roder, and R. J. P. Williams, *Naturwissenschaften* **57**, 69 (1970).
127. E. Bayer, H. A. O. Hill, A. Roder, and R. J. P. Williams, *Chem. Commun.* **1969**, 109 (1969).
128. J. O. Stern and J. Peisach, *J. Biol. Chem.* **249**, 7495 (1974).
129. J. P. Collman, T. N. Sorrell, and B. M. Hoffman, *J. Am. Chem. Soc.* **97**, 913 (1975)
130. S. Koch, S. C. Tang, R. H. Holm, R. B. Frankel, and J. A. Ibers, *J. Am. Chem. Soc.* **97**, 918 (1975).
131. J. P. Collman and T. N. Sorell, *J. Am. Chem. Soc.* **97**, 4133 (1975).
132. J. P. Collman, T. N. Sorrell, J. H. Dawson, J. Trudell, E. Bunnenberg, and C. Djerassi, *Proc. Natl. Acad. Sci. USA* **73**, 6 (1976).

133. C. K. Chang and D. Dolphin, *J. Am. Chem. Soc.* **97**, 5948 (1975).
134. S. Koch, S. C. Tang, R. H. Holm, and R. B. Frankel, *J. Am. Chem. Soc.* **97**, 914 (1975).
135. L. Y. Martin, L. J. DeHayes, J. L. Zompa, and D. H. Busch, *J. Am. Chem. Soc.* **96**, 4046 (1974).
136. J. P. Collman, E. M. Elliott, T. R. Halbert, and B. S. Tovrog, *Proc. Natl. Acad. Sci. USA* **74**, 18 (1977).
137. C. K. Chang, *J. Chem. Soc. Chem. Commun.* **1977**, 800 (1977); C. K. Chang, *Adv. Chem. Ser.*, in press (1979).
138. F. C. McElroy and J. D. Dabrowiak, *J. Am. Chem. Soc.* **98**, 7112 (1976).
139. S. G. Boxer and G. L. Closs, *J. Am. Chem. Soc.* **98**, 5406 (1976).
140. M. R. Wasielewski, M. H. Studier, and J. J. Katz, *Proc. Natl. Acad. Sci. USA* **73**, 4282 (1976).
141. H. Ogoshi, S. Sugimoto, and Z. Yoshida, *Tetrahedron Lett.* 169 (1977).
142. N. E. Kagan, D. Mauzerall, and R. B. Merrifield, *J. Am. Chem. Soc.* **99**, 5484 (1977).
143. F. P. Schwarz, M. Gouterman, Z. Muljiani, and D. H. Dolphin, *Bioinorg. Chem.* **2**, 1 (1972); J. A. Anton, J. Kwong, and P. A. Looch, *J. Heterocycl. Chem.* **13**, 717 (1976).
144. J.-M. Lehn, J. Simon, and J. Wagner, *Angew. Chem.* **85**, 621–622 (1973); *Angew. Chem. Int. Ed. Engl.* **12**, 578, 579 (1973); *Nuov. J. Chim.* **1**, 77 (1977).
145. J.-M. Lehn and J. Simon, *Helv. Chim. Acta* **60**, 141 (1977).
146. J.-M. Lehn, S. H. Pine, E. Watanabe, and A. Willard, *J. Am. Chem. Soc.* **99**, 6766 (1977).
147. C. K. Chang, *J. Am. Chem. Soc.* **99**, 2819 (1977).
148. D. A. Buckingham, M. C. Gunter, and L. N. Mander, *J. Am. Chem. Soc.* **100**, 2899 (1978).
149. C. M. Elliott, *J. Chem. Soc. Chem. Commun.*, 399 (1978).
150. A. H. Alberts, R. Annunziata, and J.-M. Lehn, *J. Am. Chem. Soc.* **99**, 8502 (1977).
151. R. Louis, Y. Agnus, and R. Weiss, *J. Am. Chem. Soc.* **100**, 3604 (1978).
152. E. L. Rickes, N. G. Brink, F. R. Koniuszy, T. R. Wood, and K. Folkers, *Science* **107**, 396 (1948).
153. E. L. Smith, *Nature* **162**, 144 (1948).
154. A. Eschenmoser, R. Scheffold, E. Bertele, M. Pesaro, and H. Gschwend, *Proc. Roy. Soc. London* **A288**, 306 (1965).
155. R. B. Woodward, *Pure Appl. Chem.* **17**, 519 (1968).
156. J. M. Pratt, in: *Inorganic Chemistry of Vitamin B₁₂* (J. M. Pratt, ed.), p. 296, Academic Press, New York (1972).
157. D. Dodd and M. D. Johnson, *J. Organometal. Chem.* **52**, 1 (1973).
158. G. Schrauzer, *Acc. Chem. Res.* **1**, 97 (1968).
159. T. S. Roche and S. F. Endicott, *Inorg. Chem.* **13**, 1575 (1974).
160. V. L. Goedken, S.-M. Peng, and Y.-A. Park, *J. Am. Chem. Soc.* **96**, 284 (1974).
161. V. L. Goedken and S.-M. Peng, *J. Chem. Soc. Chem. Commun.*, 258 (1975).
162. H. R. V. Arnstein and R. J. Wrighton, *The Cobalamins*, Churchill and Livingston, Edinburgh (1971).
163. D. R. Brown, in: *Progress in Inorganic Chemistry*, Vol. 18, p. 177 (S. J. Lippard, ed.), Interscience, New York (1973).
164. J. M. Pratt and R. G. Thorp, *Advances in Chemistry and Radiochemistry* (H. J. Emeleus and A. G. Sharp, eds.), p. 375, Academic Press, New York (1972).
165. D. H. Busch, private communication (1979).

Index